Fundamentals of Momentum, Heat, and Mass Transfer

5th Edition

BICENTENNIAL

1807

WILEY

2007

BICENTENNIAL · BICENTENNIAL · BICENTENNIAL · BICENTENNIAL

THE WILEY BICENTENNIAL—KNOWLEDGE FOR GENERATIONS

*E*ach generation has its unique needs and aspirations. When Charles Wiley first opened his small printing shop in lower Manhattan in 1807, it was a generation of boundless potential searching for an identity. And we were there, helping to define a new American literary tradition. Over half a century later, in the midst of the Second Industrial Revolution, it was a generation focused on building the future. Once again, we were there, supplying the critical scientific, technical, and engineering knowledge that helped frame the world. Throughout the 20th Century, and into the new millennium, nations began to reach out beyond their own borders and a new international community was born. Wiley was there, expanding its operations around the world to enable a global exchange of ideas, opinions, and know-how.

For 200 years, Wiley has been an integral part of each generation's journey, enabling the flow of information and understanding necessary to meet their needs and fulfill their aspirations. Today, bold new technologies are changing the way we live and learn. Wiley will be there, providing you the must-have knowledge you need to imagine new worlds, new possibilities, and new opportunities.

Generations come and go, but you can always count on Wiley to provide you the knowledge you need, when and where you need it!

WILLIAM J. PESCE
PRESIDENT AND CHIEF EXECUTIVE OFFICER

PETER BOOTH WILEY
CHAIRMAN OF THE BOARD

Fundamentals of Momentum, Heat, and Mass Transfer

5th Edition

James R. Welty

Department of Mechanical Engineering

Charles E. Wicks

Department of Chemical Engineering

Robert E. Wilson

Department of Mechanical Engineering

Gregory L. Rorrer

Department of Chemical Engineering
Oregon State University

John Wiley & Sons, Inc.

ASSOCIATE PUBLISHER	Daniel Sayre
ACQUISITIONS EDITOR	Jennifer Welter
MARKETING MANAGER	Christopher Ruel
CREATIVE DIRECTOR	Harry Nolan
DESIGNER	Michael St. Martine
SENIOR MEDIA EDITOR	Lauren Sapira
SENIOR PRODUCTION EDITOR	Patricia McFadden
PRODUCTION MANAGEMENT SERVICES	Thomson Digital

This book was set in by Thomson Digital and printed and bound by Hamilton Printing. The cover was printed by Lehigh Press, Inc.

This book is printed on acid free paper. ∞

To order books or for customer service please, call 1-800-CALL WILEY (225-5945).

ISBN-13   978-0470128688

Printed in the United States of America

10 9 8 7 6 5 4 3

Preface to the 5th Edition

The first edition of *Fundamentals of Momentum, Heat, and Mass Transfer,* published in 1969, was written to become a part of what was then known as the engineering science core of most engineering curricula. Indeed, requirements for ABET accreditation have stipulated that a significant part of all curricula be devoted to fundamental subjects.

The emphasis on engineering science has evolved over the intervening years as new subjects and technologies have entered the world of engineering. Engineering curricula have made appropriate accommodations to these new application areas. Nonetheless, the subjects of momentum transfer (fluid mechanics), heat transfer, and mass transfer remain, at least in part, important components of all engineering curricula. It is in this context that we now present the fifth edition.

This text is intended for sophomore- or junior-level engineering students whose technical interests require an understanding of transport phenomena. The most likely users would include students majoring in Chemical, Environmental and/or Biochemical engineering. Other engineering majors may also find the idea of commonality among fluid mechanics, heat and mass transfer to be useful and important. It is assumed that students using this book will have completed courses in calculus and mass and energy balances. A rudimentary knowledge of differential equations is also recommended.

Advances in computing capability have been monumental since 1969. At that time, the pocket calculator was quite new and not generally in the hands of engineering students. Now, more than 38 years since the first edition, computer competency among students is taken for granted. Students may find it useful to solve some of the homework problems from this text using numerical computing packages however the bulk of our problems require the use of fundamental mathematical tools.

Noteworthy features of this new edition include the following:

- Considerable revision has been made in the momentum transfer (fluid mechanics) portion of the text. The topics of compressibility and surface tension have been added to the introductory chapter. Turbulent flow, which has been treated in a separate chapter in previous editions, is included as part of the chapter on viscous flow. The final chapter in this section is now devoted to rotating machinery with particular emphasis on pumps and fans. Analysis of the operation of these devices and the familiar "fan laws" are presented in this new chapter.

- Mass transfer applications have been expanded to include examples and numerous additional homework problems dealing with chemical separation phenomena. Application areas include bioremediation of waste products, materials separations, chemical deposition, and membrane separations among others.

Resources for adopting instructors, including the Instructor Solutions Manual and electronics images from the book suitable for use in lecture slides, are available on the Instructor Companion Site at www.wiley.com/college/welty. Please visit the Instructor Companion Site to register for a password to access these resources.

We wish to acknowledge contributions of the numerous instructors and students who have offered invaluable feedback over the years. Specifically for 5th edition we owe our thanks to those who responded to our publisher's survey.

We are gratified that the use of this book has continued at a significant level since the first edition appeared some 38 years ago. It is our continuing belief that the transport phenomena remain essential parts of the foundation of engineering education and practice. With the modifications and modernization of this fifth edition, it is our hope that Fundamentals of Momentum, Heat, and Mass Transfer will continue to be an essential part of students' educational experiences.

Corvallis, Oregon
September 2007

J.R. Welty
C.E. Wicks
R.E. Wilson
G.L. Rorrer

Contents

Fundamentals of Momentum, Heat, and Mass Transfer

5th Edition

Chapter 1

Introduction to Momentum Transfer

Momentum transfer in a fluid involves the study of the motion of fluids and the forces that produce these motions. From Newton's second law of motion it is known that force is directly related to the time rate of change of momentum of a system. Excluding action-at-a-distance forces such as gravity, the forces acting on a fluid, such as those resulting from pressure and shear stress, may be shown to be the result of microscopic (molecular) transfer of momentum. Thus the subject under consideration, which is historically fluid mechanics, may equally be termed momentum transfer.

The history of fluid mechanics shows the skillful blending of the nineteenth- and twentieth century analytical work in hydrodynamics with the empirical knowledge in hydraulics that man has collected over the ages. The mating of these separately developed disciplines was started by Ludwig Prandtl in 1904 with his boundary-layer theory, which was verified by experiment. Modern fluid mechanics, or momentum transfer, is both analytical and experimental.

Each area of study has its phraseology and nomenclature. Momentum transfer being typical, the basic definitions and concepts will be introduced in order to provide a basis for communication.

1.1 FLUIDS AND THE CONTINUUM

A fluid is defined as a substance that deforms continuously under the action of a shear stress. An important consequence of this definition is that when a fluid is at rest, there can be no shear stresses. Both liquids and gases are fluids. Some substances such as glass are technically classified as fluids. However, the rate of deformation in glass at normal temperatures is so small as to make its consideration as a fluid impractical.

Concept of a Continuum. Fluids, like all matter, are composed of molecules whose numbers stagger the imagination. In a cubic inch of air at room conditions there are some 10^{20} molecules. Any theory that would predict the individual motions of these many molecules would be extremely complex, far beyond our present abilities.

Most engineering work is concerned with the macroscopic or bulk behavior of a fluid rather than with the microscopic or molecular behavior. In most cases it is convenient to think of a fluid as a continuous distribution of matter or a *continuum*. There are, of course, certain instances in which the concept of a continuum is not valid. Consider, for example, the number of molecules in a small volume of a gas at rest. If the volume were taken small enough, the number of molecules per unit volume would be time-dependent for the microscopic volume even though the macroscopic volume had a constant number of molecules in it. The concept of a continuum would be valid only for the latter case. The validity of the continuum approach is seen to be dependent upon the type of information desired rather than the nature of the fluid. The treatment of fluids as continua is valid whenever the smallest fluid volume of interest contains a sufficient number of molecules to make statistical averages meaningful. The macroscopic properties of a continuum are considered to vary smoothly (continuously) from point to point in the fluid. Our immediate task is to define these properties at a point.

1.2   PROPERTIES AT A POINT

When a fluid is in motion, the quantities associated with the state and the motion of the fluid will vary from point to point. The definition of some fluid variables at a point is presented below.

Density at a Point.   The density of a fluid is defined as the mass per unit volume. Under flow conditions, particularly in gases, the density may vary greatly throughout the fluid. The density, ρ, at a particular point in the fluid is defined as

$$\rho = \lim_{\Delta V \to \delta V} \frac{\Delta m}{\Delta V}$$

where Δm is the mass contained in a volume ΔV, and δV is the smallest volume surrounding the point for which statistical averages are meaningful. The limit is shown in Figure 1.1.

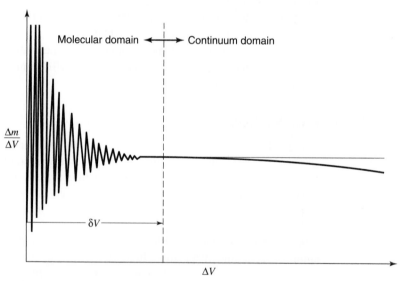

Figure 1.1 Density at a point.

     The concept of the density at a mathematical point, that is, at $\Delta V = 0$ is seen to be fictitious; however, taking $\rho = \lim_{\Delta V \to \delta V}(\Delta m/\Delta V)$ is extremely useful, as it allows us to describe fluid flow in terms of continuous functions. The density, in general, may vary from point to point in a fluid and may also vary with respect to time as in a punctured automobile tire.

Fluid Properties and Flow Properties. Some fluids, particularly liquids, have densities that remain almost constant over wide ranges of pressure and temperature. Fluids which exhibit this quality are usually treated as being incompressible. The effects of compressibility, however, are more a property of the situation than of the fluid itself. For example, the flow of air at low velocities is described by the same equations that describe the flow of water. From a static viewpoint, air is a compressible fluid and water incompressible. Instead of being classified according to the fluid, compressibility effects are considered a property of the flow. A distinction, often subtle, is made between the properties of the fluid and the properties of the flow, and the student is hereby alerted to the importance of this concept.

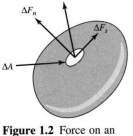

Figure 1.2 Force on an element of fluid.

Stress at a Point. Consider the force $\Delta\mathbf{F}$ acting on an element ΔA of the body shown in Figure 1.2. The force $\Delta\mathbf{F}$ is resolved into components normal and parallel to the surface of the element. The force per unit area or stress at a point is defined as the limit of $\Delta\mathbf{F}/\Delta A$ as $\Delta A \to \delta A$ where δA is the smallest area for which statistical averages are meaningful

$$\lim_{\Delta A \to \delta A} \frac{\Delta F_n}{\Delta A} = \sigma_{ii} \qquad \lim_{\Delta A \to \delta A} \frac{\Delta F_s}{\Delta A} = \tau_{ij}$$

Here σ_{ii} is the normal stress and τ_{ij} the shear stress. In this text, the double-subscript stress notation as used in solid mechanics will be employed. The student will recall that normal stress is positive in tension. The limiting process for the normal stress is illustrated in Figure 1.3.

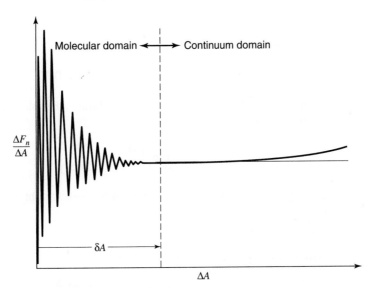

Figure 1.3 Normal stress at a point.

Forces acting on a fluid are divided into two general groups: body forces and surface forces. Body forces are those which act without physical contact, for example, gravity and electrostatic forces. On the contrary, pressure and frictional forces require physical contact for transmission. As a surface is required for the action of these forces they are called surface forces. Stress is therefore a surface force per unit area.[1]

---

[1] Mathematically, stress is classed as a tensor of second order, as it requires magnitude, direction, and orientation with respect to a plane for its determination.

Pressure at a point in a Static Fluid. For a static fluid, the normal stress at a point may be determined from the application of Newton's laws to a fluid element as the fluid element approaches zero size. It may be recalled that there can be no shearing stress in a static fluid. Thus, the only surface forces present will be those due to normal stresses. Consider the element shown in Figure 1.4. This element, while at rest, is acted upon by gravity and normal stresses. The weight of the fluid element is $\rho \mathbf{g}(\Delta x\, \Delta y\, \Delta z/2)$.

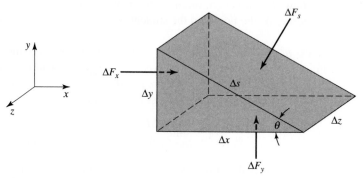

Figure 1.4 Element in a static fluid.

For a body at rest, $\Sigma \mathbf{F} = 0$. In the x direction

$$\Delta F_x - \Delta F_s \sin \theta = 0$$

Since $\sin \theta = \Delta y/\Delta s$, the above equation becomes

$$\Delta F_x - \Delta F_s \frac{\Delta y}{\Delta s} = 0$$

Dividing through by $\Delta y\, \Delta z$ and taking the limit as the volume of the element approaches zero, we obtain

$$\lim_{\Delta V \to 0} \left[\frac{\Delta F_x}{\Delta y\, \Delta z} - \frac{\Delta F_s}{\Delta s\, \Delta z} \right] = 0$$

Recalling that normal stress is positive in tension, we obtain, by evaluating the above equation

$$\sigma_{xx} = \sigma_{ss} \tag{1-1}$$

In the y direction, applying $\Sigma \mathbf{F} = 0$ yields

$$\Delta F_y - \Delta F_s \cos \theta - \rho g \frac{\Delta x\, \Delta y\, \Delta z}{2} = 0$$

Since $\cos \theta = \Delta x/\Delta s$, one has

$$\Delta F_y - \Delta F_s \frac{\Delta x}{\Delta s} - \rho g \frac{\Delta x\, \Delta y\, \Delta z}{2} = 0$$

Dividing through by $\Delta x\, \Delta z$ and taking the limit as before, we obtain

$$\lim_{\Delta V \to 0} \left[\frac{\Delta F_y}{\Delta x\, \Delta z} - \frac{\Delta F_s}{\Delta s\, \Delta z} - \frac{\rho g \Delta y}{2} \right] = 0$$

which becomes

$$-\sigma_{yy} + \sigma_{ss} - \frac{\rho g}{2}(0) = 0$$

or

$$\sigma_{yy} = \sigma_{ss} \qquad (1\text{-}2)$$

It may be noted that the angle θ does not appear in equation (1-1) or (1-2), thus the normal stress at a point in a static fluid is independent of direction, and is therefore a scalar quantity.

As the element is at rest, the only surface forces acting are those due to the normal stress. If we were to measure the force per unit area acting on a submerged element, we would observe that it acts inward or places the element in compression. The quantity measured is, of course, pressure, which in light of the preceding development, must be the negative of the normal stress. This important simplification, the reduction of stress, a tensor, to pressure, a scalar, may also be shown for the case of zero shear stress in a flowing fluid. When shearing stresses are present, the normal stress components at a point may not be equal; however, the pressure is still equal to the average normal stress; that is

$$P = -\tfrac{1}{3}(\sigma_{xx} + \sigma_{yy} + \sigma_{zz})$$

with very few exceptions, one being flow in shock waves.

Now that certain properties at a point have been discussed, let us investigate the manner in which fluid properties vary from point to point.

1.3 POINT-TO-POINT VARIATION OF PROPERTIES IN A FLUID

In the continuum approach to momentum transfer, use will be made of pressure, temperature, density, velocity, and stress fields. In previous studies, the concept of a gravitational field has been introduced. Gravity, of course, is a vector, and thus a gravitational field is a vector field. In this book, vectors will be written in boldfaced type. Weather maps illustrating the pressure variation over this country are published daily in our newspapers. As pressure is a scalar quantity, such maps are an illustration of a scalar field. Scalars in this book will be set in regular type.

In Figure 1.5, the lines drawn are the loci of points of equal pressure. The pressure varies continuously throughout the region, and one may observe the pressure levels and infer the manner in which the pressure varies by examining such a map.

Figure 1.5 Weather map—an example of a scalar field.

Of specific interest in momentum transfer is the description of the point-to-point variation in the pressure. Denoting the directions east and north in Figure 1.5 by x and y, respectively, we may represent the pressure throughout the region by the general function $P(x, y)$.

The change in P, written as dP, between two points in the region separated by the distances dx and dy is given by the total differential

$$dP = \frac{\partial P}{\partial x} dx + \frac{\partial P}{\partial y} dy \tag{1-3}$$

In equation (1-3), the partial derivatives represent the manner in which P changes along the x and y axes, respectively.

Along an arbitrary path s in the xy plane the total derivative is

$$\frac{dP}{ds} = \frac{\partial P}{\partial x} \frac{dx}{ds} + \frac{\partial P}{\partial y} \frac{dy}{ds} \tag{1-4}$$

In equation (1-4), the term dP/ds is the directional derivative, and its functional relation describes the rate of change of P in the s direction.

A small portion of the pressure field depicted in Figure 1.5 is shown in Figure 1.6. The arbitrary path s is shown, and it is easily seen that the terms dx/ds and dy/ds are the cosine and sine of the path angle, α, with respect to the x axis. The directional derivative, therefore, may be written as

$$\frac{dP}{ds} = \frac{\partial P}{\partial x} \cos \alpha + \frac{\partial P}{\partial y} \sin \alpha \tag{1-5}$$

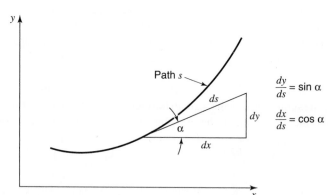

Figure 1.6 Path s in the xy plane.

There are an infinite number of paths to choose from in the xy plane; however, two particular paths are of special interest: the path for which dP/ds is zero and that for which dP/ds is maximum.

The path for which the directional derivative is zero is quite simple to find. Setting dP/ds equal to zero, we have

$$\left.\frac{\sin \alpha}{\cos \alpha}\right|_{dP/ds=0} = \tan \alpha \bigg|_{dP/ds=0} = -\frac{\partial P/\partial x}{\partial P/\partial y}$$

or, since $\tan \alpha = dy/dx$, we have

$$\left.\frac{dy}{dx}\right|_{dP/ds=0} = -\frac{\partial P/\partial x}{\partial P/\partial y} \tag{1-6}$$

Along the path whose slope is defined by equation (1-6), we have $dP = 0$, and thus P is constant. Paths along which a scalar is constant are called *isolines*.

In order to find the direction for which dP/ds is a maximum, we must have the derivative $(d/d\alpha)(dP/ds)$ equal to zero, or

$$\frac{d}{d\alpha}\frac{dP}{ds} = -\sin \alpha \frac{\partial P}{\partial x} + \cos \alpha \frac{\partial P}{\partial y} = 0$$

or

$$\left.\tan \alpha\right|_{dP/ds \text{ is max}} = \frac{\partial P/\partial y}{\partial P/\partial x} \tag{1-7}$$

Comparing relations (1-6) and (1-7), we see that the two directions defined by these equations are perpendicular. The magnitude of the directional derivative when the directional derivative is maximum is

$$\left.\frac{dP}{ds}\right|_{\max} = \frac{\partial P}{\partial x}\cos \alpha + \frac{\partial P}{\partial y}\sin \alpha$$

where $\cos \alpha$ and $\sin \alpha$ are evaluated along the path given by equation (1-7). As the cosine is related to the tangent by

$$\cos \alpha = \frac{1}{\sqrt{1 + \tan^2 \alpha}}$$

we have

$$\left.\cos \alpha\right|_{dP/ds \text{ is max}} = \frac{\partial P/\partial x}{\sqrt{(\partial P/\partial x)^2 + (\partial P/\partial y)^2}}$$

Evaluating $\sin \alpha$ in a similar manner gives

$$\left.\frac{dP}{ds}\right|_{\max} = \frac{(\partial P/\partial x)^2 + (\partial P/\partial y)^2}{\sqrt{(\partial P/\partial x)^2 + (\partial P/\partial y)^2}} = \sqrt{\left(\frac{\partial P}{\partial x}\right)^2 + \left(\frac{\partial P}{\partial y}\right)^2} \tag{1-8}$$

Equations (1-7) and (1-8) suggest that the maximum directional derivative is a vector of the form

$$\frac{\partial P}{\partial x}\mathbf{e}_x + \frac{\partial P}{\partial y}\mathbf{e}_y$$

where $\mathbf{e}_x$ and $\mathbf{e}_y$ are unit vectors in the x and y directions, respectively.

The directional derivative along the path of maximum value is frequently encountered in the anlaysis of transfer processes and is given a special name, the *gradient*. Thus the gradient of P, grad P, is

$$\text{grad } P = \frac{\partial P}{\partial x}\mathbf{e}_x + \frac{\partial P}{\partial y}\mathbf{e}_y$$

where $P = P(x, y)$. This concept can be extended to cases in which $P = P(x, y, z)$. For this more general case

$$\text{grad } P = \frac{\partial P}{\partial x}\mathbf{e}_x + \frac{\partial P}{\partial y}\mathbf{e}_y + \frac{\partial P}{\partial z}\mathbf{e}_z \tag{1-9}$$

Equation (1-9) may be written in more compact form by use of the operation ∇ (pronounced "del"), giving

$$\nabla P = \frac{\partial P}{\partial x}\mathbf{e}_x + \frac{\partial P}{\partial y}\mathbf{e}_y + \frac{\partial P}{\partial z}\mathbf{e}_z$$

where

$$\nabla = \frac{\partial}{\partial x}\mathbf{e}_x + \frac{\partial}{\partial y}\mathbf{e}_y + \frac{\partial}{\partial z}\mathbf{e}_z \tag{1-10}$$

Equation (1-10) is the defining relationship for the ∇ operator in Cartesian coordinates. This symbol indicates that differentiation is to be performed in a prescribed manner. In other coordinate systems, such as cylindrical and spherical coordinates, the gradient takes on a different form.[2] However, the geometric meaning of the gradient remains the same; it is a vector having the direction and magnitude of the maximum rate of change of the dependent variable with respect to distance.

1.4   UNITS

In addition to the International Standard (SI) system of units, there are two different English systems of units commonly used in engineering. These systems have their roots in Newton's second law of motion: force is equal to the time rate of change of momentum. In defining each term of this law, a direct relationship has been established between the four basic physical quantities used in mechanics: force, mass, length, and time. Through the arbitrary choice of fundamental dimensions, some confusion has occurred in the use of the English systems of units. Using the SI system of units has greatly reduced these difficulties.

The relationship between force and mass may be expressed by the following statement of Newton's second law of motion:

$$\mathbf{F} = \frac{m\mathbf{a}}{g_c}$$

where g_c is a conversion factor which is included to make the equation dimensionally consistent.

In the SI system, mass, length, and time are taken as basic units. The basic units are mass in kilograms (kg), length in meters (m), and time in seconds (s). The corresponding unit of force is the newton (N). One newton is the force required to accelerate a mass of one kilogram at a rate of one meter per second per second (1 m/s^2). The conversion factor, g_c, is then equal to one kilogram meter per newton per second per second (1 kg $\cdot$ m/N $\cdot$ s^2).

In engineering practice, force, length, and time have been frequently chosen as defining fundamental units. With this system, force is expressed in pounds force (lb$_f$), length in feet, and time in seconds. The corresponding unit of mass will be that which will be accelerated at the rate of 1 ft/(s)^2 by 1 lb$_f$.

---

[2] The forms of the gradient operator in rectangular, cylindrical, and spherical coordinate systems are listed in Appendix B.

This unit of mass having the dimensions of $(lb_f)(s)^2/(ft)$ is called the *slug*. The conversion factor, g_c, is then a multiplying factor to convert slugs into $(lb_f)(s)^2/(ft)$, and its value is 1 $(slug)(ft)/(lb_f)(s)^2$.

A third system encountered in engineering practice involves all four fundamental units. The unit of force is 1 lb_f, the unit of mass is 1 lb_m; length and time are given in units of feet and seconds, respectively. When 1 lb_m at sea level is allowed to fall under the influence of gravity, its acceleration will be 32.174 $(ft)/(s)^2$. The force exerted by gravity on 1 lb_m at sea level is defined as 1 lb_f. Therefore the conversion factor, g_c, for this system is 32.174 $(lb_m)(ft)/(lb_f)(s)^2$.[3]

A summary of the values of g_c is given in Table 1.1 for these three English systems of engineering units, along with the units of length, time, force, and mass.

Table 1.1

System	Length	Time	Force	Mass	g_c
1	meter	second	newton	kilogram	$1\dfrac{kg \cdot m}{N \cdot s^2}$
2	foot	second	lb_f	slug	$\dfrac{1\,(slug)(ft)}{(lb_f)(s)^2}$
3	foot	second	lb_f	lb_m	$\dfrac{32.174\,(lb_m)(ft)}{(lb_f)(s)^2}$

As all three systems are in current use in the technical literature, the student should be able to use formulas given in any particular situation. Careful checking for dimensional consistency will be required in *all* calculations. The conversion factor, g_c, will correctly relate the units corresponding to a system. There will be no attempt by the authors to incorporate the conversion factor in any equations; instead, it will be the reader's responsibility to use units that are consistent with every term in the equation.

1.5 COMPRESSIBILITY

A fluid is considered *compressible* or *incompressible* depending on whether its density is variable or constant. Liquids are generally considered to be incompressible whereas gases are certainly compressible.

The *bulk modulus of elasticity*, often referred to as simply the *bulk modulus*, is a fluid property that characterizes compressibility. It is defined according to

$$\beta \equiv \frac{dP}{dV/V} \qquad (1\text{-}11a)$$

or as

$$\beta \equiv -\frac{dP}{d\rho/\rho} \qquad (1\text{-}11b)$$

and has the dimensions N/m^2.

---

[3] In subsequent calculations in this book, g_c will be rounded off to a value of 32.2 $lb_m ft/^2\ lb_f$.

Disturbances introduced at some location in a fluid continuum will be propagated at a finite velocity. The velocity is designated the *acoustic velocity*; that is, the speed of sound in the fluid. It is symbolized C.

It can be shown that the acoustic velocity is related to changes in pressure and density according to

$$C = \left(\frac{dP}{d\rho}\right)^{\frac{1}{2}} \tag{1-12}$$

Introducing equation (1-11b) into this relationship yields

$$C = \left(-\frac{\beta}{\rho}\right)^{\frac{1}{2}} \tag{1-13}$$

For a gas, undergoing an isentropic process where $PV^k = C$, a constant, we have

$$C = \left(\frac{kP}{\rho}\right)^{\frac{1}{2}} \tag{1-14}$$

or

$$C = (kRT)^{\frac{1}{2}} \tag{1-15}$$

The question arises concerning when a gas, which is compressible, may be treated in a flow situation as incompressible, that is, when density variations are negligibly small. A common criterion for such a consideration involves the *Mach number*. The Mach number, a dimensionless parameter, is defined as the ratio of the fluid velocity, v, to the speed of sound, C, in the fluid:

$$M = \frac{v}{C} \tag{1-16}$$

A general rule of thumb is that when $M < 0.2$ the flow may be treated as incompressible with negligible error.

EXAMPLE 1

A jet aircraft is flying at an altitude of 15,500 m, where the air temperature is 239 K. Determine whether compressibility effects are significant at airspeeds of (a) 220 km/h and (b) 650 km/h.

The test for compressibility effects requires calculating the Mach number, M, which, in turn, requires that the acoustic velocity at each airspeed, be evaluated.

For air, $k = 1.4$, $R = 0.287$ kJ/kg·K, and

$$C = (kRT)^{\frac{1}{2}}$$
$$= [1.4\,(0.287\text{ kJ/kg} \cdot \text{K})(239\text{ K})(1000\text{ N} \cdot \text{m/kJ})(\text{kg} \cdot \text{m/N} \cdot \text{s}^2)]^{\frac{1}{2}}$$
$$= 310\text{ m/s}$$

(a) At $v = 220$ km/hr (61.1 m/s)

$$M = \frac{v}{C} = \frac{61.1\text{ m/s}}{310\text{ m/s}} = 0.197$$

The flow may be treated as incompressible.

(b) At $v = 650$ km/hr (180.5 m/s)

$$M = \frac{v}{C} = \frac{180.5\text{ m/s}}{310\text{ m/s}} = 0.582$$

Compressible effects must be accounted for.

1.6 SURFACE TENSION

The situation where a small amount of unconfined liquid forms a spherical drop is familiar to most of us. The phenomenon is the consequence of the attraction that exists between liquid molecules. Within a drop a molecule of liquid is completely surrounded by many others. Particles near the surface, on the contrary, will experience an imbalance of net force because of the nonuniformity in the numbers of adjacent particles. The extreme condition is the density discontinuity at the surface. Particles at the surface experience a relatively strong inwardly directed attractive force.

Given this behavior, it is evident that some work must be done when a liquid particle moves toward the surface. As more fluid is added the drop will expand creating additional surface. The work associated with creating this new surface is the *surface tension*, symbolized, σ. Quantitatively, σ is the work per unit area, N·m/m^2 or force per unit length of interface in N/m.

A surface is, in reality, an interface between two phases. Thus both phases will have the property of surface tension. The most common materials involving phase interfaces are water and air, but many others are also possible. For a given interfacial composition, the surface tension property is a function of both pressure and temperature, but a much stronger function of temperature. Table 1.2 lists values of σ for several fluids in air at 1 atm and 20°C. For water in air the surface tension is expressed as a function of temperature according to

$$\sigma = 0.123\,(1 - 0.00139\,T)\text{N/m} \tag{1-17}$$

where T is in Kelvins.

Table 1.2 Surface tensions of some fluids in air at 1atm and 20 °C

Fluid	σ(N/m)
Ammonia	0.021
Ethyl alcohol	0.028
Gasoline	0.022
Glycerin	0.063
Kerosene	0.028
Mercury	0.440
Soap solution	0.02S
SAE 30 oil	0.035

Source: Handbook of Chemistry and Physics, 62nd Ed, Chemical Rubber Publishing Co., Cleveland, OH, 1980.

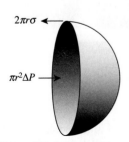

$2\pi r\sigma$

$\pi r^2 \Delta P$

Figure 1.7 A free body diagram of a hemispherical liquid droplet.

In Figure 1.7 we show a free body diagram of a hemispherical drop of liquid with the pressure and surface tension forces in balance. The condition examined is typically used for this analysis as a sphere represents the minimum surface area for a prescribed volume. The pressure difference, ΔP, between the inside and outside of the hemisphere produces a net pressure force that is balanced by the surface tension force. This force balance can be expressed as

$$\pi r^2 \Delta P = 2\pi r\sigma$$

and the pressure difference is given by

$$\Delta P = \frac{2\sigma}{r} \tag{1-18}$$

For the case of a soap bubble, which has a very thin wall, there are two interfaces and the pressure difference will be

$$\Delta P = \frac{4\sigma}{r} \tag{1-19}$$

Equations (1-18) and (1-19) indicate that the pressure difference is inversely proportional to r. The limit of this relationship is the case of a fully wetted surface where $r \cong \infty$, and the pressure difference due to surface tension is zero.

A consequence of the pressure difference resulting from surface tension is the phenomenon *of capillary action*. This effect is related to how well a liquid *wets* a solid boundary. The indicator for wetting or nonwetting is the *contact angle*, θ, defined as illustrated in Figure 1.8. With θ measured through the liquid, a nonwetting case, as shown in the figure, is associated with $\theta > 90°$. For a wetting case $\theta < 90°$. For mercury in contact with a clean

Figure 1.8 Contact angle for a non-wetting gas–liquid–solid interface.

glass tube $\theta \cong 130°$. Water in contact with a clean glass surface will completely wet the surface and, for this case, $\theta \cong 0$.

Illustrated in Figure 1.9 is the case of a small glass tube inserted into a pool of (a) water and (b) mercury. Note that water will rise in the tube and that in mercury the level in the tube is depressed.

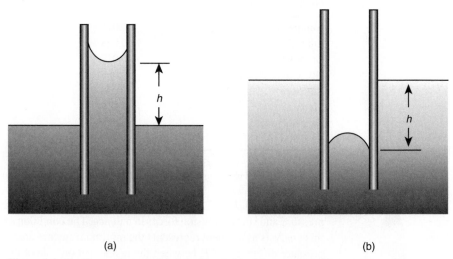

(a)   (b)

Figure 1.9 Capillary effects with a tube inserted in (a) water and (b) mercury.

For the water case, the liquid rises a distance h above the level in the pool. This is the result of attraction between the liquid molecules and the tube wall being greater than the attraction between water molecules at the liquid surface. For the mercury case, the intermolecular forces at the liquid surface are greater than the attractive forces between

$2\pi r\sigma$

$\pi r^2 h$

Figure 1.10 Free body diagram of a wetting liquid in a tube.

liquid mercury and the glass surface. The mercury is, thus, depressed a distance h below the level of the pool.

A free body diagram of the wetting liquid is shown in Figure 1.10. The upward force, due to surface tension

$$2\pi r\sigma\cos\theta$$

will be equal to the downward force due to the weight of liquid having volume $V = \pi r^2 h$.
Equating these forces we obtain

$$2\pi r\,\sigma\cos\theta = \rho g\pi r^2 h$$

and the value of h becomes

$$h = \frac{2\sigma\cos\theta}{\rho gr} \tag{1-20}$$

EXAMPLE 2

Determine the distance h that mercury will be depressed with a 4-mm-diameter glass tube inserted into a pool of mercury at 20°C (Figure 1.11).

Equation (1-20) applies, so we have

$$h = \frac{2\sigma\cos\theta}{\rho gr}$$

Recall that, for mercury and glass, $\theta = 130°$.

h

Figure 1.11 Capillary depression of mercury in a glass tube.

For mercury at 20°C $\rho = 13{,}580\,\text{kg/m}^3$, and for mercury in air $\sigma = 0.44\,\text{N/m}$ (Table 1.2) giving

$$h = \frac{2(0.44\,\text{N/m})(\cos 130°)}{(13580\,\text{kg/m}^3)(9.81\,\text{m/s}^2)(2 \times 10^{-3}\,\text{m})}$$

$$= 2.12 \times 10^{-3}\,\text{m}\quad(2.12\,\text{mm})$$

PROBLEMS

1.1 The number of molecules crossing a unit area per unit time in one direction is given by

$$N = \tfrac{1}{4}n\bar{v}$$

where n is the number of molecules per unit volume and v is the average molecular velocity. As the average molecular velocity is approximately equal to the speed of sound in a perfect gas, estimate the number of molecules crossing a circular hole 10^{-3} in. in diameter. Assume that the gas is at standard conditions. At standard conditions, there are 4×10^{20} molecules per in.^{3}.

1.2 Which of the quantities listed below are flow properties and which are fluid properties?

pressure	temperature	velocity
density	stress	speed of sound
specific heat	pressure gradient	

1.3 For a fluid of density ρ in which solid particles of density ρ_s are uniformly dispersed, show that if x is the mass fraction of solid in the mixture, the density is given by

$$\rho_{\text{mixture}} = \frac{\rho_s \rho}{\rho x + \rho_s(1-x)}$$

1.4 An equation linking water density and pressure is

$$\frac{P+B}{P_1+B} = \left(\frac{\rho}{\rho_1}\right)^7$$

where the pressure is in atmospheres and $B = 3000$ atm. Determine the pressure in psi required to increase water density by 1% above its nominal value.

1.5 What pressure change is required to change the density of air by 10% under standard conditions?

1.6 Using the information given in Problem 1.1 and the properties of the standard atmosphere given in Appendix G, estimate the number of molecules per cubic inch at an altitude of 250,000 ft.

1.7 Show that the unit vectors $\mathbf{e}_r$ and $\mathbf{e}_\theta$ in a cylindrical coordinate system are related to the unit vectors $\mathbf{e}_x$ and $\mathbf{e}_y$ by

$$\mathbf{e}_r = \mathbf{e}_x \cos\theta + \mathbf{e}_y \sin\theta$$

and

$$\mathbf{e}_\theta = -\mathbf{e}_x \sin\theta + \mathbf{e}_y \cos\theta$$

1.8 Using the results of Problem 1.7, show that $d\mathbf{e}_r/d\theta = \mathbf{e}_\theta$ and $d\mathbf{e}_\theta/d\theta = -\mathbf{e}_r$.

1.9 Using the geometric relations given below and the chain rule for differentiation, show that

$$\frac{\partial}{\partial x} = -\frac{\sin\theta}{r}\frac{\partial}{\partial\theta} + \cos\theta\frac{\partial}{\partial r}$$

and

$$\frac{\partial}{\partial y} = \frac{\cos\theta}{r}\frac{\partial}{\partial\theta} + \sin\theta\frac{\partial}{\partial r}$$

when $r^2 = x^2 + y^2$ and $\tan\theta = y/x$.

1.10 Transform the operator ∇ to cylindrical coordinates (r, θ, z), using the results of Problems 1.7 and 1.9.

1.11 Find the pressure gradient at point (a,b) when the pressure field is given by

$$P = \rho_\infty v_\infty^2 \left(\sin\frac{x}{a}\sin\frac{y}{b} + 2\frac{x}{a}\right)$$

where ρ_∞, v_∞, a, and b are constants.

1.12 Find the temperature gradient at point (a, b) at time $t = (L^2/\alpha)$ ln e when the temperature field is given by

$$T = T_0 e^{-\alpha t/4L^2} \sin\frac{x}{a}\cosh\frac{y}{b}$$

where T_0, α, a, and b are constants.

1.13 Are the fields described in Problems 1.11 and 1.12 dimensionally homogeneous? What must the units of ρ_∞ be in order that the pressure be in pounds per square foot when v_∞ is given in feet per second (problem 1.11)?

1.14 A scalar field is given by the function $\phi = 3x^2y + 4y^2$.

a. Find $\nabla\phi$ at the point $(3, 5)$.

b. Find the component of $\nabla\phi$ that makes a $-60°$ angle with the x axis at the point $(3, 5)$.

1.15 If the fluid of density ρ in Problem 1.3 obeys the perfect gas law, obtain the equation of state of the mixture, that is, $P = f(\rho_s, (RT/M), \rho_m, x)$. Will this result be valid if a liquid is present instead of a solid?

1.16 Using the expression for the gradient in polar coordinates (Appendix A), find the gradient of $\psi(r, \theta)$ when

$$\psi = A r\sin\theta\left(1 - \frac{a^2}{r^2}\right)$$

Where is the gradient maximum? The terms A and a are constant.

1.17 Given the following expression for the pressure field where x, y, and z are space coordinates, t is time, and P_0, ρ, V_∞, and L are constants, find the pressure gradient

$$P = P_0 + \tfrac{1}{2}\rho V_\infty^2\left[2\frac{xyz}{L^3} + 3\left(\frac{x}{L}\right)^2 + \frac{V_\infty t}{L}\right]$$

1.18 A vertical cylindrical tank having a base diameter of 10 m and a height of 5 m is filled to the top with water at 20°C. How much water will overflow if the water is heated to 80°C?

1.19 A liquid in a cylinder has a volume of 1200 cm^3 at 1.25 MPa and a volume of 1188 cm^3 at 2.50 MPa. Determine its bulk modulus of elasticity.

1.20 A pressure of 10 MPa is applied to 0.25 m^3 of a liquid, causing a volume reduction of 0.005 m^3. Determine the bulk modulus of elasticity.

1.21 The bulk modulus of elasticity for water is 2.205 GPa. Determine the change in pressure required to reduce a given volume by 0.75%.

1.22 Water in a container is originally at 100 kPa. The water is subjected to a pressure of 120 MPa. Determine the percentage decrease in its volume.

1.23 Determine the height to which water at 68°C will rise in a clean capillary tube having a diameter of 0.2875 cm.

1.24 Two clean and parallel glass plates, separated by a gap of 1.625 mm, are dipped in water. If $\sigma = 0.0735$ N/m, determine how high the water will rise.

1.25 A glass tube having an inside diameter of 0.25 mm and an outside diameter of 0.35 mm is inserted into a pool of mercury at 20°C such that the contact angle is 130°. Determine the upward force on the glass.

1.26 Determine the capillary rise for a water–air–glass interface at 40°C in a clean glass tube having a radius of 1 mm.

1.27 Determine the difference in pressure between the inside and outside of a soap film bubble at 20 °C if the diameter of the bubble is 4 mm.

1.28 An open, clean glass tube, having a diameter of 3 mm, is inserted vertically into a dish of mercury at 20°C. Determine how far the column of mercury in the tube will be depressed for a contact angle of 130°.

1.29 At 60°C the surface tension of water is 0.0662 N/m and that of mercury is 0.47 N/m. Determine the capillary height changes in these two fluids when they are in contact with air in a glass tube of diameter 0.55 mm. Contact angles are 0° for water and 130° for mercury.

1.30 Determine the diameter of the glass tube necessary to keep the capillary-height change of water at 30°C less than 1 mm.

Chapter **2**

Fluid Statics

$\mathbf{T}$he definition of a fluid variable at a point was considered in Chapter 1. In this chapter, the point-to-point variation of a particular variable, pressure, will be considered for the special case of a fluid at rest.

A frequently encountered static situation exists for a fluid that is stationary on Earth's surface. Although Earth has some motion of its own, we are well within normal limits of accuracy to neglect the absolute acceleration of the coordinate system that, in this situation, would be fixed with reference to Earth. Such a coordinate system is said to be an *inertial reference*. If, on the contrary, a fluid is stationary with respect to a coordinate system that has some significant absolute acceleration of its own, the reference is said to be *noninertial*. An example of this latter situation would be the fluid in a railroad tank car as it travels around a curved portion of track.

The application of Newton's second law of motion to a fixed mass of fluid reduces to the expression that the sum of the external forces is equal to the product of the mass and its acceleration. In the case of an inertial reference, we would naturally have the relation $\Sigma \mathbf{F} = 0$; whereas the more general statement $\Sigma \mathbf{F} = m\mathbf{a}$ must be used for the noninertial case.

2.1 PRESSURE VARIATION IN A STATIC FLUID

From the definition of a fluid, it is known that there can be no shear stress in a fluid at rest. This means that the only forces acting on the fluid are those due to gravity and pressure. As the sum of the forces must equal zero throughout the fluid, Newton's law may be satisfied by applying it to an arbitrary free body of fluid of differential size. The free body selected, shown in Figure 2.1, is the element of fluid $\Delta x\, \Delta y\, \Delta z$ with a corner at the point xyz. The coordinate system xyz is an inertial coordinate system.

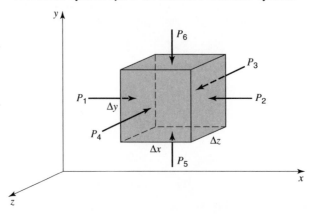

Figure 2.1 Pressure forces on a static fluid element.

The pressures that act on the various faces of the element are numbered 1 through 6. To find the sum of the forces on the element, the pressure on each face must first be evaluated.

We shall designate the pressure according to the face of the element upon which the pressure acts. For example, $P_1 = P|_x$, $P_2 = P|_{x+\Delta x}$, and so on. Evaluating the forces acting on each face, along with the force due to gravity acting on the element $\rho \mathbf{g}\, \Delta x\, \Delta y\, \Delta z$, we find that the sum of the forces is

$$\rho \mathbf{g}(\Delta x\, \Delta y\, \Delta z) + (P|_x - P|_{x+\Delta x})\, \Delta y\, \Delta z \mathbf{e}_x$$
$$+ (P|_y - P|_{y+\Delta y})\Delta x\, \Delta z \mathbf{e}_y + (P|_z - P|_{z+\Delta z})\Delta x\, \Delta y \mathbf{e}_z = 0$$

Dividing by the volume of the element $\Delta x\, \Delta y\, \Delta z$, we see that the above equation becomes

$$\rho \mathbf{g} - \frac{P|_{x+\Delta x} - P|_x}{\Delta x}\, \mathbf{e}_x - \frac{P|_{y+\Delta y} - P|_y}{\Delta y}\, \mathbf{e}_y - \frac{P|_{z+\Delta z} - P|_z}{\Delta z}\, \mathbf{e}_z = 0$$

where the order of the pressure terms has been reversed. As the size of the element approaches zero, Δx, Δy, and Δz approach zero and the element approaches the point (x, y, z). In the limit

$$\rho \mathbf{g} = \lim_{\Delta x, \Delta y, \Delta z \to 0} \left[\frac{P|_{x+\Delta x} - P|_x}{\Delta x}\, \mathbf{e}_x + \frac{P|_{y+\Delta y} - P|_y}{\Delta y}\, \mathbf{e}_y + \frac{P|_{z+\Delta z} - P|_z}{\Delta z}\, \mathbf{e}_z \right]$$

or

$$\rho \mathbf{g} = \frac{\partial P}{\partial x}\mathbf{e}_x + \frac{\partial P}{\partial y}\mathbf{e}_y + \frac{\partial P}{\partial z}\mathbf{e}_z \tag{2-1}$$

Recalling the form of the gradient, we may write equation (2-1) as

$$\rho \mathbf{g} = \nabla P \tag{2-2}$$

Equation (2-2) is the basic equation of fluid statics and states that the maximum rate of change of pressure occurs in the direction of the gravitational vector. In addition, as isolines are perpendicular to the gradient, constant pressure lines are perpendicular to the gravitational vector. The point-to-point variation in pressure may be obtained by integrating equation (2-2).

EXAMPLE 1

The manometer, a pressure measuring device, may be analyzed from the previous discussion. The simplest type of manometer is the U-tube manometer shown in Figure 2.2. The pressure in the tank at point A is to be measured. The fluid in the tank extends into the manometer to point B.

Choosing the y axis in the direction shown, we see that equation (2-2) becomes

$$\frac{dP}{dy}\mathbf{e}_y = -\rho g \mathbf{e}_y$$

Integrating between C and D in the manometer fluid, we have

$$P_{\text{atm}} - P_C = -\rho_m g d_2$$

and then integrating between B and A in the tank fluid, we obtain

$$P_A - P_B = -\rho_T g d_1$$

Figure 2.2 A U-Tube manometer.

As elevations B and C are equal, the pressures, P_B and P_C, must be the same. We may, thus, combine the above equations to obtain

$$P_A - P_{atm} = \rho_m g d_2 - \rho_T g d_1$$

The U-tube manometer measures the difference between the absolute pressure and the atmospheric pressure. This difference is called the *gage pressure* and is frequently used in pressure measurement.

EXAMPLE 2

In the fluid statics of gases, a relation between the pressure and density is required to integrate equation (2-2). The simplest case is that of the isothermal perfect gas, where $P = \rho RT/M$. Here R is the universal gas constant, M is the molecular weight of the gas, and T is the temperature, which is constant for this case. Selecting the y axis parallel to $\mathbf{g}$, we see that equation (2-2) becomes

$$\frac{dP}{dy} = -\rho g = -\frac{PMg}{RT}$$

Separating variables, we see that the above differential equation becomes

$$\frac{dP}{P} = -\frac{Mg}{RT} dy$$

Integration between $y = 0$ (where $P = P_{atm}$) and $y = y$ (where the pressure is P) yields

$$\ln \frac{P}{P_{atm}} = -\frac{Mgy}{RT}$$

or

$$\frac{P}{P_{atm}} = \exp\left\{-\frac{Mgy}{RT}\right\}$$

In the above examples, the atmospheric pressure and a model of pressure variation with elevation have appeared in the results. As performance of aircraft, rockets, and many types of industrial machinery varies with ambient pressure, temperature, and density, a standard atmosphere has been established in order to evaluate performance. At sea level, standard atmospheric conditions are

$$P = 29.92 \text{ in. Hg} = 2116.2 \text{ lb}_f/\text{ft}^2 = 14.696 \text{ lb}_f/\text{in.}^2 = 101\,325 \text{ N/m}^2$$
$$T = 519°\text{R} = 59°\text{F} = 288 \text{ K}$$
$$\rho = 0.07651 \text{ lb}_m/\text{ft}^3 = 0.002378 \text{ slug/ft}^3 = 1.226 \text{ kg/m}^3$$

A table of the standard atmospheric properties as a function of altitude is given in Appendix G.[1]

2.2   UNIFORM RECTILINEAR ACCELERATION

For the case in which the coordinate system xyz in Figure 2.1 is not an inertial coordinate system, equation (2-2) does not apply. In the case of uniform rectilinear acceleration, however, the fluid will be at rest with respect to the accelerating coordinate system. With a constant acceleration, we may apply the same analysis as in the case of the inertial coordinate system except that $\Sigma\mathbf{F} = m\mathbf{a} = \rho\,\Delta x\,\Delta y\,\Delta z\mathbf{a}$, as required by Newton's second law of motion. The result is

$$\nabla P = \rho(\mathbf{g} - \mathbf{a}) \tag{2-3}$$

The maximum rate of change of pressure is now in the $\mathbf{g} - \mathbf{a}$ direction, and lines of constant pressure are perpendicular to $\mathbf{g} - \mathbf{a}$.

The point-to-point variation in pressure is obtained from integration of equation (2-3).

EXAMPLE 3

A fuel tank is shown in Figure 2.3. If the tank is given a uniform acceleration to the right, what will be the pressure at point B?

From equation (2-3) the pressure gradient is in the $\mathbf{g} - \mathbf{a}$ direction, therefore the surface of the fluid will be perpendicular to this direction. Choosing the y axis parallel to $\mathbf{g} - \mathbf{a}$, we find that equation (2-3) may be integrated between point B and the surface. The pressure gradient becomes $dP/dy\,\mathbf{e}_y$ with the selection of the y axis parallel to $\mathbf{g} - \mathbf{a}$ as shown in Figure 2.4. Thus,

$$\frac{dP}{dy}\mathbf{e}_y = -\rho|\mathbf{g} - \mathbf{a}|\mathbf{e}_y = -\rho\sqrt{g^2 + a^2}\mathbf{e}_y$$

Figure 2.3 Fuel tank at rest.

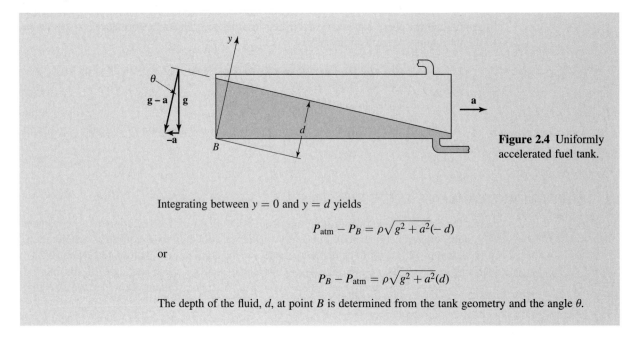

Figure 2.4 Uniformly accelerated fuel tank.

Integrating between $y = 0$ and $y = d$ yields

$$P_{atm} - P_B = \rho\sqrt{g^2 + a^2}(-d)$$

or

$$P_B - P_{atm} = \rho\sqrt{g^2 + a^2}(d)$$

The depth of the fluid, d, at point B is determined from the tank geometry and the angle θ.

2.3 FORCES ON SUBMERGED SURFACES

Determination of the force on submerged surfaces is done frequently in fluid statics. As these forces are due to pressure, use will be made of the relations describing the point-to-point variation in pressure which have been developed in the previous sections. The plane surface illustrated in Figure 2.5 is inclined at an angle α to the surface of the fluid. The area of the inclined plane is A, and the density of the fluid is ρ.

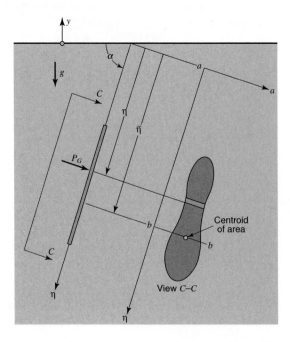

View C–C

Figure 2.5 A submerged plane surface.

The magnitude of the force on the element dA is $P_G\,dA$, where P_G is the gage pressure; $P_G = -\rho g y = \rho g \eta \sin\alpha$, giving

$$dF = \rho g \eta \sin\alpha\,dA$$

Integration over the surface of the plate yields

$$F = \rho g \sin\alpha \int_A \eta\,dA$$

The definition or the centroid of the area is

$$\bar\eta \equiv \frac{1}{A}\int_A \eta\,dA$$

and thus

$$F = \rho g \sin\alpha\,\bar\eta A \tag{2-4}$$

Thus, the force due to the pressure is equal to the pressure evaluated at the centroid of the submerged area multiplied by the submerged area. The point at which this force acts (the center of pressure) is not the centroid of the area. In order to find the center of pressure, we must find the point at which the total force on the plate must be concentrated in order to produce the same moment as the distributed pressure, or

$$F\eta_{\text{c.p.}} = \int_A \eta P_G\,dA$$

Substitution for the pressure yields

$$F\eta_{\text{c.p.}} = \int_A \rho g \sin\alpha\,\eta^2\,dA$$

and since $F = \rho g \sin\alpha\,\bar\eta A$, we have

$$\eta_{\text{c.p.}} = \frac{1}{A\bar\eta}\int_A \eta^2\,dA = \frac{I_{aa}}{A\bar\eta} \tag{2-5}$$

The moment of the area about the surface may be translated from an axis aa located at the fluid surface to an axis bb through the centroid by

$$I_{aa} = I_{bb} + \bar\eta^2 A$$

and thus

$$\eta_{\text{c.p.}} - \bar\eta = \frac{I_{bb}}{A\bar\eta} \tag{2-6}$$

The center of pressure is located below the centroid a distance $I_{bb}/A\bar\eta$.

EXAMPLE 4

A circular viewing port is to be located 1.5 ft belowthe surface of a tank as shown in Figure 2.6 the magnitude and location of the force acting Find on the window.

The force on the window is

$$F = \rho g \sin\alpha\,A\bar\eta$$

where

$$\alpha = \pi/2 \qquad \text{and} \qquad \bar\eta = 1.5\,\text{ft};$$

Figure 2.6 Submerged window.

the force is

$$F = \rho g A \bar{\eta} = \frac{(62.4\,\mathrm{lb_m/ft^3})(32.2\,\mathrm{ft/s^2})(\pi/4\,\mathrm{ft^2})(1.5\,\mathrm{ft})}{32.2\,\mathrm{lb_m\,ft/s^2\,lb_f}}$$

$$= 73.5\,\mathrm{lb_f}\,(327\,\mathrm{N})$$

The force F acts at $\bar{\eta} + \dfrac{I_{\mathrm{centroid}}}{A\bar{\eta}}$. For a circular area, $I_{\mathrm{centroid}} = \pi R^4/4$, so we obtain

$$\eta_{\mathrm{c.p.}} = 1.5 + \frac{\pi R^4}{4\pi R^2 1.5}\,\mathrm{ft} = 1.542\,\mathrm{ft}$$

EXAMPLE 5

Rainwater collects behind the concrete retaining wall shown in Figure 2.7. If the water-saturated soil (specific gravity = 2.2) acts as a fluid, determine the force and center of pressure on a 1-m width of the wall.

The force on a unit width of the wall is obtained by integrating the pressure difference between the right and left sides of the wall. Taking the origin at the top of the wall and measuring y downward, the force due to pressure is

$$F = \int (P - P_{\mathrm{atm}})(1)\,dy$$

Figure 2.7 Retaining wall.

The pressure difference in the region in contact with the water is

$$P - P_{\mathrm{atm}} = \rho_{\mathrm{H_2O}} g y$$

and the pressure difference in the region in contact with the soil is

$$P - P_{\mathrm{atm}} = \rho_{\mathrm{H_2O}} g(1) + 2.2\,\rho_{\mathrm{H_2O}} g(y - 1)$$

The force F is

$$F = \rho_{\mathrm{H_2O}} g \int_0^1 y\,dy + \rho_{\mathrm{H_2O}} g \int_0^4 [1 + 2.2(y - 1)]\,dy$$

$$F = (1000\,\mathrm{kg/m^3})(9.807\,\mathrm{m/s^2})(1\,\mathrm{m})(13.4\,\mathrm{m^2}) = 131\,414\,\mathrm{N}\,(29\,546\,\mathrm{lb_f})$$

The center of pressure of the force on the wall is obtained by taking moments about the top of the wall.

$$Fy_{\mathrm{c.p.}} = \rho_{\mathrm{H_2O}} g \left\{ \int_0^1 y^2\,dy + \int_1^4 y[1 + 2.2(y - 1)]\,dy \right\}$$

$$y_{\mathrm{c.p.}} = \frac{1}{(131\,414\,\mathrm{N})}(1000\,\mathrm{kg/m^3})(9.807\,\mathrm{m/s^2})(1\,\mathrm{m})(37.53\,\mathrm{m^3}) = 2.80\,\mathrm{m}\,(9.19\,\mathrm{ft})$$

The force on a submerged curved surface may be obtained from knowledge of the force on a plane surface and the laws of statics. Consider the curved surface BC illustrated in Figure 2.8.

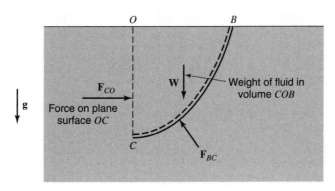

Figure 2.8 Submerged curved surface.

By consideration of the equilibrium of the fictitious body BCO, the force of the curved plate on body BCO may be evaluated. Since $\Sigma \mathbf{F} = 0$, we have

$$\mathbf{F}_{CB} = -\mathbf{W} - \mathbf{F}_{CO} \qquad (2\text{-}7)$$

The force of the fluid on the curved plate is the negative of this or $\mathbf{W} + \mathbf{F}_{CO}$. Thus, the force on a curved submerged surface may be obtained from the weight on the volume BCO and the force on a submerged plane surface.

2.4 BUOYANCY

The body shown in Figure 2.9 is submerged in a fluid with density ρ. The resultant force $\mathbf{F}$ holds the body in equilibrium.

The element of volume $h\,dA$ has gravity and pressure forces acting on it. The component of the force due to the pressure on the top of the element is $-P_2\,dS_2\cos\alpha\,\mathbf{e}_y$, where α is the angle between the plane of the element dS_2 and the xz plane. The product $dS_2\cos\alpha$ then is the projection of dS_2 onto the xz plane, or simply dA. The net

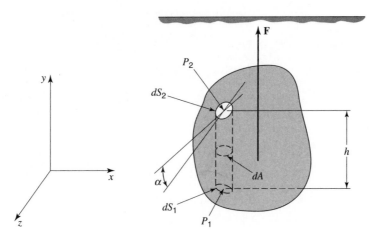

Figure 2.9 Forces on submerged volume.

pressure force on the element is $(P_1 - P_2)\,dA\,\mathbf{e}_y$, and the resultant force on the element is

$$dF = (P_1 - P_2)\,dA\,\mathbf{e}_y - \rho_B gh\,dA\,\mathbf{e}_y$$

where ρ_B is the density of the body. The difference in pressure $P_1 - P_2$ may be expressed as ρgh, so

$$dF = (\rho - \rho_B)gh\,dA\,\mathbf{e}_y$$

Integration over the volume of the body, assuming constant densities, yields

$$\mathbf{F} = (\rho - \rho_B)gV\mathbf{e}_y \qquad\qquad (2\text{-}8)$$

where V is the volume of the body. The resultant force $\mathbf{F}$ is composed of two parts, the weight $-\rho_B gV\mathbf{e}_y$ and the buoyant force $\rho gV\mathbf{e}_y$. The body experiences an upward force equal to the weight of the displaced fluid. This is the well-known principle of Archimedes. When $\rho > \rho_B$, the resultant force will cause the body to float on the surface. In the case of a floating body, the buoyant force is $\rho gV_s\mathbf{e}_y$, where V_s is the submerged volume.

EXAMPLE 6

A cube measuring 1 ft on a side is submerged so that its top face is 10 ft below the free surface of water. Determine the magnitude and direction of the applied force necessary to hold the cube in this position if it is made of

(a) cork $(\rho = 10\,\mathrm{lb_m/ft^3})$

(b) steel $(\rho = 490\,\mathrm{lb_m/ft^3})$.

The pressure forces on all lateral surfaces of the cube cancel. Those on the top and bottom do not, as they are at different depths.

Summing forces on the vertical direction, we obtain

$$\Sigma F_y = -W + P(1)|_{\text{bottom}} - P(1)|_{\text{top}} + F_y = 0$$

where F_y is the additional force required to hold the cube in position.

Expressing each of the pressures as $P_{\text{atm}} + \rho_w gh$, and W as $\rho_c gV$, we obtain, for our force balance

$$-\rho_c gV + \rho_w g(11\,\text{ft})(1\,\text{ft}^2) - \rho_w g(10\,\text{ft})(1\,\text{ft}^2) + F_y = 0$$

Solving for F_y, we have

$$F_y = -\rho_w g[(11)(1) - 10(1)] + \rho_c gV = -\rho_w gV + \rho_c gV$$

The first term is seen to be a buoyant force, equal to the weight of displaced water.

Finally, solving for F_y, we obtain

(a) $\rho_c = 10\,\mathrm{lb_m/ft^3}$

$$F_y = -\frac{(62.4\,\mathrm{lb_m/ft^3})(32.2\,\mathrm{ft/s^2})(1\,\mathrm{ft^3})}{32.2\,\mathrm{lb_m\,ft/s^2\,lb_f}} + \frac{(10\,\mathrm{lb_m\,ft^3})(32.2\,\mathrm{ft/s^2})(1\,\mathrm{ft^3})}{32.2\,\mathrm{lb_m\,ft/s^2\,lb_f}}$$

$$= -52.4\,\mathrm{lb_f}\ (\text{downward})(-233\,\mathrm{N})$$

(b) $\rho_c = 490 \, lb_m/ft^3$

$$F_y = -\frac{(62.4 \, lb_m/ft^3)(32.2 \, ft/s^2)(1 \, ft^3)}{32.2 \, lb_m ft/s^2 \ lb_f} + \frac{(490 \, lb_m ft^3)(32.2 \, ft/s^2)(1 \, ft^3)}{32.2 \, lb_m ft/s^2/lb_f}$$

$$= +427.6 \, lb_f(\text{upward})(1902 \, N)$$

In case (a), the buoyant force exceeded the weight of the cube, thus to keep it submerged 10 ft below the surface, a downward force of over 52 lb was required. In the second case, the weight exceeded the buoyant force, and an upward force was required.

2.5 CLOSURE

In this chapter the behavior of static fluids has been examined. The application of Newton's laws of motion led to the description of the point-to-point variation in fluid pressure, from which force relations were developed. Specific applications have been considered, including manometry, forces on plane and curved submerged surfaces, and the buoyancy of floating objects.

The static analyses that have been considered will later be seen as special cases of more general relations governing fluid behavior. Our next task is to examine the behavior of fluids in motion to describe the effect of that motion. Fundamental laws other than Newton's laws of motion will be necessary for this analysis.

PROBLEMS

2.1 On a certain day the barometric pressure at sea level is 30.1 in. Hg, and the temperature is 70°F. The pressure gage in an airplane in flight indicates a pressure of 10.6 psia, and the temperature gage shows the air temperature to be 46°F. Estimate as accurately as possible the altitude of the airplane above sea level.

2.2 The open end of a cylindrical tank 2 ft in diameter and 3 ft high is submerged in water as shown. If the tank weights 250 lb, to what depth h will the tank submerge? The air barometric pressure is 14.7 psia. The thickness of tank wall may be neglected. What additional force is required to bring the top of the tank flush with the water surface?

2.3 In Problem 2.2 above, find the depth at which the net force on the tank is zero.

2.4 If the density of sea water is approximated by the equation of state $\rho = \rho_0 \exp[(p - p_{atm})/\beta)]$, where β is the compressibility, determine the pressure and density at a point 32,000 ft below the surface of the sea. Assume $\beta = 300,00$ psi.

2.5 The change in density due to temperature causes the takeoff and landing speeds of a heavier-than-air craft to increase as the square root of the temperature. What effect do temperature-induced density changes have on the lifting power of a rigid lighter-than-air craft?

2.6 The practical depth limit for a suited diver is about 185 m. What is the gage pressure in sea water at that depth? The specific gravity of sea water is 1.025.

2.7 Matter is attracted to the center of Earth with a force proportional to the radial distance from the center. Using the known value of g at the surface where the radius is 6330 km, compute the pressure at Earth's center, assuming the material behaves like a liquid, and that the mean specific gravity is 5.67.

2.8 The deepest known point in the ocean is 11,034 m in the Mariana Trench in the Pacific. Assuming sea water to have a constant density of 1050 kg/m³, determine the pressure at this point in atmospheres.

2.9   Determine the depth change to cause a pressure increase of 1 atm for (a) water, (b) sea water ($SG = 1.0250$), and (c) mercury ($SG = 13.6$).

2.10   Find the pressure at point A.

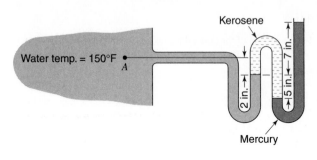

2.11   Using a liquid having a specific gravity of 1.2 and inverting a tube full of this material as shown in the figure, what is the value of h if the vapor pressure of the liquid is 3 psia?

2.12   What is the pressure p_A in the figure? The specific gravity of the oil is 0.8.

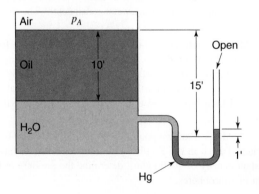

2.13   Find the difference in pressure between tanks A and B, if $d_1 = 2$ ft, $d_2 = 6$ in., $d_3 = 2.4$ in., and $d_4 = 4$ in.

2.14   What is the pressure difference between points A and B if $d_1 = 1.7$ ft, $d_2 = 1$ in., and $d_3 = 6.3$ in.?

2.15   A differential manometer is used to measure the pressure change caused by a flow constriction in a piping system as shown. Determine the pressure difference between points A and B in psi. Which section has the higher pressure?

2.16 The car shown in the figure is accelerated to the right at a uniform rate. What way will the balloon move relative to the car?

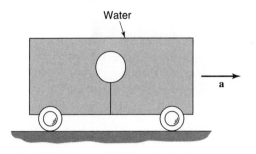

2.17 The tank is accelerated upward at a uniform rate. Does the manometer level go up or down?

2.18 Glass viewing windows are to be installed in an aquarium. Each window is to be 0.6 m in diameter and to be centered 2 m below the surface level. Find the force and location of the force acting on the window.

2.19 Find the minimum value of h for which the gate shown will rotate counterclockwise if the gate cross section is (a) rectangular, 4 ft × 4 ft; (b) triangular, 4 ft at the base × 4 ft high. Neglect bearing friction.

2.20 A circular log of radius r is to be used as a barrier as shown in the figure below. If the point of contact is at O, determine the required density of the log.

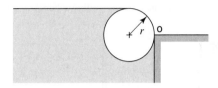

2.21 A rectangular block of concrete 3 ft × 3ft × 6 in. has its 6-in. side half buried at the bottom of a lake 23 ft deep. What force is required to lift the block free of the bottom? What force is required to maintain the block in this position? (Concrete weights 150 lb/ft³.)

2.22 A dam spillway gate holds back water of depth h. The gate weights 500 lb/ft and is hinged at A. At what depth of water will the gate rise and permit water to flow under it?

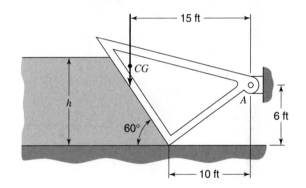

2.23 It is desired to use a 0.75-m diameter beach ball to stop a drain in a swimming pool. Obtain an expression that relates the drain diameter D and the minimum water depth h for which the ball will remain in place.

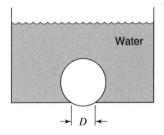

2.24 A watertight bulkhead 22 ft high forms a temporary dam for some construction work. The top 12 ft behind the bulkhead consists of sea water with a density of 2 slugs/ft³, but the bottom 10 ft begin a mixture of mud and water can be considered a fluid of density 4 slugs/ft³. Calculate the total horizontal load per unit width and the location of the center of pressure measured from the bottom.

2.25   The circular gate *ABC* has a 1 m radius and is hinged at *B*. Neglecting atmospheric pressure, determine the force *P* just sufficient to keep the gate from opening when $h = 12$ m.

2.26   The figure below shows an open triangular channel in which the two sides, *AB* and *AC*, are held together by cables, spaced 1 m apart, between *B* and *C*. Determine the cable tension.

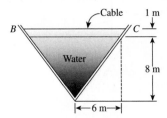

2.27   The dam shown below is 100 m wide. Determine the magnitude and location of the force on the inclined surface.

2.28   The float in a toilet tank is a sphere of radius *R* and is made of a material with density ρ. An upward buoyant force *F* is required to shut the ballcock valve. The density of water is designated ρ_w. Develop an expression for *x*, the fraction of the float submerged, in terms of *R*, ρ, *F*, *g*, and ρ_w.

2.29   A cubical piece of wood with an edge *L* in length floats in the water. The specific gravity of the wood is 0.90. What moment *M* is required to hold the cube in the position shown? The right-hand edge of the block is flush with the water.

Chapter **3**

Description of a Fluid in Motion

The development of an analytical description of fluid flow is based upon the expression of the physical laws related to fluid flow in a suitable mathematical form. Accordingly, we shall present the pertinent physical laws and discuss the methods used to describe a fluid in motion.

3.1 FUNDAMENTAL PHYSICAL LAWS

There are three fundamental physical laws that, with the exception of relativistic and nuclear phenomena, apply to each and every flow independently of the nature of the fluid under consideration. These laws are listed below with the designations of their mathematical formulations.

Law	Equation
1. The law of conservation of mass	Continuity equation
2. Newton's second law of motion	Momentum theorem
3. The first law of thermodynamics	Energy equation

The next three chapters will be devoted exclusively to the development of a suitable working form of these laws.[1]

In addition to the above laws, certain auxiliary or subsidiary relations are employed in describing a fluid. These relations depend upon the nature of the fluid under consideration. Unfortunately, most of these auxiliary relations have also been termed "laws." Already in our previous studies, Hooke's law, the perfect gas law, and others have been encountered. However accurate these "laws" may be over a restricted range, their validity is entirely dependent upon the nature of the material under consideration. Thus, while some of the auxiliary relations that will be used will be called laws, the student will be responsible for noting the difference in scope between the fundamental physical laws and the auxiliary relations.

3.2 FLUID-FLOW FIELDS: LAGRANGIAN AND EULERIAN REPRESENTATIONS

The term *field* refers to a quantity defined as a function of position and time throughout a given region. There are two different forms of representation for fields in fluid mechanics. *Lagrange's* form and *Euler's* form. The difference between these approaches lies in the manner in which the position in the field is identified.

In the Lagrangian approach, the physical variables are described for a particular element of fluid as it traverses the flow. This is the familiar approach of particle and

---

[1] The second law of thermodynamics is also fundamental to fluid-flow analysis. An analytic consideration of the second law is beyond the scope of the present treatment.

rigid-body dynamics. The coordinates (x, y, z) are the coordinates of the element of fluid and, as such, are functions of time. The coordinates (x, y, z) are therefore dependent variables in the Lagrangian form. The fluid element is identified by its position in the field at some arbitrary time, usually $t = 0$. The velocity field in this case is written in functional form as

$$\mathbf{v} = \mathbf{v}(a, b, c, t) \tag{3-1}$$

where the coordinates (a, b, c) refer to the *initial* position of the fluid element. The other fluid-flow variables, being functions of the same coordinates, may be represented in a similar manner. The Lagrangian approach is seldom used in fluid mechanics, as the type of information desired is usually the value of a particular fluid variable at a fixed point in the flow rather than the value of a fluid variable experienced by an element of fluid along its trajectory. For example, the determination of the force on a stationary body in a flow field requires that we know the pressure and shear stress at every point on the body. The Eulerian form provides us with this type of information.

The Eulerian approach gives the value of a fluid variable at a given point and at a given time. In functional form the velocity field is written as

$$\mathbf{v} = \mathbf{v}(x, y, z, t) \tag{3-2}$$

where x, y, z, and t are *all* independent variables. For a particular point (x_1, y_1, z_2) and t_1, equation (3-2) gives the velocity of the fluid at that location at time t_1. In this text the Eulerian approach will be used exclusively.

3.3   STEADY AND UNSTEADY FLOWS

In adopting the Eulerian approach, we note that the fluid flow will, in general, be a function of the four independent variables (x, y, z, t). If the flow at every point in the fluid is independent of time, the flow is termed *steady*. If the flow at a point varies with time, the flow is termed *unsteady*. It is possible in certain cases to reduce an unsteady flow to a steady flow by changing the frame of reference. Consider an airplane flying through the air at constant speed v_0, as shown in Figure 3.1. When observed from the stationary x, y, z coordinate system, the flow pattern is unsteady. The flow at the point P illustrated, for example, will vary as the vehicle approaches it.

Figure 3.1 Unsteady flow with respect to a fixed coordinate system.

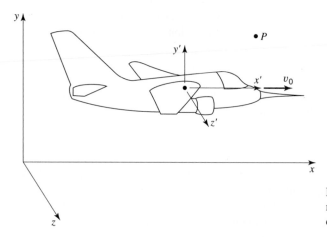

Figure 3.2 Steady flow with respect to a moving coordinate system.

Now consider the same situation when observed from the x', y', z' coordinate system which is moving at constant velocity v_0 as illustrated in Figure 3.2.

The flow conditions are now independent of time at every point in the flow field, and thus the flow is steady when viewed from the moving coordinate system. Whenever a body moves through a fluid with a constant velocity, the flow field may be transformed from an unsteady flow to a steady flow by selecting a coordinate system that is fixed with respect to the moving body.

In the wind-tunnel testing of models, use is made of this concept. Data obtained for a static model in a moving stream will be the same as the data obtained for a model moving through a static stream. The physical as well as the analytical simplifications afforded by this transformation are considerable. We shall make use of this transformation whenever applicable.

3.4 STREAMLINES

A useful concept in the description of a flowing fluid is that of a *streamline*. A streamline is defined as the line-drawn tangent to the velocity vector at each point in a flow field. Figure 3.3 shows the streamline pattern for ideal flow past a football-like object. In steady flow, as all velocity vectors are invariant with time, the path of a fluid particle follows a streamline, hence a streamline is the trajectory of an element of fluid in such a situation. In unsteady flow, streamline patterns change from instant to instant. Thus, the trajectory of a fluid element will be different from a streamline at any particular time. The actual trajectory of a fluid element as it traverses the flow is designated as a *path line*. It is obvious that path lines and streamlines are coincident only in steady flow.

Figure 3.3 Illustration of streamlines.

The streamline is useful in relating that fluid velocity components to the geometry of the flow field. For two-dimensional flow this relation is

$$\frac{v_y}{v_x} = \frac{dy}{dx} \qquad (3\text{-}3)$$

as the streamline is tangent to the velocity vector having x and y components v_x and v_y. In three dimensions this becomes

$$\frac{dx}{v_x} = \frac{dy}{v_y} = \frac{dz}{v_z} \qquad (3\text{-}4)$$

The utility of the above relations is in obtaining an analytical relation between velocity components and the streamline pattern.

Some additional discussion is provided in Chapter 10 regarding the mathematical description of streamlines around solid objects.

3.5   SYSTEMS AND CONTROL VOLUMES

The three basic physical laws listed in Section 3.1 are all stated in terms of a *system*. A system is defined as a collection of matter of fixed identity. The basic laws give the interaction of a system with its surroundings. The selection of the system for the application of these laws is quite flexible and is, in many cases, a complex problem. Any analysis utilizing a fundamental law must follow the designation of a specific system, and the difficulty of solution varies greatly depending on the choice made.

As an illustration, consider Newton's second law, $\mathbf{F} = m\mathbf{a}$. The terms represented are as follows:

$\mathbf{F} =$ the resultant force exerted by the surroundings on the system.
$m =$ the mass of the system.
$\mathbf{a} =$ the acceleration of the center of mass of the system.

Figure 3.4 An easily identifiable system.

Figure 3.5 Control volume for analysis of flow through a nozzle.

In the piston-and-cylinder arrangement shown in Figure 3.4, a convenient system to analyze, readily identified by virtue of its confinement, is the mass of material enclosed within the cylinder by the piston.

In the case of the nozzle shown in Figure 3.5, the fluid occupying the nozzle changes from instant to instant. Thus different systems occupy the nozzle at different times.

A more convenient method of analysis of the nozzle would be to consider the region bounded by the dotted line. Such a region is a *control volume*. A control volume is a region in space through which fluid flows.[2]

---

[2] A control volume may be fixed or moving uniformly (inertial), or it may be accelerating (noninertial). Primary consideration here will be given to inertial control volumes.

The extreme mobility of a fluid makes the identification of a particular system a tedious task. By developing the fundamental physical laws in a form that applies to a control volume (where the system changes from instant to instant), the analysis of fluid flow is greatly simplified. The control-volume approach circumvents the difficulty in system identification. Succeeding chapters will convert the fundamental physical laws from the system approach to a control-volume approach. The control volume selected may be either finite or infinitesimal.

Chapter 4

Conservation of Mass: Control-Volume Approach

The initial application of the fundamental laws of fluid mechanics involves the law of conservation of mass. In this chapter, we shall develop an integral relationship that expresses the law of conservation of mass for a general control volume. The integral relation thus developed will be applied to some often-encountered fluid-flow situations.

4.1 INTEGRAL RELATION

The law of conservation of mass states that mass may be neither created nor destroyed. With respect to a control volume, the law of conservation of mass may be simply stated as

$$
\left\{ \begin{array}{c} \text{rate of mass} \\ \text{efflux from} \\ \text{control} \\ \text{volume} \end{array} \right\} - \left\{ \begin{array}{c} \text{rate of mass} \\ \text{flow into} \\ \text{control} \\ \text{volume} \end{array} \right\} + \left\{ \begin{array}{c} \text{rate of} \\ \text{accumulation} \\ \text{of mass within} \\ \text{control volume} \end{array} \right\} = 0
$$

Consider now the general control volume located in a fluid flow field, as shown in Figure 4.1.

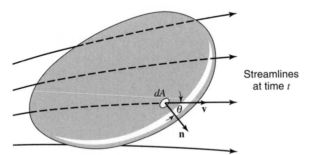

Streamlines at time t

Figure 4.1 Fluid flow through a control volume.

For the small element of area dA on the control surface, the rate of mass efflux $= (\rho v)(dA \cos \theta)$, where $dA \cos \theta$ is the projection of the area dA in a plane normal to the velocity vector, $\mathbf{v}$, and θ is the angle between the velocity vector, $\mathbf{v}$, and the *outward* directed unit normal vector, $\mathbf{n}$, to dA. From vector algebra, we recognize the product

$$
\rho v \, dA \cos \theta = \rho \, dA |\mathbf{v}| \; |\mathbf{n}| \cos \theta
$$

as the "scalar" or "dot" product

$$
\rho(\mathbf{v} \cdot \mathbf{n}) \, dA
$$

which is the form we shall use to designate the rate of mass efflux through dA. The product ρv is the mass flux, often called the mass velocity, G. Physically, this product represents the amount of mass flowing through a unit cross-sectional area per unit time.

If we now integrate this quantity over the entire control surface, we have

$$\iint_{c.s.} \rho(\mathbf{v} \cdot \mathbf{n})\, dA$$

which is the net outward flow of mass across the control surface, or the *net mass efflux* from the control volume.

Note that if mass is entering the control volume, that is, flowing inward across the control surface, the product $\mathbf{v} \cdot \mathbf{n} = |\mathbf{v}|\,|\mathbf{n}| \cos \theta$ is negative, since $\theta > 90°$, and $\cos \theta$ is therefore negative. Thus, if the integral is

positive, there is a net efflux of mass;

negative, there is a net influx of mass;

zero, the mass within the control volume is constant.

The rate of accumulation of mass within the control volume may be expressed as

$$\frac{\partial}{\partial t} \iiint_{c.v.} \rho\, dV$$

and the integral expression for the mass balance over a general control volume becomes

$$\iint_{c.s.} \rho(\mathbf{v} \cdot \mathbf{n})\, dA + \frac{\partial}{\partial t} \iiint_{c.v.} \rho\, dV = 0 \qquad (4\text{-}1)$$

4.2   SPECIFIC FORMS OF THE INTEGRAL EXPRESSION

Equation (4-1) gives the mass balance in its most general form. We now consider some frequently encountered situations where equation (4-1) may be applied.

If flow is steady relative to coordinates fixed to the control volume, the accumulation term, $\partial/\partial t \iiint_{c.v.} \rho\, dV$, will be zero. This is readily seen when one recalls that, by the definition of steady flow, the properties of a flow field are invariant with time, hence the partial derivative with respect to time is zero. Thus, for this situation the applicable form of the continuity expression is

$$\iint_{c.s.} \rho(\mathbf{v} \cdot \mathbf{n})\, dA = 0 \qquad (4\text{-}2)$$

Another important case is that of an incompressible flow with fluid filling the control volume. For incompressible flow the density, ρ, is constant, hence the accumulation term involving the partial derivative with respect to time is again zero. Additionally, the density term in the surface integral may be canceled. The conservation-of-mass expression for incompressible flow of this nature thus becomes

$$\iint_{c.s.} (\mathbf{v} \cdot \mathbf{n})\, dA = 0 \qquad (4\text{-}3)$$

The following examples illustrate the application of equation (4-1) to some cases that recur frequently in momentum transfer.

EXAMPLE 1

As our first example, let us consider the common situation of a control volume for which mass efflux and influx are steady and one dimensional. Specifically, consider the control volume indicated by dashed lines in Figure 4.2.

Figure 4.2 Steady one-dimensional flow into and out of a control volume.

Equation (4-2) applies. As mass crosses the control surface at positions (1) and (2) only, our expression is

$$\iint_{c.s.} \rho(\mathbf{v}\cdot\mathbf{n})\,dA = \iint_{A_1} \rho(\mathbf{v}\cdot\mathbf{n})\,dA + \iint_{A_2} \rho(\mathbf{v}\cdot\mathbf{n})\,dA = 0$$

The absolute value of the scalar product $(\mathbf{v}\cdot\mathbf{n})$ is equal to the magnitude of the velocity in each integral, as the velocity vectors and outwardly directed normal vectors are collinear both at (1) and (2). At (2) these vectors have the same sense, thus this product is positive, as it should be for an efflux of mass. At (1), where mass flows into the control volume, the two vectors are opposite in sense, hence the sign is negative. We may now express the continuity equation in scalar form

$$\iint_{c.s.} \rho(\mathbf{v}\cdot\mathbf{n})\,dA = -\iint_{A_1} \rho v\,dA + \iint_{A_2} \rho v\,dA = 0$$

Integration gives the familiar result

$$\rho_1 v_1 A_1 = \rho_2 v_2 A_2 \tag{4-4}$$

In obtaining equation (4-4), it is noted that the flow situation inside the control volume was unspecified. In fact, this is the beauty of the control-volume approach; the flow inside the control volume can be analyzed from information (measurements) obtained on the surface of the control volume. The box-shaped control volume illustrated in Figure 4.2 is defined for analytical purposes; the actual system contained in this box could be as simple as a pipe or as complex as a propulsion system or a distillation tower.

In solving Example 1, we assumed a constant velocity at sections (1) and (2). This situation may be approached physically, but a more general case is one in which the velocity varies over the cross-sectional area.

EXAMPLE 2

Let us consider the case of an incompressible flow, for which the flow area is circular and the velocity profile is parabolic (see Figure 4.3), varying according to the expression

$$v = v_{max}\left[1 - \left(\frac{r}{R}\right)^2\right]$$

Figure 4.3 A parabolic velocity profile in a circular flow passage.

where v_{max} is the maximum velocity which exists at the center of the circular passage (i.e., at $r = 0$), and R is the radial distance to the inside surface of the circular area considered.

The above velocity-profile expression may be obtained experimentally. It will also be derived theoretically in Chapter 8 for the case of laminar flow in a circular conduit. This expression represents the velocity at a radial distance, r, from the center of the flow section. As the average velocity is of particular interest in engineering problems, we will now consider the means of obtaining the average velocity from this expression.

At the station where this velocity profile exists, the mass rate of flow is

$$(\rho v)_{avg} A = \iint_A \rho v \, dA$$

For the present case of incompressible flow, the density is constant. Solving for the average velocity, we have

$$
v_{avg} = \frac{1}{A} \iint_A v \, dA
$$

$$
= \frac{1}{\pi R^2} \int_0^{2\pi} \int_0^R v_{max} \left[1 - \left(\frac{r}{R}\right)^2 \right] r \, dr \, d\theta
$$

$$
= \frac{v_{max}}{2}
$$

In the previous examples, we were not concerned with the composition of the fluid streams. Equation (4-1) applies to fluid streams containing more than one constituent as well as to the individual constituents alone. This type application is common to chemical processes in particular. Our final example will use the law of conservation of mass for both the total mass and for a particular species, in this case, salt.

EXAMPLE 3

Let us now examine the situation illustrated in Figure 4.4. A tank initially contains 1000 kg of brine containing 10% salt by mass. An inlet stream of brine containing 20% salt by mass flows into the tank at a rate of 20 kg/min. The mixture in the tank is kept uniform by stirring. Brine is removed from the tank via an outlet pipe at a rate of 10 kg/min. Find the amount of salt in the tank at any time t, and the elapsed time when the amount of salt in the tank is 200 kg.

Figure 4.4 A mixing process.

We first apply equation (4-1) to express the total amount of brine in the tank as a function of time. For the control volume shown

$$\iint_{c.s.} \rho(\mathbf{v} \cdot \mathbf{n}) \, dA = 10 - 20 = -10 \, \text{kg/min}$$

$$\frac{\partial}{\partial t} \iiint_{c.v.} \rho \, dV = \frac{d}{dt} \int_{1000}^{M} dM = \frac{d}{dt}(M - 1000)$$

where M is the total mass of brine in the tank at any time. Writing the complete expression, we have

$$\iint_{c.s.} \rho(\mathbf{v} \cdot \mathbf{n}) \, dA + \frac{\partial}{\partial t} \iiint_{c.v.} \rho \, dV = -10 + \frac{d}{dt}(M - 1000) = 0$$

Separating variables and solving for M gives

$$M = 1000 + 10t \quad (\text{kg})$$

We now let S be the amount of salt in the tank at any time. The concentration by weight of salt may be expressed as

$$\frac{S}{M} = \frac{S}{1000 + 10t} \quad \frac{\text{kg salt}}{\text{kg brine}}$$

Using this definition, we may now apply equation (4-1) to the salt, obtaining

$$\iint_{c.s.} \rho(\mathbf{v} \cdot \mathbf{n}) \, dA = \frac{10S}{1000 + 10t} - (0.2)(20) \quad \frac{\text{kg salt}}{\text{min}}$$

and

$$\frac{\partial}{\partial t} \iiint_{c.v.} \rho \, dV = \frac{d}{dt} \int_{S_0}^{S} dS = \frac{dS}{dt} \quad \frac{\text{kg salt}}{\text{min}}$$

The complete expression is now

$$\iint_{c.s.} \rho(\mathbf{v} \cdot \mathbf{n}) \, dA + \frac{\partial}{\partial t} \iiint_{c.v.} \rho \, dV = \frac{S}{100 + t} - 4 + \frac{dS}{dt} = 0$$

This equation may be written in the form

$$\frac{dS}{dt} + \frac{S}{100 + t} = 4$$

which we observe to be a first-order linear differential equation. The general solution is

$$S = \frac{2t(200 + t)}{100 + t} + \frac{C}{100 + t}$$

The constant of integration may be evaluated, using the initial condition that $S = 100$ at $t = 0$ to give $C = 10{,}000$. Thus the first part of the answer, expressing the amount of salt present as a function of time, is

$$S = \frac{10\,000 + 400t + 2t^2}{100 + t}$$

The elapsed time necessary for S to equal 200 kg may be evaluated to give $t = 36.6$ min.

4.3 CLOSURE

In this chapter we have considered the first of the fundamental laws of fluid flow: conservation of mass. The integral expression developed for this case was seen to be quite general in its form and use.

Similar integral expressions for conservation of energy and of momentum for a general control volume will be developed and used in subsequent chapters. The student should now develop the habit of *always* starting with the applicable integral expression and evaluating each term for a particular problem. There will be a strong temptation simply to write down an equation without considering each term in detail. Such temptations should be overcome. This approach may seem needlessly tedious at the outset, but it will always ensure a complete analysis of a problem and circumvent any errors that may otherwise result from a too-hasty consideration.

PROBLEMS

4.1 The velocity vector in a two-dimensional flow is given by the expression $\mathbf{v} = 10\mathbf{e}_x + 7x\mathbf{e}_y$ m/s when x is measured in meters. Determine the component of the velocity that makes a $-30°$ angle with the x axis at the point $(2, 2)$.

4.2 Using the velocity vector of the previous problem, determine (a) the equation of the streamline passing through point $(2, 1)$; (b) the volume of flow that crosses a plane surface connecting points $(1, 0)$ and $(2, 2)$.

4.3 Water is flowing through a large circular conduit with a velocity profile given by the equation $v = 9(1 - r^2/16)$ fps. What is the average water velocity in the 1.5-ft pipe?

4.4 Water enters a 4-in. square channel as shown at a velocity of 10 fps. The channel converges to a 2-in. square configuration as shown at the discharge end. The outlet section is cut at 30° to the vertical as shown, but the mean velocity of the discharging water remains horizontal. Find the exiting water's average velocity and total rate of flow.

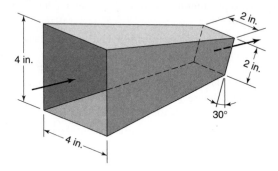

4.5 Water enters one end of a perforated pipe 0.2 m in diameter with a velocity of 6 m/s. The discharge through the pipe wall is approximated by a linear profile. If the flow is steady, find the discharge velocity.

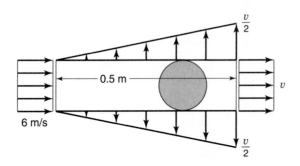

4.6 The velocities in a circular duct of 20-in. diameter were measured as follows:

Distance from center (in.)	Velocity (fps)	Distance from center (in.)	Velocity (fps)
0	7.5	7.75	5.47
3.16	7.10	8.37	5.10
4.45	6.75	8.94	4.50
5.48	6.42	9.49	3.82
6.33	6.15	10.00	2.40
7.07	5.81	...	...

Find (a) the average velocity; (b) the flow rate in cubic feet per second.

4.7 Salt water containing 1.92 lb/gal of salt flows at a fixed rate of 2 gal/min into a 100-gal tank, initially filled with fresh water. The density of the incoming solution is 71.8 lb/ft^3. The solution, kept uniform by stirring, flows out at a fixed rate of 19.2 lb/min.

(a) How many pounds of salt will there be in the tank at the end of 1 h and 40 min?

(b) What is the upper limit for the number of pounds of salt in the tank if the process continues indefinitely?

(c) How much time will elapse while the quantity of salt in the tank changes from 100 to 150 lb?

4.8 In the piston and cylinder arrangement shown below, the large piston has a velocity of 2 fps and an acceleration of 5 fps^2. Determine the velocity and acceleration of the smaller piston.

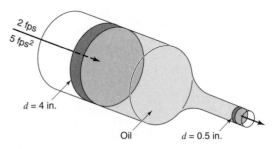

4.9 Show that in a one-dimensional steady flow the following equation is valid:

$$\frac{dA}{A} + \frac{dv}{v} + \frac{d\rho}{\rho} = 0$$

4.10 Using the symbol M for the mass in the control volume, show that equation (4-6) may be written

$$\frac{\partial M}{\partial t} + \iint_{c.s.} d\dot{m} = 0$$

4.11 A shock wave moves down a pipe as shown below. The fluid properties change across the shock, but they are not functions of time. The velocity of the shock is v_w. Write the continuity equation and obtain the relation between ρ_2, ρ_1, v_2, and v_w. The mass in the control volume at any time is $M = \rho_2 Ax + \rho_1 Ay$. Hint: Use a control volume that is moving to the right at velocity, v_w.

4.12 The velocity profile in circular pipe is given by $v = v_{max}(1 - r/R)^{1/7}$, where R is the radius of the pipe. Find the average velocity in the pipe in terms of v_{max}.

4.13 In the figure below, the x-direction velocity profiles are shown for a control volume surrounding a cylinder. If the flow is incompressible, what must the rate of flow be across the horizontal control-volume surface?

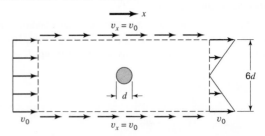

4.14 Two very long parallel plates of length $2L$ are separated a distance b. The upper plate moves downward at a constant rate V. A fluid fills the space between the plates. Fluid is squeezed out between the plates. Determine the mass flow rate and maximum velocity:

(a) If the exit velocity is uniform.
(b) If the exit velocity is parabolic.

4.15 A thin layer of liquid, draining from an inclined plane, has a velocity profile $v_x \approx v_0(2y/h - y^2/h^2)$, where v_0 is the surface velocity. If the plane has width 10 cm into the paper, determine the volume rate of flow in the film. Suppose that $h = 2$ cm and the flow rate is 2 L/min. Estimate v_0.

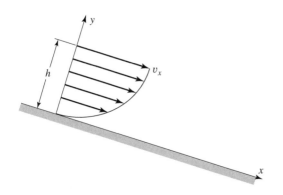

4.16 The V-shaped tank has width w into the paper and is filled from the inlet pipe at volume flow Q. Derive expressions for (a) the rate of change dh/dt and (b) the time required for the surface to rise from h_1 to h_2.

4.17 A bellows may be modeled as a deforming wedge-shaped volume. The check valve on the left (pleated) end is closed during the stroke. If w is the bellows width into the paper, derive an expression for outlet mass flow $\dot{m}_0$ as a function of stroke $\theta(t)$.

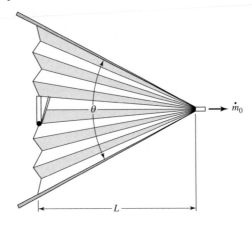

4.18 Water flows steadily through the piping junction, entering section 1 at 0.0013 m³/s. The average velocity at section 2 is 2.1 m/s. A portion of the flow is diverted through the showerhead, which contains 100 holes of 1-mm diameter. Assuming uniform shower flow, estimate the exit velocity from the showerhead jets.

4.19 The jet pump injects water at $V_1 = 40$ m/s through a 7.6 cm pipe and entrains a secondary flow of water $V_2 = 3$ m/s in the annular region around the small pipe. The two flows become fully mixed downstream, where V_3 is approximately constant. For steady incompressible flow, compute V_3.

4.20 A vertical, cylindrical tank closed at the bottom is partially filled with an incompressible liquid. A cylindrical rod of diameter d_i (less than tank diameter, d_0) is lowered into the liquid at a velocity V. Determine the average velocity of the fluid escaping between the rod and the tank walls (a) relative to the bottom of the tank and (b) relative to the advancing rod.

4.21   The hypodermic needle shown below contains a liquid serum ($\rho = 1$ g/cm^3). If the serum is to be injected steadily at 6 cm^3/s, how fast should the plunger be advanced (a) if leakage in the plunger clearance is neglected and (b) if leakage is 10% of the needle flow?

4.22   Incompressible steady flow in the inlet between parallel plates is uniform, $V_0 = 8$ cm/s, while downstream, the flow develops into the parabolic profile $v_x = az(z_0 - z)$, where a is a constant. What is the maximum value of v_x?

4.23   An incompressible fluid flows past a flat plate, as in the figure below, with a uniform inlet profile and a polynomial exit profile

$$v_x = v_0 \left(\frac{3\eta - \eta^3}{2} \right) \text{ where } \eta = \frac{y}{\delta}$$

Compute the volume flow Q across the top surface of the control volume. The plate has width b into the paper.

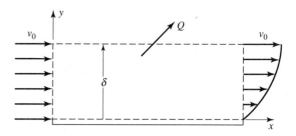

4.24   Rework Problem 4.14 if the plates are circular and have radius L.

Chapter 5

Newton's Second Law of Motion: Control-Volume Approach

The second of the fundamental physical laws upon which fluid-flow analyses are based is Newton's second law of motion. Starting with Newton's second law, we shall develop integral relations for linear and angular momentum. Applications of these expressions to physical situations will be considered.

5.1 INTEGRAL RELATION FOR LINEAR MOMENTUM

Newton's second law of motion may be stated as follows:

> *The time rate of change of momentum of a system is equal to the net force acting on the system and takes place in the direction of the net force.*

We note at the outset two very important parts of this statement: first, this law refers to a specific system, and second, it includes direction as well as magnitude and is therefore a vector expression. In order to use this law, it will be necessary to recast its statement into a form applicable to control volume which contains different fluid particles (i.e., a different system) when examined at different times.

In Figure 5.1, observe the control volume located in a fluid-flow field. The system considered is the material occupying the control volume at time t, and its position is shown both at time t and at time $t + \Delta t$.

Referring to the figure, we see that

Region I is occupied by the system only at time t.

Region II is occupied by the system at $t + \Delta t$.

Region III is common to the system both at t and at $t + \Delta t$.

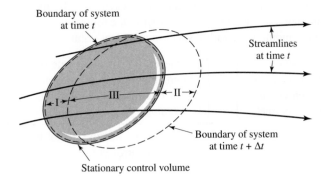

Figure 5.1 Relation between a system and a control volume in a fluid-flow field.

Writing Newton's second law for such a situation, we have

$$\Sigma\mathbf{F} = \frac{d}{dt}(m\mathbf{v}) = \frac{d}{dt}\mathbf{P} \tag{5-1}$$

where the symbols $\mathbf{F}$, m, and $\mathbf{v}$ have their usual meanings and $\mathbf{P}$ represents the total linear momentum of the system.

At time $t + \Delta t$, the linear momentum of the system that now occupies regions II and III may be expressed as

$$\mathbf{P}|_{t+\Delta t} = \mathbf{P_{II}}|_{t+\Delta t} + \mathbf{P_{III}}|_{t+\Delta t}$$

and at time t we have

$$\mathbf{P}|_t = \mathbf{P_I}|_t + \mathbf{P_{III}}|_t$$

Subtracting the second of these expressions from the first and dividing by the time interval Δt gives

$$\frac{\mathbf{P}|_{t+\Delta t} - \mathbf{P}|_t}{\Delta t} = \frac{\mathbf{P_{II}}|_{t+\Delta t} + \mathbf{P_{III}}|_{t+\Delta t} - \mathbf{P_I}|_t - \mathbf{P_{III}}|_t}{\Delta t}$$

We may rearrange the right-hand side of this expression and take the limit of the resulting equation to get

$$\lim_{\Delta t\to 0}\frac{\mathbf{P}|_{t+\Delta t} - \mathbf{P}|_t}{\Delta t} = \lim_{\Delta t\to 0}\frac{\mathbf{P_{III}}|_{t+\Delta t} - \mathbf{P_{III}}|_t}{\Delta t} + \lim_{\Delta t\to 0}\frac{\mathbf{P_{II}}|_{t+\Delta t} - \mathbf{P_I}|_t}{\Delta t} \tag{5-2}$$

Considering each of the limiting processes separately, we have, for the left-hand side

$$\lim_{\Delta t\to 0}\frac{\mathbf{P}|_{t+\Delta t} - \mathbf{P}|_t}{\Delta t} = \frac{d}{dt}\mathbf{P}$$

which is the form specified in the statement of Newton's second law, equation (5-1).

The first limit on the right-hand side of equation (5-2) may be evaluated as

$$\lim_{\Delta t\to 0}\frac{\mathbf{P_{III}}|_{t+\Delta t} - \mathbf{P_{III}}|_t}{\Delta t} = \frac{d}{dt}\mathbf{P_{III}}$$

This we see to be the rate of change of linear momentum of the control volume itself, since, as $\Delta t\to 0$, region III becomes the control volume.

The next limiting process

$$\lim_{\Delta t\to 0}\frac{\mathbf{P_{II}}|_{t+\Delta t} - \mathbf{P_I}|_t}{\Delta t}$$

expresses the net rate of momentum efflux across the control surface during the time interval Δt. As Δt approaches zero, regions I and II become coincident with the control-volume surface.

Considering the physical meaning of each of the limits in equation (5-2) and Newton's second law, equation (5-1), we may write the following word equation for the conservation of linear momentum with respect to a control volume:

$$\left\{\begin{array}{c}\text{sum of}\\\text{forces acting}\\\text{on control}\\\text{volume}\end{array}\right\} = \underbrace{\left\{\begin{array}{c}\text{rate of}\\\text{momentum}\\\text{out of control}\\\text{volume}\end{array}\right\} - \left\{\begin{array}{c}\text{rate of}\\\text{momentum}\\\text{into control}\\\text{volume}\end{array}\right\}}_{\substack{\text{net rate of momentum efflux from}\\\text{control volume}}} + \left\{\begin{array}{c}\text{rate of}\\\text{accumulation}\\\text{of momentum}\\\text{within control}\\\text{volume}\end{array}\right\}$$

$$\tag{5-3}$$

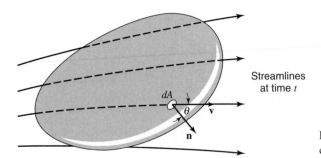

Figure 5.2 Fluid flow through a control volume.

We shall now apply equation (5-3) to a general control volume located in a fluid-flow field as shown in Figure 5.2 and evaluate the various terms.

The total force acting on the control volume consists both of surface forces due to interactions between the control-volume fluid, and its surroundings through direct contact, and of body forces resulting from the location of the control volume in a force field. The gravitational field and its resultant force are the most common examples of this latter type. We will designate the total force acting on the control volume as $\Sigma\mathbf{F}$.

If the small area dA on the control surface is considered, we may write

$$\text{rate of momentum efflux} = \mathbf{v}(\rho v)(dA \cos\theta)$$

Observe that the product $(\rho v)(dA \cos\theta)$ is the rate of mass efflux from the control volume through dA, as discussed in Chapter 4. Recall further that $dA \cos\theta$ is the area, dA projected in a direction normal to the velocity vector, $\mathbf{v}$, where θ is the angle between $\mathbf{v}$ and the outwardly directed normal vector, $\mathbf{n}$. We may then multiply the rate of mass efflux by $\mathbf{v}$ to give the rate of momentum efflux through dA. From vector algebra this product may be written as

$$\mathbf{v}(\rho v)(dA \cos\theta) = \mathbf{v}(\rho\,dA)[|\mathbf{v}|\,|\mathbf{n}|\cos\theta]$$

The term in square brackets is the scalar or dot product, $\mathbf{v}\cdot\mathbf{n}$ and the momentum efflux term becomes

$$\rho\mathbf{v}(\mathbf{v}\cdot\mathbf{n})\,dA$$

Integrating this quantity over the entire control surface, we have

$$\iint_{\text{c.s.}} \mathbf{v}\rho(\mathbf{v}\cdot\mathbf{n})\,dA$$

which is the *net momentum efflux* from the control volume.

In its integral form the momentum flux term stated above includes the rate of momentum entering the control volume as well as that leaving. If mass is entering the control volume, the sign of the product $\mathbf{v}\cdot\mathbf{n}$ is negative, and the associated momentum flux is an input. Conversely, a positive sign of the product $\mathbf{v}\cdot\mathbf{n}$ is associated with a momentum efflux from the control volume. Thus, the first two terms on the right-hand side of equation (5-3) may be written

$$\left\{ \begin{array}{c} \text{rate of momentum} \\ \text{out of control} \\ \text{volume} \end{array} \right\} - \left\{ \begin{array}{c} \text{rate of momentum} \\ \text{into control} \\ \text{volume} \end{array} \right\} = \iint_{\text{c.s.}} \mathbf{v}\rho(\mathbf{v}\cdot\mathbf{n})\,dA$$

The rate of accumulation of linear momentum within the control volume may be expressed as

$$\frac{\partial}{\partial t} \iiint_{\text{c.v.}} \mathbf{v}\rho \, dV$$

and the overall linear-momentum balance for a control volume becomes

$$\sum \mathbf{F} = \iint_{\text{c.s.}} \mathbf{v}\rho(\mathbf{v} \cdot \mathbf{n}) \, dA + \frac{\partial}{\partial t} \iiint_{\text{c.v.}} \rho \mathbf{v} \, dV \qquad (5\text{-}4)$$

This extremely important relation is often referred to in fluid mechanics as the *momentum theorem*. Note the great similarity between (5-4) and (4-1) in the form of the integral terms; observe, however, that equation (5-4) is a vector expression opposed to the scalar form of the overall mass balance considered in Chapter 4. In rectangular coordinates the single-vector equation, (5-4), may be written as three scalar equations

$$\sum F_x = \iint_{\text{c.s.}} v_x \rho(\mathbf{v} \cdot \mathbf{n}) \, dA + \frac{\partial}{\partial t} \iiint_{\text{c.v.}} \rho v_x \, dV \qquad (5\text{-}5a)$$

$$\sum F_y = \iint_{\text{c.s.}} v_y \rho(\mathbf{v} \cdot \mathbf{n}) \, dA + \frac{\partial}{\partial t} \iiint_{\text{c.v.}} \rho v_y \, dV \qquad (5\text{-}5b)$$

$$\sum F_z = \iint_{\text{c.s.}} v_z \rho(\mathbf{v} \cdot \mathbf{n}) \, dA + \frac{\partial}{\partial t} \iiint_{\text{c.v.}} \rho v_z \, dV \qquad (5\text{-}5c)$$

When applying any or all of the above equations, it must be remembered that each term has a sign with respect to the positively defined x, y, and z directions. The determination of the sign of the surface integral should be considered with special care, as both the velocity component (v_x) and the scalar product ($\mathbf{v} \cdot \mathbf{n}$) have signs. The combination of the proper sign associated with each of these terms will give the correct sense to the integral. It should also be remembered that as equations (5-5a–c) are written for the fluid in the control volume, *the forces to be employed in these equations are those acting on the fluid.*

A detailed study of the example problems to follow should aid in the understanding of, and afford facility in using, the overall momentum balance.

5.2 APPLICATIONS OF THE INTEGRAL EXPRESSION FOR LINEAR MOMENTUM

In applying equation (5-4), it is first necessary to define the control volume that will make possible the simplest and most direct solution to the problem at hand. There are no general rules to aid in this definition, but experience in handling problems of this type will enable such a choice to be made readily.

EXAMPLE 1

Consider first the problem of finding the force exerted on a reducing pipe bend resulting from a steady flow of fluid in it. A diagram of the pipe bend and the quantities significant to its analysis are shown in Figure 5.3.

The first step is the definition of the control volume. One choice for the control volume, of the several available, is all fluid in the pipe at a given time. The control volume chosen in this manner is designated in Figure 5.4, showing the external forces imposed upon it. The external

Figure 5.3 Flow in a reducing pipe bend.

forces imposed on the fluid include the pressure forces at sections (1) and (2), the body force due to the weight of fluid in the control volume, and the forces due to pressure and shear stress, P_w and τ_w, exerted on the fluid by the pipe wall. The resultant force on the fluid (due to P_w and τ_w) by the pipe is symbolized as **B**, and its x and y components as B_x and B_y, respectively.

Considering the x- and y-directional component equations, (5-5a) and (5-5b), of the overall momentum balance, the external forces acting on the fluid in the control volume are

$$\Sigma F_x = P_1 A_1 - P_2 A_2 \cos\theta + B_x$$

and

$$\Sigma F_y = P_2 A_2 \sin\theta - W + B_y$$

Each component of the unknown force **B** is assumed to have a positive sense. The actual signs for these components, when a solution is obtained, will indicate whether or not this assumption is correct.

Evaluating the surface integral in both the x and y directions, we have

$$\iint_{c.s.} v_x \rho(\mathbf{v}\cdot\mathbf{n})\, dA = (v_2 \cos\theta)(\rho_2 v_2 A_2) + (v_1)(-\rho_1 v_1 A_1)$$

$$\iint_{c.s.} v_y \rho(\mathbf{v}\cdot\mathbf{n})\, dA = (-v_2 \sin\theta)(\rho_2\, v_2 A_2)$$

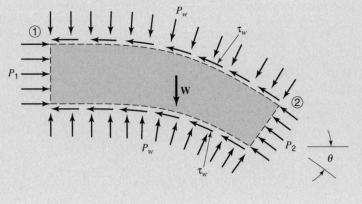

Figure 5.4 Control volume defined by pipe surface.

The complete momentum expressions in the x and y directions are

$$B_x + P_1A_1 - P_2A_2 \cos\theta = (v_2 \cos\theta)(\rho_2 v_2 A_2) + v_1(-\rho_1 v_1 A_1)$$

and

$$B_y + P_2A_2 \sin\theta - W = (-v_2 \sin\theta)(\rho_2 v_2 A_2)$$

Solving for the unknown force components B_x and B_y, we have

$$B_x = v_2^2 \rho_2 A_2 \cos\theta - v_1^2 \rho_1 A_1 - P_1A_1 + P_2A_2 \cos\theta$$

and

$$B_y = -v_2^2 \rho_2 A_2 \sin\theta - P_2A_2 \sin\theta + W$$

Recall that we were to evaluate the force exerted on the pipe rather than that on the fluid. The force sought is the reaction to **B** and has components equal in magnitude and opposite in sense to B_x and B_y. The components of the reaction force, **R**, exerted on the pipe are

$$R_x = v_2^2 \rho_2 A_2 \cos\theta + v_1^2 \rho_1 A_1 + P_1A_1 - P_2A_2 \cos\theta$$

and

$$R_y = v_2^2 \rho_2 A_2 \sin\theta + P_2A_2 \sin\theta - W$$

Some simplification in form may be achieved if the flow is steady. Applying equation (4-3), we have

$$\rho_1 v_1 A_1 = \rho_2 v_2 A_2 = \dot{m}$$

where $\dot{m}$ is the mass flow rate.

The final solution for the components of **R** may now be written as

$$R_x = \dot{m}(v_1 - v_2 \cos\theta) + P_1A_1 - P_2A_2 \cos\theta$$
$$R_y = \dot{m}v_2 \sin\theta + P_2A_2 \sin\theta - W$$

The control volume shown in Figure 5.4 for which the above solution was obtained represents only possible choice. Another is depicted in Figure 5.5. This control volume is bounded simply by the straight planes cutting through the pipe at sections (1) and (2). The fact that a control volume such as this can be used indicates the versatility of this approach, that is, that the results of complicated processes occurring internally may be analyzed quite simply by considering only those quantities of transfer across the control surface.

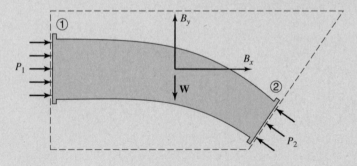

Figure 5.5 Control volume including fluid and pipe.

For this control volume, the x- and y-directional momentum equations are

$$B_x + P_1A_1 - P_2A_2 \cos\theta = (v_2 \cos\theta)(\rho_2 v_2 A_2) + v_1(-v_1\rho_1 A_1)$$

and

$$B_y + P_2A_2 \sin\theta - W = (-v_2 \sin\theta)(\rho_2 v_2 A_2)$$

where the force having components B_x and B_y is that exerted on the control volume by the section of pipe cut through at sections (1) and (2). The pressures at (1) and (2) in the above equations are gage pressures, as the atmospheric pressures acting on all surfaces cancel.

Note that the resulting equations for this control volume are identical to those obtained for the one defined previously. Thus, a correct solution may be obtained for each of several chosen control volumes so long as they are analyzed carefully and completely.

EXAMPLE 2

As our second example of the application of the control-volume expression for linear momentum (the momentum theorem), consider the steam locomotive tender schematically illustrated in Figure 5.6, which obtains water from a trough by means of a scoop. The force on the train due to the water is to be obtained.

Figure 5.6 Schematic of locomotive tender scooping water from a trough.

The logical choice for a control volume in this case is the water-tank/scoop combination. Our control-volume boundary will be selected as the *interior* of the tank and scoop. As the train is moving with a uniform velocity, there are two possible choices of coordinate systems. We may select a coordinate system either fixed in space or moving[1] with the velocity of the train, v_0. Let us first analyze the system by using a moving coordinate system.

The moving control volume is shown in Figure 5.7 with the xy coordinate system moving at velocity v_0. All velocities are determined with respect to the x and y axes.

Figure 5.7 Moving coordinate system and control volume.

---

[1] Recall that a uniformly translating coordinate system is an inertial coordinate system, hence Newton's second law and the momentum theorem may be employed directly.

The applicable expression is equation (5-5a)

$$\Sigma F_x = \iint_{c.s.} v_x \rho (\mathbf{v} \cdot \mathbf{n})\, dA + \frac{\partial}{\partial t} \iiint_{c.v.} v_x \rho\, dV$$

In Figure 5.7, ΣF_x is represented as F_x and is shown in the positive sense. As the forces due to pressure and shear are to be neglected, F_x is the total force exerted on the fluid by the train and scoop. The momentum flux term is

$$\iint_{c.s.} v_x \rho (\mathbf{v} \cdot \mathbf{n})\, dA = \rho(-v_0)(-1)(v_0)(h) \quad \text{(per unit length)}$$

and the rate of change of momentum within the control volume is zero, as the fluid in the control volume has zero velocity in the x direction.

Thus,

$$F_x = \rho v_0^2 h$$

This is the force exerted by the train on the fluid. The force exerted by the fluid on the train is the opposite of this, or $-\rho v_0^2 h$.

Now let us consider the same problem with a stationary coordinate system (see Figure 5.8). Employing once again the control-volume relation for linear momentum

$$\Sigma F_x = \iint_{c.s.} v_x \rho (\mathbf{v} \cdot \mathbf{n})\, dA + \frac{\partial}{\partial t} \iiint_{c.v.} v_x \rho\, dV$$

we obtain

$$F_x = 0 + \frac{\partial}{\partial t} \iiint_{c.v.} v_x \rho\, dV$$

where the momentum flux is zero, as the entering fluid has zero velocity. There is, of course, no fluid leaving the control volume. The terms $\partial/\partial t \iiint_{c.v.} v_x \rho\, dV$, as the velocity, $v_x = v_0 = $ constant, may be written as $v_0 \partial/\partial t \iiint_{c.v.} \rho\, dV$ or $v_0(\partial m/\partial t)$, where $\dot m$ is the mass of fluid entering the control volume at the rate $\partial m/\partial t = \rho v_0 h$ so that $F_x = \rho v_0^2 h$ as before.

Figure 5.8 Stationary coordinate system and moving control volume.

The student should note that, in the case of a stationary coordinate system and a moving control volume, care must be exercised in the interpretation of the momentum flux

$$\iint_{c.s.} \mathbf{v} \rho (\mathbf{v} \cdot \mathbf{n})\, dA$$

Regrouping the terms, we obtain

$$\iint_{c.s.} \mathbf{v} \rho (\mathbf{v} \cdot \mathbf{n})\, dA \equiv \iint_{c.s.} \mathbf{v}\, d\dot m$$

Thus, it is obvious that while $\mathbf{v}$ is the velocity relative to fixed coordinates, $\mathbf{v} \cdot \mathbf{n}$ is the velocity relative to the control-volume boundary.

EXAMPLE 3

A jet of fluid exits a nozzle and strikes a vertical plane surface as shown in Figure 5.9.

(a) Determine the force required to hole the plate stationary if the jet is composed of
   i. water
   ii. air.

(b) Determine the magnitude of the rest the restraining force for a water jet when the plate is moving to the right with a uniform velocity of 4 m/s.

Figure 5.9 A fluid jet striking a vertical plate.

The control volume to be used in this analysis is shown Figure 5.10.

The coordinates are fixed with the control volume which, for parts (a) and (b) of this example, is stationary.

Writing the x-directional scalar form of the momentum theroem, we have

$$\sum F_x = \iint_{c.s.} v_x \rho(\mathbf{v} \cdot \mathbf{n})\, dA + \frac{\partial}{\partial t} \iiint_{c.v.} v_x \rho\, dv$$

Evaluation of each term in this expression yields

$$\sum F_x = -F$$

$$\iint_{c.s.} v_x \rho(\mathbf{v} \cdot \mathbf{n})\, dA = v_j \rho(-v_j A_j)$$

$$\frac{\partial}{\partial t} \iiint_{c.v.} v_x \rho\, dv = 0$$

Figure 5.10 Control volume for Example 3.

and the governing equation is

$$F = \rho A_j v_j^2$$

We may now introduce the appropriate numerical values and solve for F. For case (a)
(i) $\rho_w = 1000\,\text{kg/m}^3$

$$F = (1000\,\text{kg/m}^3)(0.005\,\text{m}^2)(12\,\text{m/s})^2$$
$$= 720\,\text{N}$$

(ii) $\rho_w = 1.206\,\text{kg/m}^3$

$$F = (1.206\,\text{kg/m}^3)(0.005\,\text{m}^2)(12\,\text{m/s})^2$$
$$= 0.868\,\text{N}$$

For case (b), the same control volume will be used. In this case, however, the control volume and *the coordinate system* are moving to the right at a velocity of 4 m/s. from the perspective of an observer moving the control volume, the velocity of the incoming water jet is $(v_j - v_0) = 8$ m/s.

The x-directional component form of the momemtum theorem will yield the expression

$$F = \rho A_j (v_j - v_0)^2$$

Substitution of appropriate numerical values yields

$$F = (1000\,\text{kg/m}^3)(0.005\,\text{m}^2)(12 - 4\,\text{m/s})^2$$
$$= 320\,\text{N}$$

5.3 INTEGRAL RELATION FOR MOMENT OF MOMENTUM

The integral relation for the moment of momentum of a control volume is an extension of the considerations just made for linear momentum.

Starting with equation (5-1), which is a mathematical expression of Newton's second law of motion applied to a system of particles (Figure 5.11)

$$\sum \mathbf{F} = \frac{d}{dt}(m\mathbf{v}) = \frac{d}{dt}\,\mathbf{P} \tag{5-1}$$

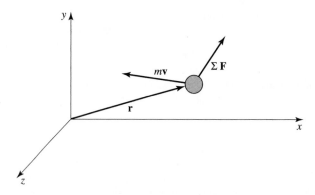

Figure 5.11 A system and its displacement vector *r*.

we take the vector or "cross" product of a position vector, **r**, with each term and get

$$\mathbf{r} \times \sum \mathbf{F} = \mathbf{r} \times \frac{d}{dt}(m\mathbf{v}) = \mathbf{r} \times \frac{d}{dt}\,\mathbf{P} \tag{5-6}$$

The quantity on the left-hand side of equation (5-6), $\mathbf{r} \times \sum \mathbf{F}$, is the resultant moment, $\sum \mathbf{M}$, about the origin as shown in Figure 5.11, due to all forces applied to the system. Clearly, we may write

$$\mathbf{r} \times \sum \mathbf{F} = \sum \mathbf{r} \times \mathbf{F} = \sum \mathbf{M}$$

where $\sum \mathbf{M}$ is, again, the total moment about the origin of all forces acting on the system.

The right-hand side of equation (5-6) is the moment of the time rate of change of linear momentum. This we can write as

$$\mathbf{r} \times \frac{d}{dt}m\mathbf{v} = \frac{d}{dt}(\mathbf{r} \times m\mathbf{v}) = \frac{d}{dt}(\mathbf{r} \times \mathbf{P}) = \frac{d}{dt}\,\mathbf{H}$$

Thus, this term is also the time rate of change of the moment of momentum of the system. We shall use the symbol **H** to designate moment of momentum. The complete expression is now

$$\sum \mathbf{M} = \frac{d}{dt}\,\mathbf{H} \tag{5-7}$$

As with its analogous expression for linear momentum, equation (5-1), equation (5-7) applies to a specific system. By the same limit process as that used for linear momentum, we may recast this expression into a form applicable to a control volume and achieve a word equation

$$
\left\{
\begin{array}{c}
\text{sum of}\\
\text{moments}\\
\text{acting on}\\
\text{control}\\
\text{volume}
\end{array}
\right\}
=
\underbrace{
\left\{
\begin{array}{c}
\text{rate of}\\
\text{moment of}\\
\text{momentum}\\
\text{out of control}\\
\text{volume}
\end{array}
\right\}
-
\left\{
\begin{array}{c}
\text{rate of}\\
\text{moment of}\\
\text{momentum into}\\
\text{control volume}
\end{array}
\right\}
}_{\substack{\text{net rate of efflux of moment of}\\ \text{momentum from control volume}}}
+
\left\{
\begin{array}{c}
\text{rate of}\\
\text{accumulation}\\
\text{of moment}\\
\text{of momen-}\\
\text{tum within}\\
\text{control}\\
\text{volume}
\end{array}
\right\}
$$

$$(5\text{-}8)$$

Equation (5-8) may be applied to a general control volume to yield the following equation:

$$
\Sigma \mathbf{M} = \iint_{\text{c.s.}} (\mathbf{r} \times \mathbf{v})\rho(\mathbf{v} \cdot \mathbf{n})\, dA + \frac{\partial}{\partial t} \iiint_{\text{c.v.}} (\mathbf{r} \times \mathbf{v})\rho\, dV
\qquad (5\text{-}9)
$$

The term on the left-hand side of equation (5-9) is the total moment of all forces acting on the control volume. The terms on the right-hand side represent the net rate of efflux of moment of momentum through the control surface and the rate of accumulation of moment of momentum within the control volume, respectively.

This single-vector equation may be expressed as three scalar equations for the orthogonal inertial coordinate directions x, y, and z as

$$
\Sigma M_x = \iint_{\text{c.s.}} (\mathbf{r} \times \mathbf{v})_x \rho(\mathbf{v} \cdot \mathbf{n})\, dA + \frac{\partial}{\partial t} \iiint_{\text{c.v.}} (\mathbf{r} \times \mathbf{v})_x \rho\, dV
\qquad (5\text{-}10\text{a})
$$

$$
\Sigma M_y = \iint_{\text{c.s.}} (\mathbf{r} \times \mathbf{v})_y \rho(\mathbf{v} \cdot \mathbf{n})\, dA + \frac{\partial}{\partial t} \iiint_{\text{c.v.}} (\mathbf{r} \times \mathbf{v})_y \rho\, dV
\qquad (5\text{-}10\text{b})
$$

and

$$
\Sigma M_z = \iint_{\text{c.s.}} (\mathbf{r} \times \mathbf{v})_z \rho(\mathbf{v} \cdot \mathbf{n})\, dA + \frac{\partial}{\partial t} \iiint_{\text{c.v.}} (\mathbf{r} \times \mathbf{v})_z \rho\, dV
\qquad (5\text{-}10\text{c})
$$

The directions associated with M_x and $(\mathbf{r} \times \mathbf{v})$ are those considered in mechanics in which the right-hand rule is used to determine the orientation of quantities having rotational sense.

5.4   APPLICATIONS TO PUMPS AND TURBINES

The moment-of-momentum expression is particularly applicable to two types of devices, generally classified as pumps and turbines. We shall, in this section, consider those having rotary motion only. If energy is derived from a fluid acting on a rotating device, it is designated a turbine, whereas a pump adds energy to a fluid. The rotating part of a turbine is called a runner and that of a pump an impeller.

The following two examples illustrate how moment-of-momentum analysis is used to generate expressions for evaluating turbine performance. Similar approaches will be used in Chapter 14 to evaluate operating characteristics of fans and pumps.

EXAMPLE 4
Let us first direct our attention to a type of turbine known as the Pelton wheel. Such a device is represented in Figure 5.12. In this turbine, a jet of fluid, usually water, is directed from a nozzle striking a system of buckets on the periphery of the runner. The buckets are shaped so that the water is diverted in such a way as to exert a force on the runner which will, in turn, cause rotation. Using the moment-of-momentum relation, we may determine the torque resulting from such a situation.

Bottom view of bucket

Figure 5.12 Pelton wheel.

Figure 5.13 Control volume for analysis of Pelton wheel.

We must initially define our control volume. The dashed lines in Figure 5.13 illustrates the control volume chosen. It encloses the entire runner and cuts the jet of water with velocity v_0 as shown. The control surface also cuts through the shaft on both sides of the runner.

The applicable scalar form of the general moment-of-momentum expression is equation (5-10c) written for the z direction. All rotation is in the xy plane, and—according to the right-hand rule—the vector representation of a quantity having angular motion, or a tendency to produce angular motion, has a sense normal to the xy plane, that is, the z direction. Recall that a positive angular sense is that conforming to the direction in which the thumb on the right hand will point when the fingers of the right hand are aligned with the direction of counterclockwise angular motion.

$$\sum M_z = \iint_{c.s.} (\mathbf{r} \times \mathbf{v})_z \rho(\mathbf{v} \cdot \mathbf{n}) \, dA + \frac{\partial}{\partial t} \iiint_{c.v.} \rho (\mathbf{r} \times \mathbf{v})_z \, dV$$

Evaluating each term separately, we have, for the external moment

$$\sum M_z = M_{shaft}$$

where M_{shaft}, the moment applied to the runner by the shaft, is the only such moment acting on the control volume.

The surface integral

$$\iint_{\text{c.s.}} (\mathbf{r} \times \mathbf{v})_z \rho(\mathbf{v} \cdot \mathbf{n}) \, dA$$

is the net rate of efflux of moment of momentum. The fluid leaving the control volume is illustrated in Figure 5.14. The x-direction component of the fluid leaving the control volume is

$$\{r\omega - (v_0 - r\omega)\cos\theta\}\mathbf{e}_x$$

Figure 5.14 Velocity vectors for turbine bucket.

Here it is assumed that the z components of the velocity are equal and opposite. The leaving velocity is the vector sum of the velocity of the turbine bucket, $r\omega$, and that of the exiting fluid relative to the bucket and leaving at an angle θ to the direction of motion of the bucket, $(v_0 - r\omega)\cos\theta$. These velocity vectors are shown in the figure. The final expression for the surface integral is now

$$\iint_{\text{c.s.}} (\mathbf{r} \times \mathbf{v})_z \rho(\mathbf{v} \cdot \mathbf{n}) \, dA = r[r\omega - (v_0 - r\omega)\cos\theta]\rho Q - rv_0\rho Q$$

The last term, $rv_0\rho Q$, is the moment of momentum of the incoming fluid stream of velocity v_0 and density ρ, with a volumetric flow rate Q.

As the problem under consideration is one in which the angular velocity, ω, of the wheel is constant, the term expressing the time derivative of moment of momentum of the control volume, $\partial/\partial t \iiint_{\text{c.v.}} (\mathbf{r} \times \mathbf{v})_z \rho \, dV = 0$. Replacing each term in the complete expression by its equivalent, we have

$$\Sigma M_z = M_{\text{shaft}} = \iint_{\text{c.s.}} (\mathbf{r} \times \mathbf{v})_z \rho(\mathbf{v} \cdot \mathbf{n}) \, dA + \frac{\partial}{\partial t} \iiint_{\text{c.v.}} \rho(\mathbf{r} \times \mathbf{v})_z \, dV$$

$$= r[r\omega - (v_0 - r\omega)\cos\theta]\rho Q - rv_0\rho Q = -r(v_0 - r\omega)(1 + \cos\theta)\rho Q$$

The torque applied to the shaft is equal in magnitude and opposite in sense to M_{shaft}. Thus our final result is

$$\text{Torque} = -M_{\text{shaft}} = r(v_0 - r\omega)(1 + \cos\theta)\rho Q$$

EXAMPLE 5

The radial-flow turbine illustrated in Figure 5.15 may be analyzed with the aid of the moment-of-momentum expression. In this device, the fluid (usually water) enters the guide vanes, which impart a tangential velocity and hence angular momentum to the fluid before it enters the revolving runner which reduces the angular momentum of the fluid while delivering torque to the runner.

Figure 5.15 Radial-flow turbine.

The control volume to be used is illustrated below in Figure 5.16. The outer boundary of the control volume is at radius r_1, and the inner boundary is at r_2. The width of the runner is h.

We will use equation (5-9) in order to determine the torque. For steady flow this equation becomes

$$\Sigma \mathbf{M} = \iint_{\text{c.s.}} (\mathbf{r} \times \mathbf{v}) \rho (\mathbf{v} \cdot \mathbf{n}) \, dA$$

Evaluating each term separately, we have, for the external moment of the runner on the fluid

$$\Sigma \mathbf{M} = M_{\text{fluid}} \mathbf{e}_z = -T \mathbf{e}_z$$

Figure 5.16 Radial-flow turbine-runner control volume.

where T is the shaft torque. The surface integral requires the evaluation of the vector product $(\mathbf{r} \times \mathbf{v})$ at the outer boundary r_1 and at the inner boundary r_2. If we express the velocity of the water in polar coordinates $v = v_r \mathbf{e}_r + v_\theta \mathbf{e}_\theta$, so that $(\mathbf{r} \times \mathbf{v}) = r \mathbf{e}_r \times (v_r \mathbf{e}_r + v_\theta \mathbf{e}_\theta) = r v_\theta \mathbf{e}_z$. Thus the surface integral, assuming uniform velocity distribution, is given by

$$\iint_{\text{c.s.}} (\mathbf{r} \times \mathbf{v}) \rho (\mathbf{v} \cdot \mathbf{n}) \, dA = \{ r_1 v_{\theta_1} \rho (-v_{r_1}) \, 2\pi r_1 h + r_2 v_{\theta_2} \rho v_{r_2} 2\pi r_2 h \} \mathbf{e}_z$$

The general result is

$$-T \mathbf{e}_z = \left(-\rho v_{r_1} v_{\theta_1} 2\pi r_1^2 h + \rho v_{r_2} v_{\theta_2} 2\pi r_2^2 h \right) \mathbf{e}_z$$

The law of conservation of mass may be used

$$\rho v_{r_1} 2\pi r_1 h = \dot{m} = \rho v_{r_2} 2\pi r_2 h$$

so that the torque is given by

$$T = \dot{m} (r_1 v_{\theta_1} - r_2 v_{\theta_2})$$

The velocity at r_1 is seen from Figures 5.15 and 5.16 to be determined by the flow rate and the guide vane angle α. The velocity at r_2, however, requires knowledge of flow conditions on the runner.

The velocity at r_2 may be determined by the following analysis. In Figure 5.17, the flow conditions at the outlet of the runner are sketched. The velocity of the water v_2 is the vector sum of the velocity with respect to the runner v_2' and the runner velocity $r_2 \omega$.

The velocity v_{θ_2}, the tangential velocity of the water leaving the runner, is given by

$$v_{\theta_2} = r_2\omega - v_0' \sin\beta$$

where β is the blade angle as shown. The fluid is assumed to flow in the same direction as the blade. The radial component of the flow may be determined from conservation of mass

$$v_{r_2} = v_2' \cos\beta = \frac{\dot{m}}{2\pi\rho r_2 h}$$

Thus,

$$T = \dot{m}\left(r_1 v_{\theta_1} - r_2\left[r_2\omega - \frac{\dot{m}\tan\beta}{2\pi\rho r_2 h}\right]\right)$$

In practice, the guide vanes are adjustable to make the relative velocity at the runner entrance tangent to the blades.

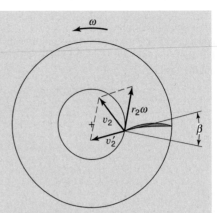

Figure 5.17 Velocity at runner exit (only one blade is shown).

5.5 CLOSURE

In this chapter, the basic relation involved has been Newton's second law of motion. This law, as written for a system, was recast so that it could apply to a control volume. The result of a consideration of a general control volume led to the integral equations for linear momentum, equation (5-4), and moment of momentum, equation (5-9). Equation (5-4) is often referred to as the momentum theorem of fluid mechanics. This equation is one of the most powerful and often-used expressions in this field.

The student is again urged to start always with the complete integral expression when working a problem. A term-by-term analysis from this basis will allow a correct solution, whereas in a hasty consideration certain terms might be evaluated incorrectly or neglected completely. As a final remark, it should be noted that the momentum theorem expression, as developed, applies to an inertial control volume only.

PROBLEMS

5.1 A two-dimensional object is placed in a 4-ft-wide water tunnel as shown. The upstream velocity, v_1, is uniform across the cross section. For the downstream velocity profile as shown, find the value of v_2.

$v_1 = 20$ fps

v_2

2 ft 4 ft

5.2 If, in the system for Problem 5.1, the total drag on the object is measured to be 800 N/m of length normal to the direction of flow, and frictional forces at the walls are neglected, find the pressure difference between inlet and outlet sections.

5.3 Rework Problem 5.1 if the exit velocity profile is given by

$$v = v_2\left(1 - \cos\frac{\pi y}{4}\right)$$

when y is measure vertically from the center line of the water tunnel.

5.4 A stationary jet engine is shown. Air with a density of 0.0805 lb_m/ft^3 enters as shown. The inlet and outlet cross-sectional areas are both 10.8 ft^2. The mass of fuel consumed is 2% of the mass of air entering the test section. For these conditions, calculate the thurst developed by the engine tested.

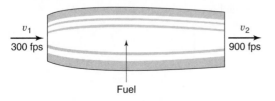

Fuel

5.5

(a) Determine the magnitude of the x and y components of the force exerted on the fixed blade shown by a $3\text{-ft}^3/\text{s}$ jet of water flowing at 25 fps.

(b) If the blade is moving to the right at 15 fps, find the magnitude and velocity of the water jet leaving the blade.

25 fps

30°

5.6 The pump in the boat shown pumps $6 \text{ ft}^3/\text{s}$ of water through a submerged water passage, which has an area of 0.25 ft^2 at the bow of the boat and 0.15 ft^2 at the stern. Determine the tension in the restraining rope, assuming that the inlet and exit pressures are equal.

30°

Pump

5.7 Oil (sp. gr. = 0.8) flows smoothly through the circular reducing section shown at $3 \text{ ft}^3/\text{s}$. If the entering and leaving velocity profiles are uniform, estimate the force which must be applied to the reducer to hold it in place.

$P = 50$ psig
$D = 12$ in.

$P = 5$ psig
$D = 2.5$ in.

5.8 At the end of a water pipe of 3-in. diameter is a nozzle that discharges a jet having a diameter of 1½ in. into the open atmosphere. The pressure in the pipe is 60 psig (pounds per square inch gage), and the rate of discharge is 400 gal/min. What are the magnitude and direction of the force necessary to hold the nozzle to the pipe?

5.9 A water jet pump has an area $A_j = 0.06 \text{ ft}^2$ and a jet velocity $v_j = 90$ fps, which entrains a secondary stream of water having a velocity $v_s = 10$ fps in a constant-area pipe of total area $A = 0.6 \text{ ft}^2$. At section 2, the water is thoroughly mixed. Assuming one-dimensional flow and neglecting wall shear

(a) find the average velocity of mixed flow at section 2;

(b) find the pressure rise $(P_2 - P_1)$, assuming the pressure of the jet and secondary stream to be the same at section 1.

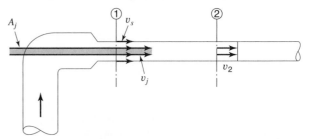

5.10 If the plate shown is inclined at an angle as shown, what are the forces F_x and F_y necessary to maintain its position? The flow is frictionless.

$v = 100$ fps

$\dot{m} = 2$ slugs/s

4
3

F_x

F_y

5.11 A steady, incompressible, frictionless, two-dimensional jet of fluid with breadth h, velocity v, and unit width impringes on a flat plate held at an angle α to its axis. Gravitational forces are to be neglected.

(a) Determine the total forceon the plate, and the breadths $a, b,$ of the two branches.

(b) Determine the distance 1 to the center of pressure (c.p.) along the plate from the point 0. (The center of pressure is the point at which the plate can be balanced without requiring an additional moment.)

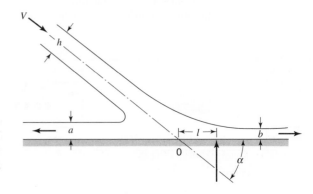

V

h

a

l

b

0

α

5.12 A plate moves perpendicularly toward a discharging jet at the rate of 5 fps. The jet discharges water at the rate of 3 ft³/s and a speed of 30 fps. Find the force of the fluid on the plate and compare it with what it would be if the plate were stationary. Assume frictionless flow.

5.13 The shock wave illustrated below is moving to the right at v_w fps. The properties in front and in back of the shock are not a function of time. By using the illustrated control volume, show that the pressure difference across the shock is

$$P_2 - P_1 = \rho_1 v_w v_2$$

5.14 If the shock-wave velocity in Problem 5.13 is approximated by the speed of sound, deterine the pressure change causing a velocity change of 10 fps in

(a) air at standard conditions;

(b) water.

5.15 Consider the differential control volume shown below. By applying the conservation of mass and the momentum theorem, show that

$$dP + \rho v\, dv + \mathbf{g}\, dy = 0$$

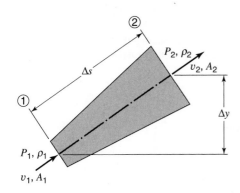

5.16 Water flows steadily through the horizontal 30° pipe bend shown below. At station 1, the diameter is 0.3 m, the velocity is 12 m/s, and the pressure is 128 kPa gage. At station

2, the diameter is 0.38 m and the pressure is 145 kPa gage. Determine the forces F_x and F_z necessary to hold the pipe bend stationary.

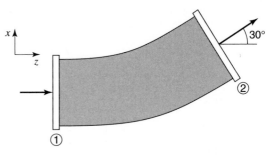

5.17 The rocket nozzle shown below consists of three welded sections. Determine the axial stress at junctions 1 and 2 when the rocket is operating at sea level. The mass flow rate is 770 lb$_m$/s.

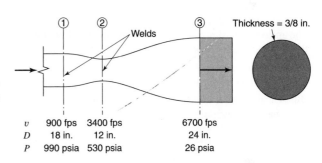

v	900 fps	3400 fps	6700 fps
D	18 in.	12 in.	24 in.
P	990 psia	530 psia	26 psia

5.18 The pressure on the control volume illustrated below is constant. The x components of velocity are as illustrated. Determine the force exerted on the cylinder by the fluid. Assume incompressible flow.

5.19 Water flows in a pipe at 3 m/s. A valve at the end of the pipe is suddenly closed. Determine the pressure rise in the pipe.

5.20 A dam discharge into a channel of constant width as shown. It is observed that a region of still water backs up behind the jet to a height H. The velocity and height of the flow in the channel are given as v and h, respectively, and the density of the water is ρ. Using the momentum theroem and the control surface indicted, determine H. Neglect the horizontal momentum of the flow that is entering the control volume from above and assume friction to be negligible. The aire pressure in the cavity below the crest of falling water is to be taken as atmospheric.

5.21 A liquid of density ρ flows through a sluice gate as shown. The upstream and downstream flows are uniform and parallel, so that the pressure variations at stations 1 and 2 may be considered hydrostatic.

(a) Determine the velocity at station 2.

(b) Determine the force per unit width, R, necessary to hold the sluice gate in place.

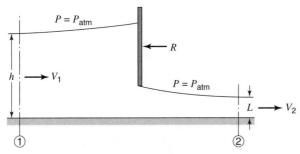

5.22 As can often be seen in a kitchen sink when the faucet is running, a high-speed channel flow (v_1, h_1) may "jump" to a low-speed, low-energy condition (v_2, h_2). The pressure at sections 1 and 2 is approximately hydrostatic, and wall friction is negligible. Use the continuity and momentum relations to find h_2 and v_2 in terms of (h_1, v_1).

5.23 For the pipe-flow-reducing section $D_1 = 8$ cm, $D_2 = 5$ cm, and $p_2 = 1$ atm. If $v_1 = 5$ m/s and the manometer reading

is $h = 58$ cm, estimate the total force resisted by the flange bolts.

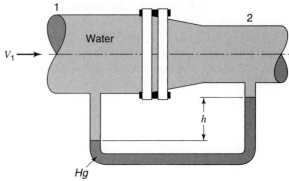

5.24 An open tank car as shown travels to the right at a uniform velocity of 4.5 m/s. At the instant shown the car passes under a jet of water issuing from a stationary 0.1-m-diameter pipe with a velocity of 20 m/s. What force is exerted on the tank by the water jet?

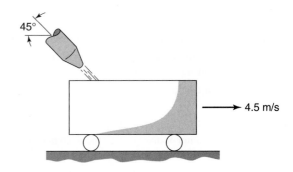

5.25 An open tank L ft long as shown below travels to the right at a velocity v_c fps. A jet of area A_j exhausts fluid of density ρ at a velocity v_j fps relative to the car. The tank car, at the same time, collects fluids from an overhead sprinkler which directs fluid downward with velocity v_s. Assuming that the sprinkler flow is uniform over the car area, A_c, determine the net force of the fluid on the tank car.

5.26 A liquid column of height h is confined in a vertical tube of cross-sectional area A by a stopper. At $t = 0$, the stopper is suddenly removed, exposing the bottom of the liquid to atmospheric pressure. Using a control-volume analysis of mass and vertical momentum, derive the differential equation for the downward motion $v(t)$ of the liquid. Assume one-dimensional, incompressible, frictionless flow.

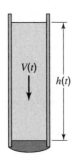

5.27 Sea water, $\rho = 64 \ \mathrm{lb_m/ft^3}$, flows through the impeller of a centrifugal pump at the rate of 800 gal/min. Determine the torque exerted on the impeller by the fluid and the power required to drive the pump. Assume that the absolute velocity of the water entering the impeller is radial. The dimensions are as follows:

$$\omega = 1180 \,\mathrm{rpm} \quad t_2 = 0.6 \,\mathrm{in.}$$
$$r_1 = 2 \,\mathrm{in.} \quad \theta_2 = 135°$$
$$r_2 = 8 \,\mathrm{in.} \quad t_1 = 0.8 \,\mathrm{in.}$$

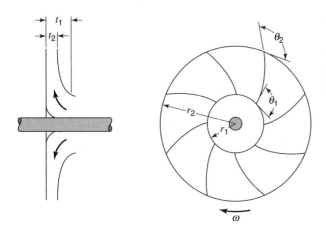

5.28 In Problem 5.27 determine

(a) the angle θ_1 such that the entering flow is parallel to the vanes;

(b) the axial load on the shaft if the shaft diameter is 1 in. and the pressure at the pump inlet is atmospheric.

5.29 A water sprinkler consists of two 1/2-in. diameter jets at the ends of a rotating hollow rod as shown. If the water

leaves at 20 fps, what torque would be necessary to hold the sprinkler in place?

5.30 A lawn sprinkler consists of two sections of curved pipe rotating about a vertical axis as shown. The sprinkler rotates with an angular velocity ω, and the effective discharge area is A, thus the water is discharged at a rate $Q = 2vv_rA$, where v_r is the velocity of the water relative to the rotating pipe. A constant friction torque M_f resists the motion of the sprinkler. Find an expression for the speed of the sprinkler in terms of the significant variables.

5.31 The pipe shown below has a slit of thickness 1/4 in. so shaped that a sheet of water of uniform thickness 1/4 in. issues out radially from the pipe. The velocity is constant along the pipe as shown and a flow rate of 8 ft³/s enters at the top. Find the moment on the tube about the axis BB from the flow of water inside the pipe system.

5.32 Water flows at 30 gal/min through the 0.75-in.-diameter double-pipe bend. The pressures are $p_1 = 30 \ \mathrm{lbf/in.^2}$ and

$p_2 = 24$ lbf/in.^{2}. Compute the torque T at point B necessary to keep the pipe from rotating.

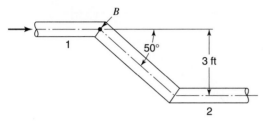

1   50°   3 ft

B

2

5.33 The illustration below shows a vane with a turning angle θ which moves with a steady speed v_c. The vane receives a jet that leaves a fixed nozzle with speed v.

(a) Assume that the vane is mounted on rails as shown in the sketch. Show that the power transmitted to the cart is maximum when $v_c/v = 1/3$

(b) Assuming that the are a large number of such vanes attached to a rotating wheel with peripheral speed, v_c, show that the power transmitted is maximum when $v_c/v = 1/2$.

v θ v_c

Chapter **6**

Conservation of Energy: Control-Volume Approach

The third fundamental law to be applied to fluid-flow analyses is the first law of thermodynamics. An integral expression for the conservation of energy applied to a control volume will be developed from the first law of thermodynamics, and examples of the application of the integral expression will be shown.

6.1 INTEGRAL RELATION FOR THE CONSERVATION OF ENERGY

The first law of thermodynamics may be stated as follows:

> *If a system is carried through a cycle, the total heat added to the system from its surroundings is proportional to the work done by the system on its surroundings.*

Note that this law is written for a specific group of particles—those comprising the defined system. The procedure will then be similar to that used in Chapter 5, that is, recasting this statement into a form applicable to a control volume which contains different fluid particles at different times. The statement of the first law of thermodynamics involves only scalar quantities however, and thus, unlike the momentum equations considered in Chapter 5, the equations resulting from the first law of thermodynamics will be scalar in form.

The statement of the first law given above may be written in equation form as

$$\oint \delta Q = \frac{1}{J} \oint \delta W \tag{6-1}$$

where the symbol $\oint$ refers to a "cyclic integral" or the integral of the quantity evaluated over a cycle. The symbols δQ and δW represent differential heat transfer and work done, respectively. The differential operator, δ, is used as both heat transfer and work are path functions and the evaluation of integrals of this type requires a knowledge of the path. The more familiar differential operator, d, is used with a "point" function. Thermodynamic properties are, by definition, point functions, and the integrals of such functions may be evaluated without a knowledge of the path by which the change in the property occurs between the initial and final states.[1] The quantity J is the so-called "mechanical equivalent of heat," numerically equal to 778.17 ft lb/Btu in engineering units. In the SI system, $J = 1$ N m/J. This factor will not be written henceforth, and the student is reminded that all equations must be dimensionally homogeneous.

---

[1] For a more complete discussion of properties, point functions and path fuctions, the reader is referred to G. N. Hatsopoulos and J. H. Keenan, *Principles of General Thermodynamics.* Wiley, New York, 1965, p. 14.

We now consider a general thermodynamic cycle, as shown in Figure 6.1. The cycle a occurs between points 1 and 2 by the paths indicated. Utilizing equation (6-1), we may write, for cycle a

$$
\int_{1a}^{2} \delta Q + \int_{2a}^{1} \delta Q \\
= \int_{1a}^{2} \delta W + \int_{2a}^{1} \delta W
$$

(6-2a)

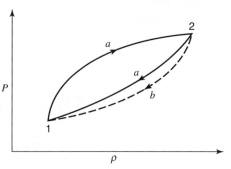

Figure 6.1 Reversible and irreversible thermodynamic cycles.

A new cycle between points 1 and 2 is postulated as follows: the path between points 1 and 2 is identical to that considered previously; however, the cycle is completed by path b between points 2 and 1, which is any path other than a between these points. Again equation (6-1) allows us to write

$$
\int_{1a}^{2} \delta Q + \int_{2b}^{1} \delta Q = \int_{1a}^{2} \delta W + \int_{2b}^{1} \delta W
$$

(6-2b)

Subtracting equation (6-2b) from equation (6-2a) gives

$$
\int_{2a}^{1} \delta Q - \int_{2b}^{1} \delta Q = \int_{2a}^{1} \delta W - \int_{2b}^{1} \delta W
$$

which may be written

$$
\int_{2a}^{1} (\delta Q - \delta W) = \int_{2b}^{1} (\delta Q - \delta W)
$$

(6-3)

As each side of equation (6-3) represents the integrand evaluated between the same two points but along different paths, it follows that the quantity, $\delta Q - \delta W$, is equal to a point function or a property. This property is designated dE, the total energy of the system. An alternate expression for the first law of thermodynamics may be written

$$
\delta Q - \delta W = dE
$$

(6-4)

The signs of δQ and δW were specified in the original statement of the first law; δQ is positive when heat is added to the system, δW is positive when work is done by the system.

For a system undergoing a process occurring in time interval dt, equation (6-4) may be written as

$$
\frac{\delta Q}{dt} - \frac{\delta W}{dt} = \frac{\delta E}{dt}
$$

(6-5)

Consider now, as in Chapter 5, a general control volume fixed in inertial space located in a fluid-flow field, as shown in Figure 6.2. The system under consideration, designated by dashed lines, occupies the control volume at time t, and its position is also shown after a period of time Δt has elapsed.

Figure 6.2 Relation between a system and a control volume in a fluid-flow field.

In this figure, region I is occupied by the system at time t, region II is occupied by the system at $t + \Delta t$, and region III is common to the system both at t and at $t + \Delta t$.

At time $t + \Delta t$ the total energy of the system may be expressed as

$$E|_{t+\Delta t} = E_{\mathrm{II}}|_{t+\Delta t} + E_{\mathrm{III}}|_{t+\Delta t}$$

and at time t

$$E|_t = E_{\mathrm{I}}|_t + E_{\mathrm{III}}|_t$$

Subtracting the second expression from the first and dividing by the elapsed time interval, Δt, we have

$$\frac{E|_{t+\Delta t} - E|_t}{\Delta t} = \frac{E_{\mathrm{III}}|_{t+\Delta t} + E_{\mathrm{II}}|_{t+\Delta t} - E_{\mathrm{III}}|_t - E_{\mathrm{I}}|_t}{\Delta t}$$

Rearranging and taking the limit as $\Delta t \rightarrow 0$ gives

$$\lim_{\Delta t \rightarrow 0} \frac{E|_{t+\Delta t} - E|_t}{\Delta t} = \lim_{\Delta t \rightarrow 0} \frac{E_{\mathrm{III}}|_{t+\Delta t} - E_{\mathrm{III}}|_t}{\Delta t} + \lim_{\Delta t \rightarrow 0} \frac{E_{\mathrm{II}}|_{t+\Delta t} - E_{\mathrm{I}}|_t}{\Delta t} \qquad (6\text{-}6)$$

Evaluating the limit of the left-hand side, we have

$$\lim_{\Delta t \rightarrow 0} \frac{E|_{t+\Delta t} - E|_t}{\Delta t} = \frac{dE}{dt}$$

which corresponds to the right-hand side of the first-law expression, equation (6-5).

On the right-hand side of equation (6-6) the first limit becomes

$$\lim_{\Delta t \rightarrow 0} \frac{E_{\mathrm{III}}|_{t+\Delta t} - E_{\mathrm{III}}|_t}{\Delta t} = \frac{dE_{\mathrm{III}}}{dt}$$

which is the rate of change of the total energy of the system, as the volume occupied by the system as $\Delta t \rightarrow 0$ is the control volume under consideration.

The second limit on the right of equation (6-6)

$$\lim_{\Delta t \rightarrow 0} \frac{E_{\mathrm{II}}|_{t+\Delta t} - E_{\mathrm{I}}|_t}{\Delta t}$$

represents the net rate of energy leaving across the control surface in the time interval Δt.

Having given physical meaning to each of the terms in equation (6-6), we may now recast the first law of thermodynamics into a form applicable to a control volume expressed by the following word equation:

$$
\left\{\begin{array}{c}\text{rate of addition}\\\text{of heat to control}\\\text{volume from}\\\text{its surroundings}\end{array}\right\} - \left\{\begin{array}{c}\text{rate of work done}\\\text{by control volume}\\\text{on its surroundings}\end{array}\right\} = \left\{\begin{array}{c}\text{rate of energy}\\\text{out of control}\\\text{volume due to}\\\text{fluid flow}\end{array}\right\}
$$
$$
- \left\{\begin{array}{c}\text{rate of energy into}\\\text{control volume due}\\\text{to fluid flow}\end{array}\right\} + \left\{\begin{array}{c}\text{rate of accumulation}\\\text{of energy within}\\\text{control volume}\end{array}\right\} \qquad (6\text{-}7)
$$

Equation (6-7) will now be applied to the general control volume shown in Figure 6.3.

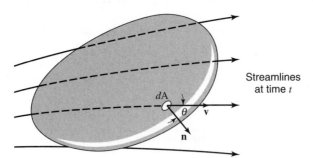

Streamlines at time t

Figure 6.3 Fluid flow through a control volume.

The rates of heat addition to and work done by the control volume will be expressed as $\delta Q/dt$ and $\delta W/dt$.

Consider now the small area dA on the control surface. The rate of energy leaving the control volume through dA may be expressed as

$$\text{rate of energy efflux} = e(\rho v)(dA \cos \theta)$$

The product $(\rho v)(dA \cos \theta)$ is the rate of mass efflux from the control volume through dA, as discussed in the previous chapters. The quantity e is the specific energy or the energy per unit mass. The specific energy includes the potential energy, gy, due to the position of the fluid continuum in the gravitational field; the kinetic energy of the fluid, $v^2/2$, due to its velocity; and the internal energy, u, of the fluid due to its thermal state.

The quantity $dA \cos \theta$ represents the area, dA, projected normal to the velocity vector, $\mathbf{v}$. Theta (θ) is the angle between $\mathbf{v}$ and the outwardly directed normal vector, $\mathbf{n}$. We may now write

$$e(\rho v)(dA \cos \theta) = e\rho \, dA[|\mathbf{v}| \, |\mathbf{n}|] \cos \theta = e\rho(\mathbf{v} \cdot \mathbf{n})dA$$

which we observe to be similar in form to the expressions previously obtained for mass and momentum. The integral of this quantity over the control surface

$$\iint_{\text{c.s.}} e\rho(\mathbf{v} \cdot \mathbf{n})dA$$

represents the *net efflux of energy* from the control volume. The sign of the scalar product, $\mathbf{v} \cdot \mathbf{n}$, accounts both for efflux and for influx of mass across the control surface as considered previously. Thus, the first two terms on the right-hand side of equation (6-7) may be evaluated as

$$\left\{ \begin{array}{c} \text{rate of energy} \\ \text{out of control} \\ \text{volume} \end{array} \right\} - \left\{ \begin{array}{c} \text{rate of energy} \\ \text{into control} \\ \text{volume} \end{array} \right\} = \iint_{\text{c.s.}} e\rho(\mathbf{v}\cdot\mathbf{n})dA$$

The rate of accumulation of energy within the control volume may be expressed as

$$\frac{\partial}{\partial t}\iiint_{\text{c.v.}} e\rho\, dV$$

Equation (6-7) may now be written as

$$\frac{\delta Q}{dt} - \frac{\delta W}{dt} = \iint_{\text{c.s.}} e\rho(\mathbf{v}\cdot\mathbf{n})\, dA + \frac{\partial}{\partial t}\iiint_{\text{c.v.}} e\rho\, dV \qquad (6\text{-}8)$$

A final form for the first-law expression may be obtained after further consideration of the work-rate or power term, $\delta W/dt$.

There are three types of work included in the work-rate term. The first is the shaft work, W_s, which is that done by the control volume on its surroundings that could cause a shaft to rotate or accomplish the raising of a weight through a distance. A second kind of work done is flow work, W_σ, which is that done on the surroundings to overcome normal stresses on the control surface where there is fluid flow. The third type of work is designated shear work, W_τ, which is performed on the surroundings to overcome shear stresses at the control surface.

Examining our control volume for flow and shear work rates, we have, as shown in Figure 6.4, another effect on the elemental portion of control surface, dA. Vector $\mathbf{S}$ is the force intensity (stress) having components σ_{ii} and τ_{ij} in the directions normal and tangential to the surface, respectively. In terms of $\mathbf{S}$, the force on dA is $\mathbf{S}\, dA$, and the rate of work done by the fluid flowing through dA is $\mathbf{S}\, dA\cdot\mathbf{v}$.

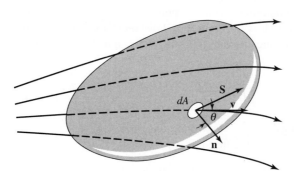

Figure 6.4 Flow and shear work for a general control volume.

The net rate of work done by the control volume on its surroundings due to the presence of $\mathbf{S}$ is

$$-\iint_{\text{c.s.}} \mathbf{v}\cdot\mathbf{S}\, dA$$

where the negative sign arises from the fact that the force per unit area *on the surroundings* is $-\mathbf{S}$.

The first-law expression, equation (6-8), may now be written as

$$\frac{\delta Q}{dt} - \frac{\delta W_s}{dt} + \iint_{c.s.} \mathbf{v} \cdot \mathbf{S} \, dA = \iint_{c.s.} e\rho(\mathbf{v} \cdot \mathbf{n}) \, dA + \frac{\partial}{\partial t} \iiint_{c.v.} e\rho \, dV \qquad (6\text{-}9)$$

where $\delta W_s/dt$ is the shaft work rate.

Writing the normal stress components of $\mathbf{S}$ as $\sigma_{ii}\mathbf{n}$, we obtain, for the net rate of work done in overcoming normal stress

$$\left(\iint_{c.s.} \mathbf{v} \cdot \mathbf{S} \, dA \right)_{normal} = \iint_{c.s.} \mathbf{v} \cdot \sigma_{ii} \mathbf{n} \, dA = \iint_{c.s.} \sigma_{ii}(\mathbf{v} \cdot \mathbf{n}) \, dA$$

The remaining part of the work to be evaluated is the part necessary to overcome shearing stresses. This portion of the required work rate, $\delta W_\tau/dt$, is transformed into a form that is unavailable to do mechanical work. This term, representing a loss of mechanical energy, is included in the derivative form given above and its analysis is included in Example 3, to follow. The work rate now becomes

$$\frac{\delta W}{dt} = \frac{\delta W_s}{dt} + \frac{\delta W_\sigma}{dt} + \frac{\delta W_\tau}{dt} = \frac{\delta W_s}{dt} - \iint_{c.s.} \sigma_{ii}(\mathbf{v} \cdot \mathbf{n}) \, dA + \frac{\delta W_\tau}{dt}$$

Substituting into equation (6-9), we have

$$\frac{\delta Q}{dt} - \frac{\delta W_s}{dt} + \iint_{c.s.} \sigma_{ii}(\mathbf{v} \cdot \mathbf{n}) dA + \frac{\delta W_\tau}{dt} = \iint_{c.s.} e\rho(\mathbf{v} \cdot \mathbf{n}) dA + \frac{\partial}{\partial t} \iiint_{c.v.} e\rho \, dV$$

The term involving normal stress must now be presented in a more usable form. A complete expression for σ_{ii} is stated in Chapter 9. For the present, we may say simply that the normal stress term is the sum of pressure effects and viscous effects. Just as with shear work, the work done to overcome the viscous portion of the normal stress is unavailable to do mechanical work. We shall thus combine the work associated with the viscous portion of the normal stress with the shear work to give a single term, $\delta W_\mu/dt$, the work rate accomplished in overcoming viscous effects at the control surface. The subscript, μ, is used to make this distinction.

The remaining part of the normal stress term, that associated with pressure, may be written in slightly different form if we recall that the bulk stress, σ_{ii}, is the negative of the thermodynamic pressure, P. The shear and flow work terms may now be written as follows:

$$\iint_{c.s.} \sigma_{ii}(\mathbf{v} \cdot \mathbf{n}) dA - \frac{\delta W_\tau}{dt} = -\iint_{c.s.} P(\mathbf{v} \cdot \mathbf{n}) \, dA - \frac{\delta W_\mu}{dt}$$

Combining this equation with the one written previously and rearranging slightly will yield the final form of the first-law expression:

$$\frac{\delta Q}{dt} - \frac{\delta W_s}{dt} = \iint_{c.s.} \left(e + \frac{P}{\rho} \right) \rho(\mathbf{v} \cdot \mathbf{n}) \, dA + \frac{\partial}{\partial t} \iiint_{c.v.} e\rho \, dV + \frac{\delta W_\mu}{dt} \qquad (6\text{-}10)$$

Equations (6-10), (4-1), and (5-4) constitute the basic relations for the analysis of fluid flow via the control-volume approach. A thorough understanding of these three equations and a mastery of their application places at the disposal of the student very powerful means of analyzing many commonly encountered problems in fluid flow.

The use of the overall energy balance will be illustrated in the following example problems.

6.2 APPLICATIONS OF THE INTEGRAL EXPRESSION

EXAMPLE 1

As a first example, let us choose a control volume as shown in Figure 6.5 under the conditions of steady fluid flow and no frictional losses.

For the specified conditions the overall energy expression, equation (6-10), becomes

$$\frac{\delta Q}{dt} - \frac{\delta W_s}{dt} = \iint_{c.s.} \rho\left(e + \frac{P}{\rho}\right)(\mathbf{v}\cdot\mathbf{n})dA + \underbrace{\frac{\partial}{\partial t}\iiint_{c.v.} e\rho\,dV}_{0-\text{steady flow}} + \underbrace{\frac{\delta W_\mu}{dt}}_{0}$$

Figure 6.5 Control volume with one-dimensional flow across boundaries.

Considering now the surface integral, we recognize the product $\rho(\mathbf{v}\cdot\mathbf{n})\,dA$ to be the mass flow rate with the sign of this product indicating whether mass flow is into or out of the control volume, dependent upon the sense of $\mathbf{v}\cdot\mathbf{n}$. The factor by which the mass-flow rate is multiplied, $e + P/\rho$, represents the types of energy that may enter or leave the control volume per mass of fluid. The specific total energy, e, may be expanded to include the kinetic, potential, and internal energy contributions, so that

$$e + \frac{P}{\rho} = gy + \frac{v^2}{2} + u + \frac{P}{\rho}$$

As mass enters the control volume only at section (1) and leaves at section (2), the surface integral becomes

$$\iint_{c.s.} \rho\left(e + \frac{P}{\rho}\right)(\mathbf{v}\cdot\mathbf{n})\,dA = \left[\frac{v_2^2}{2} + gy_2 + u_2 + \frac{P_2}{\rho_2}\right](\rho_2 v_2 A_2)$$
$$- \left[\frac{v_1^2}{2} + gy_1 + u_1 + \frac{P_1}{\rho_1}\right](\rho_1 v_1 A_1)$$

The energy expression for this example now becomes

$$\frac{\delta Q}{dt} - \frac{\delta W_s}{dt} = \left[\frac{v_2^2}{2} + gy_2 + u_2 + \frac{P_2}{\rho_2}\right](\rho_2 v_2 A_2) - \left[\frac{v_1^2}{2} + gy_1 + u_1 + \frac{P_1}{\rho_1}\right](\rho_1 v_1 A_1)$$

In Chapter 4, the mass balance for this same situation was found to be

$$\dot{m} = \rho_1 v_1 A_1 = \rho_1 v_2 A_2$$

If each term in the above expression is now divided by the mass flow rate, we have

$$\frac{q - \dot{W}_s}{\dot{m}} = \left[\frac{v_2^2}{2} + gy_2 + u_2 + \frac{P_2}{\rho_2}\right] - \left[\frac{v_1^2}{2} + gy_1 + u_1 + \frac{P_1}{\rho_1}\right]$$

or, in more familiar form

$$\frac{v_1^2}{2} + gy_1 + h_1 + \frac{q}{\dot{m}} = \frac{v_2^2}{2} + gy_2 + h_2 + \frac{\dot{W}_s}{\dot{m}}$$

where the sum of the internal energy and flow energy, $u + P/\rho$, has been replaced by the enthalpy, h, which is equal to the sum of these quantities by definition $h \equiv u + P/\rho$.

EXAMPLE 2

As a second example, consider the situation shown in Figure 6.6. If water flows under steady conditions in which the pump delivers 3 horsepower to the fluid, find the mass flow rate if frictional losses may be neglected.

Figure 6.6 A control volume for pump analysis.

Defining the control volume as shown by the dashed lines, we may evaluate equation (6-10) term by term as follows:

$$\frac{\delta Q}{dt} = 0$$

$$-\frac{\delta W_s}{dt} = (3\,\text{hp})(2545\,\text{Btu}\,/\,\text{hp--h})(778\,\text{ft--lb}_f\,/\,\text{Btu})(\text{h}\,/3600\,\text{s})$$

$$= 1650\,\text{ft}\,\text{lb}_f/\text{s}$$

$$\iint_{c.s.} \left(e + \frac{P}{\rho}\right)\rho(\mathbf{v}\cdot\mathbf{n})dA = \overset{-}{\iint_{A_2}}\left(e + \frac{P}{\rho}\right)\rho(\mathbf{v}\cdot\mathbf{n})dA + \iint_{A_1}\left(e + \frac{P}{\rho}\right)\rho(\mathbf{v}\cdot\mathbf{n})\,dA$$

$$= \left(\frac{v_2^2}{2} + gy_2 + u_2 + \frac{P_2}{\rho_2}\right)(\rho_2 v_2 A_2)$$

$$- \left(\frac{v_1^2}{2} + gy_1 + u_1 + \frac{P_1}{\rho_1}\right)(\rho_1 v_1 A_1)$$

$$= \left[\frac{v_2^2 - v_1^2}{2} + g(y_2 - y_1) + (u_2 - u_1) + \left(\frac{P_2}{\rho_2} - \frac{P_1}{\rho_1}\right)\right](\rho v A)$$

Here it may be noted that the pressure measured at station (1) is the static pressure while the pressure measured at station (2) is measured using a pressure port that is oriented normal to the oncoming flow, that is, where the velocity has been reduced to zero. Such a pressure is designated the *stagnation pressure*, which is greater than the static pressure by an amount equivalent to the change in kinetic energy of the flow. The stagnation pressure is, thus, expressed as

$$P_{\text{stagnation}} = P_0 = P_{\text{static}} + \frac{1}{2}\rho v^2$$

for incompressible flow, hence the energy flux term may be rewritten as

$$\iint_{c.s.}\left(e+\frac{P}{\rho}\right)\rho(\mathbf{v}\cdot\mathbf{n})\,dA=\left(\frac{P_{0_2}-P_1}{\rho}-\frac{v_1^2}{2}\right)(\rho vA)$$

$$=\left\{\frac{6(1-1/13.6)\,\text{in. Hg}(14.7\,\text{lb/in.}^2)(144\,\text{in.}^2/\text{ft}^2)}{(62.4\,\text{lb}_m/\text{ft}^3)(29.92\,\text{in. Hg})}\right.$$

$$\left.-\frac{v_1^2}{64.4(\text{lb}_m\,\text{ft}/\text{s}^2\,\text{lb}_f)}\right\}\{(62.4\,\text{lb}_m/\text{ft}^3)(v_1)(\pi/4\,\text{ft}^2)\}$$

$$=\left(6.30-\frac{v_1^2}{64.4}\right)(49v_1)\,\text{ft lb}_f/\text{s}$$

$$\frac{\partial}{\partial t}\iiint_{c.v.}e\rho\,dV=0$$

$$\frac{\delta W_\mu}{dt}=0$$

In the evaluation of the surface integral the choice of the control volume coincided with the location of the pressure taps at sections (1) and (2). The pressure sensed at section (1) is the static pressure, as the manometer opening is parallel to the fluid-flow direction. At section (2), however, the manometer opening is normal to the flowing fluid stream. The pressure measured by such an arrangement includes both the static fluid pressure and the pressure resulting as a fluid flowing with velocity v_2 is brought to rest. The sum of these two quantities is known as the impact or stagnation pressure.

The potential energy change is zero between sections (1) and (2) and as we consider the flow to be isothermal, the variation in internal energy is also zero. Hence, the surface integral reduces to the simple form indicated.

The flow rate of water necessary for the stated conditions to exist is achieved by solving the resulting cubic equation. The solution is

$$v_1=16.59\,\text{fps}\,(5.057\,\text{m/s})$$
$$\dot{m}=\rho Av=813\,\text{lb}_m/\text{s}\,(370\,\text{kg/s})$$

EXAMPLE 3

A shaft is rotating at constant angular velocity ω in the bearing shown in Figure 6.7. The shaft diameter is d and the shear stress acting on the shaft is τ. Find the rate at which energy must be removed from the bearing in order that the lubricating oil between the rotating shaft and the stationary bearing surface remains at constant temperature.

The shaft is assumed to be lightly loaded and concentric with the journal. The control volume selected consists of a unit length of the fluid surrounding the shaft as shown in Figure 6.7. The first law of thermodynamics for the control volume is

$$\frac{\delta Q}{dt}-\frac{\delta W_s}{dt}=\iint_{c.s.}\rho\left(e+\frac{P}{\rho}\right)(\mathbf{v}\cdot\mathbf{n})dA$$

$$+\frac{\partial}{\partial t}\iiint_{c.v.}\rho e\,dV+\frac{\delta W_\mu}{dt}$$

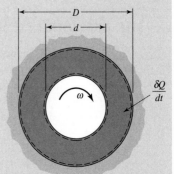

Figure 6.7 Bearing and control volume for bearing analysis.

From the figure we may observe the following:

1. No fluid crosses the control surface.

2. No shaft work crosses the control surface.

3. The flow is steady.

Thus $\delta Q/dt = \delta W_\mu/dt = \delta W_\tau/dt$. The viscous work rate must be determined. In this case all of the viscous work is done to overcome shearing stresses; thus the viscous work is $\iint_{c.s.} \tau(\mathbf{v} \cdot \mathbf{e}_t)\, dA$. At the outer boundary, $v = 0$ and at the inner boundary $\iint_{c.s.} \tau(\mathbf{v} \cdot \mathbf{e}_t)dA = -\tau(\omega d/2)A$, where $\mathbf{e}_t$ indicates the sense of the shear stress, τ, on the surroundings. The resulting sign is consistent with the concept of work being positive when done by a system on its surroundings. Thus,

$$\frac{\delta Q}{dt} = -\tau \frac{\omega d^2 \pi}{2}$$

which is the heat transfer rate required to maintain the oil at a constant temperature.

If energy is not removed from the system then $\delta Q/dt = 0$, and

$$\frac{\partial}{\partial t} \iiint_{c.v.} e\rho \, dV = -\frac{\delta W_\mu}{dt}$$

As only the internal energy of the oil will increase with respect to time

$$\frac{\partial}{\partial t} \iiint_{c.v.} e\rho \, dV = \rho\pi \left(\frac{D^2 - d^2}{4}\right) \frac{d\mu}{dt} = -\frac{\delta W_\mu}{dt} = \omega \frac{d^2 \pi}{2} \tau$$

or, with constant specific heat c

$$c\frac{dT}{dt} = \frac{2\tau\omega d^2}{\rho(D^2 - d^2)}$$

where D is the outer bearing diameter.

In this example the use of the viscous-work term has been illustrated. Note that

1. The viscous-work term involves only quantities on the surface of the control volume.

2. When the velocity on the surface of the control volume is zero, the viscous-work term is zero.

6.3 THE BERNOULLI EQUATION

Under certain flow conditions, the expression of the first law of thermodynamics applied to a control volume reduces to an extremely useful relation known as the Bernoulli equation.

If equation (6-10) is applied to a control volume as shown in Figure 6.8, in which flow is steady, incompressible, and inviscid, and in which no heat transfer or change in internal energy occurs, a term-by-term evaluation of equation (6-10) gives the following:

$$\frac{\delta Q}{dt} = 0$$

$$\frac{\delta W_s}{dt} = 0$$

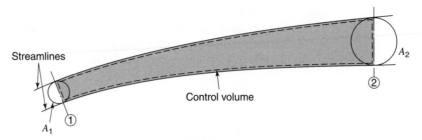

Figure 6.8 Control volume for steady, incompressible, inviscid, isothermal flow.

$$\iint_{\text{c.s.}} \rho \left(e + \frac{P}{\rho} \right) (\mathbf{v} \cdot \mathbf{n}) \, dA = \iint_{A_1} \rho \left(e + \frac{P}{\rho} \right) (\mathbf{v} \cdot \mathbf{n}) \, dA$$

$$+ \iint_{A_2} \rho \left(e + \frac{P}{\rho} \right) (\mathbf{v} \cdot \mathbf{n}) \, dA$$

$$= \left(gy_1 + \frac{v_1^2}{2} + \frac{P_1}{\rho_1} \right) (-\rho_1 v_1 A_1)$$

$$+ \left(gy_2 + \frac{v_2^2}{2} + \frac{P_2}{\rho_2} \right) (\rho_2 v_2 A_2)$$

$$\frac{\partial}{\partial t} \iiint_{\text{c.v.}} e\rho \, dV = 0$$

The first-law expression now becomes

$$0 = \left(gy_2 + \frac{v_2^2}{2} + \frac{P_2}{\rho} \right) (\rho v_2 A_2) - \left(gy_1 + \frac{v_1^2}{2} + \frac{P_1}{\rho} \right) (\rho v_1 A_1)$$

As flow is steady, the continuity equation gives

$$\rho_1 v_1 A_1 = \rho_2 v_2 A_2$$

which may be divided through to give

$$gy_1 + \frac{v_1^2}{2} + \frac{P_1}{\rho} = gy_2 + \frac{v_2^2}{2} + \frac{P_2}{\rho} \tag{6-11a}$$

Dividing through by g, we have

$$y_1 + \frac{v_1^2}{2g} + \frac{P_1}{\rho g} = y_2 + \frac{v_2^2}{2g} + \frac{P_2}{\rho g} \tag{6-11b}$$

Either of the above expressions is designated the Bernoulli equation.

Note that each term in equation (6-11b) has the unit of length. The quantities are often designated "heads" due to elevation, velocity, and pressure, respectively. These terms, both individually and collectively, indicate the quantities which may be directly converted to produce mechanical energy.

Equation (6-11) may be interpreted physically to mean that the total mechanical energy is conserved for a control volume satisfying the conditions upon which this relation is based, that is, steady, incompressible, inviscid, isothermal flow, with no heat transfer or work done. These conditions may seem overly restrictive, but they are met, or approached, in many physical systems. One such situation of practical value is for flow into and out of a stream tube. As stream tubes may vary in size, the Bernoulli equation can actually describe the variation in elevation, velocity, and pressure head from point-to-point in a fluid-flow field.

A classic example of the application of the Bernoulli equation is depicted in Figure 6.9, in which it is desired to find the velocity of the fluid exiting the tank as shown.

Figure 6.9 Control volume for Bernoulli equation analysis.

The control volume is defined as shown by dashed lines in the figure. The upper boundary of the control volume is just below the fluid surface, and thus can be considered to be at the same height as the fluid. There is fluid flow across this surface, but the surface area is large enough that the velocity of this flowing fluid may be considered negligible.

Under these conditions, the proper form of the first law of thermodynamics is equation (6-11), the Bernoulli equation. Applying equation (6-11), we have

$$y_1 + \frac{P_{atm}}{\rho g} = \frac{v_2^2}{2g} + \frac{P_{atm}}{\rho g}$$

from which the exiting velocity may be expressed in the familiar form

$$v_2 = \sqrt{2gy}$$

As a final illustration of the use of the control-volume relations, an example using all three expressions is presented below.

EXAMPLE 4

In the sudden enlargement shown below in Figure 6.10, the pressure acting at section (1) is considered uniform with value P_1. Find the change in internal energy between stations (1) and (2) for steady, incompressible flow. Neglect shear stress at the walls and express $u_2 - u_1$ in terms of v_1, A_1, and A_2. The control volume selected is indicated by the dotted line.

Figure 6.10 Flow through a sudden enlargement.

Conservation of Mass

$$\iint_{c.s.} \rho(\mathbf{v} \cdot \mathbf{n})dA + \frac{\partial}{\partial t} \iiint_{c.v.} \rho \, dV = 0$$

If we select station (2), a considerable distance downstream from the sudden enlargement, the continuity expression, for steady, incompressible flow, becomes

$$\rho_1 v_1 A_1 = \rho_2 v_2 A_2$$

or

$$v_2 = v_1 \frac{A_1}{A_2} \tag{6-12}$$

Momentum

$$\sum F = \iint_{c.s.} \rho \mathbf{v}(\mathbf{v} \cdot \mathbf{n}) dA + \frac{\partial}{\partial t} \iiint_{c.v.} \rho \mathbf{v} \, dV$$

and thus

$$P_1 A_2 - P_2 A_2 = \rho v_2^2 A_2 - \rho v_1^2 A_1$$

or

$$\frac{P_1 - P_2}{\rho} = v_2^2 - v_1^2 \left(\frac{A_1}{A_2}\right) \tag{6-13}$$

Energy

$$\frac{\delta Q}{\partial t} - \frac{\delta W_s}{dt} = \iint_{c.s.} \rho \left(e + \frac{P}{\rho}\right)(\mathbf{v} \cdot \mathbf{n}) \, dA + \frac{\partial}{\partial t} \iiint_{c.v.} \rho e \, dV + \frac{\delta W_\mu}{dt}$$

Thus,

$$\left(e_1 + \frac{P_1}{\rho}\right)(\rho v_1 A_1) = \left(e_2 + \frac{P_2}{\rho}\right)(\rho v_2 A_2)$$

or, since $\rho v_1 A_1 = \rho v_2 A_2$,

$$e_1 + \frac{P_1}{\rho} = e_2 + \frac{P_2}{\rho}$$

The specific energy is

$$e = \frac{v^2}{2} + gy + u$$

Thus our energy expression becomes

$$\frac{v_1^2}{2} + gy_1 + u_1 + \frac{P_1}{\rho} = \frac{v_2^2}{2} + gy_2 + u_2 + \frac{P_2}{\rho} \tag{6-14}$$

The three control-volume expressions may now be combined to evaluate $u_2 - u_1$. From equation (6-14), we have

$$u_2 - u_1 = \frac{P_1 - P_2}{\rho} + \frac{v_1^2 - v_2^2}{2} + g(y_1 - y_2) \tag{6-14a}$$

Substituting equation (6-13) for $(P_1 - P_2)/\rho$ and equation (6-12) for v_2 and noting that $y_1 = y_2$, we obtain

$$u_2 - u_1 = v_1^2 \left(\frac{A_1}{A_2}\right)^2 - v_1^2 \frac{A_1}{A_2} + \frac{v_1^2}{2} - \frac{v_1^2}{2} \left(\frac{A_1}{A_2}\right)^2$$

$$= \frac{v_1^2}{2} \left[1 - 2\frac{A_1}{A_2} + \left(\frac{A_1}{A_2}\right)^2\right] = \frac{v_1^2}{2} \left[1 - \frac{A_1}{A_2}\right]^2 \tag{6-15}$$

Equation (6-15) shows that the internal energy increases in a sudden enlargement. The temperature change corresponding to this change in internal energy is insignificant, but from equation (6-14a) it can be seen that the change in total head,

$$\left(\frac{P_1}{\rho} + \frac{v_1^2}{2} + gy_1\right) - \left(\frac{P_2}{\rho} + \frac{v_2^2}{2} + gy_2\right)$$

is equal to the internal energy change. Accordingly, the internal energy change in an incompressible flow is designated as the head loss, h_L, and the energy equation for steady, adiabatic, incompressible flow in a stream tube is written as

$$\frac{P_1}{\rho g} + \frac{v_1^2}{2g} + y_1 = h_L + \frac{P_2}{\rho g} + \frac{v_2^2}{2g} + y_2 \tag{6-16}$$

Note the similarity to equation (6-11).

6.4   CLOSURE

In this chapter the first law of thermodynamics, the third of the fundamental relations upon which fluid-flow analyses are based, has been used to develop an integral expression for the conservation of energy with respect to a control volume. The resulting expression, equation (6-10), is, in conjunction with equations (4–1) and (5–4), one of the fundamental expressions for the control-volume analysis of fluid-flow problems.

A special case of the integral expression for the conservation of energy is the Bernoulli equation, equation (6-11). Although simple in form and use, this expression has broad application to physical situations.

PROBLEMS

6.1 The velocity profile in the annular control volume of Example 3 is approximately linear, varying from a velocity of zero at the outer boundary to a value of $\omega d/2$ at the inner boundary. Develop an expression for the fluid velocity, $v(r)$, where r is the distance from the center of the shaft.

6.2 Sea water, $\rho = 1025\,\text{kg/m}^3$, flows through a pump at $0.21\,\text{m}^3/\text{s}$. The pump inlet is 0.25 m in diameter. At the inlet the pressure is –0.15 m of mercury. The pump outlet, 0.152 m in diameter, is 1.8 m above the inlet. The outlet pressure is 175 kPa. If the inlet and exit temperature are equal, how much power does the pump add to the fluid?

6.3 Air at 70°F, Flows into a 10-ft^3 reservoir at a velocity of 110 fps. If the reservoir pressure is 14 psig and the reservoir temperature 70°F, find the rate of temperature increase in the reservoir. Assume the incoming air is at reservoir pressure and flows through a 8-in.-diameter pipe.

6.4 Water flows through a 2-in.-diameter horizontal pipe at a flow rate of 35 gal/min. The heat transfer to the pipe can be neglected, and frictional forces result in a pressure drop of 10 psi. What is the temperature change of the water?

6.5 During the flow of 200 ft^3/s of water through the hydraulic turbine shown, the pressure indicated by gage A is 12 psig. What should gage B read if the turbine is delivering 600 hp at 82% efficiency? Gage B is designed to measure the total pressure, that is, $P + \rho v^2/2$ for an incompressible fluid.

6.6 During the test of a centrifugal pump, a Bourdon pressure gage just outside the casing of the 12-in.-diameter suction pipe reads –6 psig (i.e., vacuum). On the 10-in.-diameter discharge pipe another gage reads 40 psig. The discharge pipe is 5 ft above the suction pipe. The discharge of water through the pump is measured to be 4 ft³/s. Compute the horsepower input of the test pump.

6.7 A fan draws air from the atmosphere through a 0.30-m-diameter round duct that has a smoothly rounded entrance. A differential manometer connected to an opening in the wall of the duct shows a vacuum pressure of 2.5 cm of water. The density of air is 1.22 kg/m³. Determine the volume rate of air flow in the duct in cubic feet per second. What is the horsepower output of the fan?

6.8 Find the change in temperature between stations (1) and (2) in terms of the quantities A_1, A_3, v_1, v_3, c_v, and θ. The internal energy is given by $c_v T$. The fluid is water and $T_1 = T_3$, $P_1 = P_3$.

Top view

6.9 A liquid flows from A to B in a horizontal pipe line shown at a rate of 3 ft³/s with a friction loss of 0.45 ft of flowing fluid. For a pressure head at B of 24 in., what will be the pressure head at A?

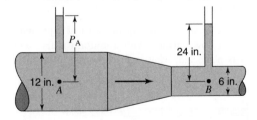

6.10 In Problem 6.26, compute the upward force on the device from water and air. Use the results of Problem 6.26 as well as any other data given in that problem that you may need. Explain why you cannot profitably use Bernoulli's equation here for a force calculation.

6.11 A Venturi meter with an inlet diameter of 0.6 m is designed to handle 6 m³/s of standard air. What is the required throat diameter if this flow is to give a reading of 0.10 m of alcohol in a differential manometer connected to the inlet and the throat? The specific gravity of alcohol may be taken as 0.8.

6.12 The pressurized tank shown has a circular cross section of 6 ft in diameter. Oil is drained through a nozzle 2 in. in diameter in the side of the tank. Assuming that the air pressure is maintained constant, how long does it take to lower the oil surface in the tank by 2 ft? The specific gravity of the oil in the tank is 0.85 and that of mercury is 13.6.

6.13 An automobile is driving into a 45-mph headwind at 40 mph. If the barometer reads 29 in. Hg and the temperature is 40°F, what is the pressure at a point on the auto where the wind velocity is 120 fps with respect to the auto?

6.14 Water is discharged from a 1.0-cm-diameter nozzle that is inclined at a 30° angle above the horizontal. If the jet strikes the ground at a horizontal distance of 3.6 m and a vertical distance of 0.6 m from the nozzle as shown, what is the rate of flow in cubic meters per second? What is the total head of the jet? (See equation (6-11b).)

6.15 The pump shown in the figure delivers water at 59°F at a rate of 550 gal/min. The inlet pipe has an inside diameter of 5.95 in. and it is 10 ft long. The inlet pipe is submerged 6 ft into the water and is vertical. Estimate the pressure inside the pipe at the pump inlet.

6.16   In the previous problem, determine the flow rate at which the pump inlet pressure is equal to the vapor pressure of the water. Assume that friction causes a head loss of 4 ft. The vapor pressure of water at 59°F is 0.247 psi.

6.17   Using the data of Problem 5.27, determine the velocity head of the fluid leaving the impeller. What pressure rise would result from such a velocity head?

6.18   In order to maneuver a large ship while docking, pumps are used to issue a jet of water perpendicular to the bow of the ship as shown in the figure. The pump inlet is located far enough away from the outlet that the inlet and outlet do not interact. The inlet is also vertical so that the net thrust of the jets on the ship is independent of the inlet velocity and pressure. Determine the pump horsepower required per pound of thrust. Assume that the inlet and outlet are at the same depth. Which will produce more thrust per horsepower, a low-volume, high-pressure pump or a high-volume, low-pressure pump?

Side

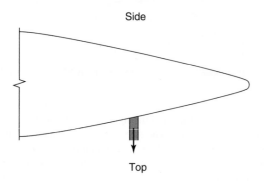

Top

6.19   Determine the head loss between stations (1) and (2) in Problem 5.7.

6.20   Multnomah Falls in Oregon has sheer drop of 165 m. Estimate the change in water temperature caused by this drop.

6.21   An "air cushion" vehicle is designed to traverse terrain while floating on a cushion of air. Air, supplied by a compressor, escapes through the clearing between the ground and the skirt of the vehicle. If the skirt has a rectangular shape 3 × 9 m, the vehicle mass is 8100 kg and the ground clearance is 3 cm, determine the airflow rate needed to maintain the cushion and the power given by the

compressor to the air. Assume that the air speeds within the cushion are very low.

6.22   The solution to Problem 5.22 is

$$h_2 = \frac{h_1}{2}\left[\left(1 + \frac{8v_1^2}{gh_1}\right)^{1/2} - 1\right]$$

Show that Bernoulli's equation applied between sections 1 and 2 does not give this result. Derive all expressions for the change in total head across a hydraulic jump.

6.23   Residential water use, exclusive of fire protection, runs about 80 gallons per person per day. If the water is delivered to a residence at 60 psig, estimate the monthly energy required to pump the water from atmospheric pressure to the delivery pressure. Neglect line losses and elevation changes. Assume the pumps are 75% efficient and are driven by electric motors with 90% efficiency.

6.24   A 1968 Volkswagen sedan is driving over a 7300-ft-high mountain pass at a speed of v m/s into a headwind of W m/s. Compute the gage pressure in mPa at a point on the auto where the velocity relative to the auto is $v - W$ m/s. The local air density is 0.984 kg/m³.

6.25   A liquid is heated in a vertical tube of constant diameter, 15 m long. The flow is upward. At the entrance the average velocity is 1 m/s, the pressure 340,000 Pa, and the density is 1001 kg/m³. If the increase in internal energy is 200,000 J/kg, find the heat added to the fluid.

6.26   Water flows steadily up the vertical pipe and is then deflected to flow outward with a uniform radial velocity. If friction is neglected, what is the flow rate of water through the pipe if the pressure at A is 10 psig?

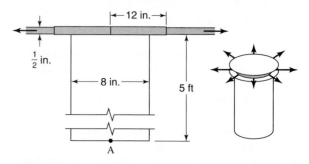

6.27   Water flows through the pipe contraction shown at a rate of 1 ft³/s. Calculate the differential manometer reading in inches of mercury, assuming no energy loss in the flow. Be sure to give the correct direction of the manometer reading.

6.28 The figure illustrates the operation of an air lift pump. Compressed air is forced into a perforated chamber to mix with the water so that the specific gravity of the air–water mixture above the air inlet is 0.5. Neglecting any pressure drop across section (1), compute the discharge velocity v of the air–water mixture. Can Bernoulli's equation be used across section (1)?

6.29 Rework Problem with the assumption that the momentum of the incoming air at section (1) is zero. Determine the exit velocity, v, and the magnitude of the pressure drop at section (1).

6.30 Air of density 1.21 kg/m^3 is flowing as shown. If $v = 15$ m/s, determine the readings on manometers (a) and (b) in the figures below.

(a)

(b)

6.31 Referring to the figure, assume the flow to be frictionless in the siphon. Find the rate of discharge in cubic feet per second,

and the pressure head at B if the pipe has a uniform diameter of 1 in. How long will it take for the water level to decrease by 3 ft? The tank diameter is 10 ft.

6.32 In Problem 6.31, find the rate of discharge if the frictional head loss in the pipe is 3.2 v^2/g where v is the flow velocity in the pipe.

6.33 Assume that the level of water in the tank remains the same and that there is no friction loss in the pipe, entrance, or nozzle. Determine

(a) the volumetric discharge rate from the nozzle;

(b) the pressure and velocity at points A, B, C, and D.

6.34 Water in an open cylindrical tank 15 ft in diameter discharges into the atmosphere through a nozzle 2 in. in diameter. Neglecting friction and the unsteadiness of the flow, find the time required for the water in the tank to drop from a level of 28 ft above the nozzle to the 4-ft level.

6.35 A fluid of density ρ_1 enters a chamber where the fluid is heated so that the density decreases to ρ_2. The fluid then escapes through a vertical chimney that has a height L. Neglecting friction and treating the flow processes as incompressible except for the heating, determine the velocity, v, in the stack. The fluid velocity entering the heating chamber may be neglected and the chimney is immersed in fluid of density ρ_1.

Heating section

6.36   Repeat the previous problem without the assumption that the velocity in the heating section is negligible. The ratio of the flow area of the heating section to the chimney flow area is R.

6.37   Consider a 4-cm pipe that runs between a tank open to the atmosphere and a station open to the atmosphere 10 m below the water surface in the tank. Assuming frictionless flow, what will be the mass flow rate? If a nozzle with a 1-cm diameter is placed at the pipe exit, what will be the mass flow rate? Repeat the problem if a head loss of 3 v^2/g occurs in the pipe where v is the flow velocity in the pipe.

6.38   The tank in the previous problem feeds two lines, a 4-cm pipe that exits 10 m below the water level in the tank and a second line, also 4 cm in diameter runs from the tank to a station 20 m below the water level in the tank. The exits of both lines are open to the atmosphere. Assuming frictionless flow, what is the mass flow rate in each line?

Chapter 7

Shear Stress in Laminar Flow

In the analysis of fluid flow thus far, shear stress has been mentioned, but it has not been related to the fluid or flow properties. We shall now investigate this relation for laminar flow. The shear stress acting on a fluid depends upon the type of flow that exists. In the so-called laminar flow, the fluid flows in smooth layers or lamina, and the shear stress is the result of the (nonobservable) microscopic action of the molecules. Turbulent flow is characterized by the large scale, observable fluctuations in fluid and flow properties, and the shear stress is the result of these fluctuations. The criteria for laminar and turbulent flows will be discussed in Chapters 12 and 13. The shear stress in turbulent flow will be discussed in Chapter 13.

7.1 NEWTON'S VISCOSITY RELATION

In a solid, the resistance to *deformation* is the modulus of elasticity. The shear modulus of an elastic solid is given by

$$\text{shear modulus} = \frac{\text{shear stress}}{\text{shear strain}} \tag{7-1}$$

Just as the shear modulus of an elastic solid is a property of the solid relating shear stress and shear strain, there exists a relation similar to (7-1), which relates the shear stress in a parallel, laminar flow to a property of the fluid. This relation is Newton's law of viscosity

$$\text{viscosity} = \frac{\text{shear stress}}{\text{rate of shear strain}} \tag{7-2}$$

Thus, the viscosity is the property of a fluid to resist the *rate* at which deformation takes place when the fluid is acted upon by shear forces. As a property of the fluid, the viscosity depends upon the temperature, composition, and pressure of the fluid, but is independent of the rate of shear strain.

The rate of deformation in a simple flow is illustrated in Figure 7.1. The flow parallel to the x axis will deform the element if the velocity at the top of the element is different than the velocity at the bottom.

The rate of shear strain at a point is defined as $-d\delta/dt$. From Figure 7.1, it may be seen that

$$-\frac{d\delta}{dt} = \lim_{\Delta x, \Delta y, \Delta t \to 0} \frac{\delta|_{t+\Delta t} - \delta|_t}{\Delta t}$$

$$= \lim_{\Delta x, \Delta y, \Delta t \to 0} \left(\frac{\{\pi/2 - \arctan\left[(v|_{y+\Delta y} - v|_y)\Delta t/\Delta y\right]\} - \pi/2}{\Delta t} \right) \tag{7-3}$$

In the limit, $-d\delta/dt = dv/dy =$ rate of shear strain.

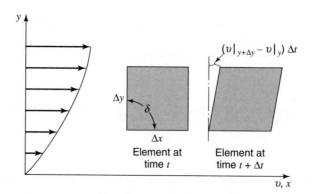

Figure 7.1 Deformation of a fluid element.

Combining equations (7-2) and (7-3) and denoting the viscosity by μ, we may write Newton's law of viscosity as

$$\tau = \mu \frac{dv}{dy} \tag{7-4}$$

The velocity profile and shear stress variation in a fluid flowing between two parallel plates is illustrated in Figure 7.2. The velocity profile[1] in this case is parabolic; as the shear stress is proportional to the derivative of the velocity, the shear stress varies in a linear manner.

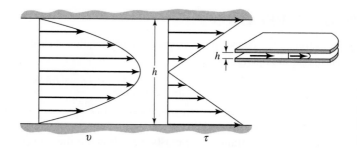

Figure 7.2 Velocity and shear stress profiles for flow between two parallel plates.

7.2 NON-NEWTONIAN FLUIDS

Newton's law of viscosity does not predict the shear stress in all fluids. Fluids are classified as Newtonian or non-Newtonian, depending upon the relation between shear stress and the rate of shearing strain. In Newtonian fluids, the relation is linear, as shown in Figure 7.3.

In non-Newtonian fluids, the shear stress depends upon the rate of shear strain. While fluids deform continuously under the action of shear stress, plastics will sustain a shear stress before deformation occurs. The "ideal plastic" has a

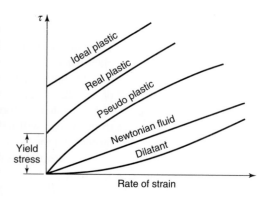

Figure 7.3 Stress rate-of-strain relation for Newtonian and non-Newtonian fluids.

---

[1] The derivation of velocity profiles is discussed in Chapter 8.

linear stress rate-of-strain relation for stresses greater than the yield stress. *Thixotropic* substances such as printer's ink have a resistance to deformation that depends upon deformation rate and time.

The No-Slip Condition

Although the substances above differ in their stress rate-of-strain relations, they are similar in their action at a boundary. In both Newtonian and non-Newtonian fluids, the layer of fluid adjacent to the boundary has zero velocity relative to the boundary. When the boundary is a stationary wall, the layer of fluid next to the wall is at rest. If the boundary or wall is moving, the layer of fluid moves at the velocity of the boundary, hence the name no-slip (boundary) condition. The no-slip condition is the result of experimental observation and fails when the fluid no longer can be treated as a continuum.

The no-slip condition is a result of the viscous nature of the fluid. In flow situations in which the viscous effects are neglected—the so called inviscid flows—only the component of the velocity normal to the boundary is zero.

7.3   VISCOSITY

The viscosity of a fluid is a measure of its resistance to deformation rate. Tar and molasses are examples of highly viscous fluids; air and water, which are the subject of frequent engineering interest, are examples of fluids with relatively low viscosities. An understanding of the existence of the viscosity requires an examination of the motion of fluid on a molecular basis.

The molecular motion of gases can be described more simply than that of liquids. The mechanism by which a gas resists deformation may be illustrated by examination of the motion of the molecules on a microscopic basis. Consider the control volume shown in Figure 7.4.

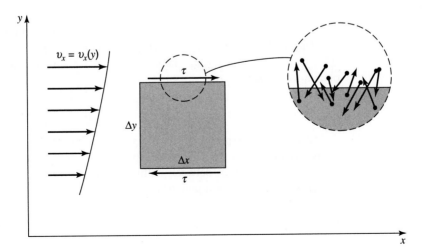

Figure 7.4 Molecular motion at the surface of a control volume.

The top of the control volume is enlarged to show that, even though the top of the element is a streamline of the flow, individual molecules cross this plane. The paths of the molecules between collisions are represented by the random arrows. Because the top of the control volume is a streamline, the net molecular flux across this surface must be zero, hence the upward molecular flux must equal the downward molecular flux. The molecules that cross the control surface in an upward direction have average velocities in the x direction corresponding to their points of origin. Denoting the y coordinate of the top of the control surface as y_0, we shall write the x-directional average velocity of the upward molecular flux as $v_x|_{y-}$, where the minus sign signifies that the average velocity is evaluated at some point below y_0. The x-directional momentum carried across the top of the control surface is then $mv_x|_{y-}$ per molecule, where m is the mass of the molecule. If Z molecules cross the plane per unit time, then the net x-directional momentum flux will be

$$\sum_{n=1}^{Z} m_n(v_x|_{y-} - v_x|_{y+}) \tag{7-5}$$

The flux of x-directional momentum on a molecular scale appears as a shear stress when the fluid is observed on a macroscopic scale. The relation between the molecular momentum flux and the shear stress may be seen from the control-volume expression for linear momentum

$$\sum \mathbf{F} = \iint_{\text{c.s.}} \rho \mathbf{v}(\mathbf{v} \cdot \mathbf{n}) \, dA + \frac{\partial}{\partial t} \iiint_{\text{c.v.}} \rho \mathbf{v} \, dV \tag{5-4}$$

The first term on the right-hand side of equation (5-4) is the momentum flux. When a control volume is analyzed on a molecular basis, this term includes both the macroscopic and molecular momentum fluxes. If the molecular portion of the total momentum flux is to be treated as a force, it must be placed on the left-hand side of equation (5-4). Thus the molecular momentum flux term changes sign. Denoting the negative of the molecular momentum flux as τ, we have

$$\tau = -\sum_{n=1}^{Z} m_n(v_x|_{y-} - v_x|_{y+}) \tag{7-6}$$

We shall treat shear stress exclusively as a force per unit area.

The bracketed term, $(v_x|_{y-} - v_x|_{y+})$ in equation (7-6), may be evaluated by noting that $v_x|_{y-} = v_x|_{y_0} - (dv_x/dy|_{y_0})\delta$, where $y- = y_0 - \delta$. Using a similar expression for $y+$, we obtain, for the shear stress

$$\tau = 2 \sum_{n=1}^{Z} m_n \frac{dv_x}{dy}\bigg|_{y_0} \delta_n$$

In the above expression δ is the y component of the distance between molecular collisions. Borrowing from the kinetic theory of gases, the concept of the mean free path, λ, as the average distance between collisions, and also noting from the same source that $\delta = 2/3\lambda$, we obtain, for a pure gas,

$$\tau = \frac{4}{3} m\lambda Z \frac{dv_x}{dy}\bigg|_{y_0} \tag{7-7}$$

as the shear stress.

Comparing equation (7-7) with Newton's law of viscosity, we see that

$$\mu = \frac{4}{3} m\lambda Z \qquad (7\text{-}8)$$

The kinetic theory gives $Z = N\overline{C}/4$, where

$$N = \text{molecules per unit volume}$$
$$\overline{C} = \text{average random molecular velocity}$$

and thus

$$\mu = \frac{1}{3} Nm\lambda\overline{C} = \frac{\rho\lambda\overline{C}}{3}$$

or, using[2]

$$\lambda = \frac{1}{\sqrt{2}\pi N d^2} \quad \text{and} \quad \overline{C} = \sqrt{\frac{8\kappa T}{\pi m}}$$

where d is the molecular diameter and κ is the Boltzmann constant, we have

$$\mu = \frac{2}{3\pi^{3/2}} \frac{\sqrt{m\kappa T}}{d^2} \qquad (7\text{-}9)$$

Equation (7-9) indicates that μ is independent of pressure for a gas. This has been shown, experimentally, to be essentially true for pressures up to approximately 10 atmospheres. Experimental evidence indicates that at low temperatures the viscosity varies more rapidly than $\sqrt{T}$. The constant-diameter rigid-sphere model for the gas molecule is responsible for the less-than-adequate viscosity-temperature relation. Even though the preceding development was somewhat crude in that an indefinite property, the molecular diameter, was introduced, the interpretation of the viscosity of a gas being due to the microscopic momentum flux is a valuable result and should not be overlooked. It is also important to note that equation (7-9) expresses the viscosity entirely in terms of fluid properties.

A more realistic molecular model utilizing a force field rather than the rigid-sphere approach will yield a viscosity-temperature relationship much more consistent with experimental data than the $\sqrt{T}$ result. The most acceptable expression for nonpolar molecules is based upon the Lennard–Jones potential energy function. This function and the development leading to the viscosity expression will not be included here. The interested reader may refer to Hirschfelder, Curtiss, and Bird[3] for the details of this approach. The expression for viscosity of a pure gas that results is

$$\mu = 2.6693 \times 10^{-6} \frac{\sqrt{MT}}{\sigma^2 \Omega_\mu} \qquad (7\text{-}10)$$

---

[2] In order of increasing complexity, the expressions for mean free path are presented in R. Resnick and D. Halliday, *Physics*, Part I, Wiley, New York, 1966, Chapter 24, and E. H. Kennard, *Kinetic Theory of Gases*, McGraw-Hill Book Company, New York, 1938, Chapter 2.

[3] J. O. Hirschfelder, C. F. Curtiss, and R. B. Bird, *Molecular Theory of Gases and Liquids*, Wiley, New York, 1954.

where μ is the viscosity, in pascal-seconds; T is absolute temperature, in K; M is the molecular weight; σ is the "collision diameter," a Lennard–Jones parameter, in Å (Angstroms); Ω_μ is the "collision integral," a Lennard–Jones parameter that varies in a relatively slow manner with the dimensionless temperature $\kappa T/\epsilon$; κ is the Boltzmann constant, $1.38 \cdot 10^{-16}$ ergs/K; and ϵ is the characteristic energy of interaction between molecules. Values of σ and ϵ for various gases are given in Appendix K, and a table of Ω_μ versus $\kappa T/\epsilon$ is also included in Appendix K.

For multicomponent gas mixtures at low density, Wilke[4] has proposed this empirical formula for the viscosity of the mixture:

$$\mu_{\text{mixture}} = \sum_{i=1}^{n} \frac{x_i \mu_i}{\sum x_j \phi_{ij}} \tag{7-11}$$

where x_i, x_j are mole-fractions of species i and j in the mixture, and

$$\phi_{ij} = \frac{1}{\sqrt{8}} \left(1 + \frac{M_i}{M_j} \right)^{-1/2} \left[1 + \left(\frac{\mu_i}{\mu_j} \right)^{1/2} \left(\frac{M_j}{M_i} \right)^{1/4} \right]^2 \tag{7-12}$$

where M_i, M_j are the molecular weights of species i and j, and μ_i, μ_j are the viscosities of species i and j. Note that when $i = j$, we have $\phi_{ij} = 1$.

Equations (7-10), (7-11), and (7-12) are for nonpolar gases and gas mixtures at low density. For polar molecules, the preceding relation must be modified.[5]

Although the kinetic theory of gases is well developed, and the more sophisticated models of molecular interaction accurately predict viscosity in a gas, the molecular theory of liquids is much less advanced. Hence, the major source of knowledge concerning the viscosity of liquids is experiment. The difficulties in the analytical treatment of a liquid are largely inherent in nature of the liquid itself. Whereas in gases the distance between molecules is so great that we consider gas molecules as interacting or colliding in pairs, the close spacing of molecules in a liquid results in the interaction of several molecules simultaneously. This situation is somewhat akin to an N-body gravitational problem. In spite of these difficulties, an approximate theory has been developed by Eyring, which illustrates the relation of the intermolecular forces to viscosity.[6] The viscosity of a liquid can be considered due to the restraint caused by intermolecular forces. As a liquid heats up, the molecules become more mobile. This results in less restraint from intermolecular forces. Experimental evidence for the viscosity of liquids shows that the viscosity decreases with temperature in agreement with the concept of intermolecular adhesive forces being the controlling factor.

Units of Viscosity

The dimensions of viscosity may be obtained from Newton's viscosity relation

$$\mu = \frac{\tau}{dv/dy}$$

---

[4] C. R. Wilke, *J. Chem. Phys.*, **18**, 517–519 (1950).

[5] J. O. Hirschfelder, C. F. Curtiss, and R. B. Bird, *Molecular Theory of Gases and Liquids*, Wiley, New York, 1954.

[6] For a description of Eyring's theory, see R. B. Bird, W. E. Stewart, and E. N. Lightfoot, *Transport Phenomena*, Wiley, New York, 1960, Chap. 1.

or, in dimensional form

$$\frac{F/L^2}{(L/t)(1/L)} = \frac{Ft}{L^2}$$

where F = force, L = length, t = time.

Using Newton's second law of motion to relate force and mass ($F = ML/t^2$), we find that the dimensions of viscosity in the mass–length–time system become M/Lt.

The ratio of the viscosity to the density occurs frequently in engineering problems. This ratio, μ/ρ, is given the name kinematic viscosity and is denoted by the symbol v. The origin of the name kinematic viscosity may be seen from the dimensions of v:

$$v \equiv \frac{\mu}{\rho} \sim \frac{M/Lt}{M/L^3} = \frac{L^2}{t}$$

The dimensions of v are those of kinematics: length and time. Either of the two names, absolute viscosity or dynamic viscosity, is frequently employed to distinguish μ from the kinematic viscosity, v.

In the SI system, dynamic viscosity is expressed in pascal-seconds (1 pascal-second = 1 N n·s/m² = 10 poise = 0.02089 slugs/ft·s = 0.02089 lb$_f$·s/ft² = 0.6720 lb$_m$/ft·s). Kinematic viscosity in the metric system is expressed in (meters)² per second (1 m²/s = 10^4 stokes = 10.76 ft²/s).

Absolute and kinematic viscosities are shown in Figure 7.5 for three common gases and two liquids as functions of temperature. A more extensive listing is contained in Appendix I.

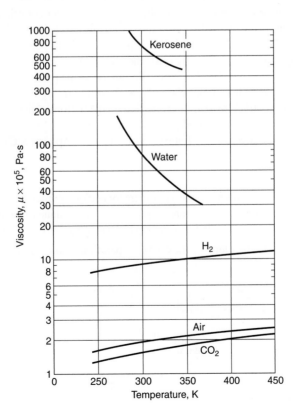

Figure 7.5 Viscosity–temperature variation for some liquids and gases.

7.4 SHEAR STRESS IN MULTIDIMENSIONAL LAMINAR FLOWS OF A NEWTONIAN FLUID

Newton's viscosity relation, discussed previously, is valid for only parallel, laminar flows. Stokes extended the concept of viscosity to three-dimensional laminar flow. The basis of Stokes' viscosity relation is equation (7-2)

$$\text{viscosity} = \frac{\text{shear stress}}{\text{rate of shear strain}} \tag{7-2}$$

where the shear stress and rate of shear strain are those of a three-dimensional element. Accordingly, we must examine shear stress and strain rate for a three-dimensional body.

Shear Stress

The shear stress is a tensor quantity requiring magnitude, direction, and orientation with respect to a plane for identification. The usual method of identification of the shear stress involves a double subscript, such as τ_{xy}. The tensor component, τ_{ij}, is identified as follows:

$$\tau = \text{magnitude}$$

first subscript $=$ direction of axis to which plane of action of shear stress

is normal

second subscript $=$ direction of action of the shear stress.

Thus τ_{xy} acts on a plane normal to the x axis (the yz plane) and acts in the y direction. In addition to the double subscript, a sense is required. The shear stresses acting on an element $\Delta x\, \Delta y\, \Delta z$, illustrated in Figure 7.6, are indicated in the positive sense. The definition of positive shear stress can be generalized for use in other coordinate systems. A shear stress component is positive when both the vector normal to the surface of action and the shear stress act in the same direction (both positive or both negative).

For example, in Figure 7.6(a), the shear stress τ_{yx} at the top of the element acts on surface $\Delta x\ \Delta z$. The vector normal to this area is in the positive y direction. The stress τ_{yx} acts in the positive x direction, hence τ_{yx} as illustrated in Figure 7.6(a) is positive. The student may apply similar reasoning to τ_{yx} acting on the bottom of the element and conclude that τ_{yx} is also positive as illustrated.

As in the mechanics of solids, $\tau_{ij} = \tau_{ji}$ (see Appendix C).

Rate of Shear Strain

The rate of shear strain for a three-dimensional element may be evaluated by determining the shear strain rate in the xy, yz, and xz planes. In the xy plane illustrated in Figure 7.7, the shear strain rate is again $-d\delta/dt$; however, the element may deform in both the x and the y directions.

Hence, as the element moves from position 1 to position 2 in time Δt

$$-\frac{d\delta}{dt} = \lim_{\Delta x, \Delta y, \Delta t \to 0} \frac{\delta|_{t+\Delta t} - \delta|_t}{\Delta t}$$

$$= \lim_{\Delta x, \Delta y, \Delta t \to 0} \left\{ \frac{\pi/2 - \arctan\{[(v_x|_{y+\Delta y} - v_x|_y)\Delta t]/\Delta y\}}{\Delta t} \right.$$

$$\left. \frac{-\arctan\{[(v_y|_{x+\Delta x} - v_y|_x)\Delta t]/\Delta x\} - \pi/2}{\Delta t} \right\}$$

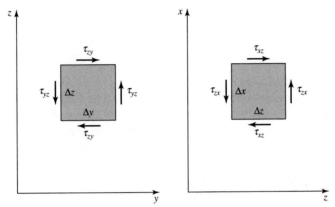

Figure 7.6 Shear stress acting in a positive sense.

As the shear strain evaluated above is in the xy plane, it will be subscripted xy. In the limit, $-d\delta_{xy}/dt = \partial v_x/\partial y + \partial v_y/\partial x$. In a similar manner, the shear strain rates in the yz and xz planes may be evaluated as

$$-\frac{d\delta_{yz}}{dt} = \frac{\partial v_y}{\partial z} + \frac{\partial v_z}{\partial y}$$

$$-\frac{d\delta_{xz}}{dt} = \frac{\partial v_x}{\partial z} + \frac{\partial v_z}{\partial x}$$

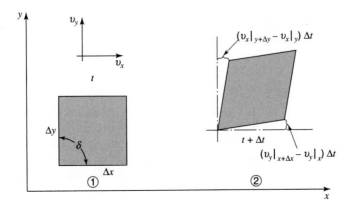

Figure 7.7 Shear strain in the xy plane.

Stokes's Viscosity Relation

(A) *Shear Stress.*   Stokes's viscosity relation for the shear-stress components in laminar flow may now be stated with the aid of the preceding developments for rate of shear strain. Using equation (7-2), we have, for the shear stresses written in rectangular coordinate form

$$\tau_{xy} = \tau_{yx} = \mu \left(\frac{\partial v_x}{\partial y} + \frac{\partial v_y}{\partial x} \right) \tag{7-13a}$$

$$\tau_{yz} = \tau_{zy} = \mu \left(\frac{\partial v_y}{\partial z} + \frac{\partial v_z}{\partial y} \right) \tag{7-13b}$$

and

$$\tau_{zx} = \tau_{xz} = \mu \left(\frac{\partial v_z}{\partial x} + \frac{\partial v_x}{\partial z} \right) \tag{7-13c}$$

(B) *Normal Stress.*   The normal stress may also be determined from a stress rate-of-strain relation; the strain rate, however, is more difficult to express than in the case of shear strain. For this reason the development of normal stress, on the basis of a generalized Hooke's law for an elastic medium, is included in detail in Appendix D, with only the result expressed below in equations (7-14a), (7-14b), and (7-14c).

The normal stress in rectangular coordinates written for a newtonian fluid is given by

$$\sigma_{xx} = \mu \left(2\frac{\partial v_x}{\partial x} - \frac{2}{3}\boldsymbol{\nabla} \cdot \mathbf{v} \right) - P \tag{7-14a}$$

$$\sigma_{yy} = \mu \left(2\frac{\partial v_y}{\partial y} - \frac{2}{3}\boldsymbol{\nabla} \cdot \mathbf{v} \right) - P \tag{7-14b}$$

and

$$\sigma_{zz} = \mu \left(2\frac{\partial v_z}{\partial z} - \frac{2}{3}\boldsymbol{\nabla} \cdot \mathbf{v} \right) - P \tag{7-14c}$$

It is to be noted that the sum of these three equations yields the previously mentioned result: the bulk stress, $\bar{\sigma} = (\sigma_{xx} + \sigma_{yy} + \sigma_{zz})/3$, is the negative of the pressure, P.

7.5   CLOSURE

The shear stress in laminar flow and its dependence upon the viscosity and kinematic derivatives has been presented for a cartesian coordinate system. The shear stress in other coordinate systems, of course, will occur frequently, and it is to be noted that equation (7-2) forms the general relation between shear stress, viscosity, and the rate of shear strain. The shear stress in other coordinate systems may be obtained from evaluating the shear-strain rate in the associated coordinate systems. Several problems of this nature are included at the end of this chapter.

PROBLEMS

7.1 Sketch the deformation of a fluid element for the following cases:

(a) $\partial v_x/\partial y$ is much larger than $\partial v_y/\partial x$;

(b) $\partial v_y/\partial x$ is much larger than $\partial v_x/\partial y$.

7.2 For a two-dimensional, incompressible flow with velocity $v_x = v_x(y)$, sketch a three-dimensional fluid element and illustrate the magnitude, direction, and surface of action of each stress component.

7.3 Show that the axial strain rate in a one-dimensional flow, $v_x = v_x(x)$, is given by $\partial v_x/\partial x$. What is the rate of volume change? Generalize for a three-dimensional element, and determine the rate of volume change.

7.4 Using a cylindrical element, show that Stokes's viscosity relation yields the following shear stress components.

$$\tau_{r\theta} = \tau_{\theta r} = \mu\left[r\frac{\partial}{\partial r}\left(\frac{v_\theta}{r}\right) + \frac{1}{r}\frac{\partial v_r}{\partial \theta}\right]$$

$$\tau_{\theta z} = \tau_{z\theta} = \mu\left[\frac{\partial v_\theta}{\partial z} + \frac{1}{r}\frac{\partial v_z}{\partial \theta}\right]$$

$$\tau_{zr} = \tau_{rz} = \mu\left[\frac{\partial v_z}{\partial r} + \frac{\partial v_r}{\partial z}\right]$$

7.5 Estimate the viscosity of nitrogen at 175 K using equation (7-10).

7.6 Calculate the viscosity of oxygen at 350°K and compare with the value given in Appendix I.

7.7 What is the percentage change in the viscosity of water when the water temperature rises from 60 to 120°F?

7.8 At what temperature is the kinematic viscosity of glycerin the same as the kinematic viscosity of helium?

7.9 According to the Hagen–Poiseuille laminar flow model, the volumetric flow rate is inversely proportional to the viscosity. What percentage change in volumetric flow rate occurs in a laminar flow as the water temperature changes from near freezing to 140°F?

7.10 Repeat the preceding problem for air.

7.11 An automobile crankshaft is 3.175 cm in diameter. A bearing on the shaft is 3.183 cm in diameter and 2.8 cm long. The bearing is lubricated with SAE 30 oil at a temperature of 365 K. Assuming that the shaft is centrally located in the bearing, determine how much heat must be removed to maintain the bearing at constant temperature. The shaft is rotating at 1700 rpm, and the viscosity of the oil is 0.01 Pa · s.

7.12 If the speed of the shaft is doubled in Problem 7.11, what will be the percentage increase in the heat transferred from the bearing? Assume that the bearing remains at constant temperature.

7.13 Two ships are traveling parallel to each other and are connected by flexible hoses. Fluid is transferred from one ship to the other for processing and then returned. If the fluid is flowing at 100 kg/s, and at a given instant the first ship is making 4 m/s whereas the second ship is making 3.1 m/s, what is the net force on ship one when the above velocities exist?

7.14 An auto lift consists of 36.02-cm-diameter ram that slides in a 36.04-cm-diameter cylinder. The annular region is filled with oil having a kinematic viscosity of 0.00037 m²/s and a specific gravity of 0.85. If the rate of travel of the ram is 0.15 m/s, estimate the frictional resistance when 3.14 m of the ram is engaged in the cylinder.

7.15 If the ram and auto rack in the previous problem together have a mass of 680 kg, estimate the maximum sinking speed of the ram and rack when gravity and viscous friction are the only forces acting. Assume 2.44 m of the ram engaged.

7.16 The conical pivot shown in the figure has angular velocity ω and rests on an oil film of uniform thickness h. Determine the frictional moment as a function of the angle α, the viscosity, the angular velocity, the gap distance, and the shaft diameter.

7.17 For water flowing in a 0.1-in.-diameter tube, the velocity distribution is parabolic (see Example 4.2). If the average velocity is 2 fps, determine the magnitude of the shear stress at the tube wall.

7.18 What pressure drop per foot of tube is caused by the shear stress in Problem 7.17?

7.19 The rate of shear work per unit volume is given by the product τv. For a parabolic velocity profile in a circular tube (see Example 4.2), determine the distance from the wall at which the shear work is maximum.

Chapter **8**

Analysis of a Differential Fluid Element in Laminar Flow

The analysis of a fluid-flow situation may follow two different paths. One type of analysis has been discussed at length in Chapters 4–6 in which the region of interest has been a definite volume, the macroscopic control volume. In analyzing a problem from the standpoint of a macroscopic control volume, one is concerned only with gross quantities of mass, momentum, and energy crossing the control surface and the total change in these quantities exhibited by the material under consideration. Changes occurring within the control volume by each differential element of fluid cannot be obtained from this type of overall analysis.

In this chapter, we shall direct our attention to elements of fluid as they approach differential size. Our goal is the estimation and description of fluid behavior from a differential point of view; the resulting expressions from such analyses will be differential equations. The solution to these differential equations will give flow information of a different nature than that achieved from a macroscopic examination. Such information may be of less interest to the engineer needing overall design information, but it can give much greater insight into the mechanisms of mass, momentum, and energy transfer.

It is possible to change from one form of analysis to the other, that is, from a differential analysis to an integral analysis by integration and vice versa, rather easily.[1]

A complete solution to the differential equations of fluid flow is possible only if the flow is laminar; for this reason only laminar-flow situations will be examined in this chapter. A more general differential approach will be discussed in Chapter 9.

8.1 FULLY DEVELOPED LAMINAR FLOW IN A CIRCULAR CONDUIT OF CONSTANT CROSS SECTION

Engineers are often confronted with flow of fluids inside circular conduits or pipes. We shall now analyze this situation for the case of incompressible laminar flow. In Figure 8.1, we have a section of pipe in which the flow is laminar and fully developed; that is, it is not influenced by entrance effects and represents a steady-flow situation. *Fully developed flow* is defined as that for which the velocity profile does not vary along the axis of flow.

We now consider the cylindrical control volume of fluid having an inside radius, r, thickness Δr, and length Δx. Applying Newton's second law to this control volume, we may

---

[1] This transformation may be accomplished by a variety of methods, among which are the methods of vector calculus. We shall use a limiting process in this text.

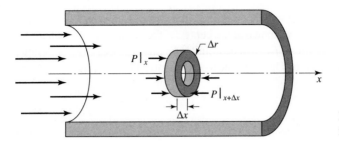

Figure 8.1 Control volume for flow in a circular conduit.

evaluate the appropriate force and momentum terms for the x direction. Starting with the control-volume expression for linear momentum in the x direction

$$\sum F_x = \iint_{\text{c.s.}} \rho v_x (\mathbf{v} \cdot \mathbf{n}) \, dA + \frac{\partial}{\partial t} \iiint_{\text{c.v.}} \rho v_x \, dV \qquad (5\text{-}5a)$$

and evaluating each term as it applies to the control volume shown, we have

$$\sum F_x = P(2\pi r \,\Delta r)\big|_x - P(2\pi r \,\Delta r)\big|_{x+\Delta x} + \tau_{rx}(2\pi r \,\Delta x)\big|_{r+\Delta r} - \tau_{rx}(2\pi r \,\Delta x)\big|_r$$

$$\iint_{\text{c.s.}} v_x \rho (\mathbf{v} \cdot \mathbf{n}) \, dA = (\rho v_x)(2\pi r \,\Delta r v_x)\big|_{x+\Delta x} - (\rho v_x)(2\pi r \,\Delta r v_x)\big|_x$$

and

$$\frac{\partial}{\partial t} \iiint_{\text{c.v.}} v_x \rho \, dV = 0$$

in steady flow.

The convective momentum flux

$$(\rho v_x)(2\pi r \,\Delta r v_x)\big|_{x+\Delta x} - (\rho v_x)(2\pi \,\Delta r v_x)\big|_x$$

is equal to zero as, by the original stipulation that flow is fully developed, all terms are independent of x. Substitution of the remaining terms into equation (5-5a) gives

$$-[P(2\pi r \,\Delta r)\big|_{x+\Delta x} - P(2\pi r \,\Delta r)\big|_x] + \tau_{rx}(2\pi r \,\Delta x)\big|_{r+\Delta r} - \tau_{rx}(2\pi r \,\Delta x)\big|_r = 0$$

Canceling terms where possible and rearranging, we find that this expression reduces to the form

$$-r \frac{P\big|_{x+\Delta x} - P\big|_x}{\Delta x} + \frac{(r\tau_{rx})\big|_{r+\Delta r} - (r\tau_{rx})\big|_r}{\Delta r} = 0$$

Evaluating this expression in the limit as the control volume approaches differential size, that is, as Δx and Δr approach zero, we have

$$-r \frac{dP}{dx} + \frac{d}{dr}(r\tau_{rx}) = 0 \qquad (8\text{-}1)$$

Note that the pressure and shear stress are functions only of x and r, respectively, and thus the derivatives formed are total rather than partial derivatives. In a region of fully developed flow, the pressure gradient, dP/dx, is constant.

The variables in equation (8-1) may be separated and integrated to give

$$\tau_{rx} = \left(\frac{dP}{dx}\right)\frac{r}{2} + \frac{C_1}{r}$$

The constant of integration C_1 may be evaluated by knowing a value of τ_{rx} at some r. Such a condition is known at the center of the conduit, $r = 0$, where for any finite value of C_1, the shear stress, τ_{rx}, will be infinite. As this is physically impossible, the only realistic value for C_1 is zero. Thus, the shear-stress distribution for the conditions and geometry specified is

$$\tau_{rx} = \left(\frac{dP}{dx}\right)\frac{r}{2} \tag{8-2}$$

We observe that the shear stress varies linearly across the conduit from a value of 0 at $r = 0$, to a maximum at $r = R$, the inside surface of the conduit.

Further information may be obtained if we substitute the Newtonian viscosity relationship, that is, assuming the fluid to be Newtonian and recalling that the flow is laminar

$$\tau_{rx} = \mu\frac{dv_x}{dr} \tag{8-3}$$

Substituting this relation into equation (8-2) gives

$$\mu\frac{dv_x}{dr} = \left(\frac{dP}{dx}\right)\frac{r}{2}$$

which becomes, upon integration,

$$v_x = \left(\frac{dP}{dx}\right)\frac{r^2}{4\mu} + C_2$$

The second constant of integration, C_2, may be evaluated, using the boundary condition that the velocity, v_x, is zero at the conduit surface (the no-slip condition), $r = R$. Thus,

$$C_2 = -\left(\frac{dP}{dx}\right)\frac{R^2}{4\mu}$$

and the velocity distribution becomes

$$v_x = -\left(\frac{dP}{dx}\right)\frac{1}{4\mu}(R^2 - r^2) \tag{8-4}$$

or

$$v_x = -\left(\frac{dP}{dx}\right)\frac{R^2}{4\mu}\left[1 - \left(\frac{r}{R}\right)^2\right] \tag{8-5}$$

Equations (8-4) and (8-5) indicate that the velocity profile is parabolic and that the maximum velocity occurs at the center of the circular conduit where $r = 0$. Thus,

$$v_{max} = -\left(\frac{dP}{dx}\right)\frac{R^2}{4\mu} \tag{8-6}$$

and equation (8-5) may be written in the form

$$v_x = v_{max}\left[1 - \left(\frac{r}{R}\right)^2\right] \tag{8-7}$$

Note that the velocity profile written in the form of equation (8-7) is identical to that used in Example 4.2. We may, therefore, use the result obtained in Example 4.2 that

$$v_{avg} = \frac{v_{max}}{2} = -\left(\frac{dP}{dx}\right)\frac{R^2}{8\mu} \tag{8-8}$$

Equation (8-8) may be rearranged to express the pressure gradient, $-dP/dx$, in terms of v_{avg}

$$-\frac{dP}{dx} = \frac{8\mu v_{\text{avg}}}{R^2} = \frac{32\mu v_{\text{avg}}}{D^2} \tag{8-9}$$

Equation (8-9) is known as the Hagen–Poiseuille equation, in honor of the two men credited with its original derivation. This expression may be integrated over a given length of conduit to find the pressure drop and associated drag force on the conduit resulting from the flow of a viscous fluid.

The conditions for which the preceding equations were derived and apply should be remembered and understood. They are as follows:

1. The fluid (a) is Newtonian,
    (b) behaves as a continuum.

2. The flow is (a) laminar,
    (b) steady,
    (c) fully developed,
    (d) incompressible.

8.2   LAMINAR FLOW OF A NEWTONIAN FLUID DOWN AN INCLINED-PLANE SURFACE

The approach used in Section 8.1 will now be applied to a slightly different situation, that of a Newtonian fluid in laminar flow down an inclined-plane surface. This configuration and associated nomenclature are depicted in Figure 8.2. We will examine the two-dimensional case, that is, we consider no significant variation in the z direction.

Figure 8.2 Laminar flow down an inclined-plane surface.

The analysis again involves the application of the control-volume expression for linear momentum in the x direction, which is

$$\sum F_x = \iint_{\text{c.s.}} v_x \rho(\mathbf{v} \cdot \mathbf{n}) \, dA + \frac{\partial}{\partial t} \iiint_{\text{c.v.}} \rho v_x \, dV \tag{5-5a}$$

Evaluating each term in this expression for the fluid element of volume $(\Delta x)(\Delta y)(1)$ as shown in the figure, we have

$$\sum F_x = P\,\Delta y|_x - P\,\Delta y|_{x+\Delta x} + \tau_{yx}\,\Delta x|_{y+\Delta y} - \tau_{yx}\,\Delta x|_y + \rho g\,\Delta x\,\Delta y \sin\theta$$
$$\iint_{\text{c.s.}} \rho v_x(\mathbf{v}\cdot\mathbf{n})\,dA = \rho v_x^2 \Delta y|_{x+\Delta x} - \rho v_x^2 \Delta y|_x$$

and

$$\frac{\partial}{\partial t} \iiint_{\text{c.v.}} \rho v_x \, dV = 0$$

Noting that the convective-momentum terms cancel for fully developed flow and that the pressure–force terms also cancel because of the presence of a free liquid surface, we see that the equation resulting from the substitution of these terms into equation (5-5a) becomes

$$\tau_{yx} \Delta x|_{y+\Delta y} - \tau_{yx} \Delta x|_y + \rho g \, \Delta x \, \Delta y \sin \theta = 0$$

Dividing by $(\Delta x)(\Delta y)(1)$, the volume of the element considered, gives

$$\frac{\tau_{yx}|_{y+\Delta y} - \tau_{yx}|_y}{\Delta y} + \rho g \, \sin\theta = 0$$

In the limit as $\Delta y \to 0$, we get the applicable differential equation

$$\frac{d}{dy} \tau_{yx} + \rho g \, \sin\theta = 0 \tag{8-10}$$

Separating the variables in this simple equation and integrating we obtain for the shear stress

$$\tau_{yx} = -\rho g \, \sin\theta \, y + C_1$$

The integration constant, C_1, may be evaluated by using the boundary condition that the shear stress, τ_{yx}, is zero at the free surface, $y = L$. Thus the shear-stress variation becomes

$$\tau_{yx} = \rho g L \, \sin\theta \left[1 - \frac{y}{L}\right] \tag{8-11}$$

The consideration of a Newtonian fluid in laminar flow enables the substitution of $\mu(dv_x/dy)$, to be made for τ_{yx}, yielding

$$\frac{dv_x}{dy} = \frac{\rho g L \, \sin\theta}{\mu} \left[1 - \frac{y}{L}\right]$$

which, upon separation of variables and integration, becomes

$$v_x = \frac{\rho g L \, \sin\theta}{\mu} \left[y - \frac{y^2}{2L}\right] + C^2$$

Using the no-slip boundary condition, that is, $v_x = 0$ at $y = 0$, the constant of integration, C_2, is seen to be zero. The final expression for the velocity profile may now be written as

$$v_x = \frac{\rho g L^2 \, \sin\theta}{\mu} \left[\frac{y}{L} - \frac{1}{2}\left(\frac{y}{L}\right)^2\right] \tag{8-12}$$

The form of this solution indicates the velocity variation to be parabolic, reaching the maximum value

$$v_{\max} = \frac{\rho g L^2 \, \sin\theta}{2\mu} \tag{8-13}$$

at the free surface, $y = L$.

Additional calculations may be performed to determine the average velocity as was indicated in Section 8.1. Note that there will be no counterpart in this case to the Hagen–Poiseuille relation, equation (8-9), for the pressure gradient. The reason for this is the presence of a free liquid surface along which the pressure is constant. Thus, for our present case, flow is not the result of a pressure gradient but rather the manifestation of the gravitational acceleration upon a fluid.

8.3 CLOSURE

The method of analysis employed in this chapter, that of applying the basic relation for linear momentum to a small control volume and allowing the control volume to shrink to differential size, enables one to find information of a sort different from that obtained previously. Velocity and shear-stress profiles are examples of this type of information. The behavior of a fluid element of differential size can give considerable insight into a given transfer process and provide an understanding available in no other type of analysis.

This method has direct counterparts in heat and mass transfer, where the element may be subjected to an energy or a mass balance.

In Chapter 9, the methods introduced in this chapter will be used to derive differential equations of fluid flow for a general control volume.

PROBLEMS

8.1 Express equation (8-9) in terms of the flow rate and the pipe diameter. If the pipe diameter is doubled at constant pressure drop, what percentage change will occur in the flow rate?

8.2 A 40-km-long pipeline delivers petroleum at a rate of 4000 barrels per day. The resulting pressure drop is 3.45×10^6 Pa. If a parallel line of the same size is laid along the last 18 km of the line, what will be the new capacity of this network? Flow in both cases is laminar and the pressure drop remains 3.45×10^6 Pa.

8.3 A 0.635-cm hydraulic line suddenly ruptures 8 m from a reservoir with a gage pressure of 207 kPa. Compare the laminar and inviscid flow rates from the ruptured line in cubic meters per second.

8.4 A common type of viscosimeter for liquids consists of a relatively large reservoir with a very slender outlet tube, the rate of outflow being determined by timing the fall in the surface level. If oil of constant density flows out of the viscosimeter shown at the rate of 0.273 cm^3/s, what is the kinematic viscosity of the fluid? The tube diameter is 0.18 cm.

8.5 Derive the expressions for the velocity distribution and for the pressure drop for a Newtonian fluid in fully developed laminar flow in the annular space between two horizontal, concentric pipes. Apply the momentum theorem to an annular fluid shell of thickness Dr and show that the analysis of such a control volume leads to

$$\frac{d}{dr}(r\tau) = r\frac{\Delta P}{L}$$

The desired expressions may then be obtained by the substitution of Newton's viscosity law and two integrations.

8.6 A thin rod of diameter d is pulled at constant velocity through a pipe of diameter D. If the wire is at the center of the pipe, find the drag per unit length of wire. The fluid filling the space between the rod and the inner pipe wall has density ρ and viscosity μ.

8.7 The viscosity of heavy liquids, such as oils, is frequently measured with a device that consists of a rotating cylinder inside a large cylinder. The annular region between these cylinders is filled with liquid and the torque required to rotate the inner cylinder at constant speed is computed, a linear velocity profile being assumed. For what ratio of cylinder diameters is the assumption of a linear profile accurate within 1% of the true profile?

8.8 Two immiscible fluids of different density and viscosity are flowing between two parallel plates. Express the boundary conditions at the interface between the two fluids.

8.9 Determine the velocity profile for fluid flowing between two parallel plates separated by a distance $2h$. The pressure drop is constant.

8.10   Fluid flows between two parallel plates, a distance h apart. The upper plate moves at velocity, v_0; the lower plate is stationary. For what value of pressure gradient will the shear stress at the lower wall be zero?

8.11   Derive the equation of motion for a one-dimensional, inviscid, unsteady compressible flow in a pipe of constant cross-sectional area neglect gravity.

8.12   A continuous belt passes upward through a chemical bath at velocity v_0 and picks up a film of liquid of thickness h, density, ρ, and viscosity μ. Gravity tends to make the liquid drain down, but the movement of the belt keeps the fluid from running off completely. Assume that the flow is a well-developed laminar flow with zero pressure gradient, and that the atmosphere produces no shear at the outer surface of the film.

(a) State clearly the boundary conditions at $y = 0$ and $y = h$ to be satisfied by the velocity.

(b) Calculate the velocity profile.

(c) Determine the rate at which fluid is being dragged up with the belt in terms of μ, ρ, h, v_0.

8.13   The device in the schematic diagram below is a viscosity pump. It consists of a rotating drum inside of a stationary case. The case and the drum are concentric. Fluid enters at A, flows through the annulus between the case and the drum, and leaves at B. The pressure at B is higher than that at A, the difference being ΔP. The length of the annulus is L. The width of the annulus h is very small compared to the diameter of the drum, so that the flow in the annulus is equivalent to the flow between two flat plates.

Assume the flow to be laminar. Find the pressure rise and efficiency as a function of the flow rate per unit depth.

8.14   Oil is supplied at the center of two long plates. The volumetric flow rate per unit length is Q and the plates remain a constant distance, b, apart. Determine the vertical force per unit length as a function of the Q, μ, L, and b.

8.15   A viscous film drains uniformly down the side of a vertical rod of radius R. At some distance down the rod, the film approaches a terminal or fully developed flow such that the film thickness, h, is constant and $v_z = f(r)$. Neglecting the shear resistance due to the atmosphere, determine the velocity distribution in the film.

8.16   Determine the maximum film velocity in Problem 8.15.

Chapter 9

Differential Equations
of Fluid Flow

The fundamental laws of fluid flow, which have been expressed in mathematical form
for an arbitrary control volume in Chapters 4–6, may also be expressed in mathematical
form for a special type of control volume, the differential element. These differential
equations of fluid flow provide a means of determining the point-to-point variation of
fluid properties. Chapter 8 involved the differential equations associated with some
one-dimensional, steady, incompressible laminar flows. In Chapter 9, we shall express
the law of conservation of mass and Newton's second law of motion in differential
form for more general cases. The basic tools used to derive these differential equations
will be the control-volume developments of Chapters 4 and 5.

9.1 THE DIFFERENTIAL CONTINUITY EQUATION

The continuity equation to be developed in this section is the law of conservation of mass
expressed in differential form. Consider the control volume $\Delta x \, \Delta y \, \Delta z$ shown in Figure 9.1.

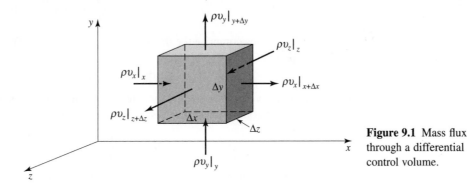

Figure 9.1 Mass flux through a differential control volume.

The control volume expression for the conservation of mass is

$$\iint \rho(\mathbf{v} \cdot \mathbf{n})dA + \frac{\partial}{\partial t} \iiint \rho \, dV = 0 \qquad (4\text{-}1)$$

which states that

$$\begin{Bmatrix} \text{net rate of mass} \\ \text{flux out of} \\ \text{control volume} \end{Bmatrix} + \begin{Bmatrix} \text{rate of accumulation} \\ \text{of mass within} \\ \text{control volume} \end{Bmatrix} = 0$$

The mass flux $\rho(\mathbf{v} \cdot \mathbf{n})$ at each face of the control volume is illustrated in Figure 9.1. The mass within the control volume is $\rho\, \Delta x\, \Delta y\, \Delta z$, and thus the time rate of change of mass within the control volume is

$$\frac{\partial}{\partial t}(\rho\, \Delta x\, \Delta y\, \Delta z)$$

The student is reminded that the density in general can vary from point to point, that is, $\rho = \rho(x, y, z, t)$.

The net mass flux out of the control volume in the x direction is

$$(\rho v_x|_{x+\Delta x} - \rho v_x|_x)\Delta y\, \Delta z$$

in the y direction

$$(\rho v_y|_{y+\Delta y} - \rho v_y|_y)\Delta x\, \Delta z$$

and in the z direction

$$(\rho v_z|_{z+\Delta z} - \rho v_z|_z)\Delta x\, \Delta y$$

The total net mass flux is the sum of the above three terms. Substituting into equation (4-1) yields

$$(\rho v_x|_{x+\Delta x} - \rho v_x|_x)\Delta y\, \Delta z + (\rho v_y|_{y+\Delta y} - \rho v_y|_y)\Delta x\, \Delta z$$
$$+ (\rho v_z|_{z+\Delta z} - \rho v_z|_z)\Delta x\, \Delta y + \frac{\partial}{\partial t}(\rho\, \Delta x\, \Delta y\, \Delta z) = 0$$

The volume does not change with time, so we may divide the above equation by $\Delta x\, \Delta y\, \Delta z$. In the limit as Δx, Δy, and Δz approach zero, we obtain

$$\frac{\partial}{\partial x}(\rho v_x) + \frac{\partial}{\partial y}(\rho v_y) + \frac{\partial}{\partial z}(\rho v_z) + \frac{\partial \rho}{\partial t} = 0 \qquad (9\text{-}1)$$

The first three terms comprise the divergence of the vector $\rho\mathbf{v}$. The divergence of a vector is the dot product with ∇

$$\text{div } \mathbf{A} \equiv \nabla \cdot \mathbf{A}$$

The student may verify that the first three terms in equation (9-1) may be written as $\nabla \cdot \rho\mathbf{v}$, and thus a more compact statement of the continuity equation becomes

$$\nabla \cdot \rho\mathbf{v} + \frac{\partial \rho}{\partial t} = 0 \qquad (9\text{-}2)$$

The continuity equation above applies to unsteady, three-dimensional flow. It is apparent that, when flow is incompressible, this equation reduces to

$$\nabla \cdot \mathbf{v} = 0 \qquad (9\text{-}3)$$

whether the flow is unsteady or not.

Equation (9-2) may be arranged in a slightly different form to illustrate the use of the *substantial derivative*. Carrying out the differentiation indicated in (9-1), we have

$$\frac{\partial \rho}{\partial t} + v_x\frac{\partial \rho}{\partial x} + v_y\frac{\partial \rho}{\partial y} + v_z\frac{\partial \rho}{\partial z} + \rho\left(\frac{\partial v_x}{\partial x} + \frac{\partial v_y}{\partial y} + \frac{\partial v_z}{\partial z}\right) = 0$$

The first four terms of the above equation comprise the substantial derivative of the density, symbolized as $D\rho/Dt$, where

$$\frac{D}{Dt} = \frac{\partial}{\partial t} + v_x \frac{\partial}{\partial x} + v_y \frac{\partial}{\partial y} + v_z \frac{\partial}{\partial z} \tag{9-4}$$

in cartesian coordinates. The continuity equation may, thus be written as

$$\frac{D\rho}{Dt} + \rho \mathbf{\nabla} \cdot \mathbf{v} = 0 \tag{9-5}$$

When considering the total differential of a quantity, three different approaches may be taken. If, for instance, we wish to evaluate the change in atmospheric pressure, P, the total differential written in rectangular coordinates is

$$dP = \frac{\partial P}{\partial t} dt + \frac{\partial P}{\partial x} dx + \frac{\partial P}{\partial y} dy + \frac{\partial P}{\partial z} dz$$

where dx, dy, and dz are arbitrary displacements in the x, y, and z directions. The rate of pressure change is obtained by dividing through by dt, giving

$$\frac{dP}{dt} = \frac{\partial P}{\partial t} + \frac{dx}{dt} \frac{\partial P}{\partial x} + \frac{dy}{dt} \frac{\partial P}{\partial y} + \frac{dz}{dt} \frac{\partial P}{\partial z} \tag{9-6}$$

As a first approach, the instrument to measure pressure is located in a weather station, which is, of course, fixed on Earth's surface. Thus, the coefficients dx/dt, dy/dt, dz/dt are all zero, and for a fixed point of observation the total derivative, dP/dt, is equal to the local derivative with respect to time $\partial P/\partial t$.

A second approach involves the pressure-measuring instrument housed in an aircraft which, at the pilot's discretion, can be made to climb or descend, or fly in any chosen x, y, z direction. In this case, the coefficients dx/dt, dy/dt, dz/dt are the x, y, and z velocities of the aircraft, and they are arbitrarily chosen, bearing only coincidental relationship to the air currents.

The third situation is one in which the pressure indicator is in a balloon that rises, falls, and drifts as influenced by the flow of air in which it is suspended. Here the coefficients dx/dt, dy/dt, dz/dt are those of the *flow* and they may be designated v_x, v_y, and v_z, respectively. This latter situation corresponds to the substantial derivative and the terms may be grouped as designated below

$$\frac{dP}{dt} = \frac{DP}{Dt} = \underbrace{\frac{\partial P}{\partial t}}_{\substack{\text{local} \\ \text{rate of} \\ \text{change of} \\ \text{pressure}}} + \underbrace{v_x \frac{\partial P}{\partial x} + v_y \frac{\partial P}{\partial y} + v_z \frac{\partial P}{\partial z}}_{\substack{\text{rate of change} \\ \text{of pressure} \\ \text{due to motion}}} \tag{9-7}$$

The derivative D/Dt may be interpreted as the time rate of change of a fluid or flow variable along the path of a fluid element. The substantial derivative will be applied to both scalar and vector variables in subsequent sections.

9.2 NAVIER–STOKES EQUATIONS

The Navier–Stokes equations are the differential form of Newton's second law of motion. Consider the differential control volume illustrated in Figure 9.1.

The basic tool we shall use in developing the Navier–Stokes equations is Newton's second law of motion for an arbitrary control volume as given in Chapter 5

$$\sum \mathbf{F} = \iint_{c.s.} \rho \mathbf{v}(\mathbf{v} \cdot \mathbf{n}) dA + \frac{\partial}{\partial t} \iiint_{c.v.} \rho \mathbf{v} dV \tag{5-4}$$

which states that

$$\left\{ \begin{array}{c} \text{sum of the external} \\ \text{forces acting on the} \\ \text{c.v.} \end{array} \right\} = \left\{ \begin{array}{c} \text{net rate of linear} \\ \text{momentum efflux} \end{array} \right\} + \left\{ \begin{array}{c} \text{time rate of change} \\ \text{of linear momentum} \\ \text{within the c.v.} \end{array} \right\}$$

As the mathematical expression for each of the above terms is rather lengthy, each will be evaluated separately and then substituted into equation (5-4).

The development may be further simplified by recalling that we have, in the prior case, divided by the volume of the control volume and taken the limit as the dimensions approach zero. Equation (5-4) can also be written

$$\lim_{\Delta x, \Delta y, \Delta z \to 0} \frac{\sum \mathbf{F}}{\Delta x\, \Delta y\, \Delta z} = \lim_{\Delta x, \Delta y, \Delta z \to 0} \frac{\iint \rho \mathbf{v}(\mathbf{v} \cdot \mathbf{n})\, dA}{\Delta x\, \Delta y\, \Delta z} + \lim_{\Delta x, \Delta y, \Delta z \to 0} \frac{\partial/\partial t \iiint \rho \mathbf{v}\, dV}{\Delta x\, \Delta y\, \Delta z} \tag{9-8}$$

$$\textcircled{1} \qquad\qquad\qquad\qquad \textcircled{2} \qquad\qquad\qquad\qquad \textcircled{3}$$

① *Sum of the external forces.*   The forces acting on the control volume are those due to the normal stress and to the shear stress, and body forces such as that due to gravity. Figure 9.2

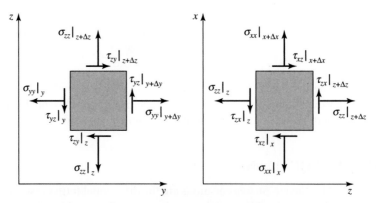

Figure 9.2 Forces acting on a differential control volume.

illustrates the various forces acting on the control volume. Summing the forces in the x direction, we obtain

$$\sum F_x = (\sigma_{xx}|_{x+\Delta x} - \sigma_{xx}|_x)\Delta y\,\Delta z + (\tau_{yx}|_{y+\Delta y} - \tau_{yx}|_y)\Delta x\,\Delta z$$
$$+ (\tau_{zx}|_{z+\Delta z} - \tau_{zx}|_z)\Delta x\,\Delta y + g_x\rho\,\Delta x\,\Delta y\,\Delta z$$

where g_x is the component of the gravitational acceleration in the x direction. In the limit as the dimensions of the element approach zero this becomes

$$\lim_{\Delta x,\Delta y,\Delta z \to 0} \frac{\sum F_x}{\Delta x\,\Delta y\,\Delta z} = \frac{\partial \sigma_{xx}}{\partial x} + \frac{\partial \tau_{yx}}{\partial y} + \frac{\partial \tau_{zx}}{\partial z} + \rho g_x \tag{9-9}$$

Similar expressions are obtained for the force summations in the y and z directions

$$\lim_{\Delta x,\Delta y,\Delta z \to 0} \frac{\sum F_y}{\Delta x\,\Delta y\,\Delta z} = \frac{\partial \tau_{xy}}{\partial x} + \frac{\partial \sigma_{yy}}{\partial y} + \frac{\partial \tau_{zy}}{\partial z} + \rho g_y \tag{9-10}$$

$$\lim_{\Delta x,\Delta y,\Delta z \to 0} \frac{\sum F_z}{\Delta x\,\Delta y\,\Delta z} = \frac{\partial \tau_{xz}}{\partial x} + \frac{\partial \tau_{yz}}{\partial y} + \frac{\partial \sigma_{zz}}{\partial z} + \rho g_z \tag{9-11}$$

② *Net momentum flux through the control volume.* The net momentum flux through the control volume illustrated in Figure 9.3 is

$$\lim_{\Delta x,\Delta y,\Delta z \to 0} \frac{\iint \rho\mathbf{v}(\mathbf{v}\cdot\mathbf{n})dA}{\Delta x\,\Delta y\,\Delta z} = \lim_{\Delta x,\Delta y,\Delta z \to 0} \left[\frac{(\rho\mathbf{v}v_x|_{x+\Delta x} - \rho\mathbf{v}v_x|_x)\Delta y\,\Delta z}{\Delta x\,\Delta y\,\Delta z} \right.$$
$$+ \frac{(\rho\mathbf{v}v_y|_{y+\Delta y} - \rho\mathbf{v}v_y|_y)\Delta x\,\Delta z}{\Delta x\,\Delta y\,\Delta z}$$
$$\left. + \frac{(\rho\mathbf{v}v_z|_{z+\Delta z} - \rho\mathbf{v}v_z|_z)\Delta x\,\Delta y}{\Delta x\,\Delta y\,\Delta z} \right] \tag{9-12}$$
$$= \frac{\partial}{\partial x}(\rho\mathbf{v}v_x) + \frac{\partial}{\partial y}(\rho\mathbf{v}v_y) + \frac{\partial}{\partial z}(\rho\mathbf{v}v_z)$$

Performing the indicated differentiation of the right-hand side of equation (9-12) yields

$$\lim_{\Delta x,\Delta y,\Delta z \to 0} \frac{\iint \rho\mathbf{v}(\mathbf{v}\cdot\mathbf{n})dA}{\Delta x\,\Delta y\,\Delta z} = \mathbf{v}\left[\frac{\partial}{\partial x}(\rho v_x) + \frac{\partial}{\partial y}(\rho v_y)\frac{\partial}{\partial z}(\rho v_z) \right]$$
$$+ \rho\left[v_x\frac{\partial \mathbf{v}}{\partial x} + v_y\frac{\partial \mathbf{v}}{\partial y} + v_z\frac{\partial \mathbf{v}}{\partial z} \right]$$

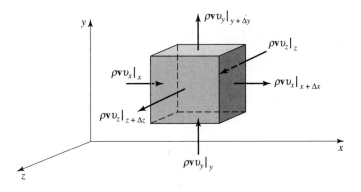

Figure 9.3 Momentum flux through a differential control volume.

The above term may be simplified with the aid of the continuity equation

$$\frac{\partial \rho}{\partial t} + \frac{\partial}{\partial x}(\rho v_x) + \frac{\partial}{\partial y}(\rho v_y) + \frac{\partial}{\partial z}(\rho v_z) = 0 \tag{9-1}$$

which, upon substitution, yields

$$\lim_{\Delta x, \Delta y, \Delta z \to 0} \frac{\iint \rho \mathbf{v}(\mathbf{v} \cdot \mathbf{n})dA}{\Delta x \, \Delta y \, \Delta z} = -\mathbf{v}\frac{\partial \rho}{\partial t} + \rho\left[v_x\frac{\partial \mathbf{v}}{\partial x} + v_y\frac{\partial \mathbf{v}}{\partial y} + vz\frac{\partial \mathbf{v}}{\partial z}\right] \tag{9-13}$$

③ *Time rate of change of momentum within the control volume.*   The time rate of change of momentum within the control volume may be evaluated directly

$$\lim_{\Delta x, \Delta y, \Delta z \to 0} \frac{\partial/\partial t \iiint \mathbf{v}\rho \, dV}{\Delta x \, \Delta y \, \Delta z} = \frac{(\partial/\partial t)\rho \mathbf{v} \, \Delta x \, \Delta y \, \Delta z}{\Delta x \, \Delta y \, \Delta z} = \frac{\partial}{\partial t}\rho \mathbf{v} = \rho\frac{\partial \mathbf{v}}{\partial t} + \mathbf{v}\frac{\partial \rho}{\partial t} \tag{9-14}$$

We have now evaluated all terms in equation 9-8

① $$\lim_{\Delta x, \Delta y, \Delta z \to 0} \frac{\sum \mathbf{F}}{\Delta x \, \Delta y \, \Delta z} = \begin{cases} \left(\dfrac{\partial \sigma_{xx}}{\partial x} + \dfrac{\partial \tau_{yx}}{\partial y} + \dfrac{\partial \tau_{zx}}{\partial z} + \rho g_x\right)\mathbf{e}_x & (9\text{-}9) \\[2em] \left(\dfrac{\partial \tau_{xy}}{\partial x} + \dfrac{\partial \sigma_{yy}}{\partial y} + \dfrac{\partial \tau_{zy}}{\partial z} + \rho g_y\right)\mathbf{e}_y & (9\text{-}10) \\[2em] \left(\dfrac{\partial \tau_{xz}}{\partial x} + \dfrac{\partial \tau_{yz}}{\partial y} + \dfrac{\partial \sigma_{zz}}{\partial z} + \rho g_z\right)\mathbf{e}_z & (9\text{-}11) \end{cases}$$

② $$\lim_{\Delta x, \Delta y, \Delta z \to 0} \frac{\iint \rho \mathbf{v}(\mathbf{v} \cdot \mathbf{n})dA}{\Delta x \, \Delta y \, \Delta z} = -\mathbf{v}\frac{\partial \rho}{\partial t} + \rho\left(v_x\frac{\partial \mathbf{v}}{\partial x} + v_y\frac{\partial \mathbf{v}}{\partial y} + v_z\frac{\partial \mathbf{v}}{\partial z}\right) \tag{9-13}$$

③ $$\lim_{\Delta x, \Delta y, \Delta z \to 0} \frac{\partial/\partial t \iiint \rho \mathbf{v}dV}{\Delta x \, \Delta y \, \Delta z} = \rho\frac{\partial \mathbf{v}}{\partial t} + \mathbf{v}\frac{\partial \rho}{\partial t} \tag{9-14}$$

It can be seen that the forces are expressed in components, whereas the rate-of-change-of-momentum terms are expressed as vectors. When the momentum terms are expressed as components, we obtain three differential equations that are the statements of Newton's second law in the x, y, and z directions

$$\rho\left(\frac{\partial v_x}{\partial t} + v_x\frac{\partial v_x}{\partial x} + v_y\frac{\partial v_x}{\partial y} + v_z\frac{\partial v_x}{\partial z}\right) = \rho g_x + \frac{\partial \sigma_{xx}}{\partial x} + \frac{\partial \tau_{yx}}{\partial y} + \frac{\partial \tau_{zx}}{\partial z} \tag{9-15a}$$

$$\rho\left(\frac{\partial v_y}{\partial t} + v_x\frac{\partial v_y}{\partial x} + v_y\frac{\partial v_y}{\partial y} + v_z\frac{\partial v_y}{\partial z}\right) = \rho g_y + \frac{\partial \tau_{xy}}{\partial x} + \frac{\partial \sigma_{yy}}{\partial y} + \frac{\partial \tau_{zy}}{\partial z} \tag{9-15b}$$

$$\rho\left(\frac{\partial v_z}{\partial t} + v_x\frac{\partial v_z}{\partial x} + v_y\frac{\partial v_z}{\partial y} + v_z\frac{\partial v_z}{\partial z}\right) = \rho g_z + \frac{\partial \tau_{xz}}{\partial x} + \frac{\partial \tau_{yz}}{\partial y} + \frac{\partial \sigma_{zz}}{\partial z} \tag{9-15c}$$

It will be noted that in equations (9-15) above, the terms on the left-hand side represent the time-rate of change of momentum, and the terms on the right-hand side represent the forces. Focusing our attention on the left-hand terms in equation (9-15a), we see that

$$\underbrace{\frac{\partial v_x}{\partial t}}_{\substack{\text{local} \\ \text{of change} \\ \text{of } v_x}} + \underbrace{v_x \frac{\partial v_x}{\partial x} + v_y \frac{\partial v_x}{\partial y} + v_z \frac{\partial v_x}{\partial z}}_{\substack{\text{rate of change in} \\ v_x \text{ due to motion}}} = \left(\frac{\partial}{\partial t} + v_x \frac{\partial}{\partial x} + v_y \frac{\partial}{\partial y} + v_z \frac{\partial}{\partial z}\right) v_x$$

The first term, $\partial v_x/\partial t$, involves the time rate of change of v_x at a point and is called the *local acceleration.* The remaining terms involve the velocity change from point to point, that is, the *convective acceleration.* The sum of these two bracketed terms is the total acceleration. The reader may verify that the terms on the left-hand side of equations (9-15) are all of the form

$$\left(\frac{\partial}{\partial t} + v_x \frac{\partial}{\partial x} + v_y \frac{\partial}{\partial y} + v_z \frac{\partial}{\partial z}\right) v_i$$

where $v_i = v_x$, v_y, or v_z. The above term is the substantial derivative of v_i.

When the substantial derivative notation is used, equations (9-15) become

$$\rho \frac{Dv_x}{Dt} = \rho g_x + \frac{\partial \sigma_{xx}}{\partial x} + \frac{\partial \tau_{yx}}{\partial y} + \frac{\partial \tau_{zx}}{\partial z} \tag{9-16a}$$

$$\rho \frac{Dv_y}{Dt} = \rho g_y + \frac{\partial \tau_{xy}}{\partial x} + \frac{\partial \sigma_{yy}}{\partial y} + \frac{\partial \tau_{zy}}{\partial z} \tag{9-16b}$$

and

$$\rho \frac{Dv_z}{Dt} = \rho g_z + \frac{\partial \tau_{xz}}{\partial x} + \frac{\partial \tau_{yz}}{\partial y} + \frac{\partial \sigma_{zz}}{\partial z} \tag{9-16c}$$

Equations (9-16) are valid for any type of fluid, regardless of the nature of the stress rate-of-strain relation. If Stokes's viscosity relations, equations (7-13) and (7-14), are used for the stress components, equations (9-16) become

$$\rho \frac{Dv_x}{Dt} = \rho g_x - \frac{\partial P}{\partial x} - \frac{\partial}{\partial x}\left(\frac{2}{3}\mu \boldsymbol{\nabla} \cdot \mathbf{v}\right) + \boldsymbol{\nabla} \cdot \left(\mu \frac{\partial \mathbf{v}}{\partial x}\right) + \boldsymbol{\nabla} \cdot (\mu \boldsymbol{\nabla} v_x) \tag{9-17a}$$

$$\rho \frac{Dv_y}{Dt} = \rho g_y - \frac{\partial P}{\partial y} - \frac{\partial}{\partial y}\left(\frac{2}{3}\mu \boldsymbol{\nabla} \cdot \mathbf{v}\right) + \boldsymbol{\nabla} \cdot \left(\mu \frac{\partial \mathbf{v}}{\partial y}\right) + \boldsymbol{\nabla} \cdot (\mu \boldsymbol{\nabla} v_y) \tag{9-17b}$$

and

$$\rho \frac{Dv_z}{Dt} = \rho g_z - \frac{\partial P}{\partial z} - \frac{\partial}{\partial z}\left(\frac{2}{3}\mu \boldsymbol{\nabla} \cdot \mathbf{v}\right) + \boldsymbol{\nabla} \cdot \left(\mu \frac{\partial \mathbf{v}}{\partial z}\right) + \boldsymbol{\nabla} \cdot (\mu \boldsymbol{\nabla} v_z) \tag{9-17c}$$

The above equations are called the Navier–Stokes[1] equations and are the differential expressions of Newton's second law of motion for a Newtonian fluid. As no assumptions

---

[1] L. M. H. Navier, Mémoire sur les Lois du Mouvements des Fluides, *Mem. de l'Acad. d. Sci.,* **6**, 398 (1822); C. G. Stokes, On the Theories of the Internal Friction of Fluids in Motion, *Trans. Cambridge Phys. Soc.,* **8** (1845).

relating to the compressibility of the fluid have been made, these equations are valid for both compressible and incompressible flows. In our study of momentum transfer we shall restrict our attention to incompressible flow with constant viscosity. In an incompressible flow, $\boldsymbol{\nabla} \cdot \mathbf{v} = 0$. Equations (9-17) thus become

$$\rho \frac{Dv_x}{Dt} = \rho g_x - \frac{\partial P}{\partial x} + \mu \left(\frac{\partial^2 v_x}{\partial x^2} + \frac{\partial^2 v_x}{\partial y^2} + \frac{\partial^2 v_x}{\partial z^2} \right) \tag{9-18a}$$

$$\rho \frac{Dv_y}{Dt} = \rho g_y - \frac{\partial P}{\partial y} + \mu \left(\frac{\partial^2 v_y}{\partial x^2} + \frac{\partial^2 v_y}{\partial y^2} + \frac{\partial^2 v_y}{\partial z^2} \right) \tag{9-18b}$$

$$\rho \frac{Dv_z}{Dt} = \rho g_z - \frac{\partial P}{\partial z} + \mu \left(\frac{\partial^2 v_z}{\partial x^2} + \frac{\partial^2 v_z}{\partial y^2} + \frac{\partial^2 v_z}{\partial z^2} \right) \tag{9-18c}$$

These equations may be expressed in a more compact form in the single vector equation

$$\rho \frac{D\mathbf{v}}{Dt} = \rho \mathbf{g} - \boldsymbol{\nabla} P + \mu \boldsymbol{\nabla}^2 \mathbf{v} \tag{9-19}$$

The above equation is the Navier–Stokes equation for an incompressible flow. The Navier–Stokes equations are written in cartesian, cylindrical, and spherical coordinate forms in Appendix E. As the development has been lengthy, let us review the assumptions and, therefore, the limitations of equation (9-19). The assumptions are

1. incompressible flow,
2. constant viscosity,
3. laminar flow.[2]

All of the above assumptions are associated with the use of the Stokes viscosity relation. If the flow is inviscid ($\mu = 0$), the Navier–Stokes equation becomes

$$\rho \frac{D\mathbf{v}}{Dt} = \rho \mathbf{g} - \boldsymbol{\nabla} P \tag{9-20}$$

which is known as Euler's equation. Euler's equation has only one limitation, that being inviscid flow.

EXAMPLE 1

Equation (9-19) may be applied to numerous flow systems to provide information regarding velocity variation, pressure gradients, and other information of the type achieved in Chapter 8. Many situations are of sufficient complexity to make the solution extremely difficult and are beyond the scope of this text. A situation for which a solution can be obtained is illustrated in Figure 9.4.

Figure 9.4 shows the situation of an incompressible fluid confined between two parallel, vertical surfaces. One surface, shown to the left, is stationary, whereas the other is moving upward at a constant velocity v_0. If we consider the fluid Newtonian and the flow laminar, the governing equation

---

[2] Strictly speaking, equation (9-19) is valid for turbulent flow, as the turbulent stress is included in the momentum flux term. This will be illustrated in Chapter 12.

of motion is the Navier–Stokes equation in the form given by equation (9-19). The reduction of each term in the vector equation into its applicable form is shown below.

$$\rho \frac{D\mathbf{v}}{Dt} = \rho \left\{ \frac{\partial \mathbf{v}}{\partial t} + v_x \frac{\partial \mathbf{v}}{\partial x} + v_y \frac{\partial \mathbf{v}}{\partial y} + v_z \frac{\partial \mathbf{v}}{\partial z} \right\} = 0$$

$$\rho \mathbf{g} = -\rho g \mathbf{e}_y$$

$$\boldsymbol{\nabla} P = \frac{dP}{dy} \mathbf{e}_y$$

where dP/dy is constant, and

$$\mu \boldsymbol{\nabla}^2 \mathbf{v} = \mu \frac{d^2 v_y}{dx^2} \mathbf{e}_y$$

The resulting equation to be solved is

$$0 = -\rho g - \frac{dP}{dy} + \mu \frac{d^2 v_y}{dx^2}$$

This differential equation is separable. The first integration yields

$$\frac{dv_y}{dx} + \frac{x}{\mu} \left\{ -\rho g - \frac{dP}{dy} \right\} = C_1$$

Integrating once more, we obtain

$$v_y + \frac{x^2}{2\mu} \left\{ -\rho g - \frac{dP}{dy} \right\} = C_1 x + C_2$$

The integration constants may be evaluated, using the boundary conditions that $v_y = 0$ at $x = 0$, and $v_y = v_0$ at $x = L$. The constants thus become

$$C_1 = \frac{v_0}{L} + \frac{L}{2\mu} \left\{ -\rho g - \frac{dP}{dy} \right\} \quad \text{and} \quad C_2 = 0$$

The velocity profile may now be expressed as

$$v_y = \underbrace{\frac{1}{2\mu} \left\{ -\rho g - \frac{dP}{dy} \right\} \left\{ Lx - x^2 \right\}}_{①} + \underbrace{v_0 \frac{x}{L}}_{②} \tag{9-21}$$

It is interesting to note, in equation (9-21), the effect of the terms labeled ① and ②, which are added. The first term is the equation for a symmetric parabola, the second for a straight line. Equation (9-21) is valid whether v_0 is upward, downward, or zero. In each case, the terms may be added to yield the complete velocity profile. These results are indicated in Figure 9.5. The resulting velocity profile obtained by superposing the two parts is shown in each case.

Figure 9.4 Fluid between two vertical plates with the one on the left stationary and the other on the right moving vertically upward with velocity v_0.

Figure 9.5 Velocity profiles for one surface moving upward, downward, or stationary.

Euler's equation may also be solved to determine velocity profiles, as will be shown in Chapter 10. The vector properties of Euler's equation are illustrated by the example below, in which the form of the velocity profile is given.

EXAMPLE 2

A rotating shaft, as illustrated in Figure 9.6, causes the fluid to move in circular streamlines with a velocity that is inversely proportional to the distance from the shaft. Find the shape of the free surface if the fluid can be considered inviscid.

Streamlines

Figure 9.6 Rotating shaft in a fluid.

As the pressure along the free surface will be constant, we may observe that the free surface is perpendicular to the pressure gradient. Determination of the pressure gradient, therefore, will enable us to evaluate the slope of the free surface.

Rearranging equation (9-20), we have

$$\boldsymbol{\nabla} P = \rho \mathbf{g} - \rho \frac{D\mathbf{v}}{Dt} \tag{9-20}$$

The velocity $\mathbf{v} = A\mathbf{e}_\theta/r$, where A is a constant, when using the coordinate system shown in Figure 9.7. Assuming that there is no slip between the fluid and the shaft at the surface of the

Figure 9.7 Cylindrical coordinate system for rotating shaft and fluid.

shaft, we have

$$v(R) = \omega R = \frac{A}{R}$$

and thus $A = \omega R^2$ and

$$\mathbf{v} = \frac{\omega R^2}{r}\mathbf{e}_\theta$$

The substantial derivative $D\mathbf{v}/Dt$ may be evaluated by taking the total derivative

$$\frac{d\mathbf{v}}{dt} = -\frac{\omega R^2}{r^2}\mathbf{e}_\theta \dot{r} + \frac{\omega R^2}{r}\frac{d\mathbf{e}_\theta}{dt}$$

where $d\mathbf{e}_\theta/dt = -\dot{\theta}\mathbf{e}_r$. The total derivative becomes

$$\frac{d\mathbf{v}}{dt} = -\frac{\omega R^2}{r^2}\dot{r}\mathbf{e}_\theta - \frac{\omega R^2}{r}\dot{\theta}\mathbf{e}_r$$

Now the fluid velocity in the r direction is zero, and $\dot{\theta}$ for the fluid is v/r, so

$$\left(\frac{d\mathbf{v}}{dt}\right)_{\text{fluid}} = \frac{D\mathbf{v}}{Dt} = -\frac{\omega R^2}{r^2}v\mathbf{e}_r = -\frac{\omega^2 R^4}{r^3}\mathbf{e}_r$$

This result could have been obtained in a more direct manner by observing that $D\mathbf{v}/Dt$ is the local fluid acceleration, which for this case is $-v^2\mathbf{e}_r/r$. The pressure gradient becomes

$$\boldsymbol{\nabla} P = -\rho g\mathbf{e}_z + \rho\frac{\omega^2 R^4 \mathbf{e}_r}{r^3}$$

From Figure 9.8, it can be seen that the free surface makes an angle β with the r axis so that

$$\tan\beta = \frac{\rho\omega^2 R^4}{r^3 \rho g}$$

$$= \frac{\omega^2 R^4}{g r^3}$$

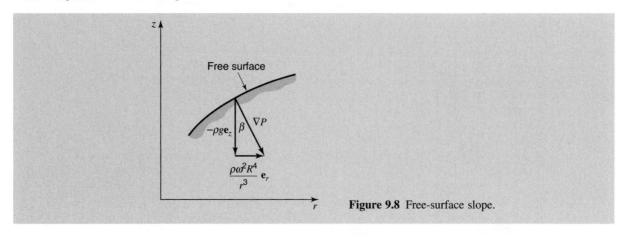

Figure 9.8 Free-surface slope.

9.3 BERNOULLI'S EQUATION

Euler's equation may be integrated directly for a particular case, flow along a streamline. In integrating Euler's equation, the use of streamline coordinates is extremely helpful. Streamline coordinates s and n are illustrated in Figure 9.9. The s direction is parallel to the streamline and the n direction is perpendicular to the streamline, directed away from the instantaneous center of curvature. The flow and fluid properties are functions of position and time. Thus, $\mathbf{v} = \mathbf{v}(s, n, t)$, and $P = P(s, n, t)$. The substantial derivatives of the velocity and pressure gradients in equation (9-20) must be expressed in terms of streamline coordinates so that equation (9-20) may be integrated.

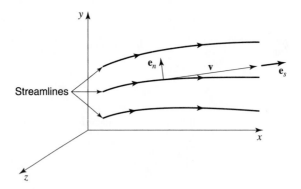

Figure 9.9 Streamline coordinates.

Following the form used in equations (9-6) to obtain the substantial derivative, we have

$$\frac{d\mathbf{v}}{dt} = \frac{\partial \mathbf{v}}{\partial t} + \dot{s}\frac{\partial \mathbf{v}}{\partial s} + \dot{n}\frac{\partial \mathbf{v}}{\partial n}$$

As the velocity of the fluid element has components $\dot{s} = v, \dot{n} = 0$, the substantial derivative of the velocity in streamline coordinates is

$$\frac{D\mathbf{v}}{Dt} = \frac{\partial \mathbf{v}}{\partial t} + v\frac{\partial \mathbf{v}}{\partial s} \tag{9-22}$$

The pressure gradient in streamline coordinates may be written as

$$\nabla P = \frac{\partial P}{\partial s}\mathbf{e}_s + \frac{\partial P}{\partial n}\mathbf{e}_n \tag{9-23}$$

Taking the dot product of equation (9-20) with $\mathbf{e}_s\,ds$, and using equations (9-22) and (9-23), we obtain

$$\rho\left(\frac{\partial \mathbf{v}}{\partial t}\cdot \mathbf{e}_s\,ds + v\frac{\partial \mathbf{v}}{\partial s}\cdot \mathbf{e}_s\,ds\right) = \rho\mathbf{g}\cdot\mathbf{e}_s\,ds - \left(\frac{\partial P}{\partial s}\mathbf{e}_s + \frac{\partial P}{\partial n}\mathbf{e}_n\right)\cdot\mathbf{e}_s\,ds$$

or, as $\partial \mathbf{v}/\partial s \cdot \mathbf{e}_s = \partial/\partial s(\mathbf{v}\cdot\mathbf{e}_s) = \partial v/\partial s$, we have

$$\rho\left(\frac{\partial \mathbf{v}}{\partial t}\cdot\mathbf{e}_s\,ds + \frac{\partial}{\partial s}\left\{\frac{v^2}{2}\right\}ds\right) = \rho\mathbf{g}\cdot\mathbf{e}_s\,ds - \frac{\partial P}{\partial s}\,ds \tag{9-24}$$

Selecting $\mathbf{g}$ to act in the $-\mathbf{y}$ direction, we have $\mathbf{g}\cdot\mathbf{e}_s\,ds = -g\,dy$. For *steady incompressible* flow, equation (9-24) may be integrated to yield

$$\frac{v^2}{2} + gy + \frac{P}{\rho} = \text{constant} \tag{9-25}$$

which is known as Bernoulli's equation. The limitations are

1. inviscid flow,
2. steady flow,
3. incompressible flow,
4. the equation applies along a streamline.

Limitation 4 will be relaxed for certain conditions to be investigated in Chapter 10.

Bernoulli's equation was also developed in Chapter 6 from energy considerations for steady incompressible flow with constant internal energy. It is interesting to note that the constant internal energy assumption and the inviscid flow assumption must be equivalent, as the other assumptions were the same. We may note, therefore, that the viscosity in some way will effect a change in internal energy.

9.4 CLOSURE

We have developed the differential equations for the conservation of mass and Newton's second law of motion. These equations may be subdivided into two special groups

$$\frac{\partial \rho}{\partial t} + \nabla \cdot \rho\mathbf{v} = 0 \tag{9-26}$$

(continuity equation)

Inviscid flow

$$\rho\frac{D\mathbf{v}}{Dt} = \rho\mathbf{g} - \nabla P \tag{9-27}$$

(Euler's equation)

Incompressible, viscous flow

$$\nabla \cdot \mathbf{v} = 0 \tag{9-28}$$

(continuity equation)

$$\rho\frac{D\mathbf{v}}{Dt} = \rho\mathbf{g} - \nabla P + \mu\nabla^2\mathbf{v} \tag{9-29}$$

(Navier–Stokes equation
for incompressible flow)

In addition, the student should note the physical meaning of the substantial derivative and appreciate the compactness of the vector representation. In component form, for example, equation (9-29) comprises some 27 terms in cartesian coordinates.

PROBLEMS

9.1 Apply the law of conservation of mass to an element in a polar coordinate system and obtain the continuity equation for a steady, two-dimensional, incompressible flow.

9.2 In cartesian coordinates, show that

$$v_x \frac{\partial}{\partial x} + v_y \frac{\partial}{\partial y} + v_z \frac{\partial}{\partial z}$$

may be written $(\mathbf{v} \cdot \nabla)$. What is the physical meaning of the term $(\mathbf{v} \cdot \nabla)$?

9.3 In an incompressible flow, the volume of the fluid is constant. Using the continuity equation, $\nabla \cdot \mathbf{v} = 0$, show that the fluid volume change is zero.

9.4 Find $D\mathbf{v}/Dt$ in polar coordinates by taking the derivative of the velocity. (Hint: $\mathbf{v} = v_r(r, \theta, t)\mathbf{e}_r + v_\theta(r, \theta, t)\mathbf{e}_\theta$. Remember that the unit vectors have derivatives.)

9.5 For flow at very low speeds and with large viscosity (the so-called creeping flows) such as occur in lubrication, it is possible to delete the inertia terms, $D\mathbf{v}/Dt$, from the Navier–Stokes equation. For flows at high velocity and small viscosity, it is not proper to delete the viscous term $v\nabla^2\mathbf{v}$. Explain this.

9.6 Using the Navier–Stokes equations and the continuity equation, obtain an expression for the velocity profile between two flat, parallel plates.

9.7 Does the velocity distribution in Example 2 satisfy continuity?

9.8 The atmospheric density may be approximated by the relation $\rho = \rho_0 \exp(-y/\beta)$, where $\beta = 22{,}000$ ft. Determine the rate at which the density changes with respect to body falling at v fps. If $v = 20{,}000$ fps at 100,000 ft, evaluate the rate of density change.

9.9 In a velocity field where $\mathbf{v} = 400[(y/L)^2\mathbf{e}_x + (x/L)^2\mathbf{e}_y]$ fps, determine the pressure gradient at the point $(L, 2L)$. The y axis is vertical, the density is $64.4\,\mathrm{lb_m/ft^3}$ and the flow may be considered inviscid.

9.10 Write equations (9-17) in component form for cartesian coordinates.

9.11 Derive equation (2-3) from equation (9-27).

9.12 In polar coordinates, the continuity equation is

$$\frac{1}{r}\frac{\partial}{\partial r}(rv_r) + \frac{1}{r}\frac{\partial v_\theta}{\partial \theta} = 0$$

Show that

(a) if $v_\theta = 0$, then $v_r = F(\theta)/r$;

(b) if $v_r = 0$, then $v_\theta = f(r)$.

9.13 Using the laws for the addition of vectors and equation (), show that in the absence of gravity,

(a) the fluid acceleration, pressure force, and viscous force all lie in the same plane;

(b) in the absence of viscous forces the fluid accelerates in the direction of decreasing pressure;

(c) a static fluid will always start to move in the direction of decreasing pressure.

9.14 Obtain the equations for a one-dimensional steady, viscous, compressible flow in the x direction from the Navier–Stokes equations. (These equations, together with an equation of state and the energy equation, may be solved for the case of weak shock waves.)

9.15 Obtain the equations for one-dimensional inviscid, unsteady, compressible flow.

9.16 Using the Navier–Stokes equations as given in Appendix E work Problems 8.17 and 8.18.

9.17 Using the Navier–Stokes equations, find the differential equation for a radial flow in which $v_z = v_\theta = 0$, and $v_r = f(r)$. Using continuity, show that the solution to the equation does not involve viscosity.

9.18 Using the Navier–Stokes equations in Appendix E, solve Problem 8.13.

9.19 For the flow described in Problem 8.13, obtain the differential equation of motion if $v_\theta = f(r, t)$.

9.20 Determine the velocity profile in a fluid situated between two coaxial rotating cylinders. Let the inner cylinder have radius R_1, and angular velocity Ω_1; let the outer cylinder have radius R_2 and angular velocity Ω_2.

Chapter 10

Inviscid Fluid Flow

An important area in momentum transfer is inviscid flow, in which, by virtue of the absence of shear stress, analytical solutions to the differential equations of fluid flow are possible.

The subject of inviscid flow has particular application in aerodynamics and hydrodynamics and general application to flow about bodies—the so called external flows. In this chapter, we shall introduce the fundamentals of inviscid flow analysis.

10.1 FLUID ROTATION AT A POINT

Consider the element of fluid shown in Figure 10.1. In time Δt the element will move in the xy plane as shown. In addition to translation, the element may also deform and rotate. We have discussed the deformation previously in Chapter 7. Now let us focus our attention on the rotation of the element. Although the element may deform, the orientation will be given by the average rotation of the line segments OB and OA or by denoting the rotation by

$$\omega_z = \frac{d}{dt}\left(\frac{\alpha + \beta}{2}\right)$$

where the counterclockwise sense is positive. From Figure 10.1, we see that

$$\omega_z = \lim_{\Delta x, \Delta y, \Delta z, \Delta t \to 0} \frac{1}{2}$$

$$\times \left(\frac{\arctan\{[(v_y|_{x+\Delta x} - v_y|_x)\Delta t]/\Delta x\}}{\Delta t} + \frac{\arctan\{-[(v_x|_{y+\Delta y} - v_x|_y)\Delta t]/\Delta y\}}{\Delta t}\right)$$

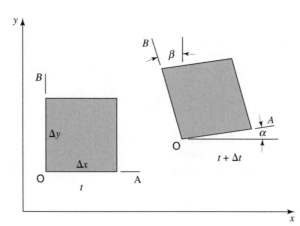

Figure 10.1 Rotation of a fluid element.

113

which becomes, in the limit,

$$\omega_z = \frac{1}{2}\left(\frac{\partial v_y}{\partial x} - \frac{\partial v_x}{\partial y}\right) \tag{10-1}$$

The subscript z indicates that the rotation is about the z axis.

In the xz and yz planes the rotation at a point is given by

$$\omega_y = \frac{1}{2}\left(\frac{\partial v_x}{\partial z} - \frac{\partial v_z}{\partial x}\right) \tag{10-2}$$

and

$$\omega_x = \frac{1}{2}\left(\frac{\partial v_z}{\partial y} - \frac{\partial v_y}{\partial z}\right) \tag{10-3}$$

The rotation at a point is related to the vector cross product of the velocity. As the student may verify

$$\nabla \times \mathbf{v} = \left(\frac{\partial v_z}{\partial y} - \frac{\partial v_y}{\partial z}\right)\mathbf{e}_x + \left(\frac{\partial v_x}{\partial z} - \frac{\partial v_z}{\partial x}\right)\mathbf{e}_y + \left(\frac{\partial v_y}{\partial x} - \frac{\partial v_x}{\partial y}\right)\mathbf{e}_z$$

and thus

$$\nabla \times \mathbf{v} = 2\omega \tag{10-4}$$

The vector $\nabla \times \mathbf{v}$ is also known as the *vorticity*. When the rotation at a point is zero the flow is said to be *irrotational*. For irrotational flow $\nabla \times \mathbf{v} = 0$, as can be seen from equation (10-4). The significance of fluid rotation at a point may be examined by a different approach. The Navier–Stokes equation for incompressible flow, equation (9-29), may also be written in the form

$$\rho\frac{D\mathbf{v}}{Dt} = -\nabla P + \rho\mathbf{g} - \mu[\nabla \times (\nabla \times \mathbf{v})] \tag{9-29}$$

It may be observed from the above equation that if viscous forces act on a fluid, the flow must be rotational.

The kinematic condition $\nabla \times \mathbf{v} = 0$ is not the first time we have encountered a kinematic relation that satisfies one of the fundamental physical laws of fluid mechanics. The law of conservation of mass for an incompressible flow, $\nabla \cdot \mathbf{v} = 0$, is also expressed as a kinematic relation. The use of this relation is the subject of the next section.

10.2   THE STREAM FUNCTION

For a two-dimensional, incompressible flow, the continuity equation is

$$\nabla \cdot \mathbf{v} = \frac{\partial v_x}{\partial x} + \frac{\partial v_y}{\partial y} = 0 \tag{9-3}$$

Equation (9-3) indicates that v_x and v_y are related in some way so that $\partial v_x/\partial x = -(\partial v_y/\partial y)$. Perhaps the easiest way to express this relation is by having v_x and v_y both related to the same function. Consider the function $F(x, y)$; if $v_x = F(x, y)$, then

$$\frac{\partial v_y}{\partial y} = -\frac{\partial F}{\partial x} \quad \text{or} \quad v_y = -\int\frac{\partial F}{\partial x}dy$$

Unfortunately, the selection of $v_x = F(x, y)$ results in an integral for v_y. We can easily remove the integral sign if we make the original $F(x,y)$ equal to the derivative of some function with respect to y. For example, if $F(x, y) = (\partial\Psi(x, y)/\partial y]$, then

$$v_x = \frac{\partial\Psi}{\partial y}$$

As $\partial v_x/\partial x = -(\partial v_y/\partial y)$, we may write

$$\frac{\partial v_y}{\partial y} = -\frac{\partial}{\partial x}\left(\frac{\partial\Psi}{\partial y}\right) \qquad \text{or} \qquad \frac{\partial}{\partial y}\left(v_y + \frac{\partial\Psi}{\partial x}\right) = 0$$

for this to be true in general

$$v_y = -\frac{\partial\Psi}{\partial x}$$

Instead of having two unknowns, v_x and v_y, we now have only one unknown, Ψ. The unknown, Ψ, is called the *stream function*. The physical significance of Ψ can be seen from the following considerations. As $\Psi = \Psi(x, y)$, the total derivative is

$$d\Psi = \frac{\partial\Psi}{\partial x}dx + \frac{\partial\Psi}{\partial y}dy$$

Also

$$\frac{\partial\Psi}{\partial x} = -v_y \qquad \text{and} \qquad \frac{\partial\Psi}{\partial y} = v_x$$

and thus

$$d\Psi = -v_y\,dx + v_x\,dy \tag{10-5}$$

Consider a path in the xy plane such that $\Psi = $ constant. Along this path, $d\Psi = 0$, and thus equation (10-5) becomes

$$\left.\frac{dy}{dx}\right|_{\Psi=\text{constant}} = \frac{v_y}{v_x} \tag{10-6}$$

The slope of the path $\Psi = $ constant is seen to be the same as the slope of a streamline as discussed in Chapter 3. The function $\Psi(x, y)$ thus represents the streamlines. Figure 10.2 illustrates the streamlines and velocity components for flow about an airfoil.

The differential equation that governs Ψ is obtained by consideration of the fluid rotation, ω, at a point. In a two-dimensional flow, $\omega_z = \frac{1}{2}[(\partial v_y/\partial x) - (\partial v_x/\partial y)]$, and thus if

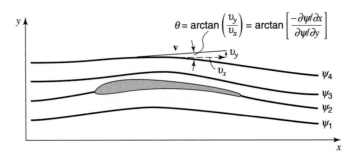

Figure 10.2 Streamlines and the stream function.

the velocity components v_y and v_x are expressed in terms of the stream function Ψ, we obtain, for an incompressible, steady flow

$$-2\omega_z = \frac{\partial^2 \Psi}{\partial x^2} + \frac{\partial^2 \Psi}{\partial y^2} \tag{10-7}$$

When the flow is irrotational, equation (10-7) becomes Laplace's equation

$$\nabla^2 \Psi = \frac{\partial^2 \Psi}{\partial x^2} + \frac{\partial^2 \Psi}{\partial y^2} = 0 \tag{10-8}$$

10.3   INVISCID, IRROTATIONAL FLOW ABOUT AN INFINITE CYLINDER

In order to illustrate the use of the stream function, the inviscid, irrotational flow pattern about a cylinder of infinite length will be examined. The physical situation is illustrated in Figure 10.3. A stationary circular cylinder of radius a is situated in uniform, parallel flow in the x direction.

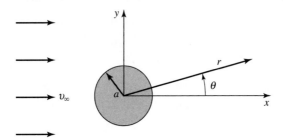

Figure 10.3 Cylinder in a uniform flow.

As there is cylindrical symmetry, polar coordinates are employed. In polar coordinates,[1] equation (10-8) becomes

$$\frac{\partial^2 \Psi}{\partial r^2} + \frac{1}{r}\frac{\partial \Psi}{\partial r} + \frac{1}{r^2}\frac{\partial^2 \Psi}{\partial \theta^2} = 0 \tag{10-9}$$

where the velocity components v_r and v_θ are given by

$$v_r = \frac{1}{r}\frac{\partial \Psi}{\partial \theta} \qquad v_\theta = -\frac{\partial \Psi}{\partial r} \tag{10-10}$$

The solution for this case must meet four boundary conditions. These are as follows:

1. The circle $r = a$ must be a streamline. As the velocity normal to a streamline is zero, $v_r|_{r=a} = 0$ or $\partial \Psi/\partial \theta|_{r=a} = 0$.
2. From symmetry, the line $\theta = 0$ must also be a streamline. Hence $v_\theta|_{\theta=0} = 0$ or $\partial \Psi/\partial r|_{\theta=0} = 0$.
3. As $r \to \infty$ the velocity must be finite.
4. The magnitude of the velocity as $r \to \infty$ is v_∞, a constant.

---

[1] The operator ∇^2 in cylindrical coordinates is developed in Appendix A.

The solution to equation (10-9) for this case is

$$\Psi(r,\theta) = v_\infty r \, \sin\theta \left[1 - \frac{a^2}{r^2} \right] \tag{10-11}$$

The velocity components v_r and v_θ are obtained from equation (10-10),

$$v_r = \frac{1}{r} \frac{\partial \Psi}{\partial \theta} = v_\infty \, \cos\theta \left[1 - \frac{a^2}{r^2} \right] \tag{10-12}$$

and

$$v_\theta = -\frac{\partial \Psi}{\partial r} = -v_\infty \, \sin\theta \left[1 + \frac{a^2}{r^2} \right] \tag{10-13}$$

By setting $r = a$ in the above equations, the velocity at the surface of the cylinder may be determined. This results in

$$v_r = 0$$

and

$$v_\theta = -2v_\infty \, \sin\theta \tag{10-14}$$

The velocity in the radial direction is, of course, zero, as the cylinder surface is a streamline. The velocity along the surface is seen to be zero at $\theta = 0$ and $\theta = 180°$. These points of zero velocity are known as *stagnation points*. The forward stagnation point is at $\theta = 180°$, and the aft or rearward stagnation point is at $\theta = 0°$. The student may verify that each of the boundary conditions for this case are satisfied.

10.4   IRROTATIONAL FLOW, THE VELOCITY POTENTIAL

In a two-dimensional irrotational flow $\nabla \times \mathbf{v} = 0$, and thus $\partial v_x/\partial y = \partial v_y/\partial x$. The similarity of this equation to the continuity equation suggests that the type of relation used to obtain the stream function may be used again. Note, however, that the order of differentiation is reversed from the continuity equation. If we let $v_x = \partial\phi(x, y)/\partial x$, we observe that

$$\frac{\partial v_x}{\partial y} = \frac{\partial^2 \phi}{\partial x \partial y} = \frac{\partial v_y}{\partial x}$$

or

$$\frac{\partial}{\partial x} \left(\frac{\partial \phi}{\partial y} - v_y \right) = 0$$

and for the general case

$$v_y = \frac{\partial \phi}{\partial y}$$

The function ϕ is called the *velocity potential*. In order for ϕ to exist, the flow must be irrotational. As the condition of irrotationality is the only condition required, the velocity potential can also exist for compressible, unsteady flows. The velocity potential is commonly used in compressible flow analysis. Additionally, the velocity potential, ϕ exists for three-dimensional flows, whereas the stream function does not.

The velocity vector is given by

$$\mathbf{v} = v_x\mathbf{e}_x + v_y\mathbf{e}_y + v_z\mathbf{e}_z = \frac{\partial\phi}{\partial x}\mathbf{e}_x + \frac{\partial\phi}{\partial y}\mathbf{e}_y + \frac{\partial\phi}{\partial z}\mathbf{e}_z$$

and thus, in vector notation

$$\mathbf{v} = \boldsymbol{\nabla}\phi \tag{10-15}$$

The differential equation defining ϕ is obtained from the continuity equation. Considering a steady incompressible flow, we have $\boldsymbol{\nabla}\cdot\mathbf{v} = 0$; thus, using equation (10-15) for $\mathbf{v}$, we obtain

$$\boldsymbol{\nabla}\cdot\boldsymbol{\nabla}\phi = \boldsymbol{\nabla}^2\phi = 0 \tag{10-16}$$

which is again Laplace's equation; this time the dependent variable is ϕ. Clearly, Ψ and ϕ must be related. This relation may be illustrated by a consideration of isolines of Ψ and ϕ. An isoline of Ψ is, of course, a streamline. Along the isolines

$$d\Psi = \frac{\partial\Psi}{\partial x}dx + \frac{\partial\Psi}{\partial y}dy$$

or

$$\left.\frac{dy}{dx}\right|_{\Psi=\text{constant}} = \frac{v_y}{v_x}$$

and

$$d\phi = \frac{\partial\phi}{\partial x}dx + \frac{\partial\phi}{\partial y}dy \qquad \left.\frac{dy}{dx}\right|_{d\phi=0} = -\frac{v_x}{v_y}$$

Accordingly

$$\left.dy/dx\right|_{\phi=\text{constant}} = -\frac{1}{dy/dx}\Bigg|_{\Psi=\text{constant}} \tag{10-17}$$

and thus Ψ and ϕ are orthogonal. The orthogonality of the stream function and the velocity potential is a useful property, particularly when graphical solutions to equations (10-8) and (10-16) are employed.

Figure 10.4 illustrates the inviscid, irrotational, steady incompressible flow about an infinite circular cylinder. Both the streamlines and constant-velocity potential lines are shown.

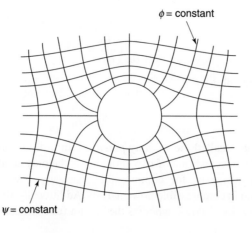

$\phi = $ constant

$\psi = $ constant

Figure 10.4 Streamlines and constant velocity potential lines for steady, incompressible, irrotational, inviscid flow about a cylinder.

10.5 TOTAL HEAD IN IRROTATIONAL FLOW

The condition of irrotationality has been shown to be of aid in obtaining analytical solutions in fluid flow. The physical meaning of irrotational flow can be illustrated by the relation between the rotation or vorticity, $\nabla \times \mathbf{v}$, and the total head, $P/\rho + v^2/2 + gy$. For an inviscid flow we may write

$$\frac{D\mathbf{v}}{Dt} = \mathbf{g} - \frac{\nabla P}{\rho} \qquad \text{(Euler's equation)}$$

and

$$\frac{D\mathbf{v}}{Dt} = \frac{\partial \mathbf{v}}{\partial t} + \nabla\left(\frac{v^2}{2}\right) - \mathbf{v} \times (\nabla \times \mathbf{v}) \qquad \text{(Vector identity)}$$

As the gradient of the potential energy is $-\mathbf{g}$, Euler's equation becomes, for incompressible flow

$$\nabla\left\{\frac{P}{\rho} + \frac{v^2}{2} + gy\right\} = \mathbf{v} \times (\nabla \times \mathbf{v}) - \frac{\partial \mathbf{v}}{\partial t}. \qquad (10\text{-}18)$$

If the flow is steady, it is seen from equation (10-18) that the gradient of the total head depends upon the vorticity, $\nabla \times \mathbf{v}$. The vector $(\nabla \times \mathbf{v})$ is perpendicular to the velocity vector; hence, the gradient of the total head has no component along a streamline. Thus, along a streamline in an incompressible, inviscid, steady flow,

$$\frac{P}{\rho} + \frac{v^2}{2} + gy = \text{constant} \qquad (10\text{-}19)$$

This is, of course, Bernoulli's equation, which was discussed in Chapters 6 and 9. If the flow is irrotational and steady, equation (10-18) yields the result that Bernoulli's equation is valid throughout the flow field. An irrotational, steady, incompressible flow, therefore, has a constant total head throughout the flow field.[2]

10.6 UTILIZATION OF POTENTIAL FLOW

Potential flow has great utility in engineering for the prediction of pressure fields, forces, and flow rates. In the field of aerodynamics, for example, potential flow solutions are used to predict force and moment distributions on wings and other bodies.

An illustration of the determination of the pressure distribution from a potential flow solution may be obtained from the solution for the flow about a circular cylinder presented in Section 10.3. From the Bernoulli equation

$$\frac{P}{\rho} + \frac{v^2}{2} = \text{constant} \qquad (10\text{-}20)$$

We have deleted the potential energy term in accordance with the original assumption of uniform velocity in the x direction. At a great distance from the cylinder the pressure is P_∞, and the velocity is v_∞, so equation (10-20) becomes[3]

$$P + \frac{\rho v^2}{2} = P_\infty + \frac{\rho v_\infty^2}{2} = P_0 \qquad (10\text{-}21)$$

---

[2] A more general result, Crocco's theorem, relates the vorticity to the entropy. Thus it can be shown that a steady, inviscid, irrotational flow, either compressible or incompressible, is isentropic.

[3] The stagnation pressure as given in equation (10-21) applies to incompressible flow only.

where P_0 is designated the *stagnation pressure* (i.e., the pressure at which the velocity is zero). In accordance with equation (10-19), the stagnation pressure is constant throughout the field in an irrotational flow. The velocity at the surface of the body is $v_\theta = -2v_\infty \sin\theta$, thus the surface pressure is

$$P = P_0 - 2\rho v_\infty^2 \sin^2\theta \qquad (10\text{-}22)$$

A plot of the potential flow pressure distribution about a cylinder is shown in Figure 10.5.

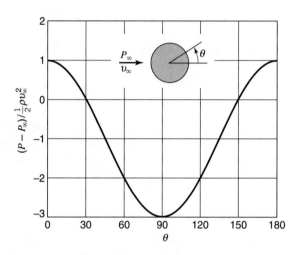

Figure 10.5 Pressure distribution on a cylinder in an inviscid, incompressible, steady flow.

10.7 POTENTIAL FLOW ANALYSIS—SIMPLE PLANE FLOW CASES

In this section, a number of cases will be considered in which solutions are achieved for two-dimensional, incompressible irrotational flow. We begin with some very straightforward flow situations.

Case 1. Uniform flow in the x direction.

For a uniform flow parallel to the x axis, with velocity $v_\infty = $ constant, the stream function and velocity potential relationships are

$$v_x = v_\infty = \frac{\partial \Psi}{\partial y} = \frac{\partial \phi}{\partial x}$$

$$v_y = 0 = \frac{\partial \Psi}{\partial x} = \frac{\partial \phi}{\partial y}$$

which integrate to yield

$$\Psi = v_\infty y$$

$$\phi = v_\infty x$$

Case 2. A line source or sink.

A line source, in two dimensions, is a flow which is radially outward from the source which is the origin in this example. The reverse, or sink flow, has the flow directed inward.

The source strength is the volume flow rate per unit depth, $Q = 2\pi r v_r$. The radial velocity associated with a source is

$$v_r = \frac{Q}{2\pi r}$$

and the azimuthal velocity is given by $v_\theta = 0$. The stream function and velocity potential are evaluated from the expressions

$$v_r = \frac{Q}{2\pi r} = \frac{1}{r}\frac{\partial \Psi}{\partial \theta} = \frac{\partial \phi}{\partial r}$$

$$v_\theta = 0 = -\frac{\partial \Psi}{\partial r} = \frac{1}{r}\frac{\partial \phi}{\partial \theta}$$

Integrating these expressions, we obtain for the line source

$$\Psi = \frac{Q}{2\pi}\theta$$

$$\phi = \frac{Q}{2\pi}\ln r$$

For sink flow, the sign of the radial velocity is negative (inward) and thus Q is negative.

The expressions for a line source or sink present a problem at $r = 0$, the origin, which is a singular point. At $r = 0$, the radial velocity approaches infinity. Physically this is unrealistic, and we use only the concept of line source or sink flow under conditions where the singularity is excluded from consideration.

Case 3. A line vortex.

Vortex flow is that which occurs in a circular fashion around a central point, such as a whirlpool. A *free vortex* is one where fluid particles are irrotational, i.e., they do not rotate as they move in concentric circles about the axis of the vortex. This would be analogous to people sitting in cabins on a ferris wheel. For an irrotational flow in polar coordinates (see Appendix B), the product rv_θ must be constant. The stream function and velocity potential can be written directly,

$$v_r = 0 = \frac{1}{r}\frac{\partial \Psi}{\partial \theta} = \frac{\partial \phi}{\partial r}$$

$$v_\theta = \frac{K}{2\pi r} = -\frac{\partial \Psi}{\partial r} = \frac{1}{r}\frac{\partial \phi}{\partial \theta}$$

which, upon integration, become

$$\Psi = -\frac{K}{2\pi}\ln r$$

$$\phi = \frac{K}{2\pi}\theta$$

where K is referred to as the *vortex strength*. When K is positive, the flow is observed to be counterclockwise about the vortex center.

10.8 POTENTIAL FLOW ANALYSIS—SUPERPOSITION

It was shown earlier that both the stream function and the velocity potential satisfy Laplace's equation for two-dimensional, irrotational, incompressible flow. As Laplace's equation is linear we can use known solutions to achieve expressions for both Ψ and φ for more complex situations using the principle of *superposition*. Superposition, simply put, is the process of adding known solutions to achieve another, i.e., if Ψ_1 and Ψ_2 are solutions to $\nabla^2\Psi = 0$, then so is $\Psi_3 = \Psi_1 + \Psi_2$ a solution.

The reader is reminded that the solutions obtained for these very specialized flow conditions are idealizations. They apply for inviscid flow which is a reasonable approximation for conditions *outside* the region, near a solid body, where viscous effects are manifested. This region, the *boundary layer*, will be considered in some depth in Chapter 12.

Some cases will now be considered where the elementary plane flows of the previous section give some interesting and useful results through the process of superposition.

Case 4. The doublet.

A useful case is achieved from considering a source-sink pair on the x axis as the separation distance, $2a$, approaches zero. Geometrically, we can note that the streamlines and velocity potential lines are circles with centers on the y and x axes but with all circles passing through the origin that is a singular point.

The strength of a double, designated λ, is defined as the finite limit of the quantity $2aQ$ as $a \to 0$. For our case, the source is placed on the x axis at $-a$ and the sink is placed on the x axis at $+a$. The resulting expressions for Ψ and ϕ in polar coordinates are

$$\Psi = -\frac{\lambda \sin\theta}{r}$$

$$\phi = \frac{\lambda \cos\theta}{r}$$

Case 5. Flow past a half body—superposition of uniform flow and a source.

The stream function and velocity potentials for uniform flow in the x direction and for a line source are added together, yielding

$$\Psi = \Psi_{\text{uniform flow}} + \Psi_{\text{source}}$$

$$= v_\infty y + \frac{Q}{2\pi}\theta = v_\infty r \sin\theta + \frac{Q}{2\pi}\theta$$

$$\phi = \phi_{\text{uniform flow}} + \phi_{\text{source}}$$

$$= v_\infty x + \frac{Q}{2\pi}\ln r = v_\infty r \cos\theta + \frac{Q}{2\pi}\ln r$$

Case 6. Flow past a cylinder—superposition of uniform flow and a doublet.

As a final illustration of the superposition method, we will consider a case of considerable utility. When the solutions for uniform flow and the doublet are superposed, the result, similar to the past case, defines a streamline pattern inside and around the outside surface of a body. In this case the body is closed and the exterior flow pattern is that of ideal flow over a cylinder. The expressions for Ψ and φ are

$$\Psi = \Psi_{\text{uniform flow}} + \Psi_{\text{doublet}}$$

$$= v_\infty y - \frac{\lambda \sin\theta}{r} = v_\infty r \sin\theta - \frac{\lambda \sin\theta}{r}$$

$$= \left[v_\infty r - \frac{\lambda}{r} \right] \sin\theta$$

$$\phi = \phi_{\text{uniform flow}} + \phi_{\text{doublet}}$$

$$= v_\infty x + \frac{\lambda \cos\theta}{r} = v_\infty r \cos\theta + \frac{\lambda \cos\theta}{r}$$

$$= \left[v_\infty r + \frac{\lambda}{r} \right] \cos\theta$$

It is useful, at this point, to examine the above expressions in more detail. First, for the stream function

$$\Psi = \left[v_\infty r - \frac{\lambda}{r} \right] \sin\theta$$

$$= v_\infty r \left[1 - \frac{\lambda/v_\infty}{r^2} \right] \sin\theta$$

where, as we recall, λ is the doublet strength. If we choose λ such that

$$\frac{\lambda}{v_\infty} = a^2$$

where a is the radius of our cylinder, we obtain

$$\Psi(r, \theta) = v_\infty r \sin\theta \left[1 - \frac{a^2}{r^2} \right]$$

which is the expression used earlier, designated as equation (10-11).

10.9   CLOSURE

In this chapter, we have examined potential flow. A short summary of the properties of the stream function and the velocity potential is given below.

Stream function

1. A stream function $\Psi(x, y)$ exists for each and every two-dimensional, steady, incompressible flow, whether viscous or inviscid.
2. Lines for which $\Psi(x, y) = $ constant are streamlines.
3. In cartesian coordinates

$$v_x = \frac{\partial \Psi}{\partial y} \qquad v_y = -\frac{\partial \Psi}{\partial x} \qquad (10\text{-}23a)$$

   and in general

$$v_s = \frac{\partial \Psi}{\partial n} \qquad (10\text{-}23b)$$

   where n is $90°$ counterclockwise from s.
4. The stream function identically satisfies the continuity equation.
5. For an irrotational, steady incompressible flow

$$\nabla^2 \Psi = 0 \qquad (10\text{-}24)$$

Velocity potential

1. The velocity potential exists if and only if the flow is irrotational. No other restrictions are required.
2. $\nabla\phi = \mathbf{v}$.
3. For irrotational, incompressible flow, $\nabla^2 \phi = 0$.
4. For steady, incompressible two-dimensional flows, lines of constant velocity potential are perpendicular to the streamlines.

PROBLEMS

10.1   In polar coordinates, show that

$$\mathbf{\nabla} \times \mathbf{v} = \frac{1}{r}\left[\frac{\partial(rv_\theta)}{\partial r} - \frac{\partial v_r}{\partial \theta}\right]\mathbf{e}_z$$

10.2   Determine the fluid rotation at a point in polar coordinates, using the method illustrated in Figure 10.1.

10.3   Find the stream function for a flow with a uniform free-stream velocity v_∞. The free-stream velocity intersects the x axis at an angle α.

10.4   In polar coordinates, the continuity equation for steady incompressible flow becomes

$$\frac{1}{r}\frac{\partial}{\partial r}(rv_r) + \frac{1}{r}\frac{\partial v_\theta}{\partial \theta} = 0$$

Derive equations (10-10), using this relation.

10.5   The velocity potential for a given two-dimensional flow field is

$$\phi = \left(\frac{5}{3}\right)x^3 - 5xy^2$$

Show that the continuity equation is satisfied and determine the corresponding stream function.

10.6   Make an analytical model of a tornado using an irrotational vortex (with velocity inversely proportional to distance from the center) outside a central core (with velocity directly proportional to distance). Assume that the core diameter is 200 ft and the static pressure at the center of the core is 38 psf below ambient pressure. Find

(a)   the maximum wind velocity;

(b)   the time it would take a tornado moving at 60 mph to lower the static pressure from -10 to -38 psfg;

(c)   the variation in stagnation pressure across the tornado. Euler's equation may be used to relate the pressure gradient in the core to the fluid acceleration.

10.7   For the flow about a cylinder, find the velocity variation along the streamline leading to the stagnation point. What is the velocity derivative $\partial v_r/\partial r$ at the stagnation point?

10.8   In Problem 10.7, explain how one could obtain $\partial v_\theta/\partial \theta$ at the stagnation point, using only r and $\partial v_r/\partial r$.

10.9   At what point on the surface of the circular cylinder in a potential flow does the pressure equal the free-stream pressure?

10.10   For the velocity potentials given below, find the stream function and sketch the streamlines

(a) $\phi = v_\infty L\left[\left(\frac{x}{L}\right)^3 - \frac{3xy^2}{L^3}\right]$

(b) $\phi = v_\infty \dfrac{xy}{L}$

(c) $\phi = \dfrac{v_\infty L}{2}\ln\left(x^2 + y^2\right).$

10.11   The stream function for an incompressible, two-dimensional flow field is

$$\psi = 2r^3 \sin 3\theta$$

For this flow field, plot several streamlines for $0 \le \theta \le \pi/3$.

10.12   For the case of a source at the origin with a uniform free-stream plot the streamline $\psi = 0$.

10.13   In Problem 10.12, how far upstream does the flow from the source reach?

10.14   Determine the pressure gradient at the stagnation point of Problem 10.10(a).

10.15   Calculate the total lift force on the Arctic hut shown below as a function of the location of the opening. The lift force results from the difference between the inside pressure and the outside pressure. Assume potential flow and the hut is in the shape of a half-cylinder.

10.16   Consider three equally spaced sources of strength m placed at $(x, y) = (-a, 0), (0, 0)$, and $(a, 0)$. Sketch the resulting streamline pattern. Are there any stagnation points?

10.17   Sketch the streamlines and potential lines of the flow due to a line source of at $(a, 0)$ plus an equivalent sink at $(-a, 0)$.

10.18   The stream function for an incompressible, two-dimensional flow field is

$$\psi = 3x^2y + y$$

For this flow field, sketch several streamlines.

10.19   A line vortex of strength K at $(x, y) = (0, a)$ is combined with opposite strength vortex at $(0, -a)$. Plot the streamline pattern and find the velocity that each vortex induces on the other vortex.

10.20   A source of strength $1.5 \text{ m}^2/\text{s}$ at the origin is combined with a uniform stream moving at 9 m/s in the x direction. For the half-body which results, find

(a)   the stagnation point;

(b)   the body height as it crosses the y axis;

(c)   the body height at large x;

(d)   the maximum surface velocity and its position (x, y).

10.21   When a doublet is added to a uniform stream so that the source part of the doublet faces the stream, a cylinder flow results. Plot the streamlines when the doublet is reversed so that the sink faces the stream.

10.22   A 2-m-diameter horizontal cylinder is formed by bolting two semicylindrical channels together on the inside. There are 12 bolts per meter of width holding the top and bottom together. The inside pressure is 60 kPa (gage). Using potential theory for the outside pressure, compute the tension force in each bolt if the free stream fluid is sea-level air and the free-stream wind speed is 25 m/s.

Chapter **11**

Dimensional Analysis and Similitude

$\mathbf{A}$n important consideration in all equations written thus far has been dimensional homogeneity. At times it has been necessary to use proper conversion factors in order that an answer be correct numerically and have the proper units. The idea of dimensional consistency can be used in another way, by a procedure known as dimensional analysis, to group the variables in a given situation into dimensionless parameters that are less numerous than the original variables. Such a procedure is very helpful in experimental work in which the very number of significant variables presents an imposing task of correlation. By combining the variables into a smaller number of dimensionless parameters, the work of experimental data reduction is considerably reduced.

This chapter will include means of evaluating dimensionless parameters both in situations in which the governing equation is known, and in those in which no equation is available. Certain dimensionless groups emerging from this analysis will be familiar, and some others will be encountered for the first time. Finally, certain aspects of similarity will be used to predict the flow behavior of equipment on the basis of experiments with scale models.

11.1 DIMENSIONS

In dimensional analysis, certain dimensions must be established as fundamental, with all others expressible in terms of these. One of these fundamental dimensions is length, symbolized L. Thus, area and volume may dimensionally be expressed as L^2 and L^3, respectively. A second fundamental dimension is time, symbolized t. The kinematic quantities, velocity and acceleration, may now be expressed as L/t and L/t^2, respectively.

Another fundamental dimension is mass, symbolized M. An example of a quantity whose dimensional expression involves mass is the density that would be expressed as M/L^3. Newton's second law of motion gives a relation between force and mass and allows force to be expressed dimensionally as $F = Ma = ML/t^2$. Some texts reverse this procedure and consider force fundamental, with mass expressed in terms of F, L, and t according to Newton's second law of motion. Here, mass will be considered a fundamental unit.

The significant quantities in momentum transfer can all be expressed dimensionally in terms of M, L, and t; thus these comprise the fundamental dimensions we shall be concerned with presently. The dimensional analysis of energy problems in Chapter 19 will require the addition of two more fundamental dimensions, heat and temperature.

Some of the more important variables in momentum transfer and their dimensional representations in terms of M, L, and t are given in Table 11.1.

Table 11.1 Important variables in momentum transfer

Variable	Symbol	Dimension
Mass	M	M
Length	L	L
Time	t	t
Velocity	v	L/t
Gravitational acceleration	g	L/t^2
Force	F	ML/t^2
Pressure	P	M/Lt^2
Density	ρ	M/L^3
Viscosity	μ	M/Lt
Surface tension	σ	M/t^2
Sonic velocity	a	L/t

11.2 DIMENSIONAL ANALYSIS OF GOVERNING DIFFERENTIAL EQUATIONS

The differential equations that describe fluid behavior as developed in Chapter 9 are powerful tools for analyzing and predicting fluid phenomena and their effects. The Navier–Stokes equations have been solved analytically for a few simple situations. For more complex applications, these relationships provide the basis for a number of sophisticated and powerful numerical codes.

In this section, we will use the differential forms of the continuity and momentum (Navier–Stokes) equations to develop some useful dimensionless parameters that will be valuable tools for subsequent analysis. This process will now be illustrated as we examine two-dimensional incompressible flow.

The governing differential equations are the following.
Continuity:

$$\frac{\partial v_x}{\partial x} + \frac{\partial v_y}{\partial y} = 0 \qquad (9\text{-}3)$$

Momentum:

$$\rho\left(\frac{\partial \mathbf{v}}{\partial t} + v_x\frac{\partial \mathbf{v}}{\partial x} + v_y\frac{\partial \mathbf{v}}{\partial y}\right) = \rho\mathbf{g} - \nabla\rho + \mu\left(\frac{\partial^2 \mathbf{v}}{\partial x^2} + \frac{\partial^2 \mathbf{v}}{\partial y^2}\right) \qquad (9\text{-}19)$$

We now stipulate the reference values for length and velocity

- reference length $\quad L$
- reference velocity $\quad v_\infty$

and, accordingly, specify nondimensional quantities for the variables in equations (9-3) and (9-19) as

$$
\begin{aligned}
x^* &= x/L & v_x^* &= v_x/v_\infty \\
y^* &= y/L & v_y^* &= v_y/v_\infty \\
t^* &= \frac{tv_\infty}{L} & \mathbf{v}^* &= \mathbf{v}/v_\infty \\
& & \nabla^* &= L\,\nabla
\end{aligned}
$$

The last quantity in this list, $\boldsymbol{\nabla}^*$, is the dimensionless gradient operator. As $\boldsymbol{\nabla}$ is composed of first derivatives with respect to space coordinates, the product $L\boldsymbol{\nabla}$ is seen to be dimensionless.

The next step is to nondimensionalize our governing equations by introducing the specified dimensionless variables. This process involves the chain rule for differentiation; for example, the two terms in equation (9-3) are transformed as follows:

$$\frac{\partial v_x}{\partial x} = \frac{\partial v_x^*}{\partial x^*}\frac{\partial v_x}{\partial v_x^*}\frac{\partial x^*}{\partial x} = \frac{\partial v_x^*}{\partial x^*}(v_\infty)(1/L) = \frac{v_\infty}{L}\frac{\partial v_x^*}{\partial x^*}$$

$$\frac{\partial v_y}{\partial y} = \frac{\partial v_y^*}{\partial y^*}\frac{\partial v_y}{\partial v_y^*}\frac{\partial y^*}{\partial y} = \frac{v_\infty}{L}\frac{v_x^*}{\partial x^*}$$

Substitution into equation (9-3) gives

$$\frac{\partial v_x^*}{\partial x^*} + \frac{\partial v_y^*}{\partial y^*} = 0 \tag{11-1}$$

and we see that the continuity equation has the same form in terms of dimensionless variables as it had originally.

Utilizing the chain rule in the same manner as just discussed, the equation of motion becomes

$$\frac{\rho v_\infty^2}{L}\left(\frac{\partial \mathbf{v}^*}{\partial t^*} + v_x^*\frac{\partial \mathbf{v}^*}{\partial x^*} + v_y^*\frac{\partial \mathbf{v}^*}{\partial y^*}\right) = \rho\mathbf{g} + \frac{1}{L}\boldsymbol{\nabla}^* P + \frac{\mu v_\infty}{L^2}\left(\frac{\partial^2 \mathbf{v}^*}{\partial x^{*2}} + \frac{\partial^2 \mathbf{v}^*}{\partial y^{*2}}\right) \tag{11-2}$$

In equation (11-2), we note that each term has the units M/L^2t^2 or F/L^3. Also, it should be observed that each term represents a certain kind of force, that is

- $\dfrac{\rho v_\infty^2}{L}$ is an inertial force

- $\dfrac{\mu v_\infty^2}{L}$ is a viscous force
- ρg is a gravitational force
- P/L is a pressure force

If we next divide through by the quantity, $\rho v_\infty^2/L$, our dimensionless equation becomes

$$\frac{\partial \mathbf{v}^*}{\partial t^*} + v_x^*\frac{\partial \mathbf{v}^*}{\partial x^*} + v_y^*\frac{\partial \mathbf{v}^*}{\partial y^*} = \mathbf{g}\frac{L}{v_\infty^2} - \frac{\boldsymbol{\nabla}^* P}{\rho v_\infty^2} + \frac{\mu}{Lv_\infty\rho}\left(\frac{\partial^2 \mathbf{v}^*}{\partial x^{*2}} + \frac{\partial^2 \mathbf{v}^*}{\partial y^{*2}}\right) \tag{11-3}$$

This resulting dimensionless equation has the same general characteristics as its original except that, as a result of its transformation into dimensionless form, each of the original force terms (those on the right-hand side) has a coefficient composed of a combination of variables. An example of these coefficients reveals that each is dimensionless. Additionally, because of the manner in which they were formed, the parameters can be interpreted as a ratio of forces.

Consideration of the first term, gL/v_∞^2, reveals that it is, indeed, dimensionless. The choice of gL/v_∞^2 or v_∞^2/gL is arbitrary; clearly both forms are dimensionless.

The conventional choice is the latter form. The *Froude number* is defined as

$$Fr \equiv v_\infty^2/gL \tag{11-4}$$

This parameter can be interpreted as a measure of the ratio of inertial to gravitational forces. The Froude number arises in analyzing flows involving a free liquid surface. It is an important parameter when dealing with open-channel flows.

The next parameter, $P/\rho v_\infty^2$, is observed to be the ratio of pressure forces to inertial forces. In this form it is designated the *Euler number*,

$$\text{Eu} \equiv P/\rho v_\infty^2 \tag{11-5}$$

A modified form of equation (11-5), also clearly dimensionless, is the coefficient of drag

$$C_D = \frac{F/A}{\rho v_\infty^2/2} \tag{11-6}$$

which, we will see directly, has application to both internal and external flows.

The third dimensionless ratio that has been generated is the *Reynolds number*, which is conventionally expressed as

$$\text{Re} \equiv L v_\infty \rho/\mu \tag{11-7}$$

In this form the Reynolds number is observed to represent the ratio of inertial forces to viscous forces. The Reynolds number is generally considered the most important dimensionless parameter in the field of fluid mechanics. It is ubiquitous in all of the transport processes. We will encounter it frequently throughout the remainder of this text.

If equation (11-3) can be solved, the results will provide the functional relationships between applicable dimensionless parameters. If direct solution is not possible then one must resort to numerical modeling or experimental determination of these functional relationships.

11.3 THE BUCKINGHAM METHOD

The procedure introduced in the previous section is, obviously, quite powerful when one knows the differential equation that pertains to a specific fluid flow process. There are, however, many situations of interest in which the governing equation is not known. In these cases, we need an alternative method for dimensional analysis. In this section, we discuss a more general approach for generating dimensionless groups of variables. This procedure was proposed by Buckingham[1] in the early part of the twentieth century. It is generally referred to as the *Buckingham method*.

The initial step in applying the Buckingham method requires the listing of the variables significant to a given problem. It is then necessary to determine the number of dimensionless parameters into which the variables may be combined. This number may be determined using the *Buckingham pi* theorem, which states

> *The number of dimensionless groups used to describe a situation involving* n *variables is equal to* n − r, *where* r *is the rank of the dimensional matrix of the variables.*

Thus,

$$i = n - r \tag{11-8}$$

where

> $i =$ the number of independent dimensionless groups
> $n =$ the number of variables involved

and

> $r =$ the rank of the dimensional matrix

---

[1] E. Buckingham, *Phys. Rev.* **2**, 345 (1914).

The dimensional matrix is simply the matrix formed by tabulating the exponents of the fundamental dimensions M, L, and t, which appear in each of the variables involved.

An example of the evaluation of r and i, as well as the application of the Buckingham method, follows.

EXAMPLE 1

Determine the dimensionless groups formed from the variables involved in the flow of fluid external to a solid body. The force exerted on the body is a function of v, ρ, μ, and L (a significant dimension of the body).

A usual first step is to construct a table of the variables and their dimensions.

Variable	Symbol	Dimensions
Force	F	ML/t^2
Velocity	v	L/t
Density	ρ	M/L^3
Viscosity	μ	M/Lt
Length	L	L

Before determining the number of dimensionless parameters to be formed, we must know r. The dimensional matrix that applies is formed from the following tabulation:

	F	v	ρ	μ	L
M	1	0	1	1	0
L	1	1	−3	−1	1
t	−2	−1	0	−1	0

The numbers in the table represent the exponents of M, L, and t in the dimensional expression for each variable involved. For example, the dimensional expression of F is ML/t^2, hence the exponents 1, 1, and −2 are tabulated versus M, L, and t, respectively, the quantities with which they are associated. The matrix is then the array of numbers shown below

$$\begin{pmatrix} 1 & 0 & 1 & 1 & 0 \\ 1 & 1 & -3 & -1 & 1 \\ -2 & -1 & 0 & -1 & 0 \end{pmatrix}$$

The rank, r, of a matrix is the number of rows (columns) in the largest nonzero determinant that can be formed from it. The rank is 3 in this case. Thus, the number of dimensionless parameters to be formed may be found by applying equation (11-4). In this example $i = 5 - 3 = 2$.

The two dimensionless parameters will be symbolized π_1 and π_2 and may be formed in several ways. Initially, a *core group* of r variables must be chosen, which will consist of those variables that will appear in each pi group and, among them, contain all of the fundamental dimensions. One way to choose a core is to exclude from it those variables whose effect one desires to isolate. In the present problem it would be desirable to have the drag force in only one dimensionless group, hence it will not be in the core. Let us arbitrarily let the viscosity be the other exclusion from the core. Our core group now consists of the remaining variables v, ρ, and L, which, we observe, include M, L, and t among them.

We now know that π_1 and π_2 both include ρ, L, and v; that one of them includes F and the other μ; and that they are both dimensionless. In order that each be dimensionless, the variables must be raised to certain exponents. Writing

$$\pi_1 = v^a \rho^b L^c F \quad \text{and} \quad \pi_2 = v^d \rho^e L^f \mu$$

we shall evaluate the exponents as follows. Considering each π group independently, we write

$$\pi_1 = v^a \rho^b L^c F$$

and dimensionally

$$M^0 L^0 t^0 = 1 = \left(\frac{L}{t}\right)^a \left(\frac{M}{L^3}\right)^b (L)^c \frac{ML}{t^2}$$

Equating exponents of M, L, and t on both sides of this expression, we have, for M

$$0 = b + 1$$

for L

$$0 = a - 3b + c + 1$$

and for t

$$0 = -a - 2$$

From these we find that $a = -2$, $b = -1$, and $c = -2$, giving

$$\pi_1 = \frac{F}{L^2 \rho v^2} = \frac{F/L^2}{\rho v^2} = \text{Eu}$$

Similarly for π_2 we have, in dimensional form

$$1 = \left(\frac{L}{t}\right)^d \left(\frac{M}{L^3}\right)^e (L)^f \frac{M}{Lt}$$

and for exponents of M

$$0 = e + 1$$

for L

$$0 = d - 3e + f - 1$$

and for t

$$0 = -d - 1$$

giving $d = -1$, $e = -1$ and $f = -1$. Thus, for our second dimensionless group we have

$$\pi_2 = \mu/\rho v L = 1/\text{Re}$$

Dimensional analysis has enabled us to relate the original five variables in terms of only two dimensionless parameters in the form

$$\text{Eu} = \phi(\text{Re}) \tag{11-9}$$

$$C_\text{D} = f(\text{Re}) \tag{11-10}$$

The two parameters, Eu and C_D, were also generated in the previous section by an alternate method. The functions $\phi(\text{Re})$ and $f(\text{Re})$ must be determined by experiment.

Table 11.2 lists several dimensionless groups that pertain to fluid flow. Similar tables will be include in later chapters that list dimensionless parameters common to heat transfer and to mass transfer.

Table 11.2 Common dimensionless parameters in momentum transfer

Name/Symbol	Dimensionless group	Physical meaning	Area of application
Reynolds number, Re	$Lv\rho/\mu$	$\dfrac{\text{Inertial force}}{\text{Viscous force}}$	Widely applicable in a host of fluid flow situations
Euler number, Eu	$P/\rho v^2$		
Coefficient of skin friction, C_f	$\dfrac{F/A}{\rho v^2/2}$	$\dfrac{\text{Pressure Force}}{\text{Inertial force}}$	Flows involving pressure differences due to frictional effects
Froude number, Fr	v^2/gL	$\dfrac{\text{Inertial force}}{\text{Gravitational force}}$	Flows involving free liquid surfaces
Weber number, We	$\dfrac{\rho v^2 L}{\sigma}$	$\dfrac{\text{Inertial force}}{\text{Surface tension force}}$	Flows with significant surface tension effects
Mach number, M	v/C	$\dfrac{\text{Inertial force}}{\text{Compressibility force}}$	Flows with significant compressibility effects

11.4 GEOMETRIC, KINEMATIC, AND DYNAMIC SIMILARITY

An important application and use of the dimensionless parameters listed in Table 11.2 is in using experimental results obtained using models to predict the performance of full-size prototypical systems. The validity of such *scaling* requires that the models and prototypes possess *similarity*. Three types of similarity are important in this regard; they are geometric, kinematic, and dynamic similarity.

Geometric similarity exists between two systems if the ratio of all significant dimensions is the same for each system. For example, if the ratio a/b for the diamond-shaped section in Figure 11.1 is equal in magnitude to the ratio a/b for the larger section, they are geometrically similar. In this example, there are only two significant dimensions. For more complex geometries, geometric similarity would be achieved when all geometric ratios between model and prototype are equal.

Figure 11.1 Two geometrically similar objects.

Kinematic similarity similarly exists when, in geometrically similar systems ① and ②, the velocities at the same locations are related according to

$$\left(\frac{v_x}{v_y}\right)_1 = \left(\frac{v_x}{v_y}\right)_2 \quad \left(\frac{v_x}{v_z}\right)_1 = \left(\frac{v_x}{v_z}\right)_2$$

The third type of similarity, *dynamic similarity*, exists when, in geometrically and kinematically similar systems, the ratios of significant forces are equal between model and prototype. These force ratios that are important in fluid flow applications include the dimensionless parameters listed in Table 11.2.

The process of scaling using these similarity requirements will be presented in Section 11.5.

11.5 MODEL THEORY

In the design and testing of large equipment involving fluid flow, it is customary to build small models geometrically similar to the larger prototypes. Experimental data achieved for the models are then scaled to predict the performance of full-sized prototypes according to the requirements of geometric, kinematic, and dynamic similarity. The following examples will illustrate the manner of utilizing model data to evaluate the conditions for a full-scale device.

EXAMPLE 2

A cylindrical mixing tank is to be scaled up to a larger size such that the volume of the larger tank is five times that of the smaller one. What will be the ratios of diameter and height between the two?

Geometric similarity between tanks a and b in Figure 11.1 requires that

$$\frac{D_a}{h_a} = \frac{D_b}{h_b}$$

or

$$\frac{h_b}{h_a} = \frac{D_b}{D_a}$$

The volumes of the two tanks are

$$V_a = \frac{\pi}{4} D_a^2 h_a \quad \text{and} \quad V_b = \frac{\pi}{4} D_b^2 h_b$$

The scaling ratio between the two is stipulated as, $\frac{V_b}{V_a} = 5$, thus,

$$\frac{V_b}{V_a} = \frac{(\pi/4)D_b^2 h_b}{(\pi/4)D_b^2 h_a} = 5$$

Figure 11.2 Cylindrical mixing tanks for Example 2.

and we get

$$\left(\frac{D_b}{D_a}\right)^2 \frac{h_b}{h_a} = 5$$

We now substitute the geometric similarity requirement that gives

$$\left(\frac{D_b}{D_a}\right)^3 = \left(\frac{L_b}{L_a}\right)^3 = 5$$

and the two ratios of interest become

$$\frac{D_b}{D_a} = \frac{L_b}{L_a} = 5^{1/3} = 1.71$$

EXAMPLE 3

Dynamic similarity may be obtained by using a cryogenic wind tunnel in which nitrogen at low temperature and high pressure is employed as the working fluid. If nitrogen at 5 atm and 183 K is used to test the low speed aerodynamics of a prototype that has a 24.38 m wing span and is to fly at standard sea-level conditions at a speed of 60 m/s, determine

(1) The scale of the model to be tested.

(2) The ratio of forces between the model and the full-scale aircraft.

Conditions of dynamic similarity should prevail. The speed of sound in nitrogen at 183 K is 275 m/s.

For dynamic similarity to exist, we know that both model and prototype must be geometrically similar and that the Reynolds number and the Mach number must be the same. A table such as the following is helpful.

	Model	Prototype
Characteristic length	L	24.38 m
Velocity	v	60 m/s
Viscosity	μ	$1.789 \cdot 10^{-5}$ Pa $\cdot$ s
Density	ρ	1.225 kg/m^3
Speed of sound	275 m/s	340 m/s

The conditions listed for the prototype have been obtained from Appendix I. Equating Mach numbers we obtain

$$M_m = M_p$$

$$v = \frac{275}{340}60 = 48.5 \,\text{m/s}$$

Equating the Reynolds numbers of the model and the prototype we obtain

$$\text{Re}_m = \text{Re}_p$$

$$\frac{\rho\, 48.5 L}{\mu} = \frac{1.225 \cdot 60 \cdot 24.38}{1.789 \cdot 10^{-5}} = 1.002 \times 10^8$$

Using equation (7-10), we may evaluate μ for nitrogen. From Appendix K, $\varepsilon/\kappa = 91.5\,\mathrm{K}$ and $\sigma = 3.681\,\text{Å}$ for nitrogen so that $\kappa T/\varepsilon = 2$ and $\Omega_\mu = 1.175$ (Appendix K). Thus,

$$\mu = 2.6693\cdot 10^{-6}\,\frac{\sqrt{28\cdot 183}}{(3.681)^2(1.175)} = 1.200\cdot 10^{-5}\,\mathrm{Pa\cdot s}$$

The density may be approximated from the perfect gas law

$$\rho = \frac{P}{P_1}\frac{M}{M_1}\frac{T_1}{T}\rho_1$$

so that

$$\rho = 5\left(\frac{28}{28.96}\right)\left(\frac{288}{183}\right)1.225 = 7.608\,\mathrm{kg/m^3}$$

Solving for the wing span of the model, we obtain

$$L = 3.26\,\mathrm{m}\,(10.7\,\mathrm{ft})$$

The ratio of the forces on the model to the forces experienced by the prototype may be determined equating values of Eu between the model and the prototype. Hence

$$\left(\frac{F}{\rho V^2 A_R}\right)_{\text{model}} = \left(\frac{F}{\rho V^2 A_R}\right)_{\text{prototype}}$$

where A_R is a suitable reference area. For an aircraft, this reference area is the projected wing area. The ratio of model force to prototype force is then given by

$$\frac{F_m}{F_p} = \frac{\rho_m}{\rho_p}\frac{V_m^2}{V_p^2}\frac{A_{R,m}}{A_{R,p}} = \frac{(\rho V^2)_m}{(\rho V^2)_p}\left(\frac{l_m}{l_p}\right)^2$$

where the ratio of reference areas can be expressed in terms of the scale ratio. Substituting numbers

$$\frac{F_m}{F_p} = \frac{7.608}{1.225}\left(\frac{48.5}{60.0}\right)^2\left(\frac{3.26}{24.38}\right)^2 = 0.0726$$

The forces on the model are seen to be 7.26% the prototype forces.

11.6 CLOSURE

The dimensional analysis of a momentum-transfer problem is simply an application of the requirement of dimensional homogeneity to a given situation. By dimensional analysis the work and time required to reduce and correlate experimental data are decreased substantially by the combination of individual variables into dimensionless π groups, which are fewer in number than the original variables. The indicated relations between dimensionless parameters are then useful in expressing the performance of the systems to which they apply.

It should be kept in mind that dimensional analysis *cannot* predict which variables are important in a given situation, nor does it give any insight into the physical transfer mechanism involved. Even with these limitations, dimensional analysis techniques are a valuable aid to the engineer.

If the equation describing a given process is known, the number of dimensionless groups is automatically determined by taking ratios of the various terms in the expression to one another. This method also gives physical meaning to the groups thus obtained.

If, on the contrary, no equation applies, an empirical method, the *Buckingham* method, may be used. This is a very general approach but gives no physical meaning to the dimensionless parameters obtained from such an analysis.

The requirements of geometric, kinematic, and dynamic similarity enable one to use model date to predict the behavior of a prototype or full-size piece of equipment. *Model theory* is thus an important application of the parameters obtained in a dimensional analysis.

PROBLEMS

11.1 The power output of a hydraulic turbine depends on the diameter D of the turbine, the density ρ of water, the height H of water surface above the turbine, the gravitational acceleration g, the angular velocity ω of the turbine wheel, the discharge Q of water through the turbine, and the efficiency η of the turbine. By dimensional analysis, generate a set of appropriate dimensionless groups.

11.2 Through a series of tests on pipe flow, *H*. Darcy derived an equation for the friction loss in pipe flow as

$$h_L = f \frac{L}{D} \frac{v^2}{2g},$$

in which f is a dimensionless coefficient that depends on (a) the average velocity v of the pipe flow; (b) the pipe diameter D; (c) the fluid density ρ; (d) the fluid viscosity μ; and (e) the average pipe wall uneveness e (length). Using the Buckingham π theorem, find a dimensionless function for the coefficient f.

11.3 The pressure rise across a pump P (this term is proportional to the head developed by the pump) may be considered to be affected by the fluid density ρ, the angular velocity ω, the impeller diameter D, the volumetric rate of flow Q, and the fluid viscosity μ. Find the pertinent dimensionless groups, choosing them so that P, Q, and μ each appear in one group only. Find similar expressions, replacing the pressure rise first by the power input to the pump, then by the efficiency of the pump.

11.4 The maximum pitching moment that is developed by the water on a flying boat as it lands is noted as c_{max} The following are the variables involved in this action:

α = angle made by flight path of plane with horizontal,
β = angle defining attitude of plane,
M = mass of plane,
L = length of hull,
ρ = density of water,
g = acceleration of gravity,
R = radius of gyration of plane about axis of pitching.

(a) According to the Buckingham π theorem, how many independent dimensionless groups should there be which characterize this problem?

(b) What is the dimensional matrix of this problem? What is its rank?

(c) Evaluate the appropriate dimensionless parameters for this problem.

11.5 The rate at which metallic ions are electroplated from a dilute electrolytic solution onto a rotating disk electrode is usually governed by the mass diffusion rate of ions to the disk. This process is believed to be controlled by the following variables:

	Dimensions
k = mass-transfer coeficient	L/t
D = diffusion coefficient	L^2/t
d = disk diameter	L
a = angular velocity	$1/t$
ρ = density	M/L^3
μ = viscosity	M/Lt

Obtain the set of dimensionless groups for these variables where k, μ, and D are kept in separate groups. How would you accumulate and present the experimental data for this system?

11.6 The performance of a journal bearing around a rotating shaft is a function of the following variables: Q, the rate of flow lubricating oil to the bearing in volume per unit time; D, the bearing diameter; N, the shaft speed in revolutions per minute; μ, the lubricant viscosity; ρ, the lubricant density; and σ, the surface tension of the lubricating oil. Suggest appropriate parameters to be used in correlating experimental data for such a system.

11.7 The mass M of drops formed by liquid discharging by gravity from a vertical tube is a function of the tube diameter D, liquid density,surface tension, and the acceleration of gravity. Determine the independent dimensionless groups that would allow the surface-tension effect to be analyzed.Neglect any effects of viscosity.

11.8 The functional frequency n of a stretched string is a function of the string length L, its diameter D, the mass density ρ, and the applied tensile force T. Suggest a set of dimensionless parameters relating these variables.

11.9 The power P required to run a compressor varies with compressor diameter D, angular velocity ω, volume flow rate Q, fluid density ρ, and fluid viscosity μ. Develop a relation between these variables by dimensional analysis, where fluid viscosity and angular velocity appear in only one dimensionless parameter.

11.10 A large amount of energy E is suddenly released in the air as in a point of explosion. Experimental evidence suggests the radius r of the high-pressure blast wave depends on time t as well as the energy E and ρ the density of the ambient air.

(a) Using the Buckingham method, find the equation for r as a function of t, ρ, and E.

(b) Show that the speed of the wave front decreases as r increases.

11.11 The size d of droplets produced by a liquid spray nozzle is thought to depend upon the nozzle diameter D, jet velocity V, and the properties of the liquid ρ, μ, and σ. Rewrite this relation in dimensionless form. Take D, ρ, and V as repeating variables.

11.12 Identify the variables associated with Problem 8.13 and find the dimensionless parameters.

11.13 A car is traveling along a road at 22.2 m/s Calculate the Reynolds number

(a) based on the length of the car,

(b) based on the diameter of the radio antenna.

The car length is 5.8 m and the antenna diameter is 6.4 mm.

11.14 In natural-convection problems, the variation of density due to the temperature difference ΔT creates an important buoyancy term in the momentum equation. If a warm gas at T_H moves through a gas at temperature T_0 and if the density change is only due to temperature changes, the equation of motion becomes

$$\rho \frac{D\mathbf{v}}{Dt} = -\nabla P + \mu \nabla^2 \mathbf{v} + \rho \mathbf{g}\left(\frac{T_H}{T_0} - 1\right)$$

Show that the ratio of gravity (buoyancy) to inertial forces acting on a fluid element is

$$\frac{Lg}{V_0^2}\left(\frac{T_H}{T_0} - 1\right)$$

where L and V_0 are reference lengths and velocity, respectively.

11.15 A 1/6-scale model of a torpedo is tested in a water tunnel to determine drag characteristics. What model velocity corresponds to a torpedo velocity of 20 knots? If the model resistance is 10 lb, what is the prototype resistance?

11.16 During the development of a 300-ft ship, it is desired to test a 10% scale model in a towing tank to determine the drag characteristics of the hull. Determine how the model is to be tested if the Froude number is to be duplicated.

11.17 A 25% scale model of an undersea vehicle that has a maximum speed of 16 m/s is to be tested in a wind tunnel with a pressure of 6 atm to determine the drag characteristics of the full-scale vehicle. The model is 3 m long. Find the air speed required to test the model and find the ratio of the model drag to the full-scale drag.

11.18 An estimate is needed on the lift provided by a hydrofoil wing section when it moves through water at 60 mph. Test data are available for this purpose from experiments in a pressurized wind tunnel with an airfoil section model geometrically similar to but twice the size of the hydrofoil. If the lift F_1 is a function of the density ρ of the fluid, the velocity v of the flow, the angle of attack θ, the chord length D, and the viscosity μ, what velocity of flow in the wind tunnel would correspond to the hydrofoil velocity for which the estimate is desired? Assume the same angle of attack in both cases, that the density of the air in the pressurized tunnel is 5.0×10^{-3} slugs/ft^3, that its kinematic viscosity is 8.0×10^{-5} ft^2/s, and that the kinematic viscosity of the water is approximately 1.0×10^{-5} ft^2/s. Take the density of water to be 1.94 slugs/ft^3.

11.19 A model of a harbor is made on the length ratio of 360:1. Storm waves of 2 m amplitude and 8 m/s velocity occur on the breakwater of the prototype harbor. Significant variables are the length scale, velocity, and g, the acceleration of gravity. The scaling of time can be made with the aid of the length scale and velocity scaling factors.

(a) Neglecting friction, what should be the size and speed of the waves in the model?

(b) If the time between tides in the prototype is 12 h, what should be the tidal period in the model?

11.20 A 40% scale model of an airplane is to be tested in a flow regime where unsteady flow effects are important. If the full-scale vehicle experiences the unsteady effects at a Mach number of 1 at an altitude of 40,000 ft, what pressure must the model be tested at to produce an equal Reynolds number? The model is to be tested in air at 70°F. What will the timescale of the flow about the model be relative to the full-scale vehicle?

11.21 A model ship propeller is to be tested in water at the same temperature that would be encountered by a full-scale propeller. Over the speed range considered, it is assumed that there is no dependence on the Reynolds or Euler numbers, but only on the Froude number (based on forward velocity V and propeller diameter d). In addition, it is thought that the ratio of forward to rotational speed of the propeller must be constant (the ratio V/Nd, where N is propeller rpm).

(a) With a model 041 m in diameter, a forward speed of 2.58 m/s and a rotational speed of 450 rpm is recorded. What are the forward and rotational speeds corresponding to a 2.45-m diameter prototype?

(b) A torque of 20 N · m is required to turn the model, and the model thrust is measured to be 245 N. What are the torque and thrust for the prototype?

Chapter 12

Viscous Flow

The concept of fluid viscosity was developed and viscosity defined in Chapter 7. Clearly, all fluids are viscous, but in certain situations and under certain conditions, a fluid may be considered ideal or inviscid, making possible an analysis by the methods of Chapter 10.

Our task in this chapter is to consider viscous fluids and the role of viscosity as it affects the flow. Of particular interest is the case of flow past solid surfaces and the interrelations between the surfaces and the flowing fluid.

12.1 REYNOLDS'S EXPERIMENT

The existence of two distinct types of viscous flow is a universally accepted phenomenon. The smoke emanating from a lighted cigarette is seen to flow smoothly and uniformly for a short distance from its source and then change abruptly into a very irregular, unstable pattern. Similar behavior may be observed for water flowing slowly from a faucet.

The well-ordered type of flow occurs when adjacent fluid layers slide smoothly over one another with mixing between layers or lamina occurring only on a molecular level. It was for this type of flow that Newton's viscosity relation was derived, and in order for us to measure the viscosity, μ, this *laminar* flow must exist.

The second flow regime, in which small packets of fluid particles are transferred between layers, giving it a fluctuating nature, is called the *turbulent* flow regime.

The existence of laminar and turbulent flow, although recognized earlier, was first described quantitatively by Reynolds in 1883. His classic experiment is illustrated in Figure 12.1. Water was allowed to flow through a transparent pipe, as shown, at a rate controlled by a valve. A dye having the same specific gravity as water was introduced at the pipe opening and its pattern observed for progressively larger flow rates of water. At low rates of flow, the dye pattern was regular and formed a single line of color as shown in Figure 12.1(a). At high flow rates, however, the dye became dispersed throughout the pipe cross section because of

(a) Re < 2300

(b) Re > 2300

Figure 12.1 Reynolds's experiment.

the very irregular fluid motion. The difference in the appearance of the dye streak was, of course, due to the orderly nature of laminar flow in the first case and to the fluctuating character of turbulent flow in the latter case.

The transition from laminar to turbulent flow in pipes is thus a function of the fluid velocity. Actually, Reynolds found that fluid velocity was the only one variable determining the nature of pipe flow, the others being pipe diameter, fluid density, and fluid viscosity. These four variables, combined into the single dimensionless parameter

$$\text{Re} \equiv \frac{D\rho v}{\mu} \tag{12-1}$$

form the Reynolds number, symbolized Re, in honor of Osborne Reynolds and his important contributions to fluid mechanics.

For flow in circular pipes, it is found that below a value for Reynolds number of 2300 the flow is *laminar*. Above this value the flow may be laminar as well, and indeed, laminar flow has been observed for Reynolds numbers as high as 40,000 in experiments wherein external disturbances were minimized. Above a Reynolds number of 2300, small disturbances will cause a transition to *turbulent* flow whereas below this value disturbances are damped out and laminar flow prevails. The *critical Reynolds number for pipe flow* thus is 2300.

12.2 DRAG

Reynolds's experiment clearly demonstrated the two different regimes of flow: laminar and turbulent. Another manner of illustrating these different flow regimes and their dependence upon Reynolds number is through the consideration of drag. A particularly illustrative case is that of external flow (i.e., flow around a body as opposed to flow inside a conduit).

The drag force due to friction is caused by the shear stresses at the surface of a solid object moving through a viscous fluid. Frictional drag is evaluated by using the expression

$$\frac{F}{A} \equiv C_f \frac{\rho v_\infty^2}{2} \tag{12-2}$$

where F is the force, A is the area of contact between the solid body and the fluid, C_f is the coefficient of skin friction, ρ is the fluid density, and v_∞ is the free-stream fluid velocity.

The coefficient of skin friction, C_f which is defined by equation (12-2), is dimensionless.

The total drag on an object may be due to pressure as well as frictional effects. In such a situation another coefficient, C_D, is defined as

$$\frac{F}{A_P} \equiv C_D \frac{\rho v_\infty^2}{2} \tag{12-3}$$

where F, ρ, and v_∞ are as described above and, additionally

$$C_D = \text{the drag coefficient}$$

and

$$A_P = \text{the projected area of the surface}$$

The value of A_P used in expressing the drag for blunt bodies is normally the maximum projected area for the body.

The quantity $\rho v_\infty^2/2$ appearing in equations (12-2) and (12-3) is frequently called the *dynamic pressure*.

Pressure drag arises from two principal sources.[1] One is induced drag, or drag due to lift. The other source is wake drag, which arises from the fact that the shear stress causes the streamlines to deviate from their inviscid flow paths, and in some cases to separate from the body altogether. This deviation in streamline pattern prevents the pressure over the rest of a body from reaching the level it would attain otherwise. As the pressure at the front of the body is now greater than that at the rear, a net rearward force develops.

In an incompressible flow, the drag coefficient depends upon the Reynolds number and the geometry of a body. A simple geometric shape that illustrates the drag dependence upon the Reynolds number is the circular cylinder. The inviscid flow about a circular cylinder was examined in Chapter 10. The inviscid flow about a cylinder of course, produced no drag, as there existed neither frictional nor pressure drag. The variation in the drag coefficient with the Reynolds number for a smooth cylinder is shown in Figure 12.2. The flow pattern about the cylinder is illustrated for several different values of Re. The flow pattern and general shape of the curve suggest that the drag variation, and hence the effects of shear stress on the flow, may be subdivided into four regimes. The features of each regime will be examined.

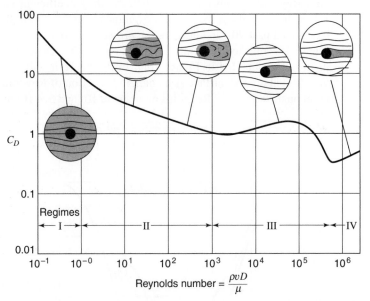

Figure 12.2 Drag coefficient for circular cylinders as a function of Reynolds number. Shaded regions indicate areas influenced by shear stress.

Regime 1

In this regime the entire flow is laminar and the Reynolds number small, being less than 1. Recalling the physical significance of the Reynolds number from Chapter 11 as the ratio of the inertia forces to the viscous forces, we may say that in regime 1 the viscous forces predominate. The flow pattern in this case is almost symmetric, the flow adheres to the body,

---

[1] A third source of pressure drag, wave drag, is associated with shock waves.

and the wake is free from oscillations. In this regime of the so-called *creeping flow*, viscous effects predominate and extend throughout the flow field.

Regime 2

Two illustrations of the flow pattern are shown in the second regime. As the Reynolds number is increased, small eddies form at the rear stagnation point of the cylinder. At higher values of the Reynolds number, these eddies grow to the point at which they separate from the body and are swept downstream into the wake. The pattern of eddies shown in regime 2 is called a von Kármán vortex trail. This change in the character of the wake from a steady to an unsteady nature is accompanied by a change in the slope of the drag curve. The paramount features of this regime are (a) the unsteady nature of the wake and (b) flow separation from the body.

Regime 3

In the third regime the point of flow separation stabilizes at a point about 80° from the forward stagnation point. The wake is no longer characterized by large eddies, although it remains unsteady. The flow on the surface of the body from the stagnation point to the point of separation is laminar, and the shear stress in this interval is appreciable only in a thin layer near the body. The drag coefficient levels out at a near-constant value of approximately 1.

Regime 4

At a Reynolds number near 5×10^5, the drag coefficient suddenly decreases to 0.3. When the flow about the body is examined, it is observed that the point of separation has moved past 90°. In addition, the pressure distribution about the cylinder (shown in Figure 12.3) up to the point of separation is fairly close to the inviscid flow pressure distribution depicted in Figure 10.5. In the figure it will be noticed that the pressure variation about the surface is a changing function of Reynolds number. The minimum point on the curves for Reynolds numbers of 10^5 and 6×10^5 are both at the point of flow

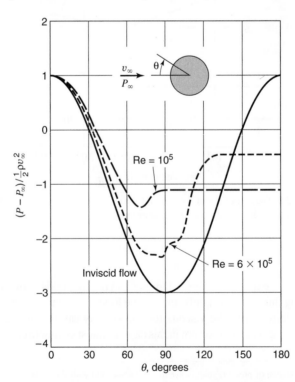

Figure 12.3 Pressure distribution on a circular cylinder at various Reynolds numbers.

separation. From this figure it is seen that separation occurs at a larger value of θ for Re $= 6 \times 10^5$ than it does for Re $= 10^5$.

The layer of flow near the surface of the cylinder is turbulent in this regime, undergoing transition from laminar flow close to the forward stagnation point. The marked decrease in drag is due to the change in the point of separation. In general, a turbulent flow resists flow separation better than a laminar flow. As the Reynolds number is large in this regime, it may be said that the inertial forces predominate over the viscous forces.

The four regimes of flow about a cylinder illustrate the decreasing realm of influence of viscous forces as the Reynolds number is increased. In regimes 3 and 4, the flow pattern over the forward part of the cylinder agrees well with the inviscid flow theory. For other geometries, a similar variation in the realm of influence of viscous forces is observed and, as might be expected, agreement with inviscid-flow predictions at a given Reynolds number increases as the slenderness of the body increases. The majority of cases of engineering interest involving external flows have flow fields similar to those of regimes 3 and 4.

Figure 12.4 shows the variation in the drag coefficient with the Reynolds number for a sphere, for infinite plates, and for circular disks and square plates. Note the similarity in form of the curve of C_D for the sphere to that for a cylinder in Figure 12.2. Specifically, one may observe the same sharp decrease in C_D to a minimum value near a Reynolds number value of 5×10^5. This is again due to the change from laminar to turbulent flow in the boundary layer.

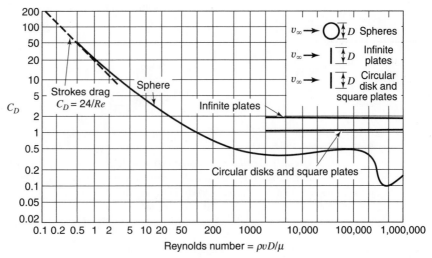

Figure 12.4 Drag coefficient versus Reynolds number for various objects.

EXAMPLE 1

Evaluate the terminal velocity of a 7.5-mm-diameter glass sphere falling freely through (a) air at 300 K; (b) water at 300 K; and (c) glycerin at 300 K. The density of glass is 2250 kg/m³.

The terminal (steady state) velocity of a falling body is reached when the force due to fluid drag matches the body's weight. In this case the weight of the glass sphere can be expressed as

$$\rho_s \frac{\pi d^3}{6} g$$

The fluid drag force is given by

$$C_D \frac{\rho_f v_\infty^2}{2} \frac{\pi d^2}{4}$$

and a force balance yields

$$C_D v_\infty^2 = \frac{4}{3} \frac{\rho_s}{\rho_f} dg$$

The drag coefficient, C_D, is plotted as a function of Reynolds number, Re_d, in Figure 12.4. As C_D is a function of v_∞, we are unable to solve explicitly for v_∞ unless $Re_d < 1$, which would permit the use of Stokes' law in expressing C_D. A trial-and-error solution is, thus, required. The conditions to be satisfied are our force balance expression and the graphical relation between C_D and Re_d in Figure 12.4.

For air at 300 K

$$\nu = 1.569 \times 10^{-5}\, \mathrm{m^2/s}$$

$$\rho = 1.177\, \mathrm{kg/m^3}$$

$$Re_d = \frac{d v_\infty}{\nu} = \frac{(7.5 \times 10^{-3}\,\mathrm{m}) v_\infty}{1.569 \times 10^{-5}\,\mathrm{m^2/s}} \tag{A}$$

$$= 478.0\, v_\infty$$

Inserting known values into our force balance expression, we have

$$C_D v_\infty^2 = \frac{4}{3} \frac{2250\,\mathrm{kg/m^3}}{1.177\,\mathrm{kg/m^3}} (7.5 \times 10^{-3}\,\mathrm{m})(9.81\,\mathrm{m/s^2}) \tag{B}$$

$$= 187.5\, \mathrm{m^2/s^2}$$

Normally the trial-and-error procedure to achieve a solution would be straightforward. In this case, however, the shape of the C_D vs. Re_d curve, given in Figure 12.4, poses a bit of a problem. Specifically, the value of C_D remains nearly uniform, that is, $0.4 < C_D < 0.5$, over a range in Re_d between $500 < Re_d < 10^5$; over 3 orders of magnitude!

In such a specific case, we will assume $C_D \cong 0.4$, and solve equation (B) for v_∞:

$$v_\infty = \left[\frac{187.5}{0}\, \mathrm{m^2/s^2}\right]^{1/2} = 21.65\, \mathrm{m/s}$$

Equation (b) then yields

$$Re_d = (478.0)(21.65) = 1.035 \times 10^4$$

These results are compatible with Figure 12.4 although the absolute accuracy is obviously not great.

Finally, for air, we determine the terminal velocity to be, approximately

$$v_\infty \cong 21.6 \, \text{m/s} \tag{a}$$

For water at 300 K

$$\nu = 0.879 \times 10^{-6} \, \text{m}^2/\text{s}$$

$$\rho = 996.1 \, \text{kg/m}^3$$

$$\text{Re}_d = \frac{(7.5 \times 10^{-3} \, \text{m})v_\infty}{0.879 \times 10^{-6} \, \text{m}^2/\text{s}} = 8530 \, v_\infty$$

$$C_D v_\infty^2 = \frac{4}{3} \frac{2250 \, \text{kg/m}^3}{996 \, \text{kg/m}^3}(7.5 \times 10^{-3} \, \text{m})(9.81 \, \text{m/s}^2)$$

$$= 0.0981$$

As in part (a), we will initially assume $C_D \cong 0.4$, and achieve the result that

$$v_\infty = 0.495 \, \text{m/s}$$

$$\text{Re}_d = 4220$$

These results, again, satisfy Figure 12.4. Thus, in water

$$v_\infty = 0.495 \, \text{m/s} \tag{b}$$

Finally, for glycerin at 300 K

$$\nu = 7.08 \times 10^{-4} \, \text{m}^2/\text{s}$$

$$\rho = 1260 \, \text{kg/m}^3$$

$$\text{Re}_d = \frac{(7.5 \times 10^{-3} \, \text{m})v_\infty}{7.08 \times 10^{-4} \, \text{m}^2/\text{s}} = 10.59 \, v_\infty$$

$$C_D v_\infty^2 = \frac{4}{3} \frac{2250 \, \text{kg/m}^3}{1260 \, \text{kg/m}^3}(7.5 \times 10^{-3} \, \text{m})(9.81 \, \text{m/s}^2)$$

$$= 0.1752$$

In this case we suspect the Reynolds number will be quite small. As an initial guess we will assume Stokes' law applies, thus $C_D = 24/\text{Re}$.

Solving for v_∞ for this case, we have

$$C_D v_\infty^2 = \frac{24\nu}{dv_\infty} v_\infty^2 = 0.1752 \, \text{m}^2/\text{s}^2$$

$$v_\infty = \frac{(0.1752 \, \text{m}^2/\text{s}^2)(7.5 \times 10^{-3} \, \text{m})}{24(7.08 \times 10^{-4} \, \text{m}^2/\text{s})}$$

$$= 0.0773 \, \text{m/s}$$

To validate the use of Stokes' law, we check the value of Reynolds number and get

$$\text{Re}_d = \frac{(7.5 \times 10^{-3} \, \text{m})(0.0773 \, \text{m/s})}{7.08 \times 10^{-4} \, \text{m}^2/\text{s}}$$

$$= 0.819$$

which is in the allowable range. The terminal velocity in glycerin thus

$$v_\infty = 0.0773 \, \text{m/s} \tag{c}$$

12.3 THE BOUNDARY-LAYER CONCEPT

The observation of a decreasing region of influence of shear stress as the Reynolds number is increased led Ludwig Prandtl to the boundary-layer concept in 1904. According to Prandtl's hypothesis, the effects of fluid friction at high Reynolds numbers are limited to a thin layer near the boundary of a body, hence the term *boundary layer*. Further, there is no significant pressure change across the boundary layer. This means that the pressure in the boundary layer is the same as the pressure in the inviscid flow outside the boundary layer. The significance of the Prandtl theory lies in the simplification that it allows in the analytical treatment of viscous flows. The pressure, for example, may be obtained from experiment or inviscid flow theory. Thus the only unknowns are the velocity components.

The boundary layer on a flat plate is shown in Figure 12.5. The thickness of the boundary layer, δ, is arbitrarily taken as the distance away from the surface where the velocity reaches 99% of the free-stream velocity. The thickness is exaggerated for clarity.

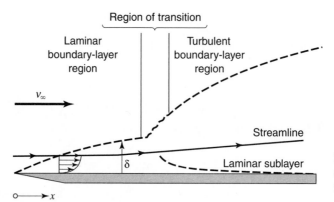

Figure 12.5 Boundary layer on a flat plate. (The thickness is exaggerated for clarity.)

Figure 12.5 illustrates how the thickness of the boundary layer increases with distance x from the leading edge. At relatively small values of x, flow within the boundary layer is laminar, and this is designated as the laminar boundary-layer region. At larger values of x the transition region is shown where fluctuations between laminar and turbulent flows occur within the boundary layer. Finally, for a certain value of x, and above, the boundary layer will always be turbulent. In the region in which the boundary layer is turbulent, there exists, as shown, a very thin film of fluid called the *laminar sublayer*, wherein flow is still laminar and large velocity gradients exist.

The criterion for the type of boundary layer present is the magnitude of Reynolds number, Re_x, known as the *local Reynolds number*, based on the distance x from the leading edge. The local Reynolds number is defined as

$$\mathrm{Re}_x \equiv \frac{x v \rho}{\mu} \tag{12-4}$$

For flow past a flat plate, as shown in Figure 12.5, experimental data indicate that for

(a) $\mathrm{Re}_x < 2 \times 10^5$ the boundary layer is laminar,

(b) $2 \times 10^5 < \mathrm{Re}_x < 3 \times 10^6$ the boundary layer may be either laminar or turbulent,

(c) $3 \times 10^6 < \mathrm{Re}_x$ the boundary layer is turbulent

12.4   THE BOUNDARY-LAYER EQUATIONS

The concept of a relatively thin boundary layer at high Reynolds numbers leads to some important simplifications of the Navier–Stokes equations. For incompressible, two-dimensional flow over a flat plate, the Navier–Stokes equations are

$$\rho\left\{\frac{\partial v_x}{\partial t} + v_x\frac{\partial v_x}{\partial x} + v_y\frac{\partial v_x}{\partial y}\right\} = \frac{\partial \sigma_{xx}}{\partial x} + \frac{\partial \tau_{yx}}{\partial y} \tag{12-5}$$

and

$$\rho\left\{\frac{\partial v_y}{\partial t} + v_x\frac{\partial v_y}{\partial x} + v_y\frac{\partial v_y}{\partial y}\right\} = \frac{\partial \tau_{xy}}{\partial x} + \frac{\partial \sigma_{yy}}{\partial y} \tag{12-6}$$

where $\tau_{xy} = \tau_{yx} = \mu(\partial v_x/\partial y + \partial v_y/\partial x)$, $\sigma_{xx} = -P + 2\mu(\partial v_x/\partial x)$ and $\sigma_{yy} = -P + 2\mu(\partial v_y/\partial y)$. The shear stress in a thin boundary layer is closely approximated by $\mu(\partial v_x/\partial y)$. This can be seen by considering the relative magnitudes of $\partial v_x/\partial y$ and $\partial v_y/\partial x$ From Figure 12.5, we may write $v_x|_\delta / v_y|_\delta \sim \mathbb{O}(x/\delta)$, where $\mathbb{O}$ signifies the order of magnitude. Then

$$\frac{\partial v_x}{\partial y} \sim \mathbb{O}\left(\frac{v_x|_\delta}{\delta}\right) \qquad \frac{\partial v_y}{\partial x} \sim \mathbb{O}\left(\frac{v_y|_\delta}{x}\right)$$

so

$$\frac{\partial v_x/\partial y}{\partial v_y/\partial x} \sim \mathbb{O}\left(\frac{x}{\delta}\right)^2$$

which, for a relatively thin boundary layer, is a large number, and thus $\partial v_x/\partial y \gg \partial v_y/\partial x$. The normal stress at a large Reynolds number is closely approximated by the negative of the pressure as $\mu(\partial v_x/\partial x) \sim \mathbb{O}(\mu v_\infty/x) = \mathbb{O}(\rho v_\infty^2/\mathrm{Re}_x)$; therefore $\sigma_{xx} \simeq \sigma_{yy} \simeq -P$. When these simplifications in the stresses are incorporated, the equations for flow over a flat plate become

$$\rho\left(\frac{\partial v_x}{\partial t} + v_x\frac{\partial v_x}{\partial x} + v_y\frac{\partial v_x}{\partial y}\right) = -\frac{\partial P}{\partial x} + \mu\frac{\partial^2 v_x}{\partial y^2} \tag{12-7}$$

and

$$\rho\left(\frac{\partial v_y}{\partial t} + v_x\frac{\partial v_y}{\partial x} + v_y\frac{\partial v_y}{\partial y}\right) = -\frac{\partial P}{\partial y} + \mu\frac{\partial^2 v_y}{\partial x^2} \tag{12-8}$$

Furthermore,[2] the terms in the second equation are much smaller than those in the first equation, and thus $\partial P/\partial y \simeq 0$; hence $\partial P/\partial x = dP/dx$, which according to Bernoulli's equation is equal to $-\rho v_\infty dv_\infty/dx$.

The final form of equation (12-7) becomes

$$\frac{\partial v_x}{\partial t} + v_x\frac{\partial v_x}{\partial x} + v_y\frac{\partial v_x}{\partial y} = v_\infty\frac{dv_\infty}{dx} + \nu\frac{\partial^2 v_x}{\partial y^2} \tag{12-9}$$

---

[2] The order of magnitude of each term may be considered as above. For example, $v_x(\partial v_y/\partial x) \sim \mathbb{O}(v_\infty(v_\infty/x)$ $(\delta/x)) = \mathbb{O}(v_\infty^2 \delta/x^2)$.

The above equation, and the continuity equation

$$\frac{\partial v_x}{\partial x} + \frac{\partial v_y}{\partial y} = 0 \qquad (12\text{-}10)$$

are known as the boundary-layer equations.

12.5 BLASIUS'S SOLUTION FOR THE LAMINAR BOUNDARY LAYER ON A FLAT PLATE

One very important case in which an analytical solution of the equations of motion has been achieved is that for the laminar boundary layer on a flat plate in steady flow.

For flow parallel to a flat surface, $v_\infty(x) = v_\infty$ and $dP/dx = 0$, according to the Bernoulli equation. The equations to be solved are now the following:

$$v_x \frac{\partial v_x}{\partial x} + v_y \frac{\partial v_x}{\partial y} = \nu \frac{\partial^2 v_x}{\partial y^2} \qquad (12\text{-}11a)$$

and

$$\frac{\partial v_x}{\partial x} + \frac{\partial v_y}{\partial y} = 0 \qquad (12\text{-}11b)$$

with boundary conditions $v_x = v_y = 0$ at $y = 0$, and $v_x = v_\infty$ at $y = \infty$.

Blasius[3] obtained a solution to the set of equations (12-11) by first introducing the stream function, Ψ, as described in Chapter 10, which automatically satisfies the two-dimensional continuity equation, equation (12-11b). This set of equations may be reduced to a single ordinary differential equation by transforming the independent variables x, y, to η and the dependent variables from $\Psi(x, y)$ to $f(\eta)$
where

$$\eta(x, y) = \frac{y}{2} \left(\frac{v_\infty}{\nu x} \right)^{1/2} \qquad (12\text{-}12)$$

and

$$f(\eta) = \frac{\Psi(x, y)}{(\nu x v_\infty)^{1/2}} \qquad (12\text{-}13)$$

The appropriate terms in equation (12-11a) may be determined from equations (12-12) and (12-13). The following expressions will result. The reader may wish to verify the mathematics involved.

$$v_x = \frac{\partial \Psi}{\partial y} = \frac{v_\infty}{2} f'(\eta) \qquad (12\text{-}14)$$

$$v_y = -\frac{\partial \Psi}{\partial x} = \frac{1}{2} \left(\frac{\nu v_\infty}{x} \right)^{1/2} (\eta f' - f) \qquad (12\text{-}15)$$

$$\frac{\partial v_x}{\partial x} = -\frac{v_\infty \eta}{4x} f'' \qquad (12\text{-}16)$$

$$\frac{\partial v_x}{\partial y} = \frac{v_\infty}{4} \left(\frac{v_\infty}{\nu x} \right)^{1/2} f'' \qquad (12\text{-}17)$$

$$\frac{\partial^2 v_x}{\partial y^2} = \frac{v_\infty}{8} \frac{v_\infty}{\nu x} f''' \qquad (12\text{-}18)$$

---

[3] H. Blasius, Grenzshichten in Flüssigkeiten mit kleiner Reibung, *Z. Math. U. Phys. Sci.*, **1**, 1908.

Substitution of (12-14) through (12-18) into equation (12-11a) and cancellation gives, as a single ordinary differential equation

$$f''' + ff'' = 0 \qquad (12\text{-}19)$$

with the appropriate boundary conditions

$$f = f' = 0 \qquad \text{at } \eta = 0$$
$$f' = 2 \qquad \text{at } \eta = \infty$$

Observe that this differential equation, although ordinary, is nonlinear and that, of the end conditions on the variable $f(\eta)$, two are initial values and the third is a boundary value. This equation was solved first by Blasius, using a series expansion to express the function, $f(\eta)$, at the origin and an asymptotic solution to match the boundary condition at $\eta = \infty$. Howarth[4] later performed essentially the same work but obtained more accurate results. Table 12.1 presents the significant numerical results of Howarth. A plot of these values is included in Figure 12.6.

A simpler way of solving equation (12-19) has been suggested in Goldstein[5] who presented a scheme whereby the boundary conditions on the function f are initial values.

Table 12.1   Values of f, f', f'', and v_x/v_∞ for laminar flow parallel to a flat plate (after Howarth)

$\eta = \dfrac{y}{2}\sqrt{\dfrac{v_\infty}{vx}}$	f	f'	f''	$\dfrac{v_x}{v_\infty}$
0	0	0	1.32824	0
0.2	0.0266	0.2655	1.3260	0.1328
0.4	0.1061	0.5294	1.3096	0.2647
0.6	0.2380	0.7876	1.2664	0.3938
0.8	0.4203	1.0336	1.1867	0.5168
1.0	0.6500	1.2596	1.0670	0.6298
1.2	0.9223	1.4580	0.9124	0.7290
1.4	1.2310	1.6230	0.7360	0.8115
1.6	1.5691	1.7522	0.5565	0.8761
1.8	1.9295	1.8466	0.3924	0.9233
2.0	2.3058	1.9110	0.2570	0.9555
2.2	2.6924	1.9518	0.1558	0.9759
2.4	3.0853	1.9756	0.0875	0.9878
2.6	3.4819	1.9885	0.0454	0.9943
2.8	3.8803	1.9950	0.0217	0.9915
3.0	4.2796	1.9980	0.0096	0.9990
3.2	4.6794	1.9992	0.0039	0.9996
3.4	5.0793	1.9998	0.0015	0.9999
3.6	5.4793	2.0000	0.0005	1.0000
3.8	5.8792	2.0000	0.0002	1.0000
4.0	6.2792	2.0000	0.0000	1.0000
5.0	8.2792	2.0000	0.0000	1.0000

[4] L. Howarth, "On the solution of the laminar boundary layer equations," *Proc. Roy. Soc. London*, **A164** 547 (1938).
[5] S. Goldstein, *Modern Developments in Fluid Dynamics*, Oxford Univ. Press, London, 1938, p. 135.

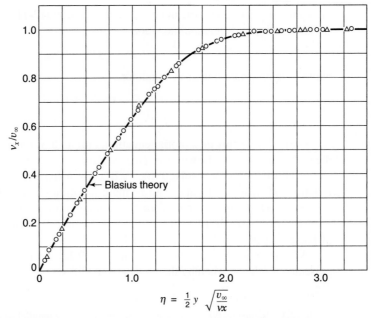

Figure 12.6 Velocity distribution in the laminar boundary layer over a flat plate. Experimental data by J. Nikuradse (monograph, Zentrale F. wiss. Berichtswesen, Berlin, 1942) for the Reynolds number range from 1.08×10^5 to 7.28×10^5.

If we define two new variables in terms of the constant, C, so that

$$\phi = f/C \tag{12-20}$$

and

$$\xi = C\eta \tag{12-21}$$

then the terms in equation (12-19) become

$$f(\eta) = C\phi(\xi) \tag{12-22}$$

$$f' = C^2\phi' \tag{12-23}$$

$$f'' = C^3\phi'' \tag{12-24}$$

and

$$f''' = C^4\phi''' \tag{12-25}$$

The resulting differential equation in $\phi(\xi)$ becomes

$$\phi''' + \phi\phi'' = 0 \tag{12-26}$$

and the initial conditions on ϕ are

$$\phi = 0 \quad \phi' = 0 \quad \phi'' = ? \quad \text{at } \xi = 0$$

The other boundary condition may be expressed as follows:

$$\phi'(\xi) = \frac{f'(\eta)}{C^2} = \frac{2}{C^2} \quad \text{at } \xi = \infty$$

An initial condition may be matched to this boundary condition if we let $f''(\eta = 0)$ equal some constant A; then $\phi''(\xi = 0) = A/C^3$. The constant A must have a certain value to satisfy the original boundary condition on f'. As an estimate we let $\phi''(\xi = 0) = 2$, giving $A = 2C^3$. Thus initial values of ϕ, ϕ', and ϕ'' are now specified. The estimate on $\phi''(0)$ requires that

$$\phi'(\infty) = \frac{2}{C^2} = 2\left(\frac{2}{A}\right)^{2/3} \tag{12-27}$$

Thus equation (12-26) may be solved as an initial-value problem with the answer scaled according to equation (12-27) to match the boundary condition at $\eta = \infty$.

The significant results of Blasius's work are the following:

(a) The boundary thickness, δ, is obtained from Table 12.1. When $\eta = 2.5$, we have $v_x/v_\infty \cong 0.99$ thus, designating $y = \delta$ at this point, we have

$$\eta = \frac{y}{2}\sqrt{\frac{v_\infty}{\nu x}} = \frac{\delta}{2}\sqrt{\frac{v_\infty}{\nu x}} = 2.5$$

and thus

$$\delta = 5\sqrt{\frac{\nu x}{v_\infty}}$$

or

$$\frac{\delta}{x} = \frac{5}{\sqrt{\dfrac{v_\infty x}{\nu}}} = \frac{5}{\sqrt{Re_x}} \tag{12-28}$$

(b) The velocity gradient at the surface is given by equation (12-27):

$$\left.\frac{\partial v_x}{\partial y}\right|_{y=0} = \frac{v_\infty}{4}\left(\frac{v_\infty}{\nu x}\right)^{1/2} f''(0) = 0.332\, v_\infty \sqrt{\frac{v_\infty}{\nu x}} \tag{12-29}$$

As the pressure does not contribute to the drag for flow over a flat plate, all the drag is viscous. The shear stress at the surface may be calculated as

$$\tau_0 = \mu \left.\frac{\partial v_x}{\partial y}\right|_{y=0}$$

Substituting equation (12-29) into this expression, we have

$$\tau_0 = \mu\, 0.332\, v_\infty \sqrt{\frac{v_\infty}{\nu x}} \tag{12-30}$$

The coefficient of skin friction may be determined by employing equation (12-2) as follows:

$$C_{fx} \equiv \frac{\tau}{\rho v_\infty^2/2} = \frac{F_d/A}{\rho v_\infty^2/2} = \frac{0.332 \mu v_\infty \sqrt{\dfrac{v_\infty}{\nu x}}}{\rho v_\infty^2/2}$$

$$= 0.664\sqrt{\frac{\nu}{x v_\infty}}$$

$$C_{fx} = \frac{0.664}{\sqrt{Re_x}} \tag{12-31}$$

Equation (12-31) is a simple expression for the coefficient of skin friction at a particular value of x. For this reason the symbol C_{fx} is used, the x subscript indicating a *local coefficient*.

While it is of interest to know values of C_{fx}, it is seldom that a local value is useful; most often one wishes to calculate the total drag resulting from viscous flow over some surface of finite size. The mean coefficient of skin friction that is helpful in this regard may be determined quite simply from C_{fx} according to

$$F_d = AC_{fL}\frac{\rho v_\infty^2}{2} = \frac{\rho v_\infty^2}{2}\int_A C_{fx}\,dA$$

or the mean coefficient, designated C_{fL}, is related to C_{fx} by

$$C_{fL} = \frac{1}{A}\int_A C_{fx}\,dA$$

For the case solved by Blasius, consider a plate of uniform width W, and length L, for which

$$C_{fL} = \frac{1}{L}\int_0^L C_{fx}\,dx = \frac{1}{L}\int_0^L 0.664\sqrt{\frac{v}{v_\infty}}x^{-1/2}dx$$

$$= 1.328\sqrt{\frac{v}{Lv_\infty}}$$

$$C_{fL} = \frac{1.328}{\sqrt{\mathrm{Re}_L}} \tag{12-32}$$

12.6 FLOW WITH A PRESSURE GRADIENT

In Blasius's solution for laminar flow over a flat plate, the pressure gradient was zero. A much more common flow situation involves flow with a pressure gradient. The pressure gradient plays a major role in flow separation, as can be seen with the aid of the boundary-layer equation (12-7). If we make use of the boundary, conditions at the wall $v_x = v_y = 0$, at $y = 0$ equation 12-7 becomes

$$\mu\frac{\partial^2 v_x}{\partial y^2}\bigg|_{y=0} = \frac{dP}{dx} \tag{12-33}$$

which relates the curvature of the velocity profile at the surface to the pressure gradient. Figure 12.7 illustrates the variation in v_x, $\partial v_x/\partial y$, and $\partial^2 v_x/\partial y^2$ across the boundary layer for the case of a zero-pressure gradient.

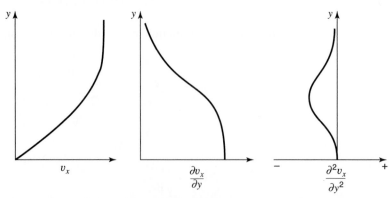

Figure 12.7 Variation in velocity and velocity derivatives across the laminar boundary layer when $dP/dx = 0$.

When $dP/dx = 0$, the second derivative of the velocity at the wall must also be zero; hence the velocity profile is linear near the wall. Further, out in the boundary layer, the velocity gradient becomes smaller and gradually approaches zero. The decrease in the velocity gradient means that the second derivative of the velocity must be negative. The derivative $\partial^2 v_x/\partial y^2$ is shown as being zero at the wall, negative within the boundary layer, and approaching zero at the outer edge of the boundary layer. It is important to note that the second derivative must approach zero from the negative side as $y \to \delta$. For values of $dP/dx \neq 0$, the variation in v_x and its derivatives is shown in Figure 12.8.

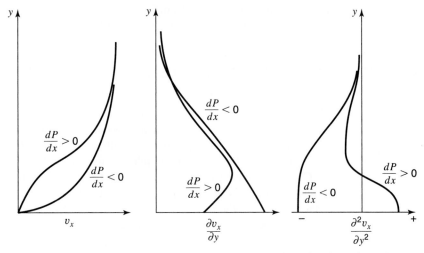

Figure 12.8 Variation in v_x and its derivatives across the boundary layer for various pressure gradients.

A negative pressure gradient is seen to produce a velocity variation somewhat similar to that of the zero-pressure-gradient case. A positive value of dP/dx, however, requires a positive value of $\partial^2 v_x/\partial y^2$ at the wall. As this derivative must approach zero from the negative side, at some point within the boundary layer the second derivative must equal zero. A zero second derivative, it will be recalled, is associated with an inflection point. The inflection point is shown in the velocity profile of Figure 12.8. We may now turn our attention to the subject of flow separation.

In order for flow separation to occur, the velocity in the layer of fluid adjacent to the wall must be zero or negative, as shown in Figure 12.9. This type of velocity profile is seen

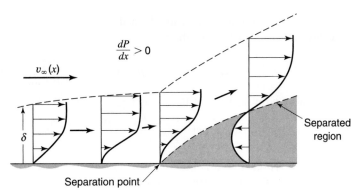

Figure 12.9 Velocity profiles in separated-flow region.

to require a point of inflection. As the only type of boundary-layer flow that has an inflection point is flow with a positive pressure gradient, it may be concluded that a positive pressure gradient is necessary for flow separation. For this reason a positive pressure gradient is called an *adverse pressure gradient*. Flow can remain unseparated with an adverse pressure gradient, thus $dP/dx > 0$ is a necessary but not a sufficient condition for separation. In contrast a negative pressure gradient, in the absence of sharp corners, cannot cause flow separation. Therefore, a negative pressure gradient is called a favorable pressure gradient.

The presence of a pressure gradient also affects the magnitude of the skin friction coefficient, as can be inferred from Figure 12.8. The velocity gradient at the wall increases as the pressure gradient becomes more favorable.

12.7 VON KÁRMÁN MOMENTUM INTEGRAL ANALYSIS

The Blasius solution is obviously quite restrictive in application, applying only to the case of a laminar boundary layer over a flat surface. Any situation of practical interest more complex than this involves analytical procedures that have, to the present time, proved inferior to experiment. An approximate method providing information for systems involving other types of flow and having other geometries will now be considered.

Consider the control volume in Figure 12.10. The control volume to be analyzed is of unit depth and is bounded in the xy plane by the x axis, here drawn tangent to the surface at point 0; the y axis, the edge of the boundary layer, and a line parallel to the y axis a distance Δx away. We shall consider the case of two-dimensional, incompressible steady flow.

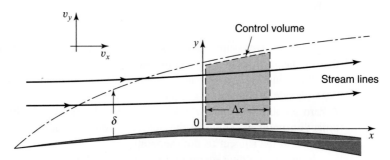

Figure 12.10 Control volume for integral analysis of the boundary layer.

A momentum analysis of the defined control volume involves the application of the x-directional scalar form of the momentum theorem

$$\Sigma F_x = \iint_{\text{c.s.}} v_x \rho (\mathbf{v} \cdot \mathbf{n}) \, dA + \frac{\partial}{\partial t} \iiint_{\text{c.v.}} v_x \rho \, dV \qquad (5\text{-}5a)$$

A term-by-term analysis of the present problem yields the following:

$$\Sigma F_x = P\delta|_x - P\delta|_{x+\Delta x} + \left(P|_x + \frac{P|_{x+\Delta x} - P|_x}{2} \right)(\delta|_{x+\Delta x} - \delta|_x) - \tau_0 \Delta x$$

where δ represents the boundary-layer thickness, and both forces are assumed negligible. The above terms represent the x-directional pressure forces on the left, right, and top sides of the control volume, and the frictional force on the bottom, respectively.

The surface integral term becomes

$$\iint_{c.s.} v_x \rho(\mathbf{v} \cdot \mathbf{n}) dA = \int_0^\delta \rho v_x^2 dy \Big|_{x+\Delta x} - \int_0^\delta \rho v_x^2 dy \Big|_x - v_\infty \dot{m}_{top}$$

and the accumulation term is

$$\frac{\partial}{\partial t} \iiint_{c.v.} v_x \rho \, dV = 0$$

as this is a steady-flow situation.

An application of the integral equation for conservation of mass will give

$$\iint_{c.s.} \rho(\mathbf{v} \cdot \mathbf{n}) \, dA + \frac{\partial}{\partial t} \iiint_{c.v.} \rho \, dV = 0 \tag{4-1}$$

$$\iint_{c.s.} \rho(\mathbf{v} \cdot \mathbf{n}) \, dA = \int_0^\delta \rho v_x \, dy \Big|_{x+\Delta x} - \int_0^\delta \rho v_x \, dy \Big|_x - \dot{m}_{top}$$

$$\frac{\partial}{\partial t} \iiint_{c.v.} \rho \, dV = 0$$

and the mass-flow rate into the top of the control volume, $\dot{m}_{top}$, may be evaluated as

$$\dot{m}_{top} = \int_0^\delta \rho v_x dy \Big|_{x+\Delta x} - \int_0^\delta \rho v_x dy \Big|_x \tag{12-34}$$

The momentum expression, including equation (12-34), now becomes

$$-(P\delta|_{x+\Delta x} - P\delta|_x) + \left(\frac{P|_{x+\Delta x} - P|_x}{2} + P|_x \right)(\delta|_{x+\Delta x} - \delta|_x) - \tau_0 \, \Delta x$$

$$= \int_0^\delta \rho v_x^2 \, dy \Big|_{x+\Delta x} - \int_0^\delta \rho v_x^2 \, dy \Big|_x - v_\infty \left(\int_0^\delta \rho v_x \, dy \Big|_{x+\Delta x} - \int_0^\delta \rho v_x \, dy \Big|_x \right)$$

Rearranging this expression and dividing through by Δx, we get

$$-\left(\frac{P|_{x+\Delta x} - P|_x}{\Delta x} \right)\delta|_{x+\Delta x} + \left(\frac{P|_{x+\Delta x} - P|_x}{2} \right)\left(\frac{\delta|_{x+\Delta x} - \delta|_x}{\Delta x} \right) + \left(\frac{P\delta|_x - P\delta|_x}{\Delta x} \right)$$

$$= \left(\frac{\int_0^\delta \rho v_x^2 dy|_{x+\Delta x} - \int_0^\delta \rho v_x^2 \, dy|_x}{\Delta x} \right) - v_\infty \left(\frac{\int_0^\delta \rho v_x dy|_{x+\Delta x} - \int_0^\delta \rho v_x \, dy|_x}{\Delta x} \right) + \tau_0$$

Taking the limit as $\Delta x \to 0$ we obtain

$$-\delta \frac{dP}{dx} = \tau_0 + \frac{d}{dx} \int_0^\delta \rho v_x^2 \, dy - v_\infty \frac{d}{dx} \int_0^\delta \rho v_x \, dy \tag{12-35}$$

The boundary-layer concept assumes inviscid flow outside the boundary layer, for which we may write Bernoulli's equation

$$\frac{dP}{dx} + \rho v_\infty \frac{dv_\infty}{dx} = 0$$

which may be rearranged to the form

$$\frac{\delta}{\rho} \frac{dP}{dx} = \frac{d}{dx}(\delta v_\infty^2) - v_\infty \frac{d}{dx}(\delta v_\infty) \tag{12-36}$$

Notice that the left-hand sides of equations (12-35) and (12-36) are similar. We may thus relate the right-hand sides and, with proper rearrangement, get the result

$$\frac{\tau_0}{\rho} = \left(\frac{d}{dx} v_\infty\right) \int_0^\delta (v_\infty - v_x)\, dy + \frac{d}{dx} \int_0^\delta v_x(v_\infty - v_x)\, dy \tag{12-37}$$

Equation (12-37) is the von Kármán momentum integral expression, named in honor of Theodore von Kármán who first developed it.

Equation (12-37) is a general expression whose solution requires a knowledge of the velocity, v_x, as a function of distance from the surface, y. The accuracy of the final result will depend on how closely the assumed velocity profile approaches the real one.

As an example of the application of equation (12-37), let us consider the case of laminar flow over a flat plate, a situation for which an exact answer is known. In this case the free-stream velocity is constant, therefore $(d/dx)v_\infty = 0$ and equation (12-36) simplifies to

$$\frac{\tau_0}{\rho} = \frac{d}{dx} \int_0^\delta v_x(v_\infty - v_x)\, dy \tag{12-38}$$

An early solution to equation (12-38) was achieved by Pohlhausen, who assumed for the velocity profile a cubic function

$$v_x = a + by + cy^2 + dy^3 \tag{12-39}$$

The constants a, b, c, and d may be evaluated if we know certain boundary conditions that must be satisfied in the boundary layer. These are

(1) $v_x = 0$ at $y = 0$

(2) $v_x = v_\infty$ at $y = \delta$

(3) $\dfrac{\partial v_x}{\partial y} = 0$ at $y = \delta$

and

(4) $\dfrac{\partial^2 v_x}{\partial y^2} = 0$ at $y = 0$

Boundary condition (4) results from equation (12-33), which states that the second derivative at the wall is equal to the pressure gradient. As the pressure is constant in this case, $\partial^2 v_x/\partial y^2 = 0$. Solving for a, b, c, and d from these conditions, we get

$$a = 0 \quad b = \frac{3}{2\delta} v_\infty \quad c = 0 \quad d = -\frac{v_\infty}{2\delta^3}$$

which, when substituted in equation (12-39), give the form of the velocity profile

$$\frac{v_x}{v_\infty} = \frac{3}{2}\left(\frac{y}{\delta}\right) - \frac{1}{2}\left(\frac{y}{\delta}\right)^3 \tag{12-40}$$

Upon substitution, equation (12-38) becomes

$$\frac{3v}{2}\frac{v_\infty}{\delta_\infty} = \frac{d}{dx}\int_0^\delta v_\infty^2 \left(\frac{3}{2}\frac{y}{\delta} - \frac{1}{2}\left(\frac{y}{\delta}\right)^3\right)\left(1 - \frac{3}{2}\frac{y}{\delta} + \frac{1}{2}\left(\frac{y}{\delta}\right)^3\right)dy$$

or, after integrating

$$\frac{3}{2}v\frac{v_\infty}{\delta} = \frac{39}{280}\frac{d}{dx}(v_\infty^2\delta)$$

As the free-stream velocity is constant, a simple ordinary differential equation in δ results

$$\delta\,d\delta = \frac{140}{13}\frac{v\,dx}{v_\infty}$$

This, upon integration, yields

$$\frac{\delta}{x} = \frac{4.64}{\sqrt{Re_x}} \tag{12-41}$$

The local skin-friction coefficient, C_{fx}, is given by

$$C_{fx} \equiv \frac{\tau_0}{\frac{1}{2}\rho v_\infty^2} = \frac{2v}{v_\infty^2}\frac{3}{2}\frac{v_\infty}{\delta} = \frac{0.646}{\sqrt{Re_x}} \tag{12-42}$$

Integration of the local skin-friction coefficient between $x = 0$ and $x = L$ as in equation (12-32) yields

$$C_{fL} = \frac{1.292}{\sqrt{Re_L}} \tag{12-43}$$

Comparing equations (12-41), (12-42), and (12-43) with exact results obtained by Blasius for the same situation, equations (12-28), (12-31), and (12-32), we observe a difference of about 7% in δ and 3% in C_f. This difference could, of course, have been smaller had the assumed velocity profile been a more accurate representation of the actual profile.

This comparison has shown the utility of the momentum integral method for the solution of the boundary layer and indicates a procedure that may be used with reasonable accuracy to obtain values for boundary-layer thickness and the coefficient of skin friction where an exact analysis is not feasible. The momentum integral method may also be used to determine the shear stress from the velocity profile.

12.8 DESCRIPTION OF TURBULENCE

Turbulent flow is the most frequently encountered type of viscous flow, yet the analytical treatment of turbulent flow is not nearly well developed as that of laminar flow. In this section, we examine the phenomenon of turbulence, particularly with respect to the mechanism of turbulent contributions to momentum transfer.

In a turbulent flow the fluid and flow variables vary with time. The instantaneous velocity vector, for example, will differ from the average velocity vector in both magnitude and direction. Figure 12.11 illustrates the type of time dependence experienced by the

Figure 12.11 Time dependence of velocity in a turbulent flow: (a) steady mean flow; (b) unsteady mean flow.

axial component of the velocity for turbulent flow in a tube. While the velocity in Figure 12.11a is seen to be steady in its mean value, small random fluctuations in velocity occur about the mean value. Accordingly, we may express the fluid and flow variables in terms of a mean value and a fluctuating value. For example, the x-directional velocity is expressed as

$$v_x = \bar{v}_x(x, y, z) + v'_x(x, y, z, t) \tag{12-44}$$

Here $\bar{v}_x(x, y, z)$ represents the time-averaged velocity at the point (x, y, z)

$$\bar{v}_x = \frac{1}{t_1} \int_0^{t_1} v_x(x, y, z, t)\, dt \tag{12-45}$$

where t_1 is a time that is very long in comparison with the duration of any fluctuation. The mean value of $v'_x(x, y, z, t)$ is zero, as expressed by

$$\bar{v}'_x = \frac{1}{t_1} \int_0^{t_1} v'_x(x, y, z, t)\, dt = 0 \tag{12-46}$$

Hereafter, $\bar{Q}$ will be used to designate the time average of the general property, Q, according to $\bar{Q} = 1/t_1 \int_0^{t_1} Q(x, y, z, t)dt$. While the mean value of the turbulent fluctuations is zero, these fluctuations contribute to the mean value of certain flow quantities. For example, the mean kinetic energy per unit volume is

$$\overline{KE} = \frac{1}{2}\rho\overline{[(\bar{v}_x + v'_x)^2 + (\bar{v}_y + v'_y)^2 + (\bar{v}_z + v'_z)^2]}$$

The average of a sum is the sum of the averages; hence the kinetic energy becomes

$$\overline{KE} = \frac{1}{2}\rho\left\{ \overline{(\bar{v}_x^2 + 2\bar{v}_x v'_x + v'^2_x)} + \overline{(\bar{v}_y^2 + 2\bar{v}_y v'_y + v'^2_y)} + \overline{(\bar{v}_z^2 + 2\bar{v}_z v'_z + v'^2_z)} \right\}$$

or, since $\overline{\bar{v}_x v'_x} = \bar{v}_x \overline{v'_x} = 0$,

$$\overline{KE} = \frac{1}{2}\rho\left(\bar{v}_x^2 + \bar{v}_y^2 + \bar{v}_z^2 + \overline{v'^2_x} + \overline{v'^2_y} + \overline{v'^2_z}\right) \tag{12-47}$$

A fraction of the total kinetic energy of a turbulent flow is seen to be associated with the magnitude of the turbulent fluctuations. It can be shown that the rms (root mean square) value of the fluctuations, $(\overline{v'^2_x} + \overline{v'^2_y} + \overline{v'^2_z})^{1/2}$ is a significant quantity. The level or *intensity of turbulence* is defined as

$$I \equiv \frac{\sqrt{(\overline{v'^2_x} + \overline{v'^2_y} + \overline{v'^2_z})/3}}{v_\infty} \tag{12-48}$$

where v_∞ is the mean velocity of the flow. The intensity of turbulence is a measure of the kinetic energy of the turbulence and is an important parameter in flow simulation. In model testing, simulation of turbulent flows requires not only duplication of Reynolds number but also duplication of the turbulent kinetic energy. Thus, the measurement of turbulence is seen to be a necessity in many applications.

The general discussion so far has indicated the fluctuating nature of turbulence. The random nature of turbulence lends itself to statistical analysis. We shall now turn our attention to the effect of the turbulent fluctuations on momentum transfer.

12.9 TURBULENT SHEARING STRESSES

In Chapter 7, the random molecular motion of the molecules was shown to result in a net momentum transfer between two adjacent layers of fluid. If the (molecular) random motions give rise to momentum transfer, it seems reasonable to expect that large-scale fluctuations, such as those present in a turbulent flow, will also result in a net transfer of momentum. Using an approach similar to that of Section 7.3, let us consider the transfer of momentum in the turbulent flow illustrated in Figure 12.12.

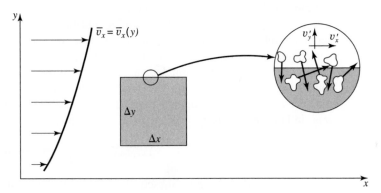

Figure 12.12 Turbulent motion at the surface of a control volume.

The relation between the macroscopic momentum flux due to the turbulent fluctuations and the shear stress may be seen from the control-volume expression for linear momentum

$$\Sigma \mathbf{F} = \iint_{c.s.} \mathbf{v}\rho(\mathbf{v}\cdot\mathbf{n})\, dA + \frac{\partial}{\partial t}\iiint_{c.v.} \mathbf{v}\rho\, dV \tag{5-4}$$

The flux of x-directional momentum across the top of the control surface is

$$\iint_{top} \mathbf{v}\rho(\mathbf{v}\cdot\mathbf{n})\, dA = \iint_{top} v_y'\rho(\bar{v}_x + v_x')\, dA \tag{12-49}$$

If the mean value of the momentum flux over a period of time is evaluated for the case of steady mean flow, the time derivative in equation (5-4) is zero; thus

$$\overline{\Sigma F_x} = \iint \overline{v_y'\rho(\bar{v}_x + v_x')dA} = \iint \overline{v_y'\rho\bar{v}_x}\, dA + \iint \overline{\rho v_y' v_x'}dA \tag{12-50}$$

The presence of the turbulent fluctuations is seen to contribute a mean x directional momentum flux of $\rho v_x' v_y' s$ per unit area. Although the turbulent fluctuations are functions of position and time, their analytical description has not been achieved, even for the simplest case. The close analogy between the molecular exchange of momentum in laminar flow and the macroscopic exchange of momentum in turbulent flow suggests that the term $\rho v_x' v_y'$ be regarded as a shear stress. Transposing this term to the left-hand side of equation (5-4) and incorporating it with the shear stress due to molecular momentum transfer, we see that the total shear stress becomes

$$\tau_{yx} = \mu \frac{d \bar{v}_x}{dy} - \overline{\rho v_x' v_y'} \tag{12-51}$$

The turbulent contribution to the shear stress is called the *Reynolds stress*. In turbulent flows it is found that the magnitude of the Reynolds stress is much greater than the molecular contribution except near the walls.

An important difference between the molecular and turbulent contributions to the shear stress is to be noted. Whereas the molecular contribution is expressed in terms of a property of the fluid and a derivative of the mean flow, the turbulent contribution is expressed solely in terms of the fluctuating properties of the flow. Further, these flow properties are not expressible in analytical terms. While Reynolds stresses exist for multidimensional flows,[6] the difficulties in analytically predicting even the one-dimensional case have proved insurmountable without the aid of experimental data. The reason for these difficulties may be seen by examining the number of equations and the number of unknowns involved. In the incompressible turbulent boundary layer, for example, there are two pertinent equations, momentum and continuity, and four unknowns, $\bar{v}_x$, $\bar{v}_y$, v_x', and v_y'.

An early attempt to formulate a theory of turbulent shear stress was made by Boussinesq.[7] By analogy with the form of Newton's viscosity relation, Boussinesq introduced the concept relating the turbulent shear stress to the shear strain rate. The shear stress in laminar flow is $\tau_{yx} = \mu(dv_x/dy)$, thus by analogy, the Reynolds stress becomes

$$(\tau_{yx})_{\text{turb}} = A_t \frac{d \bar{v}_x}{dy}$$

where A_t is the *eddy viscosity*. Subsequent refinements have led to the introduction of the *eddy diffusivity of momentum*, $\epsilon_M \equiv A_t/\rho$, and thus

$$(\tau_{yx})_{\text{turb}} = \rho \epsilon_M \frac{d \bar{v}_x}{dy} \tag{12-52}$$

The difficulties in analytical treatment still exist, however, as the eddy diffusivity, ϵ_M, is a property of the flow and not of the fluid. By analogy with the kinematic viscosity in a laminar flow, it may be observed that the units of the eddy diffusivity are L^2/t.

12.10   THE MIXING-LENGTH HYPOTHESIS

A general similarity between the mechanism of transfer of momentum in turbulent flow and that in laminar flow permits an analog to be made for turbulent shear stress. The analog to the mean free path in molecular momentum exchange for the turbulent case is the mixing length proposed by Prandtl[8] in 1925. Consider the simple turbulent flow shown in Figure 12.13.

---

[6] The existence of the Reynolds stresses may also be shown by taking the time average of the Navier–Stokes equations.

[7] J. Boussinesq, *Mem. Pre. par div. Sav.*, XXIII, (1877).

[8] L. Prandtl, *ZAMM*, **5**, 136 (1925).

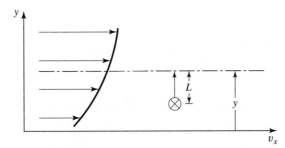

Figure 12.13 The Prandtl mixing length.

The velocity fluctuation v_x' is hypothesized as being due to the y-directional motion of a "lump" of fluid through a distance L. In undergoing translation the lump of fluid retains the mean velocity from its point of origin. Upon reaching a destination, a distance L from the point of origin, the lump of fluid will differ in mean velocity from that of the adjacent fluid by an amount $\bar{v}_x|_{y-L} - \bar{v}_x|_y$. If the lump of fluid originated at $y + L$, the velocity difference would be $\bar{v}_x|_{y+L} - \bar{v}_x|_y$. The instantaneous value of $v_x'|_y$ is then $\bar{v}_x|_{y\pm L} - \bar{v}_x|_y$, the sign of L, of course, depending on the point of origin with respect to y. Further, the mixing length, although finite, is assumed to be small enough to permit the velocity difference to be written as and thus

$$\bar{v}_x|_{y\pm L} - \bar{v}_x|_y = \pm L \frac{d\bar{v}_x}{dy}$$

and thus

$$v_x' = \pm L \frac{d\bar{v}_x}{dy} \tag{12-52}$$

The concept of the mixing length is somewhat akin to that of the mean free path of a gas molecule. The important differences are its magnitude and dependence upon flow properties rather than fluid properties. With an expression for v_x' at hand, an expression for v_y' is necessary to determine the turbulent shear stress, $-\rho \overline{v_x' v_y'}$.

Prandtl assumed that v_x' must be proportional to v_y'. If v_x' and v_y' were completely independent, then the time average of their product would be zero. Both the continuity equation and experimental data show that there is some degree of proportionality between v_x' and v_y'. Using the fact that $v_y' \sim v_x'$, Prandtl expressed the time average, $\overline{v_x' v_y'}$, as

$$\overline{v_x' v_y'} = -(\text{constant})L^2 \left| \frac{d\bar{v}_x}{dy} \right| \frac{d\bar{v}_x}{dy} \tag{12-53}$$

The constant represents the unknown proportionality between v_x' and v_y' as well as their correlation in taking the time average. The minus sign and the absolute value were introduced to make the quantity $\overline{v_x' v_y'}$ agree with experimental observations. The constant in (12-53), which is unknown, may be incorporated into the mixing length, which is also unknown, giving

$$\overline{v_x' v_y'} = -L^2 \left| \frac{d\bar{v}_x}{dy} \right| \frac{d\bar{v}_x}{dy} \tag{12-54}$$

Comparison with Boussinesq's expression for the eddy diffusivity yields

$$\epsilon_M = L^2 \left| \frac{d\bar{v}_x}{dy} \right| \tag{12-55}$$

At first glance it appears that little has been gained in going from the eddy viscosity to the mixing length. There is an advantage, however, in that assumptions regarding the nature and variation of the mixing length may be made on an easier basis than assumptions concerning the eddy viscosity.

12.11 VELOCITY DISTRIBUTION FROM THE MIXING-LENGTH THEORY

One of the important contributions of the mixing-length theory is its use in correlating velocity profiles at large Reynolds numbers. Consider a turbulent flow as illustrated in Figure 12.13. In the neighborhood of the wall the mixing length is assumed to vary directly with y, and thus $L = Ky$, where K remains a dimensionless constant to be determined via experiment. The shear stress is assumed to be entirely due to turbulence and to remain constant over the region of interest. The velocity $\bar{v}_x$ is assumed to increase in the y direction, and thus $d\bar{v}_x/dy = |d\bar{v}_x/dy|$. Using these assumptions, we may write the turbulent shear stress as

$$\tau_{yx} = \rho K^2 y^2 \left(\frac{d\bar{v}_x}{dy} \right)^2 = \tau_0 (\text{a constant})$$

or

$$\frac{d\bar{v}_x}{dy} = \frac{\sqrt{\tau_0/\rho}}{Ky}$$

The quantity $\sqrt{\tau_0/\rho}$ is observed to have units of velocity. Integration of the above equation yields

$$\bar{v}_x = \frac{\sqrt{\tau_0/\rho}}{K} \ln y + C \tag{12-56}$$

where C is a constant of integration. This constant may be evaluated by setting $\bar{v}_x = \bar{v}_{x\,\text{max}}$ at $y = h$, whereby

$$\frac{\bar{v}_{x\,\text{max}} - \bar{v}_x}{\sqrt{\tau_0/\rho}} = -\frac{1}{K} \left[\ln \frac{y}{h} \right] \tag{12-57}$$

The constant K was evaluated by Prandtl[9] and Nikuradse[10] from data on turbulent flow in tubes and found to have a value of 0.4. The agreement of experimental data for turbulent flow in smooth tubes with equation (12-57) is quite good, as can be seen from Figure 12.14.

The empirical nature of the preceding discussion cannot be overlooked. Several assumptions regarding the flow are known to be incorrect for flow in tubes, namely that the shear stress is not constant and that the geometry was treated from a two-dimensional viewpoint rather than an axisymmetric viewpoint. In view of these obvious difficulties, it is remarkable that equation (13-15) describes the velocity profile so well.

---

[9] L. Prandtl, *Proc. Intern. Congr. Appl. Mech.*, 2nd Congr., Zurich (1927), 62.
[10] J. Nikuradse, *VDI-Forschungsheft*, **356**, 1932.

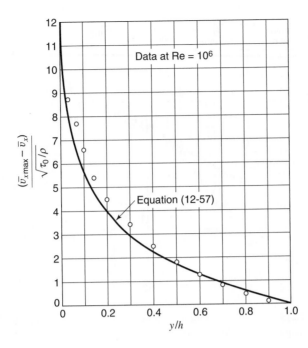

Figure 12.14 Comparison of data for flow in smooth tube with equation (12-57).

12.12 THE UNIVERSAL VELOCITY DISTRIBUTION

For turbulent flow in smooth tubes, equation (12-57) may be taken as a basis for a more general development. Recalling that the term $\sqrt{\tau_0/\rho}$ has the units of velocity, we may introduce a dimensionless velocity $\bar{v}_x/\sqrt{\tau_0/\rho}$. Defining

$$v^+ \equiv \frac{\bar{v}_x}{\sqrt{\tau_0/\rho}} \tag{12-58}$$

we may write equation (12-56) as

$$v^+ = \frac{1}{K}[\ln y] + C \tag{12-59}$$

The left-hand side of (12-59) is, of course, dimensionless; therefore the right-hand side of this equation must also be dimensionless. A pseudo-Reynolds number is found useful in this regard. Defining

$$y^+ \equiv \frac{\sqrt{\tau_0/\rho}}{\nu}y \tag{12-60}$$

we find that equation (12-59) becomes

$$v^+ = \frac{1}{K}\ln\frac{\nu y^+}{\sqrt{\tau_0/\rho}} + C = \frac{1}{K}(\ln y^+ + \ln \beta) \tag{12-61}$$

where the constant β is dimensionless.

Equation (12-61) indicates that for flow in smooth tubes $v^+ = f(y^+)$ or

$$v^+ \equiv \frac{\bar{v}_x}{\sqrt{\tau_0/\rho}} = f\left\{\ln\frac{y\sqrt{\tau_0/\rho}}{\nu}\right\} \tag{12-62}$$

The range of validity of equation (13-19) may be observed from a plot (see Figur 12.15) of v^+ versus $\ln y^+$, using the data of Nikuradse and Reichardt.

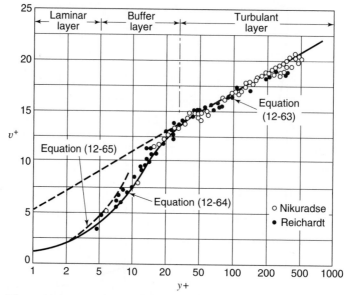

Figure 12.15 Velocity correlation for flow in circular smooth tubes at high Reynolds number (H. Reichardt, NACA TM1047, 1943).

Three distinct regions are apparent: a turbulent core, a buffer layer, and a laminar sublayer. The velocity is correlated as follows:

for turbulent core, $y^+ \geq 30$

$$v^+ = 5.5 + 2.5 \ln y^+ \tag{12-63}$$

for the buffer layer, $30 \geq y^+ \geq 5$

$$v^+ = -3.05 + 5 \ln y^+ \tag{12-64}$$

for the laminar sublayer, $5 > y^+ > 0$

$$v^+ = y^+ \tag{12-65}$$

Equations (12-63) through (12-65) define the *universal velocity distribution*. Because of the empirical nature of these equations, there are, of course, inconsistencies. The velocity gradient, for example, at the center of the tube predicted by (12-63) is not zero. In spite of this and other inconsistencies, these equations are extremely useful for describing flow in smooth tubes.

In rough tubes, the scale of the roughness e is found to affect the flow in the turbulent core, but not in the laminar sublayer. The constant β in equation (13-19) becomes $\ln \beta = 3.4 - \ln\left[(e\sqrt{\tau_0/\rho})/\nu\right]$ for rough tubes. As the wall shear stress appears in the revised expression for $\ln \beta$, it is important to note that wall roughness affects the magnitude of the shear stress in a turbulent flow.

12.13 FURTHER EMPIRICAL RELATIONS FOR TURBULENT FLOW

Two important experimental results that are helpful in studying turbulent flows are the power-law relation for velocity profiles and a turbulent-flow shear-stress relation due to Blasius. Both of these relations are valid for flow adjacent to smooth surfaces.

For flow in smooth circular tubes, it is found that over much of the cross section the velocity profile may be correlated by

$$\frac{\overline{v}_x}{\overline{v}_{x\,\text{max}}} = \left(\frac{y}{R}\right)^{1/n} \tag{12-66}$$

where R is the radius of the tube and n is a slowly varying function of Reynolds number. The exponent n is found to vary from a value of 6 at Re = 4000 to 10 at Re = 3,200,000. At Reynolds numbers of 10^5 the value of n is 7. This leads to the frequently used one-seventh-power law, $\overline{v}_x/\overline{v}_{x\,\text{max}} = (y/R)^{1/7}$. The power-law profile has also been found to represent the velocity distribution in boundary layers. For boundary layers of thickness δ, the power law is written

$$\frac{\overline{v}_x}{\overline{v}_{x\,\text{max}}} = \left(\frac{y}{\delta}\right)^{1/n} \tag{12-67}$$

The power-law profile has two obvious difficulties: the velocity gradients at the wall and those at δ are incorrect. This expression indicates that the velocity gradient at the wall is infinite and that the velocity gradient at δ is nonzero.

In spite of these inconsistencies, the power law is extremely useful in connection with the von Kármán-integral relation, as we shall see in Section 12.14.

Another useful relation is Blasius's correlation for shear stress. For pipe-flow Reynolds numbers up to 10^5 and flat-plate Reynolds numbers up to 10^7, the wall shear stress in a turbulent flow is given by

$$\tau_0 = 0.0225\rho\overline{v}_{x\,\text{max}}^2 \left(\frac{\nu}{\overline{v}_{x\,\text{max}}\ y_{\text{max}}}\right)^{1/4} \tag{12-68}$$

where $y_{\text{max}} = R$ in pipes and $y_{\text{max}} = \delta$ for flat surfaces.

12.14   THE TURBULENT BOUNDARY LAYER ON A FLAT PLATE

The variation in boundary-layer thickness for turbulent flow over a smooth flat plate may be obtained from the von Kármán momentum integral. The manner of approximation involved in a turbulent analysis differs from that used previously. In a laminar flow, a simple polynomial was assumed to represent the velocity profile. In a turbulent flow, we have seen that the velocity profile depends upon the wall shear stress and that no single function adequately represents the velocity profile over the entire region. The procedure we shall follow in using the von Kármán integral relation in a turbulent flow is to utilize a simple profile for the integration with the Blasius correlation for the shear stress. For a zero pressure gradient the von Kármán integral relation is

$$\frac{\tau_0}{\rho} = \frac{d}{dx}\int_0^\delta v_x(v_\infty - v_x)\,dy \tag{12-38}$$

Employing the one-seventh-power law for v_x and the Blasius relation, equation (12-68) for τ_0, we see that equation (12-38) becomes

$$0.0225v_\infty^2 \left(\frac{\nu}{v_\infty\delta}\right)^{1/4} = \frac{d}{dx}\int_0^\delta v_\infty^2 \left\{\left(\frac{y}{\delta}\right)^{1/7} - \left(\frac{y}{\delta}\right)^{2/7}\right\} dy \tag{12-69}$$

where the free-stream velocity, v_∞, is written in place of $\overline{v}_{x\,\text{max}}$. Performing the indicated integration and differentiation, we obtain

$$0.0225 \left(\frac{\nu}{v_\infty\delta}\right)^{1/4} = \frac{7}{72}\frac{d\delta}{dx} \tag{12-70}$$

which becomes, upon integration

$$\left(\frac{v}{v_\infty}\right)^{1/4} x = 3.45\,\delta^{5/4} + C \tag{12-71}$$

If the boundary layer is assumed to be turbulent from the leading edge, $x = 0$ (a poor assumption), the above equation may be rearranged to give

$$\frac{\delta}{x} = \frac{0.376}{\mathrm{Re}_x^{1/5}} \tag{12-72}$$

The local skin-friction coefficient may be computed from the Blasius relation for shear stress, equation (12-67), to give

$$C_{fx} = \frac{0.0576}{\mathrm{Re}_x^{1/5}} \tag{12-73}$$

Several things are to be noted about these expressions. First, they are limited to values of $\mathrm{Re}_x < 10^7$, by virtue of the Blasius relation. Second, they apply only to smooth flat plates. Last, a major assumption has been made in assuming the boundary layer to be turbulent from the leading edge. The boundary layer is known to be laminar initially and to undergo transition to turbulent flow at a value of Re_x of about 2×10^5. We shall retain the assumption of a completely turbulent boundary layer for the simplicity it affords; it is recognized, however, that this assumption introduces some error in the case of a boundary layer that is not completely turbulent.

A comparison of a laminar and a turbulent boundary layer can be made from Blasius's laminar-flow solution and equations (12-28), (12-72), and (12-73). At the same Reynolds number, the turbulent boundary layer is observed to be thicker, and is associated with a larger skin friction coefficient. While it would appear then that a laminar boundary layer is more desirable, the reverse is generally true. In most cases of engineering interest, a turbulent boundary layer is desired because it resists separation better than a laminar boundary layer. The velocity profiles in laminar and turbulent boundary layers are compared qualitatively in Figure 12.16.

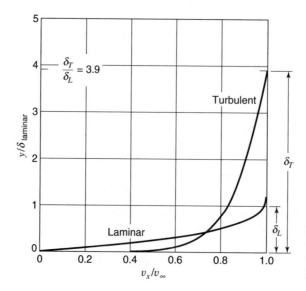

Figure 12.16 Comparison of velocity profiles in laminar and turbulent boundary layers. The Reynolds number is 500,000.

It can be seen that the turbulent boundary layer has a greater mean velocity, hence both greater momentum and energy than the laminar boundary layer. The greater momentum and energy permit the turbulent boundary layer to remain unseparated for a greater distance in the presence of an adverse pressure gradient than would be the case for a laminar boundary layer.

Consider a flat plate with transition from laminar flow to turbulent flow occurring on the plate. If transition from laminar flow to turbulent flow is assumed to occur abruptly (for computational purposes), a problem arises in how to join the laminar boundary layer to the turbulent layer at the point of transition. The prevailing procedure is to equate the momentum thicknesses, equation (12-44), at the transition point. That is, at the start of the turbulent portion of the boundary layer, the momentum thickness, θ, is equal to the momentum thickness at the end of the laminar portion of the boundary layer.

The general approach for turbulent boundary layers with a pressure gradient involves the use of the von Kármán momentum integral as given in equation (12-46). Numerical integration is required.

12.15 FACTORS AFFECTING THE TRANSITION FROM LAMINAR TO TURBULENT FLOW

The velocity profiles and momentum-transfer mechanisms have been examined for both laminar and turbulent flow regimes and found to be quite different. Laminar flow has also been seen to undergo transition to turbulent flow at certain Reynolds numbers.

So far the occurrence of transition has been expressed in terms of the Reynolds number alone, while a variety of factors other than Re actually influence transition. The Reynolds number remains, however, the principal parameter for predicting transition.

Table 12.2 indicates the influence of some of these factors on the transition Reynolds number.

Table 12.2 Factors affecting the Reynolds number of transition from laminar to turbulent flow

Factor	Influence
Pressure gradient	Favorable pressure gradient retards transition; unfavorable pressure gradient hastens it
Free-stream turbulence	Free-stream turbulence decreases transition Reynolds number
Roughness	No effect in pipes; decreases transition in external flow
Suction	Suction greatly increases transition Re
Wall curvatures	Convex curvature increases transition Re. Concave curvature decreases it
Wall temperature	Cool walls increase transition Re. Hot walls decrease it

12.16 CLOSURE

Viscous flow has been examined in this chapter for both internal and external geometries. Two approaches were employed for analyzing laminar boundary layer flows—exact analysis using boundary layer equations and approximate integral methods. For turbulent boundary layer analysis along a plane surface, an integral method was employed.

Concepts of skin friction and drag coefficients were introduced and quantitative relationships were developed for both internal and external flows.

Approaches to modeling turbulent flows were introduced, culminating in expressions for the "universal" velocity distribution. This approach considers turbulent flows to be described in three parts: the laminar sublayer, the transition or buffer layer, and the turbulent core.

The concepts developed in this chapter will be used to develop important expressions for momentum flow application in the next chapter. Similar applications will be developed in the sections related to both heat and mass transfer in later sections of this text.

PROBLEMS

12.1 If Reynolds's experiment was performed with a 38-mm-ID pipe, what flow velocity would occur at transition?

12.2 Modern subsonic aircraft have been refined to such an extent that 75% of the parasite drag (portion of total aircraft drag not directly associated with producing lift) can be attributed to friction along the external surfaces. For a typical subsonic jet, the parasite drag coefficient based on wing area is 0.011. Determine the friction drag on such an aircraft

a. at 500 mph at 35 000 ft;

b. at 200 mph at sea level.

The wing area is 2400 ft^2.

12.3 Consider the flow of air at 30 m/s along a flat plate. At what distance from the leading edge will transition occur?

12.4 Find a velocity profile for the laminar boundary layer of the form

$$\frac{v_x}{v_{x\delta}} = c_1 + c_2 y + c_3 y^2 + c_4 y^3$$

when the pressure gradient is not zero.

12.5 Evaluate and compare with the exact solution δ, C_{fx}, and C_{fL} for the laminar boundary layer over a flat plate, using the velocity profile

$$v_x = \alpha \sin by.$$

12.6 There is fluid evaporating from a surface at which $v_x|_{y=0} = 0$, but $v_x|_{y=0} \neq 0$. Derive the von Kármán momentum relation.

12.7 The drag coefficient for a smooth sphere is shown below. Determine the speed at the critical Reynolds number for a 42-mm-diameter sphere in air.

12.8 Plot a curve of drag vs. velocity for a 1.65-in.-diameter sphere in air between velocities of 50 fps and 400 fps.

12.9 For what wind velocities will a 12.7-mm-diameter cable be in the unsteady wake region of Figure 12.2?

12.10 Estimate the drag force on a 3-ft-long radio antenna with an average diameter of 0.2 in. at a speed of 60 mph.

12.11 A 2007 Toyota Prius has a drag coefficient of 0.26 at road speeds, using a reference area of 2.33 m^2. Determine the horsepower required to overcome drag at a velocity of 30 m/s. Compare this figure with the case of head and tail winds of 6 m/s.

12.12 The lift coefficient is defined as $C_L \equiv$ (lift force)/$\frac{1}{2}\rho v_x^2 A_r$. If the lift coefficient for the auto in the previous problem is 0.21, determine the lift force at a road speed of 100 mph.

12.13 The auto in Problems and showed a sensitivity to yaw angle. At a yaw angle of 20°, the lift coefficient increased to 1.0. What is the lift force at 100 mph for this case?

12.14 What diameter circular plate would have the same drag as the auto of Problem 12.11?

12.15 Estimate the normal force on a circular sign 8 ft in diameter during a hurricane wind (120 mph).

12.16 A 1998 Lexus LS400 has a drag coefficient of 0.28 and a reference area of 2.4 m^2. Determine the horsepower required to overcome drag when driving at 70 mph at sea level.

a. on a hot summer day $T \cong 100°$F.

b. on a cold winter day $T \cong 0°$F

12.17 A baseball has a circumference of $9\frac{1}{4}$ inches and a weight of $5\frac{1}{4}$ ounces. At 95 mph determine

a. the Reynolds number.

b. the drag force.

c. the type of flow (see the illustration for Problem).

12.18 Golfball "dimples" cause the drag drop (see Figure 12.4 and the illustration for Problem 12.7) to occur at a lower Reynolds number. The table below gives the drag coefficient for a rough sphere as a function of the Reynolds number. Plot the drag for 1.65-in.-diameter sphere as a function of velocity. Show several comparison points for a smooth sphere.

Re · 10^{-4}	7.5	10	15	20	25
C_D	0.48	0.38	0.22	0.12	0.10

12.19 The lift coefficient on a rotating sphere is very approximately given by

$$C_L \cong 0.24 \left(\frac{R\Omega}{V} \right) - 0.05 \quad \text{over the range of } 1.5 > \left(\frac{R\Omega}{V} \right) > 0.2.$$

Here R is the sphere radius and Ω is rotation rate of the sphere. For the baseball in Problem 12.17, determine the rotation rate for a baseball thrown at 110 mph to have the lift equal the weight. How many rotations would such a ball make in 60 ft 6 in.?

12.20 If the vertical velocity at the wall is not zero such as would be the case with suction or blowing, what modifications occur to equation (12-33)?

12.21 If the turbulence intensity is 10%, what fraction of the total kinetic energy of the flow is due to the turbulence?

12.22 In a house, water flows through a copper tube with a 0.75-in.ID, at a flow rate of 2 gpm. Determine the Reynolds number for

a. hot water ($T \cong 120°F$).

b. cold water ($T \cong 45°F$).

12.23 Plot the boundary-layer thickness along a flat plate for the flow of air at 30 m/s assuming

a. laminar flow;

b. turbulent flow.

Indicate the probable transition point.

12.24 For the fully developed flow of water in a smooth 0.15-m pipe at a rate of 0.006 m³/s, determine the thickness of

a. the laminar sublayer;

b. the buffer layer;

c. the turbulent core.

12.25 Using Blasius' correlation for shear stress (equation 12-68), develop an expression for the local skin-friction coefficient in pipes. In pipes, the average velocity is used for the friction coefficient and the Reynolds number. Use the one-seventh-power law.

12.26 For a thin a plate 6 in. wide and 3 ft long, estimate the friction force in air at a velocity of 40 fps, assuming

a. turbulent flow;

b. laminar flow.

The flow is parallel to the 6-in. dimension.

12.27 Using a sine profile the laminar flow and a one-seventh-power law for turbulent flow, make a dimensionless plot of the momentum and kinetic energy profiles in the boundary layer a Reynolds number of 10^5.

12.28 Estimate the friction drag on a wing by considering the following idealization. Consider the wing to be a rectangular flat plate, 7 ft by 40 ft, with a smooth surface. The wing is flying at 140 mph at 5000 ft. Determine the drag, assuming

a. a laminar boundary layer;

b. a turbulent boundary layer.

12.29 Compare the boundary-layer thicknesses and local skin-friction coefficients of a laminar boundary layer and a turbulent boundary layer on a smooth flat plate at a Reynolds number of 10^6. Assume both boundary layers to originate at the leading edge of the flat plate.

12.30 Use the 1/7 power-law profile and compute the drag force and boundary layer thickness on a plate 20 ft long and 10 ft wide (for one side) if it is immersed in a flow of water of 20 ft/s velocity. Assume turbulent flow to exist over the entire length of the plate. What would the drag be if laminar flow could be maintained over the entire surface?

12.31 The turbulent shear stress in a two-dimensional flow is given by

$$(\tau_{yx})_{\text{turb}} = \rho \epsilon_M \frac{\partial \bar{v}_x}{\partial y} = -\rho \overline{v_x v_y}$$

Expanding v_x' and v_y' in a Taylor series in x and y near the wall and with the aid of the continuity equation

$$\frac{\partial v_x'}{\partial x} + \frac{\partial v_y'}{\partial y} = 0$$

show that, near the wall, $\epsilon_M \sim y^3 +$ higher order terms in y. How does this compare with the mixing-length theory?

12.32 Evaluate the velocity derivative, $\partial \bar{v}_x / \partial y$, for the power-law velocity profile at $y = 0$ and $y = R$.

12.33 Using the Blasius shear-stress relation (12-68) and the power-law velocity profile, determine the boundary-layer thickness on a flat plate as a function of the Reynolds number and the exponent n.

Chapter 13

Flow in Closed Conduits

Many of the theoretical relations that have been developed in the previous chapters apply to special situations such as inviscid flow, incompressible flow, and the like. Some experimental correlations were introduced in Chapter 12 for turbulent flow in or past surfaces of simple geometry. In this chapter, an application of the material that has been developed thus far will be considered with respect to a situation of considerable engineering importance, namely fluid flow, both laminar and turbulent, through closed conduits.

13.1 DIMENSIONAL ANALYSIS OF CONDUIT FLOW

As an initial approach to conduit flow, we shall utilize dimensional analysis to obtain the significant parameters for the flow of an incompressible fluid in a straight, horizontal, circular pipe of constant cross section.

The significant variables and their dimensional expressions are as represented in the following table:

Variable	Symbol	Dimension
Pressure drop	ΔP	M/Lt^2
Velocity	v	L/t
Pipe diameter	D	L
Pipe length	L	L
Pipe roughness	e	L
Fluid viscosity	μ	M/Lt
Fluid density	ρ	M/L^3

Each of the variables is familiar, with the exception of the pipe roughness, symbolized e. The roughness is included to represent the condition of the pipe surface and may be thought of as the characteristic height of projections from the pipe wall, hence the dimension of length.

According to the Buckingham π theorem, the number of independent dimensionless groups to be formed with these variables is four. If the core group consists of the variables v, D, and ρ, then the groups to be formed are as follows:

$$\pi_1 = v^a D^b \rho^c \, \Delta P$$
$$\pi_2 = v^d D^e \rho^f L$$
$$\pi_3 = v^g D^h \rho^i e$$
$$\pi_4 = v^j D^k \rho^l \mu$$

Carrying out the procedure outlined in Chapter 11 to solve for the unknown exponents in each group, we see that the dimensionless parameters become

$$\pi_1 = \frac{\Delta P}{\rho v^2}$$

$$\pi_2 = \frac{L}{D}$$

$$\pi_3 = \frac{e}{D}$$

and

$$\pi_4 = \frac{vD\rho}{\mu}$$

The first π group is the Euler number. As the pressure drop is due to fluid friction, this parameter is often written with $\Delta P/\rho$ replaced by gh_L where h_L is the "head loss"; thus π_1 becomes

$$\frac{h_L}{v^2/g}$$

The third π group, the ratio of pipe roughness to diameter, is the so-called relative roughness. The fourth π group is the Reynolds number, Re.

A functional expression resulting from dimensional analysis may be written as

$$\frac{h_L}{v^2/g} = \phi_1\left(\frac{L}{D}, \frac{e}{D}, \text{Re}\right) \tag{13-1}$$

Experimental data have shown that the head loss in fully developed flow is directly proportional to the ratio L/D. This ratio may, then, be removed from the functional expression, giving

$$\frac{h_L}{v^2/g} = \frac{L}{D}\phi_2\left(\frac{e}{D}, \text{Re}\right) \tag{13-2}$$

The function ϕ_2, which varies with the relative roughness and Reynolds number, is designated f, the friction factor. Expressing the head loss from equation (13-2) in terms of f, we have

$$h_L = 2f_f\frac{L}{D}\frac{v^2}{g} \tag{13-3}$$

With the factor 2 inserted in the right-hand side, equation (13-3) is the defining relation for f_f, the *Fanning friction factor*. Another friction factor in common use is the *Darcy friction factor*, f_D, defined by equation 13-4.

$$h_L = f_D\frac{L}{D}\frac{v^2}{2g} \tag{13-4}$$

Quite obviously, $f_D = 4f_f$. The student should be careful to note which friction factor he is using to properly calculate frictional head loss by either equation (13-3) or (13-4). The Fanning friction factor, f_f, will be used exclusively in this text. The student may easily verify that the Fanning friction factor is the same as the skin friction coefficient C_f.

Our task now becomes that of determining suitable relations for f_f from that theory and experimental data.

13.2   FRICTION FACTORS FOR FULLY DEVELOPED LAMINAR, TURBULENT, AND TRANSITION FLOW IN CIRCULAR CONDUITS

Laminar Flow

Some analysis has been performed already for incompressible laminar flow. As fluid behavior can be described quite well in this regime according to Newton's viscosity relation, we should expect no difficulty in obtaining a functional relationship for f_f in the case of laminar flow. Recall that, for closed conduits, the flow may be considered laminar for values of the Reynolds number less than 2300.

From Chapter 8, the Hagen–Poiseuille equation was derived for incompressible, laminar, conduit flow

$$-\frac{dP}{dx} = 32\frac{\mu v_{\text{avg}}}{D^2} \tag{8-9}$$

Separating variables and integrating this expression along a length, L, of the passage, we get

$$-\int_{P_0}^{P} dP = 32\frac{\mu v_{\text{avg}}}{D^2}\int_{0}^{L} dx$$

and

$$\Delta P = 32\frac{\mu v_{\text{avg}}L}{D^2} \tag{13-5}$$

Recall that equation (8-9) held for the case of fully developed flow; thus v_{avg} does not vary along the length of the passage.

Forming an expression for frictional head loss from equation (13-5), we have

$$h_L = \frac{\Delta P}{\rho g} = 32\frac{\mu v_{\text{avg}}L}{g\rho D^2} \tag{13-6}$$

Combining this equation with equation (13-3), the defining relation for f_f

$$h_L = 32\frac{\mu v_{\text{avg}}L}{g\rho D^2} = 2f_f\frac{L}{D}\frac{v^2}{g}$$

and solving for f_f, we obtain

$$f_f = 16\frac{\mu}{D v_{\text{avg}}\rho} = \frac{16}{\text{Re}} \tag{13-7}$$

This very simple result indicates that f_f is inversely proportional to Re in the laminar flow range; the friction factor is *not* a function of pipe roughness for values of Re < 2300, but varies only with the Reynolds number.

This result has been experimentally verified and is the manifestation of the viscous effects in the fluid, damping out any irregularities in the flow caused by protrusions from a rough surface.

Turbulent Flow

In the case of turbulent flow in closed conduits or pipes, the relation for f_f is not so simply obtained or expressed as in the laminar case. No easily derived relation such as the Hagen–Poiseuille law applies; however, some use can be made of the velocity profiles expressed in Chapter 12 for turbulent flow. All development will be based on circular conduits; thus we are primarily concerned with pipes or tubes. In turbulent flow a distinction must be made between smooth- and rough-surfaced tubes.

Smooth Tubes. The velocity profile in the turbulent core has been expressed as

$$v^+ = 5.5 + 2.5 \ln y^+ \tag{12-63}$$

where the variables v^+ and y^+ are defined according to the relations

$$v^+ \equiv \frac{\overline{v}}{\sqrt{\tau_0/\rho}} \tag{12-58}$$

and

$$y^+ \equiv \frac{\sqrt{\tau_0/\rho}}{\nu} y \tag{12-60}$$

The average velocity in the turbulent core for flow in a tube of radius R can be evaluated from equation (12-63) as follows:

$$
\begin{aligned}
v_{\text{avg}} &= \frac{\int_0^A \overline{v}\, dA}{A} \\
&= \frac{\sqrt{\tau_0/\rho} \int_0^R \left(2.5 \ln\left\{ \dfrac{\sqrt{\tau_0/\rho}\, y}{\nu} \right\} + 5.5 \right) 2\pi r\, dr}{\pi R^2}
\end{aligned}
$$

Letting $y = R - r$, we obtain

$$v_{\text{avg}} = 2.5\sqrt{\tau_0/\rho}\, \ln\left\{ \frac{\sqrt{\tau_0/\rho}\, R}{\nu} \right\} + 1.75\sqrt{\tau_0/\rho} \tag{13-8}$$

The functions $\sqrt{\tau_0/\rho}$ and C_f are related according to equation (12-2). As C_f and f_f are equivalent, we may write

$$\frac{v_{\text{avg}}}{\sqrt{\tau_0/\rho}} = \frac{1}{\sqrt{f_f/2}} \tag{13-9}$$

The substitution of equation (13-9) into equation (13-8) yields

$$\frac{1}{\sqrt{f_f/2}} = 2.5 \ln\left\{ \frac{R}{\nu} v_{\text{avg}} \sqrt{f_f/2} \right\} + 1.75 \tag{13-10}$$

Rearranging the argument of the logarithm into Reynolds number form, and changing to $\log_{10}$, we see that equation (13-10) reduces to

$$\frac{1}{\sqrt{f_f}} = 4.06 \log_{10}\left\{ \text{Re} \sqrt{f_f} \right\} - 0.60 \tag{13-11}$$

This expression gives the relation for the friction factor as a function of Reynolds number for turbulent flow in smooth circular tubes. The preceding development was first performed by von Kármán.[1] Nikuradse,[2] from experimental data, obtained the equation

$$\frac{1}{\sqrt{f_f}} = 4.0 \log_{10}\left\{ \text{Re} \sqrt{f_f} \right\} - 0.40 \tag{13-12}$$

which is very similar to equation (13-11).

[1] T. von Kármán, *NACA TM* 611, 1931.
[2] J. Nikuradse, *VDI-Forschungsheft*, **356**, 1932.

Rough Tubes. By an analysis similar to that used for smooth tubes, von Kármán developed equation (13-13) for turbulent flow in rough tubes

$$\frac{1}{\sqrt{f_f}} = 4.06 \, \log_{10} \frac{D}{e} + 2.16 \tag{13-13}$$

which compares very well with the equation obtained by Nikuradse from experimental data

$$\frac{1}{\sqrt{f_f}} = 4.0 \, \log_{10} \frac{D}{e} + 2.28 \tag{13-14}$$

Nikuradse's results for fully developed pipe flow indicated that the surface condition, that is, roughness, had nothing to do with the transition from laminar to turbulent flow. Once the Reynolds number becomes large enough so that flow is fully turbulent, then either equation (13-12) or (13-14) must be used to obtain the proper value for f_f. These two equations are quite different in that equation (13-12) expresses f_f as a function of Re only and equation (13-14) gives f_f as a function only of the relative roughness. The difference is, of course, that the former equation is for smooth tubes and the latter for rough tubes. The question that naturally arises at this point is "what is 'rough'?"

It has been observed from experiment that equation (13-12) describes the variation in f_f for a range in Re, even for rough tubes. Beyond some value of Re, this variation deviates from the smooth-tube equation and achieves a constant value dictated by the tube roughness as expressed by equation (13-14). The region wherein f_f varies both with Re and e/D is called the *transition region*. An empirical equation describing the variation of f_f in the transition region has been proposed by Colebrook.[3]

$$\frac{1}{\sqrt{f_f}} = 4 \, \log_{10} \frac{D}{e} + 2.28 - 4 \log_{10}\left(4.67 \frac{D/e}{\mathrm{Re}\sqrt{f_f}} + 1 \right) \tag{13-15}$$

Equation (13-15) is applicable to the transition region above a value of $(D/e)/(\mathrm{Re}\sqrt{f_f}) = 0.01$. Below this value, the friction factor is independent of the Reynolds number, and the flow is said to be *fully turbulent*.

To summarize the development of this section, the following equations express the friction-factor variation for the surface and flow conditions specified:

For laminar flow (Re < 2300)

$$f_f = \frac{16}{\mathrm{Re}} \tag{13-7}$$

For turbulent flow (smooth pipe, Re > 3000)

$$\frac{1}{\sqrt{f_f}} = 4.0 \, \log_{10}\{\mathrm{Re}\sqrt{f_f}\} - 0.40 \tag{13-12}$$

For turbulent flow (rough pipe, (Re > 3000, $D/e)/(\mathrm{Re}\sqrt{f_f}) < 0.01$)

$$\frac{1}{\sqrt{f_f}} = 4.0 \, \log_{10} \frac{D}{e} + 2.28 \tag{13-14}$$

And for transition flow

$$\frac{1}{\sqrt{f_f}} = 4 \, \log_{10} \frac{D}{e} + 2.28 - 4 \log_{10}\left(4.67 \frac{D/e}{\mathrm{Re}\sqrt{f_f}} + 1 \right) \tag{13-15}$$

---

[3] C. F. Colebrook, *J. Inst. Civil Engr.* (London) II, **133** (1938–39).

13.3 FRICTION FACTOR AND HEAD-LOSS DETERMINATION FOR PIPE FLOW

Friction Factor

A single friction-factor plot based upon equations (13-7), (13-13), (13-14), and (13-15) has been presented by Moody.[4] Figure 13.1 is a plot of the Fanning friction factor vs. the Reynolds number for a range of values of the roughness parameter e/D.

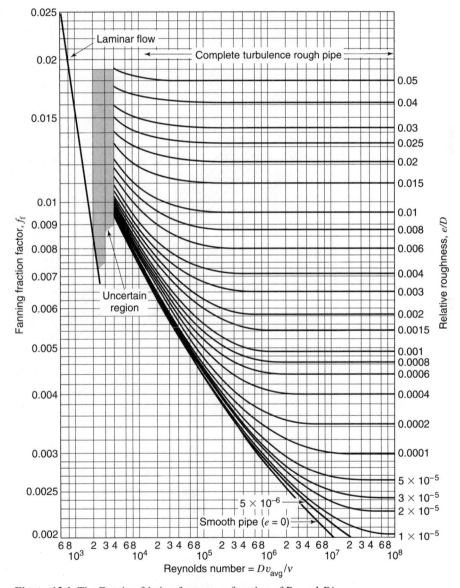

Figure 13.1 The Fanning friction factor as a function of Re and D/e.

---

[4] L. F. Moody, *Trans. ASME*, **66**, 671 (1944).

When using the friction-factor plot, Figure 13.1, it is necessary to know the value of the roughness parameter that applies to a pipe of given size and material. After a pipe or tube has been in service for some time, its roughness may change considerably, making the determination of e/D quite difficult. Moody has presented a chart, reproduced in Figure 13.2, by which a value of e/D can be determined for a given size tube or pipe constructed of a particular material.

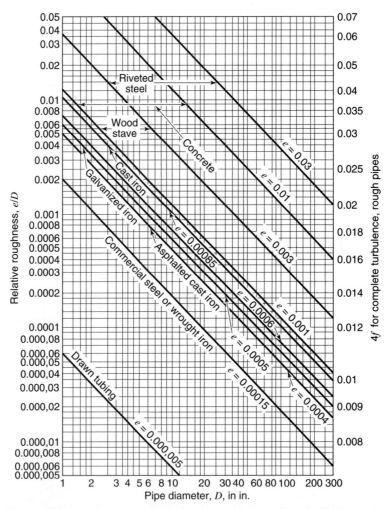

Figure 13.2 Roughness parameters for pipes and tubes. (From L. F. Moody, *Trans. ASME*, (1944).) Values of e are given in feet.

The combination of these two plots enables the frictional head loss for a length, L, of pipe having diameter D to be evaluated, using the relation

$$h_L = 2 f_f \frac{L}{D} \frac{v^2}{g} \tag{13-3}$$

Recently Haaland[5] has shown that over the range $10^8 \geq \text{Re} \geq 4 \times 10^4, 0.05 \geq e/D \geq 0$, the friction factor may be expressed (within $\pm 1.5\%$) as

$$\frac{1}{\sqrt{f_f}} = -3.6 \log_{10}\left[\frac{6.9}{\text{Re}} + \left(\frac{e}{3.7D}\right)^{10/9}\right] \tag{13-15a}$$

This expression allows explicit calculation of the friction factor.

Head Losses Due to Fittings

The frictional head loss calculated from equation (13-3) is only a part of the total head loss that must be overcome in pipe lines and other fluid-flow circuits. Other losses may occur due to the presence of valves, elbows, and any other fittings that involve a change in the direction of flow or in the size of the flow passage. The head losses resulting from such fittings are functions of the geometry of the fitting, the Reynolds number, and the roughness. As the losses in fittings, to a first approximation, have been found to be independent of the Reynolds number, the head loss may be evaluated as

$$h_L = \frac{\Delta P}{\rho} = K \frac{v^2}{2g} \tag{13-16}$$

where K is a coefficient depending upon the fitting.

An equivalent method of determining the head loss in fittings is to introduce an *equivalent length*, L_{eq}, such that

$$h_L = 2 f_f \frac{L_{\text{eq}}}{D} \frac{v^2}{g} \tag{13-17}$$

where L_{eq} is the length of pipe that produces a head loss equivalent to the head loss in a particular fitting. Equation (13-17) is seen to be in the same form as equation (13-3), and thus the total head loss for a piping system may be determined by adding the equivalent lengths for the fittings to the pipe length to obtain the total effective length of the pipe.

Table 13.1 Friction loss factors for various pipe fittings

Fitting	K	L_{eq}/D
Globe valve, wide open	7.5	350
Angle valve, wide open	3.8	170
Gate valve, wide open	0.15	7
Gate valve, $\frac{3}{4}$ open	0.85	40
Gate valve, $\frac{1}{2}$ open	4.4	200
Gate valve, $\frac{1}{4}$ open	20	900
Standard 90° elbow	0.7	32
Short-radius 90° elbow	0.9	41
Long-radius 90° elbow	0.4	20
Standard 45° elbow	0.35	15
Tee, through side outlet	1.5	67
Tee, straight through	0.4	20
180° Bend	1.6	75

---

[5] S. E. Haaland, *Trans. ASME, JFE*, **105**, 89 (1983).

Comparison of equations (13-16) and (13-17) shows that the constant K must be equal to $4f_f L_{eq}/D$. Although equation (13-17) appears to be dependent upon the Reynolds number because of the appearance of the Fanning friction factor, it is not. The assumption made in both equations (13-16) and (13-17) is that the Reynolds number is large enough so that the flow is fully turbulent. The friction coefficient for a given fitting, then, is dependent only upon the roughness of the fitting. Typical values for K and L_{eq}/D are given in Table 13.1.

Recall that the head loss due to a sudden expansion is calculated in Chapter 6, with the result given in equation (6-13).

Equivalent Diameter

Equations (13-16) and (13-17) are based upon a circular flow passage. These equations may be used to estimate the head loss in a closed conduit of any configuration if an "equivalent diameter" for a noncircular flow passage is used. An equivalent diameter is calculated according to

$$D_{eq} = 4\frac{\text{cross-sectional area of flow}}{\text{wetted perimeter}} \tag{13-18}$$

The ratio of the cross-sectional area of flow to the wetted perimeter is called the hydraulic radius.

The reader may verify that D_{eq} corresponds to D for a circular flow passage. One type of noncircular flow passage often encountered in transfer processes is the annular area between two concentric pipes. The equivalent diameter for this configuration is determined as follows:

$$\text{Cross-sectional area} = \frac{\pi}{4}(D_0^2 - D_i^2)$$

$$\text{Wetted perimeter} = \pi(D_0 + D_i)$$

yielding

$$D_{eq} = 4\frac{\pi/4}{\pi}\frac{(D_0^2 - D_i^2)}{(D_0 + D_i)} = D_0 - D_i \tag{13-19}$$

This value of D_{eq} may now be used to evaluate the Reynolds number, the friction factor, and the frictional head loss, using the relations and methods developed previously for circular conduits.

13.4 PIPE-FLOW ANALYSIS

Application of the equations and methods developed in the previous sections is common in engineering systems involving pipe networks. Such analyses are always straightforward but may vary as to the complexity of calculation. The following three example problems are typical, but by no means all-inclusive, of the types of problems found in engineering practice.

EXAMPLE 1

Water at 59°F flows through a straight section of a 6-in.-ID cast-iron pipe with an average velocity of 4 fps. The pipe is 120 ft long, and there is an increase in elevation of 2 ft from the inlet of the pipe to its exit.

Find the power required to produce this flow rate for the specified conditions.

The control volume in this case is the pipe and the water it encloses. Applying the energy equation to this control volume, we obtain

$$\frac{\delta Q}{dt} - \frac{\delta W_s}{dt} - \frac{\delta W_\mu}{dt} = \iint_{c.s.} \rho\left(e + \frac{P}{\rho}\right)(\mathbf{v} \cdot \mathbf{n})\, dA + \frac{\partial}{\partial t}\iiint_{c.v.} \rho e\, dV \tag{6-10}$$

An evaluation of each term yields

$$\frac{\delta Q}{dt} = 0 \qquad \frac{\delta W_s}{dt} = \dot{W}$$

$$\iint_{c.s.} \rho\left(e + \frac{P}{\rho}\right)(\mathbf{v \cdot n}) \, dA = \rho A v_{avg}\left(\frac{v_2^2}{2} + gy_2 + \frac{P_2}{\rho} + u_2 - \frac{v_1^2}{2} - gy_1 - \frac{P_1}{\rho} - u_1\right)$$

$$\frac{\partial}{\partial t} \iiint_{c.v.} \rho e \, dV = 0$$

and

$$\frac{\delta W_\mu}{dt} = 0$$

The applicable form of the energy equation written on a unit mass basis is now

$$\dot{W}/\dot{m} = \frac{v_1^2 - v_2^2}{2} + g(y_1 - y_2) + \frac{P_1 - P_2}{\rho} + u_1 - u_2$$

and with the internal energy change written as gh_L, the frictional head loss, the expression for w becomes

$$\dot{W}/\dot{m} = \frac{v_1^2 - v_2^2}{2} + g(y_1 - y_2) + \frac{P_1 - P_2}{\rho} - gh_L$$

Assuming the fluid at both ends of the control volume to be at atmospheric pressure, $(P_1 - P_2)/\rho = 0$, and for a pipe of constant cross section $(v_1^2 - v_2^2)/2 = 0$, giving for $\dot{W}/\dot{m}$

$$\dot{W}/\dot{m} = g(y_1 - y_2) - gh_L$$

Evaluating h_L, we have

$$\mathrm{Re} = \frac{(\frac{1}{2})(4)}{1.22 \times 10^{-5}} = 164,000$$

$$\frac{e}{D} = 0.0017 \quad \text{(from Figure 14.2)}$$

$$f_f = 0.0059 \quad \text{(from equation (14-15a))}$$

yielding

$$h_L = \frac{2(0.0059)(120\,\text{ft})(16\,\text{ft}^2/\text{s}^2)}{(0.5\,\text{ft})(32.2\,\text{ft/s}^2)} = 1.401\,\text{ft}$$

The power required to produce the specified flow conditions thus becomes

$$\dot{W} = \frac{-g((-2\,\text{ft}) - 1.401\,\text{ft})}{550\,\text{ft}\,\text{lb}_f/\text{hp-s}}\left[\frac{62.3\,\text{lb}_m/\text{ft}^3}{32.2\,\text{lb}_m\,\text{ft/s}^2\,\text{lb}_f}\left(\frac{\pi}{4}\right)\left(\frac{1}{2}\text{ft}\right)^2\left(4\frac{\text{ft}}{\text{s}}\right)\right]$$

$$= 0.300\,\text{hp}$$

EXAMPLE 2

A heat exchanger is required, which will be able to handle 0.0567 m³/s of water through a smooth pipe with an equivalent length of 122 m. The total pressure drop is 103,000 Pa. What size pipe is required for this application?

Once again, applying equation (6-10), we see that a term by term evaluation gives

$$\frac{\delta Q}{dt} = 0 \qquad \frac{\delta W_s}{dt} = 0 \qquad \frac{\delta W_\mu}{dt} = 0$$

$$\iint_{c.s.} \rho\left(e + \frac{P}{\rho}\right)(\mathbf{v} \cdot \mathbf{n})\, dA = \rho A v_{avg}\left(\frac{v_2^2}{2} + g y_2 + \frac{P_2}{\rho} + u_2 - \frac{v_1^2}{2} - g y_1 - \frac{P_1}{\rho} - u_1\right)$$

$$\frac{\partial}{\partial t}\iiint_{c.v.} \rho e\, dV = 0$$

and the applicable equation for the present problem is

$$0 = \frac{P_2 - P_1}{\rho} + g h_L$$

The quantity desired, the diameter, is included in the head-loss term but cannot be solved for directly, as the friction factor also depends on D. Inserting numerical values into the above equation and solving, we obtain

$$0 = -\frac{103\,000\,\text{Pa}}{1000\,\text{kg/m}^3} + 2 f_f \left(\frac{0.0567}{\pi D^2/4}\right)^2 \frac{\text{m}^2}{\text{s}^2} \cdot \frac{122}{D}\frac{\text{m}}{\text{m}}\frac{g}{g}$$

or

$$0 = -103 + 1.27\frac{f_f}{D^5}$$

The solution to this problem must now be obtained by trial and error. A possible procedure is the following:

1. Assume a value for f_f.
2. Using this f_f, solve the above equation for D.
3. Calculate Re with this D.
4. Using e/D and the calculated Re, check the assumed value of f_f.
5. Repeat this procedure until the assumed and calculated friction factor values agree.

Carrying out these steps for the present problem, the required pipe diameter is 0.132 m (5.2 in.).

EXAMPLE 3

An existing heat exchanger has a cross section as shown in Figure 13.3 with nine 1-in.-OD tubes inside a 5-in.-ID pipe. For a 5-ft length of heat exchanger, what flow rate of water at 60°F can be achieved in the shell side of this unit for a pressure drop of 3 psi?

An energy-equation analysis using equation (6-10) will follow the same steps as in example 13.2, yielding, as the governing equation

$$0 = \frac{P_2 - P_1}{\rho} + g h_L$$

Figure 13.3 Shell-and-tube head-exchanger configuration.

The equivalent diameter for the shell is evaluated as follows:

$$\text{Flow area} = \frac{\pi}{4}(25 - 9) = 4\pi \,\text{in.}^2$$

$$\text{Wetted perimeter} = \pi(5 + 9) = 14\pi \,\text{in.}$$

thus

$$D_{eq} = 4\frac{4\pi}{14\pi} = 1.142 \,\text{in.}$$

Substituting the proper numerical values into the energy equation for this problem reduces it to

$$0 = -\frac{3\,\text{lb}_f/\text{in.}^2\,(144\,\text{in.}^2/\text{ft}^2)}{1.94\,\text{slugs/ft}^3} + 2\,f_f v_{avg}^2 \,\text{ft}^2/\text{s}^2 \,\frac{5\,\text{ft}}{(1.142/12)\,\text{ft}}\frac{g}{g}$$

or

$$0 = -223 + 105\,f_f v_{avg}^2$$

As f_f cannot be determined without a value of Re, which is a function of v_{avg}, a simple trial-and-error procedure such as the following might be employed:

1. Assume a value for f_f.
2. Calculate v_{avg} from the above expression.
3. Evaluate Re from this value of v_{avg}.
4. Check the assumed value of f_f using equation (13-15a).
5. If the assumed and calculated values for f_f do not agree, repeat this procedure until they do.

Employing this method, we find the velocity to be 23.6 fps, giving a flow rate for this problem of 2.06 ft^3/min (0.058 m^3/s).

Notice that in each of the last two examples in which a trial-and-error approach was used, the assumption of f_f was made initially. This was not, of course, the only way to approach these problems; however, in both cases a value for f_f could be assumed within a much closer range than either D or v_{avg}.

13.5 FRICTION FACTORS FOR FLOW IN THE ENTRANCE TO A CIRCULAR CONDUIT

The development and problems in the preceding section have involved flow conditions that did not change along the axis of flow. This condition is often met, and the methods just described will be adequate to evaluate and predict the significant flow parameters.

In many real flow systems this condition is never realized. A boundary layer forms on the surface of a pipe, and its thickness increases in a similar manner to that of the boundary layer on a flat plate as described in Chapter 12. The buildup of the boundary layer in pipe flow is depicted in Figure 13.4.

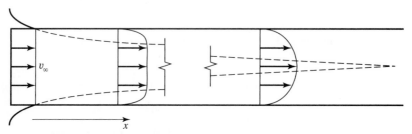

Figure 13.4 Boundary-layer buildup in a pipe.

A boundary layer forms on the inside surface and occupies a larger amount of the flow area for increasing values of x, the distance downstream from the pipe entrance. At some value of x, the boundary layer fills the flow area. The velocity profile will not change the downstream from this point, and the flow is said to be *fully developed*. The distance downstream from the pipe entrance to where flow becomes fully developed is called the entrance length, symbolized as L_e. Observe that the fluid velocity outside the boundary layer increases with x, as is required to satisfy continuity. The velocity at the center of the pipe finally reaches a value of $2v_\infty$ for fully developed laminar flow.

The entrance length required for a fully developed velocity profile to form in laminar flow has been expressed by Langhaar[6] according to

$$\frac{L_e}{D} = 0.0575\,\text{Re} \tag{13-20}$$

where D represents the inside diameter of the pipe. This relation, derived analytically, has been found to agree well with experiment.

There is no relation available to predict the entrance length for a fully developed turbulent velocity profile. An additional factor that affects the entrance length in turbulent flow is the nature of the entrance itself. The reader is referred to the work of Deissler[7] for experimentally obtained turbulent velocity profiles in the entrance region of the circular pipes. A general conclusion of the results of Deissler and others is that the turbulent velocity profile becomes fully developed after a minimum distance of 50 diameters downstream from the entrance.

The reader should realize that the entrance length for the velocity profile differs considerably from the entrance length for the velocity gradient at the wall. As the friction factor is a function of dv/dy at the pipe surface, we are also interested in this starting length.

Two conditions exist in the entrance region, which cause the friction factor to be greater than in fully developed flow. The first of these is the extremely large wall velocity gradient right at the entrance. The gradient decreases in the downstream direction, becoming constant before the velocity profile becomes fully developed. The other factor is the existence of a "core" of fluid outside the viscous layer whose velocity must increase as

---

[6] H. L. Langhaar, *Trans. ASME*, **64**, A-55 (1942).
[7] R. G. Deissler, NACA TN 2138 (1950).

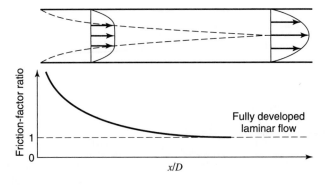

Figure 13.5 Velocity profile and friction-factor variation for laminar flow in the region near a pipe entrance.

dictated by continuity. The fluid in the core is thus being accelerated, thereby producing an additional drag force whose effect is incorporated in the friction factor.

The friction factor for laminar flow in the entrance to a pipe has been studied by Langhaar.[8] His results indicated the friction factor to be highest in the vicinity of the entrance, then to decrease smoothly to the fully developed flow value. Figure 13.5 is a qualitative representation of this variation. Table 13.2 gives the results of Langhaar for the average friction factor between the entrance and a location, a distance x from the entrance.

For turbulent flow in the entrance region, the friction factor as well as the velocity profile is difficult to express. Deissler[9] has analyzed this situation and presented his results graphically.

Even for very high free-stream velocities, there will be some portion of the entrance over which the boundary layer is laminar. The entrance configuration, as well as the Reynolds number, affects the length of the pipe over which the laminar boundary layer exists before becoming turbulent. A plot similar to Figure 13.5 is presented in Figure 13.6 for turbulent-flow friction factors in the entrance region.

Table 13.2 Average friction factor for laminar flow in the entrance to a circular pipe

$\dfrac{x/D}{Re}$	$f_f\left(\dfrac{x}{D}\right)$
0.000205	0.0530
0.000830	0.0965
0.001805	0.1413
0.003575	0.2075
0.00535	0.2605
0.00838	0.340
0.01373	0.461
0.01788	0.547
0.02368	0.659
0.0341	0.845
0.0449	1.028
0.0620	1.308
0.0760	1.538

[8] *Op cit.*
[9] R. G. Deissler, NACA TN 3016 (1953).

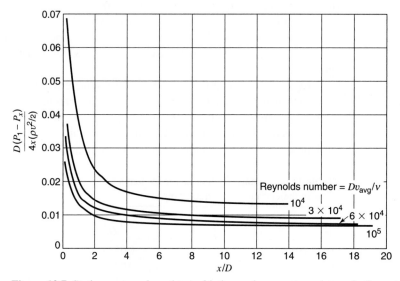

Figure 13.6 Velocity profile and friction-factor variation in turbulent flow in the region near a pipe entrance.

The foregoing description of the entrance region has been qualitative. For an accurate analytical consideration of a system involving entrance-length phenomena, Deissler's results portrayed in Figure 13.7 may be utilized.

Figure 13.7 Static pressure drop due to friction and momentum change in the entrance to a smooth, horizontal, circular tube (Deissler).

It is important to realize that in many situations flow is never fully developed; thus the friction factor will be higher than that predicted from the equations for fully developed flow or the friction-factor plot.

13.6 CLOSURE

The information and techniques presented in this chapter have included applications of the theory developed in earlier chapters supported by correlations of experimental data.

The chapters to follow will be devoted to heat and mass transfer. One specific type of transfer, momentum transfer, has been considered up to this point. The student will find that he is able to apply much of the information learned in momentum transfer to counterparts in the areas of heat and mass transfer.

PROBLEMS

13.1 An oil with kinematic viscosity of 0.08×10^{-3} ft^2/s and a density of 57 lb$_m$/ft^3 flows through a horizontal tube 0.24 in. in diameter at the rate of 10 gal/h. Determine the pressure drop in 50 ft of tube.

13.2 A lubricating line has an inside diameter of 0.1 in. and is 30 in. long. If the pressure drop is 15 psi, determine the flow rate of the oil. Use the properties given in Problem 13.1.

13.3 The pressure drop in a section of a pipe is determined from tests with water. A pressure drop of 13 psi is obtained at a flow rate of 28.3 lb$_m$/s. If the flow is fully turbulent, what will be the pressure drop when liquid oxygen ($\rho = 70$ lb$_m$/ft^3) flows through the pipe at the rate of 35 lb$_m$/s?

13.4 A 280-km-long pipeline connects two pumping stations. If 0.56 m^3/s are to be pumped through a 0.62-m-diameter line, the discharge station is 250 m lower in elevation than the upstream station, and the discharge pressure is to be maintained at 300,000 Pa, determine the power required to pump the oil. The oil has a kinematic viscosity of 4.5×10^{-6} m^2/s and a density of 810 kg/m^3. The pipe is constructed of commercial steel. The inlet pressure may be taken as atmospheric.

13.5 In the previous problem, a 10-km-long section of the pipeline is replaced during a repair process with a pipe with internal diameter of 0.42 m. Determine the total pumping power required when using the modified pipeline. The total pipeline length remains 280 km.

13.6 Oil having a kinematic viscosity of 6.7×10^{-6} m^2/s and density of 801 kg/m^3 is pumped through a pipe of 0.71 m diameter at an average velocity of 1.1 m/s. The roughness of the pipe is equivalent to that of a commercial steel pipe. If pumping stations are 320 km apart, find the head loss (in meters of oil) between the pumping stations and the power required.

13.7 The cold-water faucet in a house is fed from a water main through the following simplified piping system:

a. A 160 ft length of 3/4-in.-ID copper pipe leading from the main line to the base of the faucet.

b. Six 90° standard elbows.

c. One wide-open angle valve (with no obstruction).

d. The faucet. Consider the faucet to be made up of two parts: (1) a conventional globe valve and (2) a nozzle having a cross-sectional area of 0.10 in.^{2}.

The pressure in the main line is 60 psig (virtually independent of flow), and the velocity there is negligible. Find the maximum rate of discharge from the faucet. As a first try, assume for the pipe $f_f = 0.007$. Neglect changes in elevation throughout the system.

13.8 Water at the rate of 118 ft^3/min flows through a smooth horizontal tube 250 ft long. The pressure drop is 4.55 psi. Determine the tube diameter.

13.9 Calculate the inlet pressure to a pump 3 ft above the level of a sump. The pipe is 6 in. in diameter, 6 ft long, and made of commercial steel. The flow rate through the pump is 500 gal/min. Use the (incorrect) assumption that the flow is fully developed.

13.10 The pipe in Problem 6.33 is 35 m long and made of commercial steel. Determine the flow rate.

13.11 The siphon of Problem 6.31 is made of smooth rubber hose and is 23 ft long. Determine the flow rate and the pressure at point B.

13.12 A galvanized rectangular duct 8 in. square is 25 ft long and carries 600 ft^3/min of standard air. Determine the pressure drop in inches of water.

13.13 A cast-iron pipeline 2 m long is required to carry 3 million gal of water per day. The outlet is 175 ft higher than the inlet. The costs of three sizes of pipe when in place are as follows:

10-in. diameter	\$11.40 per ft
12-in. diameter	\$14.70 per ft
14-in. diameter	\$16.80 per ft

Power costs are estimated at \$0.07 per kilowatt hour over the 20-year life of the pipeline. If the line can be bonded with 6.0% annual interest, what is the most economical pipe diameter? The pump efficiency is 80%, and the water inlet temperature is expected to be constant at 42°F.

13.14 Estimate the flow rate of water through 50 ft of garden hose from a 40-psig source for

a. A 1/2-in.-ID hose;

b. A 3/4-in.-ID hose.

13.15 Two water reservoirs of height $h_1 = 60$ m and $h_2 = 30$ m are connected by a pipe that is 0.35 m in diameter. The exit of the pipe is submerged at distance $h_3 = 8$ m from the reservoir surface.

a. Determine the flow rate through the pipe if the pipe is 80 m long and the friction factor $f_f = 0.004$. The pipe inlet is set flush with the wall.

b. If the relative roughness $e/D = 0.004$, determine the friction factor and flow rate.

13.16   An 8-km-long, 5-m-diameter headrace tunnel at the Paute river hydroelectric project in Ecuador supplies a power station 668 m below the entrance of the tunnel. If the tunnel surface is concrete, find the pressure at the end of the tunnel if the flow rate is 90 m^3/s.

13.17   Determine the flow rate through a 0.2-m gate valve with upstream pressure of 236 kPa when the valve is

a. open;

b. 1/4 closed;

c. 1/2 closed;

d. 3/4 closed.

13.18   Water at 20°C flows through a cast-iron pipe at a velocity of 34 m/s. The pipe is 400 m long and has a diameter of 0.18 m. Determine the head loss due to friction.

13.19   A 2.20-m diameter pipe carries water at 15°C. The head loss due to friction is 0.500 m per 300 m of pipe. Determine the volumetric flow rate of the water leaving the pipe.

13.20   Water at 20°C is being drained from an open tank through a cast-iron pipe 0.6 m diameter and 30 m long. The surface of the water in the tank is at atmospheric pressure and at an elevation of 46.9 m, and the pipe discharges to the atmosphere at an elevation 30 m. Neglecting minor losses due to configuration, bends, valves, and fittings, determine the volumetric flow rate of the water leaving the pipe.

13.21   A 15-cm diameter wrought-iron pipe is to carry water at 20°C. Assuming a level pipe, determine the volumetric flow rate at the discharge if the pressure loss is not permitted to exceed 30.0 kPa per 100 m.

13.22   A level 10-m-long water pipe has a manometer at both the inlet and the outlet. The manometers indicate pressure head of 1.5 and 0.2 m, respectively. The pipe diameter is 0.2 m and the pipe roughness is 0.0004 m. Determine the mass flow rate in the pipe in kg/s.

13.23   Determine the depth of water behind the dam in the figure that will provide a flow rate of 5.675 × 10^{-4} m^3/s through a 20-m-long, 1.30 cm commercial steel pipe.

13.24   Water flows at a volumetric flow rate of 0.25 m^3/s from reservoir 1 to reservoir 2 through three concrete pipes connected in series. Pipe 1 is 900 m long and has a diameter of 0.16 m. Pipe 2 has a length of 1500 m and a diameter of 0.18 m. Pipe 3 is 800 m long and the diameter is 0.20 m. Neglecting minor losses, determine the difference in surface elevations.

13.25   A system consists of three pipes in series. The total pressure drop is 180 kPa, and the decrease in elevation is 5 m. Data for the three pipes are as follows:

Pipe	Length, m	Diameter, cm	Roughness, mm
1	125	8	0.240
2	150	6	0.120
3	100	4	0.200

Determine the total flow rate of 20°C water for the system. Neglect minor losses.

13.26   Two concrete pipes are connected in series. The flow rate of water at 20°C through the pipes is 0.18 m^3/s, with a total head loss of 18 m for both pipes. Each pipe has a length of 312.5 m and a relative roughness of 0.0035 m. Neglecting minor losses, if one pipe has a diameter of 0.30 m, determine the diameter of the other.

13.27   A 0.2-m-diameter cast-iron pipe and a 67-mm-diameter commercial steel pipe are parallel, and both run from the same pump to a reservoir. The pressure drop is 210 kPa and the lines are 150 m long. Determine the flow rate of water in each line.

13.28   A system consists of three pipes in parallel with a total head loss of 24 m. Data for the three pipes are as follows:

Pipe	Length, m	Diameter, cm	Roughness, mm
1	100	8	0.240
2	150	6	0.120
3	80	4	0.200

For water at 20°C, neglect minor losses and determine the volumetric flow rate in the system.

Chapter 14

Fluid Machinery

In this chapter we will examine the operating principles of mechanical devices that exchange fluid energy and mechanical work. A *pump* is a machine whose purpose is to apply mechanical energy to a fluid, thereby generating flow, or producing a higher pressure, or both. A *turbine* does just the opposite—producing work through the application of fluid energy.

There are two principal types of fluid machines—*positive displacement machines* and *turbomachines*. In positive displacement machines, a fluid is confined in a chamber whose volume is varied. Examples of positive-displacement-type machines are shown in Figure 14.1.

Figure 14.1 Some examples of positive-displacement configurations.

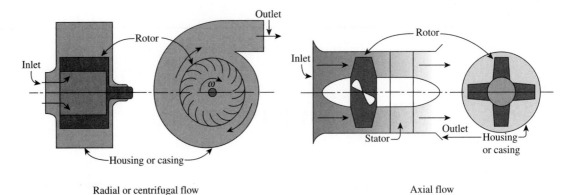

Radial or centrifugal flow   Axial flow

Figure 14.2 Turbomachines.

Turbomachines, as the name implies, involve rotary motion. Window fans and aircraft propellers are examples of *unshrouded* turbomachines. Pumps used with liquids generally have *shrouds* that confine and direct the flow. The two general types of pumps in this category are shown in Figure 14.2. The designations *radial flow* and *axial flow* refer to the direction of fluid flow relative to the axis of rotation of the rotating element.

The term *pump* is generally used when the working fluid is a liquid. If the fluid is a gas or vapor, the following terms are used:

- *Fans* are associated with relatively small pressure changes, on the order of $\Delta P \sim$ 35 cm of H_2O (0.5 psi).
- *Blowers* are of both positive and variable displacement types, with ΔP up to 2.8 m of H_2O (40 psi).
- *Compressors* are of both positive and variable configurations having delivery pressures as high as 69 MPa (10^3 psi).

Turbines, as previously stated, extract energy from high-pressure fluids. They are of two primary types, *impulse* and *reaction*, which convert fluid energy into mechanical work in different ways. In the impulse turbine, the high-energy fluid is converted, by means of a nozzle, into a high-velocity jet. This jet then strikes the turbine blades as they pass. In this configuration, the jet flow is essentially at a constant pressure. The basic analysis of these devices is examined in Chapter 5.

In reaction turbines the fluid fills the blade passages and a pressure decrease occurs as it flows through the impeller. The energy transfer in such devices involves some thermodynamic considerations beyond simple momentum analysis.

The remainder of this chapter will be devoted entirely to pumps and fans. Consideration will be given to general pump and fan performances, scaling laws, and their compatibility with piping systems.

14.1   CENTRIFUGAL PUMPS

Figure 14.3 shows two cutaway views of a typical centrifugal pump. In this configuration, fluid enters the pump casing axially. It then encounters the impeller blades that direct the flow tangentially and radially outward into the outer part of the casing and is then discharged. The fluid experiences an increase in velocity and pressure as it passes through

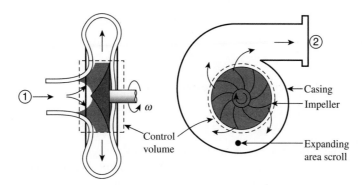

Figure 14.3 Cutaway views of a centrifugal pump.

the impeller. The discharge section, which is doughnut shaped, causes the flow to decelerate and the pressure to increase further.

The impeller blades shown have a *backward-curved* shape, which is the most common configuration.

Pump Performance Parameters

We now focus on the control volume designated in Figure 14.3 by dashed lines. Note that flow enters at section one and leaves at two.

Applying the first law of thermodynamics to this control volume we have

$$\frac{\delta Q}{dt} - \frac{\delta W_s}{dt} = \iint_{c.s.} \left(e + \frac{P}{\rho}\right)\rho(\mathbf{v} \cdot \mathbf{n})\, dA + \frac{\partial}{\partial t}\iiint_{c.v.} e\rho\, dv + \frac{\delta W_\mu}{dt} \qquad (6\text{-}10)$$

which for steady, adiabatic flow with no viscous work, becomes

$$-\frac{\delta W_s}{dt} = \dot{m}\left[h_2 - h_1 + \frac{v_2^2 - v_1^2}{2} + g(y_2 - y_1)\right]$$

It is customary to neglect the small differences in velocity and elevation between sections one and two, thus

$$v_2^2 - v_1^2 \approx 0 \quad \text{and} \quad y_2 - y_1 \approx 0$$

and the remaining expression is

$$-\frac{\delta W_s}{dt} = \dot{m}(h_2 - h_1) = \dot{m}\left(u_2 - u_1 + \frac{P_2 - P_1}{\rho}\right)$$

Recalling that the term $u_2 - u_1$ represents the loss due to friction and other irreversible effects, we write

$$u_2 - u_1 = h_L$$

The net pressure head produced in the pump is

$$\frac{P_2 - P_1}{\rho} = \frac{1}{\dot{m}}\frac{\delta W_s}{\delta t} - h_L \qquad (14\text{-}1)$$

An important performance parameter, the *efficiency*, can now be expressed in broad terms as the ratio of actual output to required input. For a centrifugal pump the efficiency, designated η, is

$$\eta = \frac{\text{power added to the fluid}}{\text{shaft power to the impeller}}$$

The power added to the fluid is given by equation (14-1)

$$\left.\frac{\delta W}{dt}\right|_{\text{fluid}} = \dot{m}\left(\frac{P_2 - P_1}{\rho}\right) \tag{14-2}$$

and the efficiency can be expressed as

$$\eta = \frac{\dot{m}(P_2 - P_1)}{\rho(\delta W_s/\delta t)_{\text{c.v.}}} \tag{14-3}$$

The difference between $\delta W_s/\delta t|_{\text{c.v.}}$ and $\delta W_s/\delta t|_{\text{liquid}}$ is clearly the head loss, h_L.

Equations (14-1), (14-2), and (14-3) provide general relationships for important pump performance parameters. To develop actual performance information for centrifugal pumps, we must examine our control volume once again from a moment of momentum perspective.

The governing equation for this analysis is

$$\sum M_z = \iint_{\text{c.s.}} (\mathbf{r} \times \mathbf{v})_z\, \rho(\mathbf{v} \cdot \mathbf{n})\, dA + \frac{\partial}{\partial t}\iiint_{\text{c.v.}} (\mathbf{r} \times \mathbf{v})_z \rho\, dv \tag{5-10c}$$

The axis of rotation of the rotor depicted in Figure 14.3 has been chosen as the z direction, hence our choice of equation (5-10c).

We now wish to solve for M_z by applying equation (5-10c) to the control volume in Figure 14.3 for one-dimensional steady flow. The coordinate system will be fixed with the z direction along the axis of rotation. Recall that the rotor contains *backward-curved* blades. In Figure 14.4, we show a detailed view of a single rotor blade. The blade is attached to the rotor hub at distance r_1 from the z axis; the outer dimension of the blade has the value r_2.

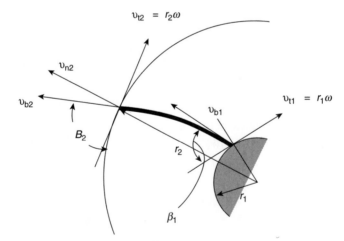

Figure 14.4 Velocity diagram for flow exiting a centrifugal pump impeller.

In this figure

v_{b1}, v_{b2} represent velocities *along the blade* at r_1 and r_2, respectively

v_{n2} is the normal velocity of the flow at r_2

v_{t2} is the tangential velocity of the flow at r_2

β_1, β_2 are the angles made between the blade and tangent directions at r_1 and r_2, respectively.

Equation (5-10c) can now be written as

$$M_z = \dot{m} \left[\begin{vmatrix} \mathbf{e}_r & \mathbf{e}_\theta & \mathbf{e}_z \\ r & 0 & 0 \\ v_r & v_\theta & v_z \end{vmatrix}_2 - \begin{vmatrix} \mathbf{e}_r & \mathbf{e}_\theta & \mathbf{e}_z \\ r & 0 & 0 \\ v_r & v_\theta & v_z \end{vmatrix}_1 \right]_z$$

which becomes

$$M_z = \dot{m}[(rv_\theta)_2 - 0]$$
$$= \rho \dot{V} r_2 v_{\theta 2}$$

(14-4)

The velocity, $v_{\theta 2}$, is the tangential component of the fluid stream exiting the rotor relative to the fixed coordinate system. The quantities shown in Figure 14.4 will be useful in evaluating $v_{\theta 2}$.

The absolute velocity of the existing flow, v_2, is the vector sum of the velocity relative to the impeller blade and the velocity of the blade tip relative to our coordinate system. For blade length, L, normal to the plane of Figure 14.4, we evaluate the following:

- the normal velocity of flow at r_2

$$v_{n2} = \frac{\dot{V}}{2\pi r_2 L}$$

(14-5)

- the velocity of flow along the blade at r_2

$$v_{b2} = \frac{v_{n2}}{\sin\beta_2}$$

(14-6)

- the blade tip velocity

$$v_{t2} = r_2 \omega$$

(14-7)

The velocity we want, $v_{\theta 2}$, can now be evaluated as

$$v_{\theta 2} = v_{t2} - v_{b2} \cos\beta_2$$

Substitution from equations (14-6) and (14-7) yields

$$v_{\theta 2} = r_2 \omega - \frac{v_{n2}}{\sin\beta_2} \cos\beta_2$$
$$= r_2 \omega - v_{n2} \cot\beta_2$$

Finally, introducing the expression for v_{n2} from equation (14-5) we have

$$v_\theta = r_2 \omega - \frac{\dot{V}}{2\pi r_2 L} \cot\beta_2$$

(14-8)

and the desired moment is

$$M_z = \rho \dot{V} r_2 \left[r_2 \omega - \frac{\dot{V}}{2\pi r_2 L} \cot\beta_2 \right]$$

(14-9)

The power delivered to the fluid is, by definition, $M_z\omega$, thus

$$\dot{W} = \frac{\delta W_s}{dt} = M_z\omega = \rho\dot{V}r_2\omega\left[r_2\omega - \frac{\dot{V}}{2\pi rL}\cot\beta_2\right] \tag{14-10}$$

Equation (14-10) expresses the power imparted to the fluid for an impeller with dimensions r_2, β_2, and L, operating at angular velocity, ω, with mass flow rate $\rho\dot{V}$.

This expression may be related to equations (14-2) and (14-3) to evaluate the imparted pressure head and the pump efficiency.

It is a standard practice to minimize friction loss at r_1, the radial location at which flow enters the impeller. This is accomplished by configuring the angle, β_1, such that inlet flow is along the blade surface. Referring to Figure 14.4, the design point for minimum losses is achieved when

$$v_{b1}\cos\beta_1 = r_1\omega$$

or, equivalently, when

$$v_{r1} = v_{b1}\sin\beta_1 = r_{1\omega}\frac{\sin\beta_1}{\cos\beta_1}$$

and, finally, when

$$v_{r1} = r_1\omega\tan\beta_1 \tag{14-11}$$

Typical performance curves, for a centrifugal pump, are shown in Figure 14.5. Pressure head, brake horsepower, and efficiency are all shown as functions of volumetric flow rate. It is reasonable to choose operating conditions at or near the flow rate where maximum efficiency is achieved.

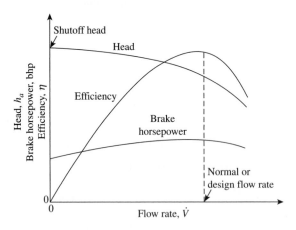

Figure 14.5 Centrifugal pump performance curves.

Example 1 illustrates how the analysis presented above relates to centrifugal pump performance.

EXAMPLE 1

Water flow is produced by a centrifugal pump with the following dimensions:

$$\begin{array}{ll} r_1 = 6\,\text{cm} & \beta_1 = 33° \\ r_2 = 10.5\,\text{cm} & \beta_2 = 21° \\ L = 4.75\,\text{cm} \end{array}$$

at a rotational speed of 1200 rpm determine

    (a) the design flow rate;

    (b) the power added to the flow;

    (c) the maximum pressure head at the pump discharge.

To experience minimum losses, equation (14-11) must be satisfied, thus

$$v_{r1} = r_1 \omega \tan\beta_1$$

$$= (0.06\,\text{m})\left(1200\frac{\text{rev}}{\text{min}}\right)\left(\frac{2\pi\,\text{rad}}{\text{rev}}\right)\left(\frac{\text{min}}{60\,\text{s}}\right)(\tan 33°) \tag{14-11}$$

$$= 4.896\,\text{m/s}$$

The corresponding flow rate is

$$\dot{V} = 2\pi r_1 L v_{r1}$$

$$= 2\pi(0.06\,\text{m})(0.0475\,\text{m})(4.896\,\text{m/s})$$

$$= 0.0877\,\text{m}^3/\text{s}\quad(1390\,\text{gpm}) \tag{a}$$

The power imparted to the flow is expressed by equation (14-10):

$$\dot{W} = \rho\dot{V}r_2\omega\left[r_2\omega - \frac{\dot{V}}{2\pi rL}\cot\beta_2\right]$$

Evaluating the following:

$$\omega = \left(1200\frac{\text{rev}}{\text{min}}\right)\left(2\pi\frac{\text{rad}}{\text{rev}}\right)\left(\frac{\text{min}}{60\,\text{s}}\right) = 125.7\,\text{rad/s}$$

$$\rho\dot{V}r_2\omega = (1000\,\text{kg/m}^3)(0.0877\,\text{m}^3/\text{s})(0.105\,\text{m})(125.7\,\text{rad/s})$$

$$= 1157\,\text{kg}\cdot\text{m/s}$$

$$\frac{\dot{V}}{2\pi r_2 L} = \frac{0.0877\,\text{m}^3/\text{s}}{2\pi(0.105\,\text{m})(0.0475\,\text{m})}$$

$$= 2.80\,\text{m/s}$$

we obtain

$$\dot{W} = (1157\,\text{kg}\cdot\text{m/s})[(0.105\,\text{m})(125.7\,\text{rad/s}) - (2.80\,\text{m/s})(\cot 21)]$$

$$= 6830\,\text{W} = 6.83\,\text{kW} \tag{b}$$

Equation (14.1) expresses the net pressure head as

$$\frac{P_2 - P_1}{\rho g} = -\frac{\dot{W}}{\dot{m}g} - h_L \tag{14-1}$$

The maximum value, with negligible loses, will be

$$\frac{P_2 - P_1}{\rho g} = \frac{6830\,\text{W}}{(1000\,\text{kg/m}^3)(0.0877\,\text{m}^3/\text{s})(9.81\,\text{m/s}^2)}$$

$$= 7.94\,\text{m H}_2\text{O gage}$$

For $P_1 = 1\,\text{atm} = 14.7\,\text{psi} = 10.33\,\text{m H}_2\text{O}$

$$P_2 = (7.94 + 10.33)\text{m H}_2\text{O} = 18.3\,\text{m H}_2\text{O}\,(26\,\text{psi}) \tag{c}$$

The actual discharge pressure will be less than this owing to friction and other irreversible losses.

Net Positive Suction Head

A major concern in pump operation is the presence of *cavitation*. Cavitation occurs when a liquid being pumped vaporizes or boils. If this occurs, the vapor bubbles that have been formed cause a decrease in efficiency and, often, structural damage to the pump that may lead to catastrophic failure. The parameter designated *net positive suction head* (NPSH) characterizes the likelihood for cavitation to occur.

At the suction side of the impeller, where pressure is lowest thus the location where cavitation will first occur, the NPSH can be expressed as

$$\text{NPSH} + \frac{P_v}{\rho g} = \frac{v_i^2}{2g} + \frac{P_i}{\rho g} \tag{14-12}$$

where v_i and P_i are evaluated at pump inlet and P_v is the liquid vapor pressure. Values of NPSH are, in general, determined experimentally over a range in flow rates, for a given pump. A typical variation of NPSH vs. $\dot{V}$ is shown in Figure 14.6.

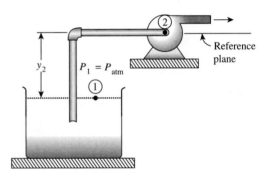

Figure 14.6 Typical variation of NPSH with V.

Figure 14.7 Pump installation at a level y above a supply reservoir.

In Figure 14.7, a representative pump installation is shown with the liquid being drawn from a reservoir located a distance, y, below the pump inlet. An energy balance between the pump inlet and the reservoir level yields

$$\frac{P_{\text{atm}}}{\rho g} = y_2 + \frac{P_2}{\rho g} + \frac{v_2^2}{2g} + \Sigma h_L \tag{14-13}$$

where the term Σh_L represents head losses between locations 1 and 2 as discussed in Chapter 13.

Combining this relationship with equation (14-12) we get

$$\begin{aligned}\text{NPSH} &= \frac{v_2^2}{2g} + \frac{P_2}{\rho g} - \frac{P_v}{\rho g} \\ &= \frac{P_{\text{atm}}}{\rho g} - y_2 - \frac{P_v}{\rho g} - \Sigma h_L\end{aligned} \tag{14-14}$$

For proper pump installation, the value of NPSH evaluated using equation (14-14) should be greater than the value obtained from a pump performance plot at the same flow rate. The principal use of these ideas is to establish a maximum value for the height, y_2. Example 2 illustrates the use of NPSH.

EXAMPLE 2

A system like the one shown in Figure 14.7 is to be assembled to pump water. The inlet pipe to the centrifugal pump is 12 cm in diameter and the desired flow rate is 0.025 m³/s. At this flow rate, the specifications for this pump show a value of NPSH of 4.2 m. The minor loss coefficient for the system may be taken as $K = 12$. Water properties are to be evaluated at 300 K. Determine the maximum value of y, the distance between pump inlet and reservoir level.

The quantity desired, y, is given by

$$y = \frac{P_{atm} - P_v}{\rho g} - \Sigma h_L - \text{NPSH} \qquad (14\text{-}14)$$

Water properties required, at 300 K, are

$$\rho = 997 \text{ kg/m}^3$$
$$P_v = 3598 \text{ Pa}$$

and we have

$$v = \frac{\dot{V}}{A} = \frac{0.0025 \text{ M}^3/\text{s}}{\frac{\pi}{4}(0.12 \text{ m})^2} = 2.21 \text{ m/s}$$

$$\Sigma h_L = K_L \frac{v^2}{2g} = \frac{12(2.21 \text{ m/s}^2)^2}{2(9.81 \text{ m/s}^2)} = 2.99 \text{ m}$$

We can now complete the solution

$$y = \frac{(101360 - 3598)P_a}{(997 \text{ kg/m}^3)(9.81 \text{ m/s}^2)} - 2.99 \text{ m} - 4.2 \text{ m}$$

$$= 2.805 \text{ m} \quad (9.2 \text{ ft})$$

Combined Pump and System Performance

As depicted in Figure 14.5, a pump has the capability of operating over a range in flow rates with its delivered head, operating efficiency, and NPSH values, all being flow-rate-dependent. An important task of the engineer is to match a given pump, with its known operating characteristics, to the performance of the system in which the pump produces flow. Piping system performance is discussed in Chapter 13.

Figure 14.8 Pumping system configuration.

A simple flow system is illustrated in Figure 14.8 where a pump is used to produce flow between two reservoirs at different elevations.

With the two reservoir surfaces designated ① and ② as shown in the figure, an energy balance between these two locations yields

$$-\frac{\dot{W}}{\dot{m}} = g(y_2 - y_1) + \frac{P_2 - P_1}{\rho} + (u_2 - u_1) \qquad (14\text{-}15)$$

Observing that $P_1 = P_2 = P_{atm}$ and expressing $u_2 - u_1 = \Sigma h_L$, we have

$$-\frac{\dot{W}}{\dot{m}g} = y_2 - y_1 + \Sigma h_L \tag{14-16}$$

From Chapter 13, we can write for the head loss

$$\Sigma h_L = \Sigma K \frac{v^2}{2g}$$

where the quantity ΣK accounts for frictional pipe loss as well as minor losses due to valves, elbows, and fittings.

The operating line for system performance is now expressed by

$$-\frac{\dot{W}}{\dot{m}g} = y_2 - y_1 + \Sigma K \frac{v^2}{2g} \tag{14-17}$$

Plotting the system operating line together with the plot of pump performance yields the combined performance diagram as shown in Figure 14.9.

Figure 14.9 Combined pump and system performance.

We note that the two operating lines intersect at a flow rate where the required head for system operation matches that which the particular pump can produce. At this operating flow rate, one can read the corresponding efficiency from the chart. A system designer would, naturally, want the system to operate at, or as near as possible to, the flow rate of maximum pump efficiency. If the operating point corresponds to an undesirable efficiency value, changes must be made either to the system, which is generally a difficult process, or to pump operating conditions.

14.2 SCALING LAWS FOR PUMPS AND FANS

The concepts of similarity and scaling are introduced in Chapter 11. The requirements of geometric, kinematic, and dynamic similarity find important applications in the scaling of rotating fluid mechanics. In this section we will develop the "fan laws" that are used to predict the effect of changing the fluid, size, or speed of rotating machines, which are in a geometrically similar family.

Dimensional Analysis of Rotating Machines

The Buckingham method of dimensional analysis, which is introduced in Chapter 11, is a useful tool in generating the dimensionless parameters that apply to rotating fluid machines.

As discussed earlier, the first step to be undertaken is to develop a table of variables that are important to our application. Table 14.1 lists the variables of interest along with their symbols and dimensional representation in the MLt system.

Table 14.1 Pump performance variables

Variable	Symbol	Dimensions
Total head	gh	L^2/t^2
Flow rate	$\dot{V}$	L^3/t
Impeller diameter	D	L
Shaft speed	ω	$1/t$
Fluid density	ρ	M/L^3
Fluid viscosity	μ	M/Lt
Power	$\dot{W}$	ML^2/t^3

Without repeating all details regarding the Buckingham method, we can establish the following:

- $i = n - r = 7 - 3 = 4$
- with a core group including the variables D, ω, ρ, the dimensionless pi groups become

$$\pi_1 = gh/D^2\omega^2$$
$$\pi_2 = \dot{V}/\omega D^3$$
$$\pi_3 = \dot{W}/\rho\omega^3 D^5$$
$$\pi_4 = \mu/D^2\omega\rho$$

The group $\pi_4 = \mu/D^2\omega\rho$ is a form of Reynolds number. The other three groups here are designated, by the pump community, as

$$\pi_1 = gh/D^2\omega^2 = C_H \quad \text{—the head coefficient} \quad (14\text{-}18)$$
$$\pi_2 = \dot{V}/\omega D^3 = C_Q \quad \text{—the flow coefficient} \quad (14\text{-}19)$$
$$\pi_3 = \dot{W}/\rho\omega^3 D^5 = C_P \quad \text{—the power coefficient} \quad (14\text{-}20)$$

Figure 14.10 is a plot of the dimensionless parameters C_H and C_P vs. the flow coefficient, C_Q, for a representative centrifugal pump family.

There is, of course, one additional dimensionless performance parameter, the efficiency. The efficiency is related to the other parameters defined above according to

$$\eta = \frac{C_H C_Q}{C_P} \quad (14\text{-}21)$$

As the parameters on the right-hand-side of the equation are functionally related to C_Q, the efficiency, η, is also a function of C_Q, and is included as one of the dependent variables in Figure 14.10.

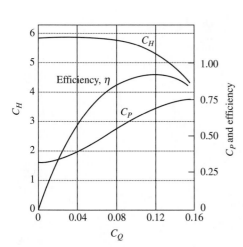

Figure 14.10 Dimensionless performance curves for a typical centrifugal pump.

The three coefficients C_H, C_Q, and C_P provide the basis for the fan laws. For similar pumps, designated 1 and 2, we may write, for C_H

$$C_{H1} = C_{H2}$$

or

$$\frac{gh_1}{\omega_1^2 D_1^2} = \frac{gh_2}{\omega_2^2 D_2^2}$$

thus

$$\frac{h_2}{h_1} = \left(\frac{\omega_2}{\omega_1}\right)^2 \left(\frac{D_2}{D_1}\right)^2 \tag{14-22}$$

Performing the same equations on C_Q and C_P, we obtain

$$\frac{\dot{V}_2}{\dot{V}_1} = \frac{\omega_2}{\omega_1} \left(\frac{D_2}{D_1}\right)^3 \tag{14-23}$$

$$\frac{P_2}{P_1} = \frac{\rho_2}{\rho_1} \left(\frac{\omega_2}{\omega_1}\right)^3 \left(\frac{D_2}{D_1}\right)^5 \tag{14-24}$$

These three equations comprise the "fan laws" or "pump laws" that are used extensively for scaling rotating machines as well as predicting their performance.

Example 3 illustrates the use of these expressions.

EXAMPLE 3

A centrifugal pump, operating at 1100 rpm against a head of 120 m H_2O produces a flow of 0.85 m³/s.

(a) For a geometrically similar pump, operating at the same speed but with an impeller diameter 30% greater than the original, what flow rate will be achieved?

(b) If the new larger pump described in part (a) is also operated at 1300 rpm, what will be the new values of flow rate and total head?

Specifying for pump 1, $D = D_1$, then for the larger pump, $D_2 = 1.3\,D_1$, thus the new flow rate will be, using equation (14-23)

$$\frac{\dot{V}_2}{\dot{V}_1} = \frac{\omega_2}{\omega_1} \left(\frac{D_2}{D_1}\right)^3 \tag{14-23}$$

$$\dot{V}_2 = 0.85\,\text{m}^3/\text{s} \left(\frac{1.3\,D_1}{D_1}\right)^3$$

$$= 1.867\,\text{m}^3/\text{s} \tag{a}$$

For the case with $D_2 = 1.3\,D_1$ and $\omega_2 = 1300$ rpm, we have, from equation (14-23)

$$\dot{V}_2 = 0.85\,\text{m}^3/\text{s} \left(\frac{1300\,\text{rpm}}{1100\,\text{rpm}}\right) \left(\frac{1.3\,D_1}{D_1}\right)^3$$

$$= 2.207\,\text{m}^3/\text{s} \tag{b}$$

The new head is determined using equation (14-22).

$$\frac{h_2}{h_1} = \left(\frac{\omega_2}{\omega_1}\right)^2 \left(\frac{D_2}{D_1}\right)^2 \tag{14-22}$$

$$= 120\,\text{m}\,_{H_2O} \left(\frac{1300\,\text{rpm}}{1100\,\text{rpm}}\right)^2 \left(\frac{1.3\,D_1}{D_1}\right)^2$$

$$= 283\,\text{m}\,_{H_2O} \tag{c}$$

14.3   AXIAL AND MIXED FLOW PUMP CONFIGURATIONS

Our examination of pumps, thus far, has focused on centrifugal pumps. The other basic configuration is axial flow. The designation centrifugal flow or axial flow relates to the direction of fluid flow in the pump. In the centrifugal case, flow is turned 90° to the axis of rotation; in the axial flow case, flow is in the direction of the axis of rotation. There is an intermediate case, designated *mixed flow*, where the flow has both normal and axial components.

The choice of centrifugal, axial, or mixed flow configurations depends on the desired values of flow rate and head needed in a specific application. The single parameter that includes both head and flow rate effects is designated N_S, the *specific speed*. It is defined as

$$N_S = \frac{C_Q^{1/2}}{C_H^{3/4}} \tag{14-25}$$

Figure 14.11 is a plot of optimum efficiencies of the three pump types as functions of N_S. The values of N_S shown in this plot correspond to the somewhat unusual units shown.

The basic message conveyed by Figure 14.11 is that higher delivery head and lower flow rate combinations dictate the use of centrifugal pumps, whereas lower heads and higher flow rates require mixed flow or axial flow pumps.

Figure 14.11 Optimum pump efficiency as a function of specific speed.

14.4   TURBINES

Analysis of turbines follows the same general steps as has been done for pumps. The reader is referred to Section 5.4 in Chapter 5 for a review of the analysis of an impulse turbine.

Turbine operation uses the energy of a fluid, emanating from a nozzle, to interact with blades attached to the rotating unit, designated the *rotor*. The momentum exchange produced as the fluid changes direction generates power at the rotor shaft.

A detailed discussion of turbine operation, other than that presented in Chapter 5, is beyond the scope of this book. Numerous treatises are available to the interested reader. A good introductory discussion, along with extensive references, is included in the text by Munson et al. (1998).

14.5   CLOSURE

This chapter has been devoted to the examination of rotating fluid machines. External power applied to pumps and fans produces higher pressure, increased flow, or both. Turbines operate in the reverse, producing power from a high-energy fluid.

Types of pumps or fans are characterized by the direction of flow through the rotor. In centrifugal pumps, the flow is turned 90° to the flow axis; flow is parallel to the flow axis in axial flow pumps. Machines with both centrifugal and axial flow components are designated mixed-flow pumps.

Standard performance plots for a family of geometrically similar pumps or fans show the head, power, efficiency, and NPSH as functions of flow rate for a designated speed of rotation.

Scaling laws were developed using parameters generated from dimensional analysis. The resulting "fan laws" that relate two similar systems are

$$\frac{h_2}{h_1} = \left(\frac{\omega_2}{\omega_1}\right)^2 \left(\frac{D_2}{D_1}\right)^2 \tag{14-22}$$

$$\frac{\dot{V}_2}{\dot{V}_1} = \frac{\omega_2}{\omega_1} \left(\frac{D_2}{D_1}\right)^3 \tag{14-23}$$

$$\frac{P_2}{P_1} = \frac{\rho_2}{\rho_1} \left(\frac{\omega_2}{\omega_1}\right)^3 \left(\frac{D_2}{D_1}\right)^5 \tag{14-24}$$

PROBLEMS

14.1 A centrifugal pump delivers 0.2 m³/s of water when operating at 850 rpm. Relevant impeller dimensions are as follows: outside diameter = 0.45 m, blade length = 50 cm, and blade exit angle = 24°. Determine (a) the torque and power required to drive the pump and (b) the maximum pressure increase across the pump.

14.2 A centrifugal pump is used with gasoline ($\rho = 680$ kg/m³). Relevant dimensions are as follows: $d_1 = 15$ cm, $d_2 = 28$ cm, $L = 9$ cm, $\beta_1 = 25°$, and $\beta_2 = 40°$. The gasoline enters the pump parallel to the pump shaft when the pump operates at 1200 rpm. Determine (a) the flow rate; (b) the power delivered to the gasoline; and (c) the head in meters.

14.3 A centrifugal pump has the following dimensions: $d_2 = 42$ cm, $L = 5$ cm, and $\beta_2 = 33°$. It rotates at 1200 rpm, and the head generated is 52 m of water. Assuming radial entry flow, determine the theoretical values for (a) the flow rate and (b) the power.

14.4 A centrifugal pump has the configuration and dimensions shown below. For water flowing at a rate of 0.0071 m³/s and an impeller speed of 1020 rpm, determine the power required to drive the pump. The inlet flow is directed radially outward, and the exiting velocity may be assumed to be tangent to the vane at its trailing edge.

14.5 A centrifugal pump is being used to pump water at a flow rate of 0.018 m³/s and the required power is measured to be 4.5 kW. If the pump efficiency is 63%, determine the head generated by the pump.

14.6 A centrifugal pump having the dimensions shown develops a flow rate of 0.032 m³/s when pumping gasoline ($\rho = 680$ kg/m³). The inlet flow may be assumed to be radial. Estimate (a) the theoretical horsepower; (b) the head increase; and (c) the proper blade angle at the impeller inlet.

14.7 A centrifugal water pump operates at 1500 rpm. The dimensions are as the following:

$$\begin{array}{ll} r_1 = 12 \text{ cm} & \beta_1 = 32° \\ r_2 = 20 \text{ cm} & \beta_2 = 20° \\ L = 4.2 \text{ cm} & \end{array}$$

Determine (a) the design point discharge rate; (b) the water horsepower; and (c) the discharge head.

14.8 The figure below represents performance, in nondimensional form, for a family of centrifugal pumps. For a pump from this family with a characteristic diameter of 0.45 m operating at maximum efficiency and pumping water at 15°C with a rotational speed of 1600 rpm, estimate (a) the head; (b) the discharge rate; (c) the pressure rise; and (d) the brake horsepower.

14.9 A pump having the characteristics described in the previous problem is to be built that will deliver water at a rate of 0.2 m³/s when operating at best efficiency and a rotational speed of 1400 rpm. Estimate (a) the impeller diameter and (b) the maximum pressure rise.

14.10 Rework Problem 14.8 for a pump diameter of 0.40 m operating at 2200 rpm.

14.11 Rework Problem 14.8 for a pump diameter of 0.35 m operating at 2400 rpm.

14.12 Rework Problem 14.9 for a desired flow rate of 0.30 m³/s at 1800 rpm.

14.13 Rework Problem 14.9 for a desired flow rate of 0.201 m³/s at 1800 rpm.

14.14 Performance curves for an operating centrifugal pump are shown below in both conventional units and in dimensionless form. The pump is used to pump water at maximum efficiency at a head of 90 m. Determine, at these new conditions, (a) the pump speed required and (b) the rate of discharge.

14.15 The pump having the characteristics shown in Problem 14.14 was used as a model for a prototype that is to be six times larger. If this prototype operates at 400 rpm, what (a) power; (b) head; and (c) discharge flow rate should be expected at maximum efficiency?

14.16 For the pump having the characteristics shown in Problem 14.14, operating at maximum efficiency with the speed increased to 1000 rpm, what will be (a) the new discharge flow rate and (b) the power required at this new speed?

14.17 The pump having the characteristics shown in Problem 14.14 is to be operated at 800 rpm. What discharge rate is to be expected if the head developed is 410 m?

14.18 If the pump having the characteristics shown in Problem 14.14 is tripled in size but halved in rotational speed, what will be the discharge rate and head when operating at maximum efficiency?

14.19 The pump having the characteristics shown in Problem 14.14 is used to pump water from one reservoir to another that is 95 m higher in elevation. The water will flow through a steel pipe that is 0.28 m in diameter and 550 m long. Determine the discharge rate.

14.20   A pump whose operating characteristics are described in Problem 14.14 is to be used in the system depicted below. Determine (a) the discharge rate and (b) power required.

14.21   For the same pump and system operation described in Problem determine (a) the discharge rate and (b) power required when the pump operates at 900 rpm.

14.22   Water at 20°C is to be pumped through the system shown. The operating data for this motor-driven pump data are as follows:

Capacity, $m^3/s \times 10^4$	Developed head, m	Efficiency, %
0	36.6	0
10	35.9	19.1
20	34.1	32.9
30	31.2	41.6
40	27.5	42.2
50	23.3	39.7

The inlet pipe to the pump is 0.06 m diameter commercial steel, 8.5 m in length. The discharge line consists of 60 m of 0.06 m diameter steel pipe. All valves are fully open globe valves. Determine the flow rate through the system.

14.23   A 0.25 m pump delivers 20°C water ($P_v = 2.34$ kPa) at 0.065 m^3/s and 2000 rpm. The pump begins to cavitate when the inlet pressure is 82.7 kPa and the inlet velocity is 6.1 m/s. Determine the corresponding NPSH.

14.24   For the pumping system described in Problem 14.23, how will the maximum elevation above the surface of the reservoir change if the water temperature is 80°C($P_v = 47.35$ kPa)?

14.25   A centrifugal pump with an impeller diameter of 0.18 m is to be used to pump water ($\rho = 1000$ kg/m^3) with the pump inlet located 3.8 m above the surface of the supply reservoir. At a flow rate of 0.760 m^3/s, the head loss between the reservoir surface and the pump inlet is 1.80 m of water. The performance curves are shown below. Would you expect cavitation to occur?

14.26   Pumps used in an aqueduct operate at 400 rpm and deliver a flow of 220 m^3/s against a head of 420 m. What types of pumps are they?

14.27   A pump is required to deliver 60,000 gpm against a head of 300 m when operating at 2000 rpm. What type of pump should be specified?

14.28   An axial flow pump has a specified specific speed of 6.0. The pump must deliver 2400 gpm against a head of 18 m. Determine the required operating rpm of the pump.

14.29   A pump operating at 520 rpm has the capability of producing 3.3 m^3/s of water flow against a head of 16 m. What type of pump is this?

14.30   A pump operating at 2400 rpm delivers 3.2 m^3/s of water against a head of 21 m. Is this pump an axial flow, mixed flow, or radial flow machine?

Chapter **15**

Fundamentals of Heat Transfer

The next nine chapters deal with the transfer of energy. Gross quantities of heat added to or rejected from a system may be evaluated by applying the control-volume expression for the first law of thermodynamics as discussed in Chapter 6. The result of a first-law analysis is only a part of the required information necessary for the complete evaluation of a process or situation that involves energy transfer. The overriding consideration is, in many instances, the rate at which energy transfer takes place. Certainly, in designing a plant in which heat must be exchanged with the surroundings, the size of heat-transfer equipment, the materials of which it is to be constructed, and the auxiliary equipment required for its utilization are all important considerations for the engineer. Not only must the equipment accomplish its required mission but it must also be economical to purchase and to operate.

Considerations of an engineering nature such as these require both a familiarity with the basic mechanisms of energy transfer and an ability to evaluate quantitatively these rates as well as the important associated quantities. Our immediate goal is to examine the basic mechanisms of energy transfer and to consider the fundamental equations for evaluating the rate of energy transfer.

There are three modes of energy transfer: conduction, convection, and radiation. All heat-transfer processes involve one or more of these modes. The remainder of this chapter will be devoted to an introductory description and discussion of these types of transfer.

15.1 CONDUCTION

Energy transfer by conduction is accomplished in two ways. The first mechanism is that of molecular interaction, in which the greater motion of a molecule at a higher energy level (temperature) imparts energy to adjacent molecules at lower energy levels. This type of transfer is present, to some degree, in all systems in which a temperature gradient exists and in which molecules of a solid, liquid, or gas are present.

The second mechanism of conduction heat transfer is by "free" electrons. The free-electron mechanism is significant primarily in pure-metallic solids; the concentration of free electrons varies considerably for alloys and becomes very low for nonmetallic solids. The ability of solids to conduct heat varies directly with the concentration of free electrons, thus it is not surprising that pure metals are the best heat conductors, as our experience has indicated.

As heat conduction is primarily a molecular phenomenon, we might expect the basic equation used to describe this process to be similar to the expression used in the molecular transfer of momentum, equation (7-4). Such an equation was first stated in 1822 by Fourier in the form

$$\frac{q_x}{A} = -k\frac{dT}{dx} \qquad (15\text{-}1)$$

where q_x is the heat-transfer rate in the x direction, in Watts or Btu/h; A is the area *normal* to the direction of heat flow, in m^2 or ft^2; dT/dx is the temperature gradient in the x direction, in K/m or °F/ft; and k is the thermal conductivity, in W/(mK) or Btu/h ft° F. The ratio q_x/A, having the dimensions of W/m^2 or Btu/h ft^2, is referred to as the heat flux in the x direction. A more general relation for the heat flux is equation (15-2)

$$\frac{\mathbf{q}}{A} = -k\mathbf{\nabla}T \qquad (15\text{-}2)$$

which expresses the heat flux as proportional to the temperature gradient. The proportionality constant is seen to be the thermal conductivity, which plays a role similar to that of the viscosity in momentum transfer. The negative sign in equation (15-2) indicates that heat flow is in the direction of a negative temperature gradient. equation (15-2) is the vector form of the *Fourier rate equation,* often referred to as Fourier's first law of heat conduction.

The thermal conductivity, k, which is defined by equation (15-1), is assumed independent of direction in equation (15-2); thus, this expression applies to an *isotropic* medium only. Most materials of engineering interest are isotropic. Wood is a good example of an *anisotropic* material where the thermal conductivity parallel to the grain may be greater than that normal to the grain by a factor of 2 or more. The thermal conductivity is a property of a conducting medium and, like the viscosity, is primarily a function of temperature, varying significantly with pressure only in the case of gases subjected to high pressures.

15.2 THERMAL CONDUCTIVITY

As the mechanism of conduction heat transfer is one of the molecular interaction, it will be illustrative to examine the motion of gas molecules from a standpoint similar to that in Section 7.3.

Considering the control volume shown in Figure 15.1, in which energy transfer in the y direction is on a molecular scale only, we may utilize the first-law analysis of Chapter 6 as follows. Mass transfer across the top of this control volume is considered to occur only on the molecular scale. This criterion is met for a gas in laminar flow.

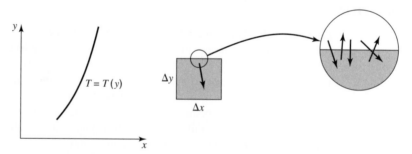

Figure 15.1 Molecular motion at the surface of a control volume.

Applying equation (6-10) and considering transfer only across the top face of the element considered

$$\frac{\delta Q}{dt} - \frac{\delta W_s}{dt} - \frac{\delta W_\mu}{dt} = \iint_{\text{c.s.}} \left(e + \frac{P}{\rho}\right)\rho(\mathbf{v} \cdot \mathbf{n})\, dA + \frac{\partial}{\partial t}\iiint_{\text{c.v.}} e\rho\, dV \qquad (6\text{-}10)$$

For Z molecules crossing the plane $\Delta x \, \Delta z$ per unit time, this equation reduces to

$$q_y = \sum_{n=1}^{Z} m_n c_p (T|_{y-} - T|_{y+}) \Delta x \, \Delta z \qquad (15\text{-}3)$$

where m_n is the mass per molecule; c_p is the molecular heat capacity of the gas; Z is the frequency with which molecules will cross area $\Delta x \, \Delta z$; and $T|_{y-}, \, - T|_{y+}$ are the temperatures of the gas slightly below and slightly above the plane considered, respectively. The right-hand term is the summation of the energy flux associated with the molecules crossing the control surface. Noting now that $T|_{y-} = T - T/y|_{y_0} \, \delta$, where $y- = y_0 - \delta$, and that a similar expression may be written for $T|_{y+}$, we may rewrite equation (15-3) in the form

$$\frac{q_y}{A} = -2 \sum_{n=1}^{Z} m_n c_p \delta \frac{T}{y}\bigg|_{y_0} \qquad (15\text{-}4)$$

where δ represents the y component of the distance between collisions. We note, as previously in Chapter 7, that $\delta = (\frac{2}{3})\lambda$, where λ is the mean free path of a molecule. Using this relation and summing over Z molecules, we have

$$\frac{q_y}{A} = -\frac{4}{3} \rho c_p Z \lambda \frac{T}{y}\bigg|_{y_0} \qquad (15\text{-}5)$$

Comparing equation (15-5) with the y component of equation (15-6)

$$\frac{q_y}{A} = -k \frac{T}{y}$$

it is apparent that the thermal conductivity, k, becomes

$$k = \frac{4}{3} \rho c_p Z \lambda$$

Utilizing further the results of the kinetic theory of gases, we may make the following substitutions:

$$Z = \frac{N\bar{C}}{4}$$

where $\bar{C}$ is the average random molecular velocity, $\bar{C} = \sqrt{8\kappa T/\pi m}$ (κ being the Boltzmann constant);

$$\lambda = \frac{1}{\sqrt{2}\pi N d^2}$$

where d is the molecular diameter; and

$$c_p = \frac{3}{2}\frac{\kappa}{N}$$

giving, finally

$$k = \frac{1}{\pi^{3/2}d^2}\sqrt{\kappa^3 T/m} \qquad (15\text{-}6)$$

This development, applying specifically to monatomic gases, is significant in that it shows the thermal conductivity of a gas to be independent of pressure, and to vary as the 1/2 power of the absolute temperature. The significance of this result should not be overlooked, even though some oversimplifications were used in its development. Some relations for thermal conductivity of gases, based upon more sophisticated molecular models, may be found in Bird, Stewart, and Lightfoot.[1]

The Chapman–Enskog theory used in Chapter 7 to predict gas viscosities at low pressures has a heat-transfer counterpart. For a monatomic gas, the recommended equation is

$$k = 0.0829\sqrt{(T/M)}/\sigma^2\Omega_k \tag{15-7}$$

where k is in W/m $\cdot$ K, σ is in Angstroms, M is the molecular weight, and Ω_k is the Lennard–Jones collision integral, identical with Ω as discussed in Section 7.3. Both σ and Ω_k may be evaluated from Appendices J and K.

The thermal conductivity of a liquid is not amenable to any simplified kinetic-theory development, as the molecular behavior of the liquid phase is not clearly understood and no universally accurate mathematical model presently exists. Some empirical correlations have met with reasonable success, but these are so specialized that they will not be included in this book. For a discussion of molecular theories related to the liquid phase and some empirical correlations of thermal conductivities of liquids, the reader is referred to Reid and Sherwood.[2] A general observation about liquid thermal conductivities is that they vary only slightly with temperature and are relatively independent of pressure. One problem in experimentally determining values of the thermal conductivity in a liquid is making sure the liquid is free of convection currents.

In the solid phase, thermal conductivity is attributed both to molecular interaction, as in other phases, and to free electrons, which are present primarily in pure metals. The solid phase is amenable to quite precise measurements of thermal conductivity, as there is no problem with convection currents. The thermal properties of most solids of engineering interest have been evaluated, and extensive tables and charts of these properties, including thermal conductivity, are available.

The free-electron mechanism of heat conduction is directly analogous to the mechanism of electrical conduction. This realization led Wiedemann and Franz, in 1853, to relate the two conductivities in a crude way; and in 1872, Lorenz[3] presented the following relation, known as the Wiedemann, Franz, Lorenz equation:

$$L = \frac{k}{k_eT} = \text{constant} \tag{15-8}$$

where k is the thermal conductivity, k_e is the electrical conductivity, T is the absolute temperature, and L is the Lorenz number.

The numerical values of the quantities in equation (15-8) are of secondary importance at this time. The significant point to note here is the simple relation between electrical and thermal conductivities and, specifically, that those materials that are good conductors of electricity are likewise good heat conductors, and vice versa.

Figure 15.2 illustrates the thermal conductivity variation with temperature of several important materials in gas, liquid, and solid phases. A more complete tabulation of thermal conductivity may be found in Appendices H and I.

The following two examples illustrate the use of the Fourier rate equation in solving simple heat-conduction problems.

---

[1] R. B. Bird, W. E. Stewart, and E. N. Lightfoot, *Transport Phenomena*, Wiley, New York, 1960, chap. 8.

[2] Reid and Sherwood, *The Properties of Gases and Liquids*, McGraw-Hill Book Company, New York, 1958, chap. 7.

[3] L. Lorenz, *Ann. Physik und Chemie* (Poggendorffs), **147**, 429 (1872).

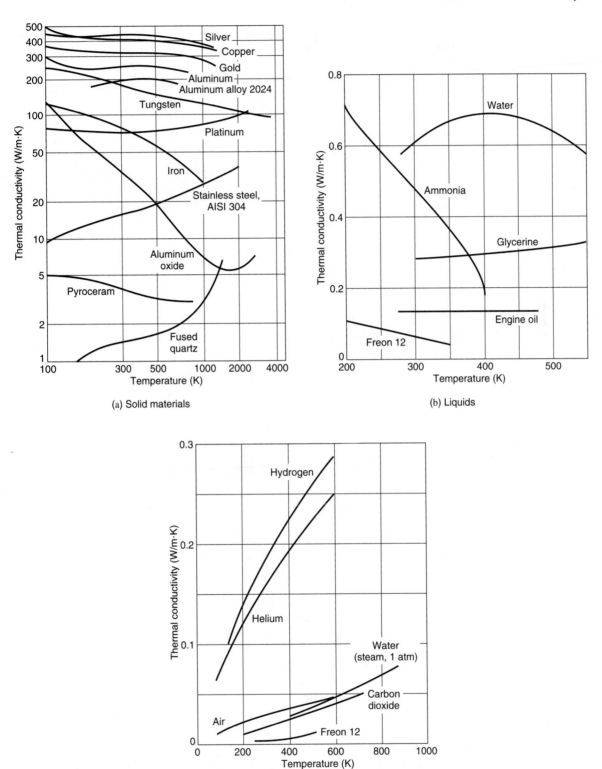

(a) Solid materials

(b) Liquids

(c) Gases and vapors

Figure 15.2 Thermal conductivity of several materials at various temperatures.

EXAMPLE 1

A steel pipe having an inside diameter of 1.88 cm and a wall thickness of 0.391 cm is subjected to inside and outside surface temperature of 367 and 344 K, respectively (see Figure 15.3). Find the heat flow rate per meter of pipe length, and also the heat flux based on both the inside and outside surface areas.

The first law of thermodynamics applied to this problem will reduce to the form $\delta Q/dt = 0$, indicating that the rate of heat transfer into the control volume is equal to the rate leaving, that is $Q = q =$ constant.

As the heat flow will be in the radial direction, the independent variable is r, and the proper form for the Fourier rate equation is

$$q_r = -kA\frac{dT}{dr}$$

Figure 15.3 Heat conduction in a radial direction with uniform surface temperatures.

Writing $A = 2\pi L$, we see that the equation becomes

$$q_r = -k(2\pi rL)\frac{dT}{dr}$$

where q_r, is constant, which may be separated and solved as follows:

$$q_r \int_{r_i}^{r_o} \frac{dr}{r} = -2\pi kL \int_{T_i}^{T_o} dT = 2\pi kL \int_{T_o}^{T_i} dT$$

$$q_r \ln \frac{r_o}{r_i} = 2\pi kL(T_i - T_o) \tag{15-9}$$

$$q_r = \frac{2\pi kL}{\ln r_o/r_i}(T_i - T_o)$$

Substituting the given numerical values, we obtain

$$q_r = \frac{2\pi(42.90\,\text{W/m} \cdot \text{K})(367 - 344)\text{K}}{\ln(2.66/1.88)}$$

$$= 17\,860\,\text{W/m}(18\,600\,\text{Btu/hr} \cdot \text{ft})$$

The inside and outside surface areas per unit length of pipe are giving

$$A_i = \pi(1.88)(10^{-2})(1) = 0.059\,\text{m}^2/\text{m}(0.194\,\text{ft}^2/\text{ft})$$

$$A_o = \pi(2.662)(10^{-2})(1) = 0.084\,\text{m}^2/\text{m}(0.275\,\text{ft}^2/\text{ft})$$

$$\frac{q_r}{A_i} = \frac{17\,860}{0.059} = 302.7\,\text{kW/m}^2(95\,900\,\text{Btu/hr} \cdot \text{ft}^2)$$

$$\frac{q_o}{A_i} = \frac{17\,860}{0.084} = 212.6\,\text{kW/m}^2(67\,400\,\text{Btu/hr} \cdot \text{ft}^2)$$

One extremely important point to be noted from the results of this example is the requirement of specifying the area upon which a heat-flux value is based. Note that for the same amount of heat flow, the fluxes based upon the inside and outside surface areas differ by approximately 42%.

EXAMPLE 2

Consider a hollow cylindrical heat-transfer medium having inside and outside radii of r_i and r_o with the corresponding surface temperatures T_i and T_o. If the thermal-conductivity variation may be described as a linear function of temperature according to

$$k = k_0(1 + \beta T)$$

calculate the steady-state heat-transfer rate in the radial direction, using the above relation for the thermal conductivity, and compare the result with that using a k value calculated at the arithmetic mean temperature.

Figure 15.3 applies. The equation to be solved is now

$$q_r = -[k_o(1 + \beta T)](2\pi r L)\frac{dT}{dr}$$

which, upon separation and integration, becomes

$$q_r \int_{r_i}^{r_o} \frac{dr}{r} = -2\pi k_o L \int_{r_i}^{r_o} (1 + \beta T)dT$$

$$= 2\pi k_o L \int_{T_o}^{T_i} (1 + \beta T)dT$$

$$q_r = \frac{2\pi k_o L}{\ln r_o/r_i}\left[T + \frac{\beta T^2}{2}\right]_{T_o}^{T_i}$$

$$q_r = \frac{2\pi k_o L}{\ln r_o/r_i}\left[1 + \frac{\beta}{2}(T_i + T_o)\right](T_i - T_o) \tag{15-10}$$

Noting that the arithmetic average value of k would be

$$k_{avg} = k_o\left[1 + \frac{\beta}{2}(T_i + T_o)\right]$$

we see that equation (15-10) could also be written as

$$q_r = \frac{2\pi k_{avg} L}{\ln r_o/r_i}(T_i - T_o)$$

Thus, the two methods give identical results.

The student may find it instructive to determine what part of the problem statement of this example is responsible for this interesting result; that is, whether a different geometrical configuration or a different thermal-conductivity expression would make the results of the two types of solutions different.

15.3 CONVECTION

Heat transfer due to convection involves the energy exchange between a surface and an adjacent fluid. A distinction must be made between *forced convection,* wherein a fluid is made to flow past a solid surface by an external agent such as a fan or pump, and *free* or *natural convection* wherein warmer (or cooler) fluid next to the solid boundary causes circulation because of the density difference resulting from the temperature variation throughout a region of the fluid.

The rate equation for convective heat transfer was first expressed by Newton in 1701, and is referred to as the *Newton rate equation* or Newton's "law" of cooling. This equation is

$$q/A = h \, \Delta T \tag{15-11}$$

where q is the rate of convective heat transfer, in W or Btu/h; A is the area normal to direction of heat flow, in m^2 or ft^2; ΔT is the temperature difference between surface and fluid, in K or °F; and h is the convective heat transfer coefficient, in $W/m^2 \cdot K$ or Btu/h ft^2 °F. Equation (15-11) is not a law but a definition of the coefficient h. A substantial portion of our work in the chapters to follow will involve the determination of this coefficient. It is, in general, a function of system geometry, fluid and flow properties, and the magnitude of ΔT.

As flow properties are so important in the evaluation of the convective heat transfer coefficient, we may expect many of the concepts and methods of analysis introduced in the preceding chapters to be of continuing importance in convective heat transfer analysis; this is indeed the case.

From our previous experience we should also recall that even when a fluid is flowing in a turbulent manner past a surface, there is still a layer, sometimes extremely thin, close to the surface where flow is laminar; also, the fluid particles next to the solid boundary are at rest. As this is always true, the mechanism of heat transfer between a solid surface and a fluid must involve conduction through the fluid layers close to the surface. This "film" of fluid often presents the controlling resistance to convective heat transfer, and the coefficient h is often referred to as the *film coefficient*.

Two types of heat transfer that differ somewhat from free or forced convection but are still treated quantitatively by equation (15-11) are the phenomena of boiling and condensation. The film coefficients associated with these two kinds of transfer are quite high. Table 15.1 represents some order-of-magnitude values of h for different convective mechanisms.

Table 15.1 Approximate values of the convective heat-transfer coefficient

Mechanism	h, Btu/h ft^2 °F	h, $W/(m^2 \cdot K)$
Free convection, air	1–10	5–50
Forced convection, air	5–50	25–250
Forced convection, water	50–3000	250–15,000
Boiling water	500–5000	2500–25,000
Condensing water vapor	1000–20,000	5000–100,000

It will also be necessary to distinguish between local heat transfer coefficients, that is, those that apply at a point, and total or average values of h that apply over a given surface area. We will designate the local coefficient h_x, according to equation (15-11)

$$dq = h_x \, \Delta T \, dA$$

Thus the average coefficient, h, is related to h_x according to the relation

$$q = \int_A h_x \, \Delta T \, dA = hA \, \Delta T \tag{15-12}$$

The values given in Table 15.1 are average convective heat-transfer coefficients.

15.4 RADIATION

Radiant heat transfer between surfaces differs from conduction and convection in that no medium is required for its propagation; indeed energy transfer by radiation is maximum when the two surfaces that are exchanging energy are separated by a perfect vacuum.

The rate of energy emission from a perfect radiator or *blackbody* is given by

$$\frac{q}{A} = \sigma T^4 \tag{15-13}$$

where q is the rate of radiant energy emission, in W or Btu/h; A is the area of the emitting surface, in m^2 or ft^2; T is the absolute temperature, in K or °R; and σ is the Stefan–Boltzmann constant, which is equal to 5.676×10^{-8} W/$m^2 \cdot K^4$ or 0.1714×10^{-8} Btu/h ft^2 °R^4. The proportionally constant relating radiant-energy flux to the fourth power of the absolute temperature is named after Stefan who, from experimental observations, proposed equation (15-13) in 1879, and Boltzmann, who derived this relation theoretically in 1884. Equation (15-13) is most often referred to as the Stefan–Boltzmann law of thermal radiation.

Certain modifications will be made in equation (15-13) to account for the *net* energy transfer between two surfaces, the degree of deviation of the emitting and receiving surfaces from blackbody behavior, and geometrical factors associated with radiant exchange between a surface and its surroundings. These considerations are discussed at length in Chapter 23.

15.5 COMBINED MECHANISMS OF HEAT TRANSFER

The three modes of heat transfer have been considered separately in Section 15.4. It is rare, in actual situations, for only one mechanism to be involved in the transfer of energy. It will be instructive to look at some situations in which heat transfer is accomplished by a combination of these mechanisms.

Consider the case depicted in Figure 15.4, that of steady-state conduction through a plane wall with its surfaces held at constant temperatures T_1 and T_2.

Writing the Fourier rate equation for the x direction, we have

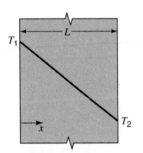

Figure 15.4 Steady-state conduction through a plane wall.

$$\frac{q_x}{A} = -k\frac{dT}{dx} \tag{15-1}$$

Solving this equation for q_x subject to the boundary conditions $T = T_1$ at $x = 0$ and $T = T_2$ at $x = L$, we obtain

$$\frac{q_x}{A}\int_0^L dx = -k\int_{T_1}^{T_2} dT = k\int_{T_2}^{T_1} dT$$

or

$$q_x = \frac{kA}{L}(T_1 - T_2) \tag{15-14}$$

Equation (15-14) bears an obvious resemblance to the Newton rate equation

$$q_x = hA\,\Delta T \tag{15-11}$$

We may utilize this similarity in form in a problem in which both types of energy transfer are involved.

Consider the composite plane wall constructed of three materials in layers with dimensions as shown in Figure 15.5. We wish to express the steady-state heat-transfer rate per unit area between a hot gas at temperature T_h on one side of this wall and a cool gas at

T_c on the other side. Temperature designations and dimensions are as shown in the figure. The following relations for q_x arise from the application of equations (15-11) and (15-14):

$$q_x = h_h A(T_h - T_1) = \frac{k_1 A}{L_1}(T_1 - T_2) = \frac{k_2 A}{L_2}(T_2 - T_3)$$

$$= \frac{k_3 A}{L_3}(T_3 - T_4) = h_c A(T_4 - T_c)$$

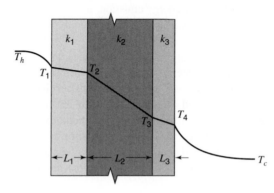

Figure 15.5 Steady-state heat transfer through a composite wall.

Each temperature difference is expressed in terms of q_x as follows:

$$T_h - T_1 = q_x(1/h_h A)$$
$$T_1 - T_2 = q_x(L_1/k_1 A)$$
$$T_2 - T_3 = q_x(L_2/k_2 A)$$
$$T_3 - T_4 = q_x(L_3/k_3 A)$$
$$T_4 - T_c = q_x(1/h_c A)$$

Adding these equations, we obtain

$$T_h - T_c = q_x \left(\frac{1}{h_h A} + \frac{L_1}{k_1 A} + \frac{L_2}{k_2 A} + \frac{L_3}{k_3 A} + \frac{1}{h_c A} \right)$$

and finally, solving for q_x, we have

$$q_x = \frac{T_h - T_c}{1/h_h A + L_1/k_1 A + L_2/k_2 A + L_3/k_3 A + 1/h_c A} \tag{15-15}$$

Note that the heat-transfer rate is expressed in terms of the *overall* temperature difference. If a series electrical circuit

is considered, we may write

$$I = \frac{\Delta V}{R_1 + R_2 + R_3 + R_4 + R_5} = \frac{\Delta V}{\sum R_i}$$

The analogous quantities in the expressions for heat flow and electrical current are apparent,

$$\Delta V \rightarrow \Delta T$$

$$I \rightarrow q_x$$

$$R_i \rightarrow 1/hA, L/kA$$

and each term in the denominator of equation (15-15) may be thought of as a thermal resistance due to convection or conduction. Equation (15-15) thus becomes a heat-transfer analog to Ohm's law, relating heat flow to the overall temperature difference divided by the total thermal resistance between the points of known temperature. Equation (15-15) may now be written simply as

$$q = \frac{\Delta T}{\sum R_{\text{thermal}}} \tag{15-16}$$

This relation applies to steady-state heat transfer in systems of other geometries as well. The thermal-resistance terms will change in form for cylindrical or spherical systems, but once evaluated, they can be utilized in the form indicated by equation (15-16). With specific reference to equation (15-9), it may be noted that the thermal resistance of a cylindrical conductor is

$$\frac{\ln(r_o/r_i)}{2\pi kL}$$

Another common way of expressing the heat-transfer rate for a situation involving a composite material or combination of mechanisms is with the *overall heat-transfer coefficient* defined as

$$U \equiv \frac{q_x}{A \, \Delta T} \tag{15-17}$$

where U is the overall heat-transfer coefficient having the same units as h, in $W/m^2 \cdot K$ or $Btu/h\,ft^2\,°F$.

EXAMPLE 3

Saturated steam at 0.276 MPa flows inside a steel pipe having an inside diameter of 2.09 cm and an outside diameter of 2.67 cm. The convective coefficients on the inner and outer pipe surfaces may be taken as 5680 and 22.7 $W/m^2 \cdot K$, respectively. The surrounding air is at 294 K. Find the heat loss per meter of bare pipe and for a pipe having a 3.8 cm thickness of 85% magnesia insulation on its outer surface.

In the case of the bare pipe there are three thermal resistances to evaluate:

$$R_1 = R_{\text{convection inside}} = 1/h_i A_i$$

$$R_2 = R_{\text{convection outside}} = 1/h_o A_{\bar{o}}$$

$$R_3 = R_{\text{conduction}} = \ln(r_{\bar{o}}/r_i)/2\pi kL$$

For conditions of this problem, these resistances have the values

$$R_1 = 1/[(5680 \, W/m^2 \cdot K)(\pi)(0.0209 \, m)(1 \, m)]$$

$$= 0.00268 \, K/W \quad \left(0.00141 \frac{h \, °R}{Btu}\right)$$

$$R_2 = 1/[(22.7 \, W/m^2 \cdot K)(\pi)(0.0267 \, m)(1 \, m)]$$

$$= 0.525 \, K/W \quad \left(0.277 \frac{h \, °R}{Btu}\right)$$

and

$$R_3 = \frac{\ln(2.67/2.09)}{2\pi(42.9\,\text{W/m}\cdot\text{K})(1\,\text{m})}$$

$$= 0.00091\,\text{K/W} \quad \left(0.00048\,\frac{\text{h}\,^\circ\text{R}}{\text{Btu}}\right)$$

The inside temperature is that of 0.276 MPa saturated steam, 404 K or 267°F. The heat transfer rate per meter of pipe may now be calculated as

$$q = \frac{\Delta T}{\sum R} = \frac{404 - 294\,\text{K}}{0.528\,\text{K/W}}$$

$$= 208\,\text{W} \quad \left(710\,\frac{\text{Btu}}{\text{h}}\right)$$

In the case of an insulated pipe, the total thermal resistance would include R_1 and R_3 evaluated above, plus additional resistances to account for the insulation. For the insulation

$$R_4 = \frac{\ln(10.27/2.67)}{2\pi(0.0675\,\text{W/m}\cdot\text{K})(1\,\text{m})}$$

$$= 3.176\,\text{K/W} \quad \left(1.675\,\frac{\text{h}\,^\circ\text{R}}{\text{Btu}}\right)$$

and for the outside surface of the insulation

$$R_5 = 1/[(22.7\,\text{W/m}^2\cdot\text{K})(\pi)(0.1027\,\text{m})(1\,\text{m})]$$

$$= 0.1365\,\text{K/W} \quad \left(0.0720\,\frac{\text{h}\,^\circ\text{R}}{\text{Btu}}\right)$$

Thus, the heat loss for the insulated pipe becomes

$$q = \frac{\Delta T}{\sum R} = \frac{\Delta T}{R_1 + R_2 + R_4 + R_5} = \frac{404 - 294\,\text{K}}{3.316\,\text{K/W}}$$

$$= 33.2\,\text{W} \quad \left(113\,\frac{\text{Btu}}{\text{h}}\right)$$

A reduction of approximately 85%!

It is apparent from this example that certain parts of the heat-transfer path offer a negligible resistance. If, for instance, in the case of the bare pipe, an increased rate of heat transfer were desired, the obvious approach would be to alter the outside convective resistance, which is almost 200 times the magnitude of the next-highest thermal-resistance value.

Example 3 could also have been worked by using an *overall heat-transfer coefficient*, which would be, in general

$$U = \frac{q_x}{A\,\Delta T} = \frac{\Delta T/\sum R}{A\,\Delta T} = \frac{1}{A\sum R}$$

or, for the specific case considered

$$U = \frac{1}{A\{1/A_i h_i + [\ln(r_o/r_i)]/2\pi kL + 1/A_o h_o\}} \tag{15-18}$$

Equation (15-18) indicates that the overall heat-transfer coefficient, U, may have a different numerical value, depending on which area it is based upon. If, for instance, U is based upon the outside surface area of the pipe, A_o, we have

$$U_o = \frac{1}{A_o/A_i h_i + [A_o \ln(r_o/r_i)]/2\pi kL + 1/h_o}$$

Thus, it is necessary, when specifying an overall coefficient, to relate it to a specific area.

One other means of evaluating heat-transfer rates is by means of the *shape factor,* symbolized as S. Considering the steady-state relations developed for plane and cylindrical shapes

$$q = \frac{kA}{L}\Delta T \tag{15-14}$$

and

$$q = \frac{2\pi kL}{\ln(r_o/r_i)}\Delta T \tag{15-9}$$

if that part of each expression having to do with the geometry is separated from the remaining terms, we have, for a plane wall,

$$q = k\left(\frac{A}{L}\right)\Delta T$$

and for a cylinder

$$q = k\left(\frac{2\pi L}{\ln(r_o/r_i)}\right)\Delta T$$

Each of the bracketed terms is the shape factor for the applicable geometry. A general relation utilizing this form is

$$q = kS\,\Delta T \tag{15-19}$$

Equation (15-19) offers some advantages when a given geometry is required because of space and configuration limitations. If this is the case, then the shape factor may be calculated and q determined for various materials displaying a range of values of k.

15.6 CLOSURE

In this chapter, the basic modes of energy transfer—conduction, convection, and radiation—have been introduced, along with the simple relations expressing the rates of energy transfer associated therewith. The transport property, thermal conductivity, has been discussed and some consideration given to energy transfer in a monatomic gas at low pressure.

The rate equations for heat transfer are as follows:

Conduction: the Fourier rate equation

$$\frac{\mathbf{q}}{A} = -k\,\nabla T$$

Convection: the Newton rate equation

$$\frac{q}{A} = h\,\Delta T$$

Radiation: the Stefan–Boltzmann law for energy emitted from a black surface

$$\frac{q}{A} = \sigma T^4$$

Combined modes of heat transfer were considered, specifically with respect to the means of calculating heat-transfer rates when several transfer modes were involved. The three ways of calculating steady-state heat-transfer rates are represented by the equations

$$q = \frac{\Delta T}{\sum R_T} \tag{15-16}$$

where $\sum R_T$ is the total thermal resistance along the transfer path;

$$q = UA \, \Delta T \tag{15-17}$$

where U is the overall heat transfer coefficient; and

$$q = kS \, \Delta T \tag{15-19}$$

where S is the shape factor.

The equations presented will be used throughout the remaining chapters dealing with energy transfer. A primary object of the chapters to follow will be the evaluation of the heat-transfer rates for special geometries or conditions of flow, or both.

Note: Effects of thermal radiation are included, along with convection, in values of surface coefficients specified in the following problems.

PROBLEMS

15.1 An asbestos pad is square in cross section, measuring 5 cm on a side at its small end increasing linearly to 10 cm on a side at the large end. The pad is 15 cm high. If the small end is held at 600 K and the large end at 300 K, what heat-flow rate will be obtained if the four sides are insulated? Assume one-dimensional heat conduction. The thermal conductivity of asbestos may be taken as 0.173 W/m·K.

15.2 Solve Problem for the case of the larger cross section exposed to the higher temperature and the smaller end held at 300 K.

15.3 Solve Problem 15.1 if, in addition to a varying cross-sectional area, the thermal conductivity varies according to $k = k_0(1 + \beta T)$, where $k_0 = 0.138$, $\beta = 1.95 \times 10^{-4}$, $T =$ temperature in Kelvin, and k is in W/m · K. Compare this result to that using a k value evaluated at the arithmetic mean temperature.

15.4 Solve Problem 15.1 if the asbestos pad has a 1.905-cm steel bolt running through its center.

15.5 A sheet of insulating material, with thermal conductivity of 0.22 W/m · K is 2 cm thick and has a surface area of 2.97 m². If 4 kW of heat are conducted through this sheet and the outer (cooler) surface temperature is measured at 55°C(328 K), what will be the temperature on the inner (hot) surface?

15.6 For the sheet of insulation specified in Problem 15.5, with a heat rate of 4 kW, evaluate the temperature at both surfaces if

the cool side is exposed to air at 30°C with a surface coefficient of 28.4 W/m² · K.

15.7 Plate glass, $k = 1.35$ W/m · K, initially at 850 K, is cooled by blowing air past both surfaces with an effective surface coefficient of 5 W/m² · K. It is necessary, in order that the glass does not crack, to limit the maximum temperature gradient in the glass to 15 K/mm during the cooling process. At the start of the cooling process, what is the lowest temperature of the cooling air that can be used?

15.8 Solve Problem 15.7 if all specified conditions remain the same but radiant energy exchange from glass to the surroundings at the air temperature is also considered.

15.9 The heat loss from a boiler is to be held at a maximum of 900 Btu/h ft² of wall area. What thickness of asbestos ($k = 0.10$ Btu/h ft °F) is required if the inner and outer surfaces of the insulation are to be 1600 and 500°F, respectively?

15.10 If, in the previous problem, a 3-in.-thick layer of kaolin brick ($k = 0.07$ Btu/h ft °F) is added to the outside of the asbestos, what heat flux will result if the outside surface of the kaolin is 250°F? What will be the temperature at the interface between the asbestos and kaolin for this condition?

15.11 A composite wall is to be constructed of 1/4-in. stainless steel ($k = 10$ Btu/h ft °F), 3 in. of corkboard ($k = 0.025$ Btu/h ft °F) and 1/2 in. of plastic ($k = 1.5$ Btu/h ft °F).

a. Draw the thermal circuit for the steady-state conduction through this wall.

b. Evaluate the individual thermal resistance of each material layer.

c. Determine the heat flux if the steel surface is maintained at 250°F and the plastic surface held at 80°F.

d. What are the temperatures on each surface of the corkboard under these conditions?

15.12 If, in the previous problem, the convective heat-transfer coefficients at the inner (steel) and outer surfaces are 40 and 5 Btu/h ft °F, respectively, determine

a. the heat flux if the gases are at 250 and 70°F, adjacent to the inner and outer surfaces;

b. the maximum temperature reached within the plastic;

c. which of the individual resistances is controlling.

15.13 A 1-in.-thick steel plate measuring 10 in. in diameter is heated from below by a hot plate, its upper surface exposed to air at 80°F. The heat-transfer coefficient on the upper surface is 5 Btu/h ft °F and k for steel is 25 Btu/h ft °F.

a. How much heat must be supplied to the lower surface of the steel if its upper surface remains at 160°F? (Include radiation.)

b. What are the relative amounts of energy dissipated from the upper surface of the steel by convection and radiation?

15.14 If, in Problem 15.13, the plate is made of asbestos, $k = 0.10$ Btu/h ft °F, what will be the temperature of the top of the asbestos if the hot plate is rated at 800 W?

15.15 A 0.20-m-thick brick wall ($k = 1.3$ W/m · K) separates the combustion zone of a furnace from its surroundings at 25°C. For an outside wall surface temperature of 100°C, with a convective heat transfer coefficient of 18 W/m² · K, what will be the inside wall surface temperature at steady-state conditions?

15.16 Solve for the inside surface temperature of the brick wall described in Problem 15.15, but with the additional consideration of radiation from the outside surface to surroundings at 25°C.

15.17 The solar radiation incident on a steel plate 2 ft square is 400 Btu/h. The plate is 1.4 in. thick and lying horizontally on an insulating surface, its upper surface being exposed to air at 90°F. If the convective heat-transfer coefficient between the top surface and the surrounding air is 4 Btu/h ft°F, what will be the steady-state temperature of the plate?

15.18 If in Problem 15.17, the lower surface of the plate is exposed to air with a convective heat transfer coefficient of 3 Btu/h ft °F, what steady-state temperature will be reached

a. if radiant emission from the plate is neglected;

b. if radiant emission from the top surface of the plate is accounted for?

15.19 The freezer compartment in a conventional refrigerator can be modeled as a rectangular cavity 0.3 m high and 0.25 m wide with a depth of 0.5 m. Determine the thickness of styrofoam insulation ($k = 0.30$ W/m · K) needed to limit the heat loss to 400 W if the inner and outer surface temperatures are −10 and 33°C, respectively.

15.20 Evaluate the required thickness of styrofoam for the freezer compartment in the previous problem when the inside wall is exposed to air at −10°C through a surface coefficient of 16 W/m² · K and the outer wall is exposed to 33°C air with a surface coefficient of 32 W/m² · K. Determine the surface temperatures for this situation.

15.21 The cross section of a storm window is shown in the sketch. How much heat will be lost through a window measuring 1.83 m by 3.66 m on a cold day when the inside and outside air temperatures are, respectively, 295 and 250 K? Convective coefficients on the inside and outside surfaces of the window are 20 and 15 W/m² · K, respectively. What temperature drop will exist across each of the glass panes? What will be the average temperature of the air between the glass panes?

Window glass 0.32 cm thick

15.22 Compare the heat loss through the storm window described in Problem 15.21 with the same conditions existing except that the window is a single pane of glass 0.32 cm thick.

15.23 The outside walls of a house are constructed of a 4-in. layer of brick, 1/2 in. of celotex, an air space 3 5/8 in. thick, and 1/4 in. of wood panelling. If the outside surface of the brick is at 30°F and the inner surface of the panelling at 75°F, what is the heat flux if

a. the air space is assumed to transfer heat by conduction only?

b. the equivalent conductance of the air space is 1.8 Btu/h ft² °F?

c. the air space is filled with glass wool?

$$k_{brick} = 0.38 \text{ Btu/h ft °F}$$
$$k_{celotex} = 0.028 \text{ Btu/h ft °F}$$
$$k_{air} = 0.015 \text{ Btu/h ft °F}$$
$$k_{wood} = 0.12 \text{ Btu/h ft °F}$$
$$k_{wool} = 0.025 \text{ Btu/h ft °F.}$$

15.24 Solve Problem 15.23 if instead of the surface temperatures being known, the air temperatures outside and inside are 30 and 75°F, and the convective heat-transfer coefficients are 7 and 2 Btu/h ft² °F, respectively.

15.25   Determine the heat-transfer rate per square meter of wall area for the case of a furnace with inside air at 1340 K. The furnace wall is composed of a 0.106-m layer of fireclay brick and a 0.635-cm thickness of mild steel on its outside surface. Heat transfer coefficients on inside and outside wall surfaces are 5110 and 45 W/m^2 · K, respectively; outside air is at 295 K. What will be the temperatures at each surface and at the brick-steel interface?

15.26   Given the furnace wall and other conditions as specified in Problem 15.25, what thickness of celotex ($k = 0.065$ W/ m · K) must be added to the furnace wall in order that the outside surface temperature of the insulation not exceed 340 K?

15.27   A 4-in.-OD pipe is to be used to transport liquid metals and will have an outside surface temperature of 1400°F under operating conditions. Insulation is 6 in. thick and has a thermal conductivity expressed as

$$k = 0.08(1 - 0.003\,T)$$

where k is in Btu/h ft °F and T is in °F, is applied to the outside surface of the pipe.

a.   What thickness of insulation would be required for the outside insulation temperature to be no higher than 300°F?

b.   What heat-flow rate will occur under these conditions?

15.28   Water at 40°F is to flow through a 1½-in. schedule 40 steel pipe. The outside surface of the pipe is to be insulated with a 1-in.-thick layer of 85% magnesia and a 1-in.-thick layer of packed glass wool, $k = 0.022$ Btu/h ft °F. The surrounding air is at 100°F.

a.   Which material should be placed next to the pipe surface to produce the maximum insulating effect?

b.   What will be the heat flux on the basis of the outside pipe surface area? The convective heat-transfer coefficients for the inner and outer surfaces are 100 and 5 Btu/h ft °F. respectively.

15.29   A 1-in.-nominal-diameter steel pipe with its outside surface at 400°F is located in air at 90°F with the convective heat-transfer coefficient between the surface of the pipe and the air equal to 1.5 Btu/h ft °F. It is proposed to add insulation having a thermal conductivity of 0.06 Btu/h ft °F to the pipe to reduce the heat loss to one half that for the bare pipe. What thickness of insulation is necessary if the surface temperature of the steel pipe and h_o remain constant?

15.30   If, for the conditions of Problem 15.29, h_o in Btu/h ft °F varies according to $h_o = 0.575/D_o^{1/4}$, where D_o is the outside diameter of the insulation in feet, determine the thickness of insulation that will reduce the heat flux to one half that of the value for the bare pipe.

15.31   Liquid nitrogen at 77 K is stored in a cylindrical container having an inside diameter of 25 cm. The cylinder is made of stainless steel and has a wall thickness of 1.2 cm. Insulation is to be added to the outside surface of the cylinder to reduce the nitrogen boil-off rate to 25% of its value without insulation. The insulation to be used has a thermal conductivity of 0.13 W/m · K. Energy loss through the top and bottom ends of the cylinder may be presumed negligible.

Neglecting radiation effects, determine the thickness of insulation when the inner surface of the cylinder is at 77 K, the convective heat-transfer coefficient at the insulation surface has a value of 12 W/m^2 · K, and the surrounding air is at 25°C.

Chapter 16

Differential Equations of Heat Transfer

$\mathbf{P}$aralleling the treatment of momentum transfer undertaken in Chapter 9, we shall now generate the fundamental equations for a differential control volume from a first-law-of-thermodynamics approach. The control-volume expression for the first law will provide our basic analytical tool. Additionally, certain differential equations already developed in previous sections will be applicable.

16.1 THE GENERAL DIFFERENTIAL EQUATION FOR ENERGY TRANSFER

Consider the control volume having dimensions Δx, Δy, and Δz as depicted in Figure 16.1. Refer to the control-volume expression for the first law of thermodynamics

$$\frac{\delta Q}{dt} - \frac{\delta W_s}{dt} - \frac{\delta W_\mu}{dt} = \iint_{\text{c.s.}} \left(e + \frac{P}{\rho}\right)\rho(\mathbf{v}\cdot\mathbf{n})dA + \frac{\partial}{\partial t}\iiint_{\text{c.v.}} e\rho\,dV \qquad (6\text{-}10)$$

The individual terms are evaluated and their meanings are discussed below.

The net rate of heat added to the control volume will include all conduction effects, the net release of thermal energy within the control volume due to volumetric effects such as a chemical reaction or induction heating, and the dissipation of electrical or nuclear energy. The generation effects will be included in the single term, $\dot{q}$, which is the volumetric rate of thermal energy generation having units W/m^3 or Btu/h ft^3. Thus, the first term may be expressed as

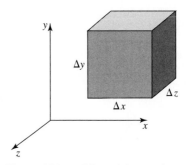

Figure 16.1 A differential control volume.

$$\frac{\delta Q}{dt} = \left[k\frac{\partial T}{\partial x}\bigg|_{x+\Delta x} - k\frac{\partial T}{\partial x}\bigg|_x\right]\Delta y\,\Delta z + \left[k\frac{\partial T}{\partial y}\bigg|_{y+\Delta y} - k\frac{\partial T}{\partial y}\bigg|_y\right]\Delta x\,\Delta z$$

$$+ \left[k\frac{\partial T}{\partial z}\bigg|_{z+\Delta z} - k\frac{\partial T}{\partial z}\bigg|_z\right]\Delta x\,\Delta y + \dot{q}\,\Delta x\,\Delta y\,\Delta z \qquad (16\text{-}1)$$

The shaft work rate or power term will be taken as zero for our present purposes. This term is specifically related to work done by some effect within the control volume that, for the differential case, is not present. The power term is thus evaluated as

$$\frac{\delta W_s}{dt} = 0 \tag{16-2}$$

The viscous work rate, occurring at the control surface, is formally evaluated by integrating the dot product of the viscous stress and the velocity over the control surface. As this operation is tedious, we shall express the viscous work rate as $\Lambda \, \Delta x \, \Delta y \, \Delta z$, where Λ is the viscous work rate per unit volume. The third term in equation (6-10) is thus written as

$$\frac{\delta W_\mu}{dt} = \Lambda \, \Delta x \, \Delta y \, \Delta z \tag{16-3}$$

The surface integral includes all energy transfer across the control surface due to fluid flow. All terms associated with the surface integral have been defined previously. The surface integral is

$$\iint_{c.s.} \left(e + \frac{P}{\rho} \right) \rho(\mathbf{v} \cdot \mathbf{n}) dA$$

$$= \left[\rho v_x \left(\frac{v^2}{2} + gy + u + \frac{P}{\rho} \right) \Big|_{x+\Delta x} - \rho v_x \left(\frac{v^2}{2} + gy + u + \frac{P}{\rho} \right) \Big|_{x} \right] \Delta y \, \Delta z$$

$$+ \left[\rho v_y \left(\frac{v^2}{2} + gy + u + \frac{P}{\rho} \right) \Big|_{y+\Delta y} - \rho v_y \left(\frac{v^2}{2} + gy + u + \frac{P}{\rho} \right) \Big|_{y} \right] \Delta x \, \Delta z$$

$$+ \left[\rho v_z \left(\frac{v^2}{2} + gy + u + \frac{P}{\rho} \right) \Big|_{z+\Delta z} - \rho v_z \left(\frac{v^2}{2} + gy + u + \frac{P}{\rho} \right) \Big|_{z} \right] \Delta x \, \Delta y \tag{16-4}$$

The energy accumulation term, relating the variation in total energy within the control volume as a function of time, is

$$\frac{\partial}{\partial t} \iiint_{c.v.} e\rho \, dV = \frac{\partial}{\partial t} \left[\frac{v^2}{2} + gy + u \right] \rho \, \Delta x \, \Delta y \, \Delta z \tag{16-5}$$

Equations (16-1) through (16-5) may now be combined as indicated by the general first-law expression, equation (6-10). Performing this combination and dividing through by the volume of the element, we have

$$\frac{k(\partial T/\partial x)|_{x+\Delta x} - k(\partial T/\partial x)|_{x}}{\Delta x} + \frac{k(\partial T/\partial y)|_{y+\Delta y} - k(\partial T/\partial y)|_{y}}{\Delta y}$$

$$+ \frac{k(\partial T/\partial z)|_{z+\Delta z} - k(\partial T/\partial z)|_{z}}{\Delta z} + \dot{q} + \Lambda$$

$$= \frac{\{\rho v_x[(v^2/2) + gy + u + (P/\rho)]|_{x+\Delta x} - \rho v_x[(v^2/2) + gy + u + (P/\rho)]|_{x}\}}{\Delta x}$$

$$+ \frac{\{\rho v_y[(v^2/2) + gy + u + (P/\rho)]|_{y+\Delta y} - \rho v_y[(v^2/2) + gy + u + (P/\rho)]|_{y}\}}{\Delta y}$$

$$+ \frac{\{\rho v_z[(v^2/2) + gy + u + (P/\rho)]|_{z+\Delta z} - \rho v_z[(v^2/2) + gy + u + (P/\rho)]|_{z}\}}{\Delta z}$$

$$+ \frac{\partial}{\partial t} \rho \left(\frac{v^2}{2} + gy + u \right)$$

Evaluated in the limit as Δx, Δy, and Δz approach zero, this equation becomes

$$\frac{\partial}{\partial x}\left(k\frac{\partial T}{\partial x}\right) + \frac{\partial}{\partial y}\left(k\frac{\partial T}{\partial y}\right) + \frac{\partial}{\partial z}\left(k\frac{\partial T}{\partial z}\right) + \dot{q} + \Lambda$$

$$= \frac{\partial}{\partial x}\left[\rho v_x\left(\frac{v^2}{2} + gy + u + \frac{P}{\rho}\right)\right] + \frac{\partial}{\partial y}\left[\rho v_y\left(\frac{v^2}{2} + gy + u + \frac{P}{\rho}\right)\right]$$

$$+ \frac{\partial}{\partial z}\left[\rho v_z\left(\frac{v^2}{2} + gy + u + \frac{P}{\rho}\right)\right] + \frac{\partial}{\partial t}\left[\rho\left(\frac{v^2}{2} + gy + u\right)\right] \qquad (16\text{-}6)$$

Equation (16-6) is completely general in application. Introducing the substantial derivative, we may write equation (16-6) as

$$\frac{\partial}{\partial x}\left(k\frac{\partial T}{\partial x}\right) + \frac{\partial}{\partial y}\left(k\frac{\partial T}{\partial y}\right) + \frac{\partial}{\partial z}\left(k\frac{\partial T}{\partial z}\right) + \dot{q} + \Lambda$$

$$= \boldsymbol{\nabla}\cdot(P\mathbf{v}) + \left(\frac{v^2}{2} + u + gy\right)\left(\boldsymbol{\nabla}\cdot\rho\mathbf{v} + \frac{\partial\rho}{\partial t}\right) + \frac{\rho}{2}\frac{Dv^2}{Dt} + \rho\frac{Du}{Dt} + \rho\frac{D(gy)}{Dt}$$

Utilizing the continuity equation, equation (9-2), we reduce this to

$$\frac{\partial}{\partial x}\left(k\frac{\partial T}{\partial x}\right) + \frac{\partial}{\partial y}\left(k\frac{\partial T}{\partial y}\right) + \frac{\partial}{\partial z}\left(k\frac{\partial T}{\partial z}\right) + \dot{q} + \Lambda$$

$$= \boldsymbol{\nabla}\cdot P\mathbf{v} + \frac{\rho}{2}\frac{Dv^2}{Dt} + \rho\frac{Du}{Dt} + \rho\frac{D(gy)}{Dt} \qquad (16\text{-}7)$$

With the aid of equation (9-19), which is valid for incompressible flow of a fluid with constant μ, the second term on the right-hand side of equation (16-7) becomes

$$\frac{\rho}{2}\frac{Dv^2}{Dt} = -\mathbf{v}\cdot\boldsymbol{\nabla}P + \mathbf{v}\cdot\rho g + \mathbf{v}\cdot\mu\,\boldsymbol{\nabla}^2\mathbf{v} \qquad (16\text{-}8)$$

Also, for incompressible flow, the first term on the right-hand side of equation (16-7) becomes

$$\boldsymbol{\nabla}\cdot P\mathbf{v} = \mathbf{v}\cdot\boldsymbol{\nabla}P \qquad (16\text{-}9)$$

Substituting equations (16-8) and (16-9) into equation (16-7), and writing the conduction terms as $\boldsymbol{\nabla}\cdot k\boldsymbol{\nabla}T$, we have

$$\boldsymbol{\nabla}\cdot k\,\boldsymbol{\nabla}T + \dot{q} + \Lambda = \rho\frac{Du}{Dt} + \rho\frac{D(gy)}{Dt} + \mathbf{v}\cdot\rho g + \mathbf{v}\cdot\mu\,\boldsymbol{\nabla}^2\mathbf{v} \qquad (16\text{-}10)$$

It will be left as an exercise for the reader to verify that equation (16-10) reduces further to the form

$$\boldsymbol{\nabla}\cdot k\boldsymbol{\nabla}T + \dot{q} + \Lambda = \rho c_v\frac{DT}{Dt} + \mathbf{v}\cdot\mu\,\boldsymbol{\nabla}^2\mathbf{v} \qquad (16\text{-}11)$$

The function Λ may be expressed in terms of the viscous portion of the normal- and shear-stress terms in equations (7-13) and (7-14). For the case of incompressible flow, it is written as

$$\Lambda = \mathbf{v}\cdot\mu\boldsymbol{\nabla}^2\,\mathbf{v} + \Phi \qquad (16\text{-}12)$$

where the "dissipation function," Φ, is given by

$$\Phi = 2\mu\left[\left(\frac{\partial v_x}{\partial x}\right)^2 + \left(\frac{\partial v_y}{\partial y}\right)^2 + \left(\frac{\partial v_z}{\partial z}\right)^2\right]$$
$$+ \mu\left[\left(\frac{\partial v_x}{\partial y} + \frac{\partial v_y}{\partial x}\right)^2 + \left(\frac{\partial v_y}{\partial z} + \frac{\partial v_z}{\partial y}\right)^2 + \left(\frac{\partial v_z}{\partial x} + \frac{\partial v_x}{\partial z}\right)^2\right]$$

Substituting for Λ in equation (16-11), we see that the energy equation becomes

$$\nabla \cdot k\nabla T + \dot{q} + \Phi = \rho c_v \frac{DT}{Dt} \qquad (16\text{-}13)$$

From equation (16-12), Φ is seen to be a function of fluid viscosity and shear-strain rates, and is positive-definite. The effect of viscous dissipation is always to increase internal energy at the expense of potential energy or stagnation pressure. The dissipation function is negligible in all cases that we will consider; its effect becomes significant in supersonic boundary layers.

16.2 SPECIAL FORMS OF THE DIFFERENTIAL ENERGY EQUATION

The applicable forms of the energy equation for some commonly encountered situations follow. In every case the dissipation term is considered negligibly small.

I. For an incompressible fluid without energy sources and with constant k

$$\rho c_v \frac{DT}{Dt} = k\nabla^2 T \qquad (16\text{-}14)$$

II. For isobaric flow without energy sources and with constant k, the energy equation is

$$\rho c_v \frac{DT}{Dt} = k\nabla^2 T \qquad (16\text{-}15)$$

Note that equations (16-14) and (16-15) are identical yet apply to completely different physical situations. The student may wish to satisfy himself at this point as to the reasons behind the unexpected result.

III. In a situation where there is no fluid motion, all heat transfer is by conduction. If this situation exists, as it most certainly does in solids where $c_v \simeq c_p$, the energy equation becomes

$$\rho c_p \frac{\partial T}{\partial t} = \nabla \cdot k\nabla T + \dot{q} \qquad (16\text{-}16)$$

Equation (16-16) applies in general to heat conduction. No assumption has been made concerning constant k. If the thermal conductivity is constant, the energy equation is

$$\frac{\partial T}{\partial t} = \alpha \nabla^2 T + \frac{\dot{q}}{\rho c_p} \qquad (16\text{-}17)$$

where the ratio $k/\rho c_p$ has been symbolized by α and is designated the *thermal diffusivity*. It is easily seen that α has the units, L^2/t; in the SI system α is expressed in m^2/s, and as ft^2/h in the English system.

If the conducting medium contains no heat sources, equation (16-17) reduces to the *Fourier field equation*

$$\frac{\partial T}{\partial t} = \alpha \nabla^2 T \qquad (16\text{-}18)$$

which is occasionally referred to as Fourier's second law of heat conduction.

For a system in which heat sources are present but there is no time variation, equation (16-17) reduces to the *Poisson equation*

$$\nabla^2 T + \frac{\dot{q}}{k} = 0 \tag{16-19}$$

The final form of the heat-conduction equation to be presented applies to a steady-state situation without heat sources. For this case, the temperature distribution must satisfy the *Laplace equation*

$$\nabla^2 T = 0 \tag{16-20}$$

Each of equations (16-17) through (16-20) has been written in general form, thus each applies to any orthogonal coordinate system. Writing the Laplacian operator, ∇^2, in the appropriate form will accomplish the transformation to the desired coordinate system. The Fourier field equation written in rectangular coordinates is

$$\frac{\partial T}{\partial t} = \alpha \left[\frac{\partial^2 T}{\partial x^2} + \frac{\partial^2 T}{\partial y^2} + \frac{\partial^2 T}{\partial z^2} \right] \tag{16-21}$$

in cylindrical coordinates

$$\frac{\partial T}{\partial t} = \alpha \left[\frac{\partial^2 T}{\partial r^2} + \frac{1}{r} \frac{\partial T}{\partial r} + \frac{1}{r^2} \frac{\partial^2 T}{\partial \theta^2} + \frac{\partial^2 T}{\partial z^2} \right] \tag{16-22}$$

and in spherical coordinates

$$\frac{\partial T}{\partial t} = \alpha \left[\frac{1}{r^2} \frac{\partial}{\partial r} \left(r^2 \frac{\partial T}{\partial r} \right) + \frac{1}{r^2 \sin\theta} \frac{\partial}{\partial \theta} \left(\sin\theta \frac{\partial T}{\partial \theta} \right) + \frac{1}{r^2 \sin^2\theta} \frac{\partial^2 T}{\partial \phi^2} \right] \tag{16-23}$$

The reader is referred to Appendix B for an illustration of the variables in cylindrical and spherical coordinate systems.

16.3 COMMONLY ENCOUNTERED BOUNDARY CONDITIONS

In solving one of the differential equations developed thus far, the existing physical situation will dictate the appropriate initial or boundary conditions, or both, which the final solutions must satisfy.

Initial conditions refer specifically to the values of T and $\mathbf{v}$ at the start of the time interval of interest. Initial conditions may be as simply specified as stating that $T|_{t=0} = T_0$ (a constant), or more complex if the temperature distribution at the start of time measurement is some function of the space variables.

Boundary conditions refer to the values of T and $\mathbf{v}$ existing at specific positions on the boundaries of a system, that is, for given values of the significant space variables. Frequently encountered boundary conditions for temperature are the case of *isothermal boundaries*, along which the temperature is constant, and *insulated boundaries*, across which no heat conduction occurs where, according to the Fourier rate equation, the temperature derivative normal to the boundary is zero. More complicated temperature functions often exist at system boundaries, and the surface temperature may also vary with time. Combinations of heat-transfer mechanisms may dictate boundary conditions as well. One situation often existing at a solid boundary is the equality between heat transfer to the surface by conduction

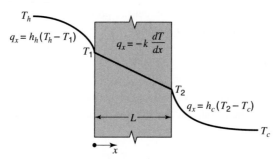

Figure 16.2 Conduction and convention at a system boundary.

and that leaving the surface by convection. This condition is illustrated in Figure 16.2. At the left-hand surface, the boundary condition is

$$h_k(T_k - T|_{x=0}) = -k\frac{\partial T}{\partial x}\bigg|_{x=0} \tag{16-24}$$

and at the right-hand surface,

$$h_c(T|_{x=L} - T_c) = -k\frac{\partial T}{\partial x}\bigg|_{x=L} \tag{16-25}$$

It is impossible at this time to foresee all the initial and boundary conditions that will be needed. The student should be aware, however, that these conditions are dictated by the physical situation. The differential equations of energy transfer are not numerous, and a specific form applying to a given situation may be found easily. It remains for the user of these equations to choose the appropriate initial and boundary conditions to make the solution meaningful.

16.4 CLOSURE

The general differential equations of energy transfer have been developed in this chapter, and some forms applying to more specific situations were presented. Some remarks concerning initial and boundary conditions have been made as well.

In the chapters to follow, analyses of energy transfer will start with the applicable differential equation. Numerous solutions will be presented and still more assigned as student exercises. The tools for heat-transfer analysis have now been developed and examined. Our remaining task is to develop a familiarity with and facility in their use.

PROBLEMS

16.1 The Fourier field equation in cylindrical coordinates is

$$\frac{\partial T}{\partial t} = \alpha\left(\frac{\partial^2 T}{\partial r^2} + \frac{1}{r}\frac{\partial T}{\partial r} + \frac{1}{r^2}\frac{\partial^2 T}{\partial \theta^2} - \frac{\partial^2 T}{\partial z^2}\right).$$

a. What form does this equation reduce to for the case of steady-state, radial heat transfer?

b. Given the boundary conditions

$$T = T_i \quad \text{at } r = r_i$$
$$T = T_o \quad \text{at } r = r_o$$

c. Generate an expression for the heat flow rate, q_r, using the result from part (b).

16.2 Perform the same operations as in parts (a), (b), and (c) of Problem 16.1 with respect to a spherical system.

16.3 Starting with the Fourier field equation in cylindrical coordinates,

a. Reduce this equation to the applicable form for steady-state heat transfer in the θ direction.

b. For the conditions depicted in the figure, that is, $T = T_o$ at $\theta = 0$, $T = T_z$ at $\theta = \pi$, the radial surfaces insulated, solve for the temperature profile.

c. Generate an expression for the heat flow rate, q_θ, using the result of part (b).

d. What is the shape factor for this configuration?

16.4 Show that equation (16-10) reduces to the form

$$\nabla \cdot k\,\nabla T + \dot{q} + \Lambda = \rho c_v \frac{DT}{Dt} + \mathbf{v} \cdot \mu \nabla^2 \mathbf{v}$$

16.5 Starting with equation (16-7), show that, for a fluid with constant thermal conductivity and no energy sources, equations (16-14) and (16-15) are obtained for incompressible and isobaric conditions, respectively. (Neglect viscous dissipation.)

16.6 Solve equation (16-19) for the temperature distribution in a plane wall if the internal heat generation per unit volume varies according to $\dot{q} = \dot{q}_0 e^{-\beta x/L}$. The boundary conditions that apply are $T = T_0$ at $x = 0$ and $T = T_L$ at $x = L$.

16.7 Solve Problem 16.6 for the same conditions, except that the boundary condition at $x = L$ is $dT/dx = 0$.

16.8 Solve Problem 16.6 for the same conditions, except that at $x = L$, $dT/dx = \xi$ (a constant).

16.9 Use the relation $T\,ds = dh - dP/\rho$ to show that the effect of the dissipation function, Φ, is to increase the entropy, S. Is the effect of heat transfer the same as the dissipation function?

16.10 In a boundary layer where the velocity profile is given by

$$\frac{v_x}{v_\infty} = \frac{3}{2}\frac{y}{\delta} - \frac{1}{2}\left(\frac{y}{\delta}\right)^3$$

where δ is the velocity boundary layer thickness, plot the dimensionless dissipation function, $\Phi\,\delta^2/\mu v_\infty^2$, vs. y/δ.

16.11 A spherical shell with inner and outer dimensions of r_i and r_o, respectively, has surface temperatures $T_i(r_i)$ and $T_o(r_o)$. Assuming constant properties and one-dimensional (radial) conduction, sketch the temperature distribution, $T(r)$. Give reasons for the shape you have sketched.

16.12 Heat is transferred by conduction (assumed to be one-dimensional) along the axial direction through the truncated conical section shown in the figure. The two base surfaces are maintained at constant temperatures: T_1 at the top, and T_2, at the bottom, where $T_1 > T_2$. Evaluate the heat transfer rate, q_x, when

a. the thermal conductivity is constant.

b. the thermal conductivity varies with temperature according to $k = k_o - aT$, where a is a constant.

16.13 Heat is generated in a radioactive plane wall according to the relationship

$$\dot{q} = \dot{q}_{max}\left[1 - \frac{x}{L}\right]$$

where $\dot{q}$ is the volumetric heat generation rate, kW/m^3, L is the half thickness of the plate, and x is measured from the plate center line.

Develop the equation that expresses the temperature difference between the plate center line and its surface.

16.14 Heat is generated in a cylindrical fuel rod in a nuclear reactor according to the relationship

$$\dot{q} = \dot{q}_{max}\left[1 - \left(\frac{r}{r_o}\right)^2\right]$$

where $\dot{q}$ is the volumetric heat generation rate, kW/m^3, and r_o is the outside cylinder radius. Develop the equation that expresses the temperature difference between the rod center line and its surface.

16.15 Heat is generated in a spherical fuel element according to the relationship

$$\dot{q} = \dot{q}_{max}\left[1 - \left(\frac{r}{r_o}\right)^3\right]$$

where $\dot{q}$ is the volumetric heat generation rate, kW/m^3, and r_o is the radius of the sphere. Develop the equation that expresses the temperature difference between the center of the sphere and its surface.

Chapter 17

Steady-State Conduction

In most equipment used in transferring heat, energy flows from one fluid to another through a solid wall. As the energy transfer through each medium is one step in the overall process, a clear understanding of the conduction mechanism of energy transfer through homogeneous solids is essential to the solutions of most heat-transfer problems.

In this chapter, we shall direct our attention to steady-state heat conduction. Steady state implies that the conditions, temperature, density, and the like at all points in the conduction region are independent of time. Our analyses will parallel the approaches used for analyzing a differential fluid element in laminar flow and those that will be used in analyzing steady-state molecular diffusion. During our discussions, two types of presentations will be used: (1) The governing differential equation will be generated by means of the control-volume concept and (2) the governing differential equation will be obtained by eliminating all irrelevant terms in the general differential equation for energy transfer.

17.1 ONE-DIMENSIONAL CONDUCTION

For steady-state conduction independent of any internal generation of energy, the general differential equation reduces to the Laplace equation

$$\nabla^2 T = 0 \tag{16-20}$$

Although this equation implies that more than one space coordinate is necessary to describe the temperature field, many problems are simpler because of the geometry of the conduction region or because of symmetries in the temperature distribution. One-dimensional cases often arise.

The one-dimensional, steady-state transfer of energy by conduction is the simplest process to describe as the condition imposed upon the temperature field is an ordinary differential equation. For one-dimensional conduction, equation (16-20) reduces to

$$\frac{d}{dx}\left(x^i \frac{dT}{dx}\right) = 0 \tag{17-1}$$

where $i = 0$ for rectangular coordinates, $i = 1$ for cylindrical coordinates, and $i = 2$ for spherical coordinates.

One-dimensional processes occur in flat planes, such as furnace walls; in cylindrical elements, such as steam pipes; and in spherical elements, such as nuclear-reactor pressure vessels. In this section, we shall consider steady-state conduction through simple systems in which the temperature and the energy flux are functions of a single space coordinate.

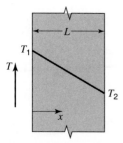

Figure 17.1 Plane wall with a one-dimensional temperature distribution.

Plane Wall. Consider the conduction of energy through a plane wall as illustrated in Figure 17.1. The one-dimensional Laplace equation is easily solved, yielding

$$T = C_1 x + C_2 \tag{17-2}$$

The two constants are obtained by applying the boundary conditions

$$\text{at } x = 0 \quad T = T_1$$

and

$$\text{at } x = L \quad T = T_2$$

These constants are

$$C_2 = T_1$$

and

$$C_1 = \frac{T_2 - T_1}{L}$$

The temperature profile becomes

$$T = \frac{T_2 - T_1}{L} x + T_1$$

or

$$T = T_1 - \frac{T_1 - T_2}{L} x \tag{17-3}$$

and is linear, as illustrated in Figure 17.1.

The energy flux is evaluated, using the Fourier rate equation

$$\frac{q_x}{A} = -k \frac{dT}{dx} \tag{15-1}$$

The temperature gradient, dT/dx, is obtained by differentiating equation (17-3) yielding

$$\frac{dT}{dx} = -\frac{T_1 - T_2}{L}$$

Substituting this term into the rate equation, we obtain for a flat wall with constant thermal conductivity

$$q_x = \frac{kA}{L}(T_1 - T_2) \tag{17-4}$$

The quantity kA/L is characteristic of a flat wall or a flat plate and is designated the *thermal conductance*. The reciprocal of the thermal conductance, L/kA, is the *thermal resistance*.

Composite Walls. The steady flow of energy through several walls in series is often encountered. A typical furnace design might include one wall for strength, an intermediate wall for insulation, and the third outer wall for appearance. This composite plane wall is illustrated in Figure 17.2.

For a solution to the system shown in this figure, the reader is referred to Section 5.

The following example illustrates the use of the composite-wall energy-rate equation for predicting the temperature distribution in walls.

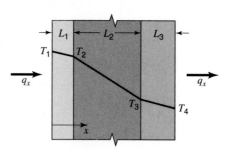

Figure 17.2 Temperature distribution for steady-state conduction of energy through a composite plane wall.

EXAMPLE 1

A furnace wall is composed of three layers, 10 cm of firebrick ($k = 1.560$ W/m·K), followed by 23 cm of kaolin insulating brick ($k = 0.073$ W/m·K), and finally 5 cm of masonry brick ($k = 1.0$ W/m·K). The temperature of the inner wall surface is 1370 K and the outer surface is at 360 K. What are the temperatures at the contacting surfaces?

The individual material thermal resistances per m² of area are

$$R_1, \text{ firebrick} = \frac{L_1}{k_1 A_1} = \frac{0.10\,\text{m}}{(1.560\,\text{W/m·K})(1\,\text{m}^2)} = 0.0641\,\text{K/W}$$

$$R_2, \text{ kaolin} = \frac{L_2}{k_2 A_2} = \frac{0.23}{(0.073)(1)} = 3.15\,\text{K/W}$$

$$R_3, \text{ masonry} = \frac{L_3}{k_3 A_3} = \frac{0.05}{(1.0)(1)} = 0.05\,\text{K/W}$$

The total resistance of the composite wall is equal to $0.0641 + 3.15 + 0.05 = 3.26$ K/W. The total temperature drop is equal to $(T_1 - T_4) = 1370 - 360 = 1010$ K.

Using equation (15-16), the energy transfer rate is

$$q = \frac{T_1 - T_4}{\Sigma R} = \frac{1010\,\text{K}}{3.26\,\text{K/W}} = 309.8\,\text{W}$$

As this is a steady-state situation, the energy transfer rate is the same for each part of the transfer path (i.e., through each wall section). The temperature at the firebrick–kaolin interface, T_2, is given by

$$T_1 - T_2 = q(R_1)$$
$$= (309.8\,\text{W})(0.0641\,\text{K/W}) = 19.9\,\text{K}$$

giving

$$T_2 = 1350.1$$

Similarly,

$$T_3 - T_4 = q(R_3)$$
$$= (309.8\,\text{W})(0.05\,\text{K/W}) = 15.5\,\text{K}$$

giving

$$T_3 = 375.5\,\text{K}$$

There are numerous situations in which a composite wall involves a combination of series and parallel energy-flow paths. An example of such a wall is illustrated in Figure 17.3, where steel is used as reinforcement for a concrete wall. The composite wall can be divided into three sections of length L_1, L_2, and L_3, and the thermal resistance for each of these lengths may be evaluated.

The intermediate layer between planes 2 and 3 consists of two separate thermal paths in parallel; the effective thermal conductance is the sum of the conductances for the two materials. For the section of the wall of height $y_1 + y_2$ and unit depth, the resistance is

$$R_2 = \frac{1}{\dfrac{k_1 y_1}{L_2} + \dfrac{k_2 y_2}{L_2}} = L_2\left(\frac{1}{k_1 y_1 + k_2 y_2}\right)$$

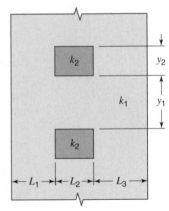

Figure 17.3 A series-parallel composite wall.

The total resistance for this wall is

$$\Sigma R_T = R_1 + R_2 + R_3$$

or

$$\Sigma R_T = \frac{L_1}{k_1(y_1 + y_2)} + L_2\left(\frac{1}{k_1 y_1 + k_2 y_2}\right) + \frac{L_3}{k_1(y_1 + y_2)}$$

The electrical circuit —WW— is an analog to the composite wall.

The rate of energy transferred from plane 1 to plane 4 is obtained by a modified form of equation (15-16).

$$q = \frac{T_1 - T_4}{\Sigma R_T} = \frac{T_1 - T_4}{\dfrac{L_1}{k_1(y_1 + y_2)} + L_2\left(\dfrac{1}{k_1 y_1 + k_2 y_2}\right) + \dfrac{L_3}{k_1(y_1 + y_2)}} \tag{17-5}$$

It important to recognize that this equation is only an approximation. Actually, there is a significant temperature distribution in the y direction close to the material that has the higher thermal conductivity.

In our discussions of composite walls, no allowance was made for a temperature drop at the contact face between two different solids. This assumption is not always valid, has there will often be vapor spaces caused by rough surfaces, or even oxide films on the surfaces of metals. These additional contact resistances must be accounted for in a precise energy-transfer equation.

Long, Hollow Cylinder. Radial energy flow by conduction through a long, hollow cylinder is another example of one-dimensional conduction. The radial heat flow for this configuration is evaluated in Example 1 of chapter 15 as

$$\frac{q_r}{L} = \frac{2\pi k}{\ln(r_o/r_i)}(T_i - T_o) \tag{17-6}$$

where r_i is the inside radius, r_o is the outside radius, T_i is the temperature on the inside surface, and T_o is the temperature on the outside surface. The resistance concept may again be used; the thermal resistance of the hollow cylinder is

$$R = \frac{\ln(r_o/r_i)}{2\pi k L} \tag{17-7}$$

The radial temperature distribution in a long, hollow cylinder may be evaluated by using equation (17-1) in cylindrical form

$$\frac{d}{dr}\left(r\frac{dT}{dr}\right) = 0 \tag{17-8}$$

Solving this equation subject to the boundary conditions

$$\text{at } r = r_i \quad T = T_i$$

and

$$\text{at } r = r_o \quad T = T_o$$

we see that temperature profile is

$$T(r) = T_i - \frac{T_i - T_o}{\ln(r_o/r_i)} \ln\frac{r}{r_i} \tag{17-9}$$

Thus, the temperature in a long, hollow cylinder is a logarithmic function of radius r, whereas for the plane wall the temperature distribution is linear.

The following example illustrates the analysis of radial energy conduction through a long, hollow cylinder.

EXAMPLE 2

A long steam pipe of outside radius r_2 is covered with thermal insulation having an outside radius of r_3. The temperature of the outer surface of the pipe, T_2, and the temperature of the surrounding air, T_∞, are fixed. The energy loss per unit area of outside surface of the insulation is described by the Newton rate equation

$$\frac{q_r}{A} = h(T_3 - T_\infty) \qquad (15\text{-}11)$$

Can the energy loss increase with an increase in the thickness of insulation? If possible, under what conditions will this situation arise? Figure 17.4 may be used to illustrate this composite cylinder.

Figure 17.4 A series composite hollow cylinder.

In Example 3 of Chapter 15, the thermal resistance of a hollow cylindrical element was shown to be

$$R = \frac{\ln(r_o/r_i)}{2\pi k L} \qquad (17\text{-}10)$$

In the present example, the total difference in temperature is $T_2 - T_\infty$ and the two resistances, due to the insulation and the surrounding air film, are

$$R_2 = \frac{\ln(r_3/r_2)}{2\pi k_2 L}$$

for the insulation, and

$$R_3 = \frac{1}{hA} = \frac{1}{h2\pi r_3 L}$$

for the air film.

Substituting these terms into the radial heat flow equation and rearranging, we obtain

$$q_r = \frac{2\pi L(T_2 - T_\infty)}{[\ln(r_3/r_2)]/k_2 + 1/hr_3} \qquad (17\text{-}11)$$

The dual effect of increasing the resistance to energy transfer by conduction and simultaneously increasing the surface area as r_3 is increased suggests that, for a pipe of given size, a particular outer radius exists for which the heat loss is maximum. As the ratio r_3/r_2 increases logarithmically, and the term $1/r_3$ decreases as r_3 increases, the relative importance of each resistance term will change as the insulation thickness is varied. In this example, $L, T_2, T_\infty, k_2 h$ and r_2 are considered constant. Differentiating equation (17-11) with respect to r_3, we obtain

$$\frac{dq_r}{dr_3} = -\frac{2\pi L(T_2 - T_\infty)\left(\dfrac{1}{k_2 r_3} - \dfrac{1}{hr_3^2}\right)}{\left[\dfrac{1}{k_2}\ln\left(\dfrac{r_3}{r_2}\right) + \dfrac{1}{hr_3}\right]^2} \qquad (17\text{-}12)$$

The radius of insulation associated with the maximum energy transfer, the *critical radius,* found by setting $dq_r/dr_3 = 0$; equation (17-12) reduces to

$$(r_3)_{\text{critical}} = \frac{k_2}{h} \tag{17-13}$$

In the case of 85% magnesia insulation ($k = 0.0692$ W/m·K) and a typical value for the heat transfer coefficient in natural convection ($h = 34$ W/m²·K), the critical radius is calculated as

$$r_{\text{crit}} = \frac{k}{h} = \frac{0.0692 \text{ W/m·K}}{34 \text{ W/m}^2\text{·K}} \begin{aligned} &= 0.0020 \text{ m} \quad (0.0067 \text{ ft}) \\ &= 0.20 \text{ cm} \quad (0.0787 \text{ in.}) \end{aligned}$$

These very small numbers indicate that the critical radius will be exceeded in any practical problem. The question then is whether the critical radius given by equation (17-13) represents a maximum or a minimum condition for q. The evaluation of the second derivative, $d^2 q_r/dr_3^2$, when $r_3 = k/h$ yields a negative result, thus r_{crit} is a maximum condition. It now follows that q_r will be decreased for any value of r_3 greater than 0.0020 m.

Hollow Sphere. Radial heat flow through a hollow sphere is another example of one-dimensional conduction. For constant thermal conductivity, the modified Fourier rate equation

$$q_r = -k \frac{dT}{dr} A$$

applies, where A = area of a sphere = $4\pi r^2$, giving

$$q_r = -4\pi k r^2 \frac{dT}{dr} \tag{17-14}$$

This relation, when integrated between the boundary conditions

$$\text{at } T = T_i \quad r = r_i$$

and

$$\text{at } T = T_o \quad r = r_o$$

yields

$$q = \frac{4\pi k(T_i - T_o)}{\dfrac{1}{r_i} - \dfrac{1}{r_0}} \tag{17-15}$$

The hyperbolic temperature distribution

$$T = T_i - \left(\frac{T_i - T_o}{1/r_i - 1/r_o} \right) \left(\frac{1}{r_i} - \frac{1}{r} \right) \tag{17-16}$$

is obtained by using the same procedure that was followed to obtain equation (17-9).

Variable Thermal Conductivity. If the thermal conductivity of the medium through which the energy is transferred varies significantly, the preceding equations in this section do not apply. As Laplace's equation involves the assumption of constant thermal conductivity, a new differential equation must be determined from the general equation for heat transfer. For steady-state conduction in the x direction without internal generation of energy, the equation that applies is

$$\frac{d}{dx}\left(k\frac{dT}{dx}\right) = 0 \qquad (17\text{-}17)$$

where k may be a function of T.

In many cases the thermal conductivity may be a linear function of temperature over a considerable range. The equation of such a straight-line function may be expressed by

$$k = k_o(1 + \beta T)$$

where k_o and $\boldsymbol{\beta}$ are constants for a particular material. In general, for materials satisfying this relation, β is negative for good conductors and positive for good insulators. Other relations for varying k have been experimentally determined for specific materials. The evaluation of the rate of energy transfer when the material has a varying thermal conductivity is illustrated in Example 2 of chapter 15.

17.2 ONE-DIMENSIONAL CONDUCTION WITH INTERNAL GENERATION OF ENERGY

In certain systems, such as electric resistance heaters and nuclear fuel rods, heat is generated within the conducting medium. As one might expect, the generation of energy within the conducting medium produces temperature profiles different than those for simple conduction.

In this section, we shall consider two simple example cases: steady-state conduction in a circular cylinder with uniform or homogeneous energy generation, and steady-state conduction in a plane wall with variable energy generation. Carslaw and Jaeger[1] and Jakob[2] have written excellent treatises dealing with more complicated problems.

Cylindrical Solid with Homogeneous Energy Generation. Consider a cylindrical solid with internal energy generation as shown in Figure 17.5. The cylinder will be considered long enough so that only radial conduction occurs. The density, ρ, the heat capacity, c_p, and the thermal conductivity of the material will

Figure 17.5 Annular element in a long, circular cylinder with internal heat generation.

---

[1] H. S. Carslaw and J. C. Jaeger, *Conduction of Heat in Solids,* 2nd Edition, Oxford Univ. Press, New York, 1959.

[2] M. Jakob, *Heat Transfer,* Vol. I, Wiley, New York, 1949.

be considered constant. The energy balance for the element shown is

$$
\left\{ \begin{array}{c} \text{rate of energy} \\ \text{conduction into} \\ \text{the element} \end{array} \right\} + \left\{ \begin{array}{c} \text{rate of energy} \\ \text{generation within} \\ \text{the element} \end{array} \right\} - \left\{ \begin{array}{c} \text{rate of energy} \\ \text{conduction out} \\ \text{of the element} \end{array} \right\}
$$

$$
= \left\{ \begin{array}{c} \text{rate of accumulation} \\ \text{of energy} \\ \text{within the element} \end{array} \right\} \tag{17-18}
$$

Applying the Fourier rate equation and letting $\dot{q}$ represent the rate of energy generated per unit volume, we may express equation (17-18) by the algebraic expression

$$
-k(2\pi rL)\frac{\partial T}{\partial r}\bigg|_{r} + \dot{q}(2\pi rL\,\Delta r) - \left[-k(2\pi rL)\frac{\partial T}{\partial r}\bigg|_{r+\Delta r} \right] = \rho c_p \frac{\partial T}{\partial t}(2\pi rL\,\Delta r)
$$

Dividing each term by $2\pi rL\,\Delta r$, we obtain

$$
\dot{q} + \frac{k[r(\partial T/\partial r)|_{r+\Delta r} - r(\partial T/\partial r)|_{r}]}{r\,\Delta r} = \rho c_p \frac{\partial T}{\partial t}
$$

In the limit as Δr approaches zero, the following differential equation is generated:

$$
\dot{q} + \frac{k}{r}\frac{\partial}{\partial r}\left(r\frac{\partial T}{\partial r} \right) = \rho c_p \frac{\partial T}{\partial t} \tag{17-19}
$$

For steady-state conditions, the accumulation term is zero; when we eliminate this term from the above expression, the differential equation for a solid cylinder with homogeneous energy generation becomes

$$
\dot{q} + \frac{k}{r}\frac{d}{dr}\left(r\frac{dT}{dr} \right) = 0 \tag{17-20}
$$

The variables in this equation may be separated and integrated to yield

$$
rk\frac{dT}{dr} + \dot{q}\frac{r^2}{2} = C_1
$$

or

$$
k\frac{dT}{dr} + \dot{q}\frac{r}{2} = \frac{C_1}{r}
$$

Because of the symmetry of the solid cylinder, a boundary condition that must be satisfied stipulates that the temperature gradient must be finite at the center of the cylinder, where $r = 0$. This can be true only if $C_1 = 0$. Accordingly, the above relation reduces to

$$
k\frac{dT}{dr} + \dot{q}\frac{r}{2} = 0 \tag{17-21}
$$

A second integration will now yield

$$
T = -\frac{\dot{q}r^2}{4k} + C_2 \tag{17-22}
$$

If the temperature T is known at any radial value, such as a surface, the second constant, C_2, may be evaluated. This, of course, provides the completed expression for the temperature profile. The energy flux in the radial direction may be obtained

from

$$\frac{q_r}{A} = -k\frac{dT}{dr}$$

by substituting equation (17-21), yielding

$$\frac{q_r}{A} = \dot{q}\frac{r}{2}$$

or

$$q_r = (2\pi rL)\dot{q}\frac{r}{2} = \pi r^2 L\dot{q} \tag{17-23}$$

Plane Wall with Variable Energy Generation. The second case associated with energy generation involves a temperature-dependent, energy-generating process. This situation develops when an electric current is passed through a conducting medium possessing an electrical resistivity that varies with temperature. In our discussion, we shall assume that the energy-generation term varies linearly with temperature, and that the conducting medium is a flat plate with temperature T_L at both surfaces. The internal energy generation is described by

$$\dot{q} = \dot{q}_L[1 + \beta(T - T_L)] \tag{17-24}$$

where $\dot{q}_L$ is the generation rate at the surface and β is a constant.

With this model for the generation function, and as both surfaces have the same temperature, the temperature distribution within the flat plate is symmetric about the midplane. The plane wall and its coordinate system are illustrated in Figure 17.6. The symmetry of the temperature distribution requires a zero temperature gradient at $x = 0$. With steady-state conditions, the differential equation may be obtained by eliminating the irrelevant terms in the general differential equation for heat transfer. Equation (16-19) for the case of steady-state conduction in the x direction in a stationary solid with constant thermal conductivity becomes

$$\frac{d^2T}{dx^2} + \frac{\dot{q}_L}{k}[1 + \beta(T - T_L)] = 0$$

The boundary conditions are

$$\text{at } x = 0 \quad \frac{dT}{dx} = 0$$

and

$$\text{at } x = \pm L \quad T = T_L$$

These relations may be expressed in terms of a new variable, $\theta = T - T_L$, by

$$\frac{d^2\theta}{dx^2} + \frac{\dot{q}_L}{k}(1 + \beta\theta) = 0$$

or

$$\frac{d^2\theta}{dx^2} + C + s\theta = 0$$

where $C = \dot{q}_L/k$ and $s = \beta\dot{q}_L/k$. The boundary conditions are

$$\text{at } x = 0 \quad \frac{d\theta}{dx} = 0$$

Figure 17.6 Flat plate with temperature-dependent energy generation.

$-L$ $2L$ x $+L$

and

$$\text{at } x = \pm L \quad \theta = 0$$

The integration of this differential equation is simplified by a second change in variables; inserting ϕ for $C + s\theta$ into the differential equation and the boundary conditions, we obtain

$$\frac{d^2\phi}{dx^2} + s\phi = 0$$

for

$$x = 0 \quad \frac{d\phi}{dx} = 0$$

and

$$x = \pm L \quad \phi = C$$

The solution is

$$\phi = C + s\theta = A\cos(x\sqrt{s}) + B\sin(x\sqrt{s})$$

or

$$\theta = A_1\cos(x\sqrt{s}) + A_2\sin(x\sqrt{s}) - \frac{C}{s}$$

The temperature distribution becomes

$$T - T_L = \frac{1}{\beta}\left[\frac{\cos(x\sqrt{s})}{\cos(L\sqrt{s})} - 1\right] \tag{17-25}$$

where $s = \beta\dot{q}_L/k$ is obtained by applying the two boundary conditions.

The cylindrical and spherical examples of one-dimensional temperature-dependent generation are more complex; solutions to these may be found in the technical literature.

17.3 HEAT TRANSFER FROM EXTENDED SURFACES

A very useful application of one-dimensional heat-conduction analysis is that of describing the effect of extended surfaces. It is possible to increase the energy transfer between a surface and an adjacent fluid by increasing the amount of surface area in contact with the fluid. This increase in area is accomplished by adding extended surfaces that may be in the forms of fins or spines of various cross sections.

The one-dimensional analysis of extended surfaces may be formulated in general terms by considering the situation depicted in Figure 17.7.

The shaded area represents a portion of the extended surface that has variable cross-sectional area, $A(x)$, and surface area, $S(x)$, which are functions of x alone. For steady-state conditions, the first law of thermodynamics, equation (6-10), reduces to the simple expression

$$\frac{\delta Q}{dt} = 0$$

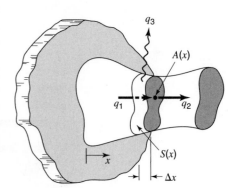

Figure 17.7 An extended surface of general configuration.

Thus, in terms of the heat flow rates designated in the figure, we may write

$$q_1 = q_2 + q_3 \tag{17-26}$$

The quantities q_1 and q_2 are conduction terms, while q_3 is a convective heat-flow rate. Evaluating each of these in the appropriate way and substituting into equation (17-26), we obtain

$$kA \frac{dT}{dx}\bigg|_{x+\Delta x} - kA \frac{dT}{dx}\bigg|_x - hS(T - T_\infty) = 0 \tag{17-27}$$

where T_∞ is the fluid temperature. Expressing the surface area, $S(x)$, in terms of the width, Δx, times the perimeter, $P(x)$, and dividing through by Δx, we obtain

$$\frac{kA(dT/dx)|_{x+\Delta x} - kA(dT/dx)|_x}{\Delta x} - hP(T - T_\infty) = 0$$

Evaluating this equation in the limit as $\Delta x \rightarrow 0$, we obtain the differential equation

$$\frac{d}{dx}\left(kA \frac{dT}{dx}\right) - hP(T - T_\infty) = 0 \tag{17-28}$$

One should note, at this point, that the temperature gradient, dT/dx, and the surface temperature, T, are expressed such that T is a function of x only. This treatment assumes the temperature to be "lumped" in the transverse direction. This is physically realistic when the cross section is thin or when the material thermal conductivity is large. Both of these conditions apply in the case of fins. More will be said about the "lumped parameter" approach in Chapter 18. This approximation in the present case leads to equation (17-28), an ordinary differential equation. If we did not make this simplifying analysis, we would have a distributed parameter problem that would require solving a partial differential equation.

A wide range of possible forms exist when equation (17-28) is applied to specific geometries. Three possible applications and the resulting equations are described in the following paragraphs.

(1) Fins or Spines of Uniform Cross Section. For either of the cases shown in Figure 17.8, the following are true: $A(x) = A$, and $P(x) = P$, both constants. If, additionally, both k and h are taken to be constant, equation (17-28) reduces to

$$\frac{d^2T}{dx^2} - \frac{hP}{kA}(T - T_\infty) = 0 \tag{17-29}$$

Figure 17.8 Two examples of extended surfaces with constant cross section.

Figure 17.9 Two examples of straight extended surfaces with variable cross section.

(2) Straight Surfaces with Linearly Varying Cross Section. Two configurations for which A and P are not constant are shown in Figure 17.9. If the area and perimeter both vary in a linear manner from the primary surface, $x = 0$, to some lesser value at the end, $x = L$, both A and P may be expressed as

$$A = A_0 - (A_0 - A_L)\frac{x}{L} \tag{17-30}$$

and

$$P = P_0 - (P_0 - P_L)\frac{x}{L} \tag{17-31}$$

In the case of the rectangular fin shown in Figure 17.9(b), the appropriate values of A and P are

$$A_0 = 2t_0 W \qquad A_L = 2t_L W$$
$$P_0 = 2[2t_0 + W] \qquad P_L = 2[2t_L + W]$$

where t_0 and t_L represent the semithickness of the fin evaluated at $x = 0$ and $x = L$, respectively, and W is the total depth of the fin.

For constant h and k, equation (17-28) applied to extended surfaces with cross-sectional area varying linearly becomes

$$\left[A_0 - (A_0 - A_L)\frac{x}{L}\right]\frac{d^2 T}{dx^2} - \frac{A_0 - A_L}{L}\frac{dT}{dx} - \frac{h}{k}\left[P_0 - (P_0 - P_L)\frac{x}{L}\right](T - T_\infty) = 0 \tag{17-32}$$

(3) Curved Surfaces of Uniform Thickness. A common type of extended surface is that of the circular fin of constant thickness as depicted in Figure 17.10. The appropriate expressions for A and P, in this case, are

$$\left.\begin{array}{l} A = 4\pi r t \\[2mm] P = 4\pi r \end{array}\right\} r_0 \leq r \leq r_L$$

When these expressions are substituted into equation (17-28), the applicable differential equation, considering k and h constant, is

$$\frac{d^2 T}{dr^2} + \frac{1}{r}\frac{dT}{dr} - \frac{h}{kt}(T - T_\infty) = 0 \tag{17-33}$$

Figure 17.10 A curved fin of constant thickness.

Equation (17-33) is a form of Bessel's equation of zero order. The solution is in terms of Bessel functions of the first kind. The description and use of these functions are beyond the mathematical scope of this text. The interested reader may consult the work of Kraus et al.[3] for a complete discussion of Bessel functions and their use.

In each of the cases considered, the thermal conductivity and convective heat-transfer coefficient were assumed constant. When the variable nature of these quantities is considered, the resulting differential equations become still more complex than those developed thus far.

Solutions for the temperature profile in the case of the straight fin of constant cross section will now be considered; equation (17-29) applies.

The general solution to equation (17-29) may be written

$$\theta = c_1 e^{mx} + c_2 e^{-mx} \tag{17-34}$$

or

$$\theta = A \cosh mx + B \sinh mx \tag{17-35}$$

where $m^2 = hP/kA$ and $\theta = T - T_\infty$. The evaluation of the constants of integration requires that two boundary conditions be known. The three sets of boundary conditions that we shall consider are as follows:

$$
\begin{aligned}
(a) \quad & T = T_0 && \text{at} \quad x = 0 \\
& T = T_L && \text{at} \quad x = L \\
(b) \quad & T = T_0 && \text{at} \quad x = 0 \\
& \frac{dT}{dx} = 0 && \text{at} \quad x = L
\end{aligned}
$$

and

$$
\begin{aligned}
(c) \quad & T = T_0 && \text{at} \quad x = 0 \\
& -k\frac{dT}{dx} = h(T - T_\infty) && \text{at} \quad x = L
\end{aligned}
$$

The first boundary condition of each set is the same and stipulates that the temperature at the base of the extended surface is equal to that of the primary surface. The second boundary condition relates the situation at a distance L from the base. In set (a) the condition is that of a known temperature at $x = L$. In set (b) the temperature gradient is zero at $x = L$. In set (c) the requirement is that heat flow to the end of an extended surface by conduction be equal to that leaving this position by convection.

The temperature profile, associated with the first set of boundary conditions, is

$$\frac{\theta}{\theta_0} = \frac{T - T_\infty}{T_0 - T_\infty} = \left(\frac{\theta_L}{\theta_0} - e^{-mL}\right)\left(\frac{e^{mx} - e^{-mx}}{e^{mL} - e^{-mL}}\right) + e^{-mx} \tag{17-36}$$

A special case of this solution applies when L becomes very large, that is, $L \rightarrow \infty$, for which equation (17-36) reduces to

$$\frac{\theta}{\theta_0} = \frac{T - T_\infty}{T_0 - T_\infty} = e^{-mx} \tag{17-37}$$

The constants, c_1 and c_2, obtained by applying set (b), yield, for the temperature profile,

$$\frac{\theta}{\theta_0} = \frac{T - T_\infty}{T_0 - T_\infty} = \frac{e^{mx}}{1 + e^{2mL}} + \frac{e^{-mx}}{1 + e^{-2mL}} \tag{17-38}$$

[3] A. D. Kraus, A. Aziz, and J. R. Welty, *Extended Surface Heat Transfer*, Wiley-Interscience, New York, 2001.

An equivalent expression to equation (17-38) but in a more compact form is

$$\frac{\theta}{\theta_0} = \frac{T - T_\infty}{T_0 - T_\infty} = \frac{\cosh[m(L - x)]}{\cosh mL} \tag{17-39}$$

Note that, in either equation (17-38) or (17-39), as $L \to \infty$ the temperature profile approaches that expressed in equation (17-37).

The application of set (c) of the boundary conditions yields, for the temperature profile,

$$\frac{\theta}{\theta_0} = \frac{T - T_\infty}{T_0 - T_\infty} = \frac{\cosh[m(L - x)] + (h/mk)\sinh[m(L - x)]}{\cosh mL + (h/mk)\sinh mL} \tag{17-40}$$

It may be noted that this expression reduces to equation (17-39) if $d\theta/dx = 0$ at $x = L$ and to equation (17-37) if $T = T_\infty$ at $L = \infty$.

The expressions for $T(x)$ that have been obtained are particularly useful in evaluating the total heat transfer from an extended surface. This total heat transfer may be determined by either of two approaches. The first is to integrate the convective heat-transfer expression over the surface according to

$$q = \int_S h[T(x) - T_\infty]dS = \int_S h\theta \, dS \tag{17-41}$$

The second method involves evaluating the energy conducted into the extended surface at the base as expressed by

2ND ODE Solutions

$$q = -kA \left.\frac{dT}{dx}\right|_{x=0} \tag{17-42}$$

The latter of   ions is easier to evaluate; accordingly, we will use this equation in th   opment.

Using equ   ind that the heat transfer rate, when set (a) of the boundary conditions ap

$$kAm\theta_0 \left[1 - 2\frac{\theta_L/\theta_0 - e^{-mL}}{e^{mL} - e^{-mL}}\right] \tag{17-43}$$

If the length   s expression becomes

$$= kAm\theta_0 = kAm(T_0 - T_\infty) \tag{17-44}$$

Substitu   9) [obtained by using set (b) of the boundary conditions] into equation   n

$$q = kAm\theta_0 \tanh mL \tag{17-45}$$

Equation (17-40), utilized in equation (17-42), yields for q the expression

$$q = kAm\theta_0 \frac{\sinh mL + (h/mk)\cosh mL}{\cosh mL + (h/mk)\sinh mL} \tag{17-46}$$

The equations for the temperature profile and total heat transfer for extended surfaces of more involved configuration have not been considered. Certain of these cases will be left as exercises for the reader.

A question that is logically asked at this point is, "What benefit is accrued by the addition of extended surfaces?" A term that aids in answering this question is the *fin efficiency*, symbolized as η_f, defined as the ratio of the actual heat transfer from an extended surface to the maximum possible heat transfer from the surface. The maximum heat transfer would occur if the temperature of the extended surface were equal to the base temperature, T_0, at all points.

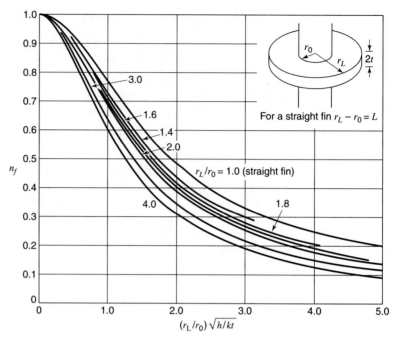

Figure 17.11 Fin efficiency for straight and circular fins of constant thickness.

Figure 17.11 is a plot of η_f as a function of a significant parameter for both straight and circular fins of constant thickness (when fin thickness is small, $t \ll r_L - r_0$).

The total heat transfer from a finned surface is

$$q_{\text{total}} = q_{\text{primary surface}} + q_{\text{fin}}$$
$$= A_0 h(T_0 - T_\infty) + A_f h(T - T_\infty) \tag{17-47}$$

The second term in equation (17-47) is the actual heat transfer from the fin surface in terms of the variable surface temperature. This may be written in terms of the fin efficiency, yielding

$$q_{\text{total}} = A_0 h(T_0 - T_\infty) + A_f h \eta_f (T_0 - T_\infty)$$

or

$$q_{\text{total}} = h(A_0 + A_f \eta_f)(T_0 - T_\infty) \tag{17-48}$$

In this expression A_0 represents the exposed area of the primary surface, A_f is the total fin surface area, and the heat transfer coefficient, h, is assumed constant.

The application of equation (17-48) as well as an idea of the effectiveness of fins is illustrated in Example 3.

EXAMPLE 3

Water and air are separated by a mild-steel plane wall. It is proposed to increase the heat-transfer rate between these fluids by adding straight rectangular fins of 1.27-mm thickness and 2.5-cm length, spaced 1.27 cm apart. The air-side and water-side heat-transfer coefficients may be assumed constant with values of 11.4 and 256 W/m²·K respectively. Determine the percent change in total heat transfer when fins are placed on (a) the water side, (b) the air side, and (c) both sides.

For a 1 m² section of the wall, the areas of the primary surface and of the fins are

$$A_o = 1\,\text{m}^2 - 79\,\text{fins}\,(1\,\text{m})\left[\frac{0.00127\,\text{m}}{\text{fin}}\right]$$
$$= 0.90\,\text{m}^2$$
$$A_f = 79\,\text{fins}\,(1\,\text{m})[(2)(0.025\,\text{m})] + 0.10\,\text{m}^2$$
$$= 4.05\,\text{m}^2$$

Values of fin efficiency can now be determined from Figure 17.11. For the air side

$$L\sqrt{h/kt} = 0.025\,\text{m}\left[\frac{11.4\,\text{W/m}^2\cdot\text{K}}{(42.9\,\text{W/m}\cdot\text{K})(0.00127\,\text{m})}\right]^{1/2}$$
$$= 0.362$$

and for the water side

$$L\sqrt{h/kT} = 0.025\,\text{m}\left[\frac{256\,\text{W/m}^2\cdot\text{K}}{(42.9\,\text{W/m}\cdot\text{K})(0.00127\,\text{m})}\right]^{1/2}$$
$$= 1.71$$

The fin efficiencies are then read from the figure as

$$\eta_{\text{air}} \cong 0.95$$
$$\eta_{\text{water}} \cong 0.55$$

The total heat transfer rates can now be evaluated. For fins on the air side

$$q = h_a\Delta T_a[A_o + \eta_{fa}\,A_f]$$
$$= 11.4\,\Delta T_a[0.90 + 0.95(4.05)]$$
$$= 54.1\,\Delta T_a$$

and on the water side

$$q = h_w\,\Delta T_w\,[A_o + \eta_{fw}\,A_f]$$
$$= 256\Delta T_w[0.90 + 0.55(4.05)]$$
$$= 801\,\Delta T_w$$

The quantities ΔT_a and ΔT_w represent the temperature differences between the steel surface at temperature T_o and the fluids.

The reciprocals of the coefficients are the thermal resistances of the finned surfaces.

Without fins the heat-transfer rate in terms of the overall temperature difference, $\Delta T = T_w - T_a$, neglecting the conductive resistance of the steel wall, is

$$q = \frac{\Delta T}{\dfrac{1}{11.4} + \dfrac{1}{256}} = 10.91\,\Delta T$$

With fins on the air side alone

$$q = \frac{\Delta T}{\dfrac{1}{54.1} + \dfrac{1}{256}} = 44.67\,\Delta T$$

an increase of 310% compared with the bare-wall case.

With fins on the water side alone

$$q = \frac{\Delta T}{\dfrac{1}{11.4} + \dfrac{1}{801}} = 11.24\,\Delta T$$

an increase of 3.0%.

With fins on both sides the heat-flow rate is

$$q = \frac{\Delta T}{\dfrac{1}{54.1} + \dfrac{1}{801}} = 50.68\,\Delta T$$

an increase of 365%.

This result indicates that adding fins is particularly beneficial where the convection coefficient has a relatively small value.

17.4 TWO- AND THREE-DIMENSIONAL SYSTEMS

In Sections 17.2 and 17.3, we discussed systems in which the temperature and the energy transfer were functions of a single-space variable. Although many problems fall into this category, there are many other systems involving complicated geometry or temperature boundary conditions, or both, for which two or even three spatial coordinates are necessary to describe the temperature field.

In this section, we shall review some of the methods for analyzing heat transfer by conduction in two- and three-dimensional systems. The problems will mainly involve two-dimensional systems, as they are less cumbersome to solve yet illustrate the techniques of analysis.

Analytical Solution. An analytical solution to any transfer problem must satisfy the differential equation describing the process as well as the prescribed boundary conditions. Many mathematical techniques have been used to obtain solutions for particular energy conduction situations in which a partial differential equation describes the temperature field. Carslaw and Jaeger[4] and Boelter et al.[5] have written excellent treatises that deal with the mathematical solutions for many of the more complex conduction problems. As most of this material is too specialized for an introductory course, a solution will be obtained to one of the first cases analyzed by Fourier[6] in the classical treatise that established the theory of energy transfer by conduction. This solution of a two-dimensional conduction medium employs the mathematical method of separation of variables.

Consider a thin, infinitely long rectangular plate that is free of heat sources, as illustrated in Figure 17.12. For a thin plate $\partial T/\partial z$ is negligible, and the temperature is a function of x and y only. The solution will be obtained for the case in which the two edges of the plate are maintained at zero temperature and the bottom is maintained at T_1 as shown. The steady-state temperature distribution in the plate of constant thermal conductivity must satisfy the differential equation

$$\frac{\partial^2 T}{\partial x^2} + \frac{\partial^2 T}{\partial y^2} = 0 \qquad (17\text{-}49)$$

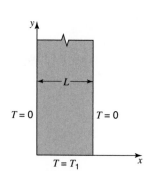

Figure 17.12 Model for two-dimensional conduction analysis.

---

[4] H. S. Carslaw and J. C. Jaeger, *Conduction of Heat in Solids,* 2nd Edition, Oxford Univ. Press, New York, 1959.

[5] L. M. K. Boelter, V. H. Cherry, H. A. Johnson, and R. C. Martinelli, *Heat Transfer Notes,* McGraw-Hill Book Company, New York, 1965.

[6] J. B. J. Fourier, *Theorie Analytique de la Chaleur,* Gauthier-Villars, Paris, 1822.

and the boundary conditions

$$T = 0 \quad \text{at} \quad x = 0 \quad \text{for all values of } y$$
$$T = 0 \quad \text{at} \quad x = L \quad \text{for all values of } y$$
$$T = T_1 \quad \text{at} \quad y = 0 \quad \text{for } 0 \leq x \leq L$$

and

$$T = 0 \quad \text{at} \quad y = \infty \quad \text{for} \quad 0 \leq x \leq L$$

Equation (17-49) is a linear, homogeneous partial differential equation. This type of equation usually can be integrated by assuming that the temperature distribution, $T(x, y)$, is of the form

$$T(x, y) = X(x)Y(y) \tag{17-50}$$

where $X(x)$ is a function of x only and $Y(y)$ is a function of y only. Substituting this equation into equation (17-49), we obtain an expression in which the variables are separated

$$-\frac{1}{X}\frac{d^2X}{dx^2} = \frac{1}{Y}\frac{d^2Y}{dy^2} \tag{17-51}$$

As the left-hand side of equation (17-51) is independent of y and the equivalent right-hand side is independent of x, it follows that both must be independent of x and y, and hence must be equal to a constant. If we designate this constant λ^2, two ordinary differential equations result

$$\frac{d^2X}{dx^2} + \lambda^2 X = 0 \tag{17-52}$$

and

$$\frac{d^2Y}{dy^2} - \lambda^2 Y = 0 \tag{17-53}$$

These differential equations may be integrated, yielding

$$X = A \cos \lambda x + B \sin \lambda x$$

and

$$Y = Ce^{\lambda y} + De^{-\lambda y}$$

According to equation (17-50), the temperature distribution is defined by the relation

$$T(x, y) = XY = (A \cos \lambda x + B \sin \lambda x)(Ce^{\lambda y} + De^{-\lambda y}) \tag{17-54}$$

where A, B, C, and D are constants to be evaluated from the four boundary conditions. The condition that $T = 0$ at $x = 0$ requires that $A = 0$. Similarly, $\sin \lambda x$ must be zero at $x = L$; accordingly, λL must be an integral multiple of π or $\lambda = n\pi/L$. Equation (17-54) is now reduced to

$$T(x, y) = B \sin \left(\frac{n\pi x}{L}\right)\left(Ce^{n\pi y/L} + De^{-n\pi y/L}\right) \tag{17-55}$$

The requirement that $T = 0$ at $y = \infty$ stipulates that C must be zero. A combination of B and D into the single constant E reduces equation (17-55) to

$$T(x, y) = Ee^{-n\pi y/L} \sin\left(\frac{n\pi x}{L}\right)$$

This expression satisfies the differential equation for any integer n greater than or equal to zero. The general solution is obtained by summing all possible solutions, giving

$$T = \sum_{n=1}^{\infty} E_n e^{-n\pi y/L} \sin\left(\frac{n\pi x}{L}\right) \tag{17-56}$$

The last boundary condition, $T = T_1$ at $y = 0$, is used to evaluate E_n according to the expression

$$T_1 = \sum_{n=1}^{\infty} E_n \sin\left(\frac{n\pi x}{L}\right) \quad \text{for } 0 \leq x \leq L$$

The constants E_n are the Fourier coefficients for such an expansion and are given by

$$E_n = \frac{4T_1}{n\pi} \quad \text{for } n = 1, 3, 5, \ldots$$

and

$$E_n = 0 \quad \text{for } n = 2, 4, 6, \ldots$$

The solution to this two-dimensional conduction problem is

$$T = \frac{4T_1}{\pi} \sum_{n=0}^{\infty} \frac{e^{[-(2n+1)\pi y]/L}}{2n+1} \sin \frac{(2n+1)\pi x}{L} \tag{17-57}$$

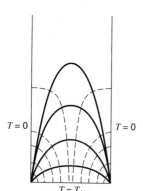

Figure 17.13 Isotherms and energy flow lines for the rectangular plate in Figure 17.12.

The isotherms and energy flow lines are plotted in Figure 17.13. The isotherms are shown in the figure as solid lines, and the dotted lines, which are orthogonal to the isotherms, are energy-flow lines. Note the similarity to the lines of constant velocity potential and stream function as discussed in momentum transfer.

The separation of variables method can be extended to three-dimensional cases by assuming T to be equal to the product $X(x)Y(y)Z(z)$ and substituting this expression for T into the applicable differential equation. When the variables are separated, three second-order ordinary differential equations are obtained, which may be integrated subject to the given boundary conditions.

Analytical solutions are useful when they can be obtained. There are, however, practical problems with complicated geometry and boundary conditions, which cannot be solved analytically. As an alternative approach, one must turn to numerical methods.

Shape Factors for Common Configurations

The shape factor, **S**, is defined and discussed briefly in Chapter 15. When a geometric case of interest involves conduction between a source and a sink, both with isothermal boundaries, a knowledge of the shape factor makes the determination of heat flow a simple calculation.

Table 17.1 lists expressions for shape factors of five configurations. In every case depicted, it is presumed that the heat transfer problem is two dimensional, that is, the dimension normal to the plane shown in very large.

Numerical Solutions

Each of the solution techniques discussed thus far for multidimensional conduction has considerable utility when conditions permit its use. Analytical solutions require relatively simple functions and geometries; the use of shape factors requires isothermal boundaries. When the situation of interest becomes sufficiently complex or when

Table 17.1 Conduction shape factors

Shape	Shape factor, S $q/L = kS(T_i - T_o)$
T_i T_b r_i r_b Concentric circular cylinders	$\dfrac{2\pi}{\ln(r_o/r_i)}$
r_i T_o r_o T_i e Eccentric circular cylinders	$\dfrac{2\pi}{\cosh^{-1}\left(\dfrac{1+\rho^2-\epsilon^2}{2\rho}\right)}$ $\rho \equiv r_i/r_o, \epsilon \equiv e/r_o$
T_b T_i r_i r_b Circular cylinder in a ⌗	$\dfrac{2\pi}{\ln(r_o/r_i) - 0.10669}$
T_i r_i r_b Circular cylinder in a ⌗	$\dfrac{2\pi}{\ln(r_o/r_i) - 0.27079}$
T_i ρ T_i r Infinite cylinder buried in s⌗	$\dfrac{2\pi}{\cosh^{-1}(\rho/r)}$

Shape Factors

boundary conditions pre⌗
simple solution techniqu⌗
to numerical solutions.

With the presence of ⌗ computers to accomplish the large number of manipulations inherent in numerical solutions rapidly and accurately, this approach is now very common. In this section we shall introduce the concepts of numerical problem formulation and solution. A more complete and detailed discussion of numerical solutions to heat conduction problems may be found in Carnahan *et al.*[7] and in Welty.[8]

Shown in Figure 17.14 is a two-dimensional representation of an element within a conducting medium. The element

Figure 17.14 Two-dimensional volume element in a conducting medium.

[7] B. Carnahan, H. A. Luther, and J. O. Wilkes, *Applied Numerical Methods,* Wiley, New York, 1969.
[8] J. R. Welty, *Engineering Heat Transfer,* Wiley, New York, 1974.

or "node" i, j is centered in the figure along with its adjacent nodes. The designation, i, j, implies a general location in a two-dimensional system where i is a general index in the x direction and j is the y index. Adjacent node indices are shown in Figure 17.14. The grid is set up with constant node width, Δx, and constant height, Δy. It may be convenient to make the grid "square," that is, $\Delta x = \Delta y$, but for now we will allow these dimensions to be different.

A direct application of equation (6-10) to node i, j yields

$$\frac{\delta Q}{dt} = \frac{\partial}{\partial t} \iiint_{\text{c.v.}} e\rho \, dV \tag{17-58}$$

The heat input term, $\delta Q/dt$, may be evaluated allowing for conduction into node i, j from the adjacent nodes and by energy generation within the medium. Evaluating $\delta Q/dt$ in this manner, we obtain

$$\frac{\delta Q}{dt} = k\frac{\Delta y}{\Delta x}(T_{i-1,j} - T_{i,j}) + k\frac{\Delta y}{\Delta x}(T_{i+1,j} - T_{i,j})$$
$$+ k\frac{\Delta x}{\Delta y}(T_{i,j-1} - T_{i,j}) + k\frac{\Delta x}{\Delta y}(T_{i,j+1} - T_{i,j}) + \dot{q}\,\Delta x\,\Delta y \tag{17-59}$$

The first two terms in this expression relate conduction in the x direction, the third and fourth express y-directional conduction, and the last is the generation term. All of these terms are positive; heat transfer is assumed positive.

The rate of energy increase within node i, j may be written simply as

$$\frac{\partial}{\partial t} \iiint_{\text{c.v.}} e\rho \, dV = \left[\frac{\rho c T|_{t+\Delta t} - \rho c T|_t}{\Delta t}\right]\Delta x\,\Delta y \tag{17-60}$$

Equation (17-58) indicates that the expressions given by equations (17-59) and (17-60) may be equated. Setting these expressions equal to each other and simplifying, we have

$$k\frac{\Delta y}{\Delta x}[T_{i-1,j} + T_{i+1,j} - 2T_{i,j}] + k\frac{\Delta x}{\Delta y}[T_{i,j-1} + T_{i,j+1} - 2T_{i,j}]$$
$$+ \dot{q}\,\Delta x\,\Delta y = \left[\frac{\rho c T_{i,j}|_{t+\Delta t} - \rho c T_{i,j}|_t}{\Delta t}\right]\Delta x\,\Delta y \tag{17-61}$$

This expression has been considered in a more complete form in the next chapter. For the present we will not consider time-variant terms; moreover, we will consider the nodes to be square, that is, $\Delta x = \Delta y$. With these simplifications equation (17-61) becomes

$$T_{i-1,j} + T_{i+1,j} + T_{i,j-1} + T_{i,j+1} - 4T_{i,j} + \dot{q}\frac{\Delta x^2}{k} = 0 \tag{17-62}$$

In the absence of internal generation, equation, (17-62) may be solved for T_{ij} to yield

$$T_{i,j} = \frac{T_{i-1,j} + T_{i+1,j} + T_{i,j-1} + T_{i,j+1}}{4} \tag{17-63}$$

or, the temperature of node i, j is the arithmetic mean of the temperatures of its adjacent nodes. A simple example showing the use of equation (17-63) in solving a two-dimensional heat conduction problem follows.

EXAMPLE 4

A hollow square duct of the configuration shown (left) has its surfaces maintained at 200 and 100 K, respectively. Determine the steady-state heat transfer rate between the hot and cold surfaces of this duct. The wall material has a thermal conductivity of 1.21 W/m·K. We may take advantage of the eightfold symmetry of this figure to lay out the simple

The grid chosen is square with $\Delta x = \Delta y = 1/2$m. Three interior node points are thus identified; their temperatures may be determined by proper application of equation (17-63). Writing the proper expressions for T_1, T_2, and T_3 using equation (17-68) as a guide, we have

$$T_1 = \frac{200 + 100 + 2T_2}{4}$$

$$T_2 = \frac{200 + 100 + T_1 + T_3}{4}$$

$$T_3 = \frac{100 + 100 + 2T_2}{4}$$

This set of three equations and three unknowns may be solved quite easily to yield the following: $T_1 = 145.83$ K, $T_2 = 141.67$ K, $T_3 = 120.83$ K.

The temperatures just obtained may now be used to find heat transfer. Implicit in the procedure of laying out a grid of the sort, we have specified is the assumption that heat flows in the x and y directions between nodes. On this basis heat transfer occurs from the hot surface to the interior only to nodes 1 and 2; heat transfer occurs to the cooler surface from nodes 1, 2, and 3. We should also recall that the section of duct that has been analyzed is one-eighth of the total thus, of the heat transfer to and from node 1, only one half should be properly considered as part of the element analyzed.

We now solve for the heat transfer rate from the hotter surface, and write

$$q = \frac{k(200 - T_1)}{2} + k(200 - T_2)$$

$$= k\left[\left(\frac{200 - 145.83}{2}\right) + (200 - 141.67)\right]$$

$$= 85.415k \quad (q \text{ in W/m}, k \text{ in W/m·K})$$

A similar accounting for the heat flow from nodes 1, 2, and 3 to the cooler surface is written as

$$q = \frac{k(T_1 - 100)}{2} + k(T_2 - 100) + k(T_3 - 100)$$

$$= k\left[\left(\frac{145.83 - 100}{2}\right) + (141.67 - 100) + (120.83 - 100)\right]$$

$$= 85.415k \quad (q \text{ in W/m}, k \text{ in W/m·K})$$

Observe that these two different means of solving for q yield identical results. This is obviously a requirement of the analysis and serves as a check on the formulation and numerical work.

The example may now be concluded. The total heat transfer per meter of duct is calculated as

$$q = 8(8.415 \text{ K})(1.21 \text{ W/m·K})$$

$$= 826.8 \text{ W/m}$$

Example 4 has illustrated, in simple fashion, the numerical approach to solving two-dimensional steady-state conduction problems. It is apparent that any added complexity in the form of more involved geometry, other types of boundary conditions such as convection, radiation, specified heat flux, among others, or simply a greater number of interior nodes, will render a problem too complex for hand calculation. Techniques for formulating such problems and some solution techniques are described by Welty.

In this section, we have considered techniques for solving two- and three-dimensional steady-state conduction problems. Each of these approaches has certain requirements that limit their use. The analytical solution is recommended for problems of simple geometrical shapes and simple boundary conditions. Numerical techniques may be used to solve complex problems involving nonuniform boundary conditions and variable physical properties.

17.5 CLOSURE

In this chapter, we have considered solutions to steady-state conduction problems. The defining differential equations were frequently established by generating the equation through the use of the control-volume expression for the conservation of energy as well as by using the general differential equation for energy transfer. It is hoped that this approach will provide the student with an insight into the various terms contained in the general differential equation and thus enable one to decide, for each solution, which terms are relevant.

One-dimensional systems with and without internal generation of energy were considered.

PROBLEMS

17.1 One-dimensional steady-state conduction, with no internal heat generation, occurs across a plane wall having a constant thermal conductivity of 30 W/m·K. The material is 30 cm thick. For each case listed in the table below, determine the unknown quantities. Show a sketch of the temperature distribution for each case.

17.2 The steady-state expression for heat conduction through a plane wall is $q = (kA/L)\Delta T$ as given by equation (17-4). For steady-state heat conduction through a hollow cylinder, an

Case	T_1	T_2	dT/dx (K/m)	q_x (W/m^2)
1	350 K	275 K		
2	300 K			-2000
3		350 K	-300	
4	250 K		200	

expression similar to equation (17-4) is

$$q = \frac{k\overline{A}}{r_o - r_i}\Delta T$$

where $\overline{A}$ is the "log-mean" area defined as

$$\overline{A} = 2\pi\frac{r_o - r_i}{\ln(r_o/r_i)}$$

a. Show that $\overline{A}$ as defined above satisfies the equations for steady-state radial heat transfer in a hollow cylindrical element.

b. If the arithmetic mean area, $\pi(r_o - r_i)$, is used rather than the logarithmic mean, calculate the resulting percent error for values of r_o/r_i of 1.5, 3, and 5.

17.3 Evaluate the appropriate "mean" area for steady-state heat conduction in a hollow sphere that satisfies an equation of the form

$$q = \frac{k\overline{A}}{r_o - r_i}\Delta T$$

Repeat part (b) of Problem 17.2 for the spherical case.

17.4 It is desired to transport liquid metal through a pipe embedded in a wall at a point where the temperature is 650 K. A 1.2-m-thick wall constructed of a material having a thermal conductivity varying with temperature according to

$k = 0.0073(1 + 0.0054\ T)$, where T is in K and k is in W/m·K, has its inside surface maintained at 925 K. The outside surface is exposed to air at 300 K with a convective heat-transfer coefficient of 23 W/m²·K. How far from the hot surface should the pipe be located? What is the heat flux for the wall?

17.5 The temperature at the inner and outer surfaces of a plane wall of thickness L are held at the constant values T_0 and T_L, respectively, where $T_0 > T_L$. The wall material has a thermal conductivity that varies linearly according to $k = k_0(1 + \beta T)$, k_0 and β being constants. At what point will the actual temperature profile differ most from that which would exist in the case of constant thermal conductivity?

17.6 Solve Problem 17.5 for the case of a hollow cylinder with boundary conditions $T = T_0$ at $r = R_0$ and $T = T_L$ at $r = R_0 + L$.

17.7 A double-pane insulated window unit consists of two 1-cm-thick pieces of glass separated by a 1.8-cm layer of air. The unit measures 4 m in width and is 3 m wide. Under conditions where the extreme outside temperature of the glass is at $-10°C$ and air, at 27°C, is adjacent to the inside glass surface, with $h_i = 12$ W/m²·K determine

a. the inside glass surface temperature.

b. the rate of heat transfer through the window unit.

The air gap between glass panes may be treated as a purely conductive layer with $k = 0.0262$ W/m·K. Thermal radiation is to be neglected.

17.8 A furnace wall is to be designed to transmit a maximum heat flux of 200 Btu/h ft² of wall area. The inside and outside wall temperatures are to be 2000 and 300°F, respectively. Determine the most economical arrangement of bricks measuring 9 by 4 ½ by 3 in. if they are made from two materials, one with a k of 0.44 Btu/h ft °F and a maximum usable temperature of 1500°F and other with a k of 0.94 Btu/h ft °F and a maximum usable temperature of 2200 °F. Bricks made of each material cost the same amount and may be laid in any manner.

17.9 A furnace wall consisting of 0.25 m of fire clay brick, 0.20 m of kaolin, and a 0.10-m outer layer of masonry brick is exposed to furnace gas at 1370 K with air at 300 K adjacent to the outside wall. The inside and outside convective heat transfer coefficients are 115 and 23 W/m²·K, respectively. Determine the heat loss per square foot of wall and the temperature of the outside wall surface under these conditions.

17.10 Given the conditions of Problem 17.9, except that the outside temperature of the masonry brick cannot exceed 325 K, by how much must the thickness of kaolin be adjusted to satisfy this requirement?

17.11 A heater composed of Nichrome wire wound back and forth and closely spaced is covered on both sides with ⅛-in. thickness of asbestos ($k = 0.15$ Btu/h ft °F) and then with a ⅛-in. thickness of stainless steel ($k = 10$ Btu/h ft °F). If the center temperature of this sandwich construction is considered constant at 1000°F and the outside convective heat-transfer

coefficient is 3 Btu/h ft² °F, how much energy must be supplied in W/ft² to the heater? What will be the outside temperature of the stainless steel?

17.12 Determine the percent increase in heat flux if, in addition to the conditions specified in Problem 17.11, there are two ¾-in.-diameter steel bolts extending through the wall per square foot of wall area (k for steel $= 22$ Btu/h ft °F).

17.13 A 2.5-cm-thick sheet of plastic ($k = 2.42$ W/m·K) is to be bonded to a 5-cm-thick aluminum plate. The glue that will accomplish the bonding is to be held at a temperature of 325 K to achieve the best adherence, and the heat to accomplish this bonding is to be provided by a radiant source. The convective heat-transfer coefficient on the outside surfaces of both the plastic and aluminum is 12 W/m²·K, and the surrounding air is at 295 K. What is the required heat flux if it is applied to the surface of (a) the plastic? (b) the aluminum?

17.14 A composite wall is to be constructed of ¼ in. of stainless steel ($k = 10$ Btu/h ft °F), 3 in. of corkboard ($k = 0.025$ Btu/h ft °F), and ½ in. of plastic ($k = 1.5$ Btu/h ft °F). Determine the thermal resistance of this wall if it is bolted together by ½-in.-diameter bolts on 6-in. centers made of

a. stainless steel;

b. aluminum ($k = 120$ Btu/h ft °F).

17.15 A cross section of a typical home ceiling is depicted below. Given the properties listed for the materials of construction, determine how much heat is transferred through the insulation and through the studs.

$$
\begin{aligned}
T_{outside} &= -10°C \\
h_o &= 20\ \text{W/m}^2 \cdot \text{K} \\
T_{inside} &= 25°C \\
h_i &= 10\ \text{W/m}^2 \cdot \text{K} \\
k_{fiberglass} &= 0.035\ \text{W/m}^2 \cdot \text{K} \\
k_{plaster} &= 0.814\ \text{W/m}^2 \cdot \text{K} \\
k_{wood} &= 0.15\ \text{W/m}^2 \cdot \text{K}
\end{aligned}
$$

17.16 A copper bus bar measuring 5 cm by 10 cm by 2.5 m long is in a room in which the air is maintained at 300 K. The bus bar is supported by two plastic pedestals to which it is attached by an adhesive. The pedestals are square in cross section, measuring 8 cm on a side. The pedestals are mounted on a wall whose temperature is 300 K. If 1 kW of energy is dissipated in the copper bar, what will be its equilibrium temperature? The convective heat-transfer coefficient for all surfaces may be taken as 23 W/m²·K. The thermal conductivity of the plastic is 2.6 W/m·K. Neglect thermal radiation.

17.17 Solve Problem 17.16 if each plastic pedestal has a 1.9-cm steel bolt running through the center.

17.18 A 2-in. schedule-40 steel pipe carries saturated steam at 60 psi through a laboratory that is 60 ft long. The pipe is insulated with 1.5 in. of 85% magnesia that costs $0.75 per foot. How long must the steam line be in service to justify the insulation cost if the heating cost for the steam is $0.68 per 10⁵ Btu? The outside-surface convective heat-transfer coefficient may be taken as 5 Btu/h ft² °F.

17.19 Saturated steam at 40 psia flows at 5 fps through a schedule-40, 1½-in. steel pipe. The convective heat-transfer coefficient by condensing steam on the inside surface may be taken as 1500 Btu/h ft² °F. The surrounding air is at 80°F, and the outside surface coefficient is 3 Btu/ft² °F. Determine the following:

a. The heat loss per 10 ft of bare pipe.

b. The heat loss per 10 ft of pipe insulated with 2 in. of 85% magnesia.

c. The mass of steam condensed in 10 ft of bare pipe.

17.20 A 10-kW heater using Nichrome wire is to be designed. The surface of the Nichrome is to be limited to a maximum temperature of 1650 K. Other design criteria for the heater are

minimum convective heat-transfer coefficient: 850 W/m²·K

minimum temperature of the surrounding medium (air): 370 K
The resistivity of Nichrome is 110 $\mu\Omega$-cm and the power to the heater is available at 12 V.

a. What size wire is required if the heater is to be in one piece 0.6 m long?

b. What length of 14-gage wire is necessary to satisfy these design criteria?

c. How will the answers to parts (a) and (b) change if $h = 1150$ W/m²·K?

17.21 Copper wire having a diameter of 3/16 in. is insulated with a 4-in. layer of material having a thermal conductivity of 0.14 Btu/h ft °F. The outer surface of the insulation is maintained at 70°F. How much current may pass through the wire if the insulation temperature is limited to a maximum of 120°F? The resistivity of copper is 1.72×10^{-6} ohm-cm.

17.22 What would be the result of Problem 17.21 if the fluid surrounding the insulated wire was maintained at 70°F with a convective heat-transfer coefficient between the insulation and the fluid of 4 Btu/h ft² °F? What would be the surface temperature of the insulation under these conditions?

17.23 Work Problem 17.21 for the case of aluminum rather than copper. The resistivity of aluminum is 2.83×10^{-6} ohm-cm.

17.24 A thin slab of material is subjected to microwave radiation that causes volumetric heating to vary according to

$$\dot{q}(x) = \dot{q}_o[1 - (x/L)]$$

where $\dot{q}_o$ has a constant value of 180 kW/m³ and the slab thickness, L, is 0.06 m. The thermal conductivity of the slab material is 0.6 W/m·K.

The boundary at $x = L$ is perfectly insulated, while the surface at $x = 0$ is maintained at a constant temperature of 320 K.

a. Determine an expression for $T(x)$ in terms of $X, L, k, \dot{q}_o$, and T_o.

b. Where, in the slab, will the maximum temperature occur?

c. What is the value of T_{max}?

17.25 Radioactive waste ($k = 20$ W/m·K) is stored in a cylindrical stainless steel ($k = 15$ W/m·K) container with inner and outer diameters of 1.0 and 1.2 m, respectively. Thermal energy is generated uniformly within the waste material at a volumetric rate of 2×10^5 W/m³. The outer container surface is exposed to water at 25°C, with a surface coefficient of 1000 W/m²·K. The ends of the cylindrical assembly are insulated so that all heat transfer occurs in the radial direction. For this situation determine

a. the steady-state temperatures at the inner and outer surfaces of the stainless steel.

b. the steady-state temperature at the center of the waste material.

17.26 A cylindrical nuclear fuel element is 10.16 cm long and 10.77 cm in diameter. The fuel generates heat uniformly at a rate of 51.7×10^3 kJ/s·m³. The fuel is placed in an environment having a temperature of 360 K with a surface coefficient of 4540 W/m²·K. The fuel material has $k = 33.9$ W/m·K. For the situation described evaluate the following at steady state:

a. The temperature profile as a function of radial position;

b. The maximum fuel temperature;

c. The surface temperature.

End effects may be neglected.

17.27 Liquid nitrogen at 77 K is stored in an insulated spherical container that is vented to the atmosphere. The container is made of a thin-walled material with an outside diameter of 0.5 m; 25 mm of insulation ($k = 0.002$ W/m·K) covers its outside surface. The latent heat of nitrogen is 200 kJ/kg; its density, in the liquid phase, is 804 kg/m^3. For surroundings at 25°C and with a convective coefficient of 18 W/m^2·K at the outside surface of the insulation, what will be the rate of liquid nitrogen boil-off?

17.28 What additional thickness of insulation will be necessary to reduce the boil-off rate of liquid nitrogen to one-half of the rate corresponding to Problem 17.27? All values and dimensions in Problem 17.27 apply.

17.29 A 1-in.-OD steel tube has its outside wall surface maintained at 250°F. It is proposed to increase the rate of heat transfer by adding fins of 3/32-in. thickness and 3/4 in. long to the outside tube surface. Compare the increase in heat transfer achieved by adding 12 longitudinal straight fins or circular fins with the same total surface area as the 12 longitudinal fins. The surrounding air is at 80°F, and the convective heat-transfer coefficient is 6 Btu/h ft^2 °F.

17.30 Solve Problem 17.29 if the convective heat-transfer coefficient is increased to 60 Btu/h ft^2 °F by forcing air past the tube surface.

17.31 A cylindrical rod 3 cm in diameter is partially inserted into a furnace with one end exposed to the surrounding air, which is at 300 K. The temperatures measured at two points 7.6 cm apart are 399 and 365 K, respectively. If the convective heat-transfer coefficient is 17 W/m^2·K, determine the thermal conductivity of the rod material.

17.32 Heat is to be transferred from water to air through an aluminum wall. It is proposed to add rectangular fins 0.05 in. thick and 3/4 in. long spaced 0.08 in. apart to the aluminum surface to aid in transferring heat. The heat-transfer coefficients on the air and water sides are 3 and 25 Btu/h ft^2 °F, respectively. Evaluate the percent increase in heat transfer if these fins are added to (a) the air side, (b) the water side, (c) and both sides. What conclusions may be reached regarding this result?

17.33 A semiconductor material with $k = 2$ W/m·K and electrical resistivity, $\rho = 2 \times 10^{-5}$ Ω-m, is used to fabricate a cylindrical rod 40 mm long with a diameter of 10 mm. The longitudinal surface of the rod is well insulated and may be considered adiabatic while the ends are maintained at temperatures of 100 and 0°C, respectively. If the rod carries a current of 10 amps, what will be its midpoint temperature? What will be the rate of heat transfer through both ends?

17.34 An iron bar used for a chimney support is exposed to hot gases at 625 K with the associated convective heat-transfer coefficient of 740 W/m^2·K. The bar is attached to two opposing chimney walls, which are at 480 K. The bar is 1.9 cm in diameter and 45 cm long. Determine the maximum temperature in the bar.

17.35 A copper rod 1/4 in. in diameter and 3 ft long runs between two bus bars, which are at 60°F. The surrounding air is at 60°F, and the convective heat-transfer coefficient is Btu/h ft^2 °F. Assuming the electrical resistivity of copper to be constant at 1.72×10^{-6} ohm-cm, determine the maximum current the copper may carry if its temperature is to remain below 150°F.

17.36 A 13 cm by 13 cm steel angle with the dimensions shown is attached to a wall with a surface temperature of 600 K. The surrounding air is at 300 K, and the convective heat-transfer coefficient between the angle surface and the air is 45 W/m^2·K.

a. Plot the temperature profile in the angle, assuming a negligible temperature drop through the side of the angle attached to the wall.

b. Determine the heat loss from the sides of the angle projecting out from the wall.

Thickness = 1.59 cm

←— 13 cm —→|

17.37 A steel I-beam with a cross-sectional area as shown has its lower and upper surfaces maintained at 700 and 370 K, respectively.

a. Assuming a negligible temperature change through both flanges, develop an expression for the temperature variation in the web as a function of the distance from the upper flange.

b. Plot the temperature profile in the web if the convective heat-transfer coefficient between the steel surface and the surrounding air is 57 W/m^2·K. The air temperature is 300 K.

c. What is the net heat transfer at the upper and lower ends of the web?

↓y

←— 0.76 cm

28 cm

17.38 Repeat Problem 17.37 for the case of an aluminum beam.

17.39 Circular fins are employed around the cylinder of a lawn mower engine to dissipate heat. The fins are made of aluminum, they are 0.3 m thick, and extend 2 cm from base to tip. The outside diameter of the engine cylinder is 0.3 m. Design operating conditions are $T_\infty = 30°C$ and $h = 12\ W/m^2 \cdot K$. The maximum allowable cylinder temperature is 300°C.

Estimate the amount of heat transfer from a single fin. How many fins are required to cool a 3-kW engine, operating at 30% thermal efficiency, if 50% of the total heat given off is transferred by the fins?

17.40   Heat from a flat wall is to be enhanced by adding straight fins, of constant thickness, made of stainless steel. The following specifications apply:

h	$= 60\ W/m^2\ K$
T_b(base)	$= 120°C$
T_∞(air)	$= 20°C$
Fin base thickness, t	$= 6\ mm$
Fin length, L	$= 20\ mm$

Determine the fin efficiency and heat loss per unit width for the finned surface.

17.41   A 2-in.-OD stainless-steel tube has 16 longitudinal fins spaced around its outside surface as shown. The fins are 1/16 in. thick and extend 1 in. from the outside surface of the tube.

a. If the outside surface of the tube wall is at 250°F, the surrounding air is at 80°F, and the convective heat-transfer coefficient is 8 Btu/h ft^2 °F, determine the percent increase in heat transfer for the finned pipe over that for the unfinned pipe.

b. Determine the same information as in part (a) for values of h of 2, 5, 15, 50, and 100 Btu/h ft^2 °F. Plot the percent increase in q vs. h. What conclusions can be reached concerning this plot?

17.42   Repeat Problem 17.41 for the case of an aluminum pipe-and-fin arrangement.

17.43   Water flows in the channels between two aluminum plates as shown in the sketch. The ribs that form the channels are also made of aluminum and are 8 mm thick. The effective surface coefficient between all surfaces and water is 300 W/m^2·K. For these conditions, how much heat is transferred at each end of each rib? How far from the lower plate

is the rib temperature a minimum? What is this minimum value?

17.44   Find the rate of heat transfer from a 3-in.-OD pipe placed eccentrically inside a 6-in.-ID cylinder with the axis of the smaller pipe displaced 1 in. from the axis of the large cylinder. The space between the cylindrical surfaces is filled with rock wool ($k = 0.023$ Btu/h ft °F). The surface temperatures at the inside and outside surfaces are 400 and 100 °F, respectively.

17.45   A cylindrical tunnel with a diameter of 2 m is dug in permafrost ($k = -0.341\ W/m^2 \cdot K$) with its axis parallel to the permafrost surface at the depth of 2.5 m.

Determine the rate of heat loss from the cylinder walls, at 280 K, to the permafrost surface at 220 K.

17.46   Determine the heat flow per foot for the configuration shown, using the numerical procedure for a grid size of 1 1/2 ft. The material has a thermal conductivity of 0.15 Btu/h ft °F. The inside and outside temperatures are at the uniform values of 200 and 0 °F, respectively.

17.47   Repeat the previous problem, using a grid size of 1 ft.

17.48   A 5-in. standard-steel angle is attached to a wall with a surface temperature of 600 °F. The angle supports a 4.375-in. by 4.375-in. section of building brine whose mean thermal conductivity may be taken as 0.38 Btu/h ft °F. The convective heat-transfer coefficient between all surfaces and the surrounding air is 8 Btu/h ft^2 °F. The air temperature is 80°F. Using numerical methods, determine

a.   the total heat loss to the surrounding air;

b. the location and value of the minimum temperature in the brick.

17.49 Saturated steam at 400 °F is transported through the 1-ft pipe shown in the figure, which may be assumed to be at the steam temperature. The pipe is centered in the 2-ft-square duct, whose surface is at 100 °F. If the space between the pipe and duct is filled with powdered 85% magnesia insulation, how much steam will condense in a 50-ft length of pipe?

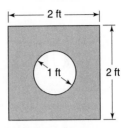

17.50 A 32.4-cm-OD pipe, 145 cm long, is buried with its centerline 1.2 m below the surface of the ground. The ground surface is at 280 K and the mean thermal conductivity of the soil is 0.66 W/m·K. If the pipe surface is at 370 K, what is the heat loss per day from the pipe?

Chapter 18

Unsteady-State Conduction

$\mathbf{T}$ransient processes, in which the temperature at a given point varies with time, will be considered in this chapter. As the transfer of energy is directly related to the temperature gradient, these processes involve an *unsteady-state* flux of energy.

Transient conduction processes are commonly encountered in engineering design. These design problems generally fall into two categories: the process that ultimately reaches steady-state conditions, and the process that is operated a relatively short time in a continually changing temperature environment. Examples of this second category would include metal stock or ingots undergoing heat treatment, missile components during reentry into Earth's atmosphere, or the thermal response of a thin laminate being bonded using a laser source.

In this chapter, we shall consider problems and their solutions that deal with unsteady-state heat transfer within systems both with and without internal energy sources.

18.1 ANALYTICAL SOLUTIONS

The solution of an unsteady-state conduction problem is, in general, more difficult than that for a steady-state problem because of the dependence of temperature on both time and position. The solution is approached by establishing the defining differential equation and the boundary conditions. In addition, the initial temperature distribution in the conducting medium must be known. By finding the solution to the partial differential equation that satisfies the initial and boundary conditions, the variation in the temperature distribution with time is established, and the flux of energy at a specific time can then be evaluated.

In heating or cooling a conducting medium, the rate of energy transfer is dependent upon both the internal and surface resistances, the limiting cases being represented either by negligible internal resistance or by negligible surface resistance. Both of these cases will be considered, as well as the more general case in which both resistances are important.

Lumped Parameter Analysis—Systems with Negligible Internal Resistance

Equation (16-17) will be the starting point for transient conduction analysis. It is repeated below for reference.

$$\frac{\partial T}{\partial t} = \alpha \, \mathbf{\nabla}^2 T + \frac{\dot{q}}{\rho c_p} \tag{16-17}$$

Recall that, in the derivation of this expression, thermal properties were taken to be independent of position and time; however, the rate of internal generation, $\dot{q}$, can vary in both.

It is frequently the case that temperature within a medium varies significantly in fewer than all three space variables. A circular cylinder, heated at one end with a fixed boundary condition, will show a temperature variation in the axial and radial directions as well as time. If the cylinder has a length that is large compared to its diameter or, if it is composed of a material with high thermal conductivity, temperature will vary with axial position and time only. If a metallic specimen, initially with uniform temperature, is suddenly exposed to surroundings at a different temperature, it may be that size, shape, and thermal conductivity may combine in such a way that the temperature within the material varies with time only, that is, is not a significant function of position. These conditions are characteristic of a "lumped" system, where the temperature of a body varies only with time; this case is the easiest of all to analyze. Because of this we will consider, as our first transient conduction case, that of a completely lumped-parameter system.

$T(0) = T_0$ (uniform)
$T(t) = T$

Figure 18.1

Shown in Figure 18.1, we have a spherical metallic specimen, initially at uniform temperature T_0 after it has been immersed in a hot oil bath at temperature T_∞ for a period of time t. It is presumed that the temperature of the metallic sphere is uniform at any given time. A first-law analysis using equation (6-10), applied to a spherical control volume coinciding with the specimen in question will reduce to

$$\frac{\delta Q}{dt} = \frac{\partial}{\partial t} \iiint_{\text{c.v.}} e\rho \, dV \tag{18-1}$$

The rate of heat addition to the control volume, $\delta Q/dt$, is due to convection from the oil and is written as

$$\frac{\delta Q}{dt} = hA(T_\infty - T) \tag{18-2}$$

The rate of energy increase within the specimen, $\partial/\partial t \iiint_{\text{c.v.}} e\rho \, dV$, with constant properties, may be expressed as

$$\frac{\partial}{\partial t} \iiint_{\text{c.v.}} e\rho \, dV = \rho V c_p \frac{dT}{dt} \tag{18-3}$$

Equating these expressions as indicated by equation (18-1) we have, with slight rearrangement

$$\frac{dT}{dt} = \frac{hA(T_\infty - T)}{\rho V c_p} \tag{18-4}$$

We may now obtain a solution for the temperature variation with time by solving equation (18-4) subject to the initial condition, $T(0) = T_0$, and obtain

$$\frac{T - T_\infty}{T_0 - T_\infty} = e^{-hAt/\rho c_p V} \tag{18-5}$$

The exponent is observed to be dimensionless. A rearrangement of terms in the exponent may be accomplished as follows:

$$\frac{hAt}{\rho c_p V} = \left(\frac{hV}{kA}\right)\left(\frac{A^2 k}{\rho V^2 c_p} t\right) = \left[\frac{hV/A}{k}\right]\left[\frac{\alpha t}{(V/A)^2}\right] \tag{18-6}$$

Each of the bracketed terms in equation (18-6) is dimensionless. The ratio, V/A, having units of length, is also seen to be a part of each of these new parametric forms. The first of the new nondimensional parameters formed is the *Biot modulus,* abbreviated Bi

$$\text{Bi} = \frac{hV/A}{k} \qquad (18\text{-}7)$$

By analogy with the concepts of thermal resistance, discussed at length earlier, the Biot modulus is seen to be the ratio of $(V/A)/k$, the conductive (internal) resistance to heat transfer, to $1/h$, the convective (external) resistance to heat transfer. The magnitude of Bi thus has some physical significance in relating where the greater resistance to heat transfer occurs. A large value of Bi indicates that the conductive resistance controls, that is, there is more capacity for heat to leave the surface by convection than to reach it by conduction. A small value for Bi represents the case where internal resistance is negligibly small and there is more capacity to transfer heat by conduction than there is by convection. In this latter case, the controlling heat transfer phenomenon is convection, and temperature gradients within the medium are quite small. An extremely small internal temperature gradient is the basic assumption in a lumped-parameter analysis.

A natural conclusion to the foregoing discussion is that the magnitude of the Biot modulus is a reasonable measure of the likely accuracy of a lumped-parameter analysis. A commonly used rule of thumb is that the error inherent in a lumped-parameter analysis will be less than 5% for a value of Bi less than 0.1. The evaluation of the Biot modulus should thus be the first thing done when analyzing an unsteady-state conduction situation.

The other bracketed term in equation (18-6) is the *Fourier modulus,* abbreviated Fo, where

$$\text{Fo} = \frac{\alpha t}{(V/A)^2} \qquad (18\text{-}8)$$

The Fourier modulus is frequently used as a nondimensional time parameter.

The lumped-parameter solution for transient conduction may now be written as

$$\frac{T - T_\infty}{T_0 - T_\infty} = e^{-\text{BiFo}} \qquad (18\text{-}9)$$

Equation (18-9) is portrayed graphically in Figure 18.2. The use of equation (18-9) is illustrated in the following example.

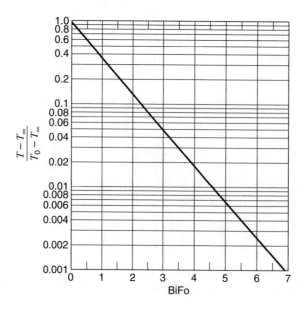

Figure 18.2 Time-temperature history of a body at initial temperature, T_0, exposed to an environment at T_∞; lumped-parameter case.

EXAMPLE 1

A long copper wire, 0.635 cm in diameter, is exposed to an air stream at a temperature, T_∞, of 310 K. After 30 s the average temperature of the wire increased from 280 to 297 K. Using this information, estimate the average surface conductance, h.

In order to determine if equation () is valid for this problem, the value of Bi must be determined. The Biot number is expressed as

$$\text{Bi} = \frac{hV/A}{k} = \frac{h\dfrac{\pi D^2 L}{4\pi DL}}{386\,\text{W/m}\cdot\text{K}} = \frac{h\dfrac{0.00635\,\text{m}}{4}}{386\,\text{W/m}\cdot\text{K}}$$

$$= 4.11 \times 10^{-6} h$$

Setting $\text{Bi} = 0.1$, which is the limiting value of Bi for a lumped-parameter analysis to be valid, and solving for h, we obtain

$$h = 0.1/4.11 \times 10^{-6} = 24300 \ \text{W/m}^2 \cdot \text{K}$$

We may conclude that a lumped-parameter solution is valid if $h < 24,300 \ \text{W/m}^2 \cdot \text{K}$, which is a near certainty.

Proceeding, we can apply equation () to yield

$$h = \frac{\rho c_p V}{tA} \ln \frac{T_0 - T_\infty}{T - T_\infty}$$

$$= \frac{(8890\,\text{kg/m}^3)(385\,\text{J/kg}\cdot\text{K})\left(\dfrac{\pi D^2 L}{4\pi DL}\right)}{(30\,\text{s})} \ln \frac{280 - 310}{297 - 310}$$

$$= 51 \ \text{W/m}^2 \cdot \text{K}$$

This result is much less than the limiting value of h indicating that a lumped-parameter solution is probably very accurate.

Heating a Body Under Conditions of Negligible Surface Resistance. A second class of time-dependent energy-transfer processes is encountered when the surface resistance is small relative to the overall resistance, that is, Bi is $\gg 0.1$. For this limiting case, the temperature of the surface, T_s is constant for all time, $t > 0$, and its value is essentially equal to the ambient temperature, T_∞.

To illustrate the analytical method of solving this class of transient heat-conduction problems, consider a large flat plate of uniform thickness L. The initial temperature distribution through the plate will be assumed to be an arbitrary function of x. The solution for the temperature history must satisfy the Fourier field equation

$$\frac{\partial T}{\partial t} = \alpha \nabla^2 T \tag{16-18}$$

For one-directional energy flow

$$\frac{\partial T}{\partial t} = \alpha \frac{\partial^2 T}{\partial x^2} \tag{18-10}$$

with initial and boundary conditions

$$T = T_0(x) \quad \text{at } t = 0 \quad \text{for } 0 \le x \le L$$
$$T = T_s \quad \ \ \text{at } x = 0 \quad \text{for } t > 0$$

and

$$T = T_s \quad \text{at } x = L \quad \text{for } t > 0$$

For convenience, let $Y = (T - T_s)/(T_0 - T_s)$, where T_0 is an arbitrarily chosen reference temperature; the partial differential equation may be rewritten in terms of the new temperature variable as

$$\frac{\partial Y}{\partial t} = \alpha \frac{\partial^2 Y}{\partial x^2} \tag{18-11}$$

and the initial and boundary conditions become

$$Y = Y_0(x) \quad \text{at } t = 0 \quad \text{for } 0 \leq x \leq L$$
$$Y = 0 \quad \quad \text{at } x = 0 \quad \text{for } t > 0$$

and

$$Y = 0 \quad \text{at } x = L \quad \text{for } t > 0$$

Solving equation (18-11) by the method of separation of variables leads to product solutions of the form

$$Y = (C_1 \cos \lambda x + C_2 \sin \lambda x) e^{-\alpha \lambda^2 t}$$

The constants C_1 and C_2 and the parameter λ are obtained by applying the initial and boundary conditions. The complete solution is

$$Y = \frac{2}{L} \sum_{n=1}^{\infty} \sin\left(\frac{n\pi}{L}\right) e^{-(n\pi/2)^2 \text{Fo}} \int_0^L Y_0(x) \sin \frac{n\pi}{L} x \, dx \tag{18-12}$$

where $\text{Fo} = \alpha t/(L/2)^2$. Equation (18-12) points out the necessity for knowing the initial temperature distribution in the conducting medium, $Y_0(x)$, before the complete temperature history may be evaluated. Consider the special case in which the conducting body has a uniform initial temperature, $Y_0(x) = Y_0$. With this temperature distribution, equation (18-12) reduces to

$$\frac{T - T_s}{T_0 - T_s} = \frac{4}{\pi} \sum_{n=1}^{\infty} \frac{1}{n} \sin\left(\frac{n\pi}{L} x\right) e^{-(n\pi/2)^2 \text{Fo}} \quad n = 1, 3, 5, \ldots \tag{18-13}$$

The temperature history at the center of the infinite plane, as well as the central temperature history in other solids, is illustrated in Figure 18.3. The central temperature history for the plane wall, infinite cylinder, and sphere is presented in Appendix F, in "Heissler charts." These charts cover a much greater range in the Fourier modulus than Figure 18.3.

The heat rate, q, at any plane in the conducting medium may be evaluated by

$$q_x = -kA \frac{\partial T}{\partial x} \tag{18-14}$$

In the case of the infinite flat plate with an initial uniform temperature distribution of T_0, the heat rate at any time t is

$$q_x = 4\left(\frac{kA}{L}\right)(T_s - T_0) \sum_{n=1}^{\infty} \cos\left(\frac{n\pi}{L} x\right) e^{-(n\pi/2)^2 \text{Fo}} \quad n = 1, 3, 5, \ldots \tag{18-15}$$

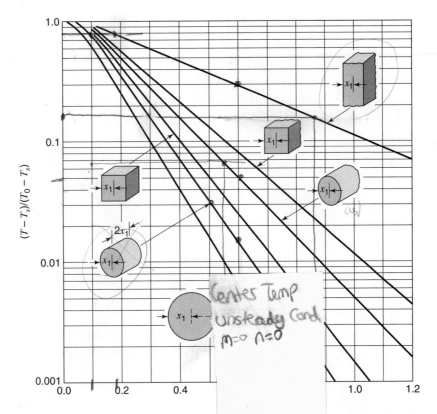

Figure 18.3 Central temperature history (...) initial uniform temperature, T_0, and constant surface temperature, T_s (From (...) uction Heat Transfer, Addison-Wesley Publishing Co., Inc., Reading Mas(...) rmission of the publishers.)

In the following example, the use of the (...) istory figure will be illustrated.

EXAMPLE 2

A concrete cylinder, 0.1 m in length and 0.1 (...) y at room temperature, 292 K. It is suspended in a steam environment where water vapor at 373 K condenses on all surfaces with an effective film coefficient, h, of 8500 W/m^2 · K. Determine the time required for the center of this stubby cylinder to reach 310 K. If the cylinder were sufficiently long so that it could be considered infinite, how long would it take?

For the first case, the finite cylinder, the Biot number is evaluated as

$$\text{Bi} = \frac{h(V/A)}{k} = \frac{h\left(\dfrac{\pi D^2 L}{4}\right)}{k\left(\pi DL + \dfrac{\pi D^2}{2}\right)} = \frac{h(DL/4)}{k(L + D/2)}$$

$$= \frac{(8500 \text{ W/m}^2 \cdot \text{K})(0.1 \text{ m})(0.1 \text{ m})/4}{1.21 \text{ W/m} \cdot \text{K}(0.1 + 0.1/2) \text{ m}}$$

$$= 117$$

For this large value, Figure 18.3 may be used. The second line from the bottom in this figure applies to a cylinder with height equal to diameter, as in this case. The ordinate is

$$\frac{T - T_s}{T_0 - T_s} = \frac{310 - 373}{292 - 373} = 0.778$$

and the corresponding abscissa value is approximately 0.11. The time required may now be determined as

$$\frac{\alpha t}{x_1^2} = 0.11$$

Thus,

$$t = 0.11 \frac{(0.05 \text{ m})^2}{5.95 \times 10^{-7} \text{ m}^2/\text{s}} = 462 \text{ s}$$
$$= 7.7 \text{ min}$$

In the case of an infinitely long cylinder, the fourth line from the bottom applies. The Biot number in this case is

$$\text{Bi} = \frac{h(V/A)}{k} = \frac{h\left(\dfrac{\pi D^2 L}{4}\right)}{k(\pi D L)} = \frac{h\dfrac{D}{4}}{k}$$
$$= \frac{(8500 \text{ W/m}^2 \cdot \text{K})(0.1 \text{ m})/4}{1.21 \text{ W/m} \cdot \text{K}} = 176$$

which is even larger than the finite cylinder case. Figure 18.3 will again be used. The ordinate value of 0.778 yields, for the abscissa, a value of approximately 0.13. The required time, in this case, is

$$t = \frac{0.13(0.05 \text{ m})^2}{5.95 \times 10^{-7} \text{ m}^2/\text{s}} = 546 \text{ s}$$
$$= 9.1 \text{ min}$$

Heating a Body with Finite Surface and Internal Resistances. The most general cases of transient heat-conduction processes involve significant values of internal and surface resistances. The solution for the temperature history without internal generation must satisfy the Fourier field equation, which may be expressed for one-dimensional heat flow by

$$\frac{\partial T}{\partial t} = \alpha \frac{\partial^2 T}{\partial x^2} \tag{18-7}$$

A case of considerable practical interest is one in which a body having a uniform temperature is placed in a new fluid environment with its surfaces suddenly and simultaneously exposed to the fluid at temperature T_∞. In this case, the temperature history must satisfy the initial, symmetry, and convective boundary conditions

$$T = T_0 \qquad \text{at } t = 0$$

$$\frac{\partial T}{\partial x} = 0 \qquad \text{at the centerline of the body}$$

and

$$-\frac{\partial T}{\partial x} = \frac{h}{k}(T - T_\infty) \quad \text{at the surface}$$

One method of solution for this class of problems involves separation of variables, which results in product solutions as previously encountered when only the internal resistance was involved.

Solutions to this case of time-dependent energy-transfer processes have been obtained for many geometries. Excellent treatises discussing these solutions have been written by Carslaw and Jaeger[1] and by Ingersoll, Zobel, and Ingersoll[2]. If we reconsider the infinite flat plate of thickness, $2x_1$, when inserted into a medium at constant temperature, T_∞, but now include a constant surface conductance, h, the following solution is obtained

$$\frac{T - T_\infty}{T_0 - T_\infty} = 2 \sum_{n=1}^{\infty} \frac{\sin \delta_n \cos (\delta_n x/x_1)}{\delta_n + \sin \delta_n \cos \delta_n} e^{-\delta_n^2 \text{Fo}} \tag{18-16}$$

where δ_n is defined by the relation

$$\delta_n \tan \delta_n = \frac{hx_1}{k} \tag{18-17}$$

The temperature history for this relatively simple geometrical shape is a function of three dimensionless quantities: $\alpha t/x_1^2$, hx_1/k, and the relative distance, x/x_1.

The complex nature of equation (18-17) has led to a number of graphical solutions for the case of one-dimensional transient conduction. The resulting plots, with dimensionless temperature as a function of other dimensionless parameters as listed above, are discussed in Section 18.2.

Heat Transfer to a Semi-Infinite Wall. An analytical solution to the one-dimensional heat-conduction equation for the case of the semi-infinite wall has some utility as a limiting case in engineering computations. Consider the situation illustrated in Figure 18.4. A large plane wall initially at a constant temperature T_0 is subjected to a surface temperature T_s, where $T_s > T_0$. The differential equation to be solved is

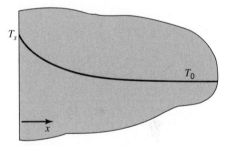

Figure 18.4 Temperature distribution in a semi-infinite wall at time t.

$$\frac{\partial T}{\partial t} = \alpha \frac{\partial^2 T}{\partial x^2} \tag{18-10}$$

and the initial and boundary conditions are

$$T = T_0 \quad \text{at } t = 0 \quad \text{for all } x$$
$$T = T_s \quad \text{at } x = 0 \quad \text{for all } t$$

and

$$T \to T_0 \quad \text{as } x \to \infty \quad \text{for all } t$$

The solution to this problem may be accomplished in a variety of ways, among which are the Laplace transformation and the Fourier transformation. We shall use an alternative procedure, which is less involved mathematically. The variables in equation (18-10)

[1] H. S. Carslaw and J. C. Jaeger, *Conduction of Heat in Solids,* Oxford University Press, 1947.

[2] L. R. Ingersoll, O. J. Zobel, and A. C. Ingersoll, *Heat Conduction (With Engineering and Geological Applications),* McGraw-Hill Book Company, New York, 1948.

may be expressed in dimensionless form by analogy with the previous case. Thus, we may write

$$\frac{T - T_0}{T_s - T_0} = f\left(\frac{x}{x_1}, \frac{\alpha t}{x_1^2}\right)$$

However, in this problem there is no finite characteristic dimension, x_1, and thus $(T - T_0)/(T_s - T_0) = f(\alpha t/x^2)$, or with equal validity, $(T - T_0)/(T_s - T_0) = f(x/\sqrt{\alpha t})$. If $\eta = x/2\sqrt{\alpha t}$ is selected as the independent variable and the dependent variable $Y = (T - T_0)/(T_s - T_0)$ is used, substitution into equation (18-10) yields the ordinary differential equation

$$d^2Y/d\eta^2 + 2\eta\, dY/d\eta = 0 \tag{18-18}$$

with the transformed boundary and initial conditions

$$Y \to 0 \quad \text{as } \eta \to \infty$$

and

$$Y = 1 \quad \text{at } \eta = 0$$

The first condition above is the same as the initial condition $T = T_0$ at $t = 0$, and the boundary condition $T \to T_0$ as $x \to \infty$. Equation (18-18) may be integrated once to yield

$$\ln \frac{dY}{d\eta} = c_1 - \eta^2$$

or

$$\frac{dY}{d\eta} = c_2 e^{-\eta^2}$$

and integrated once more to yield

$$Y = c_3 + c_2 \int e^{-\eta^2}\, d\eta \tag{18-19}$$

The integral is related to a frequently encountered form, the *error function*, designated "erf," where

$$\text{erf } \phi \equiv \frac{2}{\sqrt{\pi}} \int_0^\phi e^{-\eta^2}\, d\eta$$

and erf $(0) = 0$, erf $(\infty) = 1$, A short table of erf ϕ is given in Appendix L. Applying the boundary conditions to equation (18-19), we obtain

$$Y = 1 - \text{erf}\left(\frac{x}{2\sqrt{\alpha t}}\right)$$

or

$$\frac{T - T_0}{T_s - T_0} = 1 - \text{erf}\left(\frac{x}{2\sqrt{\alpha t}}\right)$$

or

$$\frac{T_s - T}{T_s - T_0} = \text{erf}\left(\frac{x}{2\sqrt{\alpha t}}\right) \tag{18-20}$$

This equation is extremely simple to use and quite valuable.

Consider a finite wall of thickness L subject to the surface temperature T_s. Until the temperature change at $x = L$ exceeds some nominal amount, say $(T - T_0)/(T_s - T_0)$ equal to 0.5%, the solution for the finite and infinite walls will be the same. The value of $L/(2\sqrt{\alpha t})$ corresponding to a 0.5% change in $(T - T_0)/(T_s - T_0)$ is $L/(2\sqrt{\alpha t}) \simeq 2$, so for $L/(2\sqrt{\alpha t}) > 2$, equation (18-20) may be used for finite geometry with little or no error.

For the case of finite surface resistance, the solution to equation (18-10) for a semi-infinite wall is

$$\frac{T_\infty - T}{T_\infty - T_0} = \mathrm{erf}\,\frac{x}{2\sqrt{\alpha t}} + \exp\left(\frac{hx}{k} + \frac{h^2\alpha t}{k^2}\right)\left[1 - \mathrm{erf}\left(\frac{h\sqrt{\alpha t}}{k} + \frac{x}{2\sqrt{\alpha t}}\right)\right] \qquad (18\text{-}21)$$

This equation may be used to determine the temperature distribution in finite bodies for small times in the same manner as equation (18-20). The surface temperature is particularly easy to obtain from the above equation, if we let $x = 0$, and the heat transfer rate may be determined from

$$\frac{q}{A} = h(T_s - T_\infty)$$

18.2 TEMPERATURE–TIME CHARTS FOR SIMPLE GEOMETRIC SHAPES

For unsteady-state energy transfer in several simple shapes with certain restrictive boundary conditions, the equations describing temperature profiles have been solved[3] and the results have been presented in a wide variety of charts to facilitate their use. Two forms of these charts are available in Appendix F.

Solutions are presented in Appendix F for the flat plate, sphere, and long cylinder in terms of four dimensionless ratios:

$$Y, \text{unaccomplished temperature change} = \frac{T_\infty - T}{T_\infty - T_0}$$

$$X, \text{relative time} \qquad = \frac{\alpha t}{x_1^2}$$

$$n, \text{relative position} \qquad = \frac{x}{x_1}$$

and

$$m, \text{relative resistance} \qquad = \frac{k}{hx_1}$$

where x_1 is the radius or semithickness of the conducting medium. These charts may be used to evaluate temperature profiles for cases involving transport of energy into or out of the conducting medium if the following conditions are met:

(a) Fourier's field equation describes the process; i.e., constant thermal diffusivity and no internal heat source.

(b) The conducting medium has a uniform initial temperature, T_0.

(c) The temperature of the boundary or the adjacent fluid is changed to a new value, T_∞, for $t \geq 0$.

---

[3] Equation (18-16) pertains to a plane wall of thickness, L, and boundary conditions $T(x, 0) = T_0$ and $dT/dx(0, t) = 0$.

For flat plates where the transport takes place from only one of the faces, the relative time, position, and resistance are evaluated as if the thickness were twice the true value.

Although the charts were drawn for one-dimensional transport, they may be combined to yield solutions for two- and three-dimensional problems. The following is a summary of these combined solutions:

1. For transport in a rectangular bar with insulated ends

$$Y_{bar} = Y_a Y_b \qquad (18\text{-}22)$$

where Y_a is evaluated with width $x_1 = a$, and Y_b is evaluated with thickness $x_1 = b$.

2. For transport in a rectangular parallelepiped

$$Y_{parallelepiped} = Y_a Y_b Y_c \qquad (18\text{-}23)$$

where Y_a is evaluated with width $x_1 = a$, Y_b is evaluated with thickness $x_1 = b$, and Y_c is evaluated with depth $x_1 = c$.

3. For transport in a cylinder, including both ends

$$Y_{\substack{cylinder \\ plus\ ends}} = Y_{cylinder} Y_a \qquad (18\text{-}24)$$

where Y_a is evaluated by using the flat-plate chart, and thickness $x_1 = a$.

The use of temperature–time charts is demonstrated in the following examples.

EXAMPLE 3

A flat wall of fire-clay brick, 0.5 m thick and originally at 200 K, has one of its faces suddenly exposed to a hot gas at 1200 K. If the heat-transfer coefficient on the hot side is 7.38 W/m$^2 \cdot$ K and the other face of the wall is insulated so that no heat passes out of that face, determine (a) the time necessary to raise the center of the wall to 600 K; (b) the temperature of the insulated wall face at the time evaluated in (a).

From the table of physical properties given in Appendix H, the following values are listed:

$$k = 1.125 \text{ W/m} \cdot \text{K}$$
$$c_p = 919 \text{ J/kg} \cdot \text{K}$$
$$\rho = 2310 \text{ kg/m}^3$$

and

$$\alpha = 5.30 \times 10^{-7} \text{ m}^2/\text{s}$$

The insulated face limits the energy transfer into the conducting medium to only one direction. This is equivalent to heat transfer from a 1-m-thick wall, where x is then measured from the line of symmetry, the insulated face. The relative position, x/x_1, is 1/2. The relative resistance, k/hx_1, is $1.125/[(7.38)(0.5)]$ or 0.305. The dimensionless temperature, $Y = (T_\infty - T)/(T_\infty - T_0)$, is equal to $(1200 - 600)/(1200 - 200)$, or 0.6. From Figure F.7, in Appendix F, the abscissa, $\alpha t/x_1^2$, is 0.35 under these conditions. The time required to raise the centerline to 600°F is

$$t = \frac{0.35x_1^2}{\alpha} = \frac{0.35(0.5)^2}{5.30 \times 10^{-7}} = 1.651 \times 10^5 \text{ s} \quad \text{or} \quad 45.9 \text{ h}$$

The relative resistance and the relative time for (b) will be the same as in part (a). The relative position, x/x_1, will be 0. Using these values and Figure F.1 in Appendix F, we find the dimensionless temperature, Y, to be 0.74. Using this value, the desired temperature can be evaluated by

$$\frac{T_s - T}{T_s - T_0} = \frac{1200 - T}{1200 - 200} = 0.74$$

or

$$T = 460 \text{ K} \quad (368°F)$$

EXAMPLE 4

A billet of steel 30.5 cm in diameter, 61 cm long, initially at 645 K, is immersed in an oil bath that is maintained at 310 K. if the surface conductance is 34 $W/m^2 \cdot K$ determine the center temperature of the billet after 1 h.

From Appendix H, the following average physical properties will be used:

$$k = 49.2 \, W/m \cdot K$$

$$c_p = 473 \, J/kg \cdot K$$

$$\rho = 7820 \, kg/m^3$$

$$\alpha = 1.16 \times 10^{-5} \, m^2/s$$

Equation (18-22) applies. To evaluate Y_a the following dimensionless parameters apply

$$X = \frac{\alpha t}{x_1^2} = \frac{(1.16 \times 10^{-5} \, m^2/s)(3600 \, s)}{(0.305 \, m)^2} = 0.449$$

$$n = x/x_1 = 0$$

$$m = k/hx_1 = \frac{42.9 \, W/m \cdot K}{(34 \, W/m^2 \cdot K)(0.305 \, m)} = 4.14$$

Using these values with Figure F.7 in Appendix F, the corresponding values of dimensionless temperature, Y_a, is approximately 0.95.

For the cylindrical surface the appropriate values are

$$X = \frac{\alpha t}{x_1^2} = \frac{(1.16 \times 10^{-5} \, m^2/s)(3600 \, s)}{(0.1525 \, m)^2}$$

$$= 1.80$$

$$n = \frac{x}{x_1} = 0$$

$$m = k/hx_1 = 42.9/(34)(0.1525) = 8.27$$

and, from Figure F.8 in Appendix F, we obtain

$$Y_{CL} = \left. \frac{T - T_\infty}{T_0 - T_\infty} \right|_{cyl} \cong 0.7$$

Now, for heat transfer across the cylindrical surface and both ends

$$Y|_{total} = \frac{T_{CL} - T_\infty}{T_0 - T_\infty} = Y_a Y_{CL} = (0.95)(0.7) = 0.665$$

The desired center temperature is now calculated as

$$T_{CL} = T_\infty + 0.665(T_0 - T_\infty)$$

$$= 310 \, K + 0.665(645 - 310) \, K$$

$$= 533 \, K \quad (499 \, F)$$

18.3 NUMERICAL METHODS FOR TRANSIENT CONDUCTION ANALYSIS

In many time-dependent or unsteady-state conduction processes, actual initial and/or boundary conditions do not correspond to those mentioned earlier with regard to analytical solutions. An initial temperature distribution may be nonuniform in nature; ambient temperature, surface conductance, or system geometry may be variable or quite irregular. For such complex cases, numerical techniques offer the best means to achieve solutions.

More recently, with sophisticated computing codes available, numerical solutions are being obtained for heat transfer problems of all types, and this trend will doubtlessly continue. It is likely that many users of this book will be involved in code development for such analysis.

Some numerical work is introduced in Chapter 17, dealing with two-dimensional, steady-state conduction. In this section, we will consider variation in time as well as position.

To begin our discussion, the reader is referred to equation (17-61) and the development leading up to it. For the case of no internal generation of energy, equation (17-61) reduces to

$$k \frac{\Delta y}{\Delta x}(T_{i-1, j} + T_{i+1, j} - 2T_{i, j}) + k \frac{\Delta x}{\Delta y}(T_{i, j-1} + T_{i, j+1} - 2T_{i, j})$$
$$= \left(\frac{\rho c_p T_{i, j}|_{t+\Delta t} - \rho c_p T_{i, j}|_t}{\Delta t} \right) \Delta x \, \Delta y \tag{18-25}$$

This expression applies to two dimensions; however, it can be extended easily to three dimensions.

The time-dependent term on the right of equation (18-25) is written such that the temperature at node i, j is presumed known at time t; this equation can then be solved to find T_{ij} at the end of time interval Δt. As $T_{ij|t+\Delta t}$ appears only once in this equation, it can be evaluated quite easily. This means of evaluating T_{ij} at the end of a time increment is designated an "explicit" technique. A more thorough discussion of explicit solutions is given by Carnahan.[4]

Equation (18-25) may be solved to evaluate the temperature at node i, j for all values of i, j that comprise the region of interest. For large numbers of nodes, it is clear that a great number of calculations are needed and that much information must be stored for use in subsequent computation. Digital computers obviously provide the only feasible way to accomplish solutions.

We will next consider the one-dimensional form of equation (18-25). For a space increment Δx, the simplified expression becomes

$$\frac{k}{\Delta x}(T_{i-1}|_t + T_{i+1}|_t - 2T_i|_t) = \left(\frac{\rho c_p T_i|_{t+\Delta t} - \rho c_p T_i|_t}{\Delta t} \right) \Delta x \tag{18-26}$$

where the j notation has been dropped. The absence of variation in the y direction allows several terms to be deleted. We next consider properties to be constant and represent the ratio $k/\rho c_p$ as α. Solving for $T_{i|t+\Delta t}$, we obtain

$$T_i|_{t+\Delta t} = \frac{\alpha \, \Delta t}{(\Delta x)^2}(T_{i+1}|_t + T_{i-1}|_t) + \left(1 - \frac{2\alpha \, \Delta t}{(\Delta x)^2} \right) T_i|_t \tag{18-27}$$

The ratio, $\alpha \, \Delta t / (\Delta x)^2$, a form resembling the Fourier modulus, is seen to arise naturally in this development. This grouping relates the time step, Δt, to the space increment, Δx. The magnitude of this grouping will, quite obviously, have an effect on the solution. It has been determined that equation (18-27) is numerically "stable" when

$$\frac{\alpha \, \Delta t}{(\Delta x)^2} \leq \frac{1}{2} \tag{18-28}$$

For a discussion of numerical stability the reader is referred to Carnahan et al.[4]

---

[4] B. Carnahan, H. A. Luther, and J. O. Wilkes, *Applied Numerical Methods,* Wiley, New York, 1969.

The choice of a time step involves a trade-off between solution accuracy-a smaller time step will produce greater accuracy, and computation time-a solution will be achieved more rapidly for larger values of Δt. When computing is done by machine, a small time step will likely be used without major difficulty.

An examination of equation (18-27) indicates considerable simplification to be achieved if the equality in equation (18-28) is used. For the case with $\alpha \, \Delta t/(\Delta x^2) = 1/2$, equation (18-27) becomes

$$T_{i|t+\Delta t} = \frac{T_{i+1|t} + T_{i-1|t}}{2} \tag{18-29}$$

EXAMPLE 5

A brick wall ($\alpha = 4.72 \times 10^{-7}$ m^2/s) with a thickness of 0.5 m is initially at a uniform temperature of 300 K. Determine the length of time required for its center temperature to reach 425 K if its surfaces are raised to, and maintained at 425 and 600 K, respectively.

Although relatively simple, this one-dimensional problem is not amenable to a solution using the charts because there is no axis of symmetry. Either analytical or numerical methods must therefore be employed.

An analytical solution using Laplace transform or separation-of-variables methodology is relatively straightforward. However, the solution is in terms of infinite series involving eigenvalues and the determination of a final answer is cumbersome. The simplest approach is thus numerical and we will proceed with the ideas introduced in this section.

The illustration below depicts the wall divided into 10 increments. Each of the nodes within the wall is at the center of a subvolume, having a width, Δx. The shaded subvolume at node 4 is considered to have uniform properties averaged at its center, i.e., the location of node 4. This same idea prevails for all 11 nodes; this includes the surface nodes, 0 and 10.

0 1 2 3 4 5 6 7 8 9 10

An energy balance for any internal node, having width Δx, will yield equation (18-27) as a result. This relationship includes the dimensionless ratio, $\alpha\Delta t/\Delta x^2$, which relates the time increment, Δt, to the space increment Δx. In this example we have specified $\Delta x = 0.05$ m.

The quantity, $\alpha\Delta t/\Delta x^2$, can have any value equal to or less than 0.5, which is the limit for a stable solution. If the limiting value is chosen, equation (18-27) reduces to a simple algorithmic form

$$T_{i,t+1} = \frac{T_{i-1,t} + T_{i+1,t}}{2} \tag{18-29}$$

This expression is valid for $i = 1$ to 9; however, as nodes 0 and 10 are at constant temperature for all time, the algorithms for nodes 1 and 9 can be written as

$$\begin{aligned} T_{1,t+1} &= \frac{T_{0,t} + T_{2,t}}{2} = \frac{425 + T_{2,t}}{2} \\ T_{9,t+1} &= \frac{T_{8,t} + T_{10,t}}{2} = \frac{T_{8,t} + 600}{2} \end{aligned} \tag{18-30}$$

The problem solution now proceeds as equations (18-29) and (18-30) and are solved at succeeding times to update nodal temperatures until the desired result, $T_s = 425$ K, is achieved.

Equations (18-29) and (18-30) are quite simple and easily programed to achieve a solution. In this case, a spreadsheet approach could also be used. The table below summarizes the form of the results for $T_{i,t}$.

	T_0	T_1	T_2	T_3	T_4	T_5	T_6	T_7	T_8	T_9	T_{10}
$t = 0$	425	300	300	300	300	300	300	300	300	300	600
$\vdots$											
$t = 10$	425	394.8	372.1	349.4	347.9	346.4	367.1	405.7	466.1	526.4	600
$\vdots$											
$t = 20$	425	411.4	403.3	395.3	402.3	409.4	436.2	463.1	506.5	550.0	600
$\vdots$											
$t = 22$	425	414.2	408.5	402.8	411.0	419.3	445.3	471.4	512.3	553.3	600
$t = 23$	425	416.8	408.5	409.8	411.0	428.2	445.3	478.8	512.3	556.2	600

The desired center temperature is reached between time increments 22 and 23; an interpolated value is $n = 22.6$ time increments.

As discussed earlier, the increment Δt is related to α and Δx according to the ratio

$$\frac{\alpha \, \Delta t}{\Delta x^2} = 1/2$$

or

$$\Delta t = \frac{1}{2} \frac{\Delta x^2}{\alpha} = \frac{1}{2} \frac{(0.05 \, \text{m})^2}{4.72 \times 10^{-7} \, \text{m}^2/\text{s}}$$

$$= 2648 \, \text{s}$$

$$= 0.736 \, \text{h}$$

The answer for total time elapsed is thus

$$t = 22.6(0.736) = 16.6 \, \text{h}$$

18.4 AN INTEGRAL METHOD FOR ONE-DIMENSIONAL UNSTEADY CONDUCTION

The von Kármán momentum integral approach to the hydrodynamic boundary layer has a counterpart in conduction. Figure 18.5 shows a portion of a semi-infinite wall, originally at uniform temperature T_0, exposed to a fluid at temperature T_∞, with the surface of the wall at any time at temperature T_s.

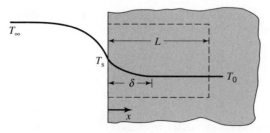

Figure 18.5 A portion of a semi-infinite wall used in integral analysis.

At any time t, heat transfer from the fluid to the wall affects the temperature profile within the wall. The "penetration distance," designated as δ, is the distance from the surface wherein this effect is manifested. At distance δ, the temperature gradient, $\partial T/\partial x$, is taken as zero.

Applying the first law of thermodynamics, equation (6-10), to a control volume extending from $x = 0$ to $x = L$, where $L > \delta$, we have

$$\frac{\delta Q}{dt} - \frac{\delta W_s}{dt} - \frac{\delta W_\mu}{dt} = \iint_{\text{c.s.}} \left(e + \frac{P}{\rho} \right) \rho(v \cdot n) \, dA + \frac{\partial}{\partial t} \iiint_{\text{c.v.}} e\rho \, dV \qquad (6\text{-}10)$$

with

$$\frac{\delta W_s}{dt} = \frac{\delta W_\mu}{dt} = \iint_{\text{c.s.}} \left(e + \frac{P}{\rho} \right) \rho(\mathbf{v} \cdot \mathbf{n}) \, dA = 0$$

The applicable form of the first law is now

$$\frac{\delta Q}{dt} = \frac{\partial}{\partial t} \iiint_{\text{c.v.}} e\rho \, dV$$

Considering all variables to be functions of x alone, we may express the heat flux as

$$\frac{q_x}{A} = \frac{d}{dt} \int_0^L \rho u \, dx = \frac{d}{dt} \int_0^L \rho c_p T \, dx \qquad (18\text{-}32)$$

The interval from 0 to L will now be divided into two increments, giving

$$\frac{q_x}{A} = \frac{d}{dt} \left[\int_0^\delta \rho c_p T \, dx + \int_\delta^L \rho c_p T_0 \, dx \right]$$

and, since T_0 is constant, this becomes

$$\frac{q_x}{A} = \frac{d}{dt} \left[\int_0^\delta \rho c_p T \, dx + \rho c_p T_0 (L - \delta) \right]$$

The integral equation to be solved is now

$$\frac{q_x}{A} = \frac{d}{dt} \int_0^\delta \rho c_p T \, dx - \rho c_p T_0 \frac{d\delta}{dt} \qquad (18\text{-}33)$$

If a temperature profile of the form $T = T(x, \delta)$ is assumed, equation (18-33) will produce a differential equation in $\delta(t)$, which may be solved, and one may use this result to express the temperature profile as $T(x, t)$.

The solution of equation (18-33) is subject to three different boundary conditions at the wall, $x = 0$, in the sections to follow.

Case 1. Constant wall temperature

The wall, initially at uniform temperature T_0, has its surface maintained at temperature T_s for $t > 0$. The temperature profile at two different times is illustrated in Figure 18.6. Assuming the temperature profile to be parabolic of the form

$$T = A + Bx + Cx^2$$

and requiring that the following boundary conditions:

$$T = T_s \quad \text{at } x = 0$$
$$T = T_0 \quad \text{at } x = \delta$$

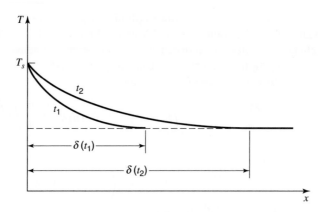

Figure 18.6 Temperature profiles at two times after the surface temperature is raised to T_s.

and

$$\frac{\partial T}{\partial x} = 0 \quad \text{at } x = \delta$$

be satisfied, we see that the expression for $T(x)$ becomes

$$\frac{T - T_0}{T_s - T_0} = \left(1 - \frac{x}{\delta}\right)^2 \tag{18-34}$$

The heat flux at the wall may now be evaluated as

$$\frac{q_x}{A} = -k \frac{\partial T}{\partial x}\bigg|_{x=0} = 2\frac{k}{\delta}(T_s - T_0) \tag{18-35}$$

which may be substituted into the integral expression along with equation (18-33), yielding

$$2\frac{k}{\delta}(T_s - T_0) = \frac{d}{dt}\int_0^\delta \rho c_p \left[T_0 + (T_s - T_0)\left(1 - \frac{x}{\delta}\right)^2\right]dx - \rho c_p T_0 \frac{d\delta}{dt}$$

and, after dividing through by ρc_p, both quantities being considered constant, we have

$$2\frac{\alpha}{\delta}(T_s - T_0) = \frac{d}{dt}\int_0^\delta \left[T_0 + (T_s - T_0)\left(1 - \frac{x}{\delta}\right)^2\right]dx - T_0 \frac{d\delta}{dt} \tag{18-36}$$

After integration, equation (18-36) becomes

$$\frac{2\alpha}{\delta}(T_s - T_0) = \frac{d}{dt}\left[(T_s - T_0)\frac{\delta}{3}\right]$$

and cancelling $(T_s - T_0)$, we obtain

$$6\alpha = \delta \frac{d\delta}{dt} \tag{18-37}$$

and thus the penetration depth becomes

$$\delta = \sqrt{12\alpha t} \tag{18-38}$$

The corresponding temperature profile may be obtained from equation (18-34) as

$$\frac{T - T_0}{T_s - T_0} = \left[1 - \frac{x}{\sqrt{3}(2\sqrt{\alpha t})}\right]^2 \tag{18-39}$$

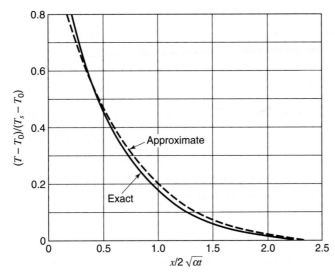

Figure 18.7 A comparison of exact and approximate results for one-dimensional conduction with a constant wall temperature.

which compares reasonably well with the exact result

$$\frac{T - T_0}{T_s - T_0} = 1 - \operatorname{erf} \frac{x}{2\sqrt{\alpha t}} \tag{18-40}$$

Figure 18.7 shows a comparison of these two results.

Case 2. A specified heat flux at the wall

In this case the appropriate boundary conditions are

$$T = T_0 \qquad \text{at } x = \delta$$

$$\frac{\partial T}{\partial x} = 0 \qquad \text{at } x = \delta$$

and

$$\frac{\partial T}{\partial x} = -\frac{F(t)}{k} \qquad \text{at } x = 0$$

where the heat flux at the wall is expressed as the general function $F(t)$.

If the parabolic temperature profile is used, the above boundary conditions yield

$$T - T_0 = \frac{[F(t)](\delta - x)^2}{2k\delta} \tag{18-41}$$

which, when substituted into equation (18-38), yields

$$\frac{d}{dt}\left(\frac{F(t)\delta^2}{6k}\right) = \frac{\alpha F(t)}{k} \tag{18-42}$$

and

$$\delta(t) = \sqrt{6\alpha}\left[\frac{1}{F(t)}\int_0^t F(t)\,dt\right]^{1/2} \tag{18-43}$$

For a constant heat flux of magnitude q_0/A the resulting expression for T_s is

$$T_s - T_0 = \frac{q_0}{Ak}\sqrt{\frac{3}{2}\alpha t} \tag{18-44}$$

which differs by approximately 8% from the exact expression

$$T_s - T_0 = \frac{1.13 q_0}{Ak}\sqrt{\alpha t} \tag{18-45}$$

Case 3. Convection at the surface

The wall temperature is a variable in this case; however, it may be easily determined. If the temperature variation within the medium is expressed generally as

$$\frac{T - T_0}{T_s - T_0} = \phi\left(\frac{x}{\delta}\right) \tag{18-46}$$

we note that the temperature gradient at the surface becomes

$$\left.\frac{\partial T}{\partial x}\right|_{x=0} = -\frac{T_s - T_0}{\delta}N \tag{18-47}$$

where N is a constant depending upon the form of $\phi(x/\delta)$.

At the surface we may write

$$\left.\frac{q}{A}\right|_{x=0} = -k\left.\frac{\partial T}{\partial x}\right|_{x=0} = h(T_\infty - T_s)$$

which becomes, upon substituting equation (18-47)

$$T_s - T_0 = \frac{h\delta}{Nk}(T_\infty - T_0) \tag{18-48}$$

or

$$T_s = \frac{T_0 + (h\delta/Nk)T_\infty}{1 + h\delta/Nk} \tag{18-49}$$

We may now write

$$\frac{T_s - T_0}{T_\infty - T_0} = \frac{h\delta/Nk}{1 + h\delta/Nk} \tag{18-50}$$

and

$$\frac{T_\infty - T_s}{T_\infty - T_0} = \frac{1}{1 + h\delta/Nk} \tag{18-51}$$

The appropriate substitutions into the integral equation and subsequent solution follow the same procedures as in cases (a) and (b); the details of this solution are left as a student exercise.

The student should recognize the marked utility of the integral solution for solving one-dimensional unsteady-state conduction problems. Temperature profile expressions more complex than a parabolic form may be assumed. However, additional boundary conditions are needed in such cases to evaluate the constants. The similarity between the penetration depth and the boundary-layer thickness from the integral analysis of Chapter 12 should also be noted.

18.5 CLOSURE

In this chapter, some of the techniques for solving transient or unsteady-state heat-conduction problems have been presented and discussed. Situations considered included cases of negligible internal resistance, negligible surface resistance, and those for which both resistances were significant.

For flat slabs, cylinders, and spheres, with a uniform initial temperature, whose surfaces are suddenly exposed to surroundings at a different temperature, charts are available for evaluating the temperature at any position and time. Numerical and integral methods were also introduced.

PROBLEMS

18.1 A household iron has a stainless-steel sole plate that weights 3 lb and has a surface area of 0.5 ft^2. The iron is rated at 500 W. If the surroundings are at a temperature of 80°F, and the convective heat-transfer coefficient between the sole plate and surroundings is 3 Btu/h ft^2 °F, how long will it take for the iron to reach 240°F after it is plugged in?

18.2 An electrical system employs fuses that are cylindrical in shape and have lengths of 0.5 cm and diameters of 0.1 mm. Air, at 30°C, surrounds a fuse with a surface coefficient of 10 W/m^2 · K. The fuse material melts at 900°C.

Assuming all heat transfer to be from the fuse surface, estimate the time it will take for the fuse to blow after a current of 3 A flows through it.

Pertinent properties of the fuse material are

$$\text{Resistance} = 0.2\,\Omega$$
$$k = 20\,\text{W/m} \cdot \text{K}$$
$$\alpha = 5 \times 10^{-5}\,\text{m}^2/\text{s}$$

18.3 Aluminum wire, having a diameter of 0.794 mm, is immersed in an oil bath that is at 25°C. Aluminum wire of this size has an electrical resistance of 0.0572 Ω/m. For conditions where an electric current of 100 A is flowing through the wire and the surface coefficient between the wire and oil bath is 550 W/m^2 · K, determine the steady state temperature of the wire.

How long, after the current is supplied, will it take for the wire to reach a temperature within 5°C of its steady-state value?

18.4 If a rectangular block of rubber (see Problem 18.12 for properties) is set out in air at 297 K to cool after being heated to a uniform temperature of 420 K, how long will it take for the rubber surface to reach 320 K? The dimensions of the block are 0.6 m high by 0.3 m long by 0.45 m wide. The block sits on one of the 0.3-m by 0.45-m bases; the adjacent surface may be considered an insulator. The effective heat-transfer coefficient at all exposed surface is 6.0 W/m^2 · K. What will the maximum temperature within the rubber block be at this time?

18.5 Cast-iron cannonballs used in the War of 1812 were occasionally heated for some extended time so that, when fired at houses or ships, they would set them afire. If one of these the so-called "hot shot" were at a uniform temperature of 2000°F, how long after being exposed to air at 0°F with an outside convective heat-transfer coefficient of 16 But/h ft^2 °F, would be required for the surface temperature to drop to 600°F? What would be the center temperature at this time? The ball diameter is 6 in. The following properties of cast iron may be used:

$$k = 23\,\text{Btu/h ft}\,°F$$
$$c_p = 0.10\,\text{Btu/lb}_m\,°F$$
$$\rho = 460\,\text{lb}_m/\text{ft}^3.$$

18.6 It is known that oranges can be exposed to freezing temperatures for short periods of time without sustaining serious damage. As a representative case, consider a 0.10-m-diameter orange, originally at a uniform temperature of 5°C, suddenly exposed to surrounding air at −5°C. For a surface coefficient, between the air and orange surface, of 15 W/m^2 · K, how long will it take for the surface of the orange to reach 0°C? Properties of the orange are the following:

$$\rho = 940\,\text{kg/m}^3$$
$$k = 0.47\,\text{W/m} \cdot \text{K}$$
$$c_p = 3.8\,\text{kJ/kg} \cdot \text{K}.$$

$\alpha = \dfrac{k}{\rho c_p}$

18.7 A copper cylinder with a diameter of 3 in. is initially at a uniform temperature of 70°F. How long after being placed in a medium at 1000°F with an associated convective heat-transfer coefficient of 4 Btu/h ft^2 °F will the temperature at the center of the cylinder reach 500°F, if the height of the cylinder is (a) 3 in.? (b) 6 in.? (c) 12 in.? (d) 24 in.? (e) 5 ft?

18.8 A cylinder 2 ft high with a diameter of 3 in. is initially at the uniform temperature of 70°F. How long after the cylinder is placed in a medium at 1000°F, with associated convective heat-transfer coefficient of 4 Btu/h ft^2 °F, will the center temperature reach 500°F if the cylinder is made from

a. copper, $k = 212\,\text{Btu/h ft}\,°F$?

b. aluminum, $k = 130\,\text{Btu/h ft}\,°F$?

c. zinc, $k = 60\,\text{Btu/h ft}\,°F$?

d. mild steel, $k = 25\,\text{Btu/h ft}\,°F$?

e. stainless steel, $k = 10.5\,\text{Btu/h ft}\,°F$?

f. asbestos, $k = 0.087\,\text{Btu/h ft}\,°F$?

18.9 Water, initially at 40°F, is contained within a thin-walled cylindrical vessel having a diameter of 18 in. Plot the temperature of the water vs. time up to 1 h if the water and container are immersed in an oil bath at a constant temperature of 300°F. Assume that the water is well stirred and that the convective heat-transfer coefficient between the oil and cylindrical surface is 40 Btu/h ft^2 °F. The cylinder is immersed to a depth of 2 ft.

18.10 A short aluminum cylinder 0.6 m in diameter and 0.6 m long is initially at 475 K. It is suddenly exposed to a convective environment at 345 K with $h = 85$ W/m^2 · K. Determine the temperature in the cylinder at a radial position of 10 cm and a distance of 10 cm from one end of the cylinder after being exposed to this environment for 1 h.

18.11 A type-304 stainless-steel billet, 6 in. in diameter, is passing through a 20-ft-long heat-treating furnace. The initial billet temperature is 200°F, and it must be raised to a minimum temperature of 1500°F before working. The heat-transfer coefficient between the furnace gases and the billet surface is 15 Btu/h ft^2 °F, and the furnace gases are at 2300°F. At what minimum velocity must the billet travel through the furnace to satisfy these conditions?

18.12 In the curing of rubber tires, the "vulcanization" process requires that a tire carcass, originally at 295 K, be heated so that its central layer reaches a minimum temperature of 410 K. This heating is accomplished by introducing steam at 435 K to both sides. Determine the time required, after introducing steam, for a 3-cm-thick tire carcass to reach the specified central temperature condition. Properties of rubber that may be used are the following: $k = 0.151$ W/m $\cdot$ K, $c_p = 200$ J/kg $\cdot$ K, $\rho = 1201$ kg/m^3, $\alpha = 6.19 \times 10^{-8}$ m^2/s.

18.13 Buckshot, 0.2 in. in diameter, is quenched in 90°F oil from an initial temperature of 400°F. The buckshot is made of lead and takes 15 s to fall from the oil surface to the bottom of the quenching bath. If the convective heat-transfer coefficient between the lead and oil is 40 Btu/h ft^2 °F, what will be the temperature of the shot as it reaches the bottom of the bath?

18.14 It is common practice to treat wooden telephone poles with tar-like materials to prevent damage by water and insects. These tars are cured into the wood at elevated temperatures and pressures.

Consider the case of a 0.3-m-diameter pole, originally at 25°C, placed in a pressurized oven. It will be removed when the tar has penetrated to a depth of 10 cm. It is known that a 10-cm depth of penetration will occur when a temperature of 100°C is achieved. For an oven temperature of 380°C and $h = 140$ W/m^2 $\cdot$ K, determine the time required for the pole to remain in the oven. Properties of the wooden pole are

$$k = 0.20 \text{ W/m} \cdot \text{K}$$

$$\alpha = 1.1 \times 10^{-7} \text{ m}^2/\text{s}$$

18.15 For an asbestos cylinder with both height and diameter of 13 cm initially at a uniform temperature of 295 K placed in a medium at 810 K with an associated convective heat-transfer coefficient of 22.8 W/m^2 $\cdot$ K, determine the time required for the center of the cylinder to reach 530 K if end effects are neglected.

18.16 A copper bus bar is initially at 400°F. The bar measures 0.2 ft by 0.5 ft and is 10 ft long. If the edges are suddenly all reduced to 100°F, how long will it take for the center to reach a temperature of 250°F?

18.17 Rework Problem 18.4 for the case when air is blown by the surfaces of the rubber block with an effective surface coefficient of 230 W/m^2 $\cdot$ K.

18.18 Consider a hot dog to have the following dimensions and properties: diameter = 20 mm, $c_p = 3.35$ kJ/kg $\cdot$ K, $\rho = 880$ kg/m^3, and $k = 0.5$ W/m $\cdot$ K. For the hot dog initially at 5°C, exposed to boiling water at 100°C, with a surface coefficient of 90 W/m^2 $\cdot$ K, what will be the cooking time if the required condition is for the center temperature to reach 80°C?

18.19 This problem involves using heat transfer principles as a guide for cooking a pork roast.

The roast is to be modeled as a cylinder, having its length and diameter equal to each other, with properties being those of water. The roast weights 2.25 kg.

Properly cooked, every portion of the meat should attain a minimum temperature of 95°C. If the meat is initially at 5°C and

the oven temperature is 190°C, with a surface coefficient of 15 W/m^2 $\cdot$ K, what is the minimum cooking time required?

18.20 Given the cylinder in Problem 18.15, construct a plot of the time for the midpoint temperature to reach 530 K as a function of H/D, where H and D are the height and diameter of the cylinder, respectively.

18.21 A rocket-engine nozzle is coated with a ceramic material having the following properties: $k = 1.73$ Btu/h ft °F, $\alpha = 0.35$ ft^2/h. The convective heat-transfer coefficient between the nozzle and the gases, which are at 3000°F, is 200 Btu/h ft^2 °F. How long after startup will it take for the temperature at the ceramic surface to reach 2700°F? What will be the temperature at a point 1/2 in. from the surface at this time? The nozzle is initially at 0°F.

18.22 One estimate of the original temperature of Earth is 7000°F. Using this value and the following properties for Earth's crust, Lord Kelvin obtained an estimate of 9.8×10^7 years for the Earth's age:

$$\alpha = 0.0456 \text{ ft}^2/\text{h}$$

$$T_2 = 0°\text{F}$$

$$\left. \frac{\partial T}{\partial y} \right|_{y=0} = 0.02°\text{F/ft, (measured)}$$

Comment on Lord Kelvin's result by considering the exact expression for unsteady-state conduction in one dimension

$$\frac{T - T_s}{T_0 - T_s} = \text{erf} \frac{x}{2\sqrt{\alpha t}}$$

18.23 After a fire starts in a room the walls are exposed to combustion products at 950°C. If the interior wall surface is made of oak, how long after exposure to the fire will the wood surface reach its combustion temperature of 400°C? Pertinent data are the following:

$$h = 30 \text{ W/m}^2 \cdot \text{K}$$

$$T_i(\text{initial}) = 21°\text{C}$$

For oak: $\rho = 545$ kg/m^3

$$k = 0.17 \text{ W/m} \cdot \text{K}$$

$$c_p = 2.385 \text{ kJ/kg} \cdot \text{K}$$

18.24 Determine an expression for the depth below the surface of a semi-infinite solid at which the rate of cooling is maximum. Substitute the information given in Problem 18.22 to estimate how far below Earth's surface this maximum cooling rate is achieved.

18.25 Soil, having a thermal diffusivity of 5.16×10^{-7} m^2/s, has its surface temperature suddenly raised and maintained at 1100 K from its initial uniform value of 280 K. Determine the temperature at a depth of 0.25 m after a period of 5 h has elapsed at this surface condition.

18.26 The convective heat-transfer coefficient between a large brick wall and air at 100°F is expressed as $h = 0.44$

$(T - T_\infty)^{1/3}$ Btu/h ft^2 °F. If the wall is initially at a uniform temperature of 1000°F, estimate the temperature of the surface after 1, 6, and 24 h.

18.27 A thick wall of oak, initially at a uniform temperature of 25°C, is suddenly exposed to combustion exhaust at 800°C. Determine the time of exposure required for the surface to reach its ignition temperature of 400°C, when the surface coefficient between the wall and combustion gas is 20 W/m^2 · K.

18.28 Air at 65°F is blown against a pane of glass 1/8 in. thick. If the glass is initially at 30°F, and has frost on the outside, estimate the length of time required for the frost to begin to melt.

18.29 How long will a 1-ft-thick concrete wall subject to a surface temperature of 1500°F on one side maintain the other side below 130°F? The wall is initially at 70°F.

18.30 A stainless-steel bar is initially at a temperature of 25°C. Its upper surface is suddenly exposed to an air stream at 200°C, with a corresponding convective coefficient of 22 W/m^2 · K. If the bar is considered semi-infinite, how long will it take for the temperature at a distance of 50 mm from the surface to reach 100°C?

18.31 A thick plate made of stainless steel is initially at a uniform temperature of 300°C. The surface is suddenly exposed to a coolant at 20°C with a convective surface coefficient of 110 W/m^2 · K. Evaluate the temperature after 3 min of elapsed time at

a. the surface;

b. a depth of 50 mm.

Work this problem both analytically and numerically.

18.32 If the heat flux into a solid is given as $F(t)$, show that the penetration depth δ for a semi-infinite solid is of the form

$$\delta = (\text{constant})\sqrt{\alpha}\left[\frac{\int_0^t F(t)dt}{F(t)}\right]^{1/2}$$

18.33 If the temperature profile through the ground is linear, increasing from 35°F at the surface by 0.5°F per foot of depth, how long will it take for a pipe buried 10 ft below the surface to reach 32°F if the outside air temperature is suddenly dropped to 0°F. The thermal diffusivity of soil may be taken as 0.02 ft^2/h, its thermal conductivity is 0.8 Btu/h ft °F, and the convective heat-transfer coefficient between the soil and the surrounding air is 1.5 Btu/h ft^2 °F.

18.34 A brick wall ($\alpha = 0.016$ ft^2/h) with a thickness of 1½ ft is initially at a uniform temperature of 80°F. How long, after the wall surfaces are raised to 300 and 600°F, respectively, will it take for the temperature at the center of the wall to reach 300°F?

18.35 A masonry brick wall 0.45 m thick has a temperature distribution at time, $t = 0$ which may be approximated by the expression $T(K) = 520 + 330 \sin \pi(x/L)$ where L is the wall width and x is the distance from either surface. How long after both surfaces of this wall are exposed to air at 280 K will the center temperature of the wall be 360 K? The convective coefficient at both surface of the wall may be taken as 14 W/m^2 · K. What will the surface temperature be at this time?

Chapter 19

Convective Heat Transfer

Heat transfer by convection is associated with energy exchange between a surface and an adjacent fluid. There are very few energy-transfer situations of practical importance in which fluid motion is not in some way involved. This effect has been eliminated as much as possible in the preceding chapters, but will now be considered in some depth.

The rate equation for convection has been expressed previously as

$$\frac{q}{A} = h \, \Delta T \qquad (15\text{-}11)$$

where the heat flux, q/A, occurs by virtue of a temperature difference. This simple equation is the defining relation for h, the convective heat-transfer coefficient. The determination of the coefficient h is, however, not at all a simple undertaking. It is related to the mechanism of fluid flow, the properties of the fluid, and the geometry of the specific system of interest.

In light of the intimate involvement between the convective heat-transfer coefficient and fluid motion, we may expect many of the considerations from the momentum transfer to be of interest. In the analyses to follow, much use will be made of the developments and concepts of Chapters 4 through 14.

19.1 FUNDAMENTAL CONSIDERATIONS IN CONVECTIVE HEAT TRANSFER

As mentioned in Chapter 12, the fluid particles immediately adjacent to a solid boundary are stationary, and a thin layer of fluid close to the surface will be in laminar flow regardless of the nature of the free stream. Thus, molecular energy exchange or conduction effects will always be present, and play a major role in any convection process. If fluid flow is laminar, then all energy transfer between a surface and contacting fluid or between adjacent fluid layers is by molecular means. If, on the contrary, flow is turbulent, then there is bulk mixing of fluid particles between regions at different temperatures, and the heat transfer rate is increased. The distinction between laminar and turbulent flow will thus be a major consideration in any convective situation.

There are two main classifications of convective heat transfer. These have to do with the driving force causing fluid to flow. *Natural* or *free convection* designates the type of process wherein fluid motion results from the heat transfer. When a fluid is heated or cooled, the associated density change and buoyant effect produce a natural circulation in which the affected fluid moves of its own accord past the solid surface, the fluid that replaces it is similarly affected by the energy transfer, and the process is repeated. *Forced convection* is the classification used to describe those convection situations in which fluid circulation is produced by an external agency such as a fan or a pump.

The hydrodynamic boundary layer, analyzed in Chapter 12, plays a major role in convective heat transfer, as one would expect. Additionally, we shall define and analyze the *thermal boundary layer,* which will also be vital to the analysis of a convective energy-transfer process.

There are four methods of evaluating the convective heat-transfer coefficient that will be discussed in this book. These are as follows:

(a) dimensional analysis, which to be useful requires experimental results;

(b) exact analysis of the boundary layer;

(c) approximate integral analysis of the boundary layer; and

(d) analogy between energy and momentum transfer.

19.2 SIGNIFICANT PARAMETERS IN CONVECTIVE HEAT TRANSFER

Certain parameters will be found useful in the correlation of convective data and in the functional relations for the convective heat-transfer coefficients. Some parameters of this type have been encountered earlier; these include the Reynolds and the Euler numbers. Several of the new parameters to be encountered in energy transfer will arise in such a manner that their physical meaning is unclear. For this reason, we shall devote a short section to the physical interpretation of two such terms.

The molecular diffusivities of momentum and energy have been defined previously as

$$\text{momentum diffusivity}: \quad \nu \equiv \frac{\mu}{\rho}$$

and

$$\text{thermal diffusivity}: \quad \alpha \equiv \frac{k}{\rho c_p}$$

That these two are designated similarly would indicate that they must also play similar roles in their specific transfer modes. This is indeed the case, as we shall see several times in the developments to follow. For the moment we should note that both have the same dimensions, those of L^2/t; thus their ratio must be dimensionless. This ratio, that of the molecular diffusivity of momentum to the molecular diffusivity of heat, is designated the *Prandtl number.*

$$\text{Pr} \equiv \frac{\nu}{\alpha} = \frac{\mu c_p}{k} \tag{19-1}$$

The Prandtl number is observed to be a combination of fluid properties; thus Pr itself may be thought of as a property. The Prandtl number is primarily a function of temperature and is tabulated in Appendix I, at various temperatures for each fluid listed.

The temperature profile for a fluid flowing past a surface is depicted in Figure 19.1. In the figure, the surface is at a higher temperature than the fluid. The temperature profile that exists is due to the energy exchange resulting from this temperature difference. For such a case the heat-transfer rate between the surface and the fluid may be written as

$$q_y = hA(T_s - T_\infty) \tag{19-2}$$

and, because heat transfer at the surface is by conduction

$$q_y = -kA \frac{\partial}{\partial y}(T - T_s)\big|_{y=0} \tag{19-3}$$

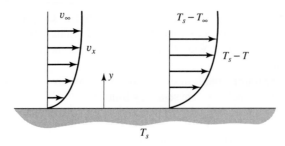

Figure 19.1 Temperature and velocity profiles for a fluid flowing past a heated plate.

These two terms must be equal; thus

$$h(T_s - T_\infty) = -k\frac{\partial}{\partial y}(T - T_s)|_{y=0}$$

which may be rearranged to give

$$\frac{h}{k} = \frac{\partial(T_s - T)/\partial y|_{y=0}}{T_s - T_\infty} \tag{19-4}$$

Equation (19-4) may be made dimensionless if a length parameter is introduced. Multiplying both sides by a representative length, L, we have

$$\frac{hL}{k} = \frac{\partial(T_s - T)/\partial y|_{y=0}}{(T_s - T_\infty)/L} \tag{19-5}$$

The right-hand side of equation (19-5) is now the ratio of the temperature gradient at the surface to an overall or reference temperature gradient. The left-hand side of this equation is written in a manner similar to that for the Biot modulus encountered in Chapter 18. It may be considered a ratio of conductive thermal resistance to the convective thermal resistance of the fluid. This ratio is referred to as the *Nusselt number*

$$\text{Nu} \equiv \frac{hL}{k} \tag{19-6}$$

where the thermal conductivity is that of the fluid as opposed to that of the solid, which was the case in the evaluation of the Biot modulus.

These two parameters, Pr and Nu, will be encountered many times in the work to follow.

19.3 DIMENSIONAL ANALYSIS OF CONVECTIVE ENERGY TRANSFER

Forced Convection. The specific forced-convection situation, which we shall now consider, is that of fluid flowing in a closed conduit at some average velocity, v, with a temperature difference existing between the fluid and the tube wall.

The important variables, their symbols, and dimensional representations are listed below. It is necessary to include two more dimensions—Q, heat, and T, temperature—to the fundamental group considered in Chapter 11; thus all variables must be expressed dimensionally as some combination of M, L, t, Q, and T. The above variables include terms descriptive of the system geometry, thermal and flow properties of the fluid, and the quantity of primary interest, h.

Variable	Symbol	Dimensions
Tube diameter	D	L
Fluid density	ρ	M/L^3
Fluid viscosity	μ	M/Lt
Fluid heat capacity	c_p	Q/MT
Fluid thermal conductivity	k	Q/tLT
Velocity	v	L/t
Heat-transfer coefficient	h	Q/tL^2T

Utilizing the Buckingham method of grouping the variables as presented in Chapter 11, the required number of dimensionless groups is found to be 3. Note that the rank of the dimensional matrix is 4, one more than the total number of fundamental dimensions.

Choosing D, k, μ, and v as the four variables comprising the core, we find that the three π groups to be formed are

$$\pi_1 = D^a k^b \mu^c v^d \rho$$

$$\pi_2 = D^e k^f \mu^g v^h c_p$$

and

$$\pi_3 = D^i k^j \mu^k v^l h$$

Writing π_1 in dimensional form

$$1 = (L)^a \left(\frac{Q}{LtT}\right)^b \left(\frac{M}{Lt}\right)^c \left(\frac{L}{t}\right)^d \frac{M}{L^3}$$

and equating the exponents of the fundamental dimensions on both sides of this equation, we have for

$$L: \quad 0 = a - b - c + d - 3$$
$$Q: \quad 0 = b$$
$$t: \quad 0 = -b - c - d$$
$$T: \quad 0 = -b$$

and

$$M: \quad 0 = c + 1$$

Solving these equations for the four unknowns yields

$$a = 1 \qquad c = -1$$
$$b = 0 \qquad d = 1$$

and π_1 becomes

$$\pi_1 = \frac{Dv\rho}{\mu}$$

which is the Reynolds number. Solving for π_2 and π_3 in the same way will give

$$\pi_2 = \frac{\mu c_p}{k} = \text{Pr} \quad \text{and} \quad \pi_3 = \frac{hD}{k} = \text{Nu}$$

The result of a dimensional analysis of forced-convection heat transfer in a circular conduit indicates that a possible relation correlating the important variables is of the form

$$\text{Nu} = f_1(\text{Re}, \text{Pr}) \tag{19-7}$$

If, in the preceding case, the core group had been chosen to include ρ, μ, c_p, and v, the analysis would have yielded the groups $Dv\rho/\mu$, $\mu c_p/k$, and $h/\rho v c_p$. The first two of these we recognize as Re and Pr. The third is the *Stanton number.*

$$\text{St} \equiv \frac{h}{\rho v c_p} \tag{19-8}$$

This parameter could also have been formed by taking the ratio Nu/(Re Pr). An alternative correlating relation for forced convection in a closed conduit is thus

$$\text{St} = f_2(\text{Re}, \text{Pr}) \tag{19-9}$$

Natural Convection.   In the case of natural-convection heat transfer from a vertical plane wall to an adjacent fluid, the variables will differ significantly from those used in the preceding case. The velocity no longer belongs in the group of variables, as it is a result of other effects associated with the energy transfer. New variables to be included in the analysis are those accounting for fluid circulation. They may be found by considering the relation for buoyant force in terms of the density difference due to the energy exchange.

The coefficient of thermal expansion, β, is given by

$$\rho = \rho_0(1 - \beta \Delta T) \tag{19-10}$$

where ρ_0 is the bulk fluid density, ρ is the fluid density inside the heated layer, and ΔT is the temperature difference between the heated fluid and the bulk value. The buoyant force per unit volume, F_{buoyant}, is

$$F_{\text{buoyant}} = (\rho_0 - \rho)g$$

which becomes, upon substituting equation (19-10)

$$F_{\text{buoyant}} = \beta g \rho_0 \Delta T \tag{19-11}$$

Equation (19-11) suggests the inclusion of the variables β, g, and ΔT into the list of those important to the natural convection situation.

The list of variables for the problem under consideration is given below.

Variable	Symbol	Dimensions
Significant length	L	L
Fluid density	ρ	M/L^3
Fluid viscosity	μ	M/Lt
Fluid heat capacity	c_p	Q/MT
Fluid thermal conductivity	k	Q/LtT
Fluid coefficient of thermal expansion	β	$1/T$
Gravitational acceleration	g	L/t^2
Temperature difference	ΔT	T
Heat-transfer coefficient	h	Q/L^2tT

The Buckingham π theorem indicates that the number of independent dimensionless parameters applicable to this problem is $9 - 5 = 4$. Choosing $L, \mu, k, g,$ and β as the core group, we see that the π groups to be formed are

$$\pi_1 = L^a \mu^b k^c \beta^d g^e c_p$$

$$\pi_2 = L^f \mu^g k^h \beta^i g^j \rho$$

$$\pi_3 = L^k \mu^l k^m \beta^n g^o \, \Delta T$$

and

$$\pi_4 = L^p \mu^q k^r \beta^s g^t h$$

Solving for the exponents in the usual way, we obtain

$$\pi_1 = \frac{\mu c_p}{k} = \text{Pr} \qquad\qquad \pi_3 = \beta \, \Delta T$$

$$\pi_2 = \frac{L^3 g \rho^2}{\mu^2} \qquad \text{and} \qquad \pi_4 = \frac{hL}{k} = \text{Nu}$$

The product of π_2 and π_3, which must be dimensionless, is $(\beta g \rho^2 L^3 \, \Delta T)/\mu^2$. This parameter, used in correlating natural-convection data, is the *Grashof number.*

$$\text{Gr} \equiv \frac{\beta g \rho^2 L^3 \, \Delta T}{\mu^2} \tag{19-12}$$

From the preceding brief dimensional-analyses considerations, we have obtained the following possible forms for correlating convection data:

(a) *Forced convection*

$$\text{Nu} = f_1(\text{Re}, \text{Pr}) \tag{19-7}$$

or

$$\text{St} = f_2(\text{Re}, \text{Pr}) \tag{19-9}$$

(b) *Natural convection*

$$\text{Nu} = f_3(\text{Gr}, \text{Pr}) \tag{19-13}$$

The similarity between the correlations of equations (19-7) and (19-13) is apparent. In equation (19-13), Gr has replaced Re in the correlation indicated by equation (19-7). It should be noted that the Stanton number can be used only in correlating forced-convection data. This becomes obvious when we observe the velocity, v, contained in the expression for St.

19.4 EXACT ANALYSIS OF THE LAMINAR BOUNDARY LAYER

An exact solution for a special case of the hydrodynamic boundary layer is discussed in Section 12.5. Blasius's solution for the laminar boundary layer on a flat plate may be extended to include the convective heat-transfer problem for the same geometry and laminar flow.

The boundary-layer equations considered previously include the two-dimensional, incompressible continuity equation

$$\frac{\partial v_x}{\partial x} + \frac{\partial v_y}{\partial y} = 0 \tag{12-10}$$

and the equation of motion in the x direction

$$\frac{\partial v_x}{\partial t} + v_x \frac{\partial v_x}{\partial x} + v_y \frac{\partial v_x}{\partial y} = v_\infty \frac{dv_\infty}{dx} + v \frac{\partial^2 v_x}{\partial y^2} \tag{12-9}$$

Recall that the y-directional equation of motion gave the result of constant pressure through the boundary layer. The proper form of the energy equation will thus be equation (16-14), for isobaric flow, written in two-dimensional form as

$$\frac{\partial T}{\partial t} + v_x \frac{\partial T}{\partial x} + v_y \frac{\partial T}{\partial y} = \alpha \left(\frac{\partial^2 T}{\partial x^2} + \frac{\partial^2 T}{\partial y^2} \right) \tag{19-14}$$

With respect to the thermal boundary layer depicted in Figure 19.2, $\partial^2 T/\partial x^2$ is much smaller in magnitude than $\partial^2 T/\partial y^2$.

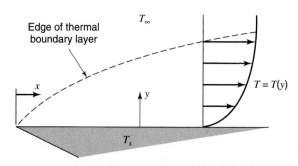

Figure 19.2 The thermal boundary layer for laminar flow past a flat surface.

In steady, incompressible, two-dimensional, isobaric flow the energy equation that applies is now

$$v_x \frac{\partial T}{\partial x} + v_y \frac{\partial T}{\partial y} = \alpha \frac{\partial^2 T}{\partial y^2} \tag{19-15}$$

From Chapter 12, the applicable equation of motion with uniform free-stream velocity is

$$v_x \frac{\partial v_x}{\partial x} + v_y \frac{\partial v_x}{\partial y} = v \frac{\partial^2 v_x}{\partial y^2} \tag{12-11a}$$

and the continuity equation

$$\frac{\partial v_x}{\partial x} + \frac{\partial v_y}{\partial y} = 0 \tag{12-11b}$$

The latter two of the above equations were originally solved by Blasius to give the results discussed in Chapter 12. The solution was based upon the boundary conditions

$$\frac{v_x}{v_\infty} = \frac{v_y}{v_\infty} = 0 \qquad \text{at } y = 0$$

and

$$\frac{v_x}{v_\infty} = 1 \qquad \text{at } y = \infty$$

The similarity in form between equations (19-15) and (12-11a) is obvious. This situation suggests the possibility of applying the Blasius solution to the energy equation. In order that this be possible, the following conditions must be satisfied:

(1) The coefficients of the second-order terms must be equal. This requires that $\nu = \alpha$ or that Pr $= 1$.

(2) The boundary conditions for temperature must be compatible with those for the velocity. This may be accomplished by changing the dependent variable from T to $(T - T_s)/(T_\infty - T_s)$. The boundary conditions now are

$$\frac{v_x}{v_\infty} = \frac{v_y}{v_\infty} = \frac{T - T_s}{T_\infty - T_s} = 0 \qquad \text{at } y = 0$$

$$\frac{v_x}{v_\infty} = \frac{T - T_s}{T_\infty - T_s} = 1 \qquad \text{at } y = \infty$$

Imposing these conditions upon the set of equations (19-15) and (12-11a), we may now write the results obtained by Blasius for the energy-transfer case. Using the nomenclature of Chapter 12,

$$f' = 2\frac{v_x}{v_\infty} = 2\frac{T - T_s}{T_\infty - T_s} \tag{19-16}$$

$$\eta = \frac{y}{2}\sqrt{\frac{v_\infty}{\nu x}} = \frac{y}{2x}\sqrt{\frac{x v_\infty}{\nu}} = \frac{y}{2x}\sqrt{\mathrm{Re}_x} \tag{19-17}$$

and applying the Blasius result, we obtain

$$\left.\frac{df'}{d\eta}\right|_{y=0} = f''(0) = \left.\frac{d[2(v_x/v_\infty)]}{d[(y/2x)\sqrt{\mathrm{Re}_x}]}\right|_{y=0}$$

$$= \left.\frac{d\{2[(T - T_s)/(T_\infty - T_s)]\}}{d[(y/2x)\sqrt{\mathrm{Re}_x}]}\right|_{y=0} = 1.328 \tag{19-18}$$

It should be noted that according to equation (19-16), the dimensionless velocity profile in the laminar boundary layer is identical with the dimensionless temperature profile. This is a consequence of having Pr $= 1$. A logical consequence of this situation is that the hydrodynamic and thermal boundary layers are of equal thickness. It is significant that the Prandtl numbers for most gases are sufficiently close to unity that the hydrodynamic and thermal boundary layers are of similar extent.

We may now obtain the temperature gradient at the surface

$$\left.\frac{\partial T}{\partial y}\right|_{y=0} = (T_\infty - T_s)\left[\frac{0.332}{x}\mathrm{Re}_x^{1/2}\right] \tag{19-19}$$

Application of the Newton and Fourier rate equations now yields

$$\frac{q_y}{A} = h_x(T_s - T_\infty) = -k\left.\frac{\partial T}{\partial y}\right|_{y=0}$$

from which

$$h_x = -\frac{k}{T_s - T_\infty}\left.\frac{\partial T}{\partial y}\right|_{y=0} = \frac{0.332k}{x}\mathrm{Re}_x^{1/2} \tag{19-20}$$

or

$$\frac{h_x x}{k} = \mathrm{Nu}_x = 0.332\,\mathrm{Re}_x^{1/2} \tag{19-21}$$

Pohlhausen[1] considered the same problem with the additional effect of a Prandtl number other than unity. He was able to show the relation between the thermal and hydrodynamic boundary layers in laminar flow to be approximately given by

$$\frac{\delta}{\delta_t} = \mathrm{Pr}^{1/3} \tag{19-22}$$

The additional factor of $\mathrm{Pr}^{1/3}$ multiplied by η allows the solution to the thermal boundary layer to be extended to Pr values other than unity. A plot of the dimensionless temperature vs. $\eta\,\mathrm{Pr}^{1/3}$ is shown in Figure 19.3. The temperature variation given in this form leads to an expression for the convective heat-transfer coefficient similar to equation (19-20). At $y = 0$, the gradient is

$$\left.\frac{\partial T}{\partial y}\right|_{y=0} = (T_\infty - T_s)\left[\frac{0.332}{x}\,\mathrm{Re}_x^{1/2}\mathrm{Pr}^{1/3}\right] \tag{19-23}$$

Figure 19.3 Temperature variation for laminar flow over a flat plate.

which, when used with the Fourier and Newton rate equations, yields

$$h_x = 0.332\frac{k}{x}\mathrm{Re}_x^{1/2}\,\mathrm{Pr}^{1/3} \tag{19-24}$$

or

$$\frac{h_x x}{k} = \mathrm{Nu}_x = 0.332\,\mathrm{Re}_x^{1/2}\,\mathrm{Pr}^{1/3} \tag{19-25}$$

The inclusion of the factor $\mathrm{Pr}^{1/3}$ in these equations extends the range of application of equations (19-20) and (19-21) to situations in which the Prandtl number differs considerably from 1.

---

[1] E. Pohlhausen, *ZAMM,* **1,** 115 (1921).

The mean heat-transfer coefficient applying over a plate of width w and length L may be obtained by integration. For a plate of these dimensions

$$q_y = hA(T_s - T_\infty) = \int_A h_x(T_s - T_\infty)\, dA$$

$$h(wL)(T_s - T_\infty) = 0.332kw\,\mathrm{Pr}^{1/3}(T_s - T_\infty)\int_0^L \frac{\mathrm{Re}_x^{1/2}}{x}\, dx$$

$$hL = 0.332k\,\mathrm{Pr}^{1/3}\left(\frac{v_\infty\rho}{\mu}\right)^{1/2}\int_0^L x^{-1/2}dx$$

$$= 0.664k\,\mathrm{Pr}^{1/3}\left(\frac{v_\infty\rho}{\mu}\right)^{1/2}L^{1/2}$$

$$= 0.664k\,\mathrm{Pr}^{1/3}\,\mathrm{Re}_L^{1/2}$$

The mean Nusselt number becomes

$$\mathrm{Nu}_L = \frac{hL}{k} = 0.664\,\mathrm{Pr}^{1/3}\,\mathrm{Re}_L^{1/2} \qquad (19\text{-}26)$$

and it is seen that

$$\mathrm{Nu}_L = 2\,\mathrm{Nu}_x \quad \text{at } x = L \qquad (19\text{-}27)$$

In applying the results of the foregoing analysis it is customary to evaluate all fluid properties at the *film temperature,* which is defined as

$$T_f = \frac{T_s + T_\infty}{2} \qquad (19\text{-}28)$$

the arithmetic mean between the wall and bulk fluid temperatures.

19.5   APPROXIMATE INTEGRAL ANALYSIS OF THE THERMAL BOUNDARY LAYER

The application of the Blasius solution to the thermal boundary layer in Section 19.4 was convenient although very limited in scope. For flow other than laminar or for a configuration other than a flat surface, another method must be utilized to estimate the convective heat-transfer coefficient. An approximate method for analysis of the thermal boundary layer employs the integral analysis as used by von Kármán for the hydrodynamic boundary layer. This approach is discussed in Chapter 12.

Consider the control volume designated by the dashed lines in Figure 19.4, applying to flow parallel to a flat surface with no pressure gradient, having width Δx, a height equal to the

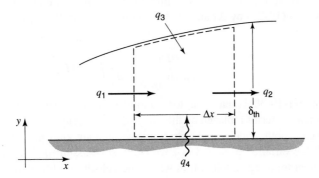

Figure 19.4 Control volume for integral energy analysis.

thickness of the thermal boundary layer, δ_t, and a unit depth. An application of the first law of thermodynamics in integral form

$$\frac{\delta Q}{dt} - \frac{\delta W_s}{dt} - \frac{\delta W_\mu}{dt} = \iint_{\text{c.s.}} (e + P/\rho)\rho(\mathbf{v} \cdot \mathbf{n}) \, dA + \frac{\partial}{\partial t} \iiint_{\text{c.v.}} e\rho \, dV \qquad (6\text{-}10)$$

yields the following under steady-state conditions:

$$\frac{\delta Q}{dt} = -k\,\Delta x \frac{\partial T}{\partial y}\bigg|_{y=0}$$

$$\frac{\delta W_s}{dt} = \frac{\delta W_\mu}{dt} = 0$$

$$\iint_{\text{c.s.}} (e + P/\rho)\rho(\mathbf{v} \cdot \mathbf{n}) \, dA = \int_0^{\delta_t} \left(\frac{v_x^2}{2} + gy + u + \frac{P}{\rho}\right)\rho v_x \, dy\bigg|_{x+\Delta x}$$

$$- \int_0^{\delta_t} \left(\frac{v_x^2}{2} + gy + u + \frac{P}{\rho}\right)\rho v_x \, dy\bigg|_{x}$$

$$- \frac{d}{dx}\int_0^{\delta_t} \left[\rho v_x\left(\frac{v_x^2}{2} + gy + u + \frac{P}{\rho}\right)\bigg|_{\delta_t}\right] dy\, \Delta x$$

and

$$\frac{\partial}{\partial t}\iiint_{\text{c.v.}} e\rho \, dV = 0$$

In the absence of significant gravitational effects, the convective-energy-flux terms become

$$\frac{v_x^2}{2} + u + \frac{P}{\rho} = h_0 \simeq c_p T_0$$

where h_0 is the stagnation enthalpy and c_p is the constant-pressure heat capacity. The stagnation temperature will now be written merely as T (without subscript) to avoid confusion. The complete energy expression is now

$$-k\,\Delta x \frac{\partial T}{\partial y}\bigg|_{y=0} = \int_0^{\delta_t} \rho v_x c_p T \, dy\bigg|_{x+\Delta x} - \int_0^{\delta_t} \rho v_x c_p T \, dy\bigg|_{x} - \rho c_p\,\Delta x \frac{d}{dx}\int_0^{\delta_t} v_x T_\infty \, dy$$

$$(19\text{-}29)$$

Equation (19-29) can also be written as $q_4 = q_2 - q_1 - q_3$ where these quantities are shown in Figure 19.4. In equation (19-29), T_∞ represents the free-stream stagnation temperature. If flow is incompressible, and an average value of c_p is used, the product ρc_p may be taken outside the integral terms in this equation. Dividing both sides of equation (19-29) by Δx and evaluating the result in the limit as by Δx approaches zero, we obtain

$$\frac{k}{\rho c_p}\frac{\partial T}{\partial y}\bigg|_{y=0} = \frac{d}{dx}\int_0^{\delta_t} v_x(T_\infty - T)\, dy \qquad (19\text{-}30)$$

Equation (19-30) is analogous to the momentum integral relation, equation (12-37), with the momentum terms replaced by their appropriate energy counterparts. This equation may be solved if both velocity and temperature profile are known. Thus, for the energy equation both the variation in v_x and in T with y must be assumed. This contrasts slightly with the momentum integral solution in which the velocity profile alone was assumed.

An assumed temperature profile must satisfy the boundary conditions

(1) $T - T_s = 0$ at $y = 0$

(2) $T - T_s = T_\infty - T_s$ at $y = \delta_t$

(3) $\dfrac{\partial}{\partial y}(T - T_s) = 0$ at $y = \delta_t$

(4) $\dfrac{\partial^2}{\partial y^2}(T - T_s) = 0$ at $y = 0$ [see equation (19-15)]

If a power-series expression for the temperature variation is assumed in the form

$$T - T_s = a + by + cy^2 + dy^3$$

the application of the boundary conditions will result in the expression for $T - T_s$

$$\frac{T - T_s}{T_\infty - T_s} = \frac{3}{2}\left(\frac{y}{\delta_t}\right) - \frac{1}{2}\left(\frac{y}{\delta_t}\right)^3 \tag{19-31}$$

If the velocity profile is assumed in the same form, then the resulting expression, as obtained in Chapter 12, is

$$\frac{v}{v_\infty} = \frac{3}{2}\frac{y}{\delta} - \frac{1}{2}\left(\frac{y}{\delta}\right)^3 \tag{12-40}$$

Substituting equations (19-31) and (12-40) into the integral expression and solving, we obtain the result

$$Nu_x = 0.36\,Re_x^{1/2}\,Pr^{1/3} \tag{19-32}$$

which is approximately 8% larger than the exact result expressed in equation (19-25).

    This result, although inexact, is sufficiently close to the known value to indicate that the integral method may be used with confidence in situations in which an exact solution is not known. It is interesting to note that equation (19-32) again involves the parameters predicted from dimensional analysis.

    A condition of considerable importance is that of an unheated starting length. Problem 19.17 at the end of the chapter deals with this situation where the wall temperature, T_s, is related to the distance from the leading edge, x, and the *unheated starting, length X*, according to

$$T_s = T_\infty \quad \text{for} \quad 0 < x < X$$
$$\text{and} \quad T_s > T_\infty \quad \text{for} \quad X < x$$

    The integral technique, as presented in this section, has proved effective in generating a modified solution for this situation. The result for $T_s = $ constant, and assuming both the hydrodynamic and temperature profiles to be cubic, is

$$Nu_x \cong 0.33\left[\frac{Pr}{1 - (X/x)^{3/4}}\right]^{\frac{1}{3}} Re_x \tag{19-33}$$

Note that this expression reduces to equations (19-25) for $X = 0$

19.6   ENERGY- AND MOMENTUM-TRANSFER ANALOGIES

Many times in our consideration of heat transfer thus far we have noted the similarities to momentum transfer both in the transfer mechanism itself and in the manner of its quantitative description. This section will deal with these analogies and use them to develop relations to describe energy transfer.

Osborne Reynolds first noted the similarities in mechanism between energy and momentum transfer in 1874.[2] In 1883, he presented[3] the results of his work on frictional resistance to fluid flow in conduits, thus making possible the quantitative analogy between the two transport phenomena.

As we have noted in the previous sections, for flow past a solid surface with a Prandtl number of unity, the dimensionless velocity and temperature gradients are related as follows:

$$\frac{d}{dy}\frac{v_x}{v_\infty}\bigg|_{y=0} = \frac{d}{dy}\left(\frac{T-T_s}{T_\infty-T_s}\right)\bigg|_{y=0} \tag{19-34}$$

For $\mathrm{Pr} = \mu c_p/k = 1$, we have $\mu c_p/k$ and we may write equation (19-34) as

$$\mu c_p \frac{d}{dy}\left(\frac{v_x}{v_\infty}\right)\bigg|_{y=0} = k\frac{d}{dy}\left(\frac{T-T_s}{T_\infty-T_s}\right)\bigg|_{y=0}$$

which may be transformed to the form

$$\frac{\mu c_p}{v_\infty}\frac{dv_x}{dy}\bigg|_{y=0} = -\frac{k}{T_s-T_\infty}\frac{d}{dy}(T-T_s)\bigg|_{y=0} \tag{19-35}$$

Recalling a previous relation for the convective heat-transfer coefficient

$$\frac{h}{k} = \frac{d}{dy}\left[\frac{(T_s-T)}{(T_s-T_\infty)}\right]\bigg|_{y=0} \tag{19-4}$$

it is seen that the entire right-hand side of equation (19-34) may be replaced by h, giving

$$h = \frac{\mu c_p}{v_\infty}\frac{dv_x}{dy}\bigg|_{y=0} \tag{19-36}$$

Introducing next the coefficient of skin friction

$$C_f \cong \frac{\tau_0}{\rho v_\infty^2/2} = \frac{2\mu}{\rho v_\infty^2}\frac{dv_x}{dy}\bigg|_{y=0}$$

we may write equation (19-36) as

$$h = \frac{C_f}{2}(\rho v_\infty c_p)$$

which, in dimensionless form, becomes

$$\frac{h}{\rho v_\infty c_p} \equiv \mathrm{St} = \frac{C_f}{2} \tag{19-37}$$

Equation (19-37) is the *Reynolds analogy* and is an excellent example of the similar nature of energy and momentum transfer. For those situations satisfying the basis for the development of equation (19-37), a knowledge of the coefficient of frictional drag will enable the convective heat-transfer coefficient to be readily evaluated.

The restrictions on the use of the Reynolds analogy should be kept in mind; they are (1) $\mathrm{Pr} = 1$ and (2) no form drag. The former of these was the starting point in the preceding development and obviously must be satisfied. The latter is sensible when one considers that, in relating two transfer mechanisms, the manner of expressing them quantitatively must

---

[2] O. Reynolds, *Proc. Manchester Lit. Phil. Soc.*, **14:7** (1874).
[3] O. Reynolds, *Trans. Roy. Soc. (London*, **174A**, 935 (1883).

remain consistent. Obviously the description of drag in terms of the coefficient of skin friction requires that the drag be wholly viscous in nature. Thus, equation (19-37) is applicable only for those situations in which form drag is not present. Some possible areas of application would be flow parallel to plane surfaces or flow in conduits. The coefficient of skin friction for conduit flow has already been shown to be equivalent to the Fanning fiction factor, which may be evaluated by using Figure 14.1.

The restriction that $Pr = 1$ makes the Reynolds analogy of limited use. Colburn[4] has suggested a simple variation of the Reynolds analogy form that allows its application to situations where the Prandtl number is other than unity. The Colburn analogy expression is

$$\text{St } Pr^{2/3} = \frac{C_f}{2} \tag{19-38}$$

which obviously reduces to the Reynolds analogy when $Pr = 1$.

Colburn applied this expression to a wide range of data for flow and geometries of different types and found it to be quite accurate for conditions where (1) no form drag exists, and (2) $0.5 < Pr < 50$. The Prandtl number range is extended to include gases, water, and several other liquids of interest. The Colburn analogy is particularly helpful for evaluating heat transfer in internal forced flows. It can be easily shown that the exact expression for a laminar boundary layer on a flat plate reduces to equation (19-38).

The Colburn analogy is often written as

$$j_H = \frac{C_f}{2} \tag{19-39}$$

where

$$j_H = \text{St } Pr^{2/3} \tag{19-40}$$

is designated the Colburn j factor for heat transfer. A mass-transfer j factor, is discussed in Chapter 28.

Note that for $Pr = 1$, the Colburn and Reynolds analogies are the same. Equation (19-38) is thus an extension of the Reynolds analogy for fluids having Prandtl numbers other than unity, within the range $0.5 - 50$ as specified above. High and low Prandtl number fluids falling outside this range would be heavy oils at one extreme and liquid metals at the other.

19.7 TURBULENT FLOW CONSIDERATIONS

The effect of the turbulent flow on energy transfer is directly analogous to the similar effects on momentum transfer as discussed in Chapter 12. Consider the temperature profile variation in Figure 19.5 to exist in turbulent flow. The distance moved by a fluid "packet" in the y direction, which is normal to the direction of bulk flow, is denoted by L, the Prandtl mixing length. The packet of fluid moving through the distance L retains the mean temperature from its point of origin, and upon reaching its destination, the packet will differ in temperature from that of the adjacent fluid by an amount $T|_{y \pm L} - T|_y$. The mixing length is assumed small enough to permit the temperature difference to be written as

$$T|_{y \pm L} - T|_y = \pm L \frac{dt}{dy}\bigg|_y \tag{19-41}$$

We now define the quantity T' as the fluctuating temperature, synonymous with the fluctuating velocity component, v'_x, described in Chapter 12. The instantaneous

---

[4] A. P. Colburn, *Trans. A.I.Ch.E.*, **29**, 174 (1933).

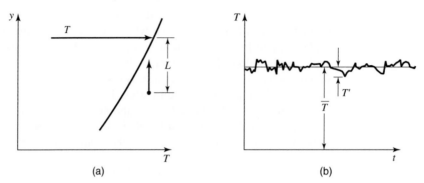

Figure 19.5 Turbulent-flow temperature variation.

temperature is the sum of the mean and fluctuating values, as indicated in Figure 19.5(b), or, in equation form

$$T = \overline{T} + T' \tag{19-42}$$

Any significant amount of energy transfer in the y direction, for bulk flow occurring in the x direction, is accomplished because of the fluctuating temperature, T'; thus, it is apparent from equations (19-41) and (19-42) that

$$T' = \pm L \frac{d\overline{T}}{dy} \tag{19-43}$$

The energy flux in the y direction may now be written as

$$\left.\frac{q_y}{A}\right|_y = \rho c_p T v'_y \tag{19-44}$$

where v'_y may be either positive or negative. Substituting for T its equivalent, according to equation (19-42)

$$\left.\frac{q_y}{A}\right|_y = \rho c_p v'_y (\overline{T} + T')$$

and taking the time average, we obtain, for the y-directional energy flux due to turbulent effects

$$\left.\frac{q_y}{A}\right|_{turb} = \rho c_p \overline{(v'_y T')} \tag{19-45}$$

or, with T' in terms of the mixing length

$$\left.\frac{q_y}{A}\right|_{turb} = \rho c_p \overline{v'_y L} \frac{d\overline{T}}{dy} \tag{19-46}$$

The total energy flux due to both microscopic and turbulent contributions may be written as

$$\frac{q_y}{A} = -\rho c_p [\alpha + \overline{|v'_y L|}] \frac{d\overline{T}}{dy} \tag{19-47}$$

As α is the molecular diffusivity of heat, the quantity $\overline{|v'_y L|}$ is the *eddy diffusivity of heat,* designated as ϵ_H. This quantity is exactly analogous to the eddy diffusivity of momentum, ϵ_M, as defined in equation (12-52). In a region of turbulent flow, $\epsilon_H \gg \alpha$ for all fluids except liquid metals.

As the Prandtl number is the ratio of the molecular diffusivities of momentum and heat, an analogous term, the *turbulent Prandtl number,* can be formed by the ratio ϵ_M/ϵ_H. Utilizing equations (19-47) and (12-55), we have

$$\text{Pr}_{\text{turb}} = \frac{\epsilon_M}{\epsilon_H} = \frac{L^2|dv_x/dy|}{|Lv_y'|} = \frac{L^2|dv_x/dy|}{L^2|dv_x/dy|} = 1 \tag{19-48}$$

Thus, in a region of fully turbulent flow the effective Prandtl number is unity, and the Reynolds analogy applies in the absence of form drag.

In terms of the eddy diffusivity of heat, the heat flux can be expressed as

$$\left.\frac{q_y}{A}\right|_{\text{turb}} = -\rho c_p \epsilon_H \frac{D\overline{T}}{dy} \tag{19-49}$$

The total heat flux, including both molecular and turbulent contributions, thus becomes

$$\frac{q_y}{A} = -\rho c_p(\alpha + \epsilon_H)\frac{d\overline{T}}{dy} \tag{19-50}$$

Equation (19-50) applies both to the region wherein flow is laminar, for which $\alpha \gg \epsilon_H$, and to that for which flow is turbulent and $\epsilon_H \gg \alpha$. It is in this latter region that the Reynolds analogy applies. Prandtl[5] achieved a solution that includes the influences of both the laminar sublayer and the turbulent core. In his analysis solutions were obtained in each region and then joined at $y = \xi$, the hypothetical distance from the wall that is assumed to be the boundary separating the two regions.

Within the laminar sublayer the momentum and heat flux equations reduce to

$$\tau = \rho v \frac{dv_x}{dy} \text{ (a constant)}$$

and

$$\frac{q_y}{A} = -\rho c_p\alpha \frac{dT}{dy}$$

Separating variables and integrating between $y = 0$ and $y = \xi$, we have, for the momentum expression

$$\int_0^{v_x|_\xi} dv_x = \frac{\tau}{\rho v}\int_0^\xi dy$$

and for the heat flux

$$\int_{T_s}^{T_\xi} dT = -\frac{q_y}{A\rho c_p\alpha}\int_0^\xi dy$$

Solving for the velocity and temperature profiles in the laminar sublayer yields

$$v_x|_\xi = \frac{\tau\xi}{\rho v} \tag{19-51}$$

and

$$T_s - T_\xi = \frac{q_y\xi}{A\rho c_p\alpha} \tag{19-52}$$

---

[5] L. Prandtl, *Zeit. Physik.,* **11,** 1072 (1910).

Eliminating the distance ξ between these two expressions gives

$$\frac{\rho v v_x|_\xi}{\tau} = \frac{\rho A c_p \alpha}{q_y}(T_s - T_\xi) \qquad (19\text{-}53)$$

Directing our attention now to the turbulent core where the Reynolds analogy applies, we may write equation (19-37)

$$\frac{h}{\rho c_p(v_\infty - v_x|_\xi)} = \frac{C_f}{2} \qquad (19\text{-}37)$$

and, expressing h and C_f in terms of their defining relations, we obtain

$$\frac{q_y/A}{\rho c_p(v_\infty - v_x|_\xi)(T_\xi - T_\infty)} = \frac{\tau}{\rho(v_\infty - v_x|_\xi)^2}$$

Simplifying and rearranging this expression, we have

$$\frac{\rho(v_\infty - v_x|_\xi)}{\tau} = \rho A c_p \frac{(T_\xi - T_\infty)}{q_y} \qquad (19\text{-}54)$$

which is a modified form of the Reynolds analogy applying from $y = \xi$ to $y = y_{max}$.

Eliminating T_ξ between equations (19-53) and (19-54), we have

$$\frac{\rho}{\tau}\left[v_\infty + v_x|_\xi\left(\frac{v}{\alpha} - 1\right)\right] = \frac{\rho A c_p}{q_y}(T_s - T_\infty) \qquad (19\text{-}55)$$

Introducing the coefficient of skin friction

$$C_f = \frac{\tau}{\rho v_\infty^2/2}$$

and the convective heat-transfer coefficient

$$h = \frac{q_y}{A(T_s - T_\infty)}$$

we may reduce equation (19-54) to

$$\frac{v_\infty + v_x|_\xi(v/\alpha - 1)}{v_\infty^2 C_f/2} = \frac{\rho c_p}{h}$$

Inverting both sides of this expression and making it dimensionless, we obtain

$$\frac{h}{\rho c_p v_\infty} \equiv \text{St} = \frac{C_f/2}{1 + (v_x|_\xi/v_\infty)[(v/\alpha) - 1]} \qquad (19\text{-}56)$$

This equation involves the ratio v/α, which has been defined previously as the Prandtl number. For a value of $\text{Pr} = 1$, equation (19-56) reduces to the Reynolds analogy. For $\text{Pr} = 1$, the Stanton number is a function of C_f, Pr, and the ratio $v_x|_\xi/v_\infty$. It would be convenient to eliminate the velocity ratio; this may be accomplished by recalling some results from Chapter 12.

At the edge of the laminar sublayer

$$v^+ = y^+ = 5$$

and by definition $v^+ = v_x/(\sqrt{\tau/\rho})$. Thus for the case at hand

$$v^+ = v_x|_\xi/(\sqrt{\tau/\rho}) = 5$$

Again introducing the coefficient of skin friction in the form

$$C_f = \frac{\tau}{\rho v_\infty^2/2}$$

we may write

$$\sqrt{\frac{\tau}{\rho}} = v_\infty \sqrt{\frac{C_f}{2}}$$

which, when combined with the previous expression given for the velocity ratio, gives

$$\frac{v_x|_\xi}{v_\infty} = 5\sqrt{\frac{C_f}{2}} \tag{19-57}$$

Substitution of equation (19-57) into (19-56) gives

$$St = \frac{C_f/2}{1 + 5\sqrt{C_f/2}(Pr - 1)} \tag{19-58}$$

which is known as the *Prandtl analogy*. This equation is written entirely in terms of measurable quantities.

von Kármán[6] extended Prandtl's work to include the effect of the transition or buffer layer in addition to the laminar sublayer and turbulent core. His result, the von Kármán analogy, is expressed as

$$St = \frac{C_f/2}{1 + 5\sqrt{C_f/2}\{Pr - 1 + \ln[1 + \frac{5}{6}(Pr - 1)]\}} \tag{19-59}$$

Note that, just as for the Prandtl analogy, equation (19-59) reduces to the Reynolds analogy for a Prandtl number of unity.

The application of the Prandtl and von Kármán analogies is, quite logically, restricted to those cases in which there is negligible form drag. These equations yield the most accurate results for Prandtl numbers greater than unity.

An illustration of the use of the four relations developed in this section is given in the example below.

EXAMPLE 1

Water at 50°F enters a heat-exchanger tube having an inside diameter of 1 in. and a length of 10 ft. The water flows at 20 gal/min. For a constant wall temperature of 210°F, estimate the exit temperature of the water using (a) the Reynolds analogy, (b) the Colburn analogy, (c) the Prandtl analogy, and (d) the von Kármán analogy. Entrance effects are to be neglected, and the properties of water may be evaluated at the arithmetic-mean bulk temperature.

Considering a portion of the heat-exchanger tube shown in Figure 19.6, we see that an application of the first law of thermodynamics to the control volume indicated will yield the result that

$$\left\{ \begin{array}{c} \text{rate of heat} \\ \text{transfer into c.v.} \\ \text{by fluid flow} \end{array} \right\} + \left\{ \begin{array}{c} \text{rate of heat} \\ \text{transfer into c.v.} \\ \text{by convection} \end{array} \right\} = \left\{ \begin{array}{c} \text{rate of heat} \\ \text{transfer out of c.v.} \\ \text{by fluid flow} \end{array} \right\}$$

---

[6] T. von Kármán, *Trans. ASME,* **61,** 705 (1939).

Figure 19.6 Analog analysis of water flowing in a circular tube.

If these heat-transfer rates are designated as q_1, q_2, and q_3, they may be evaluated as follows:

$$q_1 = \rho \frac{\pi D^2}{4} v_x c_p T|_x$$

$$q_2 = h\pi D \, \Delta_x (T_s - T)$$

and

$$q_3 = \rho \frac{\pi D^2}{4} v_x c_p T|_{x+\Delta x}$$

The substitution of these quantities into the energy balance expression gives

$$\rho \frac{\pi D^2}{4} v_x c_p [T|_{x+\Delta x} - T|_x] - h\pi D \, \Delta x (T_s - T) = 0$$

which may be simplified and rearranged into the form

$$\frac{D}{4} \frac{T|_{x+\Delta x} - T|_x}{\Delta x} + \frac{h}{\rho v_x c_p}(T - T_s) = 0 \tag{19-60}$$

Evaluated in the limit as $\Delta x \to 0$, equation (19-59) reduces to

$$\frac{dT}{dx} + \frac{h}{\rho v_x c_p} \frac{4}{D}(T - T_s) = 0 \tag{19-61}$$

Separating the variables, we have

$$\frac{dT}{T - T_s} + \frac{h}{\rho v_x c_p} \frac{4}{D} dx = 0$$

and integrating between the limits indicated, we obtain

$$\int_{T_0}^{T_L} \frac{dT}{T - T_s} + \frac{h}{\rho v_x c_p} \frac{4}{D} \int_0^L dx = 0$$

$$\ln \frac{T_L - T_s}{T_0 - T_s} + \frac{h}{\rho v_x c_p} \frac{4L}{D} = 0 \tag{19-62}$$

Equation (19-62) may now be solved for the exit temperature T_L. Observe that the coefficient of the right-hand term, $h/\rho v_x c_p$, is the Stanton number. This parameter has been achieved quite naturally from our analysis.

The coefficient of skin friction may be evaluated with the aid of Figure 14.1. The velocity is calculated as

$$v_x = 20 \, \text{gal/min} \, (\text{ft}^3/7.48 \, \text{gal})[144/(\pi/4)(1^2)]\text{ft}^2(\text{min}/60\,\text{s}) = 8.17 \, \text{fps}$$

Initially, we will assume the mean bulk temperature to be 90°F. The film temperature will then be 150°F, at which $\nu = 0.474 \times 10^{-5}$ ft²/s. The Reynolds number is

$$\text{Re} = \frac{Dv_x}{\nu} = \frac{(1/12 \, \text{ft})(8.17 \, \text{ft/s})}{0.474 \times 10^{-5} \, \text{ft}^2/\text{s}} = 144{,}000$$

At this value of Re, the friction factor, f_f, assuming smooth tubing, is 0.0042. For each of the four analogies, the Stanton number is evaluated as follows:

(a) *Reynolds analogy*

$$St = \frac{C_f}{2} = 0.0021$$

(b) *Colburn analogy*

$$St = \frac{C_f}{2} Pr^{-2/3} = 0.0021(2.72)^{-2/3} = 0.00108$$

(c) *Prandtl analogy*

$$St = \frac{C_f/2}{1 + 5\sqrt{C_f/2}(Pr - 1)}$$

$$= \frac{0.0021}{1 + 5\sqrt{0.0021}(1.72)} = 0.00151$$

(d) *von Kármán analogy*

$$St = \frac{C_f/2}{1 + 5\sqrt{C_f/2}\left\{Pr - 1 + \ln\left[1 + \frac{5}{6}(Pr - 1)\right]\right\}}$$

$$= \frac{0.0021}{1 + 5\sqrt{0.0021}\left\{2.72 - 1 + \ln\left[1 + \frac{5}{6}(2.72 - 1)\right]\right\}}$$

$$= 0.00131$$

Substituting these results into equation (19-62), we obtain, for T_L, the following results:

(a) $T_L = 152°F$

(b) $T_L = 115°F$

(c) $T_L = 132°F$

(d) $T_L = 125°F$

Some fine tuning of these results may be necessary to adjust the physical property values for the calculated film temperatures. In none of these cases is the assumed film temperature different than the calculated one by more than 6°F, so the results are not going to change much.

The Reynolds analogy value is much different from the other results obtained. This is not surprising, as the Prandtl number was considerably above a value of one. The last three analogies yielded quite consistent results. The Colburn analogy is the simplest to use and is preferable from that standpoint.

19.8 CLOSURE

The fundamental concepts of convection heat transfer have been introduced in this chapter. New parameters pertinent to convection are the Prandtl, Nusselt, Stanton, and Grashof numbers.

Four methods of analyzing a convection heat-transfer process have been discussed. These are as follows:

(1) dimensional analysis coupled with experiment;

(2) exact analysis of the boundary layer;

(3) integral analysis of the boundary layer; and

(4) analogy between momentum and energy transfer.

Several empirical equations for the prediction of convective heat-transfer coefficients will be given in the chapters to follow.

PROBLEMS

19.1 Using dimensional analysis, demonstrate that the parameters

$$\frac{T - T_\infty}{T_0 - T_\infty} \quad \frac{x}{L} \quad \frac{\alpha t}{L^2} \quad \text{and} \quad \frac{hL}{k}$$

are possible combinations of the appropriate variables in describing unsteady-state conduction in a plane wall.

19.2 Dimensional analysis has shown the following parameters to be significant for forced convection:

$$\frac{x v_\infty \rho}{\mu} \quad \frac{\mu c_p}{k} \quad \frac{hx}{k} \quad \frac{h}{\rho c_p v_\infty}$$

Evaluate each of these parameters at 340 K, for air, water, benzene, mercury, and glycerin. The distance x may be taken as 0.3 m, $v_\infty = 15$ m/s, and $h = 34$ W/m$^2 \cdot$ K.

19.3 Plot the parameters $x v_\infty \rho / \mu$, $\mu c_p / k$, hx/k, and $h/\rho c_p v_\infty$ vs. temperature for air, water, and glycerin, using the values for x, h, and v from Problem 19.2.

19.4 Using the relations from Problem 19.19, determine, for the case of air at 310 K adjacent to a vertical wall with its surface at 420 K,

a. the thickness of the boundary layer at $x = 15$ cm, 30 cm, 1.5 m,

b. the magnitude of h_x at 15 cm, 30 cm, 1.5 m.

19.5 Given the conditions specified in Problem 19.16, construct a plot of local heat-transfer coefficient vs. position along the plate for glycerin temperatures of 30°, 50°, and 80°F.

19.6 The fuel plates in a nuclear reactor are 4 ft long and stacked with a 1/2-in. gap between them. The heat flux along the plate surfaces varies sinusoidally according to the equation

$$\frac{q}{A} = \alpha + \beta \sin \frac{\pi x}{L}$$

where $\alpha = 250$ Btu/h ft^2, $\beta = 1500$ Btu/h ft^2, x is the distance from the leading edge of the plates, and L is the total plate

length. If air at 120°F, 80 psi, flowing at a mass velocity of 6000 lb$_m$/h ft^2, is used to cool the plates, prepare plots showing

a. the heat flux vs. x;

b. the mean air temperature vs. x.

19.7 Given the information in Problem 19.6, determine the total heat transferred for a stack of plates with a combined surface area of 640 ft^2, each plate being 4 ft wide.

19.8 In a thermal heat sink the heat flux variation along the axis of a cooling passage is approximated as

$$\frac{q}{A} = a + b \sin \left(\frac{\pi x}{L} \right)$$

where x is measured along the passage axis and L is its total length.

A large installation involves a stack of plates with a 3-mm air space between them. The flow passages are 1.22 m long, and the heat flux in the plates varies according to the above equation where $a = 900$ W/m^2 and $v = 2500$ W/m^2. Air enters at 100°C with a mass velocity (the product of ρV) of 7.5 kg/s $\cdot$ m^2. The surface coefficient along the flow passage can be considered constant with a value of 56 W/m$^2 \cdot$ K.

Generate a plot of heat flux, mean air temperature, and plate surface temperature as functions of x. Where does the maximum surface temperature occur and what is its value?

19.9 Determine the total heat transfer from the vertical wall described in Problem 19.4 to the surrounding air per meter of width if the wall is 2.5 m high.

19.10 Repeat Problem 19.21 for velocity and temperature profiles of the form

$$v = a + by + cy^2 \qquad T - T_s = \alpha + \beta y + \gamma y^2$$

19.11 Solve Problem 19.26 for the case of a wall heat flux varying according to

$$\frac{q}{A} = \alpha + \beta \sin \frac{\pi x}{L}$$

where $\alpha = 250\,\text{Btu/h ft}^2$, $\beta = 1500\,\text{Btu/h ft}^2$, x is the distance from the entrance, and L is the tube length.

19.12 Work Problem 19.25 for the case in which the flowing fluid is air at 15 fps.

19.13 Nitrogen at 100°F and 1 atm flows at a velocity of 100 fps. A flat plate 6 in. wide, at a temperature of 200°F, is aligned parallel to the direction of flow. At a position 4 ft from the leading edge, determine the following (a) δ; (b) δ_t; (c) C_{fx}; (d) C_{fl}; (e) h_x; (f) h; (g) total drag force; (h) total heat transfer.

19.14 A plane surface, 25 cm wide, has its temperature maintained at 80°C. Atmospheric air, at 25°C, flows parallel to the surface with a velocity of 2.8 m/s. Using the results of boundary-layer analysis, determine the following for a 1-m long plate:

a. the mean coefficient of skin friction, C_{fL};

b. the total drag force exerted on the plate by the air flow;

c. the total heat transfer rate from the plate to the air stream.

19.15 Show that, for the case of natural convection adjacent to a plane vertical wall, the appropriate integral equations for the hydrodynamic and thermal boundary layers are

$$\alpha \left.\frac{\partial T}{\partial y}\right|_{y=0} = \frac{d}{dx}\int_0^{\delta_1} v_x(T_\infty - T)\,dy$$

and

$$-v\left.\frac{dv_x}{dy}\right|_{y=0} + \beta g \int_0^{\delta_1}(T - T_\infty)\,dy = \frac{d}{dx}\int_0^{\delta} v_x^2\,dy$$

19.16 Glycerin flows parallel to a flat plate measuring 2 ft by 2 ft with a velocity of 10 fps. Determine values for the mean convective heat-transfer coefficient and the associated drag force imposed on the plate for glycerin temperatures of 350°, 50°, and 180°F. What heat flux will result, in each case, if the plate temperature is 50°F above that of the glycerin?

19.17 Shown in the figure is the case of a fluid flowing parallel to a flat plate, where for a distance X from the leading edge, the plate and fluid are at the same temperature. For values of $x > X$, the plate is maintained at a constant temperature, T_s, where $T_s > T_\infty$. Assuming a cubic profile for both the hydrodynamic and the thermal boundary layers, show that the ratio of the thickness, ξ, is expressed as

$$\xi = \frac{\delta_t}{\delta} \cong \frac{1}{\text{Pr}^{1/3}}\left[1 - \left(\frac{X}{x}\right)^{3/4}\right]^{1/3}$$

Also show that the local Nusselt number can be expressed as

$$\text{Nu}_x \cong 0.33 \left(\frac{\text{Pr}}{1 - (X/x)^{3/4}}\right)^{1/3} \text{Re}_x^{1/2}$$

19.18 Simplified relations for natural convection in air are of the form

$$h = \alpha(\Delta T/L)^\beta$$

where α, β are constants; L is a significant length, in ft; ΔT is $T_s - T_\infty$, in °F; and h is the convective heat-transfer coefficient, Btu/h ft^2 °F. Determine the values for α and β for the plane vertical wall, using the equation from Problem 19.14.

19.19 Using the integral relations from Problem 19.8, and assuming the velocity and temperature profiles of the form

$$\frac{v}{v_x} = \left(\frac{y}{\delta}\right)\left(1 - \frac{y}{\delta}\right)^2$$

and

$$\frac{T - T_\infty}{T_s - T_\infty} = \left(1 - \frac{y}{\delta}\right)^2$$

where δ is the thickness of both the hydrodynamic and thermal boundary layers, show that the solution in terms of δ and v_x from each integral equation reduce to

$$\frac{2\alpha}{\delta} = \frac{d}{dx}\left(\frac{dv_x}{30}\right)$$

and

$$-\frac{vv_x}{\delta} + \beta g\,\Delta T\,\frac{\delta}{3} = \frac{d}{dt}\left(\frac{\delta v_x^2}{105}\right)$$

Next, assuming that both δ and v_x vary with x according to

$$\delta = Ax^a \quad \text{and} \quad v_x = Bx^b$$

show that the resulting expression for δ becomes

$$\delta/x = 3.94\,\text{Pr}^{-1/2}(\text{Pr} + 0.953)^{1/4}\text{Gr}_x^{-1/4}$$

and that the local Nusselt number is

$$\text{Nu}_x = 0.508 \quad \text{Pr}^{-1/2}(\text{Pr} + 0.953)^{-1/4}\text{Gr}_x^{1/4}$$

19.20 Use the results of Problem along with those of Chapter 12 to determine δ, C_{fx}, δ_1, and h_x at a distance of 40 cm from the leading edge of a flat plane. Air with a free stream velocity of 5 m/s and $T_\infty = 300$ K flows parallel to the plate surface. The first 20 cm of the plate is unheated; the surface temperature is maintained at 400 K beyond that point.

19.21 Using the appropriate integral formulas for flow parallel to a flat surface with a constant free-stream velocity, develop expressions for the local Nusselt number in terms of Re_x and Pr for velocity and temperature profiles of the form

$$v = a + by, \quad T - T_s = \alpha + \beta y$$

19.22 Repeat Problem 19.21 for velocity and temperature profiles of the form

$$v = a \sin by, \quad T - T_s = \alpha \sin \beta u$$

19.23 For the case of a turbulent boundary layer on a flat plate, the velocity profile has been shown to follow closely the form

$$\frac{v}{v_\infty} = \left(\frac{y}{\delta}\right)^{1/7}$$

Assuming a temperature profile of the same form, that is

$$\frac{T - T_s}{T_\infty - T_s} = \left(\frac{y}{\delta_1}\right)^{1/7}$$

and assuming that $\delta = \delta_1$, use the integral relation for the boundary layer to solve for h_x and Nu_x. The temperature gradient at the surface may be considered similar to the velocity gradient at $y = 0$ given by equation (13-26).

19.24 A blacktop road surface 18.3 m wide receives solar radiation at the rate of 284 W/m² at noon and 95 W/m₂ are lost by reradiation to the atmosphere. A wind, at 300 K, flows across the road. Determine the wind velocity that will cause the road surface to be at 308 K if all energy not reradiated to the sky is removed by convection.

19.25 Water, at 60°F, enters a 1-in.-ID tube that is used to cool a nuclear reactor. The water flow rate is 30 gal/min. Determine the total heat transfer and the exiting water temperature for a 15-ft-long tube if the tube surface temperature is a constant value of 300°F. Compare the answer obtained, using the Reynolds and Colburn analogies.

19.26 Water at 60°F enters a 1-in.-ID tube that is used to cool a nuclear reactor. The water flow rate is 30 gal/min. Determine the total heat transfer, the exiting water temperature, and the wall temperature at the exit of a 15-ft long tube if the tube wall condition is one of uniform heat flux of 500 Btu/hr ft².

19.27 Work Problem 19.26 for the case in which the flowing fluid is air at 15 fps.

19.28 Work Problem 19.25 for the case in which the flowing fluid is sodium entering the tube at 200°F

19.29 Work Problem 19.26 for the case in which the flowing fluid is sodium entering the tube at 200°F.

Chapter 20

Convective Heat-Transfer Correlations

Convective heat transfer was treated from an analytical point of view in Chapter 19. Although the analytic approach is very meaningful, it may not offer a practical solution to every problem. There are many situations for which no mathematical models have as yet been successfully applied. Even in those cases for which an analytical solution is possible, it is necessary to verify the results by experiment. In this chapter, we shall present some of the most useful correlations of experimental heat-transfer data available. Most correlations are in the forms indicated by dimensional analysis.

The sections to follow include discussion and correlations for natural convection, forced convection for internal flow, and forced convection for external flow, respectively. In each case, those analytical relations that are available are presented along with the most satisfactory empirical correlations for a particular geometry and flow condition.

20.1 NATURAL CONVECTION

The mechanism of energy transfer by natural convection involves the motion of a fluid past a solid boundary, which is the result of the density differences resulting from the energy exchange. Because of this, it is quite natural that the heat-transfer coefficients and their correlating equations will vary with the geometry of a given system.

Vertical Plates. The natural convection system most amenable to analytical treatment is that of a fluid adjacent to a vertical wall.

Standard nomenclature for a two-dimensional consideration of natural convection adjacent to a vertical plane surface is indicated in Figure 20.1. The x direction is commonly taken along the wall, with y measured normal to the plane surface.

Schmidt and Beckmann[1] measured the temperature and velocity of air at different locations near a vertical plate and found a significant variation in both quantities along the direction parallel to the plate. The variations of velocity and temperature for a 12.5-cm-high vertical plate are shown in Figures 20.2 and 20.3 for the conditions $T_s = 65°C$, $T_\infty = 15°C$.

The two limiting cases for vertical plane walls are those with constant surface temperature and with constant wall heat flux. The former of these cases have been solved by Ostrach[2] and the latter by Sparrow and Gregg.[3]

---

[1] E. Schmidt and W. Beckmann, *Tech. Mech. U. Thermodynamik,* **1,** 341 and 391 (1930).
[2] S. Ostrach, NACA Report 1111, 1953.
[3] E. M. Sparrow and J. L. Gregg, *Trans. A.S.M.E.,* **78,** 435 (1956).

Figure 20.1 Coordinate system for the analysis of natural convection adjacent to a heated vertical wall.

Ostrach, employing a similarity transformation with governing equations of mass conservation, motion, and energy in a free-convection boundary layer, obtained an expression for local Nusselt number of the form

$$\mathrm{Nu}_x = f(\mathrm{Pr})\left(\frac{\mathrm{Gr}_x}{4}\right)^{1/4} \tag{20-1}$$

The coefficient, $f(\mathrm{Pr})$, varies with Prandtl number, with values given in Table 20.1.

We usually find the mean Nusselt number, Nu_L, to be of more value than Nu_x. Using an integration procedure, as discussed earlier, an expression for Nu_L may be determined

Figure 20.2 Velocity distribution in the vicinity of a vertical heated plate in air.

Figure 20.3 Temperature distribution in the vicinity of a vertical heated plate in air.

Table 20.1 Values of the coefficient $f(Pr)$ for use in equation (20-1)

Pr	0.01	0.072	1	2	10	100	1000
$f(Pr)$	0.081	0.505	0.567	0.716	1.169	2.191	3.966

using equation (20-1). The mean heat transfer coefficient for a vertical surface of height, L, is related to the local value according to

$$h_L = \frac{1}{L}\int_0^L h_x \, dx$$

Inserting equation (20-1) appropriately, we proceed to

$$h_L = \frac{k}{L} f(Pr) \left[\frac{\beta g \Delta T}{4\nu^2}\right]^{1/4} \int_0^L x^{-1/4} dx$$

$$= \left(\frac{4}{3}\right)\left(\frac{k}{L}\right) f(Pr) \left[\frac{\beta g L^3 \Delta T}{4\nu^2}\right]^{1/4}$$

and, in dimensionless form, we have

$$Nu_L = \frac{4}{3} f(Pr) \left(\frac{Gr_L}{4}\right)^{1/4} \tag{20-2}$$

Sparrow and Gregg's[4] results for the constant wall heat flux case compared within 5% to those of Ostrach for like values of Pr. Equations (20-1) and (20-2), along with coefficients from Table 20.1, may thus be used, with reasonable accuracy, to evaluate any vertical plane surface regardless of wall conditions, provided boundary layer flow is laminar.

Fluid properties, being temperature dependent, will have some effect on calculated results. It is important, therefore, that properties involved in equations (20-1) and (20-2) be evaluated at the film temperature.

$$T_f = \frac{T_s + T_\infty}{2}$$

As with forced convection, turbulent flow will also occur in free convection boundary layers. When turbulence is present, an analytical approach is quite difficult, and we must rely heavily on correlations of experimental data.

Transition from laminar to turbulent flow in natural convection boundary layers adjacent to vertical plane surfaces has been determined to occur at, or near

$$Gr_t \, Pr = Ra_t \cong 10^9 \tag{20-3}$$

where the subscript, t, indicates transition. The product, Gr Pr, is often referred to as Ra, the *Rayleigh* number.

Churchill and Chu[5] have correlated a large amount of experimental data for natural convection adjacent to vertical planes over 13 orders of magnitude of Ra. They propose a single equation for Nu_L that applies to all fluids. This powerful equation is

$$Nu_L = \left\{0.825 + \frac{0.387 \, Ra_L^{1/6}}{\left[1 + (0.492/Pr)^{9/16}\right]^{8/27}}\right\}^2 \tag{20-4}$$

[4] E. M. Sparrow and J. L. Gregg, *Trans. A.S.M.E.*, **78**, 435 (1956).
[5] S. W. Churchill and H. H. S. Chu, *Int. J. Heat & Mass Tr.*, **18**, 1323 (1975).

Churchill and Chu show this expression to provide accurate results for both laminar and turbulent flows. Some improvement was found for the laminar range $\left(\text{Ra}_L < 10^9\right)$ by using the following equation:

$$\text{Nu}_L = 0.68 + \frac{0.670 \, \text{Ra}_L^{1/4}}{\left[1 + (0.492/\text{Pr})^{9/16}\right]^{4/9}} \tag{20-5}$$

Vertical Cylinders.   For the case of cylinders with their axes vertical, the expressions presented for plane surfaces can be used provided the curvature effect is not too great. The criterion for this is expressed in equation (20-6); specifically, a vertical cylinder can be evaluated using correlations for vertical plane walls when

$$\frac{D}{L} \geq \frac{35}{\text{Gr}_L^{1/4}} \tag{20-6}$$

Physically, this represents the limit where boundary layer thickness is small relative to cylinder diameter, D.

Horizontal Plates.   The correlations suggested by McAdams[6] are well accepted for this geometry. A distinction is made regarding whether the fluid is hot or cool, relative to the adjacent surface, and whether the surface faces up or down. It is clear that the induced buoyancy will be much different for a hot surface facing up than down. McAdams's correlations are, for a hot surface facing up or cold surface facing down

$$10^5 < \text{Ra}_L < 2 \times 10^7 \quad \text{Nu}_L = 0.54 \, \text{Ra}_L^{1/4} \tag{20-7}$$

$$2 \times 10^7 < \text{Ra}_L < 3 \times 10^{10} \quad \text{Nu}_L = 0.14 \, \text{Ra}_L^{1/3} \tag{20-8}$$

and for a hot surface facing down or cold surface facing up

$$3 \times 10^5 < \text{Ra}_L < 10^{10} \quad \text{Nu}_L = 0.27 \, \text{Ra}_L^{1/4} \tag{20-9}$$

In each of these correlating equations, the film temperature, T_f, should be used for fluid property evaluation. The length scale, L, is the ratio of the plate-surface area to perimeter.

For plane surfaces inclined at an angle, θ, with the vertical, equations (20-4) and (20-5) may be used, with modification, for values of θ up to 60°. Churchill and Chu[7] suggest replacing g by $g \cos \theta$ in equation (20-5) when boundary layer flow is laminar. With turbulent flow, equation (20-4) may be used without modification.

Horizontal Cylinders.   With cylinders of sufficient length that end effects are insignificant, two correlations are recommended. Churchill and Chu[8] suggest the following correlation

$$\text{Nu}_D = \left\{ 0.60 + \frac{0.387 \, \text{Ra}_D^{1/6}}{\left[1 + (0.559/\text{Pr})^{9/16}\right]^{8/27}} \right\}^2 \tag{20-10}$$

over the Rayleigh number range $10^{-5} < \text{Ra}_D < 10^{12}$.

---

[6] W. H. McAdams, *Heat Transmission*, Third Edition, Chapter 7, McGraw-Hill Book Company, New York, 1957.
[7] S. W. Churchill and H. H. S. Chu, *Int. J. Heat & Mass Tr.*, **18**, 1323 (1975).
[8] S. W. Churchill and H. H. S. Chu, *Int. J. Heat & Mass Tr.*, **18**, 1049 (1975).

A simpler equation has been suggested by Morgan,[9] in terms of variable coefficients

$$\mathrm{Nu}_D = C\,\mathrm{Ra}_D^n \tag{20-11}$$

where values of c and n are specified as functions of Ra_D in Table 20.2.

Table 20.2 Values of constants C and n in equation (20-11)

	C	n
$10^{-10} < \mathrm{Ra}_D < 10^{-2}$	0.675	0.058
$10^{-2} < \mathrm{Ra}_D < 10^{2}$	1.02	0.148
$10^{2} < \mathrm{Ra}_D < 10^{4}$	0.850	0.188
$10^{4} < \mathrm{Ra}_D < 10^{7}$	0.480	0.250
$10^{7} < \mathrm{Ra}_D < 10^{12}$	0.125	0.333

The film temperature should be used in evaluating fluid properties in the above equations.

Spheres. The correlation suggested by Yuge[10] is recommended for the case with $\mathrm{Pr} \simeq 1$, and $1 < \mathrm{Ra}_D < 10^5$.

$$\mathrm{Nu}_D = 2 + 0.43\,\mathrm{Ra}_D^{1/4} \tag{20-12}$$

We may notice that, for the sphere, as Ra approaches zero, heat transfer from the surface to the surrounding medium is by conduction. This problem may be solved to yield a limiting value for Nu_D equal to 2. This result is obviously compatible with equation (20-12).

Rectangular Enclosures. Shown in Figure 20.4 is the configuration and nomenclature pertinent to rectangular enclosures. These cases have become much more important in recent years due to their application in solar collectors. Clearly, heat transfer will be affected

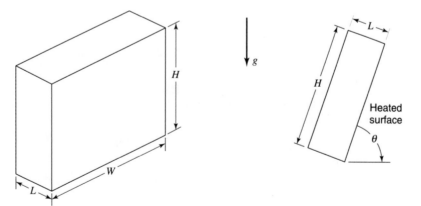

Figure 20.4 The rectangle enclosure.

[9] V. T. Morgan, *Advances in Heat Transfer,* Vol. II, T. F. Irvine and J. P. Hartnett, Eds., Academic Press, New York, 1975, pp. 199–264.

[10] T. Yuge, *J. Heat Transfer,* **82,** 214 (1960).

by the angle of tilt, θ; by the aspect ratio, H/L; and by the usual dimensionless parameters, Pr and Ra_L.

In each of the correlations to follow, the temperature of the hotter of the two large surfaces is designated as T_1, and the cooler surface is at temperature T_2. Fluid properties are evaluated at the film temperature, $T_f = (T_1/T_2)/2$. Convective heat flux is expressed as

$$\frac{q}{A} = h(T_1 - T_2) \tag{20-13}$$

Case 1. Horizontal enclosures, $\theta = 0$

With the bottom surface heated, a critical Rayleigh member has been determined by several investigators to exist. For cases where

$$Ra_L = \frac{\beta g L^3 (T_1 - T_2)}{\alpha v} > 1700$$

conditions within an enclosure are thermally unstable and natural convection will occur. A correlation for this case has been proposed by Globe and Dropkin[11] in the form

$$Nu_L = 0.069 \, Ra_L^{1/3} Pr^{0.074} \tag{20-14}$$

for the range $3 \times 10^5 < Ra_L < 7 \times 10^9$.

When $\theta = 180°$, that is, the upper surface is heated, or when $Ra_L < 1700$, heat transfer is by conduction; thus, $Nu_L = 1$.

Case 2. Vertical enclosures, $\theta = 90°$

For aspect ratios less than 10, Catton[12] suggests the use of the following correlations:

$$Nu_L = 0.18 \left(\frac{Pr}{0.2 + Pr} Ra_L \right)^{0.29} \tag{20-15}$$

when

$$1 < H/L < 2, \, 10^{-3} < Pr < 10^5, \, 10^3 < Ra_L Pr/(0.2 + Pr)$$

and

$$Nu_L = 0.22 \left(\frac{Pr}{0.22 + Pr} Ra_L \right)^{0.28} \left(\frac{H}{L} \right)^{-1/4} \tag{20-16}$$

when $2 < H/L < 10$, $Pr < 10^5$, $Ra_L < 10^{10}$.

For higher values of H/L, the correlations of MacGregor and Emery[13] are recommended. These are

$$Nu_L = 0.42 \, Ra_L^{1/4} \, Pr^{0.012} (H/L)^{-0.3} \tag{20-17}$$

for

$$10 < H/L < 40, 1 < Pr < 2 \times 10^4, 10^4 < Ra_L < 10^7$$

and

$$Nu_L = 0.046 \, Ra_L^{1/3} \tag{20-18}$$

[11] S. Globe and D. Dropkin, *J. Heat Transfer,* **81C,** 24 (1959).
[12] I. Catton, *Proc. 6th Int. Heat Tr. Conference, Toronto, Canada,* **6,** 13 (1978).
[13] P. K. MacGregor and A. P. Emery, *J. Heat Transfer,* **91,** 391 (1969).

for

$$10 < H/L < 40, 1 < \text{Pr} < 20, 10^6 < \text{Ra}_L < 10^9$$

Case 3. Tilted vertical enclosures, $0 < \theta < 90°$

Numerous publications have dealt with this configuration. Correlations for this case, when the aspect ratio is large $(H/L > 12)$, are the following:

$$\text{Nu}_L = 1 + 1.44\left[1 - \frac{1708}{\text{Ra}_L \cos\theta}\right]\left[1 - \frac{1708(\sin 1.8\theta)^{1.6}}{\text{Ra}_L \cos\theta}\right]$$
$$+ \left[\left(\frac{\text{Ra}_L \cos\theta}{5830}\right)^{1/3} - 1\right] \tag{20-19}$$

when $H/L \geq 12$, $0 < \theta < 70°$. In applying this relationship any bracketed term with a negative value should be set equal to zero. Equation (20-19) was suggested by Hollands et al.[14] With enclosures nearing the vertical, Ayyaswamy and Catton[15] suggest the relationship

$$\text{Nu}_L = \text{Nu}_{LV}(\sin\theta)^{1/4} \tag{20-20}$$

for all aspect ratios, and $70° < \theta < 90°$. The value $\theta = 70°$ is termed the "critical" tilt angle for vertical enclosures with $H/L > 12$. For smaller aspect ratios the critical angle of tilt is also smaller. A recommended review article on the subject of inclined rectangular cavities is that of Buchberg, Catton, and Edwards.[16]

EXAMPLE 1

Determine the surface temperature of a cylindrical tank measuring 0.75 m in diameter and 1.2 m high. The tank contains a transformer immersed in an oil bath that produces a uniform surface temperature condition. All heat loss from the surface may be assumed due to natural convection to surrounding air at 295 K. The heat dissipation rate from the transformer is constant at 1.5 KW.

Surface areas that apply are

$$A_{\text{top}} = A_{\text{bottom}} = \frac{\pi}{4}(0.75\,\text{m})^2 = 0.442\,\text{m}^2$$

$$A_{\text{side}} = \pi(1.2\,\text{m})(0.75\,\text{m}) = 2.83\,\text{m}^2$$

The total heat transfer is the sum of the contributions from the three surfaces. This may be written as

$$q_{\text{total}} = [h_t(0.442) + h_b(0.442) + h_s(2.83)](T - 295)$$

where subscripts t (top), b (bottom), and s (sides) apply to the surfaces in question.

Rewriting this expression in terms of Nu, we have

$$q_{\text{total}} = \left[\text{Nu}_t\frac{k}{0.75}(0.442) + \text{Nu}_b\frac{k}{0.75}(0.442) + \text{Nu}_s\frac{k}{1.2}(2.83)\right](T - 295)$$

---

[14] K. G. T. Hollands, S. E. Unny, G. D. Raithby, and L. Konicek, *J. Heat Transfer,* **98**, 189 (1976).

[15] P. S. Ayyaswamy and I. Catton, *J. Heat Transfer,* **95**, 543 (1973).

[16] H. Buchberg, I. Catton and D. K. Edwards, *J. Heat Transfer,* **98**, 182 (1976).

or

$$q_{\text{total}} = [0.589\,\text{Nu}_t + 0.589\,\text{Nu}_b + 2.36 \quad \text{Nu}_s]k(T - 295)$$

A complication exists in solving this equation because the unknown quantity is surface temperature. The procedure to be used is trial and error where, initially, a surface temperature value is assumed for property evaluation, and then solved for T. This new value for surface temperature will then be used and the procedure continued until the resulting temperature agrees with the value used in finding fluid properties.

To begin the problem, we assume that $T_{\text{surface}} = 38.5\,\text{K}$. Properties will thus be evaluated at $T_f = 340\,\text{K}$. For air at 340 K, $v = 1.955 \times 10^{-5}\,\text{m}^2/\text{s}$, $k = 0.0293\,\text{W/m·K}$, $\alpha = 2.80 \times 10^{-5}\,\text{m}^2/\text{s}$, Pr $= 0.699$, and $\beta g/v^2 = 0.750 \times 10^8\,\text{1/K·m}^3$.

For the vertical surface

$$\begin{aligned}
\text{Gr} &= \frac{\beta g}{v^2} L^3 \Delta T \\
&= (0.750 \times 10^8\,\text{1/K·m}^3)(1.2\,\text{m})^3(90\,\text{K}) \\
&= 11.7 \times 10^9
\end{aligned}$$

According to equation (20-6), the effect of curvature may be neglected if

$$\frac{D}{L} \geq \frac{35}{(11.7 \times 10^9)^{1/4}} = 0.106$$

In the present case $D/L = 0.75/1.2 = 0.625$; thus the vertical surface will be treated using equations for a plane wall.

Values must now be determined for Nu$_t$, Nu$_b$, and Nu$_s$. Equations (20-8), (20-9), and (20-4) will be employed. The three values for Nu are determined as follows:

Nu$_t$:

$$L = A/p = \frac{\pi D^2/4}{\pi D} = \frac{D}{4} = 0.1875\,\text{m}$$
$$\begin{aligned}
\text{Nu}_t &= 0.14[(0.750 \times 10^8)(0.1875)^3(90)(0.699)]^{1/3} \\
&= 44.0
\end{aligned}$$

Nu$_b$:

$$\begin{aligned}
\text{Nu}_b &= 0.27[(0.750 \times 10^8)(0.1875)^3(90)(0.699)]^{1/4} \\
&= 20.2
\end{aligned}$$

Nu$_s$:

$$\text{Nu}_s = \left\{ 0.825 + \frac{0.387[(0.750 \times 10^8)(1.2)^3(90)(0.699)]^{1/6}}{\left[1 + \left(\frac{0.492}{0.699}\right)^{9/16}\right]^{8/27}} \right\}^2$$

$$= 236$$

The solution for T is now

$$T = 295 + \frac{1500\,\text{W}}{[0.589(44) + 0.589(20.2) + 2.36(236)](0.0293)}$$

$$= 381.1\,\text{K}$$

Using this as the new estimate for T_{surface}, we have a film temperature, $T_f \cong 338$ K. Air properties at this temperature are

$$\nu = 1.936 \times 10^{-5} \text{ m}^2/\text{s} \qquad k = 0.0291 \text{ W/m·K}$$
$$\alpha = 2.77 \times 10^{-5} \text{ m}^2/\text{s} \qquad \text{Pr} = 0.699 \qquad \beta g/\nu^2 = 0.775 \times 10^8 \text{ 1/K·m}^3$$

The new values for Nu become

$$\text{Nu}_t = 0.14\big[(0.775 \times 10^8)(0.1875)^3(86)(0.699)\big]^{1/3} = 43.8$$

$$\text{Nu}_t = 0.27\big[(0.775 \times 10^8)(0.1875)^3(86)(0.699)\big]^{1/4} = 20.10$$

$$\text{Nu}_s = \left\{ 0.825 + \frac{0.387\big[0.775 \times 10^8(1.2)^3(86)(0.699)\big]^{1/6}}{\big[1 + (0.492/0.699)^{9/16}\big]^{8/27}} \right\}^2$$
$$= 235$$

The revised value for T_s is now

$$T_s = 295 + \frac{1500/0.0293}{[0.589(43.8) + 0.589(20.1) + 2.36(235)]}$$
$$= 381.4 \text{ K}$$

This result is obviously close enough and the desired result for surface temperature is

$$T_{\text{surface}} \cong 381 \text{ K}$$

20.2 FORCED CONVECTION FOR INTERNAL FLOW

Undoubtedly the most important convective heat-transfer process from an industrial point of view is that of heating or cooling a fluid that is flowing inside a closed conduit. The momentum transfer associated with this type of flow is studied in Chapter 13. Many of the concepts and terminology of that chapter will be used in this section without further discussion.

Energy transfer associated with forced convection inside closed conduits will be considered separately for laminar and turbulent flow. The reader will recall that the critical Reynolds number for conduit flow is approximately 2300.

Laminar Flow. The first analytical solution for laminar flow forced convection inside tubes was formulated by Graetz[17] in 1885. The assumptions basic to the Graetz solution are as follows:

1. The velocity profile is parabolic and fully developed before any energy exchange between the tube wall and the fluid occurs.

2. All properties of the fluid are constant.

3. The surface temperature of the tube is constant at a value T_s during the energy transfer.

Considering the system as depicted in Figure 20.5, we may write the velocity profile as

$$v_x = v_{\text{max}}\left[1 - \left(\frac{r}{R}\right)^2\right] \tag{8-7}$$

---

[17] L. Graetz, Ann. *Phys. u. Chem.*, **25**, 337 (1885).

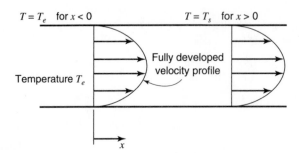

$T = T_e$ for $x < 0$ $T = T_s$ for $x > 0$

Fully developed velocity profile

Temperature T_e

x

Figure 20.5 Boundary and flow conditions for the Graetz solution.

or, recalling that $v_{\max} = 2v_{\mathrm{avg}}$, we may write

$$v_x = 2v_{\mathrm{avg}}\left[1 - \left(\frac{r}{R}\right)^2\right] \tag{20-21}$$

The applicable form of the energy equation written in cylindrical coordinates, assuming radial symmetry, and neglecting $\partial^2 T/\partial x^2$ (axial conduction) in comparison to the radial variation in temperature is

$$v_x \frac{\partial T}{\partial x} = \alpha\left[\frac{1}{r}\frac{\partial}{\partial r}\left(r\frac{\partial T}{\partial r}\right)\right] \tag{20-22}$$

Substituting equation (20-21) for v_x into equation (20-22) gives

$$2v_{\mathrm{avg}}\left[1 - \left(\frac{r}{R}\right)^2\right]\frac{\partial T}{\partial x} = \alpha\left[\frac{1}{r}\frac{\partial}{\partial r}\left(r\frac{\partial T}{\partial r}\right)\right] \tag{20-23}$$

which is the equation to be solved subject to the boundary conditions

$$T = T_e \quad \text{at } x = 0 \quad \text{for} \quad 0 \leq r \leq R$$
$$T = T_s \quad \text{at } x > 0, \qquad r = R$$

and

$$\frac{\partial T}{\partial r} = 0 \quad \text{at} \quad x > 0, \quad r = 0$$

The solution to equation (20-23) takes the form

$$\frac{T - T_e}{T_s - T_e} = \sum_{n=0}^{\infty} c_n f\left(\frac{r}{R}\right)\exp\left[-\beta_n^2 \frac{\alpha}{Rv_{\mathrm{avg}}}\frac{x}{R}\right] \tag{20-24}$$

The terms c_n, $f(r/R)$, and β_n are all coefficients to be evaluated by using appropriate boundary conditions.

The argument of the exponential, exclusive of β_n, that is, $(\alpha/Rv_{\mathrm{avg}})(x/R)$, may be rewritten as

$$\frac{4}{(2Rv_{\mathrm{avg}}/\alpha)(2R/x)} = \frac{4}{(Dv_{\mathrm{avg}}\rho/\mu)(c_p\mu/k)(D/x)}$$

or, in terms of dimensionless parameters already introduced, this becomes

$$\frac{4}{\mathrm{Re}\,\mathrm{Pr}\,D/x} = \frac{4x/D}{\mathrm{Pe}}$$

The product of Re and Pr is often referred to as the *Peclet number*, Pe. Another parameter encountered in laminar forced convection is the *Graetz number*, Gz, defined as

$$\mathrm{Gz} \equiv \frac{\pi}{4}\frac{D}{x}\mathrm{Pe}$$

Detailed solutions of equation (20-24) are found in the literature, and Knudsen and Katz[18] summarize these quite well. Figure 20.6 presents the results of the Graetz solution graphically for two different boundary conditions at the wall, these being (1) a constant wall temperature and (2) uniform heat input at the wall.

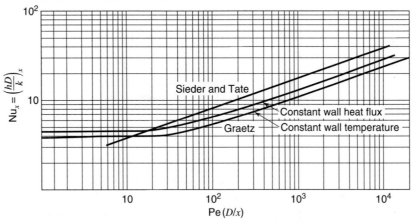

Figure 20.6 Variation in the local Nusselt number for laminar flow in tubes.

Note that, in Figure 20.6, the analytical results approach constant limiting values for large values of x. These limits are

$$\mathrm{Nu}_x = 3.658 \quad \text{for } T_{\text{wall}} = \text{constant} \tag{20-25}$$

$$\mathrm{Nu}_x = 4.364 \quad \text{for } q/A_{\text{wall}} = \text{constant} \tag{20-26}$$

Experimental data for laminar flow in tubes have been correlated by Sieder and Tate[19] by the equation

$$\mathrm{Nu}_D = 1.86\left(\mathrm{Pe}\frac{D}{L}\right)^{1/3}\left(\frac{\mu_b}{\mu_w}\right)^{0.14} \tag{20-27}$$

The Sieder–Tate relation is also shown in Figure 20.6 along with the two Graetz results. These results cannot be compared directly because the Graetz results yield local values of h_x and the Sieder–Tate equation gives mean values of the heat-transfer coefficient. The last part of equation (20-27), the ratio of the fluid viscosity at the arithmetic-mean bulk temperature to that at the temperature of the wall, takes into account the significant effect that variable fluid viscosity has on the heat-transfer rate. All properties other than μ_w are evaluated at the bulk fluid temperature.

Turbulent Flow. When considering energy exchange between a conduit surface and a fluid in turbulent flow, we must resort to correlations of experimental data as suggested by

[18] J. G. Knudsen and D. L. Katz, Fluid *Dynamics and Heat Transfer*, McGraw-Hill Book Company, New York, 1958, p. 370.
[19] F. N. Sieder and G. E. Tate, *Ind. Eng. Chem.*, **28,** 1429 (1936).

dimensional analysis. The three most-used equations of this nature and the restrictions on their use are as follows.

Dittus and Boelter[20] proposed the following equation of the type suggested earlier by dimensional analysis, equation (19-7):

$$\text{Nu}_D = 0.023\,\text{Re}_D^{0.8}\text{Pr}^n \tag{20-28}$$

where

1. $n = 0.4$ if the fluid is being heated, $n = 0.3$ if the fluid is being cooled;
2. all fluid properties are evaluated at the arithmetic-mean bulk temperature;
3. the value of Re_D should be $> 10^4$;
4. Pr is in the range $0.7 < \text{Pr} < 100$; and
5. $L/D > 60$.

Colburn[21] proposed an equation using the Stanton number, St, in place of Nu_D as related in equation (19-9). His equation is

$$\text{St} = 0.023\,\text{Re}_D^{-0.2}\text{Pr}^{-2/3} \tag{20-29}$$

where

1. Re_D and Pr are evaluated at the *film* temperature, and St is evaluated at the bulk temperature;
2. Re_D, Pr, and L/D should have values within the following limits:

$$\text{Re}_D > 10^4 \quad 0.7 < \text{Pr} < 160 \quad \text{and} \quad L/D > 60$$

To account for high Prandtl number fluids, such as oils, Sieder and Tate[22] proposed the equation

$$\text{St} = 0.023\,\text{Re}_D^{-0.2}\text{Pr}^{-2/3}\left(\frac{\mu_b}{\mu_w}\right)^{0.14} \tag{20-30}$$

where

1. All fluid properties except μ_w are evaluated at bulk temperature;
2. $\text{Re}_D > 10^4$;
3. $0.7 < \text{Pr} < 17\,000$;

and

4. $L/D > 60$.

Of the three equations presented, the first two are most often used for those fluids whose Prandtl numbers are within the specified range. The Dittus–Boelter equation is simpler to use than the Colburn equation because of the fluid property evaluation at the bulk temperature.

---

[20] F. W. Dittus and L. M. K. Boelter, University of California, *Publ. Eng.,* 2, 443 (1930).
[21] A. P. Colburn, *Trans. A.I.Ch.E.,* **29,** 174 (1933).
[22] E. N. Sieder and G. E. Tate, *Ind. Eng. Chem.,* **28,** 1429 (1936).

The following examples illustrate the use of some of the expressions presented in this section.

EXAMPLE 2

Hydraulic fluid (MIL-M-5606), in fully developed flow, flows through a 2.5 cm diameter copper tube that is 0.61 m long, with a mean velocity of 0.05 m/s. The oil enters at 295 K. steam condenses on the outside tube surface with an effective heat-transfer coefficient of $11,400$ W/m$^2 \cdot$K. find the rate of heat transfer to the oil.

To evaluate oil properties at either the film temperature or the mean bulk temperature, we need to know the exiting oil temperature. Equation (19-61) applies in this case

$$\ln \frac{T_L - T_s}{T_o - T_s} + 4\frac{L}{D}\frac{h}{\rho v c_p} = 0 \qquad (19\text{-}62)$$

If the thermal resistance of the copper tube wall is negligible, the heat transfer rate can be expressed as

$$q = \frac{A_{\text{surf}}(T_{\text{stm}} - T_{\text{oil}})}{1/h_i + 1/h_o} = \rho A v c_p (T_L - T_o)$$

To get an indication of whether the flow is laminar or turbulent, we will assume a bulk oil temperature of 300 K. The Reynolds number, at this temperature, is

$$\text{Re}_D = \frac{(0.025\,\text{m})(0.05\,\text{m/s})}{9.94 \times 10^{-6}\,\text{m}^2/\text{s}} = 126$$

and the flow is clearly laminar. The heat-transfer coefficient on the oil side can then be determined using equation (20-27)

$$h_i = \frac{k}{D}\text{Nu}_D = \frac{k}{D}1.86\left(\text{Pe}\frac{D}{L}\right)^{0.33}\left(\frac{\mu_b}{\mu_w}\right)^{0.14}$$

Initially, the bulk temperature of the oil and the wall temperature will be assumed to be 300 and 372 K, respectively. Using fluid properties at these temperatures, we have

$$h_i = \frac{(0.123\,\text{W/m}\cdot\text{K})(1.86)}{0.025\,\text{m}}\left[(126)(155)\frac{0.025}{0.61}\right]^{0.33}\left(\frac{1.036 \times 10^{-4}}{3.72 \times 10^{-3}}\right)^{0.14}$$

$$= 98.1\,\text{W/m}^2 \cdot \text{K}$$

Substituting into equation (19-61), we get

$$\ln\frac{T_s - T_L}{T_s - T_o} = -4\left(\frac{0.61\,\text{m}}{0.025\,\text{m}}\right)\frac{98.1\,\text{W/m}^2 \cdot \text{K}}{(843\,\text{kg/m}^3)(0.05\,\text{m/s})(1897\,\text{J/kg}\cdot\text{K})}$$

$$= -0.120$$

$$\frac{T_s - T_L}{T_s - T_o} = e^{-0.120} = 0.887$$

$$T_L = 372 - 0.887(372 - 295)$$

$$= 304\,\text{K}$$

With this value of T_L, the mean bulk temperature of the oil is

$$T_b = \frac{295 + 304}{2} = 299.5\,\text{K}$$

which is sufficiently close to the initial assumption that there is no need to iterate further.

With an exiting temperature of 304 K the heat-transfer rate to the oil is

$$q = \rho A v c_p (T_L - T_o)$$

$$= (843 \text{ kg/m}^3)\left(\frac{\pi}{4}\right)(0.025 \text{ m})^2(0.05 \text{ m/s})(1897 \text{ J/kg·K})(9 \text{ K})$$

$$= 353 \text{ W}$$

EXAMPLE 3

Air at 1 atmosphere and a temperature of 290 K enters a 1.27-cm-ID tube at a velocity of 24 m/s. The wall temperature is maintained at 372 K by condensing steam. Evaluate the convective heat-transfer coefficient for this situation if the tube is 1.52 m long.

As in the previous example, it will be necessary to evaluate the exiting air temperature by

$$\ln \frac{T_L - T_s}{T_o - T_s} + 4\frac{L}{D}\frac{h}{\rho v c_p} = 0 \tag{19-62}$$

To determine the type of flow, we first evaluate the Reynolds number at the tube entrance.

$$Re = \frac{Dv}{\nu} = \frac{(0.0127 \text{ m})(24 \text{ m/s})}{1.478 \times 10^{-5} \text{ Pa·s}} = 20\,600$$

Flow is clearly turbulent and Re is sufficiently large that equation (20-28), (20-29), or (20-30) may be used.

Equation (20-29) will be used. An exit temperature of 360 K will be assumed; the corresponding mean bulk temperature is 325 K and $T_f = 349$. We now have

$$St = \frac{h}{\rho v c_p} = 0.023\,Re^{-0.2}\,Pr^{-2/3}$$

$$= 0.023\left[\frac{(0.0127)(24)}{2.05 \times 10^{-5}}\right]^{-0.2}(0.697)^{-2/3}$$

$$= 0.00428$$

Substituting into equation (19-61) we have

$$\frac{T_L - T_s}{T_o - T_s} = \exp\left[-4\left(\frac{1.52 \text{ m}}{0.0127 \text{ m}}\right)(0.00428)\right]$$

$$= 0.129$$

and the calculated value of T_L is

$$T_L = 372 - (0.129)(372 - 290)$$

$$= 361 \text{ K}$$

This value agrees closely with the initially assumed value for T_L, so there is no need to perform a second calculation. The heat-transfer coefficient is now evaluated as

$$h = \rho v c_p\, St$$

$$= (1.012 \text{ kg/m}^3)(24 \text{ m/s})(1009 \text{ J/kg·K})(0.00428)$$

$$= 105 \text{ W/m}^2\text{·K}$$

For flow in short passages the correlations presented thus far must be modified to account for variable velocity and temperature profiles along the axis of flow. Deissler[23] has

---

[23] R. G. Deissler, *Trans. A.S.M.E.*, **77**, 1221 (1955).

analyzed this region extensively for the case of turbulent flow. The following equations may be used to modify the heat-transfer coefficients in passages for which $L/D < 60$:

for $2 < L/D < 20$

$$\frac{h_L}{h_\infty} = 1 + (D/L)^{0.7} \tag{20-31}$$

and for $20 < L/D < 60$

$$\frac{h_L}{h_\infty} = 1 + 6\,D/L \tag{20-32}$$

Both of these expressions are approximations relating the appropriate coefficient, h_L, in terms of h_∞, where h_∞ is the value calculated for $L/D > 60$.

20.3 FORCED CONVECTION FOR EXTERNAL FLOW

Numerous situations exist in practice in which one is interested in analyzing or describing heat transfer associated with the flow of a fluid past the exterior surface of a solid. The sphere and cylinder are the shapes of greatest engineering interest, with heat transfer between these surfaces and a fluid in crossflow frequently encountered.

The reader will recall the nature of momentum-transfer phenomena discussed in Chapter 12 relative to external flow. The analysis of such flow and of heat transfer in these situations is complicated when the phenomenon of boundary-layer separation is encountered. Separation will occur in those cases in which an adverse pressure gradient exists; such a condition will exist for most situations of engineering interest.

Flow Parallel to Plane Surfaces.　This condition is amenable to analysis and has already been discussed in Chapter 19. The significant results are repeated here for completeness.

We recall that, in this case, the boundary layer flow regimes are laminar for $Re_x < 2 \times 10^5$ and turbulent for $3 \times 10^6 < Re_x$. For the laminar range

$$Nu_x = 0.332\,Re_x^{1/2}\,Pr^{1/3} \tag{19-25}$$

and

$$Nu_L = 0.664\,Re_L^{1/2}Pr^{1/3} \tag{19-26}$$

With turbulent flow in the boundary layer, an application of the Colburn analogy

$$St_x Pr^{2/3} = \frac{C_{fx}}{2} \tag{19-37}$$

along with equation (12-73) yields

$$Nu_x = 0.0288\,Re_x^{4/5}Pr^{1/3} \tag{20-33}$$

A mean Nusselt number can be calculated using this expression for Nu_x. The resulting expression is

$$Nu_L = 0.036\,Re_L^{4/5}Pr^{1/3} \tag{20-34}$$

Fluid properties should be evaluated at the film temperature when using these equations.

Cylinders in Crossflow.　Eckert and Soehngen[24] evaluated local Nusselt numbers at various positions on a cylindrical surface past which flowed an air stream with a range

---

[24] E. R. G. Eckert and E. Soehngen, *Trans. A.S.M.E.*, **74**, 343 (1952).

in Reynolds numbers from 20 to 600. Their results are shown in Figure 20.7. A much higher Reynolds number range was investigated by Giedt[25] whose results are shown in Figure 20.8.

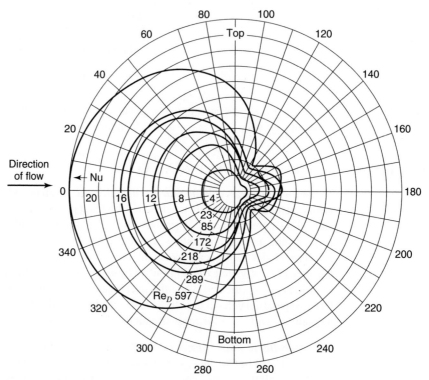

Figure 20.7 Local numbers for crossflow about a circular cylinder at low Reynolds numbers. (From E. R. G. Eckert and E. Soehngen, *Trans. A.S.M.E.*, **74**, 346 (1952). By permission of the publishers.)

Figures 20.7 and 20.8 show a smooth variation in the Nusselt number near the stagnation point. At low Reynolds numbers, the film coefficient decreases almost continuously from the stagnation point, the only departure being a slight rise in the separated-wake region of the cylinder. At higher Reynolds numbers, as illustrated in Figure 20.7, the film coefficient reaches a second maximum, which is greater than the stagnation-point value. The second peak in the Nusselt number at high Reynolds numbers is due to the fact that the boundary layer undergoes transition from laminar to turbulent flow. In the bottom curves of Figure 20.7, the laminar boundary layer separates from the cylinder near 80° from the stagnation point, and no large change in the Nusselt number occurs. The effect of higher Reynolds number is twofold. First, the separation point moves past 90° as the boundary layer becomes turbulent; thus less of the cylinder is engulfed in the wake. A second effect is that the Nusselt number reaches a value that is higher than the stagnation-point value. The increase is due to the greater conductance of the turbulent boundary layer.

It is quite apparent from the figures that the convective heat-transfer coefficient varies in an irregular, complex manner in external flow about a cylinder. It is likely, in practice, that an average h for the entire cylinder is desired. McAdams[26] has plotted the data of 13 separate

[25] W. H. Giedt, *Trans. A.S.M.E.*, **71**, 378 (1949).

[26] W. H. McAdams, *Heat Transmission*, Third Edition, McGraw-Hill Book Company, New York, 1949.

Figure 20.8 Local Nusselt numbers for crossflow about a circular cylinder at high Reynolds numbers. (From W.H. Giedt, *Trans. A.S.M.E.,* **71,** 378 (1949). By permission of the publishers.)

investigations for the flow of air normal to single cylinders and found excellent agreement when plotted as Nu_D vs. Re_D. His plot is reproduced in Figure 20.9. Note that values of Nu_D are for $Pr = 1$. For other fluids a correction factor, $Pr^{1/3}$, should be employed, that is, $Nu_D = Nu_{D(\text{figure})}Pr^{1/3}$.

A widely used correlation for these data is of the form

$$Nu_D = B\,Re^n Pr^{1/3} \qquad (20\text{-}35)$$

where the constants B and n are functions of the Reynolds number. Values for these constants are given in Table 20.3. The film temperature is appropriate for physical property evaluation.

Churchill and Bernstein[27] have recommended a single correlating equation covering conditions for which $Re_D Pr > 0.2$. This correlation is expressed in equation (20-36).

$$Nu_D = 0.3 + \frac{0.62\,Re_D^{1/2}Pr^{1/3}}{\left[1 + (0.4/Pr)^{2/3}\right]^{1/4}}\left[1 + \left(\frac{Re_D}{282,000}\right)^{5/8}\right]^{4/5} \qquad (20\text{-}36)$$

[27] S. W. Churchill and M. Bernstein, *J. Heat Transfer,* **99,** 300 (1977).

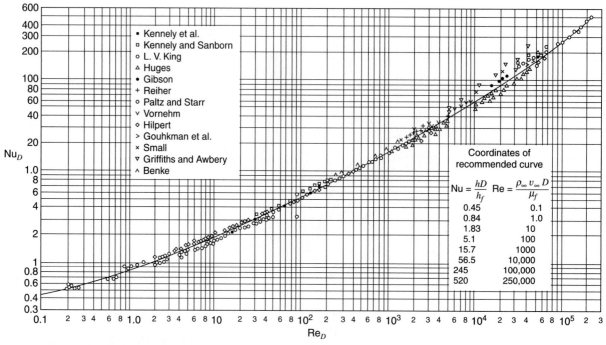

Figure 20.9 Nu vs. Re for flow normal to single cylinders.
Note: These values are strictly valid when $\Pr \cong 1$.
(From W. H. McAdams, *Heat Transmission*, Third Edition, McGraw-Hill Book Company, New York, 1954, p. 259.)

Table 20.3 Values of B and n for use in equation (20-35)

Re_D	B	n
0.4/–4	0.989	0.330
4–40	0.911	0.385
40–4000	0.683	0.466
4000–40,000	0.193	0.618
40,000–400,000	0.027	0.805

Single Spheres. Local convective heat-transfer coefficients at various positions relative to the forward stagnation point for flow past a sphere are plotted in Figure 20.10, following the work of Cary.[28]

McAdams[29] has plotted the data of several investigators relating Nu_D vs. Re_D for air flowing past spheres. His plot is duplicated in Figure 20.11.

A recent correlation proposed by Whitaker[30] is recommended for the following conditions: $0.71 < \Pr < 380$, $3.5 < Re_D < 7.6 \times 10^4$, $1.0 < \mu_\infty/\mu_s < 3.2$. All properties are

---

[28] J. R. Cary, *Trans. A.S.M.E.*, **75**, 483 (1953).
[29] W. H. McAdams, *op. cit.*
[30] S. Whitaker, *A.I.Ch.E.J.*, **18**, 361 (1972).

Figure 20.10 Local heat-transfer coefficients for flow past a sphere.
(From J. R. Cary, *Trans. A.S.M.E.*, 75, 485 (1953). By permission of the publishers.)

Figure 20.11 Nu_D vs. Re_D for air-flow past single spheres.
(From McAdams, *Heat Transmission*, Third Edition, McGraw-Hill Book Company, New York, 1954, p. 266. By permission of the publishers.)

evaluated at T_∞ except for μ_s, which is the value at the surface temperature. Whitaker's correlation is

$$\text{Nu}_D = 2 + (0.4\,\text{Re}_D^{1/2} + 0.06\,\text{Re}_D^{2/3})\text{Pr}^{0.4}(\mu_\infty/\mu_s)^{1/4} \tag{20-37}$$

An important case is that of falling liquid drops, modeled as spheres. The correlation of Ranz and Marshall,[31] for this case, is

$$\text{Nu}_D = 2 + 0.6\,\text{Re}_D^{1/2}\text{Re}^{1/3} \tag{20-38}$$

Tube Banks in Crossflow. When a number of tubes are placed together in a bank or bundle, as might be encountered in a heat exchanger, the effective heat-transfer coefficient is affected by the tube arrangement and spacing, in addition to those factors already considered for flow past single cylinders. Several investigators have made significant contributions to the analysis of these configurations.

As fluid flow through and past tube bundles involves an irregular flow path, some investigators have chosen significant lengths other than D, the tube diameter, to use in calculating Reynolds numbers. One such term is the equivalent diameter of a tube bundle D_{eq}, defined as

$$D_{\text{eq}} = \frac{4(S_L S_T - \pi D^2/4)}{\pi D} \tag{20-39}$$

where S_L is the center-to-center distance between tubes *along* the direction of flow, S_T is the center-to-center distance between tubes *normal* to the flow direction, and D is the OD of a tube.

Bergelin, Colburn, and Hull[32] studied the flow of liquids past tube bundles in the region of laminar flow with $1 < \text{Re} < 1000$. Their results, plotted as $\text{St}\,\text{Pr}^{2/3}(\mu_w/\mu_b)^{0.14}$ vs. Re for various configurations, are presented in Figure 20.12. In that figure all fluid properties except μ_w are evaluated at the average bulk temperature.

For liquids in transition flow across tube bundles, Bergelin, Brown, and Doberstein[33] extended the work just mentioned for five of the tube arrangements to include values of Re up to 10^4. Their results are presented both for energy transfer and friction factor vs. Re in Figure 20.13.

In addition to the greater Reynolds number range, Figure 20.13 involves Re calculated by using the tube diameter, D, as opposed to Figure 20.12, in which D_{eq}, defined by equation (20-37), was used.

Wind–Chill Equivalent Temperature. The reader is undoubtedly familiar with weather reports where, in addition to measured air temperatures, an indication of how cold one actually feels is expressed as the *wind–chill equivalent temperature*. This temperature indicates how cold the wind actually makes one feel; the stronger the wind blows, the colder the air feels to the human body.

The determination of wind–chill equivalent temperature is an interesting example of the combined effects of convective and conductive heat transfer between the body and the adjacent air. For a complete explanation of the modeling used in determining this quantity,

[31] W. Ranz and W. Marshall, *Chem. Engr. Progr.,* **48,** 141 (1952).
[32] O. P. Bergelin, A. P. Colburn, and H. L. Hull, Univ. Delaware, Eng. Expt. Sta. Bulletin No. 2 (1950).
[33] O. P. Bergelin, G. A. Brown, and S. C. Doberstein, *Trans. A.S.M.E.,* **74,** 953 (1952).

Figure 20.12 Convective heat-transfer exchange between liquids in laminar flow and tube bundles. (From O. P. Bergelin, A. P. Colburn, and H. L. Hull, Univ. of Delaware, Engr. Dept., Station Bulletin 10.2, 1950, p. 8. By permission of the publishers.)

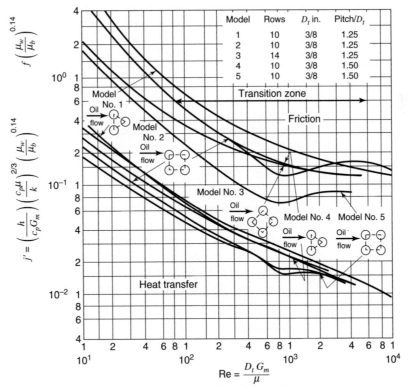

Figure 20.13 Energy transfer and frictional loss for liquids in laminar and transition flow past tube bundles.
(From O. P. Bergelin, G. A. Brown, and S. C. Doberstein, *Trans. A.S.M.E.*, **74**, 1958 (1952). By permission of the publishers.)

Table 20.4   Wind–chill equivalent temperature—english units

		35	30	25	20	15	10	5	0	−5	−10	−15	−20	−25
							Air temperature (°F)							
Wind speed (mi/hr)	5	32	27	22	16	11	6	0	−5	−10	−15	−21	−26	−31
	10	22	16	10	3	−3	−9	−15	−22	−27	−34	−40	−46	−52
	15	16	9	2	−5	−11	−18	−25	−31	−38	−45	−51	−58	−65
	20	12	4	−3	−10	−17	−24	−31	−39	−46	−53	−60	−67	−74
	25	8	1	−7	−15	−22	−29	−36	−44	−51	−59	−66	−74	−81
	30	6	−2	−10	−18	−25	−33	−41	−49	−56	−64	−71	−79	−86
	35	4	−4	−12	−20	−27	−35	−43	−52	−58	−67	−74	−82	−89
	40	3	−5	−13	−21	−29	−37	−45	−53	−60	−69	−76	−84	−92

Table 20.5   Wind–chill equivalent temperature—SI units

		8	4	0	−4	−8	−12	−16	−20	−24	−28	−32	−36	−40
							Air temperature (°C)							
Wind speed (km/hr)	0	8	4	0	−4	−8	−12	−16	−20	−24	−28	−32	−36	−40
	10	5	0	−4	−8	−13	−17	−22	−26	−31	−35	−40	−44	−49
	20	0	−5	−10	−15	−21	−26	−31	−36	−42	−47	−52	−57	−63
	30	−3	−8	−14	−20	−25	−31	−37	−43	−48	−54	−60	−65	−71
	40	−5	−11	−17	−23	−29	−35	−41	−47	−53	−59	−65	−71	−77
	50	−6	−12	−18	−25	−31	−37	−43	−49	−56	−62	−68	−74	−80
	60	−7	−13	−19	−26	−32	−39	−45	−51	−58	−64	−70	−77	−83

the reader is referred to a 1971 paper by Steadman.[34] Tables 20.4 and 20.5, provide values of the wind–chill equivalent temperature as a function of air temperature and wind speed, in English and SI units, respectively.

20.4   CLOSURE

Many of the more useful experimentally developed correlations for predicting convective heat-transfer coefficients have been presented in this chapter. Those graphs and equations presented are a small part of the information of this type available in the literature. The information included should, in any case, allow most of the more common convection heat-transfer coefficients to be predicted with some confidence.

The convection phenomena considered have included the following:

1. *Natural convection* past vertical and horizontal surfaces, plus some useful simplified expressions for air;

2. *Forced convection for internal flow,* including laminar and turbulent flow correlations; and

3. *Forced convection for external flow* with cylinders, spheres, and tube bundles being the types of surfaces so considered.

---

[34] R. G. Steadman, Indices of windchill of clothed persons, *J. App. Meteorol,* **10,** 674–683 (1971).

The reader is reminded to observe any special considerations relative to the equations and plots in this chapter. Such considerations include whether to evaluate fluid properties at the bulk or film temperature, what significant length is used in a given correlation, and what is the allowable Prandtl and Reynolds number range for a given set of data.

PROBLEMS

20.1 A 750-W immersion heater in the form of a cylinder with 3/4-in. diameter and 6 in. in length is placed in 95°F stagnant water. Calculate the surface temperature of the heater if it is oriented with its axis

a. vertical;

b. horizontal.

20.2 Repeat Problem 20.1 if the stagnant liquid is

a. bismuth at 700°F;

b. hydraulic fluid at 0°F.

20.3 An immersion heater, rated at 1000 W, is in the form of a rectangular solid with dimensions 16 cm by 10 cm by 1 cm. Determine the surface temperature of the heater if it is oriented in 295 K water with

a. the 16-cm dimension vertical,

b. the 10-cm dimension vertical.

20.4 A 2-in. copper cylinder, 6 in. in length, at a uniform temperature of 200°F, is plunged vertically into a large tank of water at 50°F.

a. How long will it take for the outside surface of the cylinder to reach 100°F?

b. How long will it take for the center of the cylinder to reach 100°F?

c. What is the surface temperature when the center temperature is 100°F? Heat transfer from ends of the cylinder may be neglected.

20.5 A fluorescent light bulb, rated at 100 W, is illuminated in air at 25°C and atmospheric pressure. Under these conditions the surface temperature of the glass is 140°C.

Determine the rate of heat transfer from the bulb by natural convection. The bulb is cylindrical, having a diameter of 35 mm and a length of 0.8 m, and is oriented horizontally.

Note that your result is not equal to 100 W. Where does the remaining energy go?

20.6 Determine the steady-state surface temperature of an electric cable, 25 cm in diameter, which is suspended horizontally in still air in which heat is dissipated by the cable at a rate of 27 W per meter of length. The air temperature is 30°C.

20.7 A copper cylinder 20.3 cm long with a diameter of 2.54 cm is being used to evaluate the surface coefficient in a laboratory experiment. When heated to a uniform temperature of 32.5°C and then plunged into a −1°C liquid bath, the center temperature of the cylinder reaches a value of 4.8°C in 3 min.

Assuming the heat exchange between the cylinder and water bath to be purely by convection, what value for the surface coefficient is indicated?

20.8 Rubber balls are molded into spheres and cured at 360 K. Following this operation they are allowed to cool in room air. What will be the elapsed time for the surface temperature of a solid rubber ball to reach 320 K when the surrounding air temperature is 295 K? Consider balls with diameters of 7.5, 5, and 1.5 cm. Properties of rubber that may be used are $k = 0.24$ W/m · K, $\rho = 1120$ kg/m^3, $c_p = 1020$ J/kg·K.

20.9 Determine the required time for the rubber balls described in Problem 20.8 to reach the condition such that the center temperature is 320 K. What will be the surface temperature when the center temperature reaches 320 K?

20.10 A 1-in., 16-BWG copper tube has its outside surface maintained at 240°F. If this tube is located in still air at 60°F, what heat flux will be achieved if the tube is oriented

a. horizontally?

b. vertically?

The tube length is 10 ft.

20.11 Solve Problem 20.10 if the medium surrounding the tube is stagnant water at 60°F.

20.12 A 0.6-m-diameter spherical tank contains liquid oxygen at 78 K. This tank is covered with 5 cm of glass wool. Determine the rate of heat gain if the tank is surrounded by air at 278 K. The tank is constructed of stainless steel 0.32 cm thick.

20.13 A "swimming-pool" nuclear reactor, consisting of 30 rectangular plates measuring 1 ft in width and 3 ft in height, spaced 2 1/2 in. apart, is immersed in water at 80°F. If 200°F is the maximum allowable plate temperature, what is the maximum power level at which the reactor may operate?

20.14 A solar energy collector measuring 20×20 ft is installed on a roof in a horizontal position. The incident solar energy flux is 200 Btu/h ft^2, and the collector surface temperature is 150°F. What fraction of incident solar energy is lost by convection to the stagnant surrounding air at a temperature of 50°F? What effect on the convective losses would result if the collector were crisscrossed with ridges spaced 1 ft apart?

20.15 Given the conditions for Problem 20.14, determine the fraction of incident solar energy lost by convection to the surrounding air at 283 K flowing parallel to the collector surface at a velocity of 6.1 m/s.

20.16   Cast-iron cannonballs used in the War of 1812 were occasionally heated for some extended time so that, when fired at houses and ships, they would set them afire. If one of these the so-called "hot shot" with a 15-cm diameter were at a uniform temperature of 1300 K, what heat flux value would exist if it were suddenly placed in still air at 270 K? The following properties of cast iron may be used:

$$k = 39.8 \text{ W/m·K}$$
$$c_p = 4.8 \text{ J/kg·K}$$
$$\rho = 7370 \text{ kg/m}^3$$

20.17   Given the information in Problem 20.16, construct a plot of convective heat-transfer coefficient vs. temperature for values of T_{surface} between 420 and 1300 K. How long would it take for the surface temperature of a cannonball to reach 600 K? What would be its center temperature at this time?

20.18   Copper wire with a diameter of 0.5 cm is covered with a 0.65-cm layer of insulating material having a thermal conductivity of 0.242 W/m·K. The air adjacent to the insulation is at 290 K. If the wire carries a current of 400 A, determine

a.   the convective heat-transfer coefficient between the insulation surface and the surrounding air;

b.   the temperatures at the insulation-copper interface and at the outside surface of the insulation.

20.19   Work Problem for an aluminum conductor of the same size (resistivity of aluminum = 2.83×10^{-6} ohm-cm).

20.20   If the steam line described in Problem 20.40 is bare and surrounded by still air at 70°F, what total heat transfer would be predicted from a 20-ft length of bare pipe? Consider the bare pipe to be a black surface and the surroundings black at 70°F.

20.21   Solve Problem 20.20 if the bare pipe is located so that 295 K air flows normal to the pipe axis at a velocity of 6.5 m/s.

20.22   Solve Problem 20.20 if 3 in. of insulation having a thermal conductivity of 0.060 Btu/h ft °F is applied to the outside of the pipe. Neglect radiation from the insulation. What will be the outside surface temperature of the insulation?

20.23   What thickness of insulation having a thermal conductivity as given in Problem 20.22 must be added to the steam pipe of Problem 20.40 in order that the outside temperature of the insulation does not exceed 250°F?

20.24   If insulation having a thermal conductivity of 0.060 Btu/h ft °F is added to the outside of the steam pipe described in Problem 20.40, what thickness will be necessary if radiation losses from the outside surface of the insulation account for no more than 15% of the total? The surroundings may be considered black at 70°F. What is the temperature at the outside surface of the insulation under these conditions?

20.25   A cooking oven has a top surface temperature of 45°C when exposed to still air. At this condition the inside oven temperature and room air temperature are 180 and 20°C, respectively, and heat is transferred from the top surface at 40 W.

To reduce the surface temperature, as required by safety regulations, room air is blown across the top with a velocity of 20 m/s. Conditions inside the oven may be considered unchanged.

a.   What will be the rate of heat loss under this new operating condition?

b.   What will be the top surface temperature?

20.26   Saturated steam at 0.1 bar condenses on the outside of a copper tube having inner and outer diameters of 16.5 and 19 mm, respectively. The surface coefficients on the inner (water) surface and outer (steam) surface are 5200 and 6800 W/m²·K, respectively.

When the mean water temperature is 28 K, estimate the rate of steam condensed per meter of tube length. The latent heat of condensation of steam may be taken as 2.390 kJ/kg.

20.27   Solve Problem 20.41 if, instead of six tubes, three tubes each 5 m in length are used to heat the oil.

20.28   Air at 60°F and atmospheric pressure flows inside a 1-in., 16-BWG copper tube whose surface is maintained at 240°F by condensing steam. Find the temperature of the air after passing through 20 ft of tubing if its entering velocity is 40 fps.

20.29   A 1-in., 16-BWG copper tube, 10 ft long, has its outside surface maintained at 240°F. Air at 60°F and atmospheric pressure is forced past this tube with a velocity of 40 fps. Determine the heat flux from the tube to the air if the flow of air is

a.   parallel to the tube;

b.   normal to the tube axis.

20.30   Solve Problem 20.29 if the medium flowing past the tube in forced convection is water at 60°F.

20.31   Solve Problem 20.29 if the medium flowing past the tube in forced convection is MIL-M-5606 hydraulic fluid.

20.32   An industrial heater is composed of a tube bundle consisting of horizontal 3/8-in.-OD tubes in a staggered array with tubes arranged in equilateral triangle fashion having a pitch-to-diameter ratio of 1.5. If water at 160°F flows at 20 ft/s past the tubes with constant surface temperature of 212°F, what will be the effective heat-transfer coefficient?

20.33   For the heater consisting of the tube bank described in Problem 20.32, evaluate the heat transferred to the water if the tube array consists of six rows of tubes in the flow direction with eight tubes per row. The tubes are 5 ft long.

20.34   Heat transfer between an electrically heated circular cylinder and water is to be examined for three different conditions. The heater has a diameter of 1.26 cm and is 7.5 cm long. The surface of the heater is at 380 K and the water is at 295 K. All heat transfer is from the lateral surface.

Evaluate the heat transfer, in W, under conditions where

a.   energy exchange is by natural convection with the heater axis oriented horizontally;

b.   energy exchange is by natural convection with heater axis oriented vertically;

c. energy exchange is by forced convection with water flowing across the lateral surface at 1.5 m/s.

20.35 Solve Problem 20.16 with all conditions as given except that the "hot shot" is traveling through the air at 270 K with a velocity of 150 m/s.

20.36 What would be the results of Problem 20.18 if a fan provided an air flow normal to the conductor axis at a velocity of 9 m/s?

20.37 An electric light bulb, rated at 60 W, has a surface temperature of 145°C when cooled by atmospheric air at 25°C. The air flows past the bulb with a velocity of 0.5 m/s. The bulb can be modeled as a sphere with a diameter of 7.5 cm. Determine the heat transfer from the bulb by the mechanism of forced convection.

20.38 A valve on a hot-water line is opened just enough to allow a flow of 0.06 fps. The water is maintained at 180°F, and the inside wall of the 1/2-in. schedule 40 water line is at 80°F. What is the total heat loss through 5 ft of water line under these conditions? What is the exit water temperature?

20.39 When the valve on the water line in Problem 20.38 is opened wide, the water velocity is 35 fps. What is the heat loss per 5 ft of water line in this case if the water and pipe temperatures are the same as specified in Problem 20.38?

20.40 Steam at 400 psi, 800°F flows through a 1-in. schedule-140 steel pipe at a rate of 10,000 lb_m/h. Estimate the value of h that applies at the inside pipe surface.

20.41 Oil at 300 K is heated by steam condensing at 372 K on the outside of steel pipes with ID = 2.09 cm, OD = 2.67 cm. The oil flow rate is 1.47 kg/s; six tubes, each 2.5 m long, are used. The properties of oil to be used are as follows:

T, K	ρ, kg/m³	c_p, J/kg·K	k, W/m·K	μ, Pa·s
300	910	1.84×10^3	0.133	0.0414
310	897	1.92×10^3	0.131	0.0228
340	870	2.00×10^3	0.130	7.89×10^{-3}
370	865	2.13×10^3	0.128	3.72×10^{-3}

Determine the rate of heat transfer to the oil.

20.42 Engine oil with properties given below flows at a rate of 136 kg per hour through a 7.5-cm-ID pipe whose inside surface is maintained at 100°C. If the oil enters at 160°C, what will its temperature be at the exit of a 15-m-long pipe?

T, K	ρ, kg/m³	c_p, J/kg·K	k, W/m·K	v, m²/s × 10³	Pr
373	842	2.219	0.137	0.0203	276
393	831	2.306	0.135	0.0124	175
413	817	2.394	0.133	0.0080	116
433	808	2.482	0.132	0.0056	84

20.43 An apparatus, used in an operating room to cool blood, consists of a coiled tube that is immersed in an ice bath. Using this apparatus, blood, flowing at 0.006 m³/h, is to be cooled from 40 to 30°C. The inside diameter of the tube is 2.5 mm, and the surface coefficient between the ice bath and outer tube surface is 500 W/m²·K. The thermal resistance of the tube wall may be neglected.

Determine the required length of tubing to accomplish the desired cooling. Properties of blood are the following:

$$\rho = 1000 \, \text{kg/m}^2$$
$$k = 0.5 \, \text{W/m·K}$$
$$c_p = 4.0 \, \text{kJ/kg·K}$$
$$v = 7 \times 10^{-7} \, \text{m}^2/\text{s}$$

Determine the total heat transfer to the oil and its temperature at the heater exit.

20.44 A 1.905-cm-diameter brass tube is used to condense steam on its outer surface at 10.13 kPa pressure. Water at 290 K is circulated through the tube. Inside and outside surface coefficients are 1700 and 8500 W/m²·K, respectively. Find the rate of steam condensed per hour per meter of the tube length under these conditions. The following information is pertinent:

$$\text{tube OD} = 1.905 \, \text{cm}$$
$$\text{tube ID} = 1.656 \, \text{cm}$$
$$\text{steam saturation temperature } = 319.5 \, \text{K}$$
$$\text{steam latent heat, } h_{fg} = 2393 \, \text{kJ/kg.}$$

20.45 A system for heating water with an inlet temperature of 25°C to an exiting temperature of 70°C involves passing the water through a thick-walled tube with inner and outer diameters of 25 and 45 mm, respectively. The outer tube surface is well insulated, and the electrical heating within the tube wall provides for a uniform generation of $\dot{q} = 1.5 \times 10^6$ W/m³.

a. For a mass flow rate of water, $\dot{m} = 0.12$ kg/s, how long must the tube be to achieve the desired outlet temperature?

b. If the inner surface of the tube at the outlet is $T_s = 110°C$, what is the local convective coefficient at this location?

20.46 Air at 25 psia is to be heated from 60 to 100°F in a smooth, 3/4-in.-ID tube whose surface is held at a constant temperature of 120°F. What is the length of the tube required for an air velocity of 25 fps? At 15 fps?

20.47 Air is transported through a rectangular duct measuring 2 ft by 4 ft. The air enters at 120°F and flows with a mass velocity of 6 lb_m/s·ft². If the duct walls are at a temperature of 80°F, how much heat is lost by the air per foot of duct length? What is the corresponding temperature decrease of the air per foot?

20.48 Cooling water flows through thin-walled tubes in a condenser with a velocity of 1.5 m/s. The tubes are 25.4 mm in diameter. The tube-wall temperature is maintained constant at 370 K by condensing steam on the outer surface. The tubes are 5 m long and the water enters at 290 K.

Estimate the exiting water temperature and the heat-transfer rate per tube.

20.49 Air, at 322 K, enters a rectangular duct with a mass velocity of 29.4 kg/s·m². The duct measures 0.61 m by 1.22 m and its walls are at 300 K.

Determine the rate of heat loss by the air per meter of duct length and the corresponding decrease in air temperature per meter.

20.50 Air at atmospheric pressure and 10°C enters a rectangular duct that is 6 m long having dimensions of 7.5 and 15 cm in cross section. The surfaces are maintained at 70°C by solar irradiation. If the exiting air temperature is to be 30°C, what is the required flow rate of the air?

20.51 A tube bank employs an in-line arrangement with $S_T = S_L = 3.2$ cm and tubes that are 1.8 cm in outside diameter. There

are 10 rows of tubes, which are held at a surface temperature of 85°C. Air at atmospheric pressure and 20°C flows normal to the tubes with a free stream velocity of 6 m/s. The tube bank is 10 rows deep, and the tubes are 1.8 m long. Determine the amount of heat transferred.

20.52 Rework Problem 20.51 for a staggered arrangement.

20.53 A tube bank employs tubes that are 1.30 cm in outside diameter at $S_T = S_L = 1.625$ cm. There are eight rows of tubes, which are held at a surface temperature of 90°C. Air, at atmospheric pressure and a bulk temperature 27°C, flows normal to the tubes with a free stream velocity of 1.25 m/s. The tube bank is eight rows deep, and the tubes are 1.8 m long. Estimate the heat-transfer coefficient.

20.54 Rework Problem 20.53 for a staggered arrangement. All other conditions remain the same.

Chapter 21

Boiling and Condensation

Energy-transfer processes associated with the phenomena of boiling and condensation may achieve relatively high heat-transfer rates, whereas the accompanying temperature differences may be quite small. These phenomena, associated with the change in phase between a liquid and a vapor, are more involved and thus more difficult to describe than the convective heat-transfer processes discussed in the preceding chapters. This is due to the additional considerations of surface tension, latent heat of vaporization, surface characteristics, and other properties of two-phase systems that were not involved in the earlier considerations. The processes of boiling and condensation deal with opposite effects relative to the change in phase between a liquid and its vapor. These phenomena will be considered separately in the following sections.

21.1 BOILING

Boiling heat transfer is associated with a change in phase from liquid to vapor. Extremely high heat fluxes may be achieved in conjunction with boiling phenomena, making the application particularly valuable where a small amount of space is available to accomplish a relatively large energy transfer. One such application is the cooling of nuclear reactors. Another is the cooling of electronic devices where space is very critical. The advent of these applications has spurred the interest in boiling, and concentrated research in this area in recent years has shed much light on the mechanism and behavior of the boiling phenomenon

There are two basic types of boiling: *pool boiling* and *flow boiling*. Pool boiling occurs on a heated surface submerged in a liquid pool that is not agitated. Flow boiling occurs in a flowing stream, and the boiling surface may itself be a portion of the flow passage. The flow of liquid and vapor associated with flow boiling is an important type of two-phase flow.

Regimes of Boiling. An electrically heated horizontal wire submerged in a pool of water at its saturation temperature is a convenient system to illustrate the regimes of boiling heat transfer. A plot of the heat flux associated with such a system as the ordinate vs. the temperature difference between the heated surface and saturated water is depicted in Figure 21.1. There are six different regimes of boiling associated with the behavior exhibited in this figure.

In regime I, the wire surface temperature is only a few degrees higher than that of the surrounding saturated liquid. Natural convection currents circulate the superheated liquid, and evaporation occurs at the free liquid surface as the superheated liquid reaches it.

An increase in wire temperature is accompanied by the formation of vapor bubbles on the wire surface. These bubbles form at certain surface sites, where vapor bubble nuclei are

323

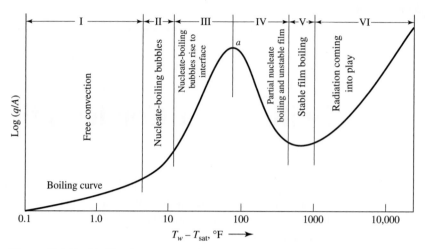

Figure 21.1 Pool boiling in water on a horizontal wire at atmospheric pressure.

present, break off, rise, and condense before reaching the free liquid surface. This is the process occurring in regime II.

At a still higher wire surface temperature, as in regime III, larger and more numerous bubbles form, break away from the wire surface, rise, and reach the free surface. Regimes II and III are associated with *nucleate boiling*.

Beyond the peak of this curve, the transition boiling regime is entered. This is region IV on the curve. In this regime, a vapor film forms around the wire, and portions of this film break off and rise, briefly exposing a portion of the wire surface. This film collapse and reformation and this unstable nature of the film is characteristic of the transition regime. When present, the vapor film provides a considerable resistance to heat transfer; thus, the heat flux decreases.

When the surface temperature reaches a value of approximately 400°F above the saturated liquid, the vapor film around the wire becomes stable. This is region V, the *stable film-boiling* regime.

For surface temperatures of 1000°F or greater above that of the saturated liquid, radiant energy transfer comes into play, and the heat flux curve rises once more. This is designated as region VI in Figure 21.1

The curve in Figure 21.1 can be achieved if the energy source is a condensing vapor. If, however, electrical heating is used, then regime IV will probably not be obtained because of wire "burnout." As the energy flux is increased, ΔT increases through regions I, II, and III. When the peak value of q/A is exceeded slightly, the required amount of energy cannot be transferred by boiling. The result is an increase in ΔT accompanied by a further decrease in the possible q/A. This condition continues until point b is reached. As ΔT at point b is extremely high, the wire will long since have reached its melting point. Point a on the curve is often referred to as the "burnout point" for these reasons.

As the mechanism of energy removal is intimately associated with buoyant forces, the magnitude of the body-force intensity will affect both the mechanism and the magnitude of boiling heat transfer. Other than normal gravitational effects are encountered in space vehicles.

Note the somewhat anomalous behavior exhibited by the heat flux associated with boiling. One normally considers a flux to be proportional to the driving force; thus, the heat flux might be expected to increase continuously as the temperature difference between the heated surface and the saturated liquid increases. This, of course, is not the case; the very high heat fluxes associated with moderate temperature differences in the nucleate-boiling regime are much higher than the heat fluxes resulting from much higher temperature differences in the film-boiling regime. The reason for this is the presence of the vapor film, which covers and insulates the heating surface in the latter case

Correlations of Boiling Heat-Transfer Data. As the fluid behavior in a boiling situation is very difficult to describe, there is no adequate analytical solution available for boiling transfer. Various correlations of experimental data have been achieved for the different boiling regimes; the most useful of these follow.

In the natural convection regime, *regime I* of Figure 21.1, the correlations presented in Chapter 20 for natural convection may be used.

Regime II, the regime of partial nucleate boiling and partial natural convection, is a combination of regimes I and III, and the results for each of these two regimes may be superposed to describe a process in regime II.

The nucleate-boiling regime, *regime III*, is of great engineering importance because of the very high heat fluxes possible with moderate temperature differences. That data for this regime are correlated by equations of the form

$$\text{Nu}_b = \phi(\text{Re}_b, \text{Pr}_L) \tag{21-1}$$

The parameter Nu_b in equation (21-1) is a Nusselt number defined as

$$\text{Nu}_b \equiv \frac{(q/A)D_b}{(T_s - T_{\text{sat}})k_L} \tag{21-2}$$

where q/A is the total heat flux, D_b is the maximum bubble diameter as it leaves the surface, $T_s - T_{\text{sat}}$ is the *excess temperature* or the difference between the surface and saturated-liquid temperatures, and k_L is the thermal conductivity of the liquid. The quantity, Pr_L, is the Prandtl number for the liquid. The bubble Reynolds number, Re_b, is defined as

$$\text{Re}_b \equiv \frac{D_b G_b}{\mu_L} \tag{21-3}$$

where G_b is the average mass velocity of the vapor leaving the surface and μ_L is the liquid viscosity

The mass velocity, G_b, may be determined from

$$G_b = \frac{q/A}{h_{fg}} \tag{21-4}$$

where h_{fg} is the latent heat of vaporization

Rohsenow[1] has used equation (21-1) to correlate Addoms's[2] pool-boiling data for a 0.024-in.-diameter platinum wire immersed in water. This correlation is shown in Figure 21.2 and is expressed in equation form as

$$\frac{q}{A} = \mu_L h_{fg} \left[\frac{g(\rho_L - \rho_v)}{\sigma} \right]^{1/2} \left[\frac{c_{pL}(T_s - T_{\text{sat}})}{C_{sf} h_{fg} \text{Pr}_L^{1.7}} \right]^3 \tag{21-5}$$

where c_{pL} is the heat capacity for the liquid and the other terms have their usual meanings.

[1] W. M. Rohsenow, *A.S.M.E. Trans.*, **74**, 969 (1952).
[2] J. N. Addoms, D.Sc. Thesis, Chemical Engineering Department, Massachusetts Institute of Technology, June 1948.

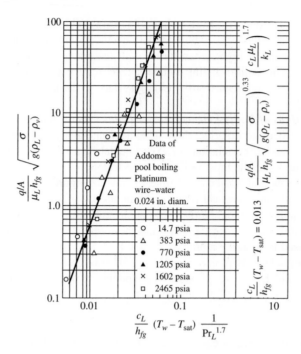

Figure 21.2 Correlation of pool-boiling data.
(From W. M. Rohsenow and H. Choi, *Heat, Mass, and Momentum Transfer*, Prentice-Hall, Inc., Englewood Cliffs, N.J., 1961, p. 224. By permission of the publishers.)

The coefficients C_{sf} in equation (21-5) vary with the surface–fluid combination. The curve drawn in Figure 21.2 is for $C_{sf} = 0.013$. A table of C_{sf} for various combinations of fluid and surface is presented by Rohsenow and Choi[3] and duplicated here as Table 21.1.

From earlier discussion it is clear that the burnout point has considerable importance. The "critical heat flux" is the value of q/A represented by point a in Figure 21.1.

Table 21.1   Values of C_{sf} for equation (21-5)

Surface/fluid combination	C_{sf}
Water/nickel	0.006
Water/platinum	0.013
Water/copper	0.013
Water/brass	0.006
CCl_4/copper	0.013
Benzene/chromium	0.010
n-Pentane/chromium	0.015
Ethyl alcohol/chromium	0.0027
Isopropyl alcohol/copper	0.0025
35% K_2CO_3/copper	0.0054
50% K_2CO_3/copper	0.0027
n-Butyl alcohol/copper	0.0030

---

[3] W. M. Rohsenow and H. Y. Choi, *Heat, Mass, and Momentum Transfer,* Prentice-Hall, Inc., Englewood Cliffs, N.J., 1961.

An analysis of conditions at burnout modified by experimental results is expressed in equation 21-6 as

$$q/A|_{\text{critical}} = 0.18 h_{fg} \rho_v \left[\frac{\sigma g(\rho_L - \rho_v)}{\rho_v^2} \right]^{1/4} \tag{21-6}$$

The interested reader is referred to the work of Zuber[4] for a discussion of this subject.

Regime IV, that of unstable film boiling, is not of great engineering interest, and no satisfactory correlation has been found for this region as yet.

The stable-film-boiling region, *regime V,* requires high surface temperatures; thus, few experimental data have been reported for this region.

Stable film boiling on the surface of horizontal tubes and vertical plates has been studied both analytically and experimentally by Bromley.[5,6] Considering conduction alone through the film on a horizontal tube, Bromley obtained the expression

$$h = 0.62 \left[\frac{k_v^3 \rho_v (\rho_L - \rho_v) g (h_{fg} + 0.4 c_{pL} \Delta T)}{D_o \mu_v (T_s - T_{\text{sat}})} \right]^{1/4} \tag{21-7}$$

where all terms are self-explanatory except D_o, which is the outside diameter of the tube.

A modification in equation (21-7) was proposed by Berenson[7] to provide a similar correlation for stable film boiling on a horizontal surface. In Berenson's correlation, the tube diameter, D_o, is replaced by the term $[\sigma/g(\rho_L - \rho_v)]^{1/2}$, and the recommended expression is

$$h = 0.425 \left[\frac{k_{vf}^3 \rho_{vf} (\rho_L - \rho_v) g (h_{fg} + 0.4 c_{pL} \Delta T)}{\mu_{vf} (T_s - T_{\text{sat}}) \sqrt{\sigma/g(\rho_L - \rho_v)}} \right]^{1/4} \tag{21-8}$$

where k_{vf}, ρ_{vf}, and μ_{vf} are to be evaluated at the film temperature as indicated.

Hsu and Westwater[8] considered film boiling for the case of a vertical tube. Their test results were correlated by the equation

$$h \left[\frac{\mu_v^2}{g \rho_v (\rho_L - \rho_v) k_v^3} \right]^{1/3} = 0.0020 \, \text{Re}^{0.6} \tag{21-9}$$

where

$$\text{Re} = \frac{4 \dot{m}}{\pi D_v \mu_v} \tag{21-10}$$

$\dot{m}$ being the flow rate of vapor in lb_m/h at the upper end of the tube and the other terms being identical to those in equation (21-7). Hsu[9] states that heat-transfer rates for film boiling are higher for vertical tubes than for horizontal tubes when all other conditions remain the same.

In *regime VI,* the correlations for film boiling still apply; however, the superimposed contribution of radiation is appreciable, becoming dominant at extremely high values of ΔT. Without any appreciable flow of liquid, the two contributions may be combined, as indicated by equation (21-11).

[4] N. Zuber, *Trans. A.S.M.E.,* **80,** 711 (1958).

[5] L. A. Bromley, *Chem. Eng Prog.,* **46,** (5), 221 (1950).

[6] L. A. Bromley, *et al., Ind. Eng Chem.,* **45,** 2639 (1953).

[7] P. Berenson, A.I.Ch.E. Paper No. 18, Heat Transfer Conference, Buffalo, N.Y., August 14–17, 960.

[8] Y. Y. Hsu and J. W. Westwater, *A.I.Ch.E. J.,* **4,** 59 (1958).

[9] S. T. Hsu, *Engineering Heat Transfer,* Van Nostrand, Princeton, N.J., 1963.

The contribution of radiation to the total heat-transfer coefficient may be expressed as

$$h = h_c \left(\frac{h_c}{h}\right)^{1/3} + h_r \tag{21-11}$$

where h is the total heat-transfer coefficient, h_c is the coefficient for the boiling phenomenon, and h_r is an effective radiant heat-transfer coefficient considering exchange between two parallel planes with the liquid between assigned a value of unity for its emissivity. This term is discussed in Chapter 23.

When there is appreciable flow of either the liquid or the vapor, the foregoing correlations are unsatisfactory. The description of *flow boiling* or *two-phase flow* will not be discussed here. The interested reader is referred to the recent literature for pertinent discussion of these phenomena. It is evident that for vertical surfaces or large-diameter horizontal tubes the density difference between liquid and vapor will produce significant local velocities. Any correlation that neglects flow contributions should, therefore, be used with caution.

21.2 CONDENSATION

Condensation occurs when a vapor contacts a surface that is at a temperature below the saturation temperature of the vapor. When the liquid condensate forms on the surface, it will flow under the influence of gravity.

Normally the liquid wets the surface, spreads out, and forms a film. Such a process is called *film condensation*. If the surface is not wetted by the liquid, then droplets form and run down the surface, coalescing as they contact other condensate droplets. This process is designated *dropwise condensation*. After a condensate film has been developed in filmwise condensation, additional condensation will occur at the liquid–vapor interface, and the associated energy transfer must occur by conduction through the condensate film. Dropwise condensation, on the contrary, always has some surface present as the condensate drop forms and runs off. Dropwise condensation is, therefore, associated with the higher heat-transfer rates of the two types of condensation phenomena. Dropwise condensation is very difficult to achieve or maintain commercially; therefore, all equipment is designed on the basis of filmwise condensation.

Film Condensation: The Nusselt Model. In 1916, Nusselt[10] achieved an analytical result for the problem of filmwise condensation of a pure vapor on a vertical wall. The meanings of the various terms in this analysis will be made clear by referring to Figure 21.3. In this figure,

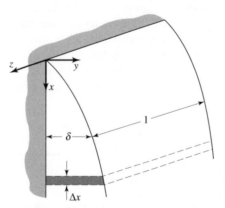

Figure 21.3 Filmwise condensation on a vertical plane wall.

---

[10] W. Nusselt, *Zeitschr. d. Ver. deutsch. Ing.*, **60**, 514 (1916).

the film thickness, δ, is seen to be zero at the top of the vertical wall, $x = 0$, and to increase with increasing values of x.

The initial assumption made by Nusselt was that of wholly laminar flow in the condensate film. Under these conditions, the velocity profile may be easily obtained from equation 8-12

$$v_x = \frac{\rho g L^2 \sin \theta}{\mu} \left[\frac{y}{L} - \frac{1}{2} \left(\frac{y}{L} \right)^2 \right] \tag{8-12}$$

For the present application, $\sin \theta = 1$ and $L = \delta$. It is also necessary to modify the density for the present case. In the derivation of equation (8-12), the density of the gas or vapor at the liquid surface was neglected. This may be true in many cases of a condensation process; however, the process may occur at a sufficiently high pressure that the vapor density, ρ_v, is significant in comparison to that of the liquid, ρ_L. To account for this possibility, the density function to be used in the present case is $\rho_L - \rho_v$ instead of simply ρ_L. The resulting expression for the velocity profile in the condensate film at a particular distance x from the top of the wall becomes

$$v_x = \frac{(\rho_L - \rho_v) g \delta^2}{\mu} \left[\frac{y}{\delta} - \frac{1}{2} \left(\frac{y}{\delta} \right)^2 \right] \tag{21-12}$$

The flow rate per unit width, Γ, at any value $x > 0$ is

$$\begin{aligned} \Gamma &= \int_0^\delta v_x \, dy \\ &= \frac{(\rho_L - \rho_v) g \delta^3}{3\mu} \end{aligned} \tag{21-13}$$

A differential change, $d\Gamma$, in the flow rate is evaluated from this expression to be

$$d\Gamma = \frac{(\rho_L - \rho_v) g \delta^2 d\delta}{\delta} \tag{21-14}$$

This result has been obtained from momentum considerations alone. We shall now, as Nusselt did originally, look at the related energy transfer.

As the flow of condensate is assumed to be laminar, it is not unreasonable to consider energy transfer through the film from the temperature at the vapor–liquid interface, T_{sat}, to the wall–liquid boundary at temperature, T_w, to be purely by conduction. On this basis, the temperature profile is linear and the heat flux to the wall is

$$\frac{q_y}{A} = k \frac{(T_{\text{sat}} - T_w)}{\delta} \tag{21-15}$$

This same amount of energy must be transferred from the vapor as it condenses and then cools to the average liquid temperature. Relating these two effects, we may write

$$\frac{q_y}{A} = k \frac{(T_{\text{sat}} - T_w)}{\delta} = \rho_L \left[h_{fg} + \frac{1}{\rho_L \Gamma} \int_0^\delta \rho_L v_x c_{pL} (T_{\text{sat}} - T) \, dy \right] \frac{d\Gamma}{dx}$$

which, if a linear temperature variation in y is utilized, becomes

$$\frac{q_y}{A} = \frac{k(T_{\text{sat}} - T_w)}{\delta} = \rho_L [h_{fg} + \tfrac{3}{8} c_{pL} (T_{\text{sat}} - T_w)] \frac{d\Gamma}{dx} \tag{21-16}$$

Solving equation () for $d\Gamma$, we have

$$d\Gamma = \frac{k(T_{sat} - T_w)\,dx}{\rho_L\delta[h_{fg} + \frac{3}{8}c_{pL}(T_{sat} - T_w)]} \tag{21-17}$$

which may now be equated to the result in equation (21-20), giving

$$\frac{(\rho_L - \rho_v)g}{\mu}\delta^2\,d\delta = \frac{k(T_{sat} - T_w)}{\rho_L\delta[h_{fg} + \frac{3}{8}c_{pL}(T_{sat} - T_w)]}\,dx$$

Simplifying this result and solving for δ, we obtain

$$\delta = \left[\frac{4k\mu(T_{sat} - T_w)x}{\rho_Lg(\rho_L - \rho_v)[h_{fg} + \frac{3}{8}c_{pL}(T_{sat} - T_w)]}\right]^{1/4} \tag{21-18}$$

We may now solve for the heat-transfer coefficient, h, from the expression

$$h = \frac{q_y/A}{T_{sat} - T_w} = \frac{k}{\delta}$$

The substitution of equation (21-18) into this expression yields

$$h_x = \left\{\frac{\rho_Lgk^3(\rho_L - \rho_v)[h_{fg} + \frac{3}{8}c_{pL}(T_{sat} - T_w)]}{4\mu(T_{sat} - T_w)x}\right\}^{1/4} \tag{21-19}$$

The average heat-transfer coefficient for a surface of length L is determined from

$$h = \frac{1}{L}\int_0^L h_x\,dx$$

which, when equation (21-19) is substituted, becomes

$$h = 0.943\left\{\frac{\rho_Lgk^3(\rho_L - \rho_v)[h_{fg} + \frac{3}{8}c_{pL}(T_{sat} - T_w)]}{L\mu(T_{sat} - T_w)}\right\}^{1/4} \tag{21-20}$$

The latent heat term, h_{fg}, in equation (21-20) and those preceding it, should be evaluated at the saturation temperature. Liquid properties should all be taken at the film temperature.

An expression similar to equation (21-20) may be achieved for a surface inclined at an angle θ from the horizontal if $\sin\theta$ is introduced into the bracketed term. This extension obviously has a limit and should not be used when θ is small, that is, when the surface is near horizontal. For such a condition, the analysis is quite simple; Example 1 illustrates such a case.

Rohsenow[11] performed a modified integral analysis of this same problem, obtaining a result that differs only in that the term $[h_{fg} + \frac{3}{8}c_{pL}(T_{sat} - T_w)]$ is replaced by $[h_{fg} + 0.68c_{pL}(T_{sat} - T_w)]$. Rohsenow's results agree well with experimental data achieved for values of $\Pr > 0.5$ and $c_{pL}(T_{sat} - T_w)/h_{fg} < 1.0$.

EXAMPLE

A square pan with its bottom surface maintained at 350 K is exposed to water vapor at 1 atm pressure and 373 K. The pan has a lip all around, so the condensate that forms cannot flow away. How deep will the condensate film be after 10 min have elapsed at this condition?

---

[11] W. M. Rohsenow, *A.S.M.E. Trans.*, **78**, 1645 (1956).

We will employ a "pseudo-steady-state" approach to solve this problem. An energy balance at the vapor–liquid interface will indicate that the heat flux and rate of mass condensed, $\dot{m}_{cond}$, are related as

$$\left.\frac{q}{A}\right|_{in} = \frac{\dot{m}_{cond}\, h_{fg}}{A}$$

The condensation rate, $\dot{m}_{cond}$, may be expressed as follows:

$$\dot{m}_{cond} = \rho \dot{V}_{cond} = \rho A \frac{d\delta}{dt}$$

where $d\delta/dt$ is the rate at which the condensate film thickness, δ, grows. The heat flux at the interface may now be expressed as

$$\left.\frac{q}{A}\right|_{in} = \rho h_{fg} \frac{d\delta}{dt}$$

This heat flux is now equated to that which must be conducted through the film to the cool pan surface. The heat flux expression that applies is

$$\left.\frac{q}{A}\right|_{out} = \frac{k_L}{\delta}(T_{sat} - T_s)$$

This is a steady-state expression; that is, we are assuming δ to be constant. If δ is not rapidly varying, this "pseudo-steady-state" approximation will give satisfactory results. Now, equating the two heat fluxes, we have

$$\rho h_{fg} = \frac{d\delta}{dt} = \frac{k_L}{\delta}(T_{sat} - T_s)$$

and, progressing, the condensate film thickness is seen to vary with time according to

$$\delta \frac{d\delta}{dt} = \frac{k_L}{\rho h_{fg}}(T_{sat} - T_s)$$

$$\int_0^\delta \delta\, d\delta = \frac{k_L}{\rho h_{fg}}(T_{sat} - T_s) \int_0^t dt$$

$$\delta = \left[\frac{2k_L}{\rho h_{fg}}(T_{sat} - T_s)\right]^{1/2} t^{1/2}$$

A quantitative answer to our example problem now yields the result

$$\delta = \left[\frac{2(0.674\,\text{W/m}\cdot\text{K})(23\,\text{K})(600\,\text{s})}{(966\,\text{kg/m}^3)(2250\,\text{kJ/kg})}\right]^{1/2}$$

$$= 2.93\,\text{mm}$$

Film Condensation: Turbulent-Flow Analysis. It is logical to expect the flow of the condensate film to become turbulent for relatively long surfaces or for high condensation rates. The criterion for turbulent flow is, as we should expect, a Reynolds number for the condensate film. In terms of an equivalent diameter, the applicable Reynolds number is

$$\text{Re} = \frac{4A}{P}\frac{\rho_L v}{\mu_f} \tag{21-21}$$

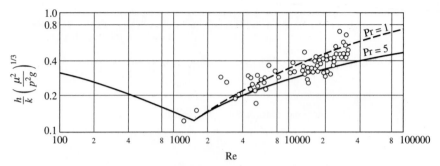

Figure 21.4 Film condensation including the regions of both laminar and turbulent flows.

where A is the condensate flow area, P is the wetted perimeter, and v is the velocity of condensate. The critical value of Re in this case is approximately 2000.

The first attempt to analyze the case of turbulent flow of a condensate film was that of Colburn,[12] who used the same j factor determined for internal pipe flow. On the basis partly of analysis and partly of experiment, Colburn formulated the plot shown in Figure 21.4. The data points shown are those of Kirkbride.[13] The correlating equations for the two regions shown are for $4\Gamma_c/\mu_f < 2000$

$$h_{\mathrm{avg}} = 1.51 \left(\frac{k^3 \rho^2 g}{\mu^2} \right)_f^{1/3} \left(\frac{4\Gamma_c}{\mu_f} \right)^{-1/3} \tag{21-22}$$

and for $4\Gamma_c/\mu_f > 2000$

$$h_{\mathrm{avg}} = 0.045 \frac{(k^3 \rho^2 g/\mu^2)_f^{1/3} (4\Gamma_c/\mu_f) \mathrm{Pr}^{1/3}}{\left[(4\Gamma_c/\mu_f)^{4/5} - 364 \right] + 576\,\mathrm{Pr}^{1/3}} \tag{21-23}$$

In these expressions, Γ_c is the mass flow rate per unit width of surface; that is, $\Gamma_c = \rho_L v_{\mathrm{avg}} \delta$, δ being film thickness and v_{avg} the average velocity. The term $4\Gamma_c/\mu_f$ is thus a Reynolds number for a condensate film on a plane vertical wall. McAdams[14] recommends a simpler expression for the turbulent range, $\mathrm{Re}_\delta > 2000$, as

$$h = 0.0077 \left[\frac{\rho_L g (\rho_L - \rho_v) k_L^3}{\mu_L^2} \right]^{1/3} \mathrm{Re}_\delta^{0.4} \tag{21-24}$$

Film Condensation: Analysis of the Horizontal Cylinder. An analysis by Nusselt[15] produced the following expression for the mean heat-transfer coefficient for a horizontal cylinder:

$$h_{\mathrm{avg}} = 0.725 \left\{ \frac{\rho_L g (\rho_L - \rho_v) k^3 \left[h_{fg} + \frac{3}{8} c_{pL}(T_{\mathrm{sat}} - T_w) \right]}{\mu D (T_{\mathrm{sat}} - T_w)} \right\}^{1/4} \tag{21-25}$$

[12] A. P. Colburn, *Ind. Eng. Chem.*, **26,** 432 (1934).

[13] C. G. Kirkbride, *Ind. Eng. Chem.*, **26,** 4 (1930).

[14] W. H. McAdams, *Heat Transmission*, 3rd edition, McGraw-Hill Book Company, New York, 1954.

[15] W. Nusselt, *Zeitschr. d. Ver. deutsch. Ing.*, **60,** 569 (1916).

The similarity between equation (21-25) for a horizontal tube and equation (21-20) for a vertical tube is marked. Combining these expressions and canceling similar terms, we obtain the result that

$$\frac{h_{\text{vert}}}{h_{\text{horiz}}} = \frac{0.943}{0.725}\left(\frac{D}{L}\right)^{1/4} = 1.3\left(\frac{D}{L}\right)^{1/4} \tag{21-26}$$

For the case of equal heat-transfer coefficients, the relation between D and L is

$$\frac{L}{D} = 2.86 \tag{21-27}$$

or, equal amounts of energy can be transferred from the same tube in either the vertical or the horizontal position if the ratio L/D is 2.86. For L/D values greater than 2.86, the horizontal position has the greater heat-transfer capability.

Film Condensation: Banks of Horizontal Tubes. For a bank of horizontal tubes there is, naturally, a different value of h for each tube, as the condensate film from one tube will drop on the next tube below it in the line. This process is depicted in Figure 21.5.

Condensate film

Figure 21.5 Condensation on a horizontal tube bank.

Nusselt also considered this situation analytically and achieved, for a vertical bank of n tubes in line, the expression

$$h_{\text{avg}} = 0.725\left\{\frac{\rho_L g(\rho_L - \rho_v)k^3\left[h_{fg} + \frac{3}{8}c_{pL}(T_{\text{sat}} - T_w)\right]}{nD\mu(T_{\text{sat}} - T_w)}\right\}^{1/4} \tag{21-28}$$

This equation yields a mean heat-transfer coefficient averaged over all n tubes.

Observing that experimental data exceeded those values predicted from equation (21-28), Chen[16] modified this expression to include the effect of condensation on the liquid layer between the tubes. His resulting equation is

$$h_{\text{avg}} = 0.725\left[1 + 0.02\frac{c_{pL}(T_{\text{sat}} - T_w)}{h_{fg}}(n - 1)\right]$$
$$\times \left\{\frac{\rho_L g(\rho_L - \rho_v)k^3\left[h_{fg} + \frac{3}{8}c_{pL}(T_{\text{sat}} - T_w)\right]}{nD\mu(T_{\text{sat}} - T_w)}\right\}^{1/4} \tag{21-29}$$

which is valid for values of $c_{pL}(T_{\text{sat}} - T_w)(n - 1)/h_{fg} > 2$. Chen's equation agrees reasonably well with experimental data for condensation on vertical banks of horizontal tubes.

---

[16] M. M. Chen, *A.S.M.E. (Trans.) Series C*, **83**, 48 (1961).

Drop Condensation.   Dropwise condensation, as mentioned earlier, is associated with higher heat-transfer coefficients than the filmwise condensation phenomenon. For dropwise condensation to occur, the surface must not be "wetted" by the condensate. Normally this requires that metal surfaces be specially treated.

Dropwise condensation is an attractive phenomenon for applications where extremely large heat-transfer rates are desired. At present, it is difficult to maintain this condition for several reasons. Because of its uncertain nature and the conservative approach of a design based on lower heat-transfer coefficients, filmwise condensation is the type predominantly used in design.

21.3   CLOSURE

The phenomena of boiling and condensation have been examined in this chapter. Each condition has a prominent place in engineering practice and both are difficult to describe analytically. Several empirical correlations for these phenomena for various surfaces oriented in different ways have been presented.

Boiling is normally described as nucleate type, film type, or a combination of the two. Very high heat-transfer rates are possible in the nucleate-boiling regime with relatively low temperature differences between the primary surface and the saturation temperature of the liquid. Film boiling is associated with a higher temperature difference yet a lower rate of heat transfer. This anomalous behavior is peculiar to the boiling phenomenon.

Condensation is categorized as either filmwise or dropwise. Dropwise condensation is associated with much higher heat-transfer coefficients than filmwise; however, it is difficult both to achieve and to maintain. Thus, filmwise condensation is of primary interest. Analytical solutions have been presented, along with empirical results, for filmwise condensation on vertical and horizontal plates and cylinders and for banks of horizontal cylinders.

PROBLEMS

The surface tension of water, a needed quantity in several of the following problems, is related to temperature according to the expression $\sigma = 0.1232[1 - 0.00146\,T]$, where σ is in N/m and T is in K. In the English system with σ given in lb_f/ft and T in °F, the surface tension may be calculated from $\sigma = (8.44 \times 10^{-3})[1 - 0.00082\,T]$.

21.1   An electrically heated square plate measuring 20 cm on a side is immersed vertically in water at atmospheric pressure. As the electrical energy supplied to the plate is increased, its surface temperature rises above that of the adjacent saturated water. At low power levels the heat-transfer mechanism is natural convection, then becoming a nucleate-boiling phenomenon at higher ΔTs. At what value of ΔT are the heat fluxes due to boiling and natural convection the same? Plot $q/A|_{convection}$, $q/A|_{boiling}$, and $q/A|_{total}$ versus ΔT values from 250 to 300 K.

21.2   Plot values of the heat-transfer coefficient for the case of pool boiling of water on horizontal metal surfaces at 1 atm total pressure and surface temperatures varying from 390 to 450 K.

Consider the following metals: (a) nickel; (b) copper; (c) platinum; (d) brass.

21.3   A cylindrical copper heating element 2 ft long and $\frac{1}{2}$ in. in diameter is immersed in water. The system pressure is maintained at 1 atm and the tube surface is held at 280°F. Determine the nucleate-boiling heat-transfer coefficient and the rate of heat dissipation for this system.

21.4   If the cylinder described in Problem 21.3 were initially heated to 500°F, how long would it take for the center of the cylinder to cool to 240°F if it were constructed of

a. copper?

b. brass?

c. nickel?

21.5   Four immersion heaters in the shape of cylinders 15 cm long and 2 cm in diameter are immersed in a water bath at 1 atm total pressure. Each heater is rated at 500 W. If the heaters operate at rated capacity, estimate the temperature of the heater surface. What is the convective heat-transfer coefficient in this case?

21.6 A horizontal circular cylinder 1 in. in diameter has its outside surface at a temperature of 1200°F. This tube is immersed in saturated water at a pressure of 40 psi. Estimate the heat flux due to film boiling that may be achieved with this configuration. At 40 psi, the temperature of saturated water is 267°F.

21.7 Estimate the heat-transfer rate per foot of length from a 0.02-in. diameter nichrome wire immersed in water at 240°F. The wire temperature is 2200°F.

21.8 Two thousand watts of electrical energy are to be dissipated through copper plates measuring 5 cm by 10 cm by 0.6 cm immersed in water at 390 K. How many plates would you recommend? Substantiate all of the design criteria used.

21.9 A steel plate is removed from a heat-treating operation at 600 K and is immediately immersed into a water bath at 373 K.

a. Construct a plot of heat flux vs. plate temperature for this system.

b. Construct a plot of convective heat-transfer coefficient vs. plate temperature.

c. For a mild-steel plate 3 cm thick and 30 cm square plot the plate temperature vs. time.

21.10 Water, flowing in a pipe, is to receive heat at a rate of 3×10^4 Btu/h·ft^2 of pipe surface. The pipe has an inside diameter of $\frac{3}{4}$ in. and is 4 ft long. If the water is to be at 212°F throughout its residence in the pipe, what rate of water flow would you suggest for safe operation? Support your results with all design criteria used.

21.11 Saturated steam at atmospheric pressure is enclosed within a vertical $\frac{1}{2}$-in.-diameter pipe whose surface is at 160°F. Construct a plot for the thickness of condensate film vs. distance from the top of the pipe.

21.12 Saturated steam at atmospheric pressure condenses on the outside surface of a 1-m-long tube with 150 mm diameter. The surface temperature is maintained at 91°C. Evaluate the condensation rate if the pipe is oriented

a. vertically;

b. horizontally.

21.13 Water flowing at a rate of 4000 kg/h through a 16.5-mm-ID tube enters at 20°C. The tube outside diameter is 19 mm. Saturated atmospheric steam condenses on the outside of the tube. For a horizontal brass tube 2 m long, evaluate

a. the convective coefficient on the water side;

b. the convective coefficient on the condensate side;

c. the exit water temperature;

d. the condensation rate.

21.14 Saturated steam at atmospheric pressure flows at a rate of 0.042 kg/s/m between two vertical surfaces maintained at 340 K that are separated by a distance of 1 cm. How tall may this configuration be if the steam velocity is not to exceed 15 m/s?

21.15 A circular pan has its bottom surface maintained at 200°F and is situated in saturated steam at 212°F. Construct a plot of condensate depth in the pan vs. time up to 1 h for this situation. The sides of the pan may be considered nonconducting.

21.16 Saturated steam at 365 K condenses on a 2-cm tube whose surface is maintained at 340 K. Determine the rate of condensation and the heat transfer coefficient for the case of a 1.5-m-long tube oriented

a. vertically;

b. horizontally.

21.17 If eight tubes of the size designated in Problem 21.14 are oriented horizontally in a vertical bank, what heat-transfer rate will occur?

21.18 Determine the heat-transfer coefficient for a horizontal $\frac{5}{8}$-in.-OD tube with its surface maintained at 100°F surrounded by steam at 200°F.

21.19 If eight tubes of the size designated in Problem 21.16 are arranged in a vertical bank and the flow is assumed laminar, determine

a. the average heat-transfer coefficient for the bank;

b. the heat-transfer coefficient for the first, third, and eighth tubes.

21.20 Given the conditions of Problem 21.16, what height of vertical wall will cause the film at the bottom of the tube to be turbulent?

21.21 A vertical flat surface 2 ft high is maintained at 60°F. If saturated ammonia at 85°F is adjacent to the surface, what heat-transfer coefficient will apply to the condensation process? What total heat transfer will occur?

21.22 A square pan measuring 40 cm on a side and having a 2-cm-high lip on all sides has its surface maintained at 350 K. If this pan is situated in saturated steam at 372 K, how long will it be before condensate spills over the lip if the pan is

a. horizontal?

b. inclined at 10° to the horizontal?

c. inclined at 30° to the horizontal?

21.23 A square pan with sides measuring 1 ft and a perpendicular lip extending 1 in. above the base is oriented with its base at an angle of 20° from the horizontal. The pan surface is kept at 180°F and it is situated in an atmosphere of 210°F steam. How long will it be before condensate spills over the lip of the pan?

Chapter 22

Heat-Transfer Equipment

$\mathbf{A}$ device whose primary purpose is the transfer of energy between two fluids is called a *heat exchanger*. Heat exchangers are usually classified into three categories:

1. regenerators;
2. open-type exchangers; and
3. closed-type exchangers or recuperators.

Regenerators are exchangers in which the hot and cold fluids flow alternately through the same space with as little physical mixing between the two streams as possible. The amount of energy transfer is dependent upon the fluid and flow properties of the fluid stream as well as the geometry and thermal properties of the surface. The required analytical tools for handling this type of heat exchanger have been developed in the preceding chapters.

Open-type heat exchangers are, as implied in their designation, devices wherein physical mixing of the two fluid streams actually occurs. Hot and cold fluids enter open-type heat exchangers and leave as a single stream. The nature of the exit stream is predicted by continuity and the first law of thermodynamics. No rate equations are necessary for the analysis of this type of exchanger.

The third type of heat exchanger, the recuperator, is the one of primary importance and the one to which we shall direct most of our attention. In the recuperator, the hot and cold fluid streams do not come into direct contact with each other but are separated by a tube wall or a surface that may be flat or curved in some manner. Energy exchange is thus accomplished from one fluid to an intermediate surface by convection, through the wall or plate by conduction, and then by convection from the surface to the second fluid. Each of these energy-transfer processes has been considered separately in the preceding chapters. We shall, in the following sections, investigate the conditions under which these three energy-transfer processes act in series with one another, resulting in a continuous change in the temperature of at least one of the fluid streams involved.

We shall be concerned with a thermal analysis of these exchangers. A complete design of such equipment involves an analysis of pressure drop, using techniques from Chapter 13, as well as material and structural considerations that are not within the scope of this text.

22.1 TYPES OF HEAT EXCHANGERS

In addition to being considered a closed-type exchanger, a recuperator is classified according to its configuration and the number of passes made by each fluid stream as it traverses the heat exchanger.

A *single-pass* heat exchanger is one in which each fluid flows through the exchanger only once. An additional descriptive term identifies the relative directions of the two streams, the terms used being *parallel flow* or *cocurrent flow* if the fluids flow in the same direction, *countercurrent flow* or simply *counterflow* if the fluids flow in opposite directions, and *crossflow* if the two fluids flow at right angles to one another. A common single-pass configuration is the double-pipe arrangement shown in Figure 22.1. A crossflow arrangement is shown in Figure 22.2.

Figure 22.1 A double-pipe heat exchanger.

Figure 22.2 A crossflow heat exchanger.

Variations on the crossflow configuration occur when one or the other, or both fluids are mixed. The arrangement shown in Figure 22.2 is one in which neither fluid is mixed. If the baffles or corrugations were not present, the fluid streams would be unseparated or mixed. In a condition such as that depicted in the figure, the fluid leaving at one end of the sandwich arrangement will have a nonuniform temperature variation from one side to the other, as each section contacts an adjacent fluid stream at a different temperature. It is normally desirable to have one or both fluids unmixed.

In order to accomplish as much transfer of energy in as little space as possible, it is desirable to utilize multiple passes of one or both fluids. A popular configuration is the *shell-and-tube* arrangement shown in Figure 22.3. In this figure, the *tube-side fluid* makes two passes, whereas the *shell-side fluid* makes one pass. Good mixing of the shell-side fluid is accomplished with the baffles shown. Without these baffles the fluid becomes stagnant in certain parts of the shell, the flow is partially channeled past these stagnant or "dead" regions, and less-than-optimum performance is achieved. Variations on the number of tube-and-shell passes are encountered in numerous applications; seldom are more than two shell-side passes used.

Figure 22.3 Shell-and-tube heat exchanger.

A number of more recent heat-transfer applications require more compact configurations than that afforded by the shell-and-tube arrangement. The subject of "compact heat exchangers" has been investigated and reported both carefully and quite thoroughly by Kays and London.[1] Typical compact arrangements are shown in Figure 22.4.

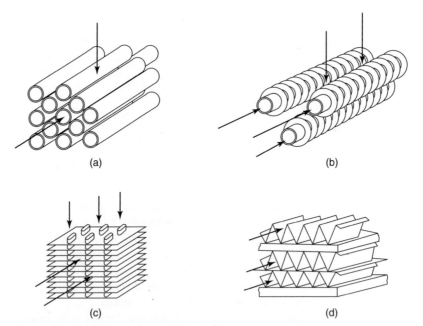

Figure 22.4 Compact heat-exchanger configurations.

The analysis of shell-and-tube, compact, or any multiple-pass heat exchanger is quite involved. As each is a composite of several single-pass arrangements, we shall initially focus our attention on the single-pass heat exchanger.

---

[1] W. M. Kays and A. L. London, *Compact Heat Exchangers,* 2nd edition, McGraw-Hill Book Company, 1964.

22.2 SINGLE-PASS HEAT-EXCHANGER ANALYSIS: THE LOG-MEAN TEMPERATURE DIFFERENCE

It is useful, when considering parallel or counterflow single-pass heat exchangers, to draw a simple sketch depicting the general temperature variation experienced by each fluid stream. There are four such profiles in this category, all of which are shown and labeled in Figure 22.5. Each of these may be found in a double-pipe arrangement.

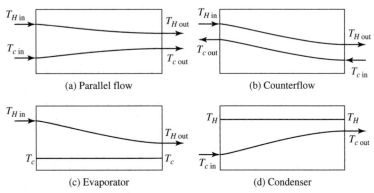

Figure 22.5 Temperature profiles for single-pass, double-pipe heat exchangers.

In Figure 22.5(c) and (d), one of the two fluids remains at constant temperature while exchanging heat with the other fluid whose temperature is changing. This situation occurs when energy transfer results in a change of phase rather than of temperature as in the cases of evaporation and condensation shown. The direction of flow of the fluid undergoing a change in phase is not depicted in the figure, as it is of no consequence to the analysis. If the situation occurs where the complete phase change such as condensation occurs within the exchanger along with some subcooling, then the diagram will appear as in Figure 22.6. In such a case, the direction of flow of the condensate stream is important. For purposes of analysis, this process may be considered the superposition of a condenser and a counterflow exchanger, as depicted in the diagram.

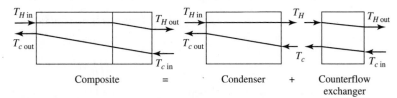

Figure 22.6 Temperature profile in a condenser with subcooling.

Also quite noticeable from Figure 22.5(a) and (b) is the significant difference in temperature profile exhibited by the parallel and counterflow arrangements. It is apparent that the exit temperatures of the hot and cold fluids in the parallel-flow case approach the same value. It is a simple exercise to show that this temperature is the one resulting if the two fluids are mixed in an open-type heat exchanger.

In the counterflow arrangement, it is possible for the hot fluid to leave the exchanger at a temperature below that at which the cold fluid leaves. This situation obviously corresponds to a case of greater total energy transfer per unit area of heat exchanger surface than would be obtained if the same fluids entered a parallel-flow configuration. The obvious conclusion to this discussion is that the counterflow configuration is the most desirable of the single-pass arrangements. It is thus the single-pass counterflow arrangement to which we shall direct our primary attention.

The detailed analysis of a single-pass counterflow heat exchanger that follows is referred to the diagram and nomenclature of Figure 22.7.

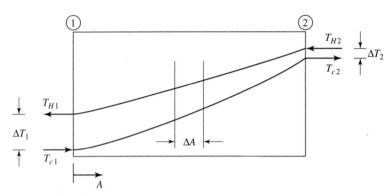

Figure 22.7 Diagram of temperature vs. contact area for single-pass counterflow analysis.

The abscissa of this figure is area. For a double-pipe arrangement, the heat-transfer area varies linearly with distance from one end of the exchanger; in the case shown, the zero reference is the end of the exchanger at which the cold fluid enters.

With reference to a general increment of area, ΔA, between the ends of this unit, a first-law-of-thermodynamics analysis of the two fluid streams will yield

$$\Delta q = (\dot{m}c_p)_c \, \Delta T_c$$

and

$$\Delta q = (\dot{m}c_p)_H \, \Delta T_H$$

As the incremental area approaches differential size, we may write

$$dq = (\dot{m}c_p)_c \, dT_c = C_c \, dT_c \qquad (22\text{-}1)$$

and

$$dq = (\dot{m}c_p)_H \, dT_H = C_H \, dT_H \qquad (22\text{-}2)$$

where the capacity coefficient, C, is introduced in place of the more cumbersome product, $\dot{m}c_p$.

Writing equation (15-17) for the energy transfer between the two fluids at this location, we have

$$dq = U \, dA(T_H - T_c) \qquad (22\text{-}3)$$

which utilizes the overall heat-transfer coefficient, U, introduced in Chapter 15. Designating $T_H - T_c$ as ΔT, we have

$$d(\Delta T) = dT_H - dT_c \qquad (22\text{-}4)$$

and substituting for dT_H and dT_c from equations (22-1) and (22-2), we obtain

$$d(\Delta T) = dq\left(\frac{1}{C_H} - \frac{1}{C_c}\right) = \frac{dq}{C_H}\left(1 - \frac{C_H}{C_c}\right) \qquad (22\text{-}5)$$

We should also note that dq is the same in each of these expressions; thus, equations (22-1) and (22-2) may be equated and integrated from one end of the exchanger to the other, yielding, for the ratio C_H/C_c

$$\frac{C_H}{C_c} = \frac{T_{c2} - T_{c1}}{T_{H2} - T_{H1}} \qquad (22\text{-}6)$$

which may be substituted into equation (22-5) and rearranged as follows:

$$d(\Delta T) = \frac{dq}{C_H}\left(1 - \frac{T_{c2} - T_{c1}}{T_{H2} - T_{H1}}\right) = \frac{dq}{C_H}\left(\frac{T_{H2} - T_{H1} - T_{c2} + T_{c1}}{T_{H2} - T_{H1}}\right)$$

$$= \frac{dq}{C_H}\left(\frac{\Delta T_2 - \Delta T_1}{T_{H2} - T_{H1}}\right) \qquad (22\text{-}7)$$

Combining equations (22-3) and (22-7), and noting that $C_H(T_{H2} - T_{H1}) = q$, we have, for constant U,

$$\int_{\Delta T_1}^{\Delta T_2} \frac{d(\Delta T)}{\Delta T} = \frac{U}{q}(\Delta T_2 - \Delta T_1)\int_0^A dA \qquad (22\text{-}8)$$

which, upon integration, becomes

$$\ln\frac{\Delta T_2}{\Delta T_1} = \frac{UA}{q}(\Delta T_2 - \Delta T_1)$$

This result is normally written as

$$q = UA\frac{\Delta T_2 - \Delta T_1}{\ln\dfrac{\Delta T_2}{\Delta T_1}} \qquad (22\text{-}9)$$

The driving force, on the right-hand side of equation (22-9), is seen to be a particular sort of mean temperature difference between the two fluid streams. This ratio, $(\Delta T_2 - \Delta T_1)/\ln(\Delta T_2/\Delta T_1)$, is designated ΔT_{lm}, the *logarithmic-mean temperature difference,* and the expression for q is written simply as

$$q = UA\,\Delta T_{lm} \qquad (22\text{-}10)$$

Even though equation (22-10) was developed for the specific case of counterflow, it is equally valid for any of the single-pass operations depicted in Figure 22.5.

It was mentioned earlier, but bears repeating, that equation (22-10) is based upon a constant value of the overall heat-transfer coefficient, U. This coefficient will not, in general, remain constant; however, calculations based upon a value of U taken midway between the ends of the exchanger are usually accurate enough. If there is considerable variation in U from one end of the exchanger to the other, then a step-by-step numerical integration is necessary, equations (22-1)–(22-3) being evaluated repeatedly over a number of small-area increments.

It is also possible that the temperature differences in equation (22-9), evaluated at either end of a counterflow exchanger, are equal. In such a case, the log-mean temperature difference is indeterminate; that is,

$$\frac{\Delta T_2 - \Delta T_1}{\ln(\Delta T_2/\Delta T_1)} = \frac{0}{0}, \qquad \text{if } \Delta T_1 = \Delta T_2$$

In such a case, L'Hôpital's rule may be applied as follows:

$$\lim_{\Delta T_2 \to \Delta T_1} \frac{\Delta T_2 - \Delta T_1}{\ln(\Delta T_2/\Delta T_1)} = \lim_{\Delta T_2/\Delta T_1 \to 1} \left[\frac{\Delta T_1\{(\Delta T_2/\Delta T_1) - 1\}}{\ln(\Delta T_2/\Delta T_1)} \right]$$

when the ratio $\Delta T_2/\Delta T_1$ is designated by the symbol F, we may write

$$= \lim_{F \to 1} \Delta T \left(\frac{F - 1}{\ln F} \right)$$

Differentiating numerator and denominator with respect to F yields the result that

$$\lim_{\Delta T_2 \to \Delta T_1} \frac{\Delta T_2 - \Delta T_1}{\ln(\Delta T_2/\Delta T_1)} = \Delta T$$

or that equation (22-10) may be used in the simple form

$$q = UA\,\Delta T \qquad (22\text{-}11)$$

From the foregoing simple analysis, it should be apparent that equation (22-11) may be used and achieve reasonable accuracy so long as ΔT_1 and ΔT_2 are not vastly different. It turns out that a simple arithmetic mean is within 1% of the logarithmic-mean temperature difference for values of $(\Delta T_2/\Delta T_1) < 1.5$.

EXAMPLE 1

Light lubricating oil ($c_p = 2090$ J/kg · K) is cooled by allowing it to exchange energy with water in a small heat exchanger. The oil enters and leaves the heat exchanger at 375 and 350 K, respectively, and flows at a rate of 0.5 kg/s. Water at 280 K is available in sufficient quantity to allow 0.201 kg/s to be used for cooling purposes. Determine the required heat-transfer area for (a) counterflow and (b) parallel-flow operations (see Figure 22.8). The overall heat-transfer coefficient may be taken as 250 W/m² · K.

The outlet water temperature is determined by applying equations (22-1) and (22-2)

$$q = (0.5 \text{ kg/s})(2090 \text{ J/kg} \cdot \text{K})(25 \text{ K}) = 26\,125 \text{ W}$$
$$= (0.201 \text{ kg/s})(4177 \text{ J/kg} \cdot \text{K})(T_{w\,\text{out}} - 280 \text{ K})$$

(a) Counterflow   (b) Parallel flow

Figure 22.8 Single-pass temperature profiles for counterflow and parallel flow.

from which we obtain

$$T_{w\,out} = 280 + \frac{(0.5)(2090)(25)}{(0.201)(4177)} = 311.1\,K \quad (100°F)$$

This result applies to both parallel flow and counterflow. For the counterflow configuration, ΔT_{lm} is calculated as

$$\Delta T_{lm} = \frac{70 - 63.9}{\ln\dfrac{70}{63.9}} = 66.9\,K \quad (120.4°F)$$

and applying equation (22-10), we see that the area required to accomplish this energy transfer is

$$A = \frac{26\,125\,W}{(250\,W/m^2 \cdot K)(66.9\,K)} = 1.562\,m^2 \quad (16.81\,ft^2)$$

Performing similar calculations for the parallel-flow situation, we obtain

$$\Delta T_{lm} = \frac{95 - 38.9}{\ln\dfrac{95}{38.9}} = 62.8\,K \quad (113°F)$$

$$A = \frac{26\,125\,W}{(250\,W/m^2 \cdot K)(62.8\,K)} = 1.66\,m^2 \quad (17.9\,ft^2)$$

The area required to transfer 26,125 W is seen to be lower for the counterflow arrangement by approximately 7%.

22.3 CROSSFLOW AND SHELL-AND-TUBE HEAT-EXCHANGER ANALYSIS

More complicated flow arrangements th   s considered in the previous sections are much more difficult to treat analytically.   1 factors to be used with equation (22-10) have been presented in chart form by   Mueller, and Nagle[2] and by the Tubular Exchanger Manufacturers Association.   2.9 and 22.10 present correction factors for six types of heat-exchanger configur   e first three are for different shell-and-tube configurations and the latter three are for different crossflow conditions.

The parameters in Figures 22.9 and 22.10 are evaluated as follows:

$$Y = \frac{T_{t\,out} - T_{t\,in}}{T_{s\,in} - T_{t\,in}} \tag{22-12}$$

$$Z = \frac{(\dot{m}c_p)_{tube}}{(\dot{m}c_p)_{shell}} = \frac{C_t}{C_s} = \frac{T_{s\,in} - T_{s\,out}}{T_{t\,out} - T_{t\,in}} \tag{22-13}$$

where the subscripts s and t refer to the shell-side and tube-side fluids, respectively. The quantity read on the ordinate of each plot, for given values of Y and Z, is F, the correction factor to be applied to equation (22-10), and thus these more complicated configurations

[2] R. A. Bowman, A. C. Mueller, and W. M. Nagle, *Trans. A.S.M.E.* **62**, 283 (1940).

[3] Tubular Exchanger Manufacturers Association, Standards, 3rd edition, TEMA, New York, 1952.

Figure 22.9 Correction factors for three shell-and-tube heat-exchanger configurations. (a) One shell pass and two or a multiple of two tube passes. (b) One shell pass and three or a multiple of three tube passes. (c) Two shell passes and two or a multiple of two tube passes.

(From R. A. Bowman, A. C. Mueller, and W. M. Nagle, *Trans. A.S.M.E.,* **62,** 284, 285 (1940). By permission of the publishers.) Correction factors, F, based on counterflow LMTD.

Correction factor plot for exchanger with one shell pass and two, four, or any multiple of tube passes

(a)

$$z = \frac{T_{H1} - T_{H2}}{T_{c2} - T_{c1}}$$

Shell fluid

Tube fluid

(b)

$$\frac{T_{H1} - T_{H2}}{T_{c2} - T_{c1}}$$

Correction factor plot for exchanger with two shell passes and four, eight, or any multiple of four tube passes

(c)

$$z = \frac{T_{H_1} - T_{H_2}}{T_{c_2} - T_{c_1}}$$

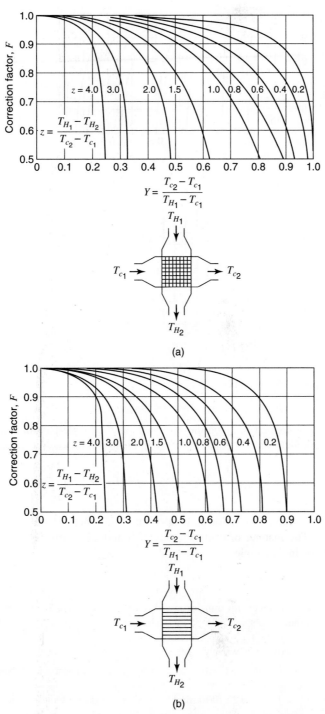

Figure 22.10 Correction factors for three crossflow heat-exchanger configurations. (a) Crossflow, single-pass, both fluids unmixed. (b) Crossflow, single-pass, one fluid unmixed. (c) Crossflow, tube passes mixed; fluid flows over first and second passes in series.

(From R. A. Bowman, A. C. Mueller, and W. M. Nagle, *Trans. A.S.M.E.,* **62,** 288–289 (1940). By permission of the publishers.)

(c)

Figure 22.10 Continued

may be treated in much the same way as the single-pass double-pipe case. The reader is cautioned to apply equation (22-10), using the factor F as in equation (22-14).

$$q = UA(F \, \Delta T_{lm}) \tag{22-14}$$

with the logarithmic-mean temperature difference calculated on the basis of *counterflow*.

The manner of using Figures 22.9 and 22.10 may be illustrated by referring to the following example.

EXAMPLE 2

In the oil–water energy transfer described in Example 1, compare the result obtained with the result that would be obtained if the heat exchanger were

(a) crossflow, water-mixed;

(b) shell-and-tube with four tube-side passes, oil being the tube-side fluid.

For part (a), Figure 22.10(b) must be used. The parameters needed to use this figure are

$$Y = \frac{T_{t\,\text{out}} - T_{t\,\text{in}}}{T_{s\,\text{in}} - T_{t\,\text{in}}} = \frac{25}{95} = 0.263$$

and

$$Z = \frac{T_{s\,in} - T_{s\,out}}{T_{t\,out} - T_{t\,in}} = \frac{31.1}{25} = 1.244$$

and from the figure we read $F = 0.96$. The required area for part (a) is thus equal to $(1.562)/(0.96) = 1.63\,\mathrm{m}^2$.

The values of Y and Z determined above are the same in part (b), yielding a value of F equal to 0.97. The area for part (b) becomes $(1.562)/(0.97) = 1.61\,\mathrm{m}^2$.

22.4   THE NUMBER-OF-TRANSFER-UNITS (NTU) METHOD OF HEAT-EXCHANGER ANALYSIS AND DESIGN

Earlier mention was made of the work of Kays and London[1] with particular reference to compact heat exchangers. The book "Compact Heat Exchangers," by Kays and London, also presents charts useful for heat-exchanger design on a different basis than discussed thus far.

Nusselt,[4] in 1930, proposed the method of analysis based upon the heat-exchanger effectiveness $\mathscr{E}$. This term is defined as the ratio of the actual heat transfer in a heat exchanger to the maximum possible heat transfer that would take place if infinite surface area were available. By referring to a temperature profile diagram for counterflow operation, as in Figure 22.11, it is seen that, in general, one fluid undergoes a greater total temperature change than the other. It is apparent that the fluid experiencing the larger change in temperature is the one having the smaller capacity coefficient, which we designate C_{min}. If $C_c = C_{min}$, as in Figure 22.11(a), and if there is infinite area available for energy transfer, the exit temperature of the cold fluid will equal the inlet temperature of the hot fluid.

(a) $C_H > C_C$, $C_C = C_{min}$ (b) $C_C > C_H$, $C_H = C_{min}$

Figure 22.11 Temperature profiles for counterflow heat exchangers.

According to the definition of effectiveness, we may write

$$\mathscr{E} = \frac{C_H(T_{H\,in} - T_{H\,out})}{C_c(T_{c\,out} - T_{c\,in})_{max}} = \frac{C_{max}(T_{H\,in} - T_{H\,out})}{C_{min}(T_{H\,in} - T_{c\,in})} \qquad (22\text{-}15)$$

If the hot fluid is the minimum fluid, as in Figure 22.11(b), the expression for $\mathscr{E}$ becomes

$$\mathscr{E} = \frac{C_c(T_{c\,out} - T_{c\,in})}{C_H(T_{H\,in} - T_{H\,out})_{max}} = \frac{C_{max}(T_{c\,out} - T_{c\,in})}{C_{min}(T_{H\,in} - T_{c\,in})} \qquad (22\text{-}16)$$

---

[4] W. Nusselt, *Tech. Mechan. Thermodyn.*, **12** (1930).

Notice that the denominators in both equations (22-15) and (22-16) are the same and that, in each case, the numerator represents the actual heat transfer. It is thus possible to write a fifth expression for q as

$$q = \mathscr{E} C_{min}(T_{H\,in} - T_{c\,in}) \tag{22-17}$$

which, along with the integrated forms of equations (22-1) and (22-2), as well as equations (22-10) and (22-14), expresses q, the rate of heat transfer, in all of its useful forms as far as heat-exchanger analysis and design are concerned. Equation (22-17) is conspicuous among these others, as the temperature difference appearing is that between the inlet streams alone. This is a definite advantage when a given heat exchanger is to be used under conditions other than those for which it was designed. The exit temperatures of the two streams are then needed quantities, and equation (22-17) is obviously the easiest means of attaining this knowledge if one can determine the value of $\mathscr{E}$.

To determine $\mathscr{E}$ for a single-pass case, we initially write equation (22-17) in the form

$$\mathscr{E} = \frac{C_H(T_{H\,in} - T_{H\,out})}{C_{min}(T_{H\,in} - T_{c\,in})} = \frac{C_c(T_{c\,out} - T_{c\,in})}{C_{min}(T_{H\,in} - T_{c\,in})} \tag{22-18}$$

The appropriate form for equation (22-18) depends on which of the two fluids has the smaller value of C. We shall consider the cold fluid to be the minimum fluid and consider the case of counterflow. For these conditions, equation (22-10) may be written as follows (numerical subscripts correspond to the situation shown in Figure 22.7):

$$q = C_c(T_{c2} - T_{c1}) = UA \frac{(T_{H1} - T_{c1}) - (T_{H2} - T_{c2})}{\ln[(T_{H1} - T_{c1})/(T_{H2} - T_{c2})]} \tag{22-19}$$

The entering temperature of the hot fluid, T_{H2}, may be written in terms of $\mathscr{E}$ by use of equation (22-18), yielding

$$T_{H2} = T_{c1} + \frac{1}{\mathscr{E}}(T_{c2} - T_{c1}) \tag{22-20}$$

and also

$$
\begin{aligned}
T_{H2} - T_{c2} &= T_{c1} - T_{c2} + \frac{1}{\mathscr{E}}(T_{c2} - T_{c1}) \\
&= \left(\frac{1}{\mathscr{E}} - 1\right)(T_{c2} - T_{c1})
\end{aligned}
\tag{22-21}
$$

From the integrated forms of equations (22-1) and (22-2), we have

$$\frac{C_c}{C_H} = \frac{T_{H2} - T_{H1}}{T_{c2} - T_{c1}}$$

which may be rearranged to the form

$$T_{H1} = T_{H2} - \frac{C_{min}}{C_{max}}(T_{c2} - T_{c1})$$

or

$$T_{H1} - T_{c1} = T_{H2} - T_{c1} - \frac{C_{min}}{C_{max}}(T_{c2} - T_{c1}) \tag{22-22}$$

Combining this expression with equation (22-20), we obtain

$$
\begin{aligned}
T_{H1} - T_{c1} &= \frac{1}{\mathscr{E}}(T_{c2} - T_{c1}) - \frac{C_{min}}{C_{max}}(T_{c2} - T_{c1}) \\
&= \left(\frac{1}{\mathscr{E}} - \frac{C_{min}}{C_{max}}\right)(T_{c2} - T_{c1})
\end{aligned}
\tag{22-23}
$$

Now substituting equations (22-21) and (22-23) into equation (22-19) and rearranging, we have

$$
\ln\frac{1/\mathscr{E} - C_{min}/C_{max}}{1/\mathscr{E} - 1} = \frac{UA}{C_{min}}\left(1 - \frac{C_{min}}{C_{max}}\right)
$$

Taking the antilog of both sides of this expression and solving for ξ we have, finally,

$$
\mathscr{E} = \frac{1 - \exp\left[-\dfrac{UA}{C_{min}}\left(1 - \dfrac{C_{min}}{C_{max}}\right)\right]}{1 - (C_{min}/C_{max})\exp\left[-\dfrac{UA}{C_{min}}\left(1 - \dfrac{C_{min}}{C_{max}}\right)\right]}
\tag{22-24}
$$

The ratio UA/C_{min} is designated the *number of transfer units,* abbreviated NTU. Equation (22-24) was derived on the basis that $C_c = C_{min}$; if we had initially considered the hot fluid to be minimum, the same result would have been achieved. Thus, equation (22-25)

$$
\mathscr{E} = \frac{1 - \exp\left[-\text{NTU}\left(1 - \dfrac{C_{min}}{C_{max}}\right)\right]}{1 - (C_{min}/C_{max})\exp\left[-\text{NTU}\left(1 - \dfrac{C_{min}}{C_{max}}\right)\right]}
\tag{22-25}
$$

is valid for counterflow operation in general. For parallel flow, an analogous development to the preceding will yield

$$
\mathscr{E} = \frac{1 - \exp\left[-\text{NTU}\left(1 + \dfrac{C_{min}}{C_{max}}\right)\right]}{1 + C_{min}/C_{max}}
\tag{22-26}
$$

Kays and London[1] have put equations (22-25) and (22-26) into chart form, along with comparable expressions for the effectiveness of several shell-and-tube and crossflow arrangements. Figures 22.12 and 22.13 are charts for $\mathscr{E}$ as functions of NTU for various values of the parameter C_{min}/C_{max}.

With the aid of these figures, equation (22-17) may be used both as an original design equation and as a means of evaluating existing equipment when it operates at other than design conditions.

The utility of the NTU approach is illustrated in the following example.

Figure 22.12 Heat-exchanger effectiveness for three shell-and-tube configurations.
(a) Counterflow. (b) Parallel flow. (c) One shell pass and two or a multiple of two tube passes.

Figure 22.12 Continued

Figure 22.13 Heat-exchanger effectiveness for three crossflow configurations. (a) Crossflow, both fluids unmixed. (b) Crossflow, one fluid mixed. (c) Crossflow, multiple pass.

(b)

(c)

Figure 22.13 Continued

EXAMPLE 3	Repeat the calculations for Examples 1 and 2 to determine the required heat-transfer area for the specified conditions if the configurations are

(a) counterflow;

(b) parallel flow;

(c) crossflow, water-mixed; and

(d) shell-and-tube with four tube-side passes.

It is first necessary to determine the capacity coefficients for the oil and water

$$C_{oil} = (\dot{m}c_p)_{oil} = (0.5\,\text{kg/s})(2090\,\text{J/kg} \cdot \text{K}) = 1045\,\text{J/s} \cdot \text{K}$$

and

$$C_{water} = (\dot{m}c_p)_w = (0.201\,\text{kg/s})(4177\,\text{J/kg} \cdot \text{K}) = 841.2\,\text{J/s} \cdot \text{K}$$

thus the water is the minimum fluid. From equation (22-16), the effectiveness is evaluated as

$$\mathscr{E} = \frac{26\,125\,\text{W}}{(841.2\,\text{J/kg} \cdot \text{s})(95\,\text{K})} = 0.327$$

By using the appropriate chart in Figures 22.12 and 22.13, the appropriate NTU values and, in turn, the required area may be evaluated for each heat-exchanger configuration.

(a) *Counterflow*

$$NTU = 0.47$$

$$A = \frac{(0.47)(841.2)}{250} = 1.581\,\text{m}^2$$

(b) *Parallel flow*

$$NTU = 0.50$$

$$A = \frac{(0.50)(841.2)}{250} = 1.682\,\text{m}^2$$

(c) *Crossflow, water-mixed*

$$NTU = 0.48$$

$$A = \frac{(0.48)(841.2)}{250} = 1.615\,\text{m}^2$$

(d) *Shell-and-tube, four tube-side passes*

$$NTU = 0.49$$

$$A = \frac{(0.49)(841.2)}{250} = 1.649\,\text{m}^2$$

These results are comparable to those obtained earlier, with some possible inaccuracies involved in reading the chart.

The NTU method offers no distinct advantage over the procedure introduced earlier, using the log-mean temperature difference, when performing calculations of the type involved in the preceding examples. In Example 4, however, the NTU approach is clearly superior.

EXAMPLE 4

In the energy exchange between water and lubricating oil as considered in the preceding examples, a crossflow heat exchanger with the shell-side fluid (water) mixed is constructed with a heat-transfer area of 1.53 m². A new pump is attached to the water supply line enabling the water flow rate to be increased to 1000 kg/h. What will be the exit temperatures of the water and oil for the new operating conditions?

If the ΔT_{lm} method were used in this problem, a trial-and-error method would be necessary as ΔT_{lm}, Y, and F are all dependent on one or both exit stream temperatures. The NTU method is, thus, a bit simpler. Using the NTU method, it is first necessary to calculate the capacity coefficients

$$C_{oil} = (0.5\,\text{kg/s})(2090\,\text{J/kg} \cdot \text{K}) = 1045\,\text{J/s} \cdot \text{K}$$
$$C_w = (1000\,\text{kg/h})(\text{h}/3600\,\text{s})(4177\,\text{J/kg} \cdot \text{K})$$
$$= 1160\,\text{J/kg} \cdot \text{K}$$

Oil is now the "minimum" fluid. With $C_{oil} = C_{min}$ we have

$$\text{NTU} = \frac{UA}{C_{min}} = \frac{(250\,\text{W/m}^2 \cdot \text{K})(1.53\,\text{m}^2)}{1045\,\text{J/s} \cdot \text{K}}$$
$$= 0.366$$

and, from Figure 22.13, the effectiveness is

$$\mathscr{E} \cong 0.29$$

Using equation (22-17) we may evaluate the heat-transfer rate as

$$q = (0.29)(1045\,\text{J/s} \cdot \text{K})(95\,\text{K})$$
$$= 28.8\,\text{K}$$

an increase of over 10%. This value may now be used in equations (22-1) and (22-2) to yield the required answers.

$$T_{oil\,out} = 375 - 28.8\,\text{kW}/(1045\,\text{W/K})$$
$$= 347.4\,\text{K}$$
$$T_{w\,out} = 280 + 28.8\,\text{kW}/(1160\,\text{W/K})$$
$$= 304.8\,\text{K}$$

22.5 ADDITIONAL CONSIDERATIONS IN HEAT-EXCHANGER DESIGN

After a heat exchanger has been in service for some time, its performance may change as a result of the buildup of scale on a heat-transfer surface or of the deterioration of the surface by a corrosive fluid. When the nature of the surface is altered in some way as to affect the heat-transfer capability, the surface is said to be "fouled."

When a fouling resistance exists, the thermal resistance is increased and a heat exchanger will transfer less energy than the design value. It is extremely difficult to predict the rate of scale buildup or the effect such buildup will have upon heat transfer. Some evaluation can be done after a heat exchanger has been in service for some time by comparing its performance with that when the surfaces were clean. The thermal resistance of the scale is determined by

$$R_{sc} = \frac{1}{U_f} - \frac{1}{U_0} \tag{22-27}$$

where U_0 is the overall heat-transfer coefficient of the clean exchanger, U_f is the overall heat-transfer coefficient of the fouled exchanger, and R_{sc} is the thermal resistance of the scale.

Fouling resistances that have been obtained from experiments may be used to roughly predict the overall heat-transfer coefficient by incorporation into an expression similar to equation (15-19). The following equation includes the fouling resistances, R_i on the inside tube surface and R_o on the outside tube surface:

$$U_0 = \frac{1}{A_0/A_i h_i + R_i + [A_0 \ln(r_o/r_i)]/2\pi k/L + R_o + 1/h_o} \tag{22-28}$$

Fouling resistances to be used in equation (22-28) have been compiled by the Tubular Exchanger Manufacturers Association.[5] Some useful values are given in Table 22.1.

Table 22.1   Heat-exchanger fouling resistances

Fluid	Fouling resistances $(m^2 \cdot K/W \times 10^5)$
Distilled water	8.8
Sea water, below 325 K	8.8
above 325 K	17.6
Boiler feed water, treated	17.6
City or well water, below 325 K	17.6
above 325 K	35.2
Refrigerating liquids	17.6
Refrigerating vapors	35.2
Liquid gasoline, organic vapors	8.8
Fuel oil	88.1
Quenching oil	70.5
Steam, non-oil-bearing	8.8
Industrial air	35.2

It is often useful to have "ball-park" figures on heat-exchanger size, flow rates, and the like. The most difficult quantity to estimate quickly is the overall heat-transfer coefficient, U. Mueller[6] has prepared the very useful table of approximate U values which is reproduced here as Table 22.2.

---

[5] Tubular Exchanger Manufacturers Association, TEMA Standards, 3rd edition, New York, 1952.
[6] A. C. Mueller, *Purdue Univ. Eng. Exp. Sta. Eng. Bull. Res. Ser.* **121** (1954).

Table 22.2 Approximate values for overall heat-transfer coefficients

Fluid combination	$U(\text{W/m}^2\cdot\text{K})$
Water to compressed air	55–165
Water to water, jacket water coolers	850–1560
Water to brine	570–1140
Water to gasoline	340–480
Water to gas oil or distillate	200–340
Water to organic solvents, alcohol	280–850
Water to condensing alcohol	250–680
Water to lubricating oil	110–340
Water to condensing oil vapors	220–570
Water to condensing or boiling Freon-12	280–850
Water to condensing ammonia	850–1350
Steam to water, instantaneous heater	2280–3400
storage-tank heater	990–1700
Steam to oil, heavy fuel	55–165
light fuel	165–340
light petroleum distillate	280–1140
Steam to aqueous solutions	570–3400
Steam to gases	28–280
Light organics to light organics	220–425
Medium organics to medium organics	110–340
Heavy organics to heavy organics	55–220
Heavy organics to light organics	55–340
Crude oil to gas oil	165–310

22.6 CLOSURE

The basic equations and procedures for heat-exchanger design are presented and developed in this chapter. All heat-exchanger design and analysis involve one or more of the following equations:

$$dq = C_c\, dT_c \tag{22-1}$$

$$dq = C_H\, dT_H \tag{22-2}$$

$$dq = U\, dA(T_H - T_c) \tag{22-3}$$

$$q = UA\, \Delta T_{lm} \tag{22-10}$$

and

$$q = \mathscr{E} C_{\min}(T_{H\,\text{in}} - T_{c\,\text{in}}) \tag{22-17}$$

Charts were presented by which single-pass techniques could be extended to include the design and analysis of crossflow and shell-and-tube configurations.

The two methods for heat-exchanger design utilize either equation (22-10) or (22-17). Either is reasonably rapid and straightforward for designing an exchanger. Equation (22-17) is a simpler and more direct approach when analyzing an exchanger that operates at other than design conditions.

PROBLEMS

22.1 A single tube-pass heat exchanger is to be designed to heat water by condensing steam in the shell. The water is to pass through the smooth horizontal tubes in turbulent flow, and the steam is to be condensed dropwise in the shell. The water flow rate, the initial and final water temperatures, the condensation temperature of the steam, and the available tube-side pressure drop (neglecting entrance and exit losses) are all specified. In order to determine the optimum exchanger design, it is desirable to know how the total required area of the exchanger varies with the tube diameter selected. Assuming that the water flow remains turbulent and that the thermal resistance of the tube wall and the steam–condensate film is negligible, determine the effect of tube diameter on the total area required in the exchanger.

22.2 One hundred thousand pounds per hour of water are to pass through a heat exchanger, which is to raise the water temperature from 140 to 200°F. Combustion products having a specific heat of 0.24 Btu/lb$_m$ °F are available at 800°F. The overall heat-transfer coefficient is 12 Btu/h ft^2 °F. If 100,000 lb$_m$/h of the combustion products are available, determine

a. the exit temperature of the flue gas;

b. the required heat-transfer area for a counterflow exchanger.

22.3 An oil having a specific heat of 1880 J/kg · K enters a single-pass counterflow heat exchanger at a rate of 2 kg/s and a temperature of 400 K. It is to be cooled to 350 K. Water is available to cool the oil at a rate of 2 kg/s and a temperature of 280 K. Determine the surface area required if the overall heat-transfer coefficient is 230 W/m^2 · K.

22.4 Air at 203 kPa and 290 K flows in a long rectangular duct with dimensions 10 cm by 20 cm. A 2.5-m length of this duct is maintained at 395 K, and the average exit air temperature from this section is 300 K. Calculate the air flow rate and the total heat transfer.

22.5 Water enters a counterflow, double-pipe heat exchanger at a rate of 150 lb$_m$/min and is heated from 60 to 140°F by an oil with a specific heat of 0.45 Btu/lb$_m$ °F. The oil enters at 240°F and leaves at 80°F. The overall heat-transfer coefficient is 150 Btu/h ft^2 °F.

a. What heat-transfer area is required?

b. What area is required if all conditions remain the same except that a shell-and-tube heat exchanger is used with the water making one shell pass and the oil making two tube passes?

c. What exit water temperature would result if, for the exchanger of part (a), the water flow rate were decreased to 120 lb$_m$/min?

22.6 A water-to-oil heat exchanger has entering and exiting temperatures of 255 and 340 K, respectively, for the water and 305 and 350 K, respectively, for the oil. What is the effectiveness of this heat exchanger?

22.7 Water at 50°F is available for cooling at a rate of 400 lb$_m$/h. It enters a double-pipe heat exchanger with a total area of 18 ft^2. Oil, with $c_p = 0.45$ Btu/lb$_m$ °F, enters the exchanger at 250°F. The exiting water temperature is limited at 212°F, and the oil must leave the exchanger at no more than 160°F. Given the value of $U = 60$ Btu/h ft^2 °F, find the maximum flow of oil that may be cooled with this unit.

22.8 A shell-and-tube heat exchanger is used for the heating of oil from 20 to 30°C; the oil flow rate is 12 kg/s ($c_{po} = 2.2$ kJ/kg · K). The heat exchanger has one shell pass and two tube passes. Hot water ($c_{pw} = 4.18$ kJ/kg · K) enters the shell at 75°C and leaves the shell at 55°C. The overall heat-transfer coefficient based on the outside surface of the tubes is estimated to be 1080 W/m^2 · K. Determine

a. the corrected logarithmic-mean temperature difference;

b. the required surface area in the exchanger.

22.9 Consider the exchanger in Problem 22.8. After 4 years of operation, the outlet of the oil reaches 28°C instead of 30°C with all other conditions remaining the same. Determine the fouling resistance on the oil side of the exchanger.

22.10 A finned-tube crossflow heat exchanger with both fluids unmixed is used to heat water ($c_{po} = 4.2$ kJ/kg · K) from 20 to 75°C. The mass flow rate of the water is 2.7 kg/s. The hot stream ($c_{pw} = 1.2$ kJ/kg · K) enters the heat exchanger at 280°C and leaves at 120°C. The overall heat-transfer coefficient is 160 W/m^2 · K. Determine

a. the mass flow rate of the heat stream;

b. the exchanger surface area.

22.11 A shell-and-tube heat exchanger is used to cool oil ($C_H = 2.2$ kJ/kg · K) from 110 to 65°C. The heat exchanger has two shell passes and four tube passes. The coolant ($C_c = 4.20$ kJ/kg · K) enters the shell at 20°C and leaves the shell at 42°C. For an overall tube-side heat-transfer coefficient of 1200 W/m^2 · K and an oil flow of 11 kg/s, determine

a. the coolant mass flow rate;

b. the required surface area in the exchanger.

22.12 A shell-and-tube exchanger having one shell pass and eight tube passes is to heat kerosene from 80 to 130°F. The kerosene enters at a rate of 2500 lb$_m$/h. Water entering at 200°F and at a rate of 900 lb$_m$/h is to flow on the shell side. The overall heat-transfer coefficient if 260 Btu/h ft^2 °F. Determine the required heat-transfer area.

22.13 If the overall heat-transfer coefficient, initial fluid temperature, and total heat-transfer area determined in Problem 22.3 remain the same, find the exit oil temperature if the configuration is changed to

a. crossflow, both fluids unmixed;

b. shell-and-tube with two tube passes and one shell pass.

22.14 A condenser unit is of a shell-and-tube configuration with steam condensing at 85°C in the shell. The coefficient on the condensate side is 10,600 W/m² · K. Water at 20°C enters the tubes, which make two passes through the single-shell unit. The water leaves the unit at a temperature of 38°C. An overall heat-transfer coefficient of 4600 W/m² · K may be assumed to apply. The heat transfer rate is 0.2×10^6 kW. What must be the required length of tubes for this case?

22.15 Determine the required heat-transfer surface area for a heat exchanger constructed from 10-cm OD tubes. A 95% ethanol solution ($c_p = 3.810$ kJ/kg · K), flowing at 6.93 kg/s is cooled from 340 to 312 K by 6.30 kg/s of water that is available at 283 K. The overall heat-transfer coefficient based on outside tube area is 568 W/m² · K. Three different exchanger configurations are of interest:

a. counterflow, single pass;

b. parallel flow, single pass;

c. crossflow with one tube pass and one shell pass, shell-side fluid mixed.

22.16 Compressed air is used in a heat-pump system to heat water, which is subsequently used to warm a house. The house demand is 95,000 Btu/h. Air enters the exchanger at 200°F and leaves at 120°F, and water enters and leaves the exchanger at 90 and 125°F, respectively. Choose from the following alternative units the one that is most compact.

a. A counterflow surface with $U = 30$ Btu/h ft² °F and a surface-to-volume ratio of 130 ft²/ft³.

b. A crossflow configuration with the water unmixed and air mixed having $U = 40$ Btu/h ft² °F and a surface-to-volume ratio of 100 ft²/ft³.

c. A crossflow unit with both fluids unmixed with $U = 50$ Btu/h ft² °F and surface-to-volume ratio of 90 ft²/ft³.

22.17 A shell-and-tube heat exchanger with two shell passes and four tube passes is used to exchange energy between two pressurized water streams. One stream flowing at 5000 lb$_m$/h is heated from 75 to 220°F. The hot stream flows at 2400 lb$_m$/h and enters at 400°F. If the overall heat-transfer coefficient is 300 W/m² · K, determine the required surface area.

22.18 For the heat exchanger described in Problem 22.17 it is observed, after a long period of operation, that the cold stream leaves at 184°F instead of at the design value of 220°F. This is for the same flow rates and entering temperatures of both streams. Evaluate the fouling factor that exists at the new conditions.

22.19 Water flowing at a rate of 10 kg/s through 50 tubes in a double-pass shell-and-tube heat exchanger heats air that flows on the shell side. The tubes are made of brass with outside diameters of 2.6 cm and are 6.7 m long. Surface coefficients on the inside and outside tube surfaces are 470 and 210 W/m² · K, respectively. Air enters the unit at 15°C with a flow rate of 16 kg/s. The entering water temperature is 350 K. Determine the following:

a. heat-exchanger effectiveness;

b. heat-transfer rate to the air;

c. exiting temperatures of the water and air streams.

If, after a long period of operation, a scale has been built up inside the tubes resulting in an added fouling resistance of 0.0021 m² · K/W, determine the new results for parts (a), (b), and (c), above.

22.20 Water flowing at a rate of 3.8 kg/s is heated from 38 to 55°C in the tubes of a shell-and-tube heat exchanger. The shell side is one-pass with water, flowing at 1.9 kg/s entering at 94°C. The overall heat-transfer coefficient is 1420 W/m² · K. The average water velocity in the 1.905-cm-ID tubes is 0.366 m/s. Because of space limitations, the tubes may not exceed 2.44 m in length. Determine the required number of tube passes, the number of tubes per pass, and the length of tubes consistent with this restriction.

22.21 Saturated steam at 373 K is to be condensed in a shell-and-tube exchanger; it is to enter as steam at 373 K and leave as condensate at approximately 373 K. If the NTU rating for the condenser is given by the manufacturer as 1.25 in this service for a circulating water flow of 0.07 kg/s, and circulating water is available at 280 K, what will be the approximate maximum flow rate of steam in kg/s that can be condensed? What will be the leaving temperature of the circulating water under these conditions? Under these conditions, the heat vaporization is 2257 kJ/kg and c_p is 4.18 kJ/kg · K.

Chapter 23

Radiation Heat Transfer

The mechanism of radiation heat transfer has no analogy in either momentum or mass transfer. Radiation heat transfer is extremely important in many phases of engineering design such as boilers, home heating, and spacecraft. In this chapter, we will concern ourselves first with understanding the nature of thermal radiation. Next, we will discuss properties of surfaces and consider how system geometry influences radiant heat transfer. Finally, we will illustrate some techniques for solving relatively simple problems where surfaces and some gases participate in radiant energy exchange.

23.1 NATURE OF RADIATION

The transfer of energy by radiation has several unique characteristics when contrasted with conduction or convection. First, matter is not required for radiant heat transfer; indeed the presence of a medium will impede radiation transfer between surfaces. Cloud cover is observed to reduce maximum daytime temperatures and to increase minimum evening temperatures, both of which are dependent upon radiant energy transfer between earth and space. A second unique aspect of radiation is that both the amount of radiation *and* the quality of the radiation depend upon temperature. In conduction and convection, the amount of heat transfer was found to depend upon the temperature difference; in radiation, the amount of heat transfer depends upon both the temperature difference between two bodies and their absolute temperatures. In addition, radiation from a hot object will be different in quality than radiation from a body at a lower temperature. The color of incandescent objects is observed to change as the temperature is changed. The changing optical properties of radiation with temperature are of paramount importance in determining the radiant-energy exchange between bodies.

Radiation travels at the speed of light, having both wave properties and particle-like properties. The electromagnetic spectrum shown in Figure 23.1 illustrates the tremendous range of frequency and wavelength over which radiation occurs.

The unit of wavelength which we shall use in discussing radiation is the micron, symbolized μ. One micron is 10^{-6} m or $3.94(10)^{-5}$ in. The frequency, v, of radiation is related to the wavelength λ, by $\lambda v = c$, where c is the speed of light. Short-wavelength radiation such as gamma rays and x-rays is associated with very high energies. To produce radiation of this type we must disturb the nucleus or the inner-shell electrons of an atom. Gamma rays and x-rays also have great penetrating ability; surfaces that are opaque to visible radiation are easily traversed by gamma and x-rays. Very-long-wavelength radiation, such as radio waves, also may pass through solids; however, the energy associated with these waves is much less than that for short-wavelength radiation. In the range from $\lambda = 0.38$ to 0.76 microns, radiation is sensed by the optical nerve of the eye and is what we call light. Radiation in the visible range is observed to have little penetrating power except in some liquids, plastics, and glasses. The radiation between wavelengths of 0.1 and 100 microns is

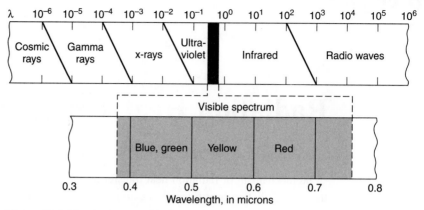

Figure 23.1 The electromagnetic spectrum.

termed as *thermal* radiation. The thermal band of the spectrum includes a portion of the ultraviolet and all of the infrared regions.

23.2 THERMAL RADIATION

Thermal radiation incident upon a surface as shown in Figure 23.2 may be either absorbed, reflected, or transmitted.

Figure 23.2 Fate of radiation incident upon a surface.

If ρ, α, and τ are the fractions of the incident radiation that are reflected, absorbed and transmitted, respectively, then

$$\rho + \alpha + \tau = 1 \tag{23-1}$$

where ρ is called the *reflectivity*, α is called the *absorptivity*, and τ is called *transmissivity*.

There are two types of reflection that can occur, specular reflection and diffuse reflection. In *specular reflection*, the angle of incidence of the radiation is equal to the angle of reflection. The reflection shown in Figure 23.2 is specular reflection. Most bodies do not reflect in a specular manner, they reflect radiation in all directions. *Diffuse reflection* is sometimes likened to a situation in which the incident thermal radiation is absorbed and then reemitted from the surface, still retaining its initial wavelength.

Absorption of thermal radiation in solids takes place in a very short distance, on the order of 1 μm in electrical conductors and about 0.05 in. in electrical nonconductors, the difference being caused by the different population of energy states in electrical conductors, which can absorb energy at thermal radiation frequencies.

For most solids, the transmissivity is zero, and thus they may be called *opaque* to thermal radiation. equation (23-1) becomes, for an opaque body, $\rho + \alpha = 1$.

The ideally absorbing body, for which $\alpha = 1$, is called a *black body*. A black body neither reflects nor transmits any thermal radiation. As we see reflected light (radiation), a so-called black body will appear black, no light being reflected from it. A small hole in a large cavity closely approaches a black body, regardless of the nature of the interior surface. Radiation incident to the hole has very little opportunity to be reflected back out of the hole. Black bodies may also be made of bright objects, as can be shown by looking at a stack of razor blades, sharp edge forward.

The *total emissive power, E*, of a surface is defined as the total rate of thermal energy emitted via radiation from a surface in all directions and at all wavelengths per unit surface area. The total emissive power is also referred to elsewhere as the emittance or the total hemispheric intensity. Closely related to the total emissive power is the emissivity. The *emissivity, ϵ*, is defined as the ratio of the total emissive power of a surface to the total emissive power of an ideally radiating surface at the same temperature. The ideal radiating surface is also called a black body, so we may write

$$\epsilon = \frac{E}{E_b} \tag{23-2}$$

where E_b is the total emissive power of a black body. As the total emissive power includes radiant-energy contributions from all wavelengths, the *monochromatic emissive power, E_λ*, may also be defined. The radiant energy E_λ contained between wavelengths λ and $\lambda + d\lambda$ is the monochromatic emissive power; thus

$$dE = E_\lambda \, d\lambda, \quad \text{or} \quad E = \int_0^\infty E_\lambda \, d\lambda$$

The monochromatic emissivity, ϵ_λ, is simply $\epsilon_\lambda = E_\lambda/E_{\lambda,b}$, where $E_{\lambda,b}$ is the monochromatic emissive power of a black body at wavelength λ at the same temperature. A monochromatic absorptivity, α_λ, may be defined in the same manner as the monochromatic emissivity. The *monochromatic absorptivity* is defined as the ratio of the incident radiation of wavelength λ that is absorbed by a surface to the incident radiation absorbed by a black surface.

A relation between the absorptivity and the emissivity is given by Kirchhoff's law. Kirchhoff's law states that, for a system in thermodynamic equilibrium, the following equality holds for each surface:

$$\epsilon_\lambda = \alpha_\lambda \tag{23-3}$$

Thermodynamic equilibrium requires that all surfaces be at the same temperature so that there is no net heat transfer. The utility of Kirchhoff's law lies in its use for situations in which the departure from equilibrium is small. In such situations the emissivity and the absorptivity may be assumed to be equal. For radiation between bodies at greatly different temperatures, such as between Earth and the sun, Kirchhoff's law does not apply. A frequent error in using Kirchhoff's law arises from confusing thermal equilibrium with steady-state conditions. Steady state means that time derivatives are zero, whereas equilibrium refers to the equality of temperatures.

23.3 THE INTENSITY OF RADIATION

In order to characterize the quantity of radiation that travels from a surface along a specified path, the concept of a single ray is not adequate. The amount of energy traveling in a given direction is determined from I, the *intensity* of radiation. With reference to Figure 23.3, we

are interested in knowing the rate at
which radiant energy is emitted from a
representative portion, dA, of the surface
shown in a prescribed direction. Our perspective will be that of an observer at
point P looking at dA. Standard spherical
coordinates will be used, these being r,
the radial coordinate; θ, the zenith angle
shown in Figure 23.3; and ϕ, the azimuthal angle, which will be discussed
shortly. If a unit area of surface, dA, emits
a total energy dq, then the *intensity of
radiation* is given by

$$I \equiv \frac{d^2q}{dA \, d\Omega \, \cos \theta} \qquad (23\text{-}4)$$

where $d\Omega$ is a differential solid angle,
that is, a portion of space. Note that with
the eye located at point P, in Figure 23.3,

Figure 23.3 The intensity of radiation.

the apparent size of the emitting area is $dA \cos \theta$. It is important to remember that the
intensity of radiation is independent of direction for a diffusely radiating surface. Rearranging equation (23-4), we see that the relation between the total emissive power, $E = dq/dA$,
and the intensity, I, is

$$\frac{dq}{dA} = E = \int I \cos \theta \, d\Omega = I \int \cos \theta \, d\Omega \qquad (23\text{-}5)$$

The relation is seen to be purely geometric for
a diffusely radiating $(I \neq I(\theta))$ surface. Consider an imaginary hemisphere of radius r covering the plane surface on which dA is located.
The solid angle $d\Omega$ intersects the shaded area
on the hemisphere as shown in Figure 23.4. A
solid angle is defined by $\Omega = A/r^2$ or $d\Omega = dA/r^2$, and thus

$$d\Omega = \frac{(r \sin \theta \, d\phi)(r \, d\theta)}{r^2} = \sin \theta \, d\theta \, d\phi$$

The total emissive power per unit area becomes

$$E = I \int \cos \theta \, d\Omega$$

$$= I \int_0^{2\pi} \int_0^{\pi/2} \cos \theta \, \sin \theta \, d\theta \, d\phi$$

or simply

Figure 23.4 Integration of intensity over
solid angles.

$$E = \pi I \qquad (23\text{-}6)$$

If the surface does not radiate diffusely, then

$$E = \int_0^{2\pi} \int_0^{\pi/2} I \cos \theta \, \sin \theta \, d\theta \, d\phi$$

The relation between the intensity of radiation, I, and the total emissive power is an important step in determining the total emissive power.

Radiation intensity is fundamental in formulating a quantitative description of radiant heat transfer but its definition, as already discussed, is cumbersome. Equation (23-6) relates intensity to emissive power that, potentially, is much easier to describe. We will now consider the means of such a description.

23.4 PLANCK'S LAW OF RADIATION

Planck[1] introduced the quantum concept in 1900 and with it the idea that radiation is emitted not in a continuous energy state but in discrete amounts or quanta. The intensity of radiation emitted by a black body, derived by Planck, is

$$I_{b,\lambda} = \frac{2c^2 h \lambda^{-5}}{\exp\left(\dfrac{ch}{\kappa \lambda T}\right) - 1}$$

where $I_{b,\lambda}$ is the intensity of radiation from a black body between wavelengths λ and $\lambda + d\lambda$, c is the speed of light, h is Planck's constant, κ is the Boltzmann constant, and T is the temperature. The total emissive power between wavelengths λ and $\lambda + d\lambda$ is then

$$E_{b,\lambda} = \frac{2\pi c^2 h \lambda^{-5}}{\exp\left(\dfrac{ch}{\kappa \lambda T}\right) - 1} \tag{23-7}$$

Figure 23.5 illustrates the spectral energy distribution of energy of a black body as given by equation (23-7).

In Figure 23.5 the area under the curve of $E_{b,\lambda}$ vs. λ (the total emitted energy) is seen to increase rapidly with temperature. The peak energy is also observed to occur at shorter and shorter wavelengths as the temperature is increased. For a black body at 5800 K (the effective temperature of solar radiation), a large part of the emitted energy is in the visible region. Equation (23-7) expresses, functionally, $E_{b,\lambda}$ as a function of wavelength and temperature. Dividing both sides of this equation by T^5, we get

$$\frac{E_{b\lambda}}{T^5} = \frac{2\pi^2 h (\lambda T)^{-5}}{\exp\left(\dfrac{ch}{\kappa \lambda T}\right) - 1} \tag{23-8}$$

where the quantity $E_{b\lambda}/T^5$ is expressed as a function of the λT product, which can be treated as a single independent variable. This functional relationship is plotted in Figure 23.6, and discrete values of $E_{b\lambda}/\sigma T^5$ are given in Table 23.1. The constant, σ, will be discussed in the next section.

The peak energy is observed to be emitted at $\lambda T = 2897.6 \ \mu\text{m K} (5215.6 \ \mu\text{m}^\circ\text{R})$, as can be determined by maximizing equation (23-8). The relation, $\lambda_{\max} T = 2897 \ \mu$ K, is called Wien's displacement law. Wien obtained this result in 1893, 7 years prior to Planck's development.

We are often interested in knowing how much emission occurs in a specific portion of the total wavelength spectrum. This is conveniently expressed as a fraction of the total emissive power. The fraction between wavelengths λ_1 and λ_2 is designated $F_{\lambda_1 - \lambda_2}$ and may be expressed as

$$F_{\lambda_1 - \lambda_2} = \frac{\int_{\lambda_1}^{\lambda_2} E_{b\lambda} \, d\lambda}{\int_0^\infty E_{b\lambda} \, d\lambda} = \frac{\int_{\lambda_1}^{\lambda_2} E_{b\lambda} \, d\lambda}{\sigma T^4} \tag{23-9}$$

---

[1] M. Planck, *Verh. d. deut. physik. Gesell.*, **2**, 237 (1900).

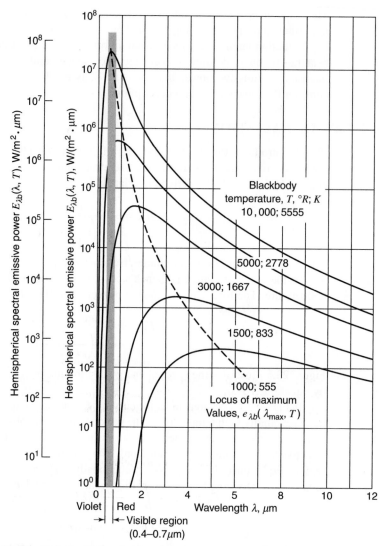

Figure 23.5 Spectral emissive power for a black body for several temperatures.
(From R. Siegel and J. R. Howell, *Thermal Radiation Heat Transfer*, Third Edition, Hemisphere Publishers, Washington, 1992. By permission of the publishers.)

Equation (23-9) is conveniently broken into two integrals as follows:

$$F_{\lambda_1 - \lambda_2} = \frac{1}{\sigma T^4} \left(\int_0^{\lambda_2} E_{b\lambda} \, d\lambda - \int_0^{\lambda_1} E_{b\lambda} \, d\lambda \right)$$

$$= F_{0-\lambda_2} - F_{0-\lambda_1} \tag{23-10}$$

So, at a given temperature, the fraction of emission between any two wavelengths can be determined by subtraction.

This process can be simplified if the temperature is eliminated as a separate variable. This may be accomplished by using the fraction $E_{b\lambda}/\sigma T^5$, as discussed. Equation (23-10)

Figure 23.6 Spectral energy distribution for a black body as a function of λT.
(From R. Siegel and J. R. Howell, *Thermal Radiation Heat Transfer*, Third Edition, Hemisphere Publishers, Washington, 1992. By permission of the publishers.)

may be modified in this manner to yield

$$F_{\lambda_1 T - \lambda_2 T} = \int_0^{\lambda_2 T} \frac{E_{b\lambda}}{\sigma T^5} \, d(\lambda T) - \int_0^{\lambda_1 T} \frac{E_{b\lambda}}{\sigma T^5} \, d(\lambda T) \qquad (23\text{-}11)$$
$$= F_{0-\lambda_2 T} - F_{0-\lambda_1 T}$$

Values of $F_{0-\lambda T}$ are given as functions of the product, λT, in Table 23.1.

23.5 STEFAN–BOLTZMANN LAW

Planck's law of radiation may be integrated over wavelengths from zero to infinity to determine the total emissive power. The result is

$$E_b = \int_0^\infty E_{b,\lambda} \, d\lambda = \frac{2\pi^5 \kappa^4 T^4}{15 c^2 h^3} = \sigma T^4 \qquad (23\text{-}12)$$

where σ is called the Stefan–Boltzmann constant and has the value $\sigma = 5.676 \times 10^{-8} \ \text{W/m}^2 \cdot \text{K}^4 (0.1714 \times 10^{-8} \ \text{Btu/h ft}^2 \ {}^\circ \text{R}^4)$. This constant is observed to be a combination of other physical constants. The Stefan–Boltzmann relation, $E_b = \sigma T^4$, was obtained prior to Planck's law, via experiment by Stefan in 1879 and via a thermodynamic derivation by Boltzmann in 1884. The exact value of the Stefan–Boltzmann constant, σ, and its relation to other physical constants were obtained after the presentation of Planck's law in 1900.

Table 23.1   Planck radiation functions

$\lambda T(\mu m\ K)$	$F_{0-\lambda_T}$	$\dfrac{E_b}{\sigma T^5}\left(\dfrac{1}{cm\ K}\right)$	$\lambda T(\mu m\ K)$	$F_{0-\lambda_T}$	$\dfrac{E_b}{\sigma T^5}\left(\dfrac{1}{cm\ K}\right)$
1000	0.0003	0.0372	5500	0.6909	1.0339
1100	0.0009	0.0855	5600	0.7010	0.9938
1200	0.0021	0.1646	5700	0.7108	0.9552
1300	0.0043	0.2774	5800	0.7201	0.9181
1400	0.0078	0.4222	5900	0.7291	0.8826
1500	0.0128	0.5933	6000	0.7378	0.8485
1600	0.0197	0.7825	6100	0.7461	0.8158
1700	0.0285	0.9809	6200	0.7541	0.7844
1800	0.0393	1.1797	6300	0.7618	0.7543
1900	0.0521	1.3713	6400	0.7692	0.7255
2000	0.0667	1.5499	6500	0.7763	0.6979
2100	0.0830	1.7111	6600	0.7832	0.6715
2200	0.1009	1.8521	6700	0.7897	0.6462
2300	0.1200	1.9717	6800	0.7961	0.6220
2400	0.1402	2.0695	6900	0.8022	0.5987
2500	0.1613	2.1462	7000	0.8081	0.5765
2600	0.1831	2.2028	7100	0.8137	0.5552
2700	0.2053	2.2409	7200	0.8192	0.5348
2800	0.2279	2.2623	7300	0.8244	0.5152
2900	0.2505	2.2688	7400	0.8295	0.4965
3000	0.2732	2.2624	7500	0.8344	0.4786
3100	0.2958	2.2447	7600	0.8391	0.4614
3200	0.3181	2.2175	7700	0.8436	0.4449
3300	0.3401	2.1824	7800	0.8480	0.4291
3400	0.3617	2.1408	7900	0.8522	0.4140
3500	0.3829	2.0939	8000	0.8562	0.3995
3600	0.4036	2.0429	8100	0.8602	0.3856
3700	0.4238	1.9888	8200	0.8640	0.3722
3800	0.4434	1.9324	8300	0.8676	0.3594
3900	0.4624	1.8745	8400	0.8712	0.3472
4000	0.4809	1.8157	8500	0.8746	0.3354
4100	0.4987	1.7565	8600	0.8779	0.3241
4200	0.5160	1.6974	8700	0.8810	0.3132
4300	0.5327	1.6387	8800	0.8841	0.3028
4400	0.5488	1.5807	8900	0.8871	0.2928
4500	0.5643	1.5238	9000	0.8900	0.2832
4600	0.5793	1.4679	9100	0.8928	0.2739
4700	0.5937	1.4135	9200	0.8955	0.2650
4800	0.6075	1.3604	9300	0.8981	0.2565
4900	0.6209	1.3089	9400	0.9006	0.2483
5000	0.6337	1.2590	9500	0.9030	0.2404
5100	0.6461	1.2107	9600	0.9054	0.2328
5200	0.6579	1.1640	9700	0.9077	0.2255
5300	0.6694	1.1190	9800	0.9099	0.2185
5400	0.6803	1.0756	9900	0.9121	0.2117

$\lambda T(\mu m\ K)$	$F_{0-\lambda_T}$	$\dfrac{E_b}{\sigma T^5}\left(\dfrac{1}{cm\ K}\right)$	$\lambda T(\mu m\ K)$	$F_{0-\lambda_T}$	$\dfrac{E_b}{\sigma T^5}\left(\dfrac{1}{cm\ K}\right)$
10,000	0.9142	0.2052	30,000	0.9953	0.0044
11,000	0.9318	0.1518	31,000	0.9957	0.0039
12,000	0.9451	0.1145	32,000	0.9961	0.0035
13,000	0.9551	0.0878	33,000	0.9964	0.0031
14,000	0.9628	0.0684	34,000	0.9967	0.0028
15,000	0.9689	0.0540	35,000	0.9970	0.0025
16,000	0.9738	0.0432	36,000	0.9972	0.0022
17,000	0.9777	0.0349	37,000	0.9974	0.0020
18,000	0.9808	0.0285	38,000	0.9976	0.0018
19,000	0.9834	0.0235	39,000	0.9978	0.0016
20,000	0.9856	0.0196	40,000	0.9979	0.0015
21,000	0.9873	0.0164	41,000	0.9981	0.0014
22,000	0.9889	0.0139	42,000	0.9982	0.0012
23,000	0.9901	0.0118	43,000	0.9983	0.0011
24,000	0.9912	0.0101	44,000	0.9984	0.0010
25,000	0.9922	0.0087	45,000	0.9985	0.0009
26,000	0.9930	0.0075	46,000	0.9986	0.0009
27,000	0.9937	0.0065	47,000	0.9987	0.0008
28,000	0.9943	0.0057	48,000	0.9988	0.0007
29,000	0.9948	0.0050	49,000	0.9988	0.0007

(From M. Q. Brewster, *Thermal Radiative Transfer and Properties*, John Wiley & Sons, New York, 1992. By permission of the publishers).

23.6   EMISSIVITY AND ABSORPTIVITY OF SOLID SURFACES

Whereas thermal conductivity, specific heat, density, and viscosity are the important physical properties of matter in heat conduction and convection, emissivity and absorptivity are the controlling properties in heat exchange by radiation.

From preceding sections it is seen that, for black-body radiation, $E_b = \sigma T^4$. For actual surfaces, $E = \epsilon E_b$, following the definition of emissivity. The emissivity of the surface, so defined, is a gross factor, as radiant energy is being sent out from a body not only in all directions but also over various wavelengths. For actual surfaces, the emissivity may vary with wavelength as well as the direction of emission. Consequently, we have to differentiate the monochromatic emissivity ϵ_λ and the directional emissivity ϵ_θ from the total emissivity ϵ.

Monochromatic Emissivity.   By definition, the monochromatic emissivity of an actual surface is the ratio of its monochromatic emissive power to that of a black surface at the same temperature. Figure 23.7 represents a typical distribution of the intensity of radiation of two

Figure 23.7 Emissivity at various wavelengths.

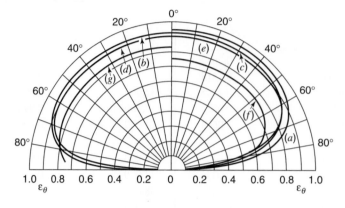

Figure 23.8 Emissivity variation with direction for nonconductors. (a) Wet ice. (b) Wood. (c) Glass. (d) Paper. (e) Clay. (f) Copper oxide. (g) Aluminum oxide.

such surfaces at the same temperature over various wavelengths. The monochromatic emissivity at a certain wavelength, λ_1, is seen to be the ratio of two ordinates such as $\overline{OQ}$ and $\overline{OP}$. That is

$$\epsilon_{\lambda_1} = \frac{\overline{OQ}}{\overline{OP}}$$

which is equal to the monochromatic absorptivity α_{λ_1} from radiation of a body at the same temperature. This is the direct consequence of Kirchhoff's law. The total emissivity of the surface is given by the ratio of the shaded area shown in Figure 23.7 to that under the curve for the black-body radiation.

Directional Emissivity. The cosine variation discussed previously, equation (23-5), is strictly applicable to radiation from a black surface but fulfilled only approximately by materials present in nature. This is due to the fact that the emissivity (averaged over all wavelengths) of actual surfaces is not a constant in all directions. The variation of emissivity of materials with the direction of emission can be conveniently represented by polar diagrams.

If the cosine law is fulfilled, the distribution curves should take the form of semicircles. Most nonconductors have much smaller emissivities for emission angles in the neighborhood of 90° (see Figure 23.8).

Deviation from the cosine law is even greater for many conductors (see Figure 23.9). The emissivity stays fairly constant in the neighborhood of the normal direction of emission; as the emission angle is increased, it first increases and then decreases as the former approaches 90°.

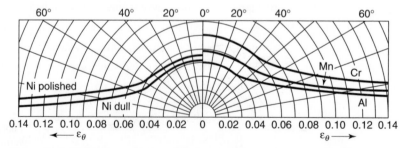

Figure 23.9 Emissivity variation with direction for conductors.

Table 23.2 The ratio ϵ/ϵ_n for bright metallic surfaces

Aluminum, bright rolled (443 K)	$\dfrac{0.049}{0.039} = 1.25$
Nickel, bright matte (373 K)	$\dfrac{0.046}{0.041} = 1.12$
Nickel, polished (373 K)	$\dfrac{0.053}{0.045} = 1.18$
Manganin, bright rolled (392 K)	$\dfrac{0.057}{0.048} = 1.19$
Chromium, polished (423 K)	$\dfrac{0.071}{0.058} = 1.22$
Iron, bright etched (423 K)	$\dfrac{0.158}{0.128} = 1.23$
Bismuth, bright (353 K)	$\dfrac{0.340}{0.336} = 1.08$

The average total emissivity may be determined by using the following expression:

$$\epsilon = \int_0^{\pi/2} \epsilon_\theta \sin 2\theta \, d\theta$$

The emissivity, ϵ, is, in general, different from the normal emissivity, ϵ_n (emissivity in the normal direction). It has been found that for most bright metallic surfaces, the total emissivity is approximately 20% higher than ϵ_n. Table 23.2 lists the ratio of ϵ/ϵ_n for a few representative bright metallic surfaces. For nonmetallic or other surfaces, the ratio ϵ/ϵ_n is slightly less than unity. Because of the inconsistency that can often be found among various sources, the normal emissivity values can be used, without appreciable error, for total emissivity (see Table 23.3).

Table 23.3 The ratio ϵ/ϵ_n for nonmetallic and other surfaces

Copper oxide (300°F)	0.96
Fire clay (183°F)	0.99
Paper (200°F)	0.97
Plywood (158°F)	0.97
Glass (200°F)	0.93
Ice (32°F)	0.95

A few generalizations may be made concerning the emissivity of surfaces:

(a) In general, emissivity depends on surface conditions.

(b) The emissivity of highly polished metallic surfaces is very low.

(c) The emissivity of all metallic surfaces increases with temperature.

(d) The formation of a thick oxide layer and roughening of the surface increase the emissivity appreciably.

(e) The ratio ϵ/ϵ_n is always greater than unity for bright metallic surfaces. The value 1.2 can be taken as a good average.

(f) The emissivities of nonmetallic surfaces are much higher than for metallic surfaces and show a decrease as temperature increases.

(g) The emissivities of colored oxides of heavy metals like Zn, Fe, and Cr are much larger than emissivities of white oxides of light metals like Ca, Mg, and Al.

Absorptivity.   The absorptivity of a surface depends on the factors affecting the emissivity and, in addition, on the quality of the incident radiation. It may be remarked once again that Kirchhoff's law holds strictly true under thermal equilibrium. That is, if a body at temperature T_1 is receiving radiation from a black body also at temperature T_1, then $\alpha = \epsilon$. For most materials, in the usual range of temperature encountered in practice (from room temperature up to about 1370 K) the simple relationship $\alpha = \epsilon$ holds with good accuracy. However, if the incident radiation is that from a very-high-temperature source, say solar radiation (~ 5800 K), the emissivity and absorptivity of ordinary surfaces may differ widely. White metal oxides usually exhibit an emissivity (and absorptivity) value of about 0.95 at ordinary temperature, but their absorptivity drops sharply to 0.15 if these oxides are exposed to solar radiation. Contrary to the above, freshly polished metallic surfaces have an emissivity value (and absorptivity under equilibrium conditions) of about 0.05. When exposed to solar radiation, their absorptivity increases to 0.2 or even 0.4.

Under these latter circumstances a double-subscript notation, $\alpha_{1,2}$, may be employed, the first subscript referring to the temperature of the receiving surface and the second subscript to the temperature of the incident radiation.

Gray surfaces.   Like emissivity, the monochromatic absorptivity, α_λ, of a surface may vary with wavelength. If α_λ is a constant and thus independent of λ, the surface is called *gray*. For a gray surface, the total average absorptivity will be independent of the spectral-energy distribution of the incident radiation. Consequently, the emissivity, ϵ, may be used in place of α, even though the temperatures of the incident radiation and the receiver are not the same. Good approximations of a gray surface are slate, tar board, and dark linoleum. Table 23.4 lists emissivities, at various temperatures, for several materials.

23.7   RADIANT HEAT TRANSFER BETWEEN BLACK BODIES

The exchange of energy between black bodies is dependent upon the temperature difference and the geometry with the geometry, in particular, playing a dominant role. Consider the two surfaces illustrated in Figure 23.10. The radiant energy emitted from a black surface at dA_1 and received at dA_2 is

$$dq_{1 \to 2} = I_{b_1} \cos \theta_1 \, d\Omega_{1-2} \, dA_1$$

Figure 23.10 Radiant heat transfer between two surfaces.

Table 23.4 Normal total emissivity of various surfaces (Compiled by H. C. Hottel)†

Surface	T, °F‡	Emissivity
A. Metals and their oxides		
Aluminum:		
Highly polished plate, 98.3% pure	440–1070	0.039–0.057
Commercial sheet	212	0.09
Oxidized at 1110°F	390–1110	0.11–0.19
Heavily oxidized	200–940	0.20–0.31
Brass:		
Polished	100–600	0.10
Oxidized by heating at 1110°F	390–1110	0.61–0.59
Chromium (see nickel alloys for Ni–Cr steels):		
Polished	100–2000	0.08–0.36
Copper		
Polished	212	0.052
Plate heated at 1110°F	390–1110	0.57
Cuprous oxide	1470–2010	0.66–0.54
Molten copper	1970–2330	0.16–0.13
Gold:		
Pure, highly polished	440–1160	0.018–0.035
Iron and steel (not including stainless):		
Metallic surfaces (or very thin oxide layer)		
Iron, polished	800–1880	0.14–0.38
Cast iron, polished	392	0.21
Wrought iron, highly polished	100–480	0.28
Oxidized surfaces		
Iron plate, completely rusted	67	0.69
Steel plate, rough	100–700	0.94–0.97
Molten surfaces		
Cast iron	2370–2550	0.29
Mild steel	2910–3270	0.28
Lead:		
Pure (99.96%), unoxidized	260–440	0.057–0.075
Gray oxidized	75	0.28
Nickel alloys:		
Chromnickel	125–1894	0.64–0.76
Copper–nickel, polished	212	0.059
Nichrome wire, bright	120–1830	0.65–0.79
Nichrome wire, oxidized	120–930	0.95–0.98
Platinum:		
Pure, polished plate	440–1160	0.054–0.104
Strip	1700–2960	0.12–0.17
Filament	80–2240	0.036–0.192
Wire	440–2510	0.073–0.182
Silver:		
Polished, pure	440–1160	0.020–0.032
Polished	100–700	0.022–0.031

(Continued)

Table 23.4 Normal total emissivity of various surfaces (Compiled by H. C. Hottel)†

Surface	T, °F‡	Emissivity
A. Metals and their oxides		
Stainless steels:		
Polished	212	0.074
Type 310 (25 Cr; 20 Ni)		
Brown, splotched, oxidized from furnace service	420–980	0.90–0.97
Tin:		
Bright tinned iron	76	0.043 and 0.064
Bright	122	0.06
Commercial tin-plated sheet iron	212	0.07, 0.08
Tungsten:		
Filament, aged	80–6000	0.032–0.35
Filament	6000	0.39
Polished coat	212	0.066
Zinc:		
Commercial 99.1% pure, polished	440–620	0.045–0.053
Oxidized by heating at 750°F	750	0.11
B. Refractories, building materials, paints, and miscellaneous		
Asbestos:		
Board	74	0.96
Paper	100–700	0.93–0.94
Brick		
Red, rough, but no gross irregularities	70	0.93
Brick, glazed	2012	0.75
Building	1832	0.45
Fireclay	1832	0.75
Carbon:		
Filament	1900–2560	0.526
Lampblack-waterglass coating	209–440	0.96–0.95
Thin layer of same on iron plate	69	0.927
Glass:		
Smooth	72	0.94
Pyrex, lead, and soda	500–1000	0.95–0.85
Gypsum, 0.02 in. thick on smooth or blackened plate	70	0.903
Magnesite refractory brick	1832	0.38
Marble, light gray, polished	72	0.93
Oak, planed	70	0.90
Paints, lacquers, varnishes:		
Snow-white enamel varnish on rough iron plate	73	0.906
Black shiny lacquer, sprayed on iron	76	0.875
Black shiny shellac on tinned iron sheet	70	0.821
Black matte shellac	170–295	0.91
Black or white lacquer	100–200	0.80–0.95
Flat black lacquer	100–200	0.96–0.98
Oil paints, 16 different, all colors	212	0.92–0.96
A1 paint, after heating to 620°F	300–600	0.35

Table 23.4   *(Continued)*

Surface	T, °F‡	Emissivity
B. Refractories, building materials, paints, and miscellaneous		
Plaster, rough lime	50–190	0.91
Roofing paper	69	0.91
Rubber:		
Hard, glossy plate	74	0.94
Soft, gray, rough (reclaimed)	76	0.86
Water	32–212	0.95–0.963

† By permission from W. H. McAdams (ed.), *Heat Transmission,* Third Edition, McGraw-Hill Book Company, 1954. Table of normal total emissivity compiled by H. C. Hottel.

‡ When temperatures and emissivities appear in pairs separated by dashes, they correspond and linear interpolation is permissible.

The reader should note that the units of temperatures in Table 23.4 are °F, in contrast to K, as has been used throughout the text thus far. Table 23.4 is presented as originally published by McAdams.

where $d\Omega_{1-2}$ is the solid angle subtended by dA_2 as seen from dA_1. Thus

$$d\Omega_{1-2} = \cos\theta_2 \frac{dA_2}{r^2}$$

and as $I_{b_1} = E_{b_1}/\pi$, the heat transfer from 1 to 2 is

$$dq_{1\rightarrow2} = E_{b_1}\, dA_1 \left\{ \frac{\cos\theta_1\,\cos\theta_2\,dA_2}{\pi r^2} \right\}$$

The bracketed term is seen to depend solely upon geometry. In exactly the same manner the energy emitted by dA_2 and captured by dA_1, may be determined. This is

$$dq_{2\rightarrow1} = E_{b_2}\, dA_2 \left\{ \frac{\cos\theta_2\,\cos\theta_1\,dA_1}{\pi r^2} \right\}$$

The net heat transfer between surfaces dA_1 and dA_2 is then simply

$$dq_{1-2\,\text{net}} = dq_{1\rightleftharpoons2} = dq_{1\rightarrow2} - dq_{2\rightarrow1}$$

or

$$dq_{1\rightleftharpoons2} = (E_{b_1} - E_{b_2}) \frac{\cos\theta_1\,\cos\theta_2\,dA_1\,dA_2}{\pi r^2}$$

Integrating over surfaces 1 and 2, we obtain

$$q_{1\rightleftharpoons2} = (E_{b_1} - E_{b_2}) \int_{A_1} \int_{A_2} \frac{\cos\theta_1\,\cos\theta_2\,dA_2\,dA_1}{\pi r^2}$$

the insertion of A_1/A_1 yields

$$q_{1\rightleftharpoons2} = (E_{b_1} - E_{b_2})A_1 \left[\frac{1}{A_1} \int_{A_1} \int_{A_2} \frac{\cos\theta_1\,\cos\theta_2\,dA_2\,dA_1}{\pi r^2} \right] \tag{23-13}$$

The bracketed term in the above equation is called the *view factor* F_{1-2}. If we had used A_2 as a reference, then the view factor would be F_{21}. Clearly, the net heat transfer is not affected by these operations, and thus $A_1 F_{12} = A_2 F_{21}$. This simple but extremely important expression is called the *reciprocity* relationship.

A physical interpretation of the view factor may be obtained from the following argument. As the total energy leaving surface A_1 is $E_b A_1$, the amount of heat that surface A_2 receives is $E_{b_1} A_1 F_{12}$. The amount of heat lost by surface A_2 is $E_{b_2} A_2$, whereas the amount that reaches A_1 is $E_{b_2} A_2 F_{21}$. The net rate of heat transfer between A_1 and A_2 is the difference or $E_{b_1} A_1 F_{12} - E_{b_2} A_2 F_{21}$. This may be arranged to yield $(E_{b_1} - E_{b_2}) A_1 F_{12}$. Thus, the view factor F_{12} can be interpreted as the fraction of black-body energy leaving A_1 which reaches A_2. Clearly the view factor cannot exceed unity.

Before some specific view factors are examined, there are several generalizations worthy of note concerning view factors.

1. The *reciprocity relation,* $A_1 F_{12} = A_2 F_{21}$, is always valid.

2. The view factor is independent of temperature. It is purely geometric.

3. For an enclosure, $F_{11} + F_{12} + F_{13} + \cdots = 1$.

In many cases the view factor may be determined without integration. An example of such a case follows.

EXAMPLE 1

Consider the view factor between a hemisphere and a plane as shown in the figure. Determine the view factors F_{11}, F_{12}, and F_{21}.

The view factor F_{21} is unity, as surface 2 sees only surface 1. For surface 1 we may write $F_{11} + F_{12} = 1$ and $A_1 F_{12} = A_2 F_{21}$. As $F_{21} = 1, A_2 = \pi r_0^2$, and $A_1 = 2\pi r_0^2$, the above relations give

Surface 1

r_0

Surface 2

$$F_{12} = F_{21} \frac{A_2}{A_1} = (1)\left(\frac{\pi r_0^2}{2\pi r_0^2}\right) = \frac{1}{2}$$

and

$$F_{11} = 1 - F_{12} = \frac{1}{2}$$

The view factor F_{12} can, in general, be determined by integration. As

$$F_{12} \equiv \frac{1}{A_1} \int_{A_1} \int_{A_2} \frac{\cos\theta_1 \cos\theta_2 \, dA_2 \, dA_1}{\pi r^2} \tag{23-14}$$

this integration process becomes quite tedious, and the view factor for a complex geometry is seen to require numerical methods. In order to illustrate the analytical evaluation of view factors, consider the view factor between the differential area dA_1 and the parallel plane A_2 shown in Figure 23.11. The view factor $F_{dA_1 A_2}$ is given by

$$F_{dA_1 A_2} = \frac{1}{dA_1} \int_{dA_1} \int_{A_2} \frac{\cos\theta_1 \cos\theta_2 \, dA_2 \, dA_1}{\pi r^2}$$

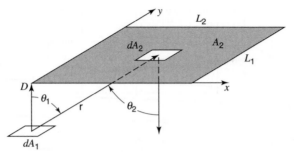

Figure 23.11 Differential area and parallel-finite area.

and as $A_2 \gg dA_1$ the view of dA_2 from dA_1 is independent of the position on dA_1, hence

$$F_{dA_1A_2} = \frac{1}{\pi} \int_{A_2} \frac{\cos \theta_1 \cos \theta_2}{r^2} dA_2$$

Also, it may be noted that $\theta_1 = \theta_2$ and $\cos \theta = D/r$, where $r^2 = D^2 + x^2 + y^2$. The resulting integral becomes

$$F_{dA_1A_2} = \frac{1}{\pi} \int_0^{L_1} \int_0^{L_2} \frac{D^2 \, dx \, dy}{(D^2 + x^2 + y^2)^2}$$

or

$$F_{dA_1A_2} = \frac{1}{2\pi} \left\{ \frac{L_1}{\sqrt{D^2 + L_1^2}} \tan^{-1} \frac{L_2}{\sqrt{D^2 + L_1^2}} + \frac{L_2}{\sqrt{D^2 + L_2^2}} \tan^{-1} \frac{L_1}{\sqrt{D^2 + L_2^2}} \right\} \quad (23\text{-}15)$$

The view factor given by equation (23-15) is shown graphically in Figure 23.12. Figures 23.13–23.15 also illustrate some view factors for simple geometries.

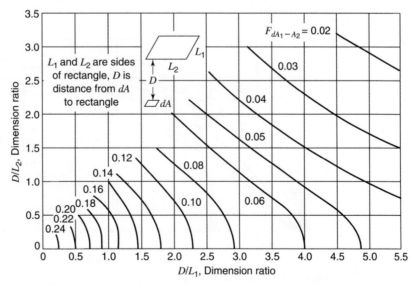

Figure 23.12 View factor for a surface element and a rectangular surface parallel to it. (From H. C. Hottel, "Radiant Heat Transmission," *Mech. Engrg.*, 52 (1930). By permission of the publishers.)

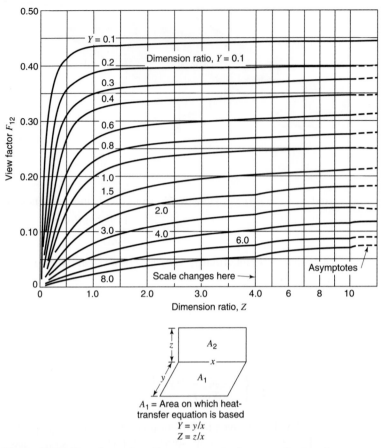

Figure 23.13 View factor for adjacent rectangles in perpendicular planes.
(From H. C. Hottel, "Radiant Heat Transmission," *Mech. Engrg.,* 52 (1930). By permission of the publishers.)

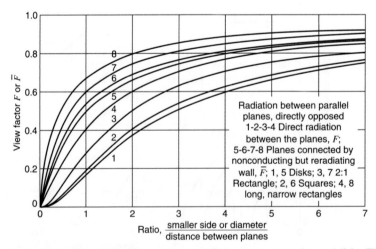

Figure 23.14 View factors for equal and parallel squares, rectangles, and disks. The curves labeled 5, 6, 7, and 8 allow for continuous variation in the sidewall temperatures from top to bottom.
(From H. C. Hottel, "Radiant Heat Transmission," *Mech. Engrg.,* 52 (1930). By permission of the publishers.)

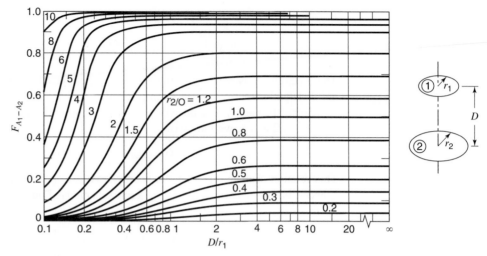

Figure 23.15 View factors for parallel opposed circular disks of unequal size.

EXAMPLE 2

Determine the view factor from a 1 m square to a parallel rectangular plane 10 m by 12 m centered 8 m above the 1 m square.

The smaller area may be considered a differential area, and Figure 23.12 may be used. The 10 m by 12 m area may be divided into four 5 m by 6 m rectangles directly over the smaller area. Thus, the total view factor is the sum of the view factors to each subdivided rectangle. Using $D = 8$, $L_1 = 6$, $L_2 = 5$, we find that the view factor from Figure 23.12 is 0.09. The total view factor is the sum of the view factors or 0.36.

View-Factor Algebra

View factors between combinations of differential- and finite-size areas have been expressed in equation form thus far. Some generalizations can be made that will be useful in evaluating radiant energy exchange in cases that, at first glance, seem quite difficult.

In an enclosure all energy leaving one surface, designated i, will be incident on the other surfaces that it can "see." If there are n surfaces in total, with j designating any surface that receives energy from i, we may write

$$\sum_{j=1}^{n} F_{ij} = 1 \tag{23-16}$$

A general form of the reciprocity relationship may be written as

$$A_i F_{ij} = A_j F_{ji} \tag{23-17}$$

these two expressions form the basis of a technique designated *view-factor algebra*.

A simplified notation will be introduced, using the symbol G_{ij}, defined as

$$G_{ij} \equiv A_i F_{ij}$$

Equations (23-16) and (23-17) may now be written as

$$\sum G_{ij} = A_i \qquad (23\text{-}18)$$
$$G_{ij} = G_{ji} \qquad (23\text{-}19)$$

The quantity G_{ij} is designated the *geometric flux*. Relations involving geometric fluxes are dictated by energy conservation principles.

Some special symbolism will now be explained. If surface 1 "sees" two surfaces, designated 2 and 3, we may write

$$G_{1-(2+3)} = G_{1-2} + G_{1-3} \qquad (23\text{-}20)$$

This relation says simply that the energy leaving surface 1 and striking both surfaces 2 and 3 is the total of that striking each separately. Equation (23-20) can be reduced further to

$$A_1 F_{1-(2+3)} = A_1 F_{12} + A_1 F_{13}$$

or

$$F_{1-(2+3)} = F_{12} + F_{13}$$

A second expression, involving four surfaces, is reduced to

$$G_{(1+2)-(3+4)} = G_{1-(3+4)} + G_{2-(3+4)}$$

which decomposes further to the form

$$G_{(1+2)-(3+4)} = G_{1-3} + G_{1-4} + G_{2-3} + G_{2-4}$$

Examples of how view-factor algebra can be used follow.

EXAMPLE 3

Determine the view factors, F_{1-2}, for the finite areas shown.

Inspection indicates that, in case (a), view factors F_{2-3} and $F_{2-(1+3)}$ can be read directly from Figure 23.13. The desired view factor, F_{1-2}, can be obtained using view-factor algebra in the following steps.

$$G_{2-(1+3)} = G_{2-1} + G_{2-3}$$

Thus,

$$G_{2-1} = G_{2-(1+3)} - G_{2-3}$$

Finally, by reciprocity, we may solve for F_{1-2} according to

$$G_{1-2} = G_{2-1} = G_{2-(1+3)} - G_{2-3}$$

$$A_1 F_{1-2} = A_2 F_{2-(1+3)} - A_2 F_{2-3}$$

$$F_{1-2} = \frac{A_2}{A_1} [F_{2-(1+3)} - F_{2-3}]$$

From Figure 23.13 we read

$$F_{2-(1+3)} = 0.15 \quad F_{2-3} = 0.10$$

Thus, for configuration (a), we obtain

$$F_{1-2} = \frac{5}{2}(0.15 - 0.10) = 0.125$$

Now, for case (b), the solution steps are

$$G_{1-2} = G_{1-(2+4)} - G_{1-4}$$

which may be written as

$$F_{1-2} = F_{1-(2+4)} - F_{1-4}$$

The result from part (a) can now be utilized to write

$$F_{1-(2+4)} = \frac{A_2 + A_4}{A_1} [F_{(2+4)-(1+3)} - F_{(2+4)-3}]$$

$$F_{1-4} = \frac{A_4}{A_1} [F_{4-(1+3)} - F_{4-3}]$$

Each of the view factors on the right side of these two expressions may be evaluated from Figure 23.13; the appropriate values are

$$F_{(2+4)-(1+3)} = 0.15 \quad F_{4-(1+3)} = 0.22$$

$$F_{(2+4)-3} = 0.10 \quad F_{4-3} = 0.165$$

Making these substitutions, we have

$$F_{1-(2+4)} = \frac{5}{2}(0.15 - 0.10) = 0.125$$

$$F_{1-4} = \frac{3}{2}(0.22 - 0.165) = 0.0825$$

The solution to case (b) now becomes

$$F_{1-2} = 0.125 - 0.0825 = 0.0425$$

23.8 RADIANT EXCHANGE IN BLACK ENCLOSURES

As pointed out earlier, a surface that views n other surfaces may be described according to

$$F_{11} + F_{12} + \cdots + F_{1i} + \cdots + F_{1n} = 1$$

or

$$\sum_{i=1}^{n} F_{1i} = 1 \tag{23-21}$$

Obviously, the inclusion of A_1 with equation (23-12) yields

$$\sum_{i=1}^{n} A_1 F_{1i} = A_1 \qquad (23\text{-}22)$$

Between any two black surfaces the radiant heat exchange rate is given by

$$q_{12} = A_1 F_{12}(E_{b_1} - E_{b_2}) = A_2 F_{21}(E_{b_1} - E_{b_2}) \qquad (23\text{-}23)$$

For surface 1 and any other surface, designated i, in a black enclosure the radiant exchange is given as

$$q_{1i} = A_1 F_{1i}(E_{b_1} - E_{b_i}) \qquad (23\text{-}24)$$

For an enclosure where surface 1 views n other surfaces, we may write for the net heat transfer with 1,

$$q_{1-\text{others}} = \sum_{i=1}^{n} q_{1i} = \sum_{i=1}^{n} A_1 F_{1i}(E_{b_1} - E_{b_i}) \qquad (23\text{-}25)$$

Equation (23-25) can be thought of as an analog to Ohm's law where the quantity of transfer, q; the potential driving force, $E_{b_1} - E_{b_i}$; and the thermal resistance, $1/A_1 F_{1i}$; have electrical counterparts I, ΔV, and R, respectively.

Figure 23.16 depicts the analogous electrical circuits for enclosures with three and four surfaces, respectively.

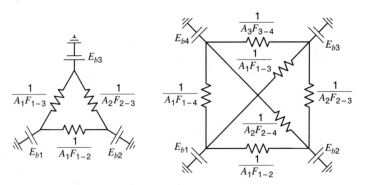

Figure 23.16 Radiation analogs.

The solution to a three-surface problem, that is, to find q_{12}, q_{13}, q_{23}, although somewhat tedious, can be accomplished in reasonable time. When analyzing enclosures with four or more surfaces, an analytical solution becomes impractical. In such situations one would resort to numerical methods.

23.9 RADIANT EXCHANGE WITH RERADIATING SURFACES PRESENT

The circuit diagrams shown in Figure 23.16 show a path to ground at each of the junctions. The thermal analog is a surface that has some external influence whereby its temperature is maintained at a certain level by the addition or rejection of energy. Such a surface is in contact with its surroundings and will conduct heat by virtue of an imposed temperature difference across it.

In radiation applications, we encounter surfaces that effectively are insulated from the surroundings. Such a surface will reemit all radiant energy that is absorbed—usually in a diffuse fashion. These surfaces thus act as reflectors and their temperatures "float" at some value that is required for the system to be in equilibrium. Figure 23.17 shows a physical situation and the corresponding electric analog for a three-surface enclosure with one being a nonabsorbing reradiating surface.

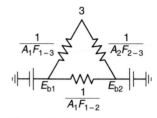

Figure 23.17

Evaluating the net heat transfer between the two black surfaces, q_{1-2}, we have

$$
\begin{aligned}
q_{12} &= \frac{E_{b_1} - E_{b_2}}{R_{\text{equiv}}} \\
&= \left[A_1 F_{12} + \frac{1}{1/A_1 F_{13} + 1/A_2 F_{23}} \right] (E_{b_1} - E_{b_2}) \\
&= A_1 \left[F_{12} + \frac{1}{1/F_{13} + A_1/A_2 F_{23}} \right] (E_{b_1} - E_{b_2}) \\
&= A_1 \overline{F}_{12} (E_{b_1} - E_{b_2})
\end{aligned}
\tag{23-26}
$$

The resulting expression, equation (23-62), contains a new term, $\overline{F}_{12}$, the *reradiating view factor*. This new factor, $\overline{F}_{12}$, is seen equivalent to the square-bracketed term in the previous expression, which includes direct exchange between surfaces 1 and 2, F_{12}, plus terms that account for the energy that is exchanged between these surfaces via the intervening reradiating surface. It is apparent that $\overline{F}_{12}$ will always be greater than F_{12}. Figure 23.14 allows reradiating view factors to be read directly for some simple geometries. In other situations where curves such as in this figure are not available, the electrical analog may be used by the simple modification that no path to ground exists at the reradiating surface.

23.10   RADIANT HEAT TRANSFER BETWEEN GRAY SURFACES

In the case of surfaces that are not black, determination of heat transfer becomes more involved. For gray bodies, that is, surfaces for which the absorptivity and emissivity are independent of wavelength, considerable simplifications can be made. The net heat transfer from the surface shown in Figure 23.18 is determined by the difference between the radiation leaving the surface and the radiation incident upon the surface. The *radiosity, J*, is defined as the rate at which radiation leaves a given surface per unit area. The *irradiation, G*, is defined as the rate at which radiation is incident on a surface per unit area. For a gray body, the radiosity, irradiation, and the total emissive power are related by

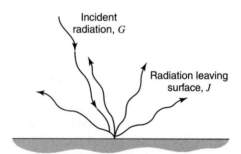

Figure 23.18 Heat transfer at a surface.

$$
J = \rho G + \epsilon E_b
\tag{23-27}
$$

where ρ is the reflectivity and ϵ is the emissivity. The net heat transfer from a surface is

$$\frac{q_{net}}{A} = J - G = \epsilon E_b + \rho G - G = \epsilon E_b - (1 - \rho)G \qquad (23\text{-}28)$$

In most cases it is useful to eliminate G from equation (23-28). This yields

$$\frac{q_{net}}{A} = \epsilon E_b - (1 - \rho)\frac{(J - \epsilon E_b)}{\rho}$$

As $\alpha + \rho = 1$ for an opaque surface

$$\frac{q_{net}}{A} = \frac{\epsilon E_b}{\rho} - \frac{\alpha J}{\rho} \qquad (23\text{-}29)$$

When the emissivity and absorptivity can be considered equal, an important simplification may be made in equation (23-29). Setting $\alpha = \epsilon$, we obtain

$$q_{net} = \frac{A\epsilon}{\rho}(E_b - J) \qquad (23\text{-}30)$$

which suggests an analogy with Ohm's law, $V = IR$, where the net heat leaving a surface can be thought of in terms of a current, the difference $E_b - J$ may be likened to a potential difference, and the quotient $\rho/\epsilon A$ may be termed a resistance. Figure 23.19 illustrates this analogy.

$q_{net} = \dfrac{(E_b - J)}{r/\epsilon A}$

Figure 23.19 Electrical analogy for radiation from a surface.

Now the net exchange of heat via radiation between two surfaces will depend upon their radiosities and their relative "views" of each other. From equation (23-17) we may write

$$q_{1 \rightleftharpoons 2} = A_1 F_{12}(J_1 - J_2) = A_2 F_{21}(J_1 - J_2)$$

We may now write the net heat exchange in terms of the different "resistances" offered by each part of the heat transfer path as follows:

Rate of heat leaving surface 1: $\qquad\qquad q = \dfrac{A_1 \epsilon_1}{\rho_1}(E_{b_1} - J_1)$

Rate of heat exchange between surfaces 1 and 2: $\qquad q = A_1 F_{12}(J_1 - J_2)$

Rate of heat received at surface 2: $\qquad\qquad q = \dfrac{A_2 \epsilon_2}{\rho_2}(J_2 - E_{b_2})$

If surfaces 1 and 2 view each other and no others then each of the qs in the previous equations is equivalent. In such a case an additional expression for q can be written in terms of the overall driving force, $E_{b_1} - E_{b_2}$. Such an expression is

$$q = \frac{E_{b_1} - E_{b_2}}{\rho_1/A_1\epsilon_1 + 1/A_1 F_{12} + \rho_2/A_2\epsilon_2} \qquad (23\text{-}31)$$

where the terms in the denominator are the equivalent resistances due to the characteristics of surface 1, geometry, and the characteristics of surface 2, respectively. The electrical analog to this equation is portrayed in Figure 23.20.

Figure 23.20 Equivalent network for gray-body relations between two surfaces.

The assumptions required to use the electrical analog approach to solve radiation problems are the following:

1. Each surface must be gray,

2. Each surface must be isothermal,

3. Kirchhoff's law must apply, that is, $\alpha = \epsilon$,

4. There is no heat absorbing medium between the participating surfaces.

Examples 4 and 5, which follow, illustrate features of gray-body problem solutions.

EXAMPLE 4

Two parallel gray surfaces maintained at temperatures T_1 and T_2 view each other. Each surface is sufficiently large that they may be considered infinite. Generate an expression for the net heat transfer between these surfaces.

A simple series electrical circuit is useful in solving this problem. The circuit and important quantities are shown here.

Utilizing Ohm's law, we obtain the expression

$$q_{12} = \frac{E_{b1} - E_{b2}}{\Sigma R} = \frac{\sigma\left(T_1^4 - T_2^4\right)}{\dfrac{\rho_1}{A_1 \epsilon_1} + \dfrac{1}{A_1 F_{12}} + \dfrac{\rho_2}{A_2 \epsilon_2}}$$

Now, noting that for infinite parallel planes $A_1 = A_2 = A$ and $F_{12} = F_{21} = 1$ and writing $\rho_1 = 1 - \epsilon_1$ and $\rho_2 = 1 - \epsilon_2$, we obtain the result

$$q_{12} = \frac{A\sigma\left(T_1^4 - T_2^4\right)}{\dfrac{1 - \epsilon_1}{\epsilon_1} + 1 + \dfrac{1 - \epsilon_2}{\epsilon_2}}$$

$$= \frac{A\sigma\left(T_1^4 - T_2^4\right)}{\dfrac{1}{\epsilon_1} + \dfrac{1}{\epsilon_2} - 1}$$

EXAMPLE 5

Two parallel planes measuring 2 m by 2 m are situated 2 m apart. Plate 1 is maintained at a temperature of 1100 K and plate 2 is maintained at 550 K. Determine the net heat transfer from the high temperature surface under the following conditions:

(a) the plates are black and the surroundings are at 0 K and totally absorbing;

(b) the plates are black and the walls connecting the plates are reradiating;

(c) the plates are gray with emissivities of 0.4 and 0.8, respectively, with black surroundings at 0 K.

Analog electrical circuits for parts (a), (b), and (c) are shown in Figure 23.21.

Figure 23.21 Equivalent circuits for example 3.

Heat flux evaluations will require evaluating the quantities F_{12}, F_{1R}, and $\overline{F}_{12}$. The appropriate values are

$$F_{12} = 0.20 \qquad \text{from Figure 23.14}$$

$$\overline{F}_{12} = 0.54 \qquad \text{from Figure 23.14}$$

and

$$F_{1R} = 1 - F_{12} = 0.80$$

Part (a). The net rate of heat leaving plate 1 is

$$q_{1\,net} = q_{12} + q_{1R}$$

$$= A_1 F_{12}(E_{b1} - E_{b2}) + A_1 F_{1R} E_{b1}$$

$$= (4\,m^2)(0.2)(5.676 \times 10^{-8}\,W/m^2 \cdot K^4)(1100^4 - 550^4)K^4$$

$$+ (4\,m^2)(0.8)(5.676 \times 10^{-8}\,W/m^2 \cdot K^4)(1100\,K)^4$$

$$= 62,300\,W + 266,000\,W$$

$$= 328.3\,kW$$

Part (b). When reradiating walls are present the heat flux becomes

$$q_{12} = (E_{b1} - E_{b2}) \left[A_1 F_{12} + \cfrac{1}{\cfrac{1}{A_1 F_{1R}} + \cfrac{1}{A_2 F_{2R}}} \right]$$

and, since $A_1 = A_2$ and $F_{1R} = F_{2R}$

$$q_{12} = (E_{b1} - E_{b2}) A_1 \left[F_{12} + \frac{F_{1R}}{2} \right]$$

Since $F_{12} + F_{1R} = 1$, the bracketed term is evaluated as

$$F_{12} + \frac{F_{1R}}{2} = 0.2 + \frac{0.8}{2} = 0.6$$

and, finally, the heat flux is

$$q_{12} = (4\,\text{m}^4)(5.678 \times 10^{-4}\,\text{W/m}^2 \cdot \text{K}^4)(1100^4 - 550^4)\text{K}^4(0.6)$$
$$= 187\,\text{kW}$$

We should note that an equivalent expression for the heat flux is

$$q_{12} = A_1 \overline{F}_{12}(E_{b1} - E_{b2})$$

and, using the value $\overline{F}_{12} = 0.54$, from Figure 23.14, the result would be

$$q_{12} = 168.3\,\text{kW}$$

This alternate result is the more accurate in that the values of $\overline{F}_{12}$ plotted in Figure 23.14 allow for the temperatures along the reradiating walls to vary from T_1 to T_2. The use of the analog circuit considers the radiating surface to be a constant temperature. Such an assumption, in this example, leads to an error of approximately 11%.

Part (c). An evaluation of the circuit shown in Figure 23.2(C) yields $q_{1,\text{out}} = 131.3\,\text{kW}$.

The concepts related to the quantities, radiosity, and irradiation are particularly useful in generalizing the analysis of radiant heat exchange in an enclosure containing any number of surfaces. The formalism to be developed in this section is directly applicable for solution by numerical methods.

For a representative surface having area, A_i, in an enclosure bounded by n surfaces, equations (20–28) and (20–30) can be written as

$$q_i = \frac{E_{bi} - J_i}{\rho_i / A_i \epsilon_i} = A_i(J_i - G_i) \tag{23-32}$$

where q_i is the net rate of heat transfer leaving surface i.

The irradiation, G_i, can be expressed as

$$A_i G_i = \sum_{j=1}^{n} J_j A_j F_{ji} \tag{23-33}$$

or, using reciprocity, as

$$A_i G_i = A_i \sum_{j=1}^{n} J_j F_{ij} \tag{23-34}$$

Combining equations (23-32) and (23-34) we obtain

$$q_i = A_i\left[J_i - \sum_{j=1}^{n} F_{ij}J_j\right] \tag{23-35}$$

$$= \frac{A_i\epsilon_i}{\rho_i} E_{bi} - \frac{A_i\epsilon_i}{\rho_i} J_i \tag{23-36}$$

We can now write the two basic expressions for a general surface in an enclosure.

If the surface heat flux is known, equation (23-35) can be expressed in the form

$$J_i - \sum_{j=1}^{n} F_{ij}J_j = \frac{q_i}{A_i}$$

or

$$J_i(1 - F_{ii}) - \sum_{\substack{j=1 \\ j \neq i}}^{n} F_{ij}J_j = \frac{q_i}{A_i} \tag{23-37}$$

and, if the temperature at surface i is known, equations (23-35) and (23-36) yield

$$\frac{A_i\epsilon_i}{\rho_i}(E_{bi} - J_i) = A_i\left[J_i - \sum_{j=1}^{n} F_{ij}J_j\right]$$

$$= A_i\left[J_i(1 - F_{ii}) - \sum_{\substack{j=1 \\ j \neq i}}^{n} F_{ij}J_j\right]$$

and, finally

$$\left(1 - F_{ii} + \frac{\epsilon_i}{\rho_i}\right)J_i - \sum_{\substack{j=1 \\ j \neq i}}^{n} F_{ij}J_j = \frac{\epsilon_i}{\rho_i} E_{bi} \tag{23-38}$$

Equations (23-37) and (23-38) comprise the algorithm for evaluating quantities of interest in a many-surface enclosure. The former applies to a surface of known heat flux; the latter is written when the surface temperature is specified.

In these two equations the terms involving the view factor, F_{ii}, have been separated out of the summation. This quantity, F_{ii} will have a nonzero value in those cases when surface i "sees"; itself, i.e., it is concave. In most cases F_{ii} will be 0.

When writing equation (23-37) or (23-38) for each surface in an enclosure a series of n simultaneous equations is generated, involving the unknowns J_i. This set of equations can be represented in matrix form as

$$[A][J] = [B] \tag{23-39}$$

where $[A]$ is the coefficient matrix, $[B]$ is a column matrix involving the right-hand sides of equations (23-37) and (23-38), and $[J]$ is a column matrix of the unknowns, J_i. The solution for the J_i then proceeds according to

$$[J] = [C][B] \tag{23-40}$$

where

$$[C] = [A]^{-1} \tag{23-41}$$

is the inverse of the coefficient matrix.

Example 6 illustrates the application of this approach.

EXAMPLE 6   Solve the problem posed in example 5 using the methods developed in this section.

For this case $n = 3$ and the problem formulation will involve 3 equations—one for each surface.

Part (a). Each of the surfaces is at a known temperature in this case, thus equation (23-38) applies. The following conditions are known:

$$
\begin{array}{lll}
T_1 = 1100\,\text{K} & T_2 = 550\,\text{K} & T_3 = 0\,\text{K} \\
F_{11} = 0 & F_{21} = 0.2 & F_{31} = 0.2 \\
F_{12} = 0.2 & F_{23} = 0 & F_{32} = 0.2 \\
F_{13} = 0.8 & F_{23} = 0.8 & F_{33} = 0.6 \\
\epsilon_1 = 1 & \epsilon_2 = 1 & \epsilon_3 = 1
\end{array}
$$

We can write the following:

$$
\left(1 + \frac{\epsilon_1}{\rho_1}\right)J_1 - [F_{12}J_2 + F_{13}J_3] = \frac{\epsilon_1}{\rho_1}E_{b1}
$$

$$
\left(1 + \frac{\epsilon_2}{\rho_2}\right)J_2 - [F_{21}J_1 + F_{23}J_3] = \frac{\epsilon_2}{\rho_2}E_{b2}
$$

$$
\left(1 - F_{33} + \frac{\epsilon_3}{\rho_3}\right)J_3 - [F_{31}J_1 + F_{32}J_2] = \frac{\epsilon_3}{\rho_3}E_{b3}
$$

which, for the given conditions, reduce to

$$
J_1 = E_{b1} = \sigma T_1^4
$$
$$
J_2 = E_{b2} = \sigma T_2^4
$$
$$
J_3 = 0
$$

The net heat leaving plate 1 is thus, according to equation (23-37), equal to

$$
q_1 = A_1[J_1 - F_{12}J_2]
$$

$$
= A_1[\sigma T_1^4 - 0.2\sigma T_2^4]
$$

$$
= 4\,\text{m}^2(5.676 \times 10^{-8}\,\text{W/m} \cdot \text{K}^4)[1100^4 - 0.2(550)^4]\text{K}^4
$$

$$
= 328.3\,\text{kW}
$$

Part (b). Values of T_i and F_{ij} remain the same. The only change from part (a) is that $\epsilon_3 = 0$. The set of equations, applying to the three surfaces are again

$$
\left(1 + \frac{\epsilon_1}{\rho_1}\right)J_1 - [F_{12}J_2 + F_{13}J_3] = \frac{\epsilon_1}{\rho_1}E_{b1}
$$

$$
\left(1 + \frac{\epsilon_2}{\rho_2}\right)J_2 - [F_{21}J_1 + F_{23}J_3] = \frac{\epsilon_2}{\rho_2}E_{b2}
$$

$$
\left(1 - F_{33} + \frac{\epsilon_3}{\rho_3}\right)J_3 - [F_{31}J_1 + F_{32}J_2] = \frac{\epsilon_3}{\rho_3}E_{b3}
$$

as before. Substituting values for T_i, F_{ij}, and ϵ_i, we have

$$
J_1 = E_{b1} = \sigma T_1^4
$$
$$
J_2 = E_{b2} = \sigma T_2^4
$$
$$
(1 - F_{33})J_3 - F_{31}J_1 - F_{32}J_2 = 0
$$

The expression for q_i is

$$q_1 = A_1[J_1 - F_{12}J_2 - F_{13}J_3] = A_1\left[J_1 - F_{12}J_2 - \frac{F_{13}}{1 - F_{33}}(F_{31}J_1 + F_{32}J_2)\right]$$

$$= A_1\left[J_1\left(1 - \frac{F_{13}F_{31}}{1 - F_{33}}\right) - J_2\left(F_{12} + \frac{F_{13}F_{32}}{1 - F_{33}}\right)\right]$$

and, with numerical values inserted, we obtain

$$q_1 = 4(5.676 \times 10^{-8})\left\{(1100)^4\left[1 - \frac{(0.8)(0.2)}{1 - 0.6}\right] - (550)^4\left[0.2 + \frac{(0.8)(0.2)}{1 - 0.6}\right]\right\} = 187.0\,\text{kW}$$

Part (c). Values of T_i and F_{ij} remain the same. Emissivities are

$$\epsilon_1 = 0.4 \quad \epsilon_2 = 0.8 \quad \epsilon_3 = 1$$

Equations for the three surfaces are, again

$$\left(1 + \frac{\epsilon_1}{\rho_1}\right)J_1 - [F_{12}J_2 + F_{13}J_3] = \frac{\epsilon_1}{\rho_1}E_{b1}$$

$$\left(1 + \frac{\epsilon_2}{\rho_2}\right)J_2 - [F_{21}J_1 + F_{23}J_3] = \frac{\epsilon_2}{\rho_2}E_{b2}$$

$$\left(1 - F_{33} + \frac{\epsilon_3}{\rho_3}\right)J_3 - [F_{31}J_1 + F_{32}J_2] = \frac{\epsilon_3}{\rho_3}E_{b3}$$

which become

$$\left(1 + \frac{0.4}{0.6}\right)J_1 - (F_{12}J_2 + F_{13}J_3) = \frac{0.4}{0.6}E_{b1}$$

$$\left(1 + \frac{0.8}{0.2}\right)J_2 - (F_{21}J_1 + F_{23}J_3) = \frac{0.8}{0.2}E_{b2}$$

$$J_3 = 0$$

We now have

$$1.67J_1 - 0.2J_2 = 0.67E_{b1}$$

$$5J_2 - 0.2J_1 = 4E_{b2}$$

Solving these two equations simultaneously for J_1 and J_2 we get

$$J_1 = 33\,900\,\text{W/m}^2$$
$$J_2 = 5510\,\text{W/m}^2$$

and the value for q_i is evaluated as

$$q_1 = \left[33\,900 - \frac{5510}{5}\right]4$$

$$= 131.2\,\text{kW}$$

23.11 RADIATION FROM GASES

So far, the interaction of radiation with gases has been neglected. Gases emit and absorb radiation in discrete energy bands dictated by the allowed energy states within the molecule. As the energy associated with, say, the vibrational or rotational motion of a molecule may have only certain values, it follows that the amount of energy emitted or absorbed by a molecule will have a

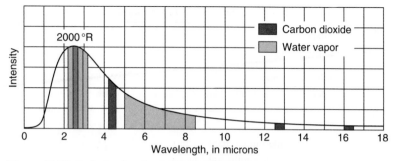

Figure 23.22 Emission bands of CO_2 and H_2O.

frequency, $\nu = \Delta E/h$, corresponding to the difference in energy ΔE between allowed states. Thus, while the energy emitted by a solid will comprise a continuous spectrum, the radiation emitted and absorbed by a gas will be restricted to bands. Figure 23.22 illustrates the emission bands of carbon dioxide and water vapor relative to black-body radiation at 1500°F.

The emission of radiation for these gases is seen to occur in the infrared region of the spectrum.

For nonluminous gases, the inert gases and diatomic gases of symmetrical composition such as O_2, N_2, and H_2 may be considered transparent to thermal radiation. Important types of media that absorb and emit radiations are polyatomic gases such as CO_2 and H_2O and unsymmetrical molecules such as CO. These gases are also associated with the products of combustion of hydrocarbons. The determination of the absorption and emission of radiation is very difficult, as it involves the temperature, composition, density, and geometry of the gas. There are several simplifications that allow estimation of radiation in gases to be made in a straightforward manner. These idealizations are as follows:

1. The gas is in thermodynamic equilibrium. The state of the gas may therefore be characterized locally by a single temperature.

2. The gas may be considered gray. This simplification allows the absorption and emission of radiation to be characterized by one parameter as $\alpha = \epsilon$ for a gray body.

In the range of temperatures associated with the products of hydrocarbon combustion, the gray gas emissivities of H_2O and CO_2 may be obtained from the results of Hottel. A hemispherical mass of gas at 1 atm pressure was used by Hottel to evaluate the emissivity. While the graphs apply strictly only to a hemispherical gas mass of radius L, other shapes can be treated by consideration of a mean beam length L as given in Table 23.5. For geometries not covered in the table, the mean beam length may be approximated by the relation $L = 3.4$ (volume)/(surface area).

Figure 23.23 gives the emissivity of a hemispherical mass of water vapor at 1 atm total pressure and near-zero partial pressure as a function of temperature and the product p_wL, where p_w is the partial pressure of the water vapor. For pressures other than atmospheric, Figure 23.24 gives the correction factor, C_w, which is the ratio of the emissivity at total pressure P to the emissivity at a total pressure of 1 atm. Figures 23.25 and 23.26 give the corresponding data for CO_2.

From Figure 23.22, it may be seen that the emission bands of CO_2 and H_2O overlap. When both carbon dioxide and water vapor are present, the total emissivity may be determined from the relation

$$\epsilon_{total} = \epsilon_{H_2O} + \epsilon_{CO_2} - \Delta\epsilon$$

where $\Delta\epsilon$ is given in Figure 23.27.

Table 23.5 Mean beam length, L, for various geometries[†]

Shape	L
Sphere	$\frac{2}{3} \times$ diameter
Infinite cylinder	$1 \times$ diameter
Space between infinite parallel planes	$1.8 \times$ distance between planes
Cube	$\frac{2}{3} \times$ side
Space outside infinite bank of tubes with centers on equilateral triangles; tube diameter equals clearance	$2.8 \times$ clearance
Same as preceding except tube diameter equals one-half clearance	$3.8 \times$ clearance

[†]From H. C. Hottel, "Radiation," Chap. IV in W. H. McAdams (ed.), *Heat Transmission,* Third Edition, McGraw-Hill Book Company, New York, 1964. By permission of the publishers.

Figure 23.23 Emissivity of water vapor at one atmosphere total pressure and near-zero partial pressure.

Figure 23.24 Correction factor for converting emissivity of H_2O at one atmosphere total pressure to emissivity at P atmospheres total pressure.

Figure 23.25 Emissivity of CO_2 at one atmosphere total pressure and near-zero partial pressure.

Figure 23.26 Correction factor for converting emissivity of CO_2 at one atmosphere total pressure to emissivity at P atmospheres total pressure.

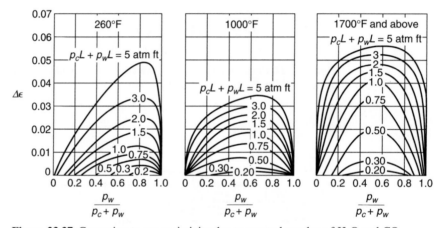

Figure 23.27 Correction to gas emissivity due to spectral overlap of H_2O and CO_2.

The results presented here for the gray gas are gross simplifications. For a more complete treatment, textbooks by Siegel and Howell,[2] Modest,[3] and Brewster[4] present the fundamentals of nongray-gas radiation, along with extensive bibliographies.

23.12 THE RADIATION HEAT-TRANSFER COEFFICIENT

Frequently in engineering analysis, convection and radiation occur simultaneously rather than as isolated phenomena. An important approximation in such cases is the linearization of the radiation contribution so that

$$h_{\text{total}} = h_{\text{convection}} + h_{\text{radiation}} \qquad (23\text{-}42)$$

[2] R. Siegel and J. R. Howell, *Thermal Radiation Heat Transfer,* 3rd Edition, Hemisphere Publishing Corp., Washington, 1992.

[3] M. F. Modest, *Radiative Heat Transfer,* McGraw-Hill, New York, 1993.

[4] M. Q. Brewster, *Thermal Radiative Transfer and Properties,* J. Wiley and Sons, New York, 1992.

where

$$h_r \equiv \frac{q_r/A_1}{(T - T_R)}$$

$$= \mathscr{F}_{1-2}\left[\frac{\sigma(T^4 - T_2^4)}{T - T_R}\right] \quad (23\text{-}43)$$

Here T_R is a reference temperature, and T_1 and T_2 are the respective surface temperatures. In effect, equation (23-43) represents a straight-line approximation to the radiant heat transfer as illustrated in Figure 23.28. The factor, $\mathscr{F}$, accounts for geometry and surface condition of the radiating and absorbing surface.

By constructing a tangent to the relation curve at $T = T_1$, the following relations are obtained for h_r and T_R:

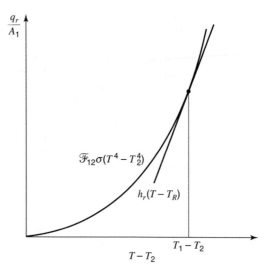

Figure 23.28 Tangent approximation for h_r.

$$h_r = 4\sigma T_1^3 \mathscr{F}_{1-2} \quad (23\text{-}44)$$

and

$$T_R = T_1 - \frac{T_1^4 - T_2^4}{4T_1^3} \quad (23\text{-}45)$$

23.13 CLOSURE

Radiation heat transfer has been considered in this chapter. Radiant energy transfer is associated with the portion of the electromagnetic spectrum between 0.1 and 100 μm, which is generally referred to as the thermal band.

The fundamental rate equation for thermal radiation, introduced in Chapter 15, is designated the Stefan–Boltzmann equation; it is expressed as

$$E_b = \sigma T^4 \quad (23\text{-}12)$$

where E_b is the black body emissive power, T is the absolute temperature, and σ is the Stefan–Boltzmann constant, having units of $W/m^2 \cdot K^4$ in the SI system.

Modifications to this relationship were made for nonblack surfaces and for geometric relationships between multiple surfaces in view of each other.

The presence of absorbing and emitting gases between surfaces was also examined. The gases of principle interest in this regard are water vapor and carbon dioxide.

PROBLEMS

23.1 The sun is approximately 93 million miles distant from Earth, and its diameter is 860,000 miles. On a clear day solar irradiation at Earth's surface has been measured at 360 Btu/h ft^2 and an additional 90 Btu/h ft^2 are absorbed by Earth's atmosphere. With this information, estimate the sun's effective surface temperature.

23.2 A greenhouse is constructed of silica glass that is known to transmit 92% of incident radiant energy between wavelengths of 0.35 and 2.7 μm. The glass may be considered opaque for wavelengths above and below these limits.

Considering the sun to emit as a black body at 5800 K, determine the percent of solar radiation that will pass through the glass.

If the plants on the inside of the greenhouse have an average temperature of 300 K, and emit as a black body, what fraction of their emitted energy will be transmitted through the glass?

23.3 A tungsten filament, radiating as a gray body, is heated to a temperature of 4000°R. At what wavelength is the emissive power maximum? What portion of the total emission lies within the visible-light range, 0.3 to 0.75 μm?

23.4 A radiation detector, oriented as shown in the sketch, is used to estimate heat loss through an opening in a furnace wall. The opening in this case is circular with a diameter of 2.5 cm. The detector has a surface area of 0.10 cm² and is located 1 m from the furnace opening. Determine the amount of radiant energy reaching the detector under two conditions:

a. the detector has a clear view of the opening;

b. the opening is covered by a semitransparent material with spectral transmissivity given by

$$\tau_\lambda = 0.8 \qquad \text{for } 0 \le \lambda \le 2\,\mu\text{m}$$
$$\tau_\lambda = 0 \qquad \text{for } 2\,\mu\text{m} < \lambda < \infty$$

23.5 The distribution of solar energy, incident on Earth, can be approximated as being from a black body at 5800 K.

Two kinds of glass, plain and tinted, are being considered for use in windows. The spectral transmissivity for these two glasses is approximated as

plain glass: $\tau_\lambda = 0$ for $0 < \lambda < 0.3\,\mu\text{m}$
              0.9 for $0.3 < \lambda < 2.5\,\mu$
              0 for $2.5\,\text{mm} < \lambda$

tinted glass: $\tau_\lambda = 0$ for $0 < \lambda < 0.5\,\mu\text{m}$
               0.9 for $0.5 < \lambda < 1.5\,\mu$
               0 for $1.5\,\text{mm} < \lambda$

Compare the fraction of incident solar energy transmitted through each material.

Compare the fraction of visible radiant energy transmitted through each.

23.6 Determine the fraction of total energy emitted by a black body, which lies in the wavelength band between 0.8 and 5.0 μm for surface temperatures of 500, 2000, 3000, and 4500 K.

23.7 The sun's temperature is approximately 5800 K and the visible light range is taken to be between 0.4 and 0.7 μm. What fraction of solar emission is visible? What fraction of solar emission lies in the ultraviolet range? The infrared range? At what wavelength is solar emissive power a maximum?

23.8 A satellite may be considered spherical with its surface properties roughly those of aluminum. Its orbit may be considered circular at a height of 500 miles above Earth. Taking the satellite diameter as 50 in., estimate the temperature of the satellite skin. Earth may be considered to be at a uniform temperature of 50°F, and the emissivity of Earth may be taken as 0.95. Solar irradiation may be taken as 450 Btu/h ft² of satellite disc area.

23.9 An opaque gray surface with $\epsilon = 0.3$ is irradiated with 1000 W/cm. For an effective convective heat-transfer coefficient of 12 W/m² · K applying, and air at 20°C adjacent to the plate, what will be the net heat flux to or from a 30°C surface?

23.10 A black solar collector, with a surface area of 60 m², is placed on the roof of a house. Incident solar energy reaches the collector with a flux of 800 W/m². The surroundings are considered black with an effective temperature of 30°C. The convective heat-transfer coefficient between the collector and the surrounding air, at 30°C, is 35 W/m² · K. Neglecting any conductive loss from the collector, determine:

a. the net radiant exchange between the collector and its surroundings;

b. the equilibrium temperature of the collector.

23.11 A 7.5-cm-diameter hole is drilled in a 10-cm-thick iron plate. If the plate temperature is 700 K and the surroundings are at 310 K, determine the energy loss through the hole. The hole sides may be considered to be black.

23.12 If the 7.5-cm-diameter hole in Problem 23.11 were drilled to a depth of 5 cm, what heat loss would result?

23.13 A sheet-metal box in the shape of a 0.70-m cube has a surface emissivity of 0.7. The box encloses electronic equipment that dissipates 1200 W of energy. If the surroundings are taken to be black at 280 K, and the top and sides of the box are considered to radiate uniformly, what will be the temperature of the box surface?

23.14 Two very large black plane surfaces are maintained at 900 and 580 K, respectively. A third large plane surface, having $\epsilon = 0.8$, is placed between these two. Determine the fractional change in radiant exchange between the two plane surfaces due to the intervening plane and evaluate the temperature of this intervening plane.

23.15 The filament of an ordinary 100 W light bulb is at 2910 K and it is presumed to be a black body. Determine (a) the wavelength of maximum emission and (b) the fraction of emission in the visible region of the spectrum.

23.16 A small circular hole is to be drilled in the surface of a large, hollow, spherical enclosure maintained at 2000 K. If 100 W of radiant energy exits through the hole, determine (a) the hole diameter, (b) the number of watts emitted in the visible range from 0.4 and 0.7 μm, (c) the ultraviolet range between 0 and 0.4 μm, and (d) the infrared range from 0.7 to 100 μm.

23.17 A large cavity with a small opening, 0.0025 m² in area, emits 8 W. Determine the wall temperature of the cavity.

23.18   Determine the wavelength of maximum emission for (a) the sun with an assumed temperature of 5790 K, (b) a light bulk filament at 2910 K, (c) a surface at 1550 K, and (d) human skin at 308 K.

23.19   A furnace that has black interior walls maintained at 1500 K contains a peephole with a diameter of 10 cm. The glass in the peephole has a transmissivity of 0.78 between 0 and 3.2 μm and 0.08 between 3.2 μm and ∞. Determine the heat lost through the peephole.

23.20   A cryogenic fluid flows in a 20-mm-diameter tube with an outer surface temperature of 75 K and an emissivity of 0.2. A larger tube, having a diameter of 50 mm, is concentric with the smaller one. This larger tube is gray, with $\epsilon = 0.05$ and its surface temperature is 300 K. The intervening space between the two tubes is evacuated.

Determine the heat gain by the cryogenic fluid, in watts per meter of tube length.

Evaluate the heat gain per meter of length if there is a thin walled radiation shield placed midway between the two tubes. The shield surfaces may be considered gray and diffuse with an emissivity of 0.04 on both sides.

23.21   A circular duct 2 ft long with a diameter of 3 in. has a thermocouple in its center with a surface area of 0.3 in.^{2}. The duct walls are at 200°F, and the thermocouple indicates 310°F. Assuming the convective heat-transfer coefficient between the thermocouple and gas in the duct to be 30 Btu/h ft^2 °F, estimate the actual temperature of the gas. The emissivity of the duct walls may be taken as 0.8 and that of the thermocouple as 0.6.

23.22   A heating element in the shape of a cylinder is maintained at 2000°F and placed at the center of a half-cylindrical reflector as shown. The rod diameter is 2 in. and that of the reflector is 18 in. The emissivity of the heater surface is 0.8, and the entire assembly is placed in a room maintained at 70°F. What is the radiant energy loss from the heater per foot of length? How does this compare to the loss from the heater without the reflector present?

23.23   A 12-ft-long, 3-in.-OD iron pipe $\epsilon = 0.7$, passes horizontally through a $12 \times 14 \times 9$ ft room whose walls are maintained at 70°F and have an emissivity of 0.8. The pipe surface is at a temperature of 205°F. Compare the radiant energy loss from the pipe with that due to convection to the surrounding air at 70°F.

23.24   The circular base of the cylindrical enclosure shown may be considered a reradiating surface. The cylindrical walls have an effective emissivity of 0.80 and are maintained at 540°F. The top of the enclosure is open to the surroundings, which are maintained at 40°F. What is the net rate of radiant transfer to the surroundings?

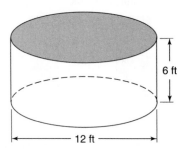

23.25   The hemispherical cavity shown in the figure has an inside surface temperature of 700 K. A plate of refractory material is placed over the cavity with a circular hole of 5 cm diameter in the center. How much energy will be lost through the hole if the cavity is

a.   black?

b.   gray with an emissivity of 0.7?

What will be the temperature of the refractory under each condition?

23.26   A room measuring 12 ft by 20 ft by 8 ft high has its floor and ceiling temperatures maintained at 85 and 65°F, respectively. Assuming the walls to be reradiating and all surfaces to have an emissivity of 0.8, determine the net energy exchange between the floor and ceiling.

23.27   A dewar flask, used to contain liquid nitrogen, is made of two concentric spheres separated by an evacuated space. The inner sphere has an outside diameter of 1 m and the outer sphere has an inside diameter of 1.3 m. These surfaces are both gray-diffuse with $\epsilon = 0.2$. Nitrogen, at 1 atmosphere, has a saturation temperature of 78 K and a latent heat of vaporization of 200 kJ/kg.

Under conditions when the inner sphere is full of liquid nitrogen and the outer sphere is at a temperature of 300 K, estimate the boil-off rate of nitrogen.

23.28   A cylindrical cavity is closed at the bottom and has an opening centered in the top surface. A cross section of this configuration is shown in the sketch. For the conditions stated below, determine the rate of radiant energy passing through the

5-mm-diameter cavity opening. What will be effective emissivity of the opening?

a. All interior surfaces are black at 600 K.

b. The bottom surface is diffuse-gray with $\epsilon = 0.6$, and has a temperature of 600 K. All other surfaces are reradiating.

c. All interior surfaces are diffuse-gray with $\epsilon = 0.6$ and are at a uniform temperature of 600 K.

23.29 A circular heater, measuring 20 cm in diameter, has its surface temperature maintained at 1000°C. The bottom of a tank, having the same diameter, is oriented parallel to the heater with a separation distance of 10 cm. The heater surface is gray ($\epsilon = 0.6$) and the tank surface is also gray ($\epsilon = 0.7$).

Determine the radiant energy reaching the bottom of the tank if

a. the surroundings are black at 27°C;

b. the space between the two cylindrical surfaces is enclosed by an adiabatic surface.

23.30 Two parallel black rectangular surfaces, whose back sides are insulated, are oriented parallel to each other with a spacing of 5 m. They measure 5 m by 10 m. The surroundings are black at 0 K. The two surfaces are maintained at 200 and 100 K, respectively. Determine the following:

a. the net radiant heat transfer between the two surfaces;

b. the net heat supplied to each surface;

c. the net heat transfer between each surface and the surroundings.

23.31 Two parallel rectangles have emissivities of 0.6 and 0.9, respectively. These rectangles are 1.2 m wide and 2.4 m high and are 0.6 m apart. The plate having $\epsilon = 0.6$ is maintained at 1000 K and the other is at 420 K. The surroundings may be considered to absorb all energy that escapes the two-plate system. Determine

a. the total energy lost from the hot plate;

b. the radiant energy interchange between the two plates.

23.32 If a third rectangular plate, with both surfaces having an emissivity of 0.8 is placed between the two plates described in Problem 23.31, how will the answer to part (a) of Problem 23.31 be affected? Draw the thermal circuit for this case.

23.33 Two disks are oriented on parallel planes separated by a distance of 10 in., as shown in the accompanying figure. The disk to the right is 4 in. in diameter and is at a temperature of 500°F. The disk to the left has an inner ring cut out such that it is annular in shape with inner and outer diameters of 2.5 and 4 in.,

respectively. The disk surface temperature is 210°F. Find the heat exchange between these disks if

a. they are black;

b. they are gray $\epsilon_1 = 0.6$, $\epsilon_2 = 0.3$.

23.34 Evaluate the net heat transfer between the disks described in Problem 23.33 if they are bases of a cylinder with the side wall considered a nonconducting, reradiating surface. How much energy will be lost through the hole?

23.35 Evaluate the heat transfer leaving disk 1 for the geometry shown in Problem 23.33. In this case the two disks comprise the bases of a cylinder with side wall at constant temperature of 350°F. Evaluate for the case where:

a. the side wall is black;

b. the side wall is gray with $\epsilon = 0.2$.

Determine the rate of heat loss through the hole in each case.

23.36 A heavily oxidized aluminum surface at 755 K is the source of energy in an enclosure, which radiantly heats the side walls of a circular cylinder surface as shown, to 395 K. The side wall is made of polished stainless steel. The top of the enclosure is made of fire clay brick and is adiabatic. For purposes of calculation, assume that all three surfaces have uniform temperatures and that they are diffuse and gray. Evaluate the heat transfer to the stainless steel surface.

23.37 A gray, diffuse circular heater with a diameter of 15 cm is placed parallel to a second gray, diffuse receiver with a spacing of 7.5 between them. The backs of both surfaces are insulated and convective effects are to be neglected. This heater–receiver assembly is placed in a large room at a temperature of 275 K. The surroundings (the room) can be considered black and the heater surface emissivity is 0.8. When the power input to the heater is 300 W, determine

a. the heater surface temperature;

b. the receiver surface temperature;

c. the net radiant exchange to the surroundings;

d. the net radiant exchange between the heater and receiver.

23.38   A small (1/4 in. diameter × 1 in. long) metal test specimen is suspended by very fine wires in a large evacuated tube. The metal is maintained at a temperature of 2500°F, at which temperature it has an emissivity of approximately 0.2. The water-cooled walls and ends of the tube are maintained at 50°F. In the upper end is a small (1/4-in.-diameter) silica glass viewing port. The inside surfaces of the steel tube are newly galvanized. Room temperature is 70°F. Estimate

a. the view factor from the specimen to the window;

b. the total net heat-transfer rate by radiation from the test specimen;

c. the energy radiated through the viewing port.

23.39   A duct with square cross section measuring 20 cm by 20 cm has water vapor at 1 atmosphere and 600 K flowing through it. One wall of the duct is held at 420 K and has an emissivity of 0.8. The other three walls may be considered refractory surfaces. Determine the rate of radiant energy transfer to the cold wall from the water vapor.

23.40   A gas of mixture at 1000 K and a pressure of 5 atm is introduced into an evacuated spherical cavity with a diameter of 3 m. The cavity walls are black and initially at a temperature of 600 K. What initial rate of heat transfer will occur between the gas and spherical walls if the gas contains 15% CO_2 with the remainder of the gas being nonradiating?

23.41   A gas consisting of 20% CO_2 and 80% oxygen and nitrogen leaves a lime kiln at 2000°F and enters a square duct measuring 6 in. by 6 in. in cross section. The specific heat of the gas is 0.28 Btu/lb$_m$ °F, and it is to be cooled to 1000°F in the duct, whose inside surface is maintained at 800°F, and whose walls have an emissivity of 0.9. The mass velocity of the kiln gas is 0.4 lb$_m$/ft^2·s and the convective heat-transfer coefficient between the gas and duct walls is 1.5 Btu/h ft^2 °F.

a. Determine the required length of duct to cool the gas to 1000°F.

b. Determine the ratio of radiant energy transfer to that by convection.

c. At what temperature would the gas leave the duct if the length of the duct were twice the value determined in part (a)?

(Courtesy of the American Institute of Chemical Engineers.)

Hint. As the response of the gas to emission and absorption of radiant energy differs, an approximation for the radiant energy exchange between the enclosure and gas contained within an arbitrary control volume is given by $A_w F_{w-g} \sigma \epsilon_w (\epsilon_g T_g^4 - \alpha_g T_w^4)$.

Chapter 24

Fundamentals of Mass Transfer

The previous chapters dealing with the transport phenomena of momentum and heat transfer have dealt with one-component phases that possessed a natural tendency to reach equilibrium conditions. When a system contains two or more components whose concentrations vary from point to point, there is a natural tendency for mass to be transferred, minimizing the concentration differences within the system. The transport of one constituent from a region of higher concentration to that of a lower concentration is called *mass transfer.*

Many of our day-to-day experiences involve mass transfer. A lump of sugar added to a cup of black coffee eventually dissolves and then diffuses uniformly throughout the coffee. Water evaporates from ponds to increase the humidity of the passing air stream. Perfume presents a pleasant fragrance that is imparted throughout the surrounding atmosphere.

Mass transfer is the basis for many biological and chemical processes. Biological processes include the oxygenation of blood and the transport of ions across membranes within the kidney. Chemical processes include the chemical vapor deposition (CVD) of silane (SiH_4) onto a silicon wafer, the doping of a silicon wafer to form a semiconducting thin film, the aeration of wastewater, and the purification of ores and isotopes. Mass transfer underlies the various chemical separation processes where one or more components migrate from one phase to the interface between the two phases in contact. For example, in adsorption or crystallization processes, the components remain at the interface, whereas in gas absorption and liquid–liquid extraction processes, the components penetrate the interface and then transfer into the bulk of the second phase.

If we consider the lump of sugar added to the cup of black coffee, experience teaches us that the length of time required to distribute the sugar will depend upon whether the liquid is quiescent or whether it is mechanically agitated by a spoon. The mechanism of mass transfer, as we have also observed in heat transfer, depends upon the dynamics of the system in which it occurs. Mass can be transferred by random molecular motion in quiescent fluids, or it can be transferred from a surface into a moving fluid, aided by the dynamic characteristics of the flow. These two distinct modes of transport, molecular mass transfer and convective mass transfer, are analogous to conduction heat transfer and convective heat transfer. Each of these modes of mass transfer will be described and analyzed. As in the case of heat transfer, we should immediately realize that the two mechanisms often act simultaneously. However, in the confluence of the two modes of mass transfer, one mechanism can dominate quantitatively so that approximate solutions involving only the dominant mode need be used.

24.1 MOLECULAR MASS TRANSFER

As early as 1815, Parrot observed qualitatively that whenever a gas mixture contains two or more molecular species, whose relative concentrations vary from point to point, an apparently natural process results, which tends to diminish any inequalities of composition. This macroscopic transport of mass, independent of any convection within the system, is defined as *molecular diffusion.*

In the specific case of gaseous mixtures, a logical explanation of this transport phenomenon can be deduced from the kinetic theory of gases. At temperatures above absolute zero, individual molecules are in a state of continual yet random motion. Within dilute gas mixtures, each solute molecule behaves independently of the other solute molecules, as it seldom encounters them. Collisions between the solute and the solvent molecules are continually occurring. As a result of the collisions, the solute molecules move along a zigzag path, sometimes toward a region of higher concentration, sometimes toward a lower concentration.

Let us consider a hypothetical section passing normal to the concentration gradient within an isothermal, isobaric gaseous mixture containing solute and solvent molecules. The two thin, equal elements of volume above and below the section will contain the same number of molecules, as stipulated by Avogadro's law. Although it is not possible to state which way any particular molecule will travel in a given interval of time, a definite number of the molecules in the lower element of the volume will cross the hypothetical section from below, and the same number of molecules will leave the upper element and cross the section from above. With the existence of a concentration gradient, there are more solute molecules in one of the elements of volume than in the other; accordingly, an overall net transfer from a region of higher concentration to one of lower concentration will result. The net flow of each molecular species occurs in the direction of a negative concentration gradient.

As pointed out in Chapters 7 and 15, the molecular transport of momentum and the transport of energy by conduction are also due to random molecular motion. Accordingly, one should expect that the three transport phenomena will depend upon many of the same characteristic properties, such as mean free path, and that the theoretical analyses of all three phenomena will have much in common.

The Fick Rate Equation

The laws of mass transfer show the relation between the flux of the diffusing substance and the concentration gradient responsible for this mass transfer. Unfortunately, the quantitative description of molecular diffusion is considerably more complex than the analogous descriptions for the molecular transfer of momentum and energy that occur in a one-component phase. As *mass transfer,* or *diffusion,* as it is also called, occurs only in mixtures, its evaluation must involve an examination of the effect of each component. For example, we will often desire to know the diffusion rate of a specific component relative to the velocity of the mixture in which it is moving. As each component may possess a different mobility, the mixture velocity must be evaluated by averaging the velocities of all of the components present.

In order to establish a common basis for future discussions, let us first consider definitions and relations that are often used to explain the role of components within a mixture.

Molecule of species A

Figure 24.1 Elemental volume containing a multicomponent mixture.

Concentrations. In a multicomponent mixture, the concentration of a molecular species can be expressed in many ways. Figure 24.1 shows an elemental volume dV that contains a mixture of components, including species A. As each molecule of each species has a mass, a *mass concentration* for each species, as well as for the mixture, can be defined. For species A, *mass* concentration, ρ_A, is defined as the mass of A per unit volume of the mixture. The total mass concentration or *density,* ρ, is the total mass of the mixture contained in the unit volume; that is,

$$\rho = \sum_{i=1}^{n} \rho_i \tag{24-1}$$

where n is the number of species in the mixture. The *mass fraction,* ω_A, is the mass concentration of species A divided by the total mass density

$$\omega_A = \frac{\rho_A}{\sum_{i}^{n} \rho_i} = \frac{\rho_A}{\rho} \tag{24-2}$$

The sum of the mass fractions, by definition, must be 1:

$$\sum_{i=1}^{n} \omega_i = 1 \tag{24-3}$$

The *molecular concentration* of species A, c_A, is defined as the number of moles of A present per unit volume of the mixture. By definition, one mole of any species contains a mass equivalent to its molecular weight; the mass concentration and molar concentration terms are related by the following relation:

$$c_A = \frac{\rho_A}{M_A} \tag{24-4}$$

where M_A is the molecular weight of species A. When dealing with a gas phase, concentrations are often expressed in terms of partial pressures. Under conditions in which the ideal gas law, $p_A V = n_A RT$, applies, the molar concentration is

$$c_A = \frac{n_A}{V} = \frac{p_A}{RT} \tag{24-5}$$

where p_A is the partial pressure of the species A in the mixture, n_A is the number of moles of species A, V is the gas volume, T is the absolute temperature, and R is the gas constant. The total molar concentration, c, is the total moles of the mixture contained in the unit volume; that is,

$$c = \sum_{i=1}^{n} c_i \tag{24-6}$$

or for a gaseous mixture that obeys the ideal gas law, $c = n_{total}/V = P/RT$, where P is the total pressure. The mole fraction for liquid or solid mixtures, x_A, and for gaseous mixtures, y_A, are the molar concentrations of species A divided by the total molar density

$$x_A = \frac{c_A}{c} \text{ (liquids and solids)}$$
$$y_A = \frac{c_A}{c} \text{ (gases)} \tag{24-7}$$

For a gaseous mixture that obeys the ideal gas law, the mole fraction, y_A, can be written in terms of pressures

$$y_A = \frac{c_A}{c} = \frac{p_A/RT}{P/RT} = \frac{p_A}{P} \tag{24-8}$$

Equation (24-8) is an algebraic representation of Dalton's law for gas mixtures. The sum of the mole fractions, by definition, must be 1:

$$\sum_{i=1}^{n} x_i = 1$$
$$\sum_{i=1}^{n} y_i = 1$$
(24-9)

A summary of the various concentration terms and of the interrelations for a binary system containing species A and B is given in Table 24.1.

Table 24.1   Concentrations in a binary mixture of A and B

Mass concentrations

ρ = total mass density of the mixture
ρ_A = mass density of species A
ρ_B = mass density of species B
ω_A = mass fraction of species A = ρ_A/ρ
ω_B = mass fraction of species B = ρ_B/ρ
$\rho = \rho_A + \rho_B$
$1 = \omega_A + \omega_B$

Molar concentrations

Liquid or solid mixture

c = molar density of mixture = n/V
c_A = molar density of species $A = n_A/V$
c_B = molar density of species $B = n_B/V$
x_A = mole fraction of species $A = c_A/c = n_A/n$
x_B = mole fraction of species B = $c_B/c = n_B/n$
$c = c_A + c_B$

$1 = x_A + x_B$

Gas mixture

$c = n/V = P/RT$
$c_A = n_A/V = p_A/RT$
$c_B = n_B/V = p_B/RT$
$y_A = c_A/c = n_A/n = p_A/p$
$y_B = c_B/c = n_B/n = p_B/p$

$c = c_A + c_B = \dfrac{p_A}{RT} + \dfrac{p_B}{RT} = \dfrac{P}{RT}$

$1 = y_A + y_B$

Interrelations

$\rho_A = c_A M_A$

x_A or $y_A = \dfrac{\omega_A/M_A}{\omega_A/M_A + \omega_B/M_B}$ (24-10)

$\omega_A = \dfrac{x_A M_A}{x_A M_A + x_A M_A}$ or $\dfrac{y_A M_A}{y_A M_A + y_B M_B}$ (24-11)

EXAMPLE 1

The composition of air is often given in terms of only the two principal species in the gas mixture

$$\text{oxygen,} \quad O_2, \quad y_{O_2} = 0.21$$
$$\text{nitrogen,} \quad N_2, \quad y_{N_2} = 0.79$$

Determine the mass fraction of both oxygen and nitrogen and the mean molecular weight of the air when it is maintained at 25°C (298 K) and 1 atm (1.013×10^5 Pa). The molecular weight of oxygen is 0.032 kg/mol and of nitrogen is 0.028 kg/mol.

As a basis for our calculations, consider 1 mol of the gas mixture

$$\text{oxygen present} = (1\,\text{mol})(0.21) = 0.21\,\text{mol}$$

$$= (0.21\,\text{mol})\frac{(0.032\,\text{kg})}{\text{mol}} = 0.00672\,\text{kg}$$

$$\text{nitrogen present} = (1\,\text{mol})(0.79) = 0.79\,\text{mol}$$

$$= (0.79\,\text{mol})\frac{(0.028\,\text{kg})}{\text{mol}} = 0.0221\,\text{kg}$$

$$\text{total mass present} = 0.00672 + 0.0221 = 0.0288\,\text{kg}$$

$$\omega_{O_2} = \frac{0.00672\,\text{kg}}{0.0288\,\text{kg}} = 0.23$$

$$\omega_{N_2} = \frac{0.0221\,\text{kg}}{0.0288\,\text{kg}} = 0.77$$

As 1 mol of the gas mixture has a mass of 0.0288 kg, the mean molecular weight of the air must be 0.0288. When one takes into account the other constituents that are present in air, the mean molecular weight of air is often rounded off to 0.029 kg/mol.

This problem could also be solved using the ideal gas law, $PV = nRT$. At ideal conditions, 0°C or 273 K and 1 atm of 1.013×10^5 Pa pressure, the gas constant is evaluated to be

$$R = \frac{PV}{nT} = \frac{(1.013 \times 10^5\,\text{Pa})(22.4\,\text{m}^3)}{(1\,\text{kg mol})(273\,\text{K})} = 8.314\frac{\text{Pa} \cdot \text{m}^3}{\text{mol} \cdot \text{K}} \tag{24-12}$$

The volume of the gas mixture, at 298 K, is

$$V = \frac{nRT}{P} = \frac{(1\,\text{mol})\left(8.314\dfrac{\text{Pa} \cdot \text{m}^3}{\text{mol} \cdot \text{K}}\right)(298\,\text{K})}{1.013 \times 10^5\,\text{Pa}}$$

$$= 0.0245\,\text{m}^3$$

The concentrations are

$$c_{O_2} = \frac{0.21\,\text{mol}}{0.0245\,\text{m}^3} = 8.57\frac{\text{mol}\,O_2}{\text{m}^3}$$

$$c_{N_2} = \frac{0.79\,\text{mol}}{0.0245\,\text{m}^3} = 32.3\frac{\text{mol}\,N_2}{\text{m}^3}$$

$$c = \sum_{i=1}^{n} c_i = 8.57 + 32.3 = 40.9\,\text{mol/m}^3$$

The total density, ρ, is

$$\rho = \frac{0.0288\,\text{kg}}{0.0245\,\text{m}^3} = 1.180\,\text{kg/m}^3$$

and the mean molecular weight of the mixture is

$$M = \frac{\rho}{c} = \frac{1.180\,\text{kg/m}^3}{40.9\,\text{mol/m}^3} = 0.0288\,\text{kg/mol}$$

Velocities. In a multicomponent system the various species will normally move at different velocities; accordingly, an evaluation of a velocity for the gas mixture requires the averaging of the velocities of each species present.

The *mass-average velocity* for a multicomponent mixture is defined in terms of the mass densities and velocities of all components by

$$\mathbf{v} = \frac{\sum\limits_{i=1}^{n} \rho_i \mathbf{v}_i}{\sum\limits_{i=1}^{n} \rho_i} = \frac{\sum\limits_{i=1}^{n} \rho_i \mathbf{v}_i}{\rho} \qquad (24\text{-}13)$$

where $\mathbf{v}_i$ denotes the absolute velocity of species i relative to stationary coordinate axes. This is the velocity that would be measured by a pitot tube and is the velocity that was previously encountered in the equations of momentum transfer. The *molar-average velocity* for a multicomponent mixture is defined in terms of the molar concentrations of all components by

$$\mathbf{V} = \frac{\sum\limits_{i=1}^{n} c_i \mathbf{v}_i}{\sum\limits_{i=1}^{n} c_i} = \frac{\sum\limits_{i=1}^{n} c_i \mathbf{v}_i}{c} \qquad (24\text{-}14)$$

The velocity of a particular species relative to the mass-average or molar-average velocity is termed a *diffusion velocity*. We can define two different diffusion velocities

$\mathbf{v}_i - \mathbf{v}$, the diffusion velocity of species i relative to the mass-average velocity

and

$\mathbf{v}_i - \mathbf{V}$, the diffusion velocity of species i relative to the molar-velocity average

According to Fick's law, a species can have a velocity relative to the mass- or molar-average velocity only if gradients in the concentration exist.

Fluxes. The mass (or molar) flux of a given species is a vector quantity denoting the amount of the particular species, in either mass or molar units, that passes per given increment of time through a unit area normal to the vector. The flux may be defined with reference to coordinates that are fixed in space, coordinates that are moving with the mass-average velocity, or coordinates that are moving with the molar-average velocity.

The basic relation for molecular diffusion defines the molar flux relative to the molar-average velocity, $\mathbf{J}_A$. An empirical relation for this molar flux, first postulated by Fick[1] and, accordingly, often referred to as Fick's first law, defines the diffusion of component A in an isothermal, isobaric system:

$$\mathbf{J}_A = -D_{AB} \nabla c_A$$

For diffusion in only the z direction, the Fick rate equation is

$$J_{A,z} = -D_{AB} \frac{dc_A}{dz} \qquad (24\text{-}15)$$

where $J_{A,z}$ is the molar flux in the z direction relative to the molar-average velocity, dc_A/dz is the concentration gradient in the z direction, and D_{AB}, the proportionality factor, is the *mass diffusivity* or *diffusion coefficient* for component A diffusing through component B.

A more general flux relation that is not restricted to isothermal, isobaric systems was proposed by de Groot[2] who chose to write

$$\text{flux} = -\begin{pmatrix} \text{overall} \\ \text{density} \end{pmatrix} \begin{pmatrix} \text{diffusion} \\ \text{coefficient} \end{pmatrix} \begin{pmatrix} \text{concentration} \\ \text{gradient} \end{pmatrix}$$

---

[1] A. Fick, *Ann. Physik.*, **94,** 59 (1855).

[2] S. R. de Groot, *Thermodynamics of Irreversible Processes,* North-Holland, Amsterdam, 1951.

or

$$J_{A,z} = -cD_{AB}\frac{dy_A}{dz} \tag{24-16}$$

As the total concentration c is constant under isothermal, isobaric conditions, equation (24-15) is a special form of the more general relation (24-16). An equivalent expression for $j_{A,z}$, the mass flux in the z direction relative to the mass-average velocity, is

$$j_{A,z} = -\rho D_{AB}\frac{d\omega_A}{dz} \tag{24-17}$$

where $d\omega_A/dz$ is the concentration gradient in terms of the mass fraction. When the density is constant, this relation simplifies to

$$j_{A,z} = -D_{AB}\frac{d\rho_A}{dz}$$

Initial experimental investigations of molecular diffusion were unable to verify Fick's law of diffusion. This was apparently due to the fact that mass is often transferred simultaneously by two possible means: (1) as a result of the concentration differences as postulated by Fick and (2) by convection differences induced by the density differences that resulted from the concentration variation. Steffan (1872) and Maxwell (1877), using the kinetic theory of gases, proved that the mass flux relative to a fixed coordinate was a result of two contributions: the concentration gradient contribution and the bulk motion contribution.

For a binary system with a constant average velocity in the z direction, the molar flux in the z direction relative to the molar-average velocity may also be expressed by

$$J_{A,z} = c_A(v_{A,z} - V_z) \tag{24-18}$$

Equating expressions (24-16) and (24-18), we obtain

$$J_{A,z} = c_A(v_{A,z} - V_z) = -cD_{AB}\frac{dy_A}{dz}$$

which, upon rearrangement, yields

$$c_A v_{A,z} = -cD_{AB}\frac{dy_A}{dz} + c_A V_z$$

For this binary system, V_z can be evaluated by equation (24-14) as

$$V_z = \frac{1}{c}(c_A v_{A,z} + c_B v_{A,z})$$

or

$$c_A V_z = y_A(c_A v_{A,z} + c_B v_{B,z})$$

Substituting this expression into our relation, we obtain

$$c_A v_{A,z} = -cD_{AB}\frac{dy_B}{dz} + y_A(c_A v_{A,z} + c_B v_{B,z}) \tag{24-19}$$

As the component velocities, $v_{A,z}$ and $v_{B,z}$, are velocities relative to the fixed z axis, the quantities $c_A v_{A,z}$ and $c_B v_{B,z}$ are fluxes of components A and B relative to a fixed z coordinate; accordingly, we symbolize this new type of flux that is relative to a set of stationary axes by

$$\mathbf{N}_A = c_A\mathbf{v}_A$$

and

$$\mathbf{N}_B = c_B \mathbf{v}_B$$

Substituting these symbols into equation (24-19), we obtain a relation for the flux of component A relative to the z axis

$$N_{A,z} = -cD_{AB}\frac{dy_A}{dz} + y_A(\mathbf{N}_{A,z} + \mathbf{N}_{B,z}) \tag{24-20}$$

This relation may be generalized and written in vector form as

$$\mathbf{N}_A = -cD_{AB}\mathbf{\nabla} y_A + y_A(\mathbf{N}_A + \mathbf{N}_B) \tag{24-21}$$

It is important to note that the molar flux, $\mathbf{N}_A$, is a resultant of the two vector quantities:

$-cD_{AB}\mathbf{\nabla}y_A$ the molar flux, $\mathbf{J}_A$, resulting from the concentration gradient. This term is referred to as the *concentration gradient contribution*;

and

$y_A(\mathbf{N}_A + \mathbf{N}_B) = c_A\mathbf{V}$ the molar flux resulting as component A is carried in the bulk flow of the fluid. This flux term is designated the *bulk motion contribution*.

Either or both quantities can be a significant part of the total molar flux, $\mathbf{N}_A$. Whenever equation (24-21) is applied to describe molar diffusion, the vector nature of the individual fluxes, $\mathbf{N}_A$ and $\mathbf{N}_B$, must be considered and then, in turn, the direction of each of two vector quantities must be evaluated.

If species A were diffusing in a multicomponent mixture, the expression equivalent to equation (24-21) would be

$$\mathbf{N}_A = -cD_{AM}\mathbf{\nabla} y_A + y_A \sum_{i=1}^{n} \mathbf{N}_i$$

where D_{AM} is the diffusion coefficient of A in the mixture.

The mass flux, $\mathbf{n}_A$, relative to a fixed spatial coordinate system, is defined for a binary system in terms of mass density and mass fraction by

$$\mathbf{n}_A = -\rho D_{AB}\mathbf{\nabla}\omega_A + \omega_A(\mathbf{n}_A + \mathbf{n}_B) \tag{24-22}$$

where

$$\mathbf{n}_A = \rho_A \mathbf{v}_A$$

and

$$\mathbf{n}_B = \rho_B \mathbf{v}_B$$

Under isothermal, isobaric conditions, this relation simplifies to

$$\mathbf{n}_A = -D_{AB}\mathbf{\nabla}\rho_A + \omega_A(\mathbf{n}_A + \mathbf{n}_B)$$

As previously noted, the flux is a resultant of two vector quantities:

$-D_{AB}\mathbf{\nabla}\rho_A,$ the mass flux, $\mathbf{j}_A$, resulting from a concentration gradient; the *concentration gradient contribution*.

$\omega_A(\mathbf{n}_A + \mathbf{n}_B) = \rho_A\mathbf{v},$ the mass flux resulting as component A is carried in the bulk flow of the fluid; the *bulk motion contribution*.

Table 24.2 Equivalent forms of the mass flux equation for binary system A and B

Flux	Gradient	Fick rate equation	Restrictions
$\mathbf{n}_A$	$\nabla \omega_A$	$\mathbf{n}_A = -\rho D_{AB} \nabla \omega_A + \omega_A(\mathbf{n}_A + \mathbf{n}_B)$	
	$\nabla \rho_A$	$\mathbf{n}_A = -D_{AB} \nabla \rho_A + \omega_A(\mathbf{n}_A + \mathbf{n}_B)$	Constant ρ
$\mathbf{N}_A$	∇y_A	$\mathbf{N}_A = -c D_{AB} \nabla y_A + y_A(\mathbf{N}_A + \mathbf{N}_B)$	
	∇c_A	$\mathbf{N}_A = -D_{AB} \nabla c_A + y_A(\mathbf{N}_A + \mathbf{N}_B)$	Constant c
$\mathbf{j}_A$	$\nabla \omega_A$	$\mathbf{j}_A = -\rho D_{AB} \nabla \omega_A$	
	$\nabla \rho_A$	$\mathbf{j}_A = -D_{AB} \nabla \rho_A$	Constant ρ
$\mathbf{J}_A$	∇y_A	$\mathbf{J}_A = -c D_{AB} \nabla y_A$	
	∇c_A	$\mathbf{J}_A = -D_{AB} \nabla c_A$	Constant c

If a balloon, filled with a color dye, is dropped into a large lake, the dye will diffuse radially as a concentration gradient contribution. When a stick is dropped into a moving stream, it will float downstream by the bulk motion contribution. If the dye-filled balloon were dropped into the moving stream, the dye would diffuse radially while being carried downstream; thus both contributions participate simultaneously in the mass transfer.

The four equations defining the fluxes, $\mathbf{J}_A$, $\mathbf{j}_A$, $\mathbf{N}_A$, and $\mathbf{n}_A$ are equivalent statements of the Fick rate equation. The diffusion coefficient, D_{AB}, is identical in all four equations. Any one of these equations is adequate to describe molecular diffusion; however, certain fluxes are easier to use for specific cases. The mass fluxes, $\mathbf{n}_A$ and $\mathbf{j}_A$, are used when the Navier–Stokes equations are also required to describe the process. Since chemical reactions are described in terms of moles of the participating reactants, the molar fluxes, $\mathbf{J}_A$ and $\mathbf{N}_A$, are used to describe mass-transfer operations in which chemical reactions are involved. The fluxes relative to coordinates fixes in space, $\mathbf{n}_A$ and $\mathbf{N}_A$, are often used to describe engineering operations within process equipment. The fluxes $\mathbf{J}_A$ and $\mathbf{j}_A$ are used to describe the mass transfer in diffusion cells used for measuring the diffusion coefficient. Table 24.2 summarizes the equivalent forms of the Fick rate equation.

Related Types of Molecular Mass Transfer

According to the second law of thermodynamics, systems not in equilibrium will tend to move toward equilibrium with time. A generalized driving force in chemical thermo-dynamic terms is $-d\mu_c/dz$ where μ_c is the *chemical potential*. The molar diffusion velocity of component A is defined in terms of the chemical potential by

$$v_{A,z} - V_z = u_A \frac{d\mu_c}{dz} = -\frac{D_{AB}}{RT} \frac{d\mu_c}{dz} \tag{24-23}$$

where u_A is the "mobility" of component A, or the resultant velocity of the molecule while under the influence of a unit driving force. Equation (24-23) is known as the Nernst–Einstein relation. The molar flux of A becomes

$$J_{A,z} = c_A(v_{A,z} - V_z) = -c_A \frac{D_{AB}}{RT} \frac{d\mu_c}{dz} \tag{24-24}$$

Equation (24-24) may be used to define all molecular mass-transfer phenomena. As an example, consider the conditions specified for equation (24-15); the chemical potential of a component in a homogeneous ideal solution at constant temperature and pressure is defined by

$$\mu_c = \mu^0 + RT \ln c_A \tag{24-25}$$

where μ^0 is a constant, the chemical potential of the standard state. When we substitute this relation into equation (24-24), the Fick rate equation for a homogeneous phase is obtained

$$J_{A,z} = -D_{AB}\frac{dc_A}{dz} \qquad (24\text{-}15)$$

There are a number of other physical conditions, in addition to differences in concentration, which will produce a chemical potential gradient: temperature differences, pressure differences, and differences in the forces created by external fields, such as gravity, magnetic, and electrical fields. We can, for example, obtain mass transfer by applying a temperature gradient to a multicomponent system. This transport phenomenon, the *Soret effect* or *thermal diffusion,* although normally small relative to other diffusion effects, is used successfully in the separation of isotopes. Components in a liquid mixture can be separated with a centrifuge by *pressure diffusion.* There are many well-known examples of mass fluxes being induced in a mixture subjected to an external force field: separation by sedimentation under the influence of gravity, electrolytic precipitation due to an electrostatic force field, and magnetic separation of mineral mixtures through the action of a magnetic force field. Although these mass-transfer phenomena are important, they are very specific processes.

The molecular mass transfer, resulting from concentration differences and described by Fick's law, results from the random molecular motion over small mean free paths, independent of any containment walls. The diffusion of fast neutrons and molecules in extremely small pores or at very low gas density cannot be described by this relationship.

Neutrons, produced in a nuclear fission process, initially possess high kinetic energies and are termed *fast neutrons* because of their high velocities; that is, up to 15 million meters per second. At these high velocities, neutrons pass through the electronic shells of other atoms or molecules with little hindrance. To be deflected, the fast neutrons must collide with a nucleus, which is a very small target compared to the volume of most atoms and molecules. The mean free path of fast neutrons is approximately one million times greater than the free paths of gases at ordinary pressures. After the fast neutrons are slowed down through elastic-scattering collisions between the neutrons and the nuclei of the reactor's moderator, these slower moving neutrons, *thermal neutrons,* migrate from positions of higher concentration to positions of lower concentration, and their migration is described by Fick's law of diffusion.

24.2   THE DIFFUSION COEFFICIENT

Fick's law of proportionality, D_{AB}, is known as the diffusion coefficient. Its fundamental dimensions, which may be obtained from equation (24-15)

$$D_{AB} = \frac{-J_{A,z}}{dc_A/dz} = \left(\frac{M}{L^2 t}\right)\left(\frac{1}{M/L^3 \cdot 1/L}\right) = \frac{L^2}{t}$$

are identical to the fundamental dimensions of the other transport properties: kinematic viscosity, ν, and thermal diffusivity, α, or its equivalent ratio, $k/\rho c_p$. The mass diffusivity has been reported in cm^2/s; the SI units are m^2/s, which is a factor 10^{-4} smaller. In the English system ft^2/h is commonly used. Conversion between these systems involves the simple relations

$$\frac{D_{AB}(\text{cm}^2/\text{s})}{D_{AB}(\text{m}^2/\text{s})} = 10^4$$

$$\frac{D_{AB}(\text{ft}^2/\text{h})}{D_{AB}(\text{cm}^2/\text{s})} = 3.87 \qquad (24\text{-}26)$$

The diffusion coefficient depends upon the pressure, temperature, and composition of the system. Experimental values for the diffusivities of gases, liquids, and solids are tabulated in Appendix Tables J.1, J.2, and J.3, respectively. As one might expect from the consideration of the mobility of the molecules, the diffusion coefficients are generally higher for gases (in the range of 5×10^{-6} to $1 \times 10^{-5} \, \text{m}^2/\text{s}$), than for liquids (in the range of 10^{-10} to $10^{-9} \, \text{m}^2/\text{s}$), which are higher than the values reported for solids (in the range of 10^{-14} to $10^{-10} \, \text{m}^2/\text{s}$).

In the absence of experimental data, semitheoretical expressions have been developed which give approximations, sometimes as valid as experimental values due to the difficulties encountered in their measurement.

Gas Mass Diffusivity

Theoretical expressions for the diffusion coefficient in low-density gaseous mixtures as a function of the system's molecular properties were derived by Sutherland,[3] Jeans,[4] and Chapman and Cowling,[5] based upon the kinetic theory of gases. In the simplest model of gas dynamics, the molecules are regarded as rigid spheres that exert no intermolecular forces. Collisions between these rigid molecules are considered to be completely elastic. With these assumptions, a simplified model for an ideal gas mixture of species A diffusing through its isotope A^* yields an equation for the *self-diffusion coefficient,* defined as

$$D_{AA^*} = \frac{1}{3}\lambda u \tag{24-27}$$

and λ is the mean free path of length of species A, given by

$$\lambda = \frac{\kappa T}{\sqrt{2}\pi\sigma_A^2 P} \tag{24-28}$$

where u is the mean speed of species A with respect to the molar-average velocity

$$u = \sqrt{\frac{8\kappa NT}{\pi M_A}} \tag{24-29}$$

Insertion of equations (24-28) and (24-29) into equation (24-27) results in

$$D_{AA^*} = \frac{2T^{3/2}}{3\pi^{3/2}\sigma_A^2 P}\left(\frac{\kappa^3 N}{M_A}\right)^{1/2} \tag{24-30}$$

where M_A is the molecular weight of the diffusing species A, (g/mol), N is Avogadro's number $(6.022 \times 10^{23} \text{ molecules/mol})$, P is the system pressure, T is the absolute temperature (K), κ is the Boltzmann constant $(1.38 \times 10^{-16} \text{ ergs/K})$, and σ_{AB} is the Lennard–Jones diameter of the spherical molecules.

Using a similar kinetic theory of gases approach for a binary mixture of species A and B composed of rigid spheres of unequal diameters, the gas-phase diffusion coefficient is shown to be

$$D_{AB} = \frac{2}{3}\left(\frac{\text{K}}{\pi}\right)^{3/2} N^{1/2}T^{3/2}\frac{\left(\dfrac{1}{2M_A}+\dfrac{1}{2M_B}\right)^{1/2}}{P\left(\dfrac{\sigma_A+\sigma_B}{2}\right)^2} \tag{24-31}$$

---

[3] W. Sutherland, *Phil. Mag.,* **36,** 507; **38,** 1 (1894).

[4] J. Jeans, *Dynamical Theory of Gases,* Cambridge University Press, London, 1921.

[5] S. Chapman and T. G. Cowling, *Mathematical Theory of Non-Uniform Gases,* Cambridge University Press, London, 1959.

Unlike the other two molecular transport coefficients for gases, the viscosity and thermal conductivity, the gas-phase diffusion coefficient is dependent on the pressure and the temperature. Specifically, the gas-phase diffusion coefficient is

- an inverse function of total system pressure

$$D_{AB} \propto \frac{1}{P}$$

- a 3/2 power-law function of the absolute temperature

$$D_{AB} \propto T^{3/2}$$

As equation (24-31) reveals, and as one of the problems at the end of this chapter points out, the diffusion coefficients for gases $D_{AB} = D_{BA}$. This is not the case for liquid diffusion coefficients.

Modern versions of the kinetic theory have been attempted to account for forces of attraction and repulsion between the molecules. Hirschfelder et al. (1949),[6] using the Lennard–Jones potential to evaluate the influence of the molecular forces, presented an equation for the diffusion coefficient for gas pairs of nonpolar, nonreacting molecules:

$$D_{AB} = \frac{0.001858T^{3/2}\left[\dfrac{1}{M_A} + \dfrac{1}{M_B}\right]^{1/2}}{P\sigma_{AB}^2\Omega_D} \tag{24-33}$$

where D_{AB} is the mass diffusivity of A through B, in cm^2/s; T is the absolute temperature, in K; M_A, M_B are the molecular weights of A and B, respectively; P is the absolute pressure, in atmospheres; σ_{AB} is the "collision diameter," a Lennard–Jones parameter, in Å; and Ω_D is the "collision integral" for molecular diffusion, a dimensionless function of the temperature and of the intermolecular potential field for one molecule of A and one molecule of B. Appendix Table K.1 lists Ω_D as a function of $\kappa T/\varepsilon_{AB}$, k is the Boltzmann constant, which is 1.38×10^{-16} ergs/K, and ε_{AB} is the energy of molecular interaction for the binary system A and B, a Lennard–Jones parameter, in ergs, see equation (24-31). Unlike the other two molecular transport coefficients, viscosity and thermal conductivity, the diffusion coefficient is dependent on pressure as well as on a higher order of the absolute temperature. When the transport process in a single component phase was examined, we did not find any composition dependency in equation (24-30) or in the similar equations for viscosity and thermal conductivity. Figure 24.2 presents the graphical dependency of the "collision integral," Ω_D, on the dimensionless temperature, $\kappa T/\varepsilon_{AB}$.

The Lennard–Jones parameters, ρ and ε_D, are usually obtained from viscosity data. Unfortunately, this information is available for only a very few pure gases. Appendix Table K.2 tabulates these values. In the absence of experimental data, the values for pure components may be estimated from the following empirical relations:

$$\sigma = 1.18 V_b^{1/3} \tag{24-34}$$

$$\sigma = 0.841 V_c^{1/3} \tag{24-35}$$

$$\sigma = 2.44 \left(\frac{T_c}{P_c}\right)^{1/3} \tag{24-36}$$

$$\varepsilon_A/\kappa = 0.77 T_c \tag{24-37}$$

---

[6] J. O. Hirschfelder, R. B. Bird, and E. L. Spotz, *Chem. Rev.*, **44**, 205 (1949).

Figure 24.2 Binary gas-phase Lennard–Jones "collision integral."

and

$$\varepsilon_A/\kappa = 1.15\, T_b \tag{24-38}$$

where V_b is the molecular volume at the normal boiling point, in $(cm)^3/g$ mol (this is evaluated by using Table 24.3); V_c is the critical molecular volume, in $(cm)^3/g$ mol; T_c is the critical temperature, in K; T_b is the normal boiling temperature, in K; and P_c is the critical pressure, in atmospheres.

Table 24.3 Atomic diffusion volumes for use in estimating D_{AB} by method of Fuller, Schettler, and Giddings

Atomic and structure diffusion-volume increments, v			
C	16.5	Cl	19.5
H	1.98	S	17.0
O	5.48	Aromatic ring	−20.2
N	5.69	Heterocyclic ring	−20.2

Diffusion volumes for simple molecules, v					
H_2	7.07	Ar	16.1	H_2O	12.7
D_2	6.70	Kr	22.8	$CClF_2$	114.8
He	2.88	CO	18.9	SF_6	69.7
N_2	17.9	CO_2	26.9	Cl_2	37.7
O_2	16.6	N_2O	35.9	Br_2	67.2
Air	20.1	NH_3	14.9	SO_2	41.1

For a binary system composed of nonpolar molecular pairs, the Lennard–Jones parameters of the pure component may be combined empirically by the following relations:

$$\sigma_{AB} = \frac{\sigma_A + \sigma_B}{2} \tag{24-39}$$

and

$$\varepsilon_{AB} = \sqrt{\varepsilon_A \varepsilon_B} \tag{24-40}$$

These relations must be modified for polar–polar and polar–nonpolar molecular pairs; the proposed modifications are discussed by Hirschfelder, Curtiss, and Bird.[7]

The Hirschfelder equation (24-33) is often used to extrapolate experimental data. For moderate ranges of pressure, up to 25 atm, the diffusion coefficient varies inversely with the pressure. Higher pressures apparently require dense gas corrections; unfortunately, no satisfactory correlation is available for high pressures. Equation () also states that the diffusion coefficient varies with the temperature as $T^{3/2}/\Omega_D$ varies. Simplifying equation (), we can predict the diffusion coefficient at any temperature and at any pressure below 25 atm from a known experimental value by

$$D_{AB_{T_2,P_1}} = D_{AB_{T_1,P_1}} \left(\frac{P_1}{P_2}\right) \left(\frac{T_2}{T_1}\right)^{3/2} \frac{\Omega_D|_{T_1}}{\Omega_D|_{T_2}} \tag{24-41}$$

In Appendix Table J.1, experimental values of the product $D_{AB}P$ are listed for several gas pairs at a particular temperature. Using equation (24-41), we may extend these values to other temperatures.

EXAMPLE 2

Evaluate the diffusion coefficient of carbon dioxide in air at 20°C and atmospheric pressure. Compare this value with the experimental value reported in appendix table J.1.

From Appendix Table K.2, the values of σ and (ε_A/κ) are obtained

	σ, in Å	ε_A/κ, in K
Carbon dioxide	3.996	190
Air	3.617	97

The various parameters for equation (24-33) may be evaluated as follows:

$$\sigma_{AB} = \frac{\sigma_A + \sigma_B}{2} = \frac{3.996 + 3.617}{2} = 3.806 \,\overset{\circ}{A}$$

$$\varepsilon_{AB}/\kappa = \sqrt{(\varepsilon_A/\kappa)(\varepsilon_B/\kappa)} = \sqrt{(190)(97)} = 136$$

$$T = 20 + 273 = 293 \,\text{K}$$

$$P = 1 \,\text{atm}$$

$$\frac{\varepsilon_{AB}}{\kappa T} = \frac{136}{293} = 0.463$$

$$\frac{\kappa T}{\varepsilon_{AB}} = 2.16$$

$$\Omega_D \,(\text{Table K.1}) = 1.047$$

$$M_{CO_2} = 44$$

and

$$M_{Air} = 29$$

Substituting these values into equation (24-33), we obtain

$$D_{AB} = \frac{0.001858 \, T^{3/2}(1/M_A + 1/M_B)^{1/2}}{P\sigma_{AB}^2 \Omega_D}$$

$$= \frac{(0.001858)(293)^{3/2}(1/44 + 1/29)^{1/2}}{(1)(3.806)^2(1.047)} = 0.147 \,\text{cm}^2/\text{s}$$

---

[7] J. O. Hirschfelder, C. F. Curtiss, and R. B. Bird, *Molecular Theory of Gases and Liquids,* John Wiley & Sons, Inc., New York, 1954.

From Appendix Table J.1 for CO_2 in air at 273 K, 1 atm, we have

$$D_{AB} = 0.136 \text{ cm}^2/\text{s}$$

Equation (24-42) will be used to correct for the differences in temperature

$$\frac{D_{AB,T_1}}{D_{AB,T_2}} = \left(\frac{T_1}{T_2}\right)^{3/2} \left(\frac{\Omega_D|T_2}{\Omega_D|T_1}\right)$$

Values for Ω_D may be evaluated as follows:

$$\text{at } T_2 = 273 \quad \varepsilon_{AB}/\kappa T = \frac{136}{273} = 0.498 \quad \Omega_D|_{T_2} = 1.074$$

$$\text{at } T_1 = 293 \quad \Omega_D|_{T_1} = 1.047 \quad \text{(previous calculations)}$$

The corrected value for the diffusion coefficient at 20°C is

$$D_{AB,T_1} = \left(\frac{293}{273}\right)^{3/2} \left(\frac{1.074}{1.047}\right)(0.136) = 0.155 \text{ cm}^2/\text{s} \quad (1.55 \times 10^{-5} \text{ m}^2/\text{s})$$

We readily see that the temperature dependency of the "collision integral" is very small. Accordingly, most scaling of diffusivities relative to temperature include only the ratio $(T_1/T_2)^{3/2}$.

Equation (24-33) was developed for dilute gases consisting of nonpolar, spherical monatomic molecules. However, this equation gives good results for most nonpolar, binary gas systems over a wide range of temperatures.[8] Other empirical equations have been proposed[9] for estimating the diffusion coefficient for nonpolar, binary gas systems at low pressures. The empirical correlation recommended by Fuller, Schettler, and Giddings permits the evaluation of the diffusivity when reliable Lennard–Jones parameters, σ_i and ε_i, are unavailable. The Fuller correlation is

$$D_{AB} = \frac{10^{-3} T^{1.75} \left(\frac{1}{M_A} + \frac{1}{M_B}\right)^{1/2}}{P\left[(\Sigma v)_A^{1/3} + (\Sigma v)_B^{1/3}\right]^2} \tag{24-42}$$

where D_{AB} is in cm²/s, T is in K, and P is in atmospheres. To determine the v terms, the authors recommend the addition of the atomic and structural diffusion-volume increments v reported in Table 24.3.

Danner and Daubert[10] have recommended the atomic and structure diffusion-volume increments for C to be corrected to 15.9 and for H to 2.31 and the diffusion volumes for H_2 to be corrected to 6.12 and for air to 19.7.

---

[8] R. C. Reid, J. M. Prausnitz, and T. K. Sherwood, *The Properties of Gases and Liquids,* Third Edition, McGraw-Hill Book Company, New York, 1977, Chapter 11.

[9] J. H. Arnold, *J. Am. Chem. Soc.,* **52,** 3937 (1930). E. R. Gilliland, *Ind. Eng. Chem.,* **26,** 681 (1934). J. C. Slattery and R. B. Bird, *A.I.Ch.E. J.,* **4,** 137 (1958). D. F. Othmer and H. T. Chen, *Ind. Eng. Chem. Process Des. Dev.,* **1,** 249 (1962). R. G. Bailey, *Chem. Engr.,* **82**(6), 86, (1975). E. N. Fuller, P. D. Schettler, and J. C. Giddings, *Ind. Eng. Chem.,* **58**(5), 18 (1966).

[10] R. P. Danner, and T. E. Daubert, *Manual for Predicting Chemical Process Design Data,* A.I.Ch.E. (1983).

EXAMPLE 3

Reevaluate the diffusion coefficient of carbon dioxide in air at 20°C and atmospheric pressure using the Fuller, Schettler, and Giddings equation and compare the new value with the one reported in example 2.

$$D_{AB} = \frac{10^{-3}T^{1.75}\left(\frac{1}{M_A}+\frac{1}{M_B}\right)^{1/2}}{P\left[(\Sigma v)_A^{1/3}+(\Sigma v)_B^{1/3}\right]^2}$$

$$= \frac{10^{-3}(293)^{1.75}\left(\frac{1}{44}+\frac{1}{29}\right)^{1/2}}{(1)[(26.9)^{1/3}+(20.1)^{1/3}]^2}$$

$$= 0.152\,\text{cm}^2/\text{s}$$

This value compares very favorably to the value evaluated with Hirschfelder equation, $0.155\,\text{cm}^2/\text{s}$, and its determination was easily accomplished.

Brokaw[11] has suggested a method for estimating diffusion coefficient for binary gas mixtures containing polar compounds. The Hirschfelder equation (24-33) is still used; however, the collision integral is evaluated by

$$\Omega_D = \Omega_{D_0} + \frac{0.196\delta_{AB}^2}{T^*} \tag{24-43}$$

where

$$\delta_{AB} = (\delta_A\delta_B)^{1/2}$$
$$\delta = \frac{1.94\times10^3\mu_p^2}{V_DT_D} \tag{24-44}$$

μ_p = dipole moment, Debye

V_b = liquid molar volume of the specific compound at its boiling point, $\text{cm}^3/\text{g mol}$

T_b = normal boiling point, K

and

$$T^* = \kappa T/\varepsilon_{AB}$$

where

$$\frac{\varepsilon_{AB}}{\kappa} = \left(\frac{\varepsilon_A}{\kappa}\frac{\varepsilon_B}{\kappa}\right)^{1/2} \tag{24-45}$$
$$\varepsilon/\kappa = 1.18(1+1.3\,\delta^2)T_b$$

δ is evaluated with (24–44). And

$$\Omega_{D_0} = \frac{A}{(T^*)^B} + \frac{C}{\exp(DT^*)} + \frac{E}{\exp(FT^*)} + \frac{G}{\exp(HT^*)} \tag{24-46}$$

---

[11] R. S. Brokaw, *Ind. Engr. Chem. Process Des. Dev.*, **8**, 240 (1969).

with

$$A = 1.060,36 \quad E = 1.035,87$$
$$B = 0.156,10 \quad F = 1.529,96$$
$$C = 0.193,00 \quad G = 1.764,74$$
$$D = 0.476,35 \quad H = 3.894,11$$

The collision diameter, σ_{AB}, is evaluated with

$$\sigma_{AB} = (\sigma_A \sigma_B)^{1/2} \tag{24-47}$$

with each component's characteristic length evaluated by

$$\sigma = \left(\frac{1.585\, V_D}{1 + 1.3\, \delta^2} \right)^{1/3} \tag{24-48}$$

Reid, Prausnitz, and Sherwood[12] noted that the Brokaw equation is fairly reliable, permitting the evaluation of the diffusion coefficients for gases involving polar compounds with errors less than 15%.

Mass transfer in gas mixtures of several components can be described by theoretical equations involving the diffusion coefficients for the various binary pairs involved in the mixture. Hirschfelder, Curtiss, and Bird[13] present an expression in its most general form. Wilke[14] has simplified the theory and has shown that a close approximation to the correct form is given by the relation

$$D_{1-\text{mixture}} = \frac{1}{y_2'/D_{1-2} + y_3'/D_{1-3} + \cdots + y_n'/D_{1-n}} \tag{24-49}$$

where $D_{1-\text{mixture}}$ is the mass diffusivity for component 1 in the gas mixture; D_{1-n} is the mass diffusivity for the binary pair, component 1 diffusing through component n; and y_n' is the mole fraction of component n in the gas mixture evaluated on a component-1-free basis, that is

$$y_2' = \frac{y_2}{y_2 + y_3 + \cdots y_n} = \frac{y_2}{1 - y_1}$$

In Problem 24.7 at the end of this chapter, equation (24-49) is developed by using Wilke's approach for extending the Stefan and Maxwell theory in order to explain the diffusion of species A through a gas mixture of several components.

EXAMPLE 4

In the chemical vapor deposition of silane (SiH_4) on a silicon wafer, a process gas stream rich in an inert nitrogen (N_2) carrier gas has the following composition:

$$y_{SiH_4} = 0.0075, \quad y_{H_2} = 0.015, \quad y_{N_2} = 0.9775$$

The gas mixture is maintained at 900 K and 100 Pa total system pressure. Determine the diffusivity of silane through the gas mixture. The Lennard–Jones constants for silane are $\varepsilon_A/\kappa = 207.6\,\text{K}$ and $\sigma_A = 4.08\,\text{Å}$

[12] R. C. Reid, J. M. Prausnitz, and T. K. Sherwood, *The Properties of Gases and Liquids,* Third Edition, McGraw-Hill Book Company, New York, 1977, Chapter 11.

[13] J. O. Hirschfelder, C. F. Curtiss, and R. B. Bird, *Molecular Theory of Gases and Liquids,* Wiley, New York, p. 718.

[14] C. R. Wilke, *Chem. Engr. Prog.,* **46**, 95–104 (1950).

The binary diffusion coefficients at 900 K and 100 Pa total system pressure estimated by the Hirschfelder equation (24-33) are

$$D_{SiH_4-N_2} = 1.09 \times 10^3 \, cm^2/s \quad \text{and} \quad D_{SiH_4-H_2} = 4.06 \times 10^3 \, cm^2/s$$

The binary diffusion coefficients are relatively high because the temperature is high and the total system pressure is low. The composition of nitrogen and hydrogen on a silane-free basis are

$$y'_{N_2} = \frac{0.9775}{1 - 0.0075} = 0.9849 \quad \text{and} \quad y'_{H_2} = \frac{0.015}{1 - 0.0075} = 0.0151$$

Upon substituting these values into the Wilke equation (24-49), we obtain

$$D_{SiH_4-mixture} = \frac{1}{\dfrac{y'_{N_2}}{D_{SiH_4-N_2}} + \dfrac{y'_{H_2}}{D_{SiH_4-H_2}}} = \frac{1}{\dfrac{0.9849}{1.09 \times 10^3} + \dfrac{0.0151}{4.06 \times 10^3}} = 1.10 \times 10^3 \, \frac{cm^2}{s}$$

This example verifies that for a dilute multicomponent gas mixture, the diffusion coefficient of the diffusing species in the gas mixture is approximated by the binary diffusion coefficient of the diffusing species in the carrier gas.

Liquid-Mass Diffusivity

In contrast to the case for gases, where we have available an advanced kinetic theory for explaining molecular motion, theories of the structure of liquids and their transport characteristics are still inadequate to permit a rigorous treatment. Inspection of published experimental values for liquid diffusion coefficients in Appendix J.2 reveals that they are several orders of magnitude smaller than gas diffusion coefficients and that they depend on concentration due to the changes in viscosity with concentration and changes in the degree of ideality of the solution.

Certain molecules diffuse as molecules, while others that are designated as electrolytes ionize in solutions and diffuse as ions. For example, sodium chloride, NaCl, diffuses in water as the ions Na^+ and Cl^-. Though each ion has a different mobility, the electrical neutrality of the solution indicates that the ions must diffuse at the same rate; accordingly, it is possible to speak of a diffusion coefficient for molecular electrolytes such as NaCl. However, if several ions are present, the diffusion rates of the individual cations and anions must be considered, and molecular diffusion coefficients have no meaning. Needless to say, separate correlations for predicting the relation between the liquid mass diffusivities and the properties of the liquid solution will be required for electrolytes and nonelectrolytes.

Two theories, the Eyring "hole" theory and the hydrodynamical theory, have been postulated as possible explanations for diffusion of nonelectrolyte solutes in low-concentration solutions. In the Eyring concept, the ideal liquid is treated as a quasi-crystalline lattice model interspersed with holes. The transport phenomenon is then described by a unimolecular rate process involving the jumping of solute molecules into the holes within the lattice model. These jumps are empirically related to Eyring's theory of reaction rate.[15] The hydrodynamical theory states that the liquid diffusion coefficient is related to the solute molecule's mobility; that is, to the net velocity of the molecule while under the influence of a unit driving force. The laws of hydrodynamics provide

---

[15] S. Glasstone, K. J. Laidler, and H. Eyring, *Theory of Rate Processes,* McGraw-Hill Book Company, New York, 1941, Chap. IX.

relations between the force and the velocity. An equation that has been developed from the hydrodynamical theory is the Stokes–Einstein equation

$$D_{AB} = \frac{\kappa T}{6\pi r \mu_B} \tag{24-50}$$

where D_{AB} is the diffusivity of A in dilute solution in D, κ is the Boltzmann constant, T is the absolute temperature, r is the solute particle radius, and μ_B is the solvent viscosity. This equation has been fairly successful in describing the diffusion of colloidal particles or large round molecules through a solvent that behaves as a continuum relative to the diffusing species.

The results of the two theories can be rearranged into the general form

$$\frac{D_{AB}\mu_B}{\kappa T} = f(V) \tag{24-51}$$

in which $f(V)$ is a function of the molecular volume of the diffusing solute. Empirical correlations, using the general form of equation (24-51), have been developed, which attempt to predict the liquid diffusion coefficient in terms of the solute and solvent properties. Wilke and Chang[16] have proposed the following correlation for nonelectrolytes in an infinitely dilute solution:

$$\frac{D_{AB}\mu_B}{T} = \frac{7.4 \times 10^{-8}(\Phi_B M_B)^{1/2}}{V_A^{0.6}} \tag{24-52}$$

where D_{AB} is the mass diffusivity of A diffusing through liquid solvent B, in cm^2/s; μ_B is the viscosity of the solution, in centipoises; T is absolute temperature, in K; M_B is the molecular weight of the solvent; V_A is the molal volume of solute at normal boiling point, in cm^3/g mol; and Φ_B is the "association" parameter for solvent B.

Molecular volumes at normal boiling points, V_A, for some commonly encountered compounds, are tabulated in Table 24.4. For other compounds, the atomic volumes of each element present are added together as per the molecular formulas. Table 24.5 lists the contributions for each of the constituent atoms. When certain ring structures are involved, corrections must be made to account for the specific ring configuration; the following

Table 24.4   Molecular volumes at normal boiling point for some commonly encountered compounds

Compound	Molecular volume, in cm^3/g mol	Compound	Molecular volume, in cm^3/g mol
Hydrogen, H$_2$	14.3	Nitric oxide, NO	23.6
Oxygen, O$_2$	25.6	Nitrous oxide, N$_2$O	36.4
Nitrogen, N$_2$	31.2	Ammonia, NH$_3$	25.8
Air	29.9	Water, H$_2$O	18.9
Carbon monoxide, CO	30.7	Hydrogen sulfide, H$_2$S	32.9
Carbon dioxide, CO$_2$	34.0	Bromine, Br$_2$	53.2
Carbonyl sulfide, COS	51.5	Chlorine, Cl$_2$	48.4
Sulfur dioxide, SO$_2$	44.8	Iodine, I$_2$	71.5

[16] C. R. Wilke and P. Chang, *A.I.Ch.E.J.*, **1**, 264 (1955).

Table 24.5 Atomic volumes for complex molecular volumes for simple substances[†]

Element	Atomic volume, in $cm^3/g\,mol$	Element	Atomic volume, in $cm^3/g\,mol$
Bromine	27.0	Oxygen, except as noted below	7.4
Carbon	14.8	Oxygen, in methyl esters	9.1
Chlorine	21.6	Oxygen, in methyl ethers	9.9
Hydrogen	3.7	Oxygen, in higher ethers	
Iodine	37.0	and other esters	11.0
Nitrogen, double bond	15.6	Oxygen, in acids	12.0
Nitrogen, in primary amines	10.5	Sulfur	25.6
Nitrogen, in secondary amines	12.0		

[†]G. Le Bas, *The Molecular Volumes of Liquid Chemical Compounds,* Longmans, Green & Company, Ltd., London, 1915.

corrections are recommended:

for three-membered ring, as ethylene oxide	deduct 6
for four-membered ring, as cyclobutane	deduct 8.5
for five-membered ring, as furan	deduct 11.5
for pyridine	deduct 15
for benzene ring	deduct 15
for naphthalene ring	deduct 30
for anthracene ring	deduct 47.5

Recommended values of the association parameter, Φ_B, are given below for a few common solvents.

Solvent	Φ_B
Water	2.26[17]
Methanol	1.9
Ethanol	1.5
Benzene, ether, heptane, and other unassociated solvents	1.0

If data for computing the molar volume of solute at its normal boiling point, V_A, are not available, Tyn and Calus[18] recommend the correlation

$$V_A = 0.285 V_c^{1.048}$$

where V_c is the critical volume of species A in $cm^3/g.\,mol$. Values of V_c are tabulated in Reid, Prausnitz, and Sherwood.[19]

[17] The correction of Φ_B is recommended by R. C. Reid, J. M. Prausnitz, and T. K. Sherwood, *The Properties of Gases and Liquids,* Third Edition, McGraw-Hill Book Company, New York, 1977, p. 578.
[18] Tyn, M.T. and W.F. Calus, *Processing,* 21, (4): 16 (1975).
[19] R.C. Reid, J.M. Prausnitz and, T.K. Sherwood, *The Properties of Gases and Liquids*, Third Edition, McGraw-Hill Book Company, New York, 1977, Appendix A.

EXAMPLE 5

Estimate the liquid diffusion coefficient of ethanol, C_2H_5OH, in a dilute solution of water at 10°C. The molecular volume of ethanol may be evaluated by using values from Table 24.5 as follows:

$$V_{C_2H_5OH} = 2V_C + 6V_H + V_O$$

$$V_{C_2H_5OH} = 2(14.8) + 6(3.7) + 7.4 = 59.2 \, \text{cm}^3/\text{mol}$$

At 10°C, the viscosity of a solution containing 0.05 mol of alcohol/liter of water is 1.45 centipoises; the remaining parameters to be used are

$$T = 283 \, \text{K}$$
$$\Phi_B \text{ for water} = 2.26$$

and

$$M_B \text{ for water} = 18$$

Substituting these values into equation (24-52), we obtain

$$D_{C_2H_5OH-H_2O} = \left(\frac{7.4 \times 10^{-8}(2.26 \times 18)^{1/2}}{(59.2)^{0.6}} \right) \left(\frac{283}{1.45} \right)$$

$$= 7.96 \times 10^{-6} \, \text{cm}^2/\text{s} \qquad (7.96 \times 10^{-10} \, \text{m}^2/\text{s})$$

This value is in good agreement with the experimental value of $8.3 \times 10^{-10} \text{m}^2/\text{s}$ reported in Appendix J.

Let us compare this value of the liquid diffusivity of ethanol in a dilute solution of water at 10°C, $7.96 \times 10^{-6} \text{cm}^2/\text{s}$, with the value of the gas diffusivity of ethanol in air at 10°C and 1 atm pressure, $0.118 \, \text{cm}^2/\text{s}$. This emphasizes the order of magnitude difference between the values of the liquid and gas diffusivities.

Performing a similar calculation, the liquid diffusion coefficient of water in an infinite dilute solution of ethanol at the same 10°C temperature predicts that the diffusion coefficient D_{BA} is equal to $1.18 \times 10^{-5} \text{cm}^2/\text{s}$. It is important to note that liquid diffusivities D_{AB_L} and D_{BA_L} are not equal as were the gas diffusivities at the same temperature and pressure.

Hayduk and Laudie[20] have proposed a much simpler equation for evaluating infinite dilution diffusion coefficients of nonelectrolytes *in water*

$$D_{AB} = 13.26 \times 10^{-5} \mu_B^{-1.14} V_A^{-0.589} \qquad (24\text{-}53)$$

where D_{AB} is the mass diffusivity of A through liquid B, in cm^2/s; μ_B is the viscosity of water, in centipoises; and V_A is the molal volume of the solute at normal boiling point, in $\text{cm}^3/\text{g} \cdot \text{mol}$. This relation is much simpler to use and gives similar results to the Wilke–Chang equation. If we substitute the values used in example 4 into the Hayduk and Laudie relationship, we would obtain a diffusion coefficient for ethanol in a dilute water solution of $7.85 \times 10^{-6} \text{cm}^2/\text{s}$; this value is essentially the same value obtained using the Wilke–Chang equation.

Scheibel[21] has proposed that the Wilke–Chang relation be modified to eliminate the association factor, Φ_B, yielding

$$\frac{D_{AB} \mu_B}{T} = \frac{K}{V_A^{1/3}} \qquad (24\text{-}54)$$

---

[20] W. Hayduk and H. Laudie, *A.I.Ch.E. J.*, **20**, 611 (1974).
[21] E. G. Scheibel, *Ind. Eng. Chem.*, **46**, 2007 (1954).

where K is determined by

$$K = (8.2 \times 10^{-8})\left[1 + \left(\frac{3V_B}{V_A}\right)^{2/3}\right]$$

except

1. For benzene as a solvent, if $V_A < 2V_B$, use $K = 18.9 \times 10^{-8}$.
2. For other organic solvents, if $V_A < 2.5V_B$, use $K = 17.5 \times 10^{-8}$.

Reid, Prausnitz, and Sherwood[22] recommend this equation for solutes diffusing into organic solvents; however, they noted that this equation might evaluate values that had errors up to 20%.

Most methods for predicting the liquid diffusion coefficients in concentration solutions have combined the infinite dilution coefficients, D_{AB} and D_{BA}, in a simple function of composition. Vignes[23] recommended the following relationship:

$$D_{AB} = (D_{AB})^{x_B}(D_{BA})^{x_A}$$

where D_{AB} is the infinitely dilute diffusion coefficient of A in solvent B, D_{BA} is the infinitely dilute diffusion coefficient of B in solvent A, and x_A and x_B are the molar fraction composition of A and B. This Vignes equation has been less successful for mixtures containing an associating compound, such as an alcohol. A modification for this type of concentrated solution has been proposed by Leffler and Cullinan[24]

$$D_{AB}\mu = (D_{AB}\mu_B)^{x_B}(D_{BA}\mu_A)^{x_A}$$

As the values of liquid diffusion coefficients reported in the literature were obtained in the neighborhood of the ambient temperature, Tyne[25] recommended the following equation for extrapolating to higher temperatures

$$\frac{(D_{AB\,T_1})}{(D_{AB\,T_2})} = \left(\frac{T_c - T_2}{T_c - T_1}\right)^n \tag{24-55}$$

where T_1 and T_2 are in K, T_c is the critical temperature of solvent B in K, and n is the exponent related to the latent heat of vaporization of solvent, ΔH_v, at its normal boiling point temperature. This exponent may be evaluated from the following table:

ΔH_v, (kJ/kmol)	7,900–30,000	30,000–39,000	39,000–46,000	46,000–50,000	>50,000
n	3	4	6	8	10

The properties of electrically conducting solutions have been studied intensively for more than 75 years. Even so, the known relations between electrical conductance and the liquid diffusion coefficient are valid only for dilute solutions of salts in water.

[25] M. J. Tyne, *Trans. I. Chem. E.*, **9**, 112 (1981).

[22] R. C. Reid, J. M. Prausnitz, and T. K. Sherwood, *The Properties of Gases and Liquids,* Third Edition, McGraw-Hill Book Company, New York, 1977, Chapter 11.

[23] A. Vignes, *Ind. Eng. Chem. Fundam.*, **5**, 189 (1966).

[24] J. Leffler and H. T. Cullinan, *Ind. Eng. Chem.*, **9**, 84 (1970).

[25] M. J. Tyne, *Trans. I. Chem. E.*, **9**, 112 (1981).

The diffusion coefficient of a univalent salt in dilute solution is given by the Nernst equation

$$D_{AB} = \frac{2RT}{(1/\lambda_+^0 + 1/\lambda_-^0)\mathscr{F}}$$

(24-56)

where D_{AB} is the diffusion coefficient based on the molecular concentration of A, in cm^2/s; R is the gas constant, 8.316 joules/(K)(g mol); T is absolute temperature, in K, λ_+^0, λ_-^0 are the limiting (zero concentration) ionic conductances in (amp/cm^2) (volt/cm) (g equivalent/cm^3), and $\mathscr{F}$ is Faraday's constant, 96,500 coulombs/g equivalent. This equation has been extended to polyvalent ions by replacing the numerical constant 2 by $(1/n^+ + 1/n^-)$, where n^+ and n^- are the valences of the cation and anion, respectively.

Pore Diffusivity

There are many instances where molecular diffusion occurs inside the pores of porous solids. For example, many catalysts are porous solid pellets containing catalytically active sites on the pore walls. The porous catalyst possesses a high internal surface area to promote chemical reactions at the catalytic surface. The separation of solutes from dilute solution by the process of adsorption is another example. In an adsorption process, the solute sticks to a feature on the solid surface that is attractive to the solute. Many adsorbent materials are porous to provide a high internal surface area for solute adsorption. In both examples, the molecules must diffuse through a gas or liquid phase residing inside the pores. As the pore diameter approaches the diameter of the diffusing molecule, the diffusing molecule can interact with the wall of the pore. Below, we describe two types of pore diffusion: the Knudsen diffusion of gases in cylindrical pores and the hindered diffusion of solutes in solvent-filled cylindrical pores.

Knudsen diffusion. Consider the diffusion of gas molecules through very small capillary pores. If the pore diameter is smaller than the mean free path of the diffusing gas molecules and the density of the gas is low, the gas molecules will collide with the pore walls more frequently than with each other. This process is known as Knudsen flow or *Knudsen diffusion*. The gas flux is reduced by the wall collisions.

The Knudsen number, Kn, given by

$$Kn = \frac{\lambda}{d_{pore}} = \frac{\text{mean free path length of the diffusing species}}{\text{pore diameter}}$$

is a good measure of the relative importance of Knudsen diffusion. If the Kn number is much greater than one, then Knudsen diffusion can be important. At a given pore diameter, the Kn number goes up as the total system pressure P decreases and absolute temperature T increases. In practice, Knudsen diffusion applies only to gases because the mean free path for molecules in the liquid state is very small, typically near the molecular diameter of the molecule itself. Consequently, Kn for liquids is very small. The diffusivity for Knudsen diffusion is obtained from the self-diffusion coefficient derived from the kinetic theory of gases

$$D_{AA^*} = \frac{\lambda u}{3} = \frac{\lambda}{3}\sqrt{\frac{8\kappa NT}{\pi M_A}}$$

(24-57)

For Knudsen diffusion, we replace path length λ with pore diameter d_{pore}, as species A is now more likely to collide with the pore wall as opposed to another molecule. In this instance, the Knudsen diffusivity for diffusing species A, D_{KA}, is

$$D_{KA} = \frac{d_{\text{pore}}}{3} u = \frac{d_{\text{pore}}}{3} \sqrt{\frac{8\kappa NT}{\pi M_A}} \qquad (24\text{-}58)$$

$$D_{KA} = \frac{d_{\text{pore}}}{3} \sqrt{\frac{8}{\pi} \left(1.38 \cdot 10^{-16} \frac{\text{g} \cdot \text{cm}}{\text{s}^2 K}\right) \left(6.023 \cdot 10^{23} \frac{molecules}{mol}\right)} \sqrt{\frac{T}{M_A}}$$

$$= 4850 d_{\text{pore}} \sqrt{\frac{T}{M_A}}$$

This simplified equation requires that d_{pore} has units of cm, M_A has units of g/mol, and temperature T has units of K. The Knudsen diffusivity, D_{KA}, is dependent on the pore diameter, species A molecular weight, and temperature. We can make two comparisons of D_{KA} to the binary gas phase diffusivity, D_{AB}. First, it is not a function of absolute pressure P, or the presence of species B in the binary gas mixture. Second, the temperature dependence for the Knudsen diffusivity is $D_{KA} \propto T^{1/2}$, vs. $D_{AB} \propto T^{3/2}$ for the binary gas phase diffusivity.

Generally, the Knudsen process is significant only at low pressure and small pore diameter. However, there are instances where both Knudsen diffusion and molecular diffusion (D_{AB}) can be important. If we consider that Knudsen diffusion and molecular diffusion compete with one another by a "resistances in series" approach, then the effective diffusivity of species A in a binary mixture of A and B, D_{Ae}, is determined by

$$\frac{1}{D_{Ae}} = \frac{1 - \alpha y_A}{D_{AB}} + \frac{1}{D_{KA}} \qquad (24\text{-}59)$$

with

$$\alpha = 1 + \frac{N_B}{N_A}$$

For cases where $\alpha = 0$ ($N_A = -N_B$), or where y_A is close to zero, equation (24-59) reduces to

$$\frac{1}{D_{Ae}} = \frac{1}{D_{AB}} + \frac{1}{D_{KA}} \qquad (24\text{-}60)$$

The above relationships for the effective diffusion coefficient are based on diffusion within straight, cylindrical pores aligned in a parallel array. However, in most porous materials, pores of various diameters are twisted and interconnected with one another, and the path for diffusion of the gas molecule within the pores is "tortuous." For these materials, if an average pore diameter is assumed, a reasonable approximation for the effective diffusion coefficient in random pores is

$$D'_{Ae} = \varepsilon^2 D_{Ae} \qquad (24\text{-}61)$$

$$\text{where } \varepsilon = \frac{\text{the volume occupied by pores within the porous solid}}{\text{total volume of porous solid (solid + pores)}}$$

ε is the volume void fraction of the porous volume within the porous material. This "void fraction" is usually experimentally determined for a specific material.

The four possible types of pore diffusion are illustrated in Figure 24.3, each with their respective diffusivity correlation. The first three, pure molecular diffusion, pure Knudsen

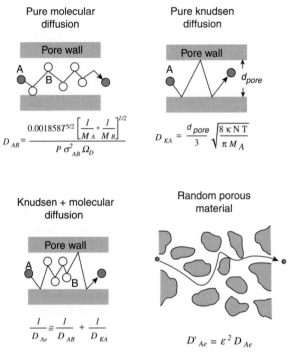

Pure molecular diffusion

$$D_{AB} = \frac{0.001858 T^{3/2} \left[\frac{1}{M_A} + \frac{1}{M_B} \right]^{1/2}}{P \, \sigma_{AB}^2 \, \Omega_D}$$

Pure knudsen diffusion

$$D_{KA} = \frac{d_{pore}}{3} \sqrt{\frac{8 \kappa N T}{\pi M_A}}$$

Knudsen + molecular diffusion

$$\frac{1}{D_{Ae}} \cong \frac{1}{D_{AB}} + \frac{1}{D_{KA}}$$

Random porous material

$$D'_{Ae} = \varepsilon^2 D_{Ae}$$

Figure 24.3 Types of porous diffusion. Shaded areas represent nonporous solids.

diffusion, and Knudsen and molecular combined diffusion, are based on diffusion within straight, cylindrical pores that are aligned in parallel array. The fourth involves diffusion via "tortuous paths" that exist within the compacted solid.

EXAMPLE 6

One step in the manufacture of optical fibers is the chemical vapor deposition of silane (SiH_4) on the inside surface of a hollow glass fiber to form a very thin cladding of solid silicon by the reaction

$$SiH_4(g) \rightarrow Si(s) + 2H_2(g)$$

as shown in Figure 24.4. Typically, the process is carried out at high temperature and very low total system pressure. Optical fibers for high bandwidth data transmission have very small inner pore diameters, typically less than $20 \, \mu m$ ($1 \, \mu m = 1 \times 10^{-6}$ m). If the inner diameter of the Si-coated hollow glass fiber is $10 \, \mu m$, assess the importance of Knudsen diffusion for SiH_4 inside the fiber lumen at 900 K and 100 Pa (0.1 kPa) total system pressure. Silane is diluted to 1.0 mol % in the inert carrier gas helium (He). The binary gas phase diffusivity of silane in helium at 25°C (298 K) and 1.0 atm (101.3 kPa) total system pressure is 0.571 cm²/s, with $\sigma_{SiH_4} = 4.08$ Å and $\varepsilon_{SiH_4}/\kappa = 207.6$ K. The molecular weight of silane is 32 g/mol.

Figure 24.4 Optical fiber.

The gas-phase molecular diffusivity of SiH_4–He, Knudsen diffusivity for SiH_4, and effective diffusivity for SiH_4 at 900 K and 100 Pa total system pressure must be calculated. The gas-phase molecular diffusivity of silane in helium is scaled to process temperature and pressure using the Hirschfelder extrapolation, equation (24-41)

$$D_{SiH_4-He}\left|\begin{matrix}900\,K\\0.1\,kPa\end{matrix}\right. = 0.571\frac{cm^2}{s}\left(\frac{900\,K}{298\,K}\right)^{1.5}\left(\frac{101.3\,kPa}{0.1\,kPa}\right)\left(\frac{0.802}{0.668}\right) = 3.32\times10^3\,\frac{cm^2}{s}$$

It is left to the reader to show that the collision integral Ω_D is equal to 0.802 at 298 K and 0.668 at 900 K for gaseous SiH_4–He mixtures. Note that the gas phase molecular diffusivity is high due to high temperature and very low system pressure. The Knudsen diffusivity of SiH_4 inside the optical fiber is calculated using equation (24-58), with $d_{pore} = 1\times10^{-3}$ cm (10 μm)

$$D_{K,SiH_4} = 4850\,d_{pore}\sqrt{\frac{T}{M_{SiH_4}}} = 4850(1\times10^{-3})\sqrt{\frac{900}{32}} = 25.7\,\frac{cm^2}{s}$$

As the SiH_4 is significantly diluted in He, the process is dilute with respect to SiH_4 and so equation (24-60) can be used to estimate the effective diffusivity

$$D_{SiH_4,e} = \frac{1}{\dfrac{1}{D_{SiH_4-He}}+\dfrac{1}{D_{K,SiH_4}}} = \frac{1}{\dfrac{1}{3.32\times10^3}+\dfrac{1}{25.7}} = 25.5\,\frac{cm^2}{s}$$

The effective diffusivity for SiH_4 is smaller than its Knudsen diffusivity, reflecting the resistance in series approach. Finally, we calculate the Knudsen number for SiH_4

$$\lambda = \frac{\kappa T}{\sqrt{2}\pi\sigma_A^2\,P} = \frac{1.38\times10^{-16}\dfrac{erg}{K}\dfrac{1\,N\,m}{10^7\,erg}900\,K}{\sqrt{2}\pi\left(0.408\,nm\dfrac{1\,m}{10^9\,nm}\right)^2100\dfrac{N}{m^2}} = 1.68\times10^{-4}\,m = 168\,\mu m$$

$$Kn = \frac{\lambda}{d_{pore}} = \frac{168\,\mu m}{10\,\mu m} = 16.8$$

As $Kn \gg 1$ and the effective diffusivity is close to the Knudsen diffusivity, then Knudsen diffusion controls the silane transport inside the optical fiber if no external bulk transport is supplied.

Hindered solute diffusion in solvent-filled pores.   Consider the diffusion of a solute molecule through a tiny capillary pore filled with liquid solvent. As the molecular diameter of the solute approaches the diameter of the pore, the diffusive transport of the solute through the solvent is hindered by the presence of the pore and the pore wall. General models for diffusion coefficients describing the "hindered diffusion" of solutes in solvent-filled pores assume the form of

$$D_{Ae} = D_{AB}^{\circ}F_1(\varphi)F_2(\varphi) \tag{24-62}$$

The molecular diffusion coefficient of solute A in the solvent B at infinite dilution, D_{Ae}°, is reduced by two correction factors, $F_1(\varphi)$, and $F_2(\varphi)$, both of which are theoretically bounded by 0 and 1. Furthermore, both correction factors are functions of the reduced pore diameter φ

$$\varphi = \frac{d_s}{d_{pore}} = \frac{\text{solute molecular diameter}}{\text{pore diameter}} \tag{24-63}$$

If $\varphi > 1$, then the solute is too large to enter the pore. This phenomena is known as solute exclusion, and is used to separate large biomolecules such as proteins from dilute aqueous mixtures containing solutes of much smaller diameter. As φ approaches 1, both $F_1(\varphi)$ and $F_2(\varphi)$ decrease asymptotically toward zero so at $\varphi = 1$, the effective diffusion coefficient is zero.

The correction factor $F_1(\varphi)$, the stearic partition coefficient, is based on simple geometric arguments for stearic exclusion, that is,

$$F_1(\varphi) = \frac{\text{flux area available to solute}}{\text{total flux area}} = \frac{\pi(d_{\text{pore}} - d_s)^2}{\pi d_{\text{pore}}^2} = (1 - \varphi)^2 \qquad (24\text{-}64)$$

and holds for $0 \leq F_1(\varphi) \leq 1.0$

The correction factor $F_2(\varphi)$, the hydrodynamic hindrance factor, is based on the complicated hydrodynamic calculations involving the hindered Brownian motion of the solute within the solvent-filled pore. Equations for $F_2(\varphi)$, assuming diffusion of a rigid spherical solute in a straight cylindrical pore, have been developed. The analytical models are generally asymptotic solutions over a limited range of φ, and ignore electrostatic or other energetic solute–solvent-pore wall intereactions, polydisperity of solute diameters, and noncircular pore cross sections. The most common equation, developed by Renkin,[26] is reasonable for $0 \leq \varphi \leq 0.6$

$$F_2(\varphi) = 1 - 2.104\varphi + 2.09\varphi^3 - 0.95\varphi^5 \qquad (24\text{-}65)$$

EXAMPLE 7

It is desired to separate a mixture of two industrial enzymes, lysozyme and catalase, in a dilute, aqueous solution by a gel filtration membrane. A mesoporous membrane with cylindrical pores of 30 nm diameter is available (Figure 24.5). The following separation factor (α) for the process is proposed

$$\alpha = \frac{D_{Ae}}{D_{Be}}$$

$d_{s,B} = 10.44$ nm

$d_{s,A} = 4.12$ nm

Bulk solvent

$d_{\text{pore}} = 30$ nm

Figure 24.5 Hindered diffusion of solutes in solvent-filled pores.

Determine the separation factor for this process. The properties of each enzyme as reported by Tanford[27] are given below.

[26] E. M. Renkin, *J. Gen. Physiol.,* **38,** 225 (1954).

[27] C. Tanford, *Physical Chemistry of Macromolecules,* John Wiley & Sons, New York, 1961.

Lysozyme (species A)

$M_A = 14\,100\ \text{g/g mol}$

$d_{s,A} = 4.12\ \text{nm}$

$D^o_{A-H_2O} = 1.04 \times 10^{-6}\ \text{cm}^2/\text{s}$

Catalase (species B)

$M_B = 250\,000\ \text{g/g mol}$

$d_{s,B} = 10.44\ \text{nm}$

$D^o_{B-H_2O} = 4.10 \times 10^{-7}\ \text{cm}^2/\text{s}$

The transport of large enzyme molecules through pores filled with liquid water represents a hindered diffusion process. The reduced pore diameters for lysozyme and catalase are

$$\varphi_A = \frac{d_{s,A}}{d_{\text{pore}}} = \frac{4.12\ \text{nm}}{30.0\ \text{nm}} = 0.137 \quad \text{and} \quad \varphi_B = \frac{d_{s,B}}{d_{\text{pore}}} = \frac{10.44\ \text{nm}}{30.0\ \text{nm}} = 0.348$$

For lysozyme, $F_1(\varphi_A)$ by equation (24-64) and $F_2(\varphi_A)$ by the Renkin equation (24-65) are

$$F_1(\varphi_A) = (1 - \varphi_A)^2 = (1 - 0.137)^2 = 0.744$$

$$F_2(\varphi_A) = 1 - 2.104\varphi_A + 2.09\varphi_A^3 - 0.95\varphi_A^5$$

$$= 1 - 2.104(0.137) + 2.09(0.137)^3 - 0.95(0.137)^5 = 0.716$$

The effective diffusivity of lysozyme in the pore, D_{Ae} is estimated by equation (24-62)

$$D_{Ae} = D^o_{A-H_2O}F_1(\varphi_A)F_2(\varphi_A) = 1.04 \times 10^{-6}\ \frac{\text{cm}^2}{\text{s}}(0.744)(0.716) = 5.54 \times 10^{-7}\ \frac{\text{cm}^2}{\text{s}}$$

Likewise, for catalase $F_1(\varphi_B) = 0.425$, $F_2(\varphi_B) = 0.351$, and $D_{Be} = 6.12 \times 10^{-8}\ \text{cm}^2/\text{s}$. Finally, the separation factor is

$$\alpha = \frac{D_{Ae}}{D_{Be}} = \frac{5.54 \times 10^{-7}\ \text{cm}^2/\text{s}}{6.12 \times 10^{-8}\ \text{cm}^2/\text{s}} = 9.06$$

It is interesting to compare the value above with α', the ratio of molecular diffusivities at infinite dilution

$$\alpha' = \frac{D^o_{A-H_2O}}{D^o_{B-H_2O}} = \frac{1.04 \times 10^{-6}\ \text{cm}^2/\text{s}}{4.1 \times 10^{-7}\ \text{cm}^2/\text{s}} = 1.75$$

The small pore diameter enhances the value for α because the diffusion of the large catalase molecule is significantly hindered inside the pore relative to the smaller lysozyme molecule.

Solid Mass Diffusivity

The diffusion of atoms within solids underlies the synthesis of many engineering materials. In semiconductor manufacturing processes, "impurity atoms," commonly called *dopants,* are introduced into solid silicon to control the conductivity in a semiconductor device. The hardening of steel results from the diffusion of carbon and other elements through iron. Vacancy diffusion and interstitial diffusion are the two most frequently encountered solid diffusion mechanisms.

In *vacancy diffusion,* the transported atom "jumps" from a lattice position of the solid into a neighboring unoccupied lattice site or vacancy, as illustrated in Figure 24.6. The atom continues to diffuse through the solid by a series of jumps into other neighboring vacancies that appear to it from time to time. This normally requires a distortion of the lattice. This mechanism has been mathematically described by assuming a unimolecular rate process and applying Eyring's "activated state" concept, as discussed in the "hole" theory for liquid diffusion. The resulting equation is a complex equation relating the diffusivity in terms of the geometric relations between the lattice positions, the length of the jump path, and the energy of activation associated with the jump.

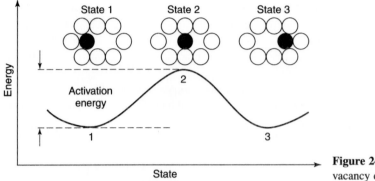

Figure 24.6 Solid-state vacancy diffusion.

An atom moves in *interstitial diffusion* by jumping from one interstitial site to a neighboring one, as illustrated in Figure 24.7. This normally involves a dilation or distortion of the lattice. This mechanism is also mathematically described by Eyring's unimolecular rate theory. Excellent references are available for a more detailed discussion on the diffusion characteristics of atoms in solids (Barrer; Shewmon; Middleman and Hochberg; Kou.[28]

Appendix Table J.3 lists a few values of binary diffusivities in solids. Figure 24.8 illustrates the dependence of solid-phase diffusion coefficients on temperature, specifically for the diffusion of common dopants in solid silicon. The solid-phase diffusion coefficient has been observed to increase with increasing temperature according to an Arrhenius equation of the form

$$D_{AB} = D_o e^{-Q/RT} \tag{24-66}$$

or

$$\ln(D_{AB}) = -\frac{Q}{R}\frac{1}{T} + \ln(D_o) \tag{24-67}$$

where D_{AB} is solid diffusion coefficient for the diffusing species A within solid B, D_o is a proportionality constant of units consistent with D_{AB}, Q is the activation energy (J/mol), R is the thermodynamic constant (8.314 J/mol · K), and T is the absolute temperature (K).

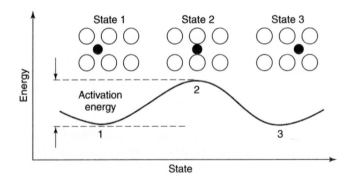

Figure 24.7 Solid-state interstitial diffusion.

[28] R. M. Barrer, *Diffusion In and Through Solids*, Cambridge University Press, London, 1941; P. G. Shewmon, *Diffusion of Solids*, McGraw-Hill Inc., New York, 1963; S. Middleman and A. K. Hochberg, *Process Engineering Analysis in Semiconductor Device Fabrication*, McGraw-Hill Inc., New York, 1993; S. Kou, *Transport Phenomena and Materials Processing*, John Wiley & Sons Inc., New York, 1996.

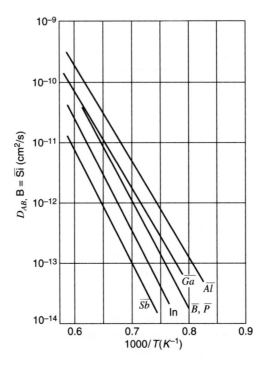

Figure 24.8 Diffusion coefficients of substitutional dopants in crystalline silicon.

Data from Figure 24.8 can be used to estimate Q for a given dopant in silicon using equation (24-67). Tables 24.6 and 24.7 provide the diffusion data needed to evaluate D_{AB} by equation (24-66) for self-diffusion in pure metals and interstitial solutes in iron. These tables point out the significant energy barrier that must be surpassed when an atom jumps between two lattice sites by vacancy diffusion (Table 24.6) and a significantly smaller energy barrier encountered in interstitial diffusion (Table 24.7).

Table 24.6 Data for self-diffusion in pure metals

Structure	Metal	D_o (mm²/s)	Q (kJ/mole)
fcc	Au	10.7	176.9
fcc	Cu	31	200.3
fcc	Ni	190	279.7
fcc	Fe(γ)	49	284.1
bcc	Fe(α)	200	239.7
bcc	Fe(δ)	1980	238.5

Diffusion coefficients and solubilities of solutes in polymers are reported by Rogers,[29] and by Crank and Park.[30] Diffusivities of solutes in dilute biological gels are reported by Friedman and Kramer[31] and by Spalding.[32]

[29] C. E. Rogers, *Engineering Design for Plastics,* Reinhold Press, New York, 1964.
[30] J. Crank and G. S. Park, *Diffusion in Polymers,* Academic Press, New York, 1968.
[31] L. Friedman and E. O. Kramer, *J. Am. Chem. Soc.,* **52,** 1311 (1930).
[32] G. E. Spalding, *J. Phys. Chem.,* 3380 (1969).

Table 24.7 Diffusion parameters for interstitial solutes in iron

Structure	Solute	D_o (mm^2/s)	Q (kJ/mole)
bcc	C	2.0	84.1
bcc	N	0.3	76.1
bcc	H	0.1	13.4
fcc	C	2.5	144.2

24.3 CONVECTIVE MASS TRANSFER

Mass transfer between a moving fluid and a surface or between immiscible moving fluids separated by a mobile interface (as in a gas/liquid or liquid/liquid contactor) is often aided by the dynamic characteristics of the moving fluid. This mode of transfer is called *convective mass transfer*, with the transfer always going from a higher to a lower concentration of the species being transferred. Convective transfer depends on both the transport properties and the dynamic characteristics of the flowing fluid.

As in the case of convective heat transfer, a distinction must be made between two types of flow. When an external pump or similar device causes the fluid motion, the process is called *forced convection*. If the fluid motion is due to a density difference, the process is called *free* or *natural convection*.

The rate equation for convective mass transfer, generalized in a manner analogous to Newton's "law" of cooling, equation 15.11 is

$$N_A = k_c \, \Delta c_A \tag{24-68}$$

where N_A is the molar mass transfer of species A measured relative to fixed spatial coordinates, Δc_A is the concentration difference between the boundary surface concentration and the average concentration of the fluid stream of the diffusing species A, and k_c is the convective mass-transfer coefficient.

As in the case of molecular mass transfer, convective mass transfer occurs in the direction of a decreasing concentration. Equation (24-68) defines the coefficient k_c in terms of the mass flux and the concentration difference from the beginning to the end of the mass-transfer path. The reciprocal of the coefficient, $1/k_c$, represents the resistance to the transfer through the moving fluid. Chapters 28 and 30 consider the methods of determining this coefficient. It is, in general, a function of system geometry, fluid and flow properties, and the concentration difference Δc_A.

From our experiences in dealing with a fluid flowing past a surface, we can recall that there is always a layer, sometimes extremely thin, close to the surface where the fluid is laminar, and that fluid particles next to the solid boundary are at rest. As this is always true, the mechanism of mass transfer between a surface and a fluid must involve molecular mass transfer through the stagnant and laminar flowing fluid layers. The controlling resistance to convective mass transfer is often the result of this "film" of fluid and the coefficient, k_c, is accordingly referred to as the *film mass-transfer coefficient*.

It is important for the student to recognize the close similarity between the convective mass-transfer coefficient and the convective heat-transfer coefficient. This immediately suggests that the techniques developed for evaluating the convective heat-transfer coefficient may be repeated for convective mass transfer. A complete discussion of convective mass-transfer coefficients and their evaluation is given in Chapters 28 and 30.

EXAMPLE 8   A pure nitrogen carrier gas flows parallel to the $0.6\ m^2$ surface of a liquid acetone in an open tank. The acetone temperature is maintained at 290 K. If the average mass-transfer coefficient, k_c, for the mass transfer of acetone into the nitrogen stream is 0.0324 m/s, determine the total rate of acetone release in units of kg.mol/s.

The total molar rate of acetone transfer from the liquid to the gas phase can be evaluated by

$$W_A = N_A A = k_c A(c_{As} - c_{A\infty})$$

The mass transfer area is specified as $0.6\ m^2$. At 290 K, acetone exerts a vapor pressure of 161 mmHg or 2.148×10^4 Pa. Therefore, the concentration of acetone in the gas phase at the acetone surface is

$$c_{As} = \frac{P_A}{RT} = \frac{2.148 \times 10^4\ \text{Pa}}{\left(8.314 \dfrac{\text{Pa} \times \text{m}^3}{\text{kg mol} \times \text{K}}\right)(290\,\text{K})} = 8.91 \frac{\text{kg mol}}{\text{m}^3}$$

and the concentration of acetone in the nitrogen carrier gas is near zero because the molar flowrate of the carrier gas is in a large excess relative to the rate of acetone transfer. Thus

$$W_A = k_c A(c_{As} - c_{A\infty}) = \left(0.0324 \frac{\text{m}}{\text{s}}\right)(0.6\,\text{m}^2)\left(8.91\frac{\text{kg.mol}}{\text{m}^3} - 0\right) = 0.1732 \frac{\text{kg.mol}}{\text{s}}$$

24.4 CLOSURE

In this chapter, the two modes of mass transport, molecular and convective mass transfer, have been introduced. As diffusion of mass involves a multicomponent mixture, fundamental relations were presented for concentrations and velocities of the individual species as well as for the mixture. The molecular transport property, D_{AB}, the diffusion coefficient or mass diffusivity in gas, liquid, and solid systems, has been discussed and correlating equations presented.

The rate equations for the mass transfer of species A in a binary mixture are as follows:

molecular mass transfer:

$\mathbf{J}_A = -cD_{AB}\nabla y_A$	molar flux relative to the molar-average velocity
$\mathbf{j}_A = -\rho D_{AB}\nabla\omega_A$	mass flux relative to the mass-average velocity
$\mathbf{N}_A = -cD_{AB}\nabla y_A + y_A(\mathbf{N}_A + \mathbf{N}_B)$	molar flux relative to fixed spatial coordinates
$\mathbf{n}_A = -\rho D_{AB}\nabla\omega_A + \omega_A(\mathbf{n}_A + \mathbf{n}_B)$	mass flux relative to fixed spatial coordinates

convective mass transfer:

$N_A = k_c\,\Delta c_A$

PROBLEMS

24.1 Liquified natural gas, LNG, is to be shipped from the Alaskan Kenai Peninsula by an ocean carrier to processing plant on Yaquina Bay, Oregon. The molar composition of the commercial LNG is

methane, CH_4	93.5 mol %
ethane, C_2H_6	4.6%
Propane, C_5H_8	1.2%
Carbon dioxide, CO_2	0.7%

determine

a. the weight fraction of ethane;

b. the average molecular weight of the LNG mixture;

c. the density of the gas mixture when heated to 207 K and at 1.4×10^5 Pa;

d. the partial pressure of methane when the total pressure is 1.4×10^5 Pa;

e. the mass fraction of carbon dioxide in parts per million by weight.

24.2 In the manufacture of microelectronic devices, a thin film of solid silicon (Si) is uniformly deposited on a wafer surface by the chemical decomposition of silane (SiH_4) in the presence of H_2 gas. If the gas composition is maintained at 40 mol % SiH_4 and 60 mol % H_2, determine

a. the weight fraction of these species;

b. the average molecular weight of the gas mixture;

c. the molar concentration, c_A, of SiH_4 if the feed gas is maintained at 900 K and a system pressure of 60 torr.

24.3 Air is contained in a $30\,m^3$ container at 400 K and 1.013×10^5 Pa. Determine the following properties of the gas mixture:

a. mole fraction of O_2;

b. volume fraction of O_2;

c. weight of the mixture;

d. mass density of O_2;

e. mass density of N_2;

f. mass density of the air;

g. average molecular weight of the gas mixture.

24.4 Starting with Fick's equation for the diffusion of A through a binary mixture of species A and B as given by $N_{Az} = -cD_{AB}\frac{dy_A}{dz} + y_A(N_{Az} + N_{Bz})$ and Fick's equation for the diffusion of B through the same binary mixture given by $N_{Bz} = -cD_{BA}\frac{dy_B}{dz} + y_B(N_{Bz} + N_{Az})$, prove the two gas diffusivities, D_{AB} and D_{BA}, are equal. Does the Hirshfelder equation for gas evaluating gas diffusivities verify this same equality?

24.5 Starting with the Fick's equation for the diffusion of A through a binary mixture of components A and B

$$\mathbf{N_A} = -cD_{AB}\nabla y_A + y_A(\mathbf{N_A} + \mathbf{N_B})$$

derive the following relations, stating the assumptions made in the derivations:

a. $\mathbf{n_A} = -D_{AB}\nabla\rho_A + w_A(\mathbf{n_A} + \mathbf{n_B})$

b. $\mathbf{J_A} = -D_{AB}\nabla c_A$

24.6 Starting with Fick's equation for the diffusion of A through a binary mixture of A and B, prove

a. $\mathbf{N_A} + \mathbf{N_B} = c\mathbf{V}$;

b. $\mathbf{n_A} + \mathbf{n_B} = \rho\mathbf{v}$

c. $\mathbf{j_A} + \mathbf{j_B} = 0$.

24.7 Stefan and Maxwell explained the diffusion of A through B in terms of the driving force dc_A, the resistances that must overcome the molecular mass transfer, and a proportionality constant, β. The following equation expresses mathematically the resistances for an isothermal, isobaric gaseous system:

$$-dc_A = \beta\frac{\rho_A}{M_A}\frac{\rho_B}{M_B}(v_{Az} - v_{Bz})dz$$

Wilke[33] extended this theory to a multicomponent gas mixture.

---

[33] C. Wilke, *Chem. Eng. Prog.*, **46**, 95 (1950).

The appropriate form of the Maxwell-type equation was assumed to be

$$-\frac{dc_A}{dz} = \beta_{AB}\frac{\rho_A}{M_A}\frac{\rho_B}{M_B}(v_{Az} - v_{Bz}) + \beta_{AC}\frac{\rho_A}{M_A}\frac{\rho_C}{M_C}(v_{Az} - v_{Cz})$$
$$+ \beta_{AD}\frac{\rho_A}{M_A}\frac{\rho_D}{M_D}(v_{Az} - v_{0z}) + \cdots$$

Using this relation, verify equation (24-49).

24.8 Determine the value of the following gas diffusivites using the Hirschfelder equation:

a. carbon dioxide/air at 310 K and 1.5×10^5 Pa

b. ethanol/air at 325 K and 2.0×10^5 Pa

c. carbon monoxide/air at 310 K and 1.5×10^5 Pa

d. carbon tetrachloride/air at 298 K and 1.913×10^5 Pa

24.9 The isomerization of *n*-butane to *iso*-butane is carried out on a catalyst surface at 2.0 atm and 400°C. What is the gas-phase molecular diffusion coefficient of *n*-butane in *iso*-butane? Compare values obtained from both the Hirschfelder and Fuller–Schettler–Giddings equations.

24.10 Determine the diffusivity of methane in air using (a) the Hirschfelder equation and (b) the Wilke equation for a gas mixture. The air is at 373 K and 1.5×10^5 *Pa*.

24.11 An absorption tower is proposed to remove selectively ammonia from an exhaust gas stream. Estimate the diffusivity of ammonia in air at 1.013×10^5 Pa and 373 K using the Brokaw equation (24-43). The dipole moment for ammonia is 1.46 debye. Compare the evaluated value with the experimental value reported in Appendix Table J.1.

24.12 Highly purified tetrachlorosilane ($SiCl_4$) gas is reacted with hydrogen gas (H_2) to produce electronic-grade polycrystalline silicon at 800°C and 1.5×10^5 Pa according to the equction:

$$SiCl_4(g) + 2H_2(g) \rightarrow Si(s) + 4HCl(g).$$

There are concerns that the reaction experiences diffusional limitations at the growing Si solid surface. Estimate the molecular diffusion coefficient for (a) $SiCl_4$ in H_2 and (b) $SiCl_4$ in a gas phase mixture containing 40 mol % $SiCl_4$, 40 mol % H_2, and 20 mol % HCl. The Lennard–Jones parameters for $SiCl_4$ (species A) are $\varepsilon_A/\kappa = 358\,K$, $\sigma_A = 5.08$ Å

24.13 An absorption tower has been proposed to remove selectively two pollutants, hydrogen sulfide (H_2S) and sulfur dioxide (SO_2), from an exhaust gas stream containing

H_2S	3 vol %
SO_2	5 vol %
N_2	92 vol %

Estimate the diffusivity of hydrogen sulfide in the gas mixture at 350 K and 1.013×10^5 Pa. The critical temperature (T_C) of H_2S is 373.2 K and the critical volume (V_C) of H_2S is 98.5 cm^3/mol.

24.14 The Stokes–Einstein equation is often used to estimate the molecular diameter of large spherical molecules from the molecular diffusion coefficient. The measured molecular diffusion coefficient of the serum albumin (an important blood protein) in water at infinite dilution is $5.94 \times 10^{-7} \text{ cm}^2/\text{s}$ at 293 K. Estimate the mean diameter of a serum albumin molecule. The known value is 7.22 nm.

24.15 Estimate the liquid diffusivity of the following solutes that are transferred through dilute solutions:

a. oxygen in ethanol at 293 K;

b. methanol in water at 283 K;

c. water in methanol at 288 K;

d. n-butanol in water at 288 K.

Compare this value with experimental value reported in Appendix J.2.

24.16 Water supplies are often treated with chlorine as one of the processing steps in treating wastewater. Determine the liquid diffusion coefficient of chlorine in an infinitely dilute solution of water at 289 K using (a) the Wilke–Chang equation and (b) the Hayduk–Laudie equation. Compare the results with the experimental value reported in Appendix J.2.

24.17 Benzene (species A) is often added to ethanol to denature the ethanol (species B). Estimate the liquid-phase diffusion of benzene in ethanol and ethanol in benzene at 288 K by (a) the Wilke–Chang equation and (b) the Scheibel equation. Does $D_{AB} = D_{BA}$?

24.18 The aeration of water is an important industrial operation. Determine the liquid diffusion coefficient of oxygen in an infinitely dilute solution of water at 288 K using (a) the Wilke–Chang equation and (b) the Hayduk–Laudie equation.

24.19 A silicon wafer is doped with phosphorus. From Figure 24.8, the nominal value of the diffusion coefficient for phosphorus in silicon at 1316 K is $1 \times 10^{-13} \text{ cm}^2/\text{s}$ and at 1408 K is $1 \times 10^{-12} \text{ cm}^2/\text{s}$. Determine the value of the diffusion coefficient at 1373 K.

24.20 The case-hardening of mild steel involves the diffusion of carbon into iron. Estimate the diffusion coefficient for carbon diffusing into fcc iron and bcc iron at 1000 K. Learn about the structures of fcc and bcc iron in a materials science textbook, and then explain why the diffusion coefficients are different.

24.21 Determine the effective diffusion coefficient for hydrogen gas (H_2, species A) diffusing into nitrogen gas (N_2, species B) at 100°C and 1.0 atm within the following materials:

a. Straight 100 Å pores in parallel array;

b. Random pores 100 Å in diameter with void fraction of 0.4;

c. Random pores 1000 Å in diameter with void fraction of 0.4;

d. Straight 20,000 Å pores in parallel array.

24.22 Researchers are proposing the development of a "nanochannel reactor" for steam reforming of methane (CH_4) to fuel-cell hydrogen gas to power microscale devices.

Gas phase diffusion in nanochannel

As each channel diameter is so small, the gas flow is likely to be very small within a given channel. Hence, gas diffusion processes may play a role in the operation of this device, particularly during the mixing and heating steps. We are specifically interested in evaluating the effective diffusion coefficient of methane gas (species A, $M_A = 16 \text{ g/g. mol}$) in water vapor (species B, $M_B = 18 \text{ g/g. mol}$) at 300°C and 0.5 atm total system pressure. The diameter of the channel is 200 nm ($1 \times 10^9 \text{ nm} = 10 \text{ m}$). A feed gas containing 20 mol % CH_4 in water vapor is fed to the nanochannel with a flux ratio $N_A/N_B = 0.25$. What is effective diffusion coefficient of CH_4 in the nanochannel at the feed gas conditions? Is Knudsen diffusion important?

24.23 Diffusion experiments were conducted with a binary mixture of synthesis gas containing H_2 (species A) diluted in a large excess of CO (species B) at 2.0 atm total system pressure and 80°C in a porous material of monodispersed pore size distribution and void volume fraction of 0.3. From the measurements, the measured effective diffusion coefficient of hydrogen was $0.036 \text{ cm}^2/\text{s}$. What is the mean pore size (d_{pore}) of this material?

24.24 A mixture 1.0 mol % O_2 (species A) in a helium carrler gas (species B) enters the microscale chamber. The chamber consists of a channel that is $5.0 \,\mu\text{m}$ (microns) in diameter. The total system pressure within the chamber is 300 Pa, which is very small. The chamber temperature is maintained at 100°C. The molecular weight of oxygen is 32 g/g. mol and helium is 4.0 g/g. mol.

a. What is the molar concentration of oxygen gas at the entrance to the microchamber?

b. What is the effective diffusion coefficent of O_2 ($D_{A_{ez}}$) within the microchamber?

24.25 Consider a single, porous, spherical, inert mineral particle. The pores inside the particle are filled with liquid water (species B). We are interested in analyzing the molecular diffusion of the contaminant benzene C_6H_6, species A within the water-filled pores of the particle. The average diameter of the pores is 150 nm and the void fraction is 0.40. The benzene solute does not adsorb onto the intersurfaces of the pores. Benzene is very sparingly soluble in water and has a molecular diameter of 0.15 nm. The process is isothermal at 298 K. The concentration of dissolved benzene in the water surrounding the particle, C_{A_c}, is constant with time. Initially, there is no dissolved benzene inside the water-filled pores. The critical volume (V_c) of benzene is 259 cm^3/g.mol.

What is the effective diffusion coefficient of benzene inside the porous particle? Is pore diffusion important?

24.26 Steam reforming of hydrocarbons is one way to make hydrogen gas (H$_2$) for fuel cells. However, the product gas contains H$_2$ contaminated with carbon monoxide (CO), which must be further enriched in H$_2$ in order for the fuel cell to work better. It is desired to separate a mixture of CO and H$_2$ using a microporous ceramic membrane. The average pore diameter of the porous membrane material is 15 nm, and the void fraction $\varepsilon = 0.10$. The system operates at 5.0 atm total system pressure and 400°C.

Determine the effective diffusion coefficient of CO in the gas mixture within the microporous membrane. Is Knudsen diffusion important? You may assume that the CO is dilute in the gas mixture.

24.27 As part of a bioseparation process, glucose (solute A) in aqueous solution is diffusing across a microporous membrane, as shown in the following figure.

Diffusion of glucose across a microporus polymer membrane

The thickness of the membrane is 2.0 mm, and the pores running through the membrane consists of parallel channels of 3.0 nm diameter. The temperature is 30°C. The mean diameter of a single glucose molecule is 0.86 nm (nanometers). Assume that the molecular diffusion coefficient of glucose in water is described by Stokes–Einstein relationship. What is the effective diffusion coefficient, D_{A_z}, of glucose through the membrane?

24.28 Protein mixtures in aqueous solution are commonly separated by molecular sieve chromatography. An important aspect of this separation process is the diffusion of the protein into the porous matrix of the chromatography support used to affect the separation. Estimate the effective diffusion coefficient of the enzyme urease in a silica gel support with 100 nm diameter pores. The molecular diffusion coefficient of urease in water at infinite dilution is 3.46×10^{-7} cm^2/s at 298° K, and the diameter of the molecule is 12.38 nm.

24.29 The diffusion rate of the enzyme ribonuclease into a porous chromatography support was measured at 298 K, and an effective diffusion coefficient of 5.0×10^{-7} cm^2/s was backed out of the data. Estimate the mean pore diameter of the support. The molecular diffusion coefficient of ribonuclease in water is 1.19×10^{-6} cm^2/s at 298 K, and the diameter of the molecule is 3.6 nm.

Chapter 25

Differential Equations of Mass Transfer

$\mathbf{I}$n Chapter 9, the general differential equations for momentum transfer are derived by the use of a differential control volume concept. By an analogous treatment, the general differential equations for heat transfer are generated in Chapter 16. Once again, we shall use this approach to develop the differential equations for mass transfer. By making a mass balance over a differential control volume, we shall establish the equation of continuity for a given species.

Additional differential equations will be obtained when we insert, into the continuity equation, mass flux relationships developed in the previous chapter.

25.1 THE DIFFERENTIAL EQUATION FOR MASS TRANSFER

Consider the control volume, $\Delta x\,\Delta y\,\Delta z$, through which a mixture including component A is flowing, as shown in Figure 25.1. The control-volume expression for the conservation of mass is

$$\iint_{\text{c.s.}} \rho(\mathbf{v}\cdot\mathbf{n})dA + \frac{\partial}{\partial t}\iiint_{\text{c.v.}} \rho\,dV = 0 \qquad (4\text{-}1)$$

which may be stated in words as

$$\left\{\begin{array}{c}\text{net rate of mass}\\ \text{efflux from}\\ \text{control volume}\end{array}\right\} + \left\{\begin{array}{c}\text{net rate of accumulation}\\ \text{of mass within control}\\ \text{volume}\end{array}\right\} = 0$$

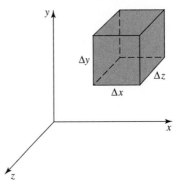

Figure 25.1 A differential control volume.

If we consider the conservation of a given species A, this relation should also include a term that accounts for the production or disappearance of A by chemical reaction within the volume. The general relation for a mass balance of species A for our control volume may be stated as

$$\left\{\begin{array}{c}\text{net rate of mass}\\ \text{efflux of } A \text{ from}\\ \text{control volume}\end{array}\right\} + \left\{\begin{array}{c}\text{net rate of accum}\\ \text{ulation of } A \text{ within}\\ \text{control volume}\end{array}\right\} - \left\{\begin{array}{c}\text{rate of chemical}\\ \text{production of } A\\ \text{within the control}\\ \text{volume}\end{array}\right\} = 0 \qquad (25\text{-}1)$$

The individual terms will be evaluated for constituent A, and a discussion of their meanings will be given below.

The net rate of mass efflux from the control volume may be evaluated by considering the mass transferred across control surfaces. For example, the mass of A transferred across the area $\Delta y \, \Delta z$ at x will be $\rho_A v_{Ax} \Delta y \, \Delta z|_x$, or in terms of the flux vector, $\mathbf{n}_A = \rho_A \mathbf{v}_A$, it would be $n_{A,x} \, \Delta y \, \Delta z|_x$. The net rate of mass efflux of constituent A will be

$$\text{in the } x \text{ direction: } n_{A,x} \Delta y \, \Delta z|_{x+\Delta x} - n_{A,x} \Delta y \, \Delta z|_x$$

$$\text{in the } y \text{ direction: } n_{A,y} \Delta x \, \Delta z|_{y+\Delta y} - n_{A,y} \Delta x \, \Delta z|_y$$

and

$$\text{in the } z \text{ direction: } n_{A,z} \Delta x \, \Delta y|_{z+\Delta z} - n_{A,z} \Delta x \, \Delta y|_z$$

The rate of accumulation of A in the control volume is

$$\frac{\partial \rho_A}{\partial t} \Delta x \, \Delta y \, \Delta z$$

If A is produced within the control volume by a chemical reaction at a rate r_A, where r_A has the units (mass of A produced)/(volume)(time), the rate of production of A is

$$r_A \Delta x \, \Delta y \, \Delta z$$

This production term is analogous to the energy generation term that appeared in the differential equation for energy transfer, as discussed in Chapter 16.

Substituting each term in equation (25-1), we obtain

$$n_{A,x} \Delta y \, \Delta z|_{x+\Delta x} - n_{A,x} \Delta y \, \Delta z|_x + n_{A,y} \Delta x \, \Delta z|_{y+\Delta y}$$
$$- n_{A,y} \Delta x \, \Delta z|_y + n_{A,z} \Delta x \, \Delta y|_{z+\Delta z} - n_{A,z} \Delta x \, \Delta y|_z \qquad (25\text{-}2)$$
$$+ \frac{\partial \rho_A}{\partial t} \Delta x \, \Delta y \, \Delta z - r_A \Delta x \, \Delta y \, \Delta z = 0$$

Dividing through by the volume, $\Delta x \, \Delta y \, \Delta z$, and canceling terms, we have

$$\frac{n_{A,x}|_{x+\Delta x} - n_{A,x}|_x}{\Delta x} + \frac{n_{A,y}|_{y+\Delta y} - n_{A,y}|_y}{\Delta y} + \frac{n_{A,z}|_{z+\Delta z} - n_{A,z}|_z}{\Delta z} + \frac{\partial \rho_A}{\partial t} - r_A = 0 \qquad (25\text{-}3)$$

Evaluated in the limit as Δx, Δy, and Δz approach zero, this yields

$$\frac{\partial}{\partial x} n_{A,x} + \frac{\partial}{\partial y} n_{A,y} + \frac{\partial}{\partial z} n_{A,z} + \frac{\partial \rho_A}{\partial t} - r_A = 0 \qquad (25\text{-}4)$$

Equation (25-4) is the *equation of continuity for component A*. As $n_{A,x}$, $n_{A,y}$, and $n_{A,z}$ are the rectangular components of the mass flux vector, $\mathbf{n}_A$, equation (25-4) may be written

$$\nabla \cdot \mathbf{n}_A + \frac{\partial \rho_A}{\partial t} - r_A = 0 \qquad (25\text{-}5)$$

A similar equation of continuity may be developed for a second constituent B in the same manner. The differential equations are

$$\frac{\partial}{\partial x} n_{B,x} + \frac{\partial}{\partial y} n_{B,y} + \frac{\partial}{\partial z} n_{B,z} + \frac{\partial \rho_B}{\partial t} - r_B = 0 \qquad (25\text{-}6)$$

and

$$\boldsymbol{\nabla} \cdot \mathbf{n}_B + \frac{\partial \rho_B}{\partial t} - r_B = 0 \tag{25-7}$$

where r_B is the rate at which B will be produced within the control volume by a chemical reaction. Adding equations (25-5) and (25-7), we obtain

$$\boldsymbol{\nabla} \cdot (\mathbf{n}_A + \mathbf{n}_B) + \frac{\partial(\rho_A + \rho_B)}{\partial t} - (r_A + r_B) = 0 \tag{25-8}$$

For a binary mixture of A and B, we have

$$\mathbf{n}_A + \mathbf{n}_B = \rho_A \mathbf{v}_A + \rho_B \mathbf{v}_B = \rho \mathbf{v}$$
$$\rho_A + \rho_B = \rho$$

and

$$r_A = -r_B$$

by the law of conservation of mass. Substituting these relations into (25-8), we obtain

$$\boldsymbol{\nabla} \cdot \rho \mathbf{v} + \frac{\partial \rho}{\partial t} = 0 \tag{25-9}$$

This is the *equation of continuity for the mixture*. Equation (25-9) is identical to the equation of continuity (9-2) for a homogeneous fluid.

The equation of continuity for the mixture and for a given species can be written in terms of the substantial derivative. As shown in Chapter 9, the continuity equation for the mixture can be rearranged and written

$$\frac{D\rho}{Dt} + \rho \boldsymbol{\nabla} \cdot \mathbf{v} = 0 \tag{9-5}$$

Through similar mathematical manipulations, the equation of continuity for species A in terms of the substantial derivative may be derived. This equation is

$$\frac{\rho D\omega_A}{Dt} + \boldsymbol{\nabla} \cdot \mathbf{j}_A - r_A = 0 \tag{25-10}$$

We could follow the same development in terms of molar units. If R_A represents the rate of molar production of A per unit volume, and R_B represents the rate of molar production of B per unit volume, the molar-equivalent equations are

for component A

$$\boldsymbol{\nabla} \cdot \mathbf{N}_A + \frac{\partial c_A}{\partial t} - R_A = 0 \tag{25-11}$$

for component B

$$\boldsymbol{\nabla} \cdot \mathbf{N}_B + \frac{\partial c_B}{\partial t} - R_B = 0 \tag{25-12}$$

and for the mixture

$$\boldsymbol{\nabla} \cdot (\mathbf{N}_A + \mathbf{N}_B) + \frac{\partial(c_A + c_B)}{\partial t} - (R_A + R_B) = 0 \tag{25-13}$$

For the binary mixture of A and B, we have

$$\mathbf{N}_A + \mathbf{N}_B = c_A v_A + c_B v_B = c\mathbf{V}$$

and

$$c_A + c_B = c$$

However, only when the stoichiometry of the reaction is

$$A \rightleftharpoons B$$

which stipulates that one molecule of B is produced for each mole of A disappearing, can we stipulate that $R_A = -R_B$. In general, the equation of continuity for the mixture in molar units is

$$\nabla \cdot c\mathbf{V} + \frac{\partial c}{\partial t} - (R_A + R_B) = 0 \tag{25-14}$$

25.2 SPECIAL FORMS OF THE DIFFERENTIAL MASS-TRANSFER EQUATION

Special forms of the equation of continuity applicable to commonly encountered situations follow. In order to use the equations for evaluating the concentration profiles, we replace the fluxes, $\mathbf{n}_A$ and $\mathbf{N}_A$, by the appropriate expressions developed in Chapter 24. These expressions are

$$\mathbf{N}_A = -cD_{AB}\nabla y_A + y_A(\mathbf{N}_A + \mathbf{N}_B) \tag{24-21}$$

or its equivalent

$$\mathbf{N}_A = -cD_{AB}\nabla y_A + c_A\mathbf{V}$$

and

$$\mathbf{n}_A = -\rho D_{AB}\nabla \omega_A + \omega_A(\mathbf{n}_A + \mathbf{n}_B) \tag{24-22}$$

or its equivalent

$$\mathbf{n}_A = -\rho D_{AB}\nabla \omega_A + \rho_A \mathbf{v}$$

Substituting equation (24-22) into equation (25-5), we obtain

$$-\nabla \cdot \rho D_{AB}\nabla \omega_A + \nabla \cdot \rho_A \mathbf{v} + \frac{\partial \rho_A}{\partial t} - r_A = 0 \tag{25-15}$$

and substituting equation (24-21) into equation (25-11), we obtain

$$-\nabla \cdot cD_{AB}\nabla y_A + \nabla \cdot c_A\mathbf{V} + \frac{\partial c_A}{\partial t} - R_A = 0 \tag{25-16}$$

Either equation (25-15) or (25-16) may be used to describe concentration profiles within a diffusing system. Both equations are completely general; however, they are relatively unwieldy. These equations can be simplified by making restrictive assumptions. Important forms of the equation of continuity, with their qualifying assumptions, include:

(i) If the density, ρ, and the diffusion coefficient, D_{AB} can be assumed constant, equation (25-15) becomes

$$-D_{AB}\nabla^2 \rho_A + \overset{0}{\cancel{\rho_A \nabla \cdot \mathbf{v}}} + \mathbf{v} \cdot \nabla \rho_A + \frac{\partial \rho_A}{\partial t} - r_A = 0$$

Dividing each term by the molecular weight of A and rearranging, we obtain

$$\mathbf{v} \cdot \nabla c_A + \frac{\partial c_A}{\partial t} = D_{AB} \nabla^2 c_A + R_A \tag{25-17}$$

(ii) If there is no production term, $R_A = 0$, and if the density and diffusion coefficient are assumed constant, equation (25-17) reduces to

$$\frac{\partial c_A}{\partial t} + \mathbf{v} \cdot \nabla c_A = D_{AB} \nabla^2 c_A \tag{25-18}$$

We recognize that $(\partial c_A / \partial t) + \mathbf{v} \cdot \nabla c_A$ is the substantial derivative of c_A; rewriting the left-hand side of equation (25-18), we obtain

$$\frac{D c_A}{Dt} = D_{AB} \nabla^2 c_A \tag{25-19}$$

which is analogous to equation (16-14) from heat transfer

$$\frac{DT}{Dt} = \frac{k}{\rho c_P} \nabla^2 T \tag{16-14}$$

or

$$\frac{DT}{Dt} = \alpha \nabla^2 T$$

where α is the thermal diffusivity. The similarity between these two equations is the basis for the analogies drawn between heat and mass transfer.

(iii) In a situation in which there is no fluid motion, $\mathbf{v} = 0$, no production term, $R_A = 0$, and no variation in the diffusivity or density, equation (25-18) reduces to

$$\frac{\partial c_A}{\partial t} = D_{AB} \nabla^2 c_A \tag{25-20}$$

Equation (25-20) is commonly referred to as *Fick's second "law" of diffusion*. The assumption of no fluid motion restricts its applicability to diffusion in solids, or stationary liquids, and for binary systems of gases or liquids, where $\mathbf{N}_A$ is equal in magnitude, but acting in the opposite direction to $\mathbf{N}_B$; that is, the case of equimolar counterdiffusion. Equation (25-20) is analogous to Fourier's second "law" of heat conduction

$$\frac{\partial T}{\partial t} = \alpha \nabla^2 T \tag{16-18}$$

(iv) Equations (25-17), (25-18), and (25-20) may be simplified further when the process to be defined is a steady-state process; that is, $\partial c_A / \partial t = 0$. For constant density and a constant-diffusion coefficient, the equation becomes

$$\mathbf{v} \cdot \nabla c_A = D_{AB} \nabla^2 c_A + R_A \tag{25-21}$$

For constant density, constant diffusivity, and no chemical production, $R_A = 0$, we obtain

$$\mathbf{v} \cdot \nabla c_A = D_{AB} \nabla^2 c_A \tag{25-22}$$

If additionally, $\mathbf{v} = 0$ the equation reduces to

$$\nabla^2 c_A = 0 \tag{25-23}$$

Equation (25-23) is the *Laplace equation* in terms of molar concentration.

Each of the equations (25-15) through (25-23) has been written in vector form, thus each applies to any orthogonal coordinate system. By writing the Laplacian operator, ∇^2, in the appropriate form, the transformation of the equation to the desired coordinate system is accomplished. Fick's second "law" of diffusion written in rectangular coordinates is

$$\frac{\partial c_A}{\partial t} = D_{AB}\left[\frac{\partial^2 c_A}{\partial x^2} + \frac{\partial^2 c_A}{\partial y^2} + \frac{\partial^2 c_A}{\partial z^2}\right] \tag{25-24}$$

in cylindrical coordinates is

$$\frac{\partial c_A}{\partial t} = D_{AB}\left[\frac{\partial^2 c_A}{\partial r^2} + \frac{1}{r}\frac{\partial c_A}{\partial r} + \frac{1}{r^2}\frac{\partial^2 c_A}{\partial \theta^2} + \frac{\partial^2 c_A}{\partial z^2}\right] \tag{25-25}$$

and in spherical coordinates is

$$\frac{\partial c_A}{\partial t} = D_{AB}\left[\frac{1}{r^2}\frac{\partial}{\partial r}\left(r^2\frac{\partial c_A}{\partial r}\right) + \frac{1}{r^2\sin\theta}\frac{\partial}{\partial\theta}\left(\sin\theta\frac{\partial c_A}{\partial\theta}\right) + \frac{1}{r^2\sin^2\theta}\frac{\partial^2 c_A}{\partial\phi^2}\right] \tag{25-26}$$

The general differential equation for mass transfer of component A, or the equation of continuity of A, written in rectangular coordinates is

$$\frac{\partial c_A}{\partial t} + \left[\frac{\partial N_{A,x}}{\partial x} + \frac{\partial N_{A,y}}{\partial y} + \frac{\partial N_{A,z}}{\partial z}\right] = R_A \tag{25-27}$$

in cylindrical coordinates is

$$\frac{\partial c_A}{\partial t} + \left[\frac{1}{r}\frac{\partial}{\partial r}(rN_{A,r}) + \frac{1}{r}\frac{\partial N_{A,\theta}}{\partial\theta} + \frac{\partial N_{A,z}}{\partial z}\right] = R_A \tag{25-28}$$

and in spherical coordinates is

$$\frac{\partial c_A}{\partial t} + \left[\frac{1}{r^2}\frac{\partial}{\partial r}(r^2 N_{A,r}) + \frac{1}{r\sin\theta}\frac{\theta}{\partial\theta}(N_{A,\theta}\sin\theta) + \frac{1}{r\sin\theta}\frac{\partial N_{A,\phi}}{\partial\phi}\right] = R_A \tag{25-29}$$

25.3 COMMONLY ENCOUNTERED BOUNDARY CONDITIONS

A mass-transfer process is fully described by the differential equations of mass transfer only if the initial boundary and initial conditions are specified. Typically, initial and boundary conditions are used to specify limits of integration or to determine integration constants associated with the mathematical solution of the differential equations for mass transfer. The initial and boundary conditions used for mass transfer are very similar to those used in Section 16.3 for energy transfer. The reader may wish to refer to that section for further discussion of initial and boundary conditions.

The initial condition in mass transfer processes is the concentration of the diffusing species at the start of the time interval of interest expressed in either mass or molar concentration units. The concentration may be simply equal to a constant, for example

$$\text{at } t = 0, \ c_A = c_{Ao} \ \text{(molar units)}$$
$$\text{at } t = 0, \ \rho_A = \rho_{Ao} \ \text{(mass units)}$$

or may be more complex if the initial concentration distribution within the control volume for diffusion is specified. Initial conditions are associated only with unsteady-state or pseudo-steady-state processes.

Four types of boundary conditions are commonly encountered in mass transfer.

(1) *The concentration of the transferring species A at a boundary surface is specified.* Surface concentration can assume a variety of units, for example, molar concentration c_{As}, mass concentration ρ_{As}, gas mole fraction y_{As}, liquid mole fraction x_{As}, etc. When the boundary surface is defined by a pure component in one phase and a mixture in the second phase, then the concentration of transferring species A in the mixture at the interface is usually at thermodynamic saturation conditions. Specifically, for a gas mixture in contact with a pure volatile liquid A or pure volatile solid A, the partial pressure of species A in the gas at the surface is saturation vapor pressure, P_A, so that $p_{As} = P_A$. For a liquid mixture in contact with a pure solid A, the concentration of species A in the liquid at the surface is the solubility limit of A in the liquid, c_A^* so that $c_{As} = c_A^*$.

For a contacting gas and liquid where transferring species A is present in both phases, there are two ways to specify the concentration at the gas–liquid interface. First, if both of the species in the liquid phase are volatile, then the boundary condition at the gas–liquid surface is defined for an ideal liquid mixture by Raoult's law

$$p_{As} = x_A P_A$$

where x_A is the mole fraction in the liquid, P_A is the vapor pressure of species A evaluated at the temperature of the liquid, and P_{As} is the partial pressure of species A in the gas. The partial pressure of species A [at the sur]face is related to surface mole fraction y_{As} by Dalton's law

$$= \frac{p_{As}}{P}$$

or to surface co[ncentration] | Gas law

$$= \frac{p_{As}}{RT}$$

Second, for solut[es that are only] weakly soluble in the liquid, Henry's law may be used to re[late the concentration of A] in the liquid to the partial pressure of A in the gas

$$= H \cdot x_A$$

where coefficient [H is the Henry's law co]nstant. Values of H in pressure units for selected gaseous so[lutes dissolved in aqueous] solution are listed in Table 25.1. A similar equation may also [be used to describe] the boundary conditions at a *gas–solid* interface

$$c_{A,solid} = S \cdot p_A$$

Table 25.1 Henry's constant for various gases in aqueous solutions (H in bars)

T (K)	NH₃	Cl₂	H₂S	SO₂	CO₂	CH₄	O₂	H₂
273	21	265	260	165	710	22,800	25,500	58,000
280	23	365	335	210	960	27,800	30,500	61,500
290	26	480	450	315	1300	35,200	37,600	66,500
300	30	615	570	440	1730	42,800	45,700	71,600
310		755	700	600	2175	50,000	52,500	76,000
320		860	835	800	2650	56,300	56,800	78,600

Table 25.2 Solubility constants for selected gas–solid combinations (1 bar = 10^5 Pa)

Gas	Solid	T (K)	$S = c_{A,solid}/P_A$ (kg mol/m³ bar)
O_2	Natural rubber	298	3.12×10^{-3}
N_2	Natural rubber	298	1.56×10^{-3}
CO_2	Natural rubber	298	40.15×10^{-3}
He	Silicon	293	0.45×10^{-3}
H_2	Ni	358	9.01×10^{-3}

where $c_{A,solid}$ is the molar concentration of A within the *solid* at the interface in units of kg mol/m³ and p_A is the partial pressure of gas phase species A over the solid in units of Pa. The partition coefficient S, also known as the solubility constant, has units of kg mol/m³ · Pa. Values of S for several gas–solid pairs reported by Barrer[1] are listed in Table 25.2.

(2) *A reacting surface boundary is specified.* There are three common situations, all dealing with heterogeneous surface reactions. First, the flux of one species may be related to the flux of another species by chemical reaction stoichiometry. For example, consider the generic chemical reaction at the boundary surface $A + 2B \rightarrow 3C$, where reactants A and B diffuse to the surface, and product C diffuses away from the surface. The fluxes for A and B move in the *opposite direction* to the flux for C. Consequently, the flux N_A is related to the flux of the other species by $N_B = +2 N_A$ or $N_C = -3 N_A$. Second, a finite rate of chemical reaction might exist at the surface, which in turn sets the flux at the surface. For example, if component A is consumed by a first-order on a surface at $z = 0$, *and* the positive z direction is opposite to the direction of flux of A along z, then

$$N_A\big|_{z=0} = -k_c c_{As}$$

where k_s is a surface reaction rate constant with units of m/s. Third, the reaction may be so rapid that $c_{As} = 0$ if species A is the limiting reagent in the chemical reaction.

(3) *The flux of the transferring species is zero at a boundary or at a centerline of symmetry.* This situation can arise at an impermeable boundary, or at the centerline of symmetry of the control volume, where the net flux is equal to zero. In either case, for a one-dimensional flux along z

$$N_A\big|_{z=0} = -D_{AB} \frac{\partial c_A}{\partial z}\bigg|_{z=0} = 0 \quad \text{or} \quad \frac{\partial c_A}{\partial z}\bigg|_{z=0} = 0$$

where the impermeable boundary or the centerline of symmetry is located at $z = 0$.

(4) *The convective mass transfer flux at the boundary surface is specified.* When a fluid flows over the boundary, the flux can be defined by convection. For example, at some surface located at $z = 0$, the convective mass transfer flux across the fluid boundary layer is

$$N_A\big|_{z=0} = k_c(c_{As} - c_{A\infty})$$

where $c_{A\infty}$ is the bulk concentration of A the flowing fluid, c_{As} is the surface concentration of A at $z = 0$ and k_c is the convection mass-transfer coefficient defined in Section 24.3.

---

[1] R. M. Barrer, *Diffusion In and Through Solids*, Macmillan Press, New York, 1941.

25.4   STEPS FOR MODELING PROCESSES INVOLVING MOLECULAR DIFFUSION

Processes involving molecular diffusion can be modeled by the appropriate simplifications to Fick's equation and the general differential equation for mass transfer. In general, most molecular diffusion problems involve working through the following five steps:

Step 1:   Draw a picture of the physical system. Label the important features, including the system boundaries. Decide where the source and the sink of mass transfer are located.

Step 2:   Make a "list of assumptions" based on your consideration of the physical system. As appropriate, make a "list of nomenclature" and update the list as you add more terms to the model development.

Step 3:   Pick the coordinate system that best describes the geometry of the physical system: rectilinear (x, y, z), cylindrical (r, z, θ), or spherical (r, θ, ϕ). Then formulate differential material balances to describe the mass transfer within a volume element of the process based on the geometry of the physical system and the assumptions proposed, making use of Fick's law and the general differential equation for mass transfer. Two approaches may be used to simplify the general differential equation for mass transfer. In the first approach, simply reduce or eliminate the terms that do not apply to the physical system. For example:

(a) If the process is steady state, then $\dfrac{\partial c_A}{\partial t} = 0$.

(b) If no chemical reaction occurs uniformly within the control volume for diffusion, then $R_A = 0$.

(c) If the molecular mass transfer process of species A is one-dimensional in the z direction,

$$\nabla \cdot N_A = \frac{\partial N_{Az}}{\partial z}$$

by cylindrical geometry in the r and z directions,

$$\nabla \cdot N_A = \frac{\partial N_{Az}}{\partial z} + \frac{1}{r}\frac{\partial (r N_{Ar})}{\partial r}$$

for radial symmetry in spherical coordinates,

$$\nabla \cdot N_A = \frac{1}{r^2}\frac{\partial (r^2 N_{Ar})}{\partial r}$$

In the second approach, perform a "shell balance" for the component of interest on a differential volume element of the process. Both of these approaches are discussed and illustrated in Chapter 26. Next, Fick's law is simplified by establishing the relationship between the fluxes in the bulk-contribution term. For example, recall the one-dimensional flux of a binary mixture of components A and B

$$N_{Az} = -c D_{AB}\frac{dy_A}{dz} + y_A(N_{Az} + N_{Bz})$$

If $N_{Az} = -N_{Bz}$, then $y_A(N_{Az} + N_{Bz}) = 0$. If $y_A(N_{Az} + N_{Bz})$ does not equal 0, then N_A is always equal to $c_A V_z$ and reduces to $c_A v_z$ for low concentrations of A in the

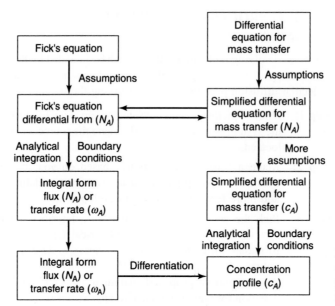

Figure 25.2 Model development pathways for processes involving molecular diffusion.

mixture. If a differential equation for the concentration profile is desired, then the simplified form of Fick's law must be substituted into the simplified form of the general differential equation for mass transfer. Figure 25.2 illustrates this process.

Step 4: Recognize and specify the boundary conditions and initial conditions. For example

(a) Known concentration of species A at a surface or interface at $z = 0$ e.g., $c_A = c_{Ao}$. This concentration can be specified or known by equilibrium relationships such as Henry's law.

(b) Symmetry condition at a centerline of the control volume for diffusion, or no net diffusive flux of species A at a surface or interface at $z = 0$,

$$N_{Az}|_{z=0} = 0 = dc_A/dz.$$

(c) Convective flux of species A at a surface or interface, e.g., $N_A = k_c(c_{As} - c_{A\infty})$.

(d) Known flux of species A at a surface or interface, e.g., at $z = 0$, $N_{Az}|_{z=0} = N_{Ao}$.

(e) Known chemical reaction at a surface or interface. For the rapid disappearance of species A at the surface or interface, e.g., at $z = 0$, $c_{As} = 0$. For a slower chemical reaction at the surface or interface with finite c_{As} at $z = 0$, e.g., $N_{Az} = k'c_{As}$, where k' is a first-order chemical reaction rate constant.

Step 5: Solve the differential equations resulting from the differential material balances and the boundary/initial conditions described to get the concentration profile, the flux, or other parameters of engineering interest. If appropriate, consider asymptotic solutions or limiting cases to more difficult problems first.

The following examples illustrate how physical and chemical processes involving molecular diffusion can be modeled by the appropriate simplifications of Fick's equation and the general differential equation for mass transfer. The examples cover many of typically encountered boundary conditions in both rectilinear and cylindrical geometry. The examples emphasize the first four steps of model development outlined, and the final model equations are generally left in differential-equation form. Chapters 26 and 27 provide analytical solution techniques for steady-state and unsteady-state diffusion processes. We have taken extra time at the beginning of each example to describe the interesting technology behind the process.

EXAMPLE 1

Microelectronic devices are fabricated by forming many layers of thin films onto a silicon wafer. Each film has unique chemical and electrical properties. For example, a thin film of solid silicon (Si) serves as a semiconductor. Silicon thin films are commonly formed by the *chemical vapor deposition*, or CVD, of silane vapor (SiH_4) onto the surface of the wafer. The chemical reaction is

$$SiH_4(g) \rightarrow Si(s) + 2H_2(g)$$

This surface reaction is usually carried out at very low pressure (100 Pa) and high temperature (900 K). In many CVD reactors, the gas phase over the Si film is not mixed. Furthermore, at high temperatures, the surface reaction is very rapid. Consequently, the molecular diffusion of the SiH_4 vapor to the surface often controls the rate of Si film formation. Consider the very simplified CVD reactor shown in Figure 25.3. A mixture of silane and hydrogen gas flows into the reactor. A diffuser provides a quiescent gas space over the growing Si film. Develop a differential model for this process, including statements of assumptions and boundary conditions.

Figure 25.3 Chemical vapor deposition of silicon hydride.

The silane in the feed gas serves as the source for mass transfer, whereas the Si film serves as the sink for silane mass transfer. In contrast, the formation of H_2 at the Si film surface serves as the source for H_2 mass transfer, whereas the feed gas serves as the sink for H_2 mass transfer. The physical system possesses rectilinear geometry, and the major assumptions for model development are listed here.

(1) The reaction occurs only at the surface of growing Si thin film. Consequently, there is no homogeneous reaction of silane within the diffusion zone, so that $R_A = 0$. In this context, the surface reaction is the sink for silane mass transfer. **(2)** The gas space in the "diffusion zone" is not externally mixed, so that molecular diffusion dominates. **(3)** The feed gas provides silane in high excess relative to that consumed by reaction, so the silane concentration in the gas space at the diffusion-zone boundary is constant. **(4)** The flux of silane is one-dimensional along z, as the source and sink for silane mass transfer are aligned at the boundaries along the z direction. **(5)** The thickness of the Si film is very thin relative to δ, the diffusion path length along the z direction. Therefore, δ is essentially constant. **(6)** The mass transfer process within the diffusion zone is at steady state.

The assumptions are used to reduce the general forms of the differential equation for mass transfer and Fick's equation. The general differential equation for mass transfer in terms of rectilinear coordinates is

$$-\left(\frac{\partial N_{Ax}}{\partial x} + \frac{\partial N_{Ay}}{\partial y} + \frac{\partial N_{Az}}{\partial z}\right) + R_A = \frac{\partial c_A}{\partial t}$$

For steady-state one-dimensional flux along the z direction with no homogeneous chemical reaction ($R_A = 0$), the general differential equation for mass transfer reduces to

$$\frac{dN_{Az}}{dz} = 0$$

which shows that the flux is constant along the z direction. As the diffusion flux is with respect to only one dimension, the partial derivative becomes an ordinary derivative. Fick's equation for the one-dimensional flux silane through a binary mixture in the gas phase is

$$N_{Az} = -cD_{AB}\frac{dy_{Az}}{dz} + y_A(N_{Az} + N_{Bz})$$

where species A represents silane vapor (SiH_4) reactant and species B represents the hydrogen gas (H_2) product. The flux of the gaseous reactant is opposite in direction to the flux of the gaseous product. From the reaction stoichiometry and Figure 25.3, N_{Az} is related to N_{Bz} as follows:

$$\frac{N_{Az}}{N_{Bz}} = \frac{-1 \text{ mol } SiH_4 \text{ reacted}}{+2 \text{ mol } H_2 \text{ formed}} = -\frac{1}{2}$$

Therefore, $N_{Bz} = -2N_{Az}$ and Fick's equation further reduces to

$$N_{Az} = -cD_{AB}\frac{dy_A}{dz} + y_A(N_{Az} - 2N_{Az}) = -\frac{cD_{AB}}{1 + y_A}\frac{dy_A}{dz}$$

It is interesting to note that increasing y_A decreases the flux. Two boundary conditions must be specified. At the surface of the Si film, the reaction is so rapid that the concentration of silane vapor is zero. Furthermore, the concentration of silane in the feed gas is constant.

At the Si film surface, $z = \delta$, $y_A = y_{As}$, and $y_{Bs} = 1$ ($y_A + y_B = 1$ for binary mixture).

At the diffusion screen, $z = 0$, $y_A = y_{Ao}$, and $y_B = y_{Bo} = 1 - y_{Ao}$.

The differential model is now specified. Although the analytical solution was not asked for in the problem statement, it is easy to obtain. We first recognize that for this particular system, N_{Az} is a constant along z. If N_{Az} is a constant, then Fick's equation can be integrated by separation of dependent variable y_A from independent variable z, with integration limits defined by the boundary conditions

$$N_{Az}\int_0^\delta dz = \int_{y_{Ao}}^{y_{As}} \frac{-cD_{AB}}{1 + y_A}\frac{dy_A}{dz}$$

If the system temperature T and total system pressure P are constant, then the total molar concentration of the gas, $c = P/RT$, is also constant. Likewise, the binary gas phase diffusion coefficient of silane vapor in hydrogen gas, D_{AB}, is also constant. The final integrated equation is

$$N_{Az} = \frac{cD_{AB}}{\delta}\ln\left(\frac{1 + y_{Ao}}{1 + y_{As}}\right)$$

If y_{As} is specified, then N_{Az} can be determined. With the silane flux N_{Az} known, parameters of engineering interest, such as the Si film formation rate, can be easily determined. These questions are considered in a problem exercise at the end of Chapter 26.

EXAMPLE 2

The formation of a tungsten thin film on unmasked surfaces of a silicon wafer is an important step in the fabrication of solid-state microelectronic devices. The tungsten metal serves as conductor for current flow between devices on the wafer. In one typical process, the tungsten thin film is formed by the chemical vapor deposition of tungsten hexafluoride (WF_6) onto the surface of the wafer in the presence of hydrogen (H_2) gas and an inert helium carrier gas

$$3H_2(g) + WF_6(g) \rightarrow W(s) + 6HF(g)$$

as shown in Figure 25.4. What is the differential form of Fick's law for the flux of WF_6 gas onto the surface?

Figure 25.4 Chemical vapor deposition of tungsten hexafluoride.

The assumptions for analysis are similar to those made in example 1. The flux of WF_6 (species A) is one dimensional in the z direction. As there are four components in the gas phase mixture, Fick's equation is

$$N_{Az} = -cD_{A-\text{mixture}} \frac{dy_A}{dz} + y_A(N_{Az} + N_{Bz} + N_{Cz} + N_{Dz})$$

where $D_{A-\text{mixture}}$ is the diffusion coefficient of WF_6 in the mixture of H_2 gas (species B), HF gas (species C), and inert He gas (species D). The flux of the gaseous reactants (WF_6, H_2) is opposite in direction to the flux of the gaseous product (HF). The reaction stoichiometry at the surface of the tungsten film relate the fluxes of all diffusing species to WF_6

$$\frac{N_{Az}}{N_{Bz}} = \frac{-1 \text{ mol } WF_6 \text{ reacted}}{-3 \text{ mol } H_2 \text{ reacted}} = \frac{1}{3} \quad or \quad N_{Bz} = +3N_{Az}$$

$$\frac{N_{Az}}{N_{Cz}} = \frac{-1 \text{ mol } WF_6 \text{ reacted}}{+6 \text{ mol HF formed}} = -\frac{1}{6} \quad or \quad N_{Cz} = -6N_{Az}$$

The net flux of He, N_{Dz}, is zero because it has no sink for mass transfer. Accordingly, Fick's equation for WF_6 reduces to

$$N_{Az} = -cD_{A-\text{mixture}} \frac{dy_A}{dz} + y_A(N_{Az} + 3N_{Az} - 6N_{Az} + 0)$$

or

$$N_{Az} = \frac{-cD_{A-\text{mixture}}}{1 + 2y_A} \frac{dy_A}{dz}$$

The flux of WF_6 to the surface is hindered by the flux of the HF product gas from the surface since the denominator term $(1 + 2y_A)$ is greater than one. The differential form of Fick's equation cannot be integrated analytically unless a mean value for the mixture-based diffusion coefficient is taken.

EXAMPLE 3

An emerging area of biotechnology called "tissue engineering" develops new processes to grow organized living tissues of human or animal origin. A typical configuration is the engineered tissue bundle. Engineered tissue bundles have several potential biomedical applications, including the production of replacement body tissue (skin, bone marrow, etc.) for transplantation into the human body, or in the future, may serve as artificial organs for direct implantation into the human body.

Living tissues require oxygen to stay alive. The mass transport of oxygen (O_2) to the tissue is an important design consideration. One potential system is schematically illustrated in Figure 25.5. Thin tubes arranged on a triangular pitch pass longitudinally through the tissue bundle. The tubes serve as a "scaffold" for supporting the living tissue matrix and supply oxygen and nutrients to the tissue at the same time. Let us focus on a single O_2 delivery tube with tissue surrounding it, as illustrated in Figure 25.5. Pure oxygen (O_2) gas flows through the tube. The tube wall is extremely permeable to O_2, and the O_2 partial pressure through the porous tube wall can be taken as the O_2 partial

Figure 25.5 Oxygen transport within an engineered tissue bundle.

pressure inside the tube. Oxygen is only sparingly soluble in the tissue, which is mostly water. The concentration of dissolved O_2 at $r = R_1$, is

$$c_{As} = \frac{p_A}{H}$$

where H is the Henry's law constant for the dissolution of O_2 in living tissue at the process temperature, and p_A is the partial pressure of O_2 in the tube. The dissolved O_2 diffuses through the tissue and is metabolically consumed. The metabolic consumption of dissolved O_2 is described by a kinetic rate equation of the form

$$R_A = -\frac{R_{A,\,max}\, c_A}{K_A + c_A}$$

A key parameter in the design of the engineered tissue bundle is the spacing between the tubes. If the tube spacing is too wide, the dissolved O_2 concentration will go to near zero and starve the tissue. Therefore, it is important to know the radial concentration profile, $c_A(r)$ of dissolved O_2. Develop a differential model to predict $c_A(r)$.

The physical system possesses cylindrical geometry, and the following assumptions for model development are listed here. (1) The source for O_2 mass transfer is the pure O_2 gas inside the tube, and the sink for mass transfer is the metabolic consumption of dissolved oxygen by the tissue. If the O_2 partial pressure p_A is maintained constant inside the tube along longitudinal coordinate z, then the flux of oxygen through the tissue is one dimensional along the radial (r) direction. (2) Tissue remains viable and maintains constant physical properties. (3) The O_2 transfer process is at steady state. (4) The tissue is stationary, and the dissolved O_2 concentration is dilute. (5) At $r = R_1$, the tube material is thin and highly permeable to O_2 so that the dissolved O_2 concentration in the tissue is in equilibrium with the O_2 partial pressure in the tube. (6) At $r = R_2$, there is no net flux of O_2.

The general differential equation for mass transfer in cylindrical coordinates is

$$-\left(\frac{1}{r}\frac{\partial}{\partial r}(rN_{Ar}) + \frac{1}{r}\frac{\partial N_{A\theta}}{\partial \theta} + \frac{\partial N_{Az}}{\partial z}\right) + R_A = \frac{\partial c_A}{\partial t}$$

For steady-state one-dimensional flux along the r direction, the general equation for mass transfer reduces to

$$-\frac{1}{r}\frac{\partial}{\partial r}(rN_{Ar}) + R_A = 0$$

For a one-dimensional system, the partial derivatives can be replaced with ordinary derivatives.

Alternatively, we can perform a material balance for dissolved O_2 on the differential element of volume $2\pi Lr\,\Delta r$ shown in Figure 25.5 and get the same result. Specifically, for steady-state one-dimensional flux along the r direction with a homogeneous reaction R_A within the differential volume element, we have

$$2\pi Lr\,N_{Ar}|_{r=r} - 2\pi Lr\,N_{Ar}|_{r=r+\Delta r} + R_A \cdot 2\pi Lr \cdot \Delta r = 0$$

Diving through by $2\pi L\,\Delta r$, and rearranging, we get

$$-\left(\frac{rN_{Ar}|_{r=r+\Delta r} - rN_{Ar}|_{r=r}}{\Delta r}\right) + R_A\,r = 0$$

Finally, taking the limits as $\Delta r \rightarrow 0$ yields

$$-\frac{1}{r}\frac{d}{dr}(rN_{Ar}) + R_A = 0$$

For one-dimensional flux of dissolved O_2 through the stagnant tissue in cylindrical coordinates along the r direction, Fick's equation reduces to

$$N_{Ar} = -D_{AB}\frac{dc_A}{dr} + \frac{c_A}{c}(N_{Ar}) \cong -D_{AB}\frac{dc_A}{dr}$$

because O_2 is only sparingly soluble in the tissue so that $c_A \ll c$, where c is the total molar concentration of the tissue, which approximates the molar concentration of water. In cylindrical geometry, N_{Ar} is not constant along diffusion path r, because (a) cross-sectional area for flux is increasing along r and (b) the R_A term is present. As a result, the flux equation cannot be integrated, as was the case in example 1. It is now necessary to combine Fick's equation and the differential equation for mass transfer in order to get the concentration profile

$$-\frac{1}{r}\frac{d}{dr}\left(-rD_{AB}\frac{dc_A}{dr}\right) + R_A = 0$$

or

$$D_{AB}\left[\frac{d^2c_A}{dr^2} + \frac{1}{r}\frac{dc_A}{dr}\right] - \frac{R_{A,\max}\,c_A}{K_A + c_A} = 0$$

The concentration profile $c_A(r)$ is now expressed as a second-order differential equation. Therefore, two boundary conditions on $c_A(r)$ must be specified:

$$r = R_1, \frac{dc_A}{dr} = 0 \quad (\text{net flux } N_A = 0 \text{ at } r = R_1)$$

$$r = R_2, c_A = c_{As} = \frac{p_A}{H}$$

The analytical solution for $c_A(r)$ and its extension to predicting the overall rate of oxygen consumption in the tissue bundle has been left as a problem exercise in Chapter 26.

25.5 CLOSURE

The general differential equation for mass transfer was developed to describe the mass balances associated with a diffusing component in a mixture. Special forms of the general differential equation for mass transfer that apply to specific situations were presented. Commonly encountered boundary conditions for molecular diffusion processes were also listed. From this theoretical framework, a five-step method for mathematically modeling processes involving molecular diffusion was proposed. Three examples illustrated how the differential form of Fick's equation presented in Chapter 24, and the general differential equation for mass transfer presented in this chapter, are reduced to simple differential equations that describe the molecular diffusion aspects of a specific process. The approaches presented in this chapter serve as the basis for problem solving in Chapters 26 and 27.

PROBLEMS

25.1 Derive equation (25-11) for component A in terms of molar units, starting with the control-volume expression for the conservation of mass.

25.2 Show that equation (25-11) may be written in the form

$$\frac{\partial \rho_A}{\partial t} + (\nabla \cdot \rho_A \mathbf{v}) - D_{AB} \nabla^2 \rho_A = r_A$$

25.3 The following sketch illustrates the gas diffusion in the neighborhood of a catalytic surface. Hot gases of heavy hydrocarbons diffuse to the catalytic surface where they are cracked into lighter compounds by the reaction: $H \rightarrow 2L$, the light products diffuse back into the gas stream.

a. Reduce the general differential equation for mass transfer to write the specific differential equation that will describe this steady-state transfer process if the catalyst is considered a flat surface. List all of the assumptions you have made in simplifying the general differential equation.

b. Determine the Fick's law relationship in terms of only compound H and insert it into the differential equation you obtained in part (a).

25.4 A hemispherical droplet of liquid water, lying on a flat surface, evaporates by molecular diffusion through still air surrounding the droplet. The droplet initially has a radius R. As the liquid water slowly evaporates, the droplet shrinks slowly with time, but the flux of the water vapor is at a nominal steady state. The temperature of the droplet and the surrounding still air are kept constant. The air contains water vapor at an infinitely long distance from the droplet's surface.

a. After drawing a picture of the physical process, select a coordinate system that will best describe this diffusion

process, list at least five reasonable assumptions for the mass-transfer aspects of the water-evaporation process and simplify the general differential equation for mass transfer in terms of the flux N_A.

b. What is the simplified differential form of Fick's equation for water vapor (species A)?

25.5 A large deep lake, which initially had a uniform oxygen concentration of 1kg/m^3, has its surface concentration suddenly raised and maintained at 9 kg/m^3 concentration level.

Reduce the general differential equation for mass transfer to write the specific differential equation for

a. the transfer of oxygen into the lake without the presence of a chemical reaction;

b. the transfer of oxygen into the lake that occurs with the simultaneous disappearance of oxygen by a first-order biological reaction.

25.6 The moisture in hot, humid, stagnant air surrounding a cold-water pipeline continually diffuses to the cold surface where it condenses. The condensed water forms a liquid film around the pipe, and then continuously drops off the pipe to the ground below. At a distance of 10 cm from the surface of the pipe, the moisture content of the air is constant. Close to the pipe, the moisture content approaches the vapor pressure of water evaluated at the temperature of the pipe.

a. Draw a picture of the physical system, select the coordinate system that best describes the transfer process and state at least five reasonable assumptions of the mass-transfer aspects of the water condensation process.

b. What is the simplified form of the general differential equation for mass transfer in terms of the flux of water vapor, N_A?

c. What is the simplified differential form of Fick's equation for water vapor, N_A?

d. What is the simplified form of the general differential equation for mass transfer in terms of the concentration of water vapor, c_A?

25.7 A liquid flows over a thin, flat sheet of a slightly soluble solid. Over the region in which diffusion is occurring, the liquid velocity may be assumed to be parallel to the plate and to be given by $v = ay$, where y is the vertical distance from the plate and a is a constant. Show that the equation governing the mass transfer, with certain simplifying assumptions, is

$$D_{AB}\left(\frac{\partial^2 c_A}{\partial x^2} + \frac{\partial^2 c_A}{\partial y^2}\right) = ay\frac{\partial c_A}{\partial x}$$

List the simplifying assumptions, and propose reasonable boundary conditions.

25.8 Consider one of the cylindrical channels that run through an isomerization catalyst as shown below. A catalyst coats the inner walls of each channel. This catalyst promotes the isomerization of n-butane $(n - C_4H_{10})$ species A to isobutene $(i - C_4H_{10})$ species B.

$$n - C_4H_{10}(g) \rightarrow i - C_4H_{10}(g)$$

The gas phase above the channels contains mixture of A and B maintained at a constant composition of 60 mol $\% n - C_4H_{10}$ (A) and 40 mol $\% i - C_4H_{10}$ (B). Gas phase species A diffuses down a straight channel of diameter $d = 0.1$ cm and length $L = 2.0$ cm. The base of each channel is sealed. This is rapid reaction so that the production rate of B is diffusion limited. The quiescent gas space in the channel consists of only species A and B.

a. State three relevant assumptions for the mass transfer process.

b. Based on your assumptions, simplify the general differential equation for the mass transfer of species A, leaving the equation in terms of the flux N_A.

c. Using equations for the flux of A in your determined equation, express the general differential equation in terms of the concentration c_A.

d. Specify relevant boundary conditions for the gas phase concentration c_A.

25.9 An early mass-transfer study of oxygen transport in human tissue won a Nobel prize for August Krough. By considering a tissue cylinder surrounding each blood vessel, he proposed the diffusion of oxygen away from the blood vessel into the annular tissue was accompanied by a zero-order reaction, that is, $R_A = -m$, where m is a constant. This reaction was necessary to explain the metabolic consumption of the oxygen to produce carbon dioxide.

Use the general differential equation for mass transfer of oxygen to write the specific differential equation that will describe the diffusion of oxygen in the human tissue. What would be the form of Ficks relationship written in terms of only the diffusing oxygen?

25.10 A fluidized coal reactor has been proposed for a new power plant. If the coal can be assumed to be spherical, reduce the general differential equation for mass transfer to obtain a specific differential equation for describing the steady-state diffusion of oxygen to the surface of the coal particle.

Determine the Fick's law relationship for the flux of oxygen from the surrounding air environment if

a. only carbon monoxide, CO, is produced at the surface of the carbon particle;

b. only carbon dioxide, CO_2, is produced at the surface of the carbon particle.

If the reaction at the surface of the carbon particle is instantenous, give two boundary conditions that might be used in solving the differential equation.

25.11 In the manufacture of semiconducting thin films, a thin film of solid arsenic laid down on the surface of a silicon water by the diffusion-limited chemical vapor deposition of arsine, AsH_3.

$$2AsH_3(g) \rightarrow 2As(s) + 3H_2(g)$$

The gas head space, 5 cm above the surface of the wafer, is stagnant. Arsenic atoms deposited on the surface then diffuse into the solid silicon to "dope" the wafer and impart semiconducting properties to the silicon, as shown in the figure below.

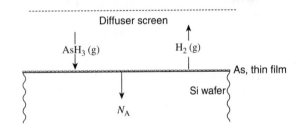

The process temperature is 1050°C. The diffusion coefficient of aresenic in silicon is 5×10^{-13} cm/s at this temperature and the maximum solubility of aresenic in silicon is 2×10^{21} atoms/cm^3. The density of solid silicon is 5×10^{22} atoms/cm^3. As the diffusion coefficient is so small, the aresenic atoms do not "penetrate" very far into the silicon solid, usually less than a few microns. Consequently, a relatively thin silicon water can be considered as a "semi-infinite" medium for diffusion.

a. State at least five reasonable assumptions for the mass transfer of aresenic in this doping process.

b. What is the simplified form of the general differential equation for the mass transfer of the aresenic concentration within the silicon? Purpose reasonable boundary and initial conditions.

25.12 A device has been proposed that will serve as a "blood oxygenator" for a heart–lung bypass machine. In this process, blood (which is mostly water, species B) containing no dissolved oxygen (O_2 species A) enters the top of the chamber and then falls vertically down as a liquid film of uniform thickness, along a surface designed to appropriately wet blood. Contacting the liquid surface is a 100% O_2 gas phase.

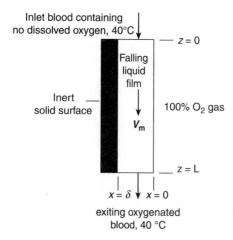

Inlet blood containing
no dissolved oxygen, 40°C

— z = 0

Falling
liquid
film

Inert
solid surface

100% O_2 gas

V_m

— z = L

$x = \delta$ $x = 0$

exiting oxygenated
blood, 40 °C

Oxygen is soluble in blood, with the equilibrium solubility described by Henry's law, $c_A^* = p_A/H$, where p_A is the partial pressure of oxygen (atm), H is the Henry's law constant, and c_A^* is the solubility concentration limit of oxygen dissolved in blood (mmol/L) at p_A.

In analyzing the mass transport of dissolved oxygen into the falling film, you may assume the following: (1) the process is dilute with respect to dissolved oxygen in the fluid; (2) the falling liquid film has a flat velocity profile with velocity v_{max}; (3) the gas space always contains 100% oxygen; (5) the width of the liquid film, W, is much larger than the length of the liquid film, L.

a. Simplify the general differential equation for O_2 transfer. If your analysis suggests more than one dimension for flux, provide a simplified flux equation for each coordinate of interest.

b. Provide one simplified differential equation in terms of the fluxes and another simplified differential equation in terms of the oxygen concentration c_A.

c. List boundary conditions associated with the oxygen mass transfer process.

25.13 One way to deliver a timed dosage within the human body is to ingest a capsule and allow it to settle in the gastro-intestinal system. Once inside the body, the capsule slowly releases the drug to the body by a diffusion-limited process. A suitable drug carrier is a spherical bead of a nontoxic gela-tinous material that can pass through the gastrointestinal system

without disintegrating. A water-soluble drug (solute A) is uni-formly dissolved within the gel, has an initial concentration, c_{Ao} of 50 mg/cm^3. The drug loaded within the spherical gel capsule is the sink for mass transfer. Consider a limiting case where the drug is immediately consumed or swept away once it reaches the surface, i.e., @ R, $c_A = 0$.

a. In analyzing the process, choose a coordinate system and simplify the general differential equation for the mass transfer of the drug in terms of the flux.

b. What reasonable assumptions were used in your simplifying of the general differential equation.

c. Simplify Fick's equation for the drug species and obtain a differential equation in terms of concentration, c_A.

25.14 Consider a single, porous, spherical, inert mineral par-ticle. The pores inside the particle are filled with liquid water (species B). We are interested in analyzing the molecular diffu-sion of the contaminant benzene C_6H_6 species A within the water-filled pores of the particle. The average diameter of the pores is 150 nm and the void fraction (porosity) is 0.40. The benzene solute does not adsorb onto the inter surfaces of the pores. Benzene is very sparingly soluble in water, and has a molecular diameter of 0.15 nm. The process is isothermal at 298 K. The concentration of dissolved benzene in the water sur-rounding the particle, c_{Ao}, is constant with time. The critical volume (V_c) of benzene is 259 cm^3/gmol. The effective diffu-sion coefficient of benzene inside the porous particle was to be calculated in Problem 24.25.

Starting with the general differential equation for mass trans-fer of benzene, develop a differential model to describe the concentration profile of benzene, species A within the single, porous, spherical, inert mineral particle. State reasonable assumptions and boundary/initial conditions for the process.

25.15 A large tank truck overturns and spills a herbicide over a field. The fluid remains on the soil 30 min before evaporating into the atmosphere. Simplify the general differential equation for the mass transfer of the herbicide to write the following:

a. the steady-state differential equation that will describe the evaporation of the herbicide into the air;

b. the differential equation that will describe the diffusion of the herbicide into the soil.

25.16 Consider the drug treatment system shown below. A hemisphere cluster of unhealthy cells is surrounded by a larger hemisphere of stagnant dead tissue (species B), which is in turn surrounded by a flowing fluid. The bulk, well-mixed fluid contains a drug compound (species A) of constant but dilute concentration c_{Ao}. Drug A is also soluble in the unhealthy tissue but does not preferentially partition into it relative to the fluid. The drug (species A) enters the dead tissue and homes in on the unhealthy cells. At the unhealthy cell boundary ($r = R_1$) the flux of A to the unhealthy cells is diffusion limited. All metabolites of drug A produced by

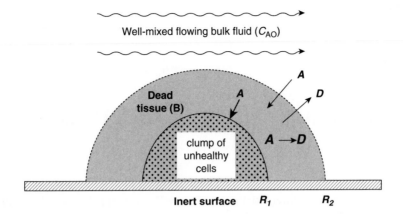

Well-mixed flowing bulk fluid (C_{AO})

Dead tissue (B)

clump of unhealthy cells

Inert surface R_1 R_2

the unhealthy cells stay within the unhealthy cells. However, drug A can also degrade to inert metabolite D by the first-order reaction on c_A, i.e., $A \rightarrow D$, that occurs only within the stagnant dead tissue.

Simplify the general differential equation in terms for mass transfer for drug A. Specify the final differential equation in terms of N_A and again in terms of c_A. State all assumptions as necessary.

Chapter 26

Steady-State Molecular Diffusion

In this chapter, we will direct our attention to describing the steady-state transfer of mass from a differential point of view. To accomplish this task, the differential equation and the boundary conditions that describe the physical situation must be established. The approach will parallel those previously used in Chapter 8 for the analysis of a differential fluid element in laminar flow and in Chapter 17 for the analysis of a differential volume element of a quiescent material for steady-state heat conduction.

During our discussion of steady-state diffusion, two approaches will be used to simplify the differential equations of mass transfer as recommended in Section 24.4. First, Fick's equation and the general differential equation for mass transfer can be simplified by eliminating the terms that do not apply to the physical situation. Second, a material balance can be performed on a differential volume element of the control volume for mass transfer. In using both approaches, the student will become more familiar with the various terms in the general differential equation for mass transfer

$$\mathbf{\nabla} \cdot \mathbf{N}_A + \frac{\partial c_A}{\partial t} - R_A = 0 \qquad (25\text{-}11)$$

To gain confidence in treating mass-transfer processes, we will initially treat the simplest case, steady-state diffusion in only one direction, which is free of any chemical production occurring uniformly throughout the process (i.e., $R_A = 0$). We will then obtain solutions for increasingly more complex mass-transfer operations.

26.1 ONE-DIMENSIONAL MASS TRANSFER INDEPENDENT OF CHEMICAL REACTION

In this section, steady-state molecular mass transfer through simple systems in which the concentration and the mass flux are functions of a single space coordinate will be considered. Although all four fluxes, $\mathbf{N}_A$, $\mathbf{n}_A$, $\mathbf{J}_A$, and $\mathbf{j}_A$, may be used to describe mass-transfer operations, only the molar flux relative to a set of axes fixed in space, $\mathbf{N}_A$, will be used in the following discussions. In a binary system, the z component of this flux is expressed by equation (24-20)

$$N_{A,z} = -cD_{AB} \frac{dy_A}{dz} + y_A(N_{A,z} + N_{B,z}) \qquad (24\text{-}20)$$

Unimolecular Diffusion

The diffusion coefficient or mass diffusivity for a gas may be experimentally measured in an Arnold diffusion cell. This cell is illustrated schematically in Figure 26.1. The narrow tube,

Flow of gas B

$z = z_2$

$N_{Az}|_{z+\Delta z}$

Δz

$N_{Az}|_z$

$z = z_1$

Pure liquid A

Figure 26.1 Arnold diffusion cell.

which is partially filled with pure liquid A, is maintained at a constant temperature and pressure. Gas B, which flows across the open end of the tube, has a negligible solubility in liquid A and is also chemically inert to A. Component A vaporizes and diffuses into the gas phase; the rate of vaporization may be physically measured and may also be mathematically expressed in terms of the molar mass flux.

Recall that the general differential equation for mass transfer is given by

$$\nabla \cdot \mathbf{N}_A + \frac{\partial c_A}{\partial t} - R_A = 0 \tag{25-11}$$

In rectilinear coordinates, this equation is

$$\frac{\partial N_{Ax}}{\partial x} + \frac{\partial N_{Ay}}{\partial y} + \frac{\partial N_{Az}}{\partial z} + \frac{\partial c_A}{\partial t} - R_A = 0 \tag{25-27}$$

Assume that (1) the mass-transfer process is at steady state with $\partial c_A/\partial t = 0$; (2) there is no chemical production of A in the diffusion path so that $R_A = 0$; and (3) the diffusion is only in the z direction, so that we are only concerned with the z component of the mass flux vector, N_A. For this physical situation, equation (25-11) reduces to

$$\frac{dN_{Az}}{dz} = 0 \tag{26-1}$$

We can also generate this governing differential equation by considering the mass transfer occurring in the differential control volume of $S\,\Delta z$, where S is the uniform cross-sectional area of the control volume and Δz is the depth of the control volume. A mass balance over this control volume for a steady-state operation, free of any chemical production of A, yields

$$SN_{Az}|_{z+\Delta z} - SN_{Az}|_z = 0$$

Dividing through by the control volume, $S\,\Delta z$, and taking limit as Δz approaches zero, we once again obtain equation (26-1).

A similar differential equation could also be generated for component B

$$\frac{d}{dz}N_{B,z} = 0 \tag{26-2}$$

and, accordingly, the molar flux of B is also constant over the entire diffusion path from z_1 to z_2. Considering only the plane at z_1 and the restriction that gas B is insoluble in liquid A, we realize that N_{Bz} at plane z_1 is zero and conclude that N_{Bz}, the net flux of B, is zero throughout the diffusion path; accordingly, component B is a *stagnant* gas.

The constant molar flux of A was described in Chapter 24 by the equation

$$N_{A,z} = -cD_{AB}\frac{dy_A}{dz} + y_A(N_{A,z} + N_{B,z}) \tag{24-20}$$

this equation reduces, when $N_{B,z} = 0$, to

$$N_{A,z} = -\frac{cD_{AB}}{1 - y_A}\frac{dy_A}{dz} \tag{26-3}$$

This equation may be integrated between the two boundary conditions:

$$\text{at } z = z_1 \quad y_A = y_{A_1}$$

and

$$\text{at } z = z_2 \quad y_A = y_{A_2}$$

Assuming the diffusion coefficient to be independent of concentration, and realizing from equation (26-1) that $N_{A,z}$ is constant along the diffusion path, we obtain, by integrating,

$$N_{A,z} \int_{z_1}^{z_2} dz = cD_{AB} \int_{y_{A_1}}^{y_{A_2}} -\frac{dy_A}{1-y_A} \tag{26-4}$$

Solving for $N_{A,z}$, we obtain

$$N_{A,z} = \frac{cD_{AB}}{(z_2 - z_1)} \ln \frac{(1 - y_{A_2})}{(1 - y_{A_1})} \tag{26-5}$$

The log-mean average concentration of component B is defined as

$$y_{B,lm} = \frac{y_{B_2} - y_{B_1}}{\ln(y_{B_2}/y_{B_1})}$$

or, in the case of a binary mixture, this equation may be expressed in terms of component A as follows:

$$y_{B,lm} = \frac{(1 - y_{A_2}) - (1 - y_{A_1})}{\ln[(1 - y_{A_2})/(1 - y_{A_1})]} = \frac{y_{A_1} - y_{A_2}}{\ln[(1 - y_{A_2})/(1 - y_{A_1})]} \tag{26-6}$$

Inserting equation (26-6) into equation (26-5), we obtain

$$N_{A,z} = \frac{cD_{AB}}{z_2 - z_1} \frac{(y_{A_1} - y_{A_2})}{y_{B,lm}} \tag{26-7}$$

Equation (26-7) may also be written in terms of pressures. For an ideal gas

$$c = \frac{n}{V} = \frac{P}{RT}$$

and

$$y_A = \frac{p_A}{P}$$

The equation equivalent to equation (26-7) is

$$N_{A,z} = \frac{D_{AB}P}{RT(z_2 - z_1)} \frac{(p_{A_1} - p_{A_2})}{p_{B,lm}} \tag{26-8}$$

Equations (26-7) and (26-8) are commonly referred to as equations for *steady-state diffusion of one gas through a second stagnant gas*. Many mass-transfer operations involve the diffusion of one gas component through another nondiffusing component; *absorption* and *humidification* are typical operations defined by these two equations.

Equation (26-8) has also been used to describe the convective mass-transfer coefficients by the "*film concept*" or *film theory*. In Figure 26.2, the flow of gas over a liquid surface is

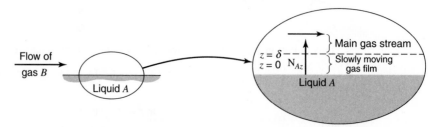

Figure 26.2 Film model for mass transfer of component A into a moving gas stream.

illustrated. The "film concept" is based upon a model in which the entire resistance to diffusion from the liquid surface to the main gas stream is assumed to occur in a stagnant or laminar film of constant thickness δ. In other words, for this model, δ is a fictitious length which represents the thickness of a fluid layer offering the same resistance to molecular diffusion as is encountered in the combined process of molecular diffusion and diffusion due to the mixing by the moving fluid. If this model is accurate, the convective mass-transfer coefficient may be expressed in terms of the gas diffusion coefficient. If $z_2 - z_1$ is set equal to δ, equation (26-8) becomes

$$N_{A,z} = \frac{D_{AB}P}{RTp_{B,lm}\delta}(p_{A_1} - p_{A_2})$$

and from equation (25-30), we have

$$N_{A,z} = k_c(c_{A_1} - c_{A_2})$$

and

$$N_{A,z} = \frac{k_c}{RT}(p_{A_1} - p_{A_2})$$

Comparison reveals that the film coefficient is expressed as

$$k_c = \frac{D_{AB}P}{p_{B,lm}\delta} \tag{26-9}$$

when the diffusing component is transported through a nondiffusing gas. Although this model is physically unrealistic, the "film concept" has had educational value in supplying a simple picture of a complicated process. The film concept has proved frequently misleading in suggesting that the convective mass-transfer coefficient is always directly proportional to the mass diffusivity. Other models for the convective coefficient will be discussed in this chapter and in Chapter 28. At that time we will find that k_c is a function of the diffusion coefficient raised to an exponent varying from 0.5 to 1.0.

Frequently, in order to complete the description of the physical operation in which mass is being transported, it is necessary to express the concentration profile. Recalling equation (26-1)

$$\frac{d}{dz}N_{A,z} = 0 \tag{26-1}$$

and equation (26-3)

$$N_{A,z} = -\frac{cD_{AB}}{1-y_A}\frac{dy_A}{dz} \tag{26-3}$$

we can obtain the differential equation that describes the variation in concentration along the diffusing path. This equation is

$$\frac{d}{dz}\left(-\frac{cD_{AB}}{1-y_A}\frac{dy_A}{dz}\right) = 0 \tag{26-10}$$

As c and D_{AB} are constant under isothermal and isobaric conditions, the equation reduces to

$$\frac{d}{dz}\left(\frac{1}{1-y_A}\frac{dy_A}{dz}\right) = 0 \tag{26-11}$$

This second-order equation may be integrated twice with respect to z to yield

$$-\ln(1 - y_A) = c_1 z + c_2 \tag{26-12}$$

The two constants of integration are evaluated, using the boundary conditions

$$\text{at } z = z_1 \qquad y_A = y_{A_1}$$

and

$$\text{at } z = z_2 \qquad y_A = y_{A_2}$$

Substituting the resulting constants into equation (26-12), we obtain the following expression for the concentration profile of component A:

$$\left(\frac{1 - y_A}{1 - y_{A_1}}\right) = \left(\frac{1 - y_{A_2}}{1 - y_{A_1}}\right)^{(z-z_1)/(z_2-z_1)} \tag{26-13}$$

or, as $y_A + y_B = 1$

$$\left(\frac{y_B}{y_{B_1}}\right) = \left(\frac{y_{B_2}}{y_{B_1}}\right)^{(z-z_1)/(z_2-z_1)} \tag{26-14}$$

Equations (26-13) and (26-14) describe logarithmic concentration profiles for both species. The average concentration of one of the species along the diffusion path may be evaluated, as an example for species B, by

$$\bar{y}_B = \frac{\displaystyle\int_{z_1}^{z_2} y_B \, dz}{\displaystyle\int_{z_1}^{z_2} dz} \tag{26-15}$$

Upon substitution of equation (26-14) into equation (26-15), we obtain

$$\bar{y}_B = y_{B_1} \frac{\displaystyle\int_{z_1}^{z_2} \left(\frac{y_{B_2}}{y_{B_1}}\right)^{(z-z_1)/(z_2-z_1)} dz}{z_2 - z_1}$$

$$= \frac{(y_{B_2} - y_{B_1})(z_2 - z_1)}{\ln(y_{B_2}/y_{B_1})(z_2 - z_1)} = \frac{y_{B_2} - y_{B_1}}{\ln(y_{B_2}/y_{B_1})}$$

$$= y_{B,lm} \tag{26-6}$$

The following example problem illustrates the application of the foregoing analysis to a mass-transfer situation.

EXAMPLE 1

Vapor degreasers like the one shown in Figure 26.3 are widely used for cleaning metal parts. Liquid solvent rests at the bottom of the degreaser tank. A heating coil immersed in the solvent vaporizes a small portion of the solvent and maintains a constant temperature, so that the solvent exerts a constant vapor pressure. The cold parts to be cleaned are suspended in the solvent vapor zone where the concentration of solvent vapors is highest. The solvent condenses on the part, dissolves the grease, and then drips back down into the tank, thereby cleaning the part. Vapor degreasers are often left open to the atmosphere for ease of dipping and removing parts and because covering them might release an explosive mixture. When the degreaser is not in use, molecular diffusion of the solvent vapor through the stagnant air inside the headspace can result in significant solvent emissions, because the surrounding atmosphere serves as an infinite sink for the mass-transfer process. As the amount

Atmosphere
$z = z_2 = 5\,\text{m}$
$y_{A2} = 0$

Still air

TCE vapors

Metal part

Vapor zone

$z = z_1 = 0.2\,\text{m}$

Heater

TCE liq. 35°C

$y_{A1} = P_A^*/P = 0.152$

Vapor degreaser tank
2 m diameter

Figure 26.3 TCE emissions from a vapor degreaser.

of solvent in the degreaser tank is large relative to the amount of vapor emitted, a steady-state diffusion process with a constant diffusion path length takes place.

At present, a cylindrical degreaser tank with a diameter of 2 m and total height of 5 m is in operation, and the solvent level height is kept constant at 0.2 m. The temperatures of the solvent and headspace of the degreaser are both constant at 35°C. The solvent used for vapor degreasing is trichloroethylene (TCE). Current regulations require that the degreaser cannot emit more than 1.0 kg TCE per day. Does the estimated emission rate of the degreaser exceed this limit? TCE has a molecular weight of 131.4 g/mol and a vapor pressure of 115.5 mmHg at 35°C. The binary-diffusion coefficient TCE in air is 0.088 cm²/s at 35°C, as determined by the Fuller–Schettler–Giddings correlation.

The source of TCE mass transfer is the liquid solvent at the bottom of the tank, and the sink for mass transfer is the surrounding atmosphere outside the tank. The steady-state molecular diffusion flux of TCE vapor through the stagnant gas headspace of the degreaser in the z direction is described by

$$N_{Az} = \frac{cD_{AB}}{z_2 - z_1} \ln\left(\frac{1 - y_{A_2}}{1 - y_{A_1}}\right)$$

with species A representing TCE vapor and species B representing air. The total molar concentration of the gas, c, is determined from the ideal gas law.

$$c = \frac{P}{RT} = \frac{1.0\,\text{atm}}{\dfrac{0.08206\,\text{m}^3\,\text{atm}}{\text{kg mol} \cdot \text{K}}(273 + 35)\text{K}} = 0.0396\,\frac{\text{kg mol}}{\text{m}^3}$$

The mole fraction of TCE vapor at the solvent surface (y_{A_1}) is determined from the vapor pressure of the solvent at 35°C.

$$y_{A_1} = \frac{P_A}{P} = \frac{115.1\,\text{mmHg}}{1.0\,\text{atm}} \frac{1.0\,\text{atm}}{760\,\text{mmHg}} = 0.152$$

The mole fraction of TCE vapor at the exit of the degreaser tank is taken as zero ($y_{A_2} = 0$), as the surrounding atmosphere serves as an infinite sink for mass transfer. The path length for diffusion is simply the difference between the solvent level height and the top of the degreaser tank

$$z_2 - z_1 = 5.0\,\text{m} - 0.2\,\text{m} = 4.8\,\text{m}$$

From these input values, the flux of TCE vapor from the degreaser is

$$N_{Az} = \frac{\left(0.0396\,\dfrac{\text{kg mol}}{\text{m}^3}\right)\left(0.088\,\dfrac{\text{cm}^2}{\text{s}}\dfrac{1\,\text{m}^2}{(100\,\text{cm})^2}\right)}{4.8\,\text{m}} \ln\left(\frac{1 - 0}{1 - 0.152}\right)$$

$$= 1.197 \times 10^{-8}\,\frac{\text{kg mol TCE}}{\text{m}^2 \cdot \text{s}}$$

The TCE emissions rate (W_A) is the product of the flux and the cross-sectional area of the degreaser tank of diameter D

$$W_A = N_{Az}\frac{\pi D^2}{4} = 1.197 \times 10^{-8}\,\frac{\text{kg mol TCE}}{\text{m}^2 \cdot \text{s}}\frac{\pi(2.0\,\text{m})^2}{4}$$

$$\left(\frac{131.4\,\text{kg TCE}}{\text{kg mol TCE}}\right)\left(\frac{3600\,\text{s}}{1\,\text{h}}\frac{24\,\text{h}}{\text{day}}\right) = 0.423\,\frac{\text{kg TCE}}{\text{day}}$$

The estimated TCE vapor emission rate is below the current regulatory limit of 1.0 kg TCE per day. In a real degreaser, it may be difficult to ensure a completely still gas space, as local air currents induced from a variety of sources may occur. The air currents would increase the mass-transfer flux by convection. Consequently, this analysis considers only the limiting case for the minimum vapor emissions from a diffusion-limited process.

Pseudo-Steady-State Diffusion

In many mass-transfer operations, one of the boundaries may move with time. If the length of the diffusion path changes a small amount over a long period of time, a pseudo-steady-state diffusion model may be used. When this condition exists, equation (26-7) describes the mass flux in the stagnant gas film. Reconsider Figure 26.1, with a moving liquid surface as illustrated in Figure 26.4. Two surface levels are shown, one at time t_0 and the other at time t_1. If the difference in the level of liquid A over the time interval considered is only a small fraction of the total diffusion path, and $t_1 - t_0$ is a relatively long period of time, at any instant in that period the molar flux in the gas phase may be evaluated by

Figure 26.4 Arnold diffusion cell with moving liquid surface.

$$N_{A,z} = \frac{cD_{AB}(y_{A_1} - y_{A_2})}{zy_{B,lm}}$$
(26-7)

where z is the length of the diffusion path at time t.

The molar flux $N_{A,Z}$ is related to the amount of A leaving the liquid by

$$N_{A,z} = \frac{\rho_{A,L}}{M_A} \frac{dz}{dt}$$
(26-16)

where $\rho_{A,L}/M_A$ is the molar density of A in the liquid phase. Under pseudo-steady-state conditions, equations (26-7) and (26-16) may be combined to give

$$\frac{\rho_{A,L}}{M_A} \frac{dz}{dt} = \frac{cD_{AB}(y_{A_1} - y_{A_2})}{zy_{B,lm}}$$
(26-17)

Equation (26-17) may be integrated from $t = 0$ to $t = t$ from $z = z_{t_0}$ to $z = z_t$ as follows:

$$\int_{t=0}^{t} dt = \frac{\rho_{A,L}y_{B,lm}/M_A}{cD_{AB}(y_{A_1} - y_{A_2})} \int_{z_{t_0}}^{z_t} z\, dz$$

This yields

$$t = \frac{\rho_{A,L}y_{B,lm}/M_A}{cD_{A,B}(y_{A_1} - y_{A_2})} \left(\frac{z_t^2 - z_{t_0}^2}{2}\right)$$
(26-18)

Therefore, for the pseudo-steady-state diffusion process, a plot of $z_t^2 - z_{t_0}^2$ vs. time t should be linear. Rearranging this expression, we obtain the equation commonly used to evaluate the gas-diffusion coefficient from Arnold cell experimental data. This equation is

$$D_{AB} = \frac{\rho_{A,L} y_{B,lm}/M_A}{c(y_{A_1} - y_{A_2})t} \left(\frac{z_t^2 - z_{t_0}^2}{2} \right) \tag{26-19}$$

As illustrated by the Arnold diffusion cell above, pseudo-steady-state diffusion processes usually involve the slow depletion of the source or sink for the mass-transfer process with time. Below, we consider another process that is modeled by pseudo-steady-state diffusion, the thermal oxidation of a silicon wafer.

EXAMPLE 2

The formation of a silicon oxide (SiO_2) thin film on a silicon (Si) wafer surface is an important step in the fabrication of solid-state microelectronic devices. A thin film of SiO_2 serves as a barrier to dopant diffusion or as a dielectric insulator to isolate various devices being formed on the wafer. In one common process, silicon is oxidized by exposure to oxygen (O_2) gas at temperatures above 700°C

$$Si(s) + O_2(g) \rightarrow SiO_2(s)$$

Molecular O_2 dissolves into the SiO_2 solid, diffuses through the SiO_2 film, and then reacts with Si at the Si/SiO_2 interface, as shown in Figure 26.5. Assuming that the diffusion of O_2 through the SiO_2 film limits the oxidation process, develop a model to predict the thickness of the SiO_2 layer (δ) as a function of time at 1000°C. The density of solid SiO_2 (ρ_B) is 2.27 g/cm³, and the molecular weight of SiO_2 (M_B) is 60 g/mol. The molecular diffusion coefficient of O_2 in SiO_2 (D_{AB}) is 2.7×10^{-9} cm²/s at 1000°C, and the maximum solubility of O_2 in SiO_2 (c_{As}) is 9.6×10^{-8} mol O_2/cm³ solid at 1000°C and 1 atm O_2 partial pressure, using data provided by Norton.[1]

Figure 26.5 Thermal oxidation of a silicon wafer.

The physical system is represented in the rectilinear coordinate system. The model development follows the approach outlined earlier in Section 25.4. The assumptions for model development are listed here. (1) The oxidation of Si to SiO_2 occurs only at the Si/SiO_2 interface. The unreacted Si at the interface serves as the sink for molecular mass transfer of O_2 through the film. (2) The O_2 in the gas phase above the wafer represents an infinite source for O_2 transfer. The O_2 molecules "dissolve" into the nonporous SiO_2 solid at the gas/solid interface. (3) The rate of SiO_2 formation is controlled by the rate of molecular diffusion of O_2 (species A) through the solid SiO_2 layer (species B) to the unreacted Si layer. The reaction is very rapid, so that the concentration of molecular O_2 at the interface is zero, that is, $c_{A,\delta} = 0$. Furthermore, there are no mass-transfer resistances

[1] F. J. Norton, *Nature*, **191**, 701 (1961).

in the gas film above the wafer surface, as O_2 is a pure component in the gas phase. (4) The flux of O_2 through the SiO_2 layer is one dimensional along coordinate z. (5) The rate of SiO_2 film formation is slow enough so that at a given film thickness δ, there is no accumulation of reactants or products within the SiO_2 film. However, the thickness of the film will still increase with time. Consequently, this is a "pseudo-steady-state" process. (6) The overall thickness of the wafer does not change as the result of the formation of the SiO_2 layer. (7) The process is isothermal.

Based on the previous assumptions, the general differential equation for mass transfer reduces to

$$\frac{dN_{Az}}{dz} = 0$$

and Fick's equation for one-dimensional diffusion of O_2 (species A) through crystalline solid SiO_2 (species B) is

$$N_{Az} = -D_{AB}\frac{dc_A}{dz} + \frac{c_A}{c}(N_{Az} + N_{Bz}) = -D_{AB}\frac{dc_A}{dz} + \frac{c_A}{c}N_{Az}$$

Usually, the concentration of molecular O_2 in the SiO_2 layer is dilute enough so that c_A/c term is very small in magnitude relative to the other terms. Therefore, Fick's equation reduces to

$$N_{Az} = -D_{AB}\frac{dc_A}{dz}$$

It is interesting to note here that unimolecular diffusion (UMD) flux mathematically simplifies to the equimolar counter diffusion (EMCD) flux at dilute concentration of the diffusing species. As N_A is constant along z, the differential flux equation can be integrated directly following separation of dependent variable c_A from independent variable z

$$\int_0^\delta N_{Az}\, dz = -D_{AB}\int_{c_{As}}^0 dc_A$$

or simply

$$N_{Az} = \frac{D_{AB}c_{As}}{\delta}$$

which describes the flux of O_2 through the SiO_2 layer of thickness δ. The surface concentration c_{As} refers to the concentration of O_2 dissolved in solid phase SiO_2 (mol O_2/cm^3 solid).

We know that δ increases slowly with time, even though there is no accumulation term for O_2 in the SiO_2 layer. In other words, the process operates under the pseudo-steady-state assumption. In order to discover how δ increases with time, consider an unsteady-state material balance for SiO_2 within the wafer

(molar rate of SiO_2 formation) = (molar rate of accumulation of SiO_2)

or

$$\text{(molar rate of accumulation of } SiO_2) = \frac{d\left(\frac{\rho_B S\delta}{M_B}\right)}{dt}$$

where ρ_B is the density of solid SiO_2 (2.27 g/cm^3), M_B is the molecular weight of the SiO_2 layer (60 g/mol), and S is the surface area of the wafer. Given the stoichiometry of the reaction, one mole of SiO_2 is formed for every mole of O_2 consumed. Therefore,

$$\text{(rate of } SiO_2 \text{ formation)} = N_{Az}S = \frac{D_{AB}c_{As}}{\delta}S$$

or

$$\frac{\rho_B}{M_B}\frac{d\delta}{dt} = \frac{D_{AB}c_{As}}{\delta}$$

Separation of dependent variable δ from the independent variable t, followed by integration at $t = 0$, $\delta = 0$ to $t = t$, $\delta = \delta$ gives

$$\int_0^\delta \delta \, d\delta = \frac{M_B D_{AB} c_{As}}{\rho_B} \int_0^t dt$$

or

$$\delta = \sqrt{\frac{2M_B D_{AB} c_{As}}{\rho_B} t}$$

The above equation predicts that the thickness of the SiO_2 thin film is proportional to the square root of time. Recall that the molecular diffusion coefficient of O_2 in SiO_2 (D_{AB}) is 2.7×10^{-9} cm^2/s at 1000°C, and the solubility of O_2 in SiO_2 (c_{As}) is 9.6×10^{-8} mol O_2/cm^3 solid at 1000°C. Figure 26.6 compares the predicted film thickness δ vs. time to process data provided by Hess[2] for 1 atm O_2 at 1000°C. As one can see, the model adequately predicts the data trend. The film is very thin, less than 0.5 μm, in part because the value for $D_{AB}c_{As}$ is so small.

Figure 26.6 SiO_2 film thickness vs. time at 1000°C.

This example illustrates how a chemical reaction at a boundary surface can serve as the driving force for molecular diffusion. This concept is formally presented in Section 26.2.

[2] D. W. Hess, *Chem. Eng. Educ.*, **24**, 34 (1990).

Equimolar Counterdiffusion

A physical situation that is encountered in the *distillation* of two constituents whose molar latent heats of vaporization are essentially equal stipulates that the flux of one gaseous component is equal to but acting in the opposite direction from the other gaseous component; that is, $N_{A,z} = -N_{B,z}$. Equation (25-11)

$$\nabla \cdot \mathbf{N}_A + \frac{\partial c_A}{\partial t} - R_A = 0 \tag{25-11}$$

for the case of steady-state mass transfer without chemical reaction may be reduced to

$$\nabla \cdot N_A = 0$$

For the transfer in the z direction, this equation reduces to

$$\frac{d}{dz} N_{A,z} = 0$$

This relation stipulates that $N_{A,z}$ is constant along the path of transfer. The molar flux, $N_{A,z}$, for a binary system at constant temperature and pressure is described by

$$N_{A,z} = -D_{AB} \frac{dc_A}{dz} + y_A(N_{A,z} + N_{B,z}) \tag{24-20}$$

The substitution of the restriction, $N_{A,z} = -N_{B,z}$, into the above equation gives an equation describing the flux of A when *equimolar-counterdiffusion* conditions exist

$$N_{A,z} = -D_{AB} \frac{dc_A}{dz} \tag{26-20}$$

Equation (26-20) may be integrated, using the boundary conditions

$$\text{at } z = z_1 \qquad c_A = c_{A_1}$$

and

$$\text{at } z = z_2 \qquad c_A = c_{A_2}$$

giving

$$N_{A,z} \int_{z_1}^{z_2} dz = -D_{AB} \int_{c_{A_1}}^{c_{A_2}} dc_A$$

from which we obtain

$$N_{A,z} = \frac{D_{AB}}{(z_2 - z_1)}(c_{A_1} - c_{A_2}) \tag{26-21}$$

When the ideal gas law is obeyed, the molar concentration of A is related to the partial pressure of A by

$$c_A = \frac{n_A}{V} = \frac{p_A}{RT}$$

Substituting this expression for c_A into equation (26-21), we obtain

$$N_{A,z} = \frac{D_{AB}}{RT(z_2 - z_1)}(p_{A_1} - p_{A_2}) \tag{26-22}$$

Equations (26-21) and (26-22) are commonly referred to as the *equations for steady-state equimolar counterdiffusion.*

The concentration profile for equimolar-counterdiffusion processes may be obtained by substituting equation (26-20) into the differential equation which describes transfer in the z direction

$$\frac{d}{dz} N_{A,z} = 0$$

or

$$\frac{d^2 c_A}{dz^2} = 0$$

This second-order equation may be integrated twice with respect to z to yield

$$c_A = C_1 z + C_2$$

The two constants of integration are evaluated, using the boundary conditions

$$\text{at } z = z_1 \qquad c_A = c_{A_1}$$
$$\text{at } z = z_2 \qquad c_A = c_{A_2}$$

to obtain the linear concentration profile

$$\frac{c_A - c_{A_1}}{c_{A_1} - c_{A_2}} = \frac{z - z_1}{z_1 - z_2} \tag{26-23}$$

Equations (26-21) and (26-23) may be used to describe any process where the bulk-contribution term is zero. Besides the equimolar-counterdiffusion phenomenon, a negligible bulk-contribution term is also encountered when a solute diffuses into or through a solid as both the mole fraction, x_A, and the flux of the diffusing species, N_{Az}, are very small. Accordingly, their resulting product is therefore negligible.

It is interesting to note that when we consider the "film concept" for mass transfer with equimolar counterdiffusion, the definition of the convective mass-transfer coefficient is different from that for diffusion in a stagnant gas film. In the case of equimolar counterdiffusion

$$k^0 = \frac{D_{AB}}{\delta} \tag{26-24}$$

The superscript on the mass-transfer coefficient is used to designate that there is no net molar transfer into the film due to the equimolar counterdiffusion. Comparing equation (26-24) with equation (26–9), we realize that these two defining equations yield the same results only when the concentration of A is very small and $p_{B,lm}$ is essentially equal to P.

26.2   ONE-DIMENSIONAL SYSTEMS ASSOCIATED WITH CHEMICAL REACTION

Many diffusional operations involve the simultaneous diffusion of a molecular species and the disappearance or appearance of the species through a chemical reaction either within or at the boundary of the phase of interest. We distinguish between the two types of chemical reactions, defining the reaction that occurs uniformly throughout a given phase as a *homogeneous reaction* and the reaction that takes place in a restricted region within or at a boundary of the phase as a *heterogeneous reaction*.

The rate of appearance of species A by a homogeneous reaction appears in the general differential equation of mass transfer as the source term, R_A

$$\nabla \cdot \mathbf{N}_A + \frac{\partial c_A}{\partial t} - R_A = 0 \tag{25-11}$$

Examples of the source term, R_A, include the first-order conversion of reactant A to product P, thus $R_A = -k_1 c_A$, where k_1 is the first-order rate constant in 1/s, and the second-order reaction of reactants A and B to form the product P with $R_A = k_2 c_A c_B$, where k_2 is the second-order rate constant in units $cm^2/mol \cdot s$.

The rate of disappearance of A by a heterogeneous reaction on a surface or at an interface does not appear in the general differential equation as R_A involves only reactions within the control volume. A heterogeneous reaction enters the analysis as a boundary condition and provides information on the fluxes of the species involved in the reaction; for example, if the surface reaction is $O_2(g) + C(s) \rightarrow CO_2(g)$, the flux of $CO_2(g)$ will be the same as the flux of $O_2(g)$, leaving in the opposite direction.

In this section, we shall consider two simple cases involving both types of chemical reactions. For a treatment of more complicated problems, the student is referred to the two excellent treatises by Crank[3] and Jost.[4]

Simultaneous Diffusion and Heterogeneous, First-Order Chemical Reaction: Diffusion with Varying Area

Many industrial processes involve the diffusion of a reactant to a surface where a chemical reaction occurs. As both diffusion and reaction steps are involved in the overall process, the relative rates of each step are important. When the reaction rate is instantaneous relative to the rate of diffusion, then the process is *diffusion controlled*. In contrast, when the reaction rate of the transferring species at the surface limits the mass-transfer rate, then the process is *reaction controlled*.

In many power plants, pulverized coal particles are fluidized within a hot combustion chamber, where oxygen in the air reacts with coal to produce carbon monoxide and/or carbon dioxide gas. This process, which produces energy by the heat of combustion, is an example of a simultaneous diffusion and a heterogeneous reaction process that is diffusion controlled.

Let us consider the steady-state, one-dimensional diffusion of oxygen to the surface of a spherical particle of coal along the r coordinate. At the surface of the particle, oxygen gas (O_2) reacts with solid carbon (C) in the coal to form carbon monoxide gas (CO) and carbon dioxide (CO_2) gas according to the heterogeneous reaction equation

$$3C(s) + 2.5O_2(g) \rightarrow 2CO_2(g) + CO(g)$$

as illustrated in Figure 26.7. No homogeneous chemical reaction occurs along the diffusion path so that $R_{O_2} = 0$. As the coal particle is oxidized, the particle shrinks with time as the carbon is converted to carbon monoxide and carbon dioxide. It is desired to predict the size of the particle with time.

Based on the above physical situation, the general differential

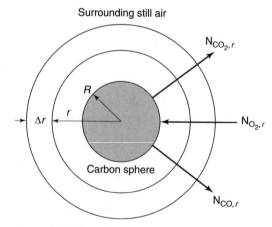

Figure 26.7 Diffusion through a spherical film.

[3] J. Crank, *The Mathematics of Diffusion*, Oxford University Press, London, 1957.

[4] W. Jost, *Diffusion in Solids, Liquids and Gases*, Academic Press, New York, 1952.

equation for mass transfer (25-29) reduces from

$$\frac{\partial c_A}{\partial t} + \frac{1}{r^2}\frac{\partial(r^2 N_{Ar})}{\partial r} + \frac{1}{r\sin\theta}\frac{\partial(N_{A\theta}\sin\theta)}{\partial\theta} + \frac{1}{r\sin\theta}\frac{\partial N_{A\phi}}{\partial\phi} - R_A = 0 \qquad (25\text{-}29)$$

to

$$\frac{1}{r^2}\frac{d(r^2 N_{Ar})}{dr} = 0$$

or

$$\frac{d(r^2 N_{Ar})}{dr} = 0 \qquad (26\text{-}25)$$

where A represents O_2, the transferring species.

As this is the first time we have encountered diffusion through a varying area, let us also derive equation (26-25) by making a mass balance in terms of moles of oxygen per time over the control volume bounded by the spherical surfaces at r and $r + \Delta r$

$$N_{O_2 r}\,4\pi r\,r^2\big|_r - N_{O_2 r}\,4\pi r^2\big|_{r+\Delta r} = 0 \qquad (26\text{-}26)$$

Dividing equation (26-26) by $4\pi\,\Delta r$ and evaluating the limit as Δr approaches zero, we obtain

$$\frac{d(r^2 N_{O_2 r})}{dr} = 0 \qquad (26\text{-}25)$$

This equation specifies that $r^2 N_{O_2 r}$ is constant over the diffusion path in the r direction, so that

$$r^2 N_{O_2 r}\big|_r = R^2 N_{O_2 r}\big|_R \qquad (26\text{-}27)$$

When we compare equation (26-25) with equation (26-1), we observe that for spherical coordinates, $r^2 N_{Ar}$ is constant along the r direction, whereas for rectilinear coordinates, N_{Az} is constant along the z direction.

Fick's equation (24-21) can be simplified once we recognize the relationship among the fluxes of the involved species. From the stoichiometry of the surface reaction, we recognize that 2.5 moles of oxygen are transferred to the surface, whereas 2 moles of carbon dioxide per 1 mole of carbon monoxide are transferred away from the surface. Therefore,

$$N_{O_2 r} = -2.5\,N_{CO r} \quad \text{and} \quad N_{O_2 r} = -1.25\,N_{CO_2 r}$$

There is no net transfer of the nitrogen in the air, as it is an inert. Therefore, $N_{N_2} = 0$. Fick's equation can now be written in terms of only oxygen

$$N_{O_2 r} = -c D_{O_2\text{-mix}}\frac{dy_{O_2}}{dr} + y_{O_2}(N_{O_2 r} + N_{CO r} + N_{CO_2 r} + N_{N_2})$$

or

$$N_{O_2 r} = -c D_{O_2\text{-mix}}\frac{dy_{O_2}}{dr} + y_{O_2}\left(N_{O_2 r} - \frac{1}{2.5}N_{O_2 r} - \frac{1}{1.25}N_{O_2 r} + 0\right)$$

Consequently, Fick's equation reduces to

$$N_{O_2 r} = -c D_{O_2\text{-mix}}\frac{dy_{O_2}}{dr} - 0.2 y_{O_2} N_{O_2 r}$$

or

$$N_{O_2 r} = -\frac{cD_{O_2\text{-mix}}}{1 + 0.2\, y_{O_2}} \frac{dy_{O_2}}{dr} \tag{26-28}$$

Both sides of equation (26-28) must be multiplied by r^2 to obtain the constant $r^2 N_{O_2 r}$ term. The diffusivity, $D_{O_2\text{-mix}}$, and the total molar concentration, c, can be treated as constants if evaluated at an average temperature and an average composition. Two boundary conditions are also needed. At the surface of the sphere, the reaction is instantaneous so that the oxygen concentration is zero

$$r = R, \quad y_{O_2} = 0$$

However, a long r-distance away from the sphere

$$r = \infty, \quad y_{O_2} = 0.21$$

Finally, equation (26-28) becomes

$$(r^2 N_{O_2 r}) \int_R^\infty \frac{dr}{r^2} = -\frac{cD_{O_2\text{-mix}}}{0.2} \int_0^{0.21} \frac{0.2\, dy_{O_2}}{1 + 0.2\, y_{O_2}}$$

which upon integration yields

$$(r^2 N_{O_2 r})\left(\frac{1}{R}\right) = \frac{cD_{O_2\text{-mix}}}{0.2} \ln\left(\frac{1}{1.042}\right)$$

The moles of oxygen transferred per time is the product of the oxygen flux and the cross-sectional area $4\pi r^2$

$$W_{O_2} = 4\pi r^2 N_{O_2 r} = -4\pi R \frac{cD_{O_2\text{-mix}}}{0.2} \ln(1.042) \tag{26-29}$$

It is important to recognize that this equation predicts a negative value for the rate of oxygen being transferred. The reason is that the direction of oxygen flux from the bulk gas to the particle surface is opposite to the increasing r direction from $r = R$ to $r = \infty$.

The pure carbon particle is the source for the CO_2 flux and the sink for O_2 flux. As the coal particle is oxidized, there will be an output of carbon as stipulated by the stoichiometry of the reaction. Although the spherical particle will decrease in size with time, the diffusion path from $r = R$ to $r = \infty$ will be essentially constant with time. This allows us to use a pseudo-steady-state approach for describing the material balance on the carbon particle. The material balance for carbon stipulates that

(input carbon rate) − (output carbon rate) = (rate of carbon accumulation)

The output rate of carbon from the spherical particle in mol C/time is related to output rate of CO_2 that is in turn related to the input mass transfer rate of O_2

$$W_C = \frac{3}{2} W_{CO_2} = -\frac{3}{2.5} W_{O_2} = +\frac{3}{2.5}\, 4\pi R\, \frac{cD_{O_2\text{-mix}}}{0.2} \ln(1.042) \tag{26-30}$$

The carbon accumulation rate in the spherical particle in mol C/time is

$$\frac{\rho_C}{M_C} \frac{dV}{dt} = \frac{\rho_C}{M_C}\, 4\pi R^2 \frac{dR}{dt}$$

where ρ_C is the density of solid carbon, M_C is the molecular weight of carbon, and V is the total volume of the solid carbon sphere. By substituting these terms into the carbon balance, we obtain

$$0 - \frac{3}{2.5} 4\pi R \frac{cD_{O_2\text{-mix}}}{0.2} \ln(1.042) = \frac{\rho_C}{M_C} 4\pi R^2 \frac{dR}{dt}$$

Rearrangement of this equation followed by separation of variables R and t gives

$$-6\,cD_{O_2\text{-mix}}\ln(1.042) \int_0^\theta dt = \frac{\rho_C}{M_C} \int_{R_i}^{R_f} R\,dR$$

where R_f is the final radius of the spherical particle, R_i is the initial radius of the particle, and θ is the time required for the particle to shrink from the initial to the final radius. Following integration, we obtain the final expression for estimating θ

$$\theta = \frac{\dfrac{\rho_C}{M_C}(R_i^2 - R_f^2)}{12\,cD_{O_2\text{-mix}}\ln(1.042)} \tag{26-31}$$

Alternative reaction equations may be proposed for the combustion process. For example, if only carbon dioxide were produced by the reaction

$$C(s) + O_2(g) \rightarrow CO_2(g) \tag{26-32}$$

then bulk contribution term to Fick's equation is zero as $N_{O_2 r} = -N_{CO_2 r}$. Therefore, Fick's equation reduces to

$$N_{O_2 r} = -cD_{O_2\text{-mix}} \frac{dy_{O_2}}{dr}$$

where the gas mixture consists of CO_2, O_2, and N_2. It is not difficult to show that the moles of oxygen transferred per time is

$$W_{O_2} = -4\pi R c D_{O_2\text{-mix}}\, y_{O_2\infty} \tag{26-33}$$

For heterogeneous reactions, information on the rate of the chemical reaction can also provide an important boundary condition

$$N_A|_{r=R} = -k_s c_{As}$$

where k_s is the first-order reaction-rate constant for a surface reaction, in units of m/s. The negative sign indicates that species A is disappearing at the surface. If the chemical reaction is instantaneous relative to the diffusion step, then the concentration of the reacting species A at the surface is essentially zero; that is, $c_{As} = 0$, as we assumed above. However, if the reaction is not instantaneous at the surface, then c_{As} will be finite.

Consider now that the reaction described by equation (26-32) is not instantaneous, and its heterogeneous rate equation can be expressed as a first-order surface reaction with respect to O_2. Consequently, the O_2 concentration at the surface is not equal to zero. In this context, it is easy to show that equation (26-33) becomes

$$W_{O_2} = -4\pi R c D_{O_2\text{-mix}}(y_{O_2\infty} - y_{O_2 s}) \tag{26-34}$$

where $y_{O_2\infty}$ is the mole fraction of O_2 in the bulk gas and $y_{O_2 s}$ is the mole fraction of O_2 at the surface $(r = R)$. For a first-order surface reaction $N_{As}|_R = -k_s c_{As}$, the mole fraction of O_2 at the surface can be expressed as

$$y_{O_2s} = \frac{c_{O_2s}}{c} = -\frac{N_{O_2R}}{k_s c} \tag{26-35}$$

The minus sign indicates that the direction of O_2 flux is opposite to increasing r. Substitution of equation (26-35) into equation (26-34) yields

$$W_{O_2} = -4\pi R c D_{O_2\text{-mix}} \left(y_{O_2\infty} + \frac{N_{O_2R}}{k_s c} \right) \tag{26-36}$$

Now recall

$$W_{O_2} = 4\pi R^2 N_{O_2R} = 4\pi R^2 N_{O_2r} \tag{26-27}$$

Combination of equations (26-36) and (26-27) to eliminate N_{O_2R} results in

$$r^2 N_{O_2r} \left(1 + \frac{D_{O_2\text{-mix}}}{k_s R} \right) = -R c D_{O_2\text{-mix}} y_{O_2\infty}$$

Finally, the oxygen-transfer rate for the combined diffusion and reaction process is

$$W_{O_2} = -\frac{4\pi R c D_{O_2\text{-mix}} y_{O_2\infty}}{1 + \frac{D_{O_2\text{-mix}}}{k_s R}} \tag{26-37}$$

Note that as k_s gets very large, equation (26-37) is approximated by equation (26-33).

EXAMPLE 3

A fluidized coal reactor has been proposed for a new power plant. If operated at 1145 K, the process will be limited by the diffusion of oxygen countercurrent to the carbon dioxide, CO_2, formed at the particle surface. Assume that the coal is pure solid carbon with a density of 1.28×10^3 kg/m^3 and that the particle is spherical with an initial diameter of 1.5×10^{-4} m (150 μm). Air (21% O_2 and 79% N_2) exists several diameters away from the sphere. Under the conditions of the combustion process, the diffusivity of oxygen in the gas mixture at 1145 K is 1.3×10^{-4} m^2/s. If a steady-state process is assumed, calculate the time necessary to reduce the diameter of the carbon particle to 5×10^{-5} m (50 μm).

The surrounding air serves as an infinite source for O_2 transfer, whereas the oxidation of the carbon at the surface of the particle is the sink for O_2 mass transfer. The reaction at the surface is

$$C(s) + O_2(g) \rightarrow CO_2(g)$$

Note that the reaction establishes a EMCD process where the flux of O_2 to the particle is equal to but opposite in the direction of the CO_2 flux away from the particle, that is,

$$N_{O_2r} = -N_{CO_2r}$$

At the surface of the coal particle, the reaction is so rapid that the concentration of oxygen is zero. Under this assumption, the instantaneous mass transfer of oxygen to the surface of the coal particle is

$$W_{O_2} = -4\pi R c D_{O_2\text{-mix}} y_{O_2\infty} \tag{26-33}$$

The stoichiometry of the surface reaction stipulates that 1 atom of carbon will disappear per each mole of oxygen reacting at the surface. Therefore,

$$W_C = W_{CO_2} = -W_{O_2} = +4\pi R c D_{O_2\text{-mix}} y_{O_2\infty}$$

The total carbon balance can be written as

$$0 - 4\pi R c D_{O_2\text{-mix}} y_{O_2\infty} = \frac{\rho_C}{M_C} 4\pi R^2 \frac{dR}{dt}$$

which simplifies to

$$dt = -\frac{\rho_C}{M_C} \frac{R\,dR}{c D_{O_2\text{-mix}} y_{O_2\infty}}$$

This equation can be integrated between the following limits

$$t = 0, \quad R = R_i = 7.5 \times 10^{-5}\,\text{m}\,(75\,\mu\text{m})$$
$$t = \theta, \quad R = R_f = 2.5 \times 10^{-5}\,\text{m}\,(25\,\mu\text{m})$$

to give

$$\theta = \frac{\rho_C(R_i^2 - R_f^2)}{2 M_C c D_{O_2\text{-mix}} y_{O_2\infty}}$$

The total gas molar concentration, c, is obtained by the ideal gas law

$$c = \frac{P}{RT} = \frac{1.0\,\text{atm}}{\dfrac{0.08206\,\text{m}^3\,\text{atm}}{\text{kg mol K}}1145\,\text{K}} = 0.0106\,\frac{\text{kg mol}}{\text{m}^3}$$

Finally,

$$\theta = \frac{\left(1.28 \times 10^3\,\dfrac{\text{kg}}{\text{m}^3}\right)((7.5 \times 10^{-5}\,\text{m})^2 - (2.5 \times 10^{-5}\,\text{m})^2)}{2\left(\dfrac{12\,\text{kg}}{\text{kg mol}}\right)\left(0.0106\,\dfrac{\text{kg mol}}{\text{m}^3}\right)\left(1.3 \times 10^{-4}\,\dfrac{\text{m}^2}{\text{s}}\right)(0.21)} = 0.92\,s$$

Diffusion with a Homogeneous, First-Order Chemical Reaction

In the unit operation of absorption, one of the constituents of a gas mixture is preferentially dissolved in a contacting liquid. Depending upon the chemical nature of the involved molecules, the absorption may or may not involve chemical reactions. When there is a production or disappearance of the diffusing component, equation (25-11) may be used to analyze the mass transfer within the liquid phase. The following analysis illustrates mass transfer that is accompanied by a homogeneous chemical reaction.

Consider a layer of the absorbing medium as illustrated in Figure 26.8. At the liquid surface, the composition of A is c_{A_0}. The thickness of the film, δ, is defined so that beyond this film the concentration of A is always zero; that is, $c_{A_\delta} = 0$. If there is very little fluid motion within the film, and if the concentration of A in the film is assumed small, the molar flux within the film is described by

$$N_{A,z} = -D_{AB}\frac{dc_A}{dz} \qquad (26\text{-}38)$$

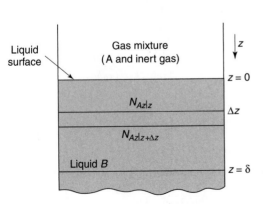

Figure 26.8 Absorption with homogeneous chemical reaction.

For one-directional steady-state mass transfer, the general differential equation of mass transfer reduces to

$$\frac{d}{dz} N_{A,z} - R_A = 0 \tag{26-39}$$

The disappearance of component A by a first-order reaction is defined by

$$-R_A = k_1 c_A \tag{26-40}$$

where k_1 is the chemical reaction rate constant. Substitution of equations (26-38) and (26-40) into equation (26-39) gives a second-order differential equation that describes simultaneous mass transfer accompanied by a first-order chemical reaction

$$-\frac{d}{dz}\left(D_{AB}\frac{dc_A}{dz}\right) + k_1 c_A = 0 \tag{26-41}$$

or with a constant diffusion coefficient, this reduces to

$$-D_{AB}\frac{d^2 c_A}{dz^2} + k_1 c_A = 0 \tag{26-42}$$

The general solution to equation (26-42) is

$$c_A = c_1 \cosh \sqrt{k_1/D_{AB}}z + c_2 \sinh \sqrt{k_1/D_{AB}}z \tag{26-43}$$

The boundary conditions

$$\text{at } z = 0 \qquad c_A = c_{A_0}$$

and

$$\text{at } z = \delta \qquad c_A = 0$$

permit the evaluation of the two constants of integration. The constant c_1 is equal to c_{A_0}, and c_2 is equal to $-(c_{A_0})/(\tanh \sqrt{k_1/D_{AB}}\delta)$, where δ is the thickness of the liquid film. Substituting these constants into equation (26-43), we obtain an equation for the concentration profile

$$c_A = c_{A_0} \cosh \sqrt{k_1/D_{AB}}z - \frac{c_{A_0}\sinh \sqrt{k_1/D_{AB}}z}{\tanh \sqrt{k_1/D_{AB}}\delta} \tag{26-44}$$

The molar mass flux at the liquid surface can be determined by differentiating equation (26-44) and evaluating the derivative, $(dc_A/dz)|_{z=0}$. The derivative of c_A with respect to z is

$$\frac{dc_A}{dz} = +c_{A_0}\sqrt{k_1/D_{AB}}\sinh \sqrt{k_1/D_{AB}}z - \frac{c_{A_0}\sqrt{k_1/D_{AB}}\cosh\sqrt{k_1/D_{AB}}z}{\tanh\sqrt{k_1/D_{AB}}\delta}$$

which, when z equals zero, becomes

$$\frac{dc_A}{dz}\bigg|_{z=0} = 0 - \frac{c_{A_0}\sqrt{k_1/D_{AB}}}{\tanh\sqrt{k_1/D_{AB}}\delta} = -\frac{c_{A_0}\sqrt{k_1/D_{AB}}}{\tanh\sqrt{k_1/D_{AB}}\delta} \tag{26-45}$$

Substituting equation (26-45) into equation (26-38) and multiplying by δ/δ, we obtain

$$N_{A,z}\big|_{z=0} = \frac{D_{AB}c_{A_0}}{\delta}\left[\frac{\sqrt{k_1/D_{AB}}\delta}{\tanh \sqrt{k_1/D_{AB}}\delta}\right] \tag{26-46}$$

It is interesting to consider the simpler mass-transfer operation involving the absorption of A into liquid B without an accompanying chemical reaction. The molar flux of A is easily determined by integrating equation (26-38) between the two boundary conditions, giving

$$N_{A,z} = \frac{D_{AB}c_{A_0}}{\delta} \tag{26-47}$$

It is apparent by comparing the two equations that the term $[(\sqrt{k_1/D_{AB}}\delta)/(\tanh\sqrt{k_1/D_{AB}}\delta)]$ shows the influence of the chemical reaction. This term is a dimensionless quantity, often called the *Hatta number.*[5]

As the rate of the chemical reaction increases, the reaction rate constant, k_1, increases and the hyperbolic tangent term, $\tanh\sqrt{k_1/D_{AB}}\delta$, approaches the value of 1.0. Accordingly, equation (26-46) reduces to

$$N_{A,z}|_{z=0} = \sqrt{D_{AB}k_1}(c_{A_0} - 0)$$

A comparison of this equation with equation (25-30)

$$N_{A,z} = k_c(c_{A_1} - c_{A_2}) \tag{25-30}$$

reveals that the film coefficient, k_c, is proportional to the diffusion coefficient raised to the $\frac{1}{2}$ power. With a relatively rapid chemical reaction, component A will disappear after penetrating only a short distance into the absorbing medium; thus, a second model for convective mass transfer has been proposed, the *penetration theory model*, in which k_c is considered a function of D_{AB} raised to the $\frac{1}{2}$ power. In our earlier discussion of another model for convective mass transfer, the film theory model, the mass-transfer coefficient was a function of the diffusion coefficient raised to the first power. We shall reconsider the penetration model in Section 26.4 and also in Chapter 28, when we discuss convective mass-transfer coefficients.

The following example considers diffusion with a homogeneous first-order chemical reaction under a different set of boundary conditions.

EXAMPLE 4

Dilute concentrations of toxic organic solutes can often be degraded by a "biofilm" attached to an inert, nonporous solid surface. A biofilm consists of living cells immobilized in a gelatinous matrix. Biofilms are not very thick, usually less than a few millimeters. A toxic organic solute (species A) diffuses into the biofilm and is degraded to harmless products, hopefully CO_2 and water, by the cells within the biofilm. For engineering applications, the biofilm can be approximated as a homogeneous substance (i.e., species B). The rate of degradation of the toxic solute per unit volume of the biofilm is described by a kinetic rate equation of the form

$$R_A = -\frac{R_{A,\max}c_A}{K_A + c_A}$$

where $R_{A,\max}$ is the maximum possible degradation rate of species A in the biofilm and K_A (mol/cm^3) is the half-saturation constant for the degradation of species A within the biofilm at hand.

Consider the simple "rotating disk" process unit shown in Figure 26.9 for the treatment of phenol (species A) in wastewater. The biofilm contains a microorganism rich in the enzyme peroxidase that oxidatively degrades phenol. The concentration of species A in the bulk-fluid phase over the biofilm is constant if the fluid phase is well mixed. However, the concentration of A within the biofilm will decrease along the depth of the biofilm z as species A is degraded. There are no

[5] S. Hatta, *Technol. Rep. Tohoku Imp. Univ.*, **10**, 199 (1932).

Figure 26.9 Biofilm for wastewater treatment.

resistances to convective mass transfer across the fluid boundary layer between the bulk fluid and the biofilm surface. Furthermore, phenol is equally soluble in both water and the biofilm, and the density difference between the biofilm and water can be neglected, so that the surface concentration of phenol in the aqueous phase equals the surface concentration of phenol in the gel phase just inside the biofilm, that is, at $z = 0$, $c_{As} = c_{A_0}$.

It is desired to treat 0.1 m^3 per hour of wastewater containing 0.1 mol/m^3 of the toxic substance phenol. If the biofilm thickness is 2.0 mm (0.002 m), what is the required surface area of the biofilm necessary to achieve the desired outlet concentration of 0.02 mol/m^3? The kinetic and mass-transport properties for the biofilm at hand are $K_A = 0.3$ mol/m^3, $R_{A,\text{max}} = 5.7 \times 10^{-3}$ mol/m$^3 \cdot$ s, and $D_{AB} = 2.0 \times 10^{-10}$ m^2/s at the process temperature of 25°C.

The source for phenol mass transfer is the inlet wastewater stream, whereas the sink for mass transfer is the steady-state consumption of phenol within the biofilm. First, a steady-state process material balance is performed on the contactor to determine the required rate of phenol degradation. A material balance on species A on the process unit is

(rate of phenol degraded) = (rate of phenol added to process) − (rate of phenol exiting process)

or

$$W_A = v_i c_{Ai} - v_o c_{Ao} \cong v_o(c_{Ai} - c_{Ao}) = \frac{0.1\,\text{m}^3}{\text{h}}\,(0.1 - 0.02)\frac{\text{mol}}{\text{m}^3} = 8.0 \times 10^{-3}\frac{\text{mol}}{\text{h}}$$

where the subscript "i" represents the inlet stream and subscript "o" represents the outlet stream. Note that c_{Ao} is the bulk concentration of phenol inside the contactor. The biofilm possesses slab geometry best described by rectilinear coordinates. The rate of degradation W_A is proportional to the flux of species A into the biofilm at $z = 0$

$$W_A = S \cdot N_{Az} = S \cdot \left(-D_{AB}\,\frac{dc_A}{dz}\bigg|_{z=0}\right)$$

where S is the required surface area of the biofilm. At low concentrations where $K_A \gg c_A$, the above rate equation approximates a first-order process with respect to c_A

$$R_A = -\frac{R_{A,\text{max}} c_A}{K_A + c_A} \cong -\frac{R_{A,\text{max}}}{K_A} c_A = -k_1 c_A$$

with k_1 equal to

$$k_1 = \frac{R_{A,max}}{K_A} = \frac{5.7 \times 10^{-3} \frac{mol}{m^3 \cdot s}}{0.3 \frac{mol}{m^3}} = 1.9 \times 10^{-2} s^{-1}$$

The flux can be obtained from the concentration profile. Recall equation (26-42) for one-dimensional, steady-state diffusion with a homogeneous first-order chemical reaction

$$D_{AB} \frac{d^2 c_A}{dz^2} - k_1 c_A = 0 \tag{26-42}$$

Recall also that this homogeneous second-order differential equation has a general solution of the form

$$c_A(z) = c_1 \cosh(z\sqrt{k_1/D_{AB}}) + c_2 \sinh(z\sqrt{k_1/D_{AB}}) \tag{26-43}$$

where c_1 and c_2 are integration constants to be determined by application of the boundary conditions. The biofilm is immobilized onto a nonporous solid surface. Therefore, the flux at $z = \delta$ is zero. Consequently, the boundary conditions are

$$z = \delta, \quad \frac{dc_A}{dz} = 0$$

$$z = 0, \quad c_A = c_{As} = c_{Ao}$$

Note that the boundary conditions previously discussed are different from the ones described earlier to develop equations (26-44) and (26-46). With a little math and a few handy identities for hyperbolic functions that can be found in any calculus text, the concentration profile is

$$c_A(z) = \frac{c_{Ao}\cosh((\delta - z)\sqrt{k_1/D_{AB}})}{\cosh(\delta\sqrt{k_1/D_{AB}})} \tag{26-48}$$

and

$$\left.\frac{dc_A}{dz}\right|_{z=0} = -c_{Ao}\sqrt{k_1/D_{AB}}\tanh(\delta\sqrt{k_1/D_{AB}})$$

From this, the flux of phenol into the biofilm is

$$N_A|_{z=0} = \frac{D_{AB}\,c_{Ao}}{\delta}(\delta\sqrt{k_1/D_{AB}})\tanh(\delta\sqrt{k_1/D_{AB}}) \tag{26-49}$$

It is useful first to calculate the dimensionless parameter, Φ, the Thiele modulus:

$$\Phi = \delta\sqrt{\frac{k_1}{D_{AB}}} = 0.002\,m\sqrt{\frac{1.9 \times 10^{-2}\frac{1}{s}}{2 \times 10^{-10}\frac{m^2}{s}}} = 19.49$$

This parameter represents the ratio of reaction rate to diffusion rate. When Φ is less than 0.1, $\tanh(\Phi)$ essentially equals zero and

$$N_A|_{z=0} = \frac{D_{AB}c_{Ao}}{\delta}\left(\delta\sqrt{k_1 D_{AB}}\right)^2$$

the reaction rate is very rapid (controlling) relative to the rate of diffusion. When Φ is greater than 5, $\tanh(\Phi) \approx 1.0$ and

$$N_{Az}|_{z=0} = \frac{D_{AB}c_{Ao}}{\delta}\left(\delta\sqrt{k_1 D_{AB}}\right)$$

Figure 26.10 Phenol concentration profile in biofilm.

The rate of diffusion is said to control. In our example, the molecular diffusion flux of phenol through the biofilm very strongly influences the overall phenol degeneration rate. The flux of phenol into the biofilm is

$$N_{Az} = \frac{\left(2 \times 10^{-10}\, \frac{m^2}{s}\right)\left(0.02\, \frac{mol}{m^3}\right)}{0.002\, m}(19.49)\tanh(19.49) = 3.9 \times 10^{-8}\, \frac{mol}{m^2 \cdot s}$$

Finally, the required surface area of the biofilm is backed out from the required degradation rate and the flux

$$S = \frac{W_A}{N_{Az}} = \frac{8.0 \times 10^{-3}\, \frac{mol}{h}\, \frac{1\, h}{3600\, s}}{3.9 \times 10^{-8}\, \frac{mol}{m^2 \cdot s}} = 57.0\, m^2$$

The steady-state concentration profile $c_A(z)$ within the biofilm is shown in Figure 26.10. It is interesting to note that concentration profile rapidly goes to zero within the first millimeter of the biofilm, again illustrating a strong diffusional resistance to the phenol degradation reaction.

26.3 TWO- AND THREE-DIMENSIONAL SYSTEMS

In Sections 26.1 and 26.2, we have discussed problems in which the concentration and the mass transfer were functions of a single-space variable. Although many problems fall into this category, there are systems involving irregular boundaries or nonuniform concentrations along the boundary for which the one-dimensional treatment may not apply. In such cases, the concentration profile may be a function of two or even three spatial coordinates.

In this section, we shall review some of the methods for analyzing molecular mass transfer in two- and three-dimensional systems. As the transfer of heat by conduction is analogous to molecular mass transfer, we shall find the analytical, analogical, and numerical techniques described in Chapter 17 to be directly applicable.

An analytical solution to any transfer problem must satisfy the general differential equation describing the transfer as well as the boundary conditions specified by the physical situation. A complete treatment of the analytical solutions for two- and three-dimensional systems requires a prior knowledge of partial differential equation and complex variable theory. As most of this material is too advanced for an introductory course, we shall limit our discussions to a relatively simple two-dimensional example. Crank[6] has written an excellent treatise dealing exclusively with mathematical solutions for more complex diffusion problems.

The classical approach to the exact solution of the Laplace equation is the separation of variables technique. We will illustrate this approach by considering the steady-state diffusion in a two-dimensional cloth strip that is used to deliver an insecticide to the surrounding atmosphere. There is no chemical reaction within the strip.

Consider a thin rectangular cloth strip, W units wide and L units long. The top of the strip is immersed in an insecticide source at $y = L$. Component A diffuses from this source down the strip to the three surfaces where it leaves by convective mass transfer. Figure 26.11 illustrates the strip and the assumed boundary conditions. The concentration at the surface, $y = L$, will be expressed functionally as $c_A(x)$, although it physically is often a constant concentration, c_{Ao}. The convective mass transfer from the three surfaces maintains zero surface concentrations. Due to the low concentration and low flux of A within the solid strip, the bulk contribution term in the Fick's equation is negligible.

The general differential equation for mass transfer (25-27) can be simplified to produce the describing partial differential equation.

$$\frac{\partial N_{Ax}}{\partial x}+\frac{\partial N_{Ay}}{\partial y}+ \overset{0}{\cancel{\frac{\partial N_{Az}}{\partial z}}} + \overset{0}{\cancel{\frac{\partial c_A}{\partial t}}} - \overset{0}{\cancel{R_A}} = 0$$

or

$$\frac{\partial N_{Ax}}{\partial x}+\frac{\partial N_{Ay}}{\partial y}=0 \qquad (26\text{-}50)$$

With negligible bulk contribution terms, the two Fick's equations are

$$N_{Ax} = -D_{AB}\frac{dc_A}{dx}$$

and

$$N_{Ay} = -D_{AB}\frac{dc_A}{dy}$$

Upon substitution of these fluxes into the partial differential equation and with the assumption of constant diffusivity, one obtains

$$\frac{\partial^2 c_A}{\partial x^2}+\frac{\partial^2 c_A}{\partial y^2}=0 \qquad (26\text{-}51)$$

Equation (26-51) is a linear, homogeneous partial differential equation. This type of equation can often be integrated by assuming a product solution of the form

$$c_A(x, y) = X(x)Y(y) \qquad (26\text{-}52)$$

Figure 26.11 Two-dimensional model for insecticide transport.

$c_A = c_A(x)$

L

$c_A = 0$

$c_A = 0$

$0 \quad c_A = 0 \quad W$

---

[6] J. Crank, *The Mathematics of Diffusion*, Oxford University Press, London, 1957.

where $X(x)$ is a function of x only and $Y(y)$ is a function of y only. Upon substituting equation (26-52) into equation (25-51), we obtain an expression in which the variables are separated

$$-\frac{1}{X}\frac{d^2X}{dx^2} = \frac{1}{Y}\frac{d^2Y}{dy^2} \tag{26-53}$$

The left-hand side of this equation is a function of x only, whereas the right-hand side is a function of y only. As neither side can change as x and y vary, both must be equal to a constant, say λ^2. We have, therefore, two ordinary differential equations

$$\frac{d^2X}{dx^2} + \lambda^2 X = 0 \tag{26-54}$$

and

$$\frac{d^2Y}{dy^2} - \lambda^2 Y = 0 \tag{26-55}$$

The general solution to equation (26-54) is

$$X = A\cos\lambda x + B\sin\lambda x \tag{26-56}$$

and the general solution to equation (26-55) is

$$Y = De^{-\lambda y} + Ee^{\lambda y} \tag{26-57}$$

According to the equation (26-48), the concentration is defined in terms of the product XY; consequently,

$$c_A = (A\cos\lambda x + B\sin\lambda x)(De^{-\lambda y} + Ee^{\lambda y}) \tag{26-58}$$

where A, B, D, and E are constants to be evaluated from the four boundary conditions

$$
\begin{array}{lll}
\text{at } x = 0 & c_A = 0 \\
\text{at } x = W & c_A = 0 \\
\text{at } y = 0 & c_A = 0
\end{array}
$$

and

$$\text{at } y = L \quad c_A = c_A(x)$$

The constants in equation (26-58) may be evaluated by the following substitutions: for the first condition at $x = 0$

$$A(De^{-\lambda y} + Ee^{\lambda y}) = 0$$

for the second condition at $x = W$

$$(A\cos\lambda W + B\sin\lambda W)(De^{-\lambda y} + Ee^{\lambda y}) = 0$$

and for the third condition, at $y = 0$

$$(A\cos\lambda x + B\sin\lambda x)(D + E) = 0$$

The third condition can be satisfied only if $D = -E$, and the first condition only if $A = 0$. Using these results, the second condition simplifies to

$$DB\sin\lambda W(e^{-\lambda y} - e^{\lambda y}) = 2DB\sin\lambda W\sinh\lambda y = 0 \tag{26-59}$$

Neither B nor D can be zero if a solution other than the trivial solution $c_A = 0$ throughout the passage is desired. As this expression is true for all y values, the condition specified by equation (26-59) can only be satisfied if $\sin \lambda W$ is zero; that is, $\lambda = n\pi/W$, where $n = 1, 2, 3, \ldots$. There exists a different solution for each integer n and each solution has a separate integration constant A_n. Summing these solutions, we obtain

$$c_A = \sum_{n=1}^{\infty} A_n \sin \frac{n\pi x}{W} \sinh \frac{n\pi y}{W} \tag{26-60}$$

The last boundary condition, at $y = L$, stipulates

$$c_A = c_A(x) = \sum_{n=1}^{\infty} A_n \sin \frac{n\pi x}{W} \sinh \frac{n\pi L}{W} \tag{26-61}$$

The constant A_n can be evaluated from equation (26-61), once the profile of $c_A(x)$ is given at the surface, $y = L$. An equation describing the variation of c_A with x and y can be obtained, after substituting the value of A_n into equation (26-56).

The separation-of-variables method can be extended to three-dimensional cases by assuming that c_A is equal to the product $X(x)Y(y)Z(z)$, and substituting this expression for c_A into the differential equation. If the variables can be separated, three second-order ordinary differential equations are obtained which may be integrated by using the given boundary conditions.

Another example of a two-dimensional steady-state diffusion process is encountered in the deposition of a silicon dioxide (SiO$_2$) thin film onto a wafer surface. One manufacturer uses the decomposition of tetraethoxysliane (TEOS) vapor by the following reaction:

$$\text{Si(OC}_2\text{H}_5)_4(g) \rightarrow 4C_2H_4(g) + 2H_2O(g) + SiO_2(s)$$

In this process, TEOS is carried to the diffusion furnace in a dilute (1 mol% TEOS) helium carrier gas from which the TEOS diffuses to vertically stacked silicon wafers, circular in shape, which are held in place by a thin support. The region between the two nonporous silicon wafers is shown in the following figure.

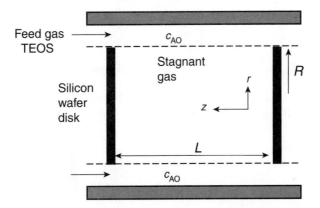

The concentration of TEOS in the gas space just outside of the vertically stacked wafer, c_{Ao}, is assumed to be constant. The gas space between the wafers is stagnant. TEOS vapors diffuse down into the gas space and then react on the wafer surface. As the temperature of the diffusion is maintained very high, the surface reaction is extremely rapid and the overall rate of SiO$_2$ formation is limited by the diffusion of the TEOS to the wafer surface.

Our analysis will focus on this "interwafer" region. The system is described in cylindrical coordinates, where the r-direction transfer is parallel to the wafer surface and the z-direction transfer is normal to the wafer surface. The general differential equation for the mass transfer of TEOS (species A) is

$$\frac{\partial c_A}{\partial t} + \left[\frac{1}{r}\frac{\partial(rN_{Ar})}{\partial r} + \frac{1}{r}\frac{\partial N_{A\theta}}{\partial \theta} + \frac{\partial N_{Az}}{\partial z}\right] = R_A$$

Due to the slow growth of the silicon dioxide film, this is a pseudo-steady-state process, $\partial c_A/\partial t = 0$; there is no homogeneous reaction within the diffusion path, $R_A = 0$; and the diffusion occurs only in the r and z directions. The equation simplifies to

$$\frac{1}{r}\frac{\partial(rN_{Ar})}{\partial r} + \frac{\partial N_{Az}}{\partial z} = 0$$

Recognizing that the gas space between the wafers is stagnant and that the gas is very dilute in TEOS, we will assume that there is negligible contribution to either of the two Fick's equations:

$$N_{Ar} = -cD_{AB}\frac{dy_A}{dr} \quad \text{and} \quad N_{Az} = -cD_{AB}\frac{dy_A}{dz}$$

These two equations can be further simplified upon the assumption that at the furnace's temperature and pressure, both c and D_{AB} are constant. By using these assumptions, the general differential equation for the mass transfer of TEOS can be simplified in terms of the concentration of TEOS, c_A.

$$D_{AB}\left[\frac{\partial^2 c_A}{\partial r^2} + \frac{1}{r}\frac{\partial c_A}{\partial r} + \frac{\partial^2 c_A}{\partial z^2}\right] = 0$$

To solve this differential equation, we need four boundary conditions:

$$@\, r = R \qquad c_A(R, z) = c_{Ao}$$

$$@\, r = 0 \qquad \frac{\partial c_A(0, z)}{\partial r} = 0$$

$$@\, z = 0 \qquad c_A(r, 0) = 0 \qquad \text{due to the rapid reaction on wafer surface}$$

and

$$@\, z = \frac{L}{2} \qquad \frac{\partial c_A\left(r, \frac{L}{2}\right)}{\partial r} = 0 \qquad \text{midpoint between two wafers}$$

A sample solution evaluated on the OSU FEMLAB, using a partial differential equation solver, is illustrated in the figure below.

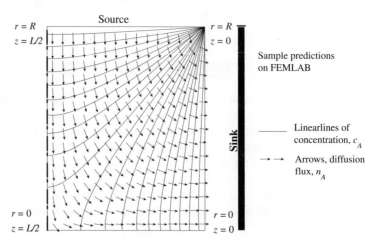

26.4 SIMULTANEOUS MOMENTUM, HEAT, AND MASS TRANSFER

In previous sections, we have considered steady-state mass transfer independent of the other transport phenomena. Many physical situations involve the simultaneous transfer of mass and either energy or momentum, and in a few cases, the simultaneous transfer of mass, energy, and momentum. The drying of a wet surface by a hot, dry gas is an excellent example in which all three transport phenomena are involved. Energy is transferred to the cooler surface by convection and radiation; mass and its associated enthalpy are transferred back into the moving gas stream. The simultaneous transport processes are more complex, requiring the simultaneous treatment of each transport phenomenon involved.

In this section, we consider two examples involving the simultaneous transfer of mass and a second transport phenomenon.

Simultaneous Heat and Mass Transfer

Generally, a diffusion process is accompanied by the transport of energy, even within an isothermal system. As each diffusing species carries its own individual enthalpy, a heat flux at a given plane is described by

$$\frac{\mathbf{q}_D}{A} = \sum_{i=1}^{n} \mathbf{N}_i \overline{H}_i \tag{26-62}$$

where $\mathbf{q}_D/A$ is the heat flux due to the diffusion of mass past the given plane and $\overline{H}_i$ is the partial molar enthalpy of species i in the mixture. When a temperature difference exists, energy will also be transported by one of the three heat-transfer mechanisms. For example, the equation for total energy transport by conduction and molecular diffusion becomes

$$\frac{\mathbf{q}}{A} = -k\boldsymbol{\nabla}T + \sum_{i=1}^{n} \mathbf{N}_i \overline{H}_i \tag{26-63}$$

If the heat transfer is by convection, the first energy-transport term in equation (26-63) would be replaced by the product of the convective heat-transfer coefficient and a ΔT driving force.

A process important in many engineering processes as well as in day-to-day events involves the condensation of a vapor upon a cold surface. Examples of this process include the "sweating" on cold water pipes and the condensation of moist vapor on a cold window pane. Figure 26.12 illustrates the process that involves a film of condensed liquid flowing down a cold surface and a film of gas through which the condensate is transferred by molecular diffusion. This process involves the simultaneous transfer of heat and mass.

The following conditions will be stipulated for this particular steady-state physical situation. Pure component A will condense from a binary gas mixture. By psychrometry, the composition, y_A, and the temperature, T_1, are known at the plane z_1. The temperature of the condensing surface, T_3, is also known. By heat-transfer considerations, the convective heat-transfer coefficients for the condensate liquid film and the gas film can be calculated from equations given in Chapter 20. For example, in the gas phase, when the carrier gas is air and the vapor content of the diffusing species is relatively low, the heat-transfer coefficient for natural convection can be estimated by equation (20-5)

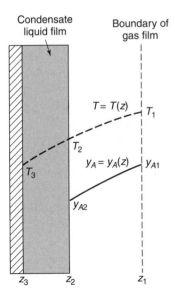

Figure 26.12 Vapor condensation on a cold surface.

$$\mathrm{Nu}_L = 0.68 + \frac{0.670\mathrm{Ra}_L^{1/4}}{[1 + (0.492/\mathrm{Pr})^{9/16}]^{4/9}} \tag{20-5}$$

Using the general differential equation for mass transfer, equation (25-11), we see that the differential equation that describes the mass transfer in the gas phase is

$$\frac{d}{dz} N_{A,z} = 0 \tag{26-64}$$

Equation (26-64) stipulates that the mass flux in the z direction is constant over the diffusion path. To complete the description of the process, the proper form of Fick's law must be chosen. If component A is diffusing through a stagnant gas, the flux is defined by equation (26-3)

$$N_{A,z} = \frac{-cD_{AB}}{1 - y_A} \frac{dy_A}{dz} \tag{26-3}$$

As a temperature profile exists within the film, and the diffusion coefficient and total gas concentration vary with temperature, this variation with z must often be considered. Needless to say, this complicates the problem and requires additional information before equation (26-3) can be integrated.

When the temperature profile is known or can be approximated, the variation in the diffusion coefficient can be treated. For example, if the temperature profile is of the form

$$\frac{T}{T_1} = \left(\frac{z}{z_1}\right)^n \tag{26-65}$$

the relation between the diffusion coefficient and the length parameter may be determined by using equation (24-41) as follows:

$$D_{AB} = D_{AB}|_{T_1} \left(\frac{T}{T_1}\right)^{3/2} = D_{AB}|_{T_1} \left(\frac{z}{z_1}\right)^{3n/2} \tag{26-66}$$

The variation in the total concentration due to the temperature variation can be evaluated by

$$c = \frac{P}{RT} = \frac{P}{RT_1 (z/z_1)^n}$$

The flux equation now becomes

$$N_{A,z} = \frac{-PD_{AB}|_{T_1}}{RT_1(1 - y_A)} \left(\frac{z}{z_1}\right)^{n/2} \frac{dy_A}{dz} \tag{26-67}$$

This is the same approach used in Example 15.2, which discussed heat transfer by conduction when the thermal conductivity was a variable.

Over a small temperature range, an average diffusion coefficient and the total molar concentration may be used. With this assumption, equation (26-3) simplifies to

$$N_{A,z} = -\frac{(cD_{AB})_{\text{avg}}}{(1 - y_A)} \frac{dy_A}{dz} \tag{26-68}$$

Integrating this equation between the boundary conditions

$$\text{at } z = z_1 \qquad y_A = y_{A_1}$$

and

$$\text{at } z = z_2 \qquad y_A = y_{A_2}$$

we obtain the relation

$$N_{A,z} = \frac{(cD_{AB})_{\text{avg}}(y_{A_1} - y_{A_2})}{(z_2 - z_1)y_{B,lm}} \tag{26-69}$$

The temperature, T_2, is needed for evaluating $(cD_{AB})_{\text{avg}}$, the temperature difference between the liquid surface and the adjacent vapor, and the vapor pressure of species A at the liquid surface. This temperature may be evaluated from heat-transfer considerations. The total energy flux through the liquid surface also passes through the liquid film. This can be expressed by

$$\frac{q_z}{A} = h_{\text{liquid}}(T_2 - T_3) = h_c(T_1 - T_2) + N_{A,z}M_A(H_1 - H_2) \tag{26-70}$$

where h_{liquid} is the convective heat-transfer coefficient in the liquid film, h_c is the natural convective heat-transfer coefficient in the gas film, M_A is the molecular weight of A, and H_1 and H_2 are the enthalpies of the vapor at plane 1 and the liquid at plane 2, respectively, for species A per unit mass. It is important to realize that there are two contributions to the energy flux entering the liquid surface from the gas film, convective heat transfer and the energy carried by the condensing species.

To solve equation (26-70), a trial-and-error solution is required. If a value for the temperature of the liquid surface is assumed, T_2, h_c, and $(cD_{AB})_{\text{avg}}$ may be calculated. The equilibrium composition, y_{A_2}, can be determined from thermodynamic relations. For example, if Raoult's law holds

$$p_{A_2} = x_A P_A$$

where x_A for a pure liquid is 1.0, and the partial pressure of A above the liquid surface is equal to the vapor pressure P_A. By Dalton's law, the mole fraction of A in the gas immediately above the liquid is

$$y_{A_2} = \frac{p_{A_2}}{P} \quad \text{or} \quad \frac{P_A}{P}$$

where P is the total pressure of the system and P_A is the vapor pressure of A at the assumed temperature T_2. Knowing $(cD_{AB})_{\text{avg}}$ and y_{A_2}, we can evaluate N_{Az} by equation (26-69). The liquid-film heat-transfer coefficients can be evaluated, using equations presented in Chapter 20. A value is now known for each term in equation (26-70). When the left- and right-hand sides of the equation are equal, the correct temperature of the liquid surface has been assumed. If the initially assumed temperature does not yield an equality, additional values must be assumed until equation (26-70) is satisfied.

There are several industrial-unit operations in which heat and mass transfer between gas and liquid phases occur simultaneously. Distillation, humidification or dehumidification of air, and water cooling are such operations. In early space exploration, the cooling of the reentry vehicles by sublimation of ablative material is another example where simultaneous transfer played an important engineering role.

In the following example, consideration of the simultaneous transfer of mass and heat is required to predict the flux relations described by Fick's law.

EXAMPLE 5

An ethanol/water vapor mixture is being distilled by contact with an ethanol/water liquid solution. The ethanol is transferred from the liquid to the vapor phase and the water is transferred in the opposite direction. The condensation of water vapor provides the energy for vaporization of ethanol. Both components are diffusing through a gas film 0.1 mm thick. The temperature is 368 K and the pressure is 1.013×10^5 Pa. At these conditions, the pure component enthalpie of vaporization of the ethanol and water are 840 and 2300 kJ/kg, respectively. Develop the flux equation for ethanol vapor. Then develop the flux equation assuming that the components have equimolar heats of vaporization.

We will assume a one-dimensional, adiabatic molecular mass-transfer process across a gas film of thickness δ, as illustrated in Figure 26.13. For one-dimensional, steady-state mass transfer, the general differential equation for mass transfer of ethanol simplifies to

$$\frac{dN_{EtOH,z}}{dz} = 0$$

For a binary gas-phase mixture, Fick's equation is

Figure 26.13 Adiabatic rectification of an ethanol/water mixture.

$$N_{EtOH,z} = -cD_{EtOH-H_2O} \frac{dy_{EtOH}}{dz} + y_{EtOH}(N_{EtOH,z} + N_{H_2O,z})$$

We perform an energy balance to relate the flux of ethanol vapor to the flux of water vapor. If the distillation operation is adiabatic, then all of the energy released when the water condenses must equal the energy used to produce the alcohol vapor. The energy balance is

$$N_{EtOH,z} \Delta H_{v,EtOH} = -N_{H_2O,z} \Delta H_{v,H_2O}$$

or

$$-N_{EtOH,z} \left(840 \frac{kJ}{kg}\right) \left(46 \frac{kg}{kg\,mol}\right) = N_{H_2O,z} \left(2300 \frac{kJ}{kg}\right) \left(18 \frac{kg}{kg\,mol}\right)$$

$$N_{EtOH,z} = -1.071\,N_{H_2O,z}$$

We recognize that flux is a vector quantity, and that according to Figure 26.13, the flux of ethanol is opposite in direction to the flux of water. Substituting this relationship into Fick's equation, we obtain

$$N_{EtOH,z} = -\frac{cD_{EtOH-H_2O}}{1 + 0.071y_{EtOH}} \frac{dy_{EtOH}}{dz}$$

As the flux is constant along the z direction, this equation can be integrated directly to obtain

$$N_{EtOH,z} \int_0^\delta dz = -cD_{EtOH-H_2O} \int_{y_{EtOH,1}}^{y_{EtOH,2}} \frac{dy_{EtOH}}{1 + 0.071y_{EtOH}}$$

or

$$N_{EtOH,z} = \frac{cD_{EtOH-H_2O}}{0.071\delta} \ln\left(\frac{1 + 0.071y_{EtOH,1}}{1 + 0.071y_{EtOH,2}}\right)$$

Now consider a simplified case where the *molar* heats of vaporization are essentially equal, i.e., $\Delta H_{v,EtOH} \cong \Delta H_{v,H_2O}$. Then from the adiabatic energy balance, it is easy to show that $-N_{EtOH} = N_{H_2O}$. Then

$$N_{EtOH,z} = -cD_{EtOH-H_2O} \frac{dy_{EtOH}}{dz} + y_{EtOH}(N_{EtOH,z} + N_{H_2O,z})$$

reduces to

$$N_{\text{EtOH}, z} = -cD_{\text{EtOH}-\text{H}_2\text{O}} \frac{dy_{\text{EtOH}}}{dz}$$

which upon integration yields

$$N_{\text{EtOH}, z} = \frac{cD_{\text{EtOH}-\text{H}_2\text{O}}}{\delta} \left(y_{\text{EtOH}, 1} - y_{\text{EtOH}, 2} \right)$$

So we see that equimolar heats of vaporization result in an equimolar-counterdiffusion flux for an adiabatic-distillation process.

Simultaneous Momentum and Mass Transfer

In several mass-transfer operations, mass is exchanged between two phases. An important example that we have previously encountered is *absorption,* the selective dissolution of one of the components of a gas mixture by a liquid. A wetted-wall column, as illustrated in Figure 26.14, is commonly used to study the mechanism of this mass-transfer operation, as it provides a well-defined area of contact between the two phases. In this operation, a thin liquid film flows along the wall of the column while in contact with a gas mixture. The time of contact between the two phases is relatively short during normal operation. As only a small quantity of mass is absorbed, the properties of the liquid are assumed to be unaltered. The velocity of the falling film will thus be virtually unaffected by the diffusion process.

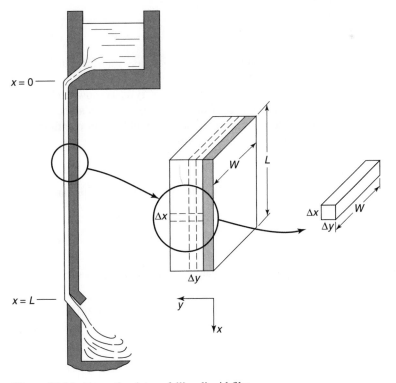

Figure 26.14 Absorption into a falling-liquid film.

The process involves both momentum and mass transfer. In Chapter 8, the laminar flow of a fluid down an inclined plane was discussed. When the angle of inclination is 90°, the results obtained in Section 8.2 can be used to describe the falling-film velocity profile. With this substitution, the differential equation for momentum transfer becomes

$$\frac{d\tau_{yx}}{dy} + \rho g = 0$$

and the boundary conditions that must be satisfied are

$$\text{at } y = 0 \qquad v_x = 0$$

and

$$\text{at } y = \delta \qquad \frac{\partial v_x}{\partial y} = 0$$

The final expression for the velocity profile is given by

$$v_x = \frac{\rho g \delta^2}{\mu} \left[\frac{y}{\delta} - \frac{1}{2} \left(\frac{y}{\delta} \right)^2 \right]$$

The maximum velocity will be at the edge of the film where $y = \delta$; its value is

$$v_{\max} = \frac{\rho g \delta^2}{2\mu}$$

Substituting this result into the velocity profile, we obtain another form of the expression for v_x

$$v_x = 2 v_{\max} \left[\frac{y}{\delta} - \frac{1}{2} \left(\frac{y}{\delta} \right)^2 \right] \tag{26-71}$$

The differential equation for mass transfer can be obtained by using the general differential equation of mass transfer and eliminating the irrelevant terms or by making a balance over the control volume, $\Delta x \, \Delta y W$, as shown in Figure 26.14. It is important to note that the y component of the mass flux, $N_{A,y}$, is associated with the negative y direction, according to the axes previously established in our fluid-flow considerations. The mass balance over the control volume is

$$N_{A,x}|_{x+\Delta x} W \, \Delta y - N_{A,x}|_x W \, \Delta y + N_{A,y}|_{y+\Delta y} W \, \Delta x - N_{A,y}|_y W \, \Delta x = 0$$

Dividing by $W \Delta x \, \Delta y$ and letting Δx and Δy approach zero, we obtain the differential equation

$$\frac{\partial N_{A,x}}{\partial x} + \frac{\partial N_{A,y}}{\partial y} = 0 \tag{26-72}$$

The one-directional molar fluxes are defined by

$$N_{A,x} = -D_{AB} \frac{\partial c_A}{\partial x} + x_A (N_{A,x} + N_{B,x}) \tag{26-73}$$

and

$$N_{A,y} = -D_{AB} \frac{\partial c_A}{\partial y} + x_A (N_{A,y} + N_{B,y}) \tag{26-74}$$

As previously mentioned, the time of contact between the vapor and liquid is relatively short; thus, a negligible concentration gradient will develop in the x direction, and equation (26-73) will reduce to

$$N_{A,x} = x_A(N_{A,x} + N_{B,x}) = c_A v_x \tag{26-75}$$

The convective transport term in the negative y direction, $x_A(N_{A,y} + N_{B,y})$, involves multiplying two extremely small values and is negligible; thus, equation (26-74) becomes

$$N_{A,y} = -D_{AB}\frac{\partial c_A}{\partial y} \tag{26-76}$$

Substituting equations (26-75) and (26-76) into equation (26-72), we obtain

$$\frac{\partial(c_A v_x)}{\partial x} - D_{AB}\frac{\partial^2 c_A}{\partial y^2} = 0 \tag{26-77}$$

or, as v_x is dependent upon y only

$$v_x\frac{\partial c_A}{\partial x} - D_{AB}\frac{\partial^2 c_A}{\partial y^2} = 0 \tag{26-78}$$

The velocity profile, as defined by equation (26-71), may be substituted into equation (26-78) yielding

$$2v_{\max}\left[\frac{y}{\delta} - \frac{1}{2}\left(\frac{y}{\delta}\right)^2\right]\frac{\partial c_A}{\partial x} = D_{AB}\frac{\partial^2 c_A}{\partial y^2} \tag{26-79}$$

The boundary conditions for mass transfer into the falling film are

$$\text{at } x = 0 \qquad c_A = 0$$

$$\text{at } y = 0 \qquad \frac{\partial c_A}{\partial y} = 0$$

and

$$\text{at } y = \delta \qquad c_A = c_{A_0}$$

Johnstone and Pigford[7] solved equation (26-79) and obtained, for the dimensionless concentration profile, the expression

$$\frac{c_A|_{x=L} - c_A|_{y=\delta}}{c_A|_{x=0} - c_A|_{y=\delta}} = 0.7857e^{-5.1213n} + 0.1001e^{-39.318n}$$
$$+ 0.03500e^{-105.64n} \tag{26-80}$$
$$+ 0.01811e^{-204.75n}$$
$$+ \cdots$$

where $c_A|_{x=L}$ is the concentration of solute at the bottom of the column, $c_A|_{y=\delta}$ is the concentration of the solute at the gas–liquid interface, $c_A|_{x=0}$ is the concentration of the solute at the top of the column, n is the ratio $D_{AB}L/\delta^2 v_{\max}$, L is the height of the column, δ is the film thickness, $v_{\max}$ is the maximum velocity in the film, located at the film surface, and D_{AB} is the diffusion coefficient of the solute in the liquid.

The specific case in which solute A penetrates only a short distance into the liquid film because of a slow rate of diffusion or a short time of exposure can be treated by the

---

[7] H. F. Johnstone and R. L. Pigford, *Trans. AIChE*, **38**, 25 (1942).

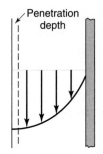

Figure 26.15
Penetration depth in
a falling film.

penetration theory model developed by Higbie.[8] As solute A is transferred into the film at $y = \delta$, the effect of the falling film on the diffusing species is such that the fluid may be considered to be flowing at the uniform velocity, v_{max}. Figure 26.15 illustrates the penetration depth. Solute A will not be affected by the presence of the wall; thus, the fluid may be considered to be of infinite depth. With these simplifications, equation (26-79) reduces to

$$v_{max} \frac{\partial c_A}{\partial x} = D_{AB} \frac{\partial^2 c_A}{\partial y^2} \tag{26-81}$$

with the boundary conditions

$$\text{at } x = 0 \quad c_A = 0$$
$$\text{at } y = \delta \quad c_A = c_{A_0}$$

and

$$\text{at } y = -\infty \quad c_A = 0$$

Equation (26-81) can be transformed into a form commonly encountered in unsteady-state mass transfer. If ξ is set equal to $\delta - y$, the transformed equation and boundary conditions become

$$v_{max} \frac{\partial c_A}{\partial x} = D_{AB} \frac{\partial^2 c_A}{\partial \xi^2} \tag{26-82}$$

and

$$\text{at } x = 0 \quad c_A = 0$$
$$\text{at } \xi = 0 \quad c_A = c_{A_0}$$

and

$$\text{at } \xi = \infty \quad c_A = 0$$

This partial differential equation can be solved by using Laplace transforms. On applying the transforms in the x direction, we obtain an ordinary differential equation in the s domain

$$v_{max} s \bar{c}_A - 0 = D_{AB} \frac{d^2 \bar{c}_A(\xi, s)}{d\xi^2}$$

or

$$\frac{d^2 \bar{c}_A}{d\xi^2} - \frac{v_{max} s \bar{c}_A}{D_{AB}} = 0 \tag{26-83}$$

This ordinary equation is readily solved to give

$$\bar{c}_A = A_1 \exp\left(\sqrt{\frac{v_{max} s}{D_{AB}}} \xi\right) + B_1 \exp\left(-\sqrt{\frac{v_{max} s}{D_{AB}}} \xi\right) \tag{26-84}$$

---

[8] R. Higbie, *Trans. AIChE*, **31**, 368–389 (1935).

The constants A_1 and B_1 are evaluated, using the two transformed boundary conditions

$$\text{at } \xi = 0 \qquad \overline{c}_A(0, s) = \frac{c_{A0}}{s}$$
$$\text{at } \xi = \infty \qquad \overline{c}_A(\infty, s) = 0$$

yielding the solution

$$\overline{c}_A = \frac{c_{A0}}{s} \exp\left(-\sqrt{\frac{v_{max} s}{D_{AB}}} \xi\right) \tag{26-85}$$

Equation (26-85) can be transformed back to the x domain by taking the inverse Laplacian, yielding

$$c_A(x, \xi) = c_{A0}\left[1 - \text{erf}\left(\frac{\xi}{\sqrt{\frac{4D_{AB}x}{v_{max}}}}\right)\right] \tag{26-86}$$

$$c_A(x, \xi) = c_{A0}\left[1 - \text{erf}\left(\frac{\xi}{\sqrt{4D_{AB}t_{exp}}}\right)\right]$$

where the time of exposure is defined by $t_{exp} = x/v_{max}$.

The error function, a mathematical form that is commonly encountered in transient problems, was discussed in Chapter 18. Similar to other mathematical functions, tables have been prepared of the error function and one of these tables is presented in Appendix L.

The local mass flux at the surface, where $\xi = 0$ or $y = \delta$, is obtained by differentiating equation (26-86) with respect to ξ and then inserting the derivative into equation (26-76).

$$N_{A,y}|_{\xi=0} = N_{A,y}|_{y=\delta} = -D_{AB}\frac{\partial c_A}{\partial y}\Big|_{y=\delta}$$

The unidirectional flux becomes

$$N_{A,y}|_{y=\delta} = c_{A_0}\sqrt{\frac{D_{AB}v_{max}}{\pi x}} \tag{26-87}$$

or

$$N_{Ay}|_{y=\delta} = c_{A_o}\sqrt{\frac{D_{AB}}{\pi t_{exp}}} \tag{26-88}$$

As the concentration driving force in the falling-liquid film

$$\Delta c_A = c_{A1} - c_{A2} = c_{A_o} - 0$$

equation (26-88) can be written

$$N_{Ay}|_{y=\delta} = \sqrt{\frac{D_{AB}}{\pi t_{exp}}}(c_{A1} - c_{A2})$$

A comparison of this equation with the convective mass-transfer equation

$$N_{Ay} = k_c(c_{A1} - c_{A2}) \tag{26-68}$$

reveals that

$$k_c = \sqrt{\frac{D_{AB}}{\pi t_{\text{exp}}}} \tag{26-89}$$

In equation (26-89), we see that the convective mass-transfer coefficient, k_c, is proportional to the diffusion coefficient raised to the $\frac{1}{2}$ power. This dependency was also shown earlier in Section 26.2 for the diffusion of a solute into a liquid accompanied by a rapid chemical reaction. *Penetration theory* considers that the solute only penetrates a short distance into the liquid phase due to a short residence time of exposure of the solute with the liquid, or because the solute rapidly disappears by a chemical reaction within the liquid. Consequently, penetration theory proposes that liquid-phase mass-transfer coefficient for transferring species A will assume the form of equation (26-89).

26.5 CLOSURE

In this chapter, we have considered solutions to steady-state molecular mass-transfer problems. The defining differential equations were established by simplifying the general differential equation for mass transfer or through the use of a control volume expression for the conservation of mass. It is hoped that this two-pronged attack will provide the student with an insight into the various terms contained in the general differential equation, and thus enable the reader to decide whether the terms are relevant or irrelevant to any specific situation.

One-directional systems both with and without chemical production were considered. Two models of convective mass transfer, film theory and penetration theory, were introduced. These models will be used in Chapter 28 to evaluate and explain convective mass-transfer coefficients.

PROBLEMS

26.1 An Arnold cell is to be operated as a pseudo-steady-state cell to determine the gas diffusivity of benzene in air at 308 K and 1.0 atm. The 20-cm-long tube, with an inner diameter of 1.0 cm, is initially loaded with liquid benzene to a depth of 1.0 cm from the bottom of the tube. The tube and the liquid are maintained at a constant temperature of 308 K. At this temperature, benzene exerts a vapor pressure of 0.195 atm. Air is continually blown over the top of the tube, removing any of the vaporized benzene vapor; the gas space within the tube is essentially stagnant. At 308 K, liquid benzene's density is 0.85 g/cm^3.

a. It was determined that 72.0 h were required to completely evaporate the benzene initially loaded into the tube. Estimate the binary gas-phase diffusion coefficient for benzene in air using these data.

b. Compare your evaluated diffusivity with the value reported in Appendix J.1. What is a possible reason for any difference in these values?

26.2 Helium gas is seperated from other components of a gas mixture by its selective diffusion through the wall of a Pyrex glass tube. Under steady-state conditions, the partial pressures of helium at the inner and outer surfaces of the Pyrex tube are 1.5 and 1.0 bar, respectively. If the wall thickness of the Pyrex tubing is 3 mm, determine

a. the flux of He through a tube having an inside diameter of 1.0 cm;

b. the concentration profile, $c_A(r)$, of helium within the wall.

26.3 A tank with its top open to the atmosphere contains liquid methanol (MeOH, molecular weight 32g/mol) at the bottom of the tank. The tank is maintained at 30°C. The diameter of the cylindrical tank is 1.0 m, the total height of the tank is 3.0 m, and the liquid level at the bottom of the tank is maintained at 0.5 m. The gas space inside the tank is stagnant and the MeOH vapors are immediately dispersed once they exit the tank. At 30°C, the vapor pressure exerted by liquid MeOH is 163 mmHg and at 40°C the MeOH vapor pressure is 265 mmHg. We are concerned that this open tank may be emitting a considerable amount of MeOH vapor.

a. What is the emission rate of MeOH vapor from the tank in units of kg MeOH/day when the tank is at a temperature of 30°C? State all assumptions and boundary conditions.

b. If the temperature of the tank is raised to 40°C, what is the new methanol emission rate?

26.4 Ethanol is diffusing through a 4-mm stagnant film of water. The ethanol concentrations of the entrance and the existing planes are maintained at 0.1 and 0.02 mol/m^3, respectively. If the water film temperature is 283 K, determine the steady-state molar flux of the ethanol and the concentration profile as a function of the position z within the liquid film. Compare these results with a 4-mm stagnant film of air at 283 K and 1 atm at the same entrance and exit ethanol concentrations.

26.5 Consider the process shown in the figure below. A slab contains parallel linear channels running through a nonporous slab of thickness 2.0 cm. The gas space over the slab contains a mixture of A and B maintained at a constant composition. Gas-phase species A diffuses down a straight, 1.0-mm-diameter channel. At the base of the slab is a catalytic surface that promotes the isomerization reaction A(g) → B(g). This reaction occurs very rapidly so that the production of B is diffusion limited. The quiescent gas space in the channel consists of only species A and B. The process is isothermal at 100°C and isobaric at 2.0 atm total system pressure. The bulk composition over the slab is maintained at 40 mol% A and 60 mol% B. The molecular weight of species A and its isomer B is 58 g/mol.

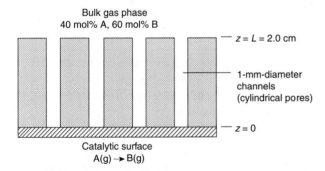

Monolith catalyst

Bulk gas phase
40 mol% A, 60 mol% B

$z = L = 2.0$ cm

1-mm-diameter channels (cylindrical pores)

$z = 0$

Catalytic surface
A(g) → B(g)

a. Listing all of your assumptions, simplify the general mass-transfer equation for species A.

b. Develop a final integrated equation for the flux of product B. Be sure to specify your boundary conditions.

c. The binary gas-phase molecular diffusion coefficient of species A in species B is 0.1 cm^2/s at 25°C and 1.0 atm. What is a reasonable estimate for the molecular flux of species B in species A under the conditions of the operation?

d. If the total production rate, W_A, is 0.01 mol B/min, what is the required number of 1.0-mm-diameter channels necessary to accomplish this production rate?

26.6 The following illustrated spherical capsule is used for long-term, sustained drug release. A saturated liquid solution containing the dissolved drug (solute A) is encapsulated within a rigid gel-like shell. The saturated solution contains a lump of solid A, which keeps the dissolved concentration of A saturated within the liquid core of the capsule. Solute A then diffuses through the gel-like shell (the gel phase) to the surroundings. Eventually, the source for A is depleted, and the amount of solute A within the liquid core goes down with time. However, as long as the lump of solid A exists within the core, the source solution is saturated in A and the concentration c_A is constant. The diffusion coefficient of solute A in the gel phase B is $D_{AB} = 1.5 \times 10^{-5}$ cm^2/s. The concentrations of solute A in the gel phase at the boundaries of the shell are

@ $r = R_1 = 0.2$ cm, $c_A = c_{As} = c_A = 0.01$ g mol/cm^3, the solubility limit of A in the gel.

@ $r = R_0 = 0.35$ cm, $c_A = c_{Ao}$.

Sustained drug-release spherical capsule

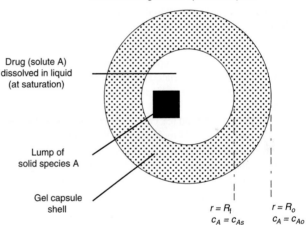

Drug (solute A) dissolved in liquid (at saturation)

Lump of solid species A

Gel capsule shell

$r = R_f$
$c_A = c_{As}$

$r = R_o$
$c_A = c_{Ao}$

a. State the differential forms of the flux equation and the differential equation for mass transfer for this diffusion process.

b. Develop the final analytical, integrated equation to determine the total rate of drug release, W_A, from the capsule under the condition where the concentration of A within the liquid core of the capsule remains constant.

c. What is the maximum possible rate of drug release from the capsule, in units of gram moles of A per hour?

26.7 A spherical ball of solid, nonporous naphthalene, a "moth ball", is suspended in still air. The naphthalene ball slowly sublimes, releasing the naphthalene into the surrounding air by molecular diffusion-limited process. Estimate the time required to reduce the diameter from 2 to 0.5 cm when the surrounding air is at 347 K and 1.013×10^5 Pa. Naphthalene has a molecular weight of 128 g/mol, a solid density of 1.145 g/cm^3, a diffusivity in air of 8.19×10^{-6} m^2/s, and exerts a vapor pressure of 5 Torr (666 Pa) at 347 K.

26.8   Consider the low-pressure chemical vapor deposition (LPCVD) diffusion reactor as shown below. It is desired to lay a thin film of the semiconductor gallium, Ga, onto a silicon wafer surface. Gallium metal is not volatile, but trimethyl gallium (TMG, Ga(CH$_3$)$_3$, 114.72 g/g mol) is volatile. In the presence of an H$_2$ gas, at high temperature, TMG will decompose to solid Ga on a surface by the following reaction:

$$Ga(CH_3)_3(g) + \tfrac{3}{2}H_2(g) \rightarrow Ga(s) + 3CH_4(g)$$

At 700°C, this surface reaction is diffusion limited. Let gas-phase species A = TMG, B = H$_2$ and C = CH$_4$.

a.   Develop an integral model to predict the flux of TMG to the wafer surface. Keep your final model in algebraic form. Provide appropriate assumptions and boundary conditions. At this point, you may not assume that the process is dilute.

For parts (b) and (c), consider a process where the feed gas consists of 99.98 mol% H$_2$ and 0.02 mol% TMG. The temperature and total system pressure are 700°C and 0.20 atm, respectively. The binary diffusion coefficient of TMG in H$_2$ at 750°C and 1.0 atm is known to be 2.0 cm^2/s.

b.   What is the simplified form of the model previously developed in part (a)?

c.   What is the value of an appropriate diffusion coefficient for this process?

26.9   Microelectronic devices are fabricated by forming many layers of thin films onto a silicon wafer. For example, thin films of arsenic as a common dopant for silicon are commonly deposited onto silicon by chemical vapor deposition, or CVD, of arsine (AsH$_3$) vapor onto the surface of the wafer. The chemical reaction for this CVD process is

$$AsH_3(g) \rightarrow As(s) + \tfrac{3}{2}H_2(g)$$

It is proposed to allow this process to take place at a pressure of 303.9 Pa and a temperature of 600°C. In many CVD reactors, the gas phase over the thin film is not mixed. Furthermore, at high temperatures, the surface reaction is very rapid. Consequently, the molecular diffusion of AsH$_3$ vapor to the surface often controls the rate of As(s) formation. Consider the very simplified CVD reactor illustrated in Chapter 25, Problem 25.11.

A mixture of arsine and hydrogen gas continuously flows into the reactor. The mixture contains 20 mol% arsine. A diffuser provides a quiescent gas space over the growing As film. The distance from the diffuser to the film surface is 6.0 cm. The gas mixture may be assumed to behave as an ideal gas. The molecular weight of arsenic, As, is 75 g/g mol. The Lennard–Jones parameters for SiH$_4$ are $\sigma_A = 4.08\,\mathring{A}$ and $\varepsilon_{A/\kappa} = 207.6$.

a.   What is the molecular diffusion coefficient of arsine vapor in hydrogen gas at 600°C and 303.9 Pa?

b.   Develop a model to predict the diffusion-limited flux of arsine to the surface of the wafer. State at least three assumptions relevant to this process.

c.   The diameter of the wafer is 15 cm. Estimate the initial deposition rate of arsenic onto the surface of the wafer, in units of grams of As per minute.

26.10   Consider a hemispherical droplet of liquid water residing on a flat surface. Still air surrounds the droplet. At an infinitely long distance from the gas film, the concentration of water vapor is effectively zero. At a constant temperature of 30°C and 1.0 atm total pressure, the evaporation rate of the droplet is controlled by the rate of the molecular diffusion through the still air. Determine the time it will take for the water droplet to completely evaporate at 30°C and 1.0 atm total system pressure if the initial droplet radius is 5 mm. The vapor pressure of water at 30°C can be found in the steam tables.

26.11   Consider the metallurgical refining process illustrated in the following figure.

Iron oxide, FeO, ore is being reduced to iron, Fe, by hydrogen gas according to the reaction

$$FeO(s) + H_2(g) \rightarrow Fe(s) + H_2O(g)$$

The FeO layer is nonporous, but the Fe layer is porous. Pure hydrogen gas flows over the surface of the slab. As FeO is reduced to Fe, the path length for diffusion of H$_2$ gas through the porous slab from the surface to the FeO/Fe boundary increases with time. In making your model, you may assume that (1) the process operates under constant conditions of 400 K and 1.0 atm; (2) the reaction is very rapid so that the reduction of FeO is limited by the diffusion of H$_2$ to the FeO/Fe boundary; (3) the diffusion process is pseudo-steady state along the path length; and (4) the effective gas-phase diffusion coefficient of water in hydrogen is 1.7 cm^2/s within the porous solid under the conditions of the process. The bulk density of the FeO solid is 2.5 g/cm^3, and the molecular weight of FeO is 71.85 g/g mol.

a. Reduce the general differential equation for mass transfer to describe the diffusion of H_2.

b. Determine the molar flux of H_2 to the FeO/Fe boundary when $\delta = 1$ cm.

c. Determine the minimum time necessary for FeO to be converted to Fe from $\delta = 1$ to $\delta = 2.0$ cm.

26.12 Condsider the timed drug-release pill illustrated below.

The pill is ingested into the stomach. The pill is a slab, 0.36 cm per side, that has an array of 16 cylindrical pores in it. Each pore is 0.4 mm in diameter and 2.0 mm deep. Pure solid drug A is loaded into each pore to a depth of 1.2 mm, which provides a total initial drug loading of 2.65 mg in all of the pores. The density of the solid drug A is 1.10 g/cm^3. The drug dissolves into the fluid inside the stomach, which approximates the properties of water (component B). The maximum solubility of drug A in water is 2.0×10^{-4} $g\,mol/cm^3$ (i.e., not very soluble) and the diffusion coefficient of the drug is 2.0×10^{-5} cm^2/s at body temperature of 37°C. The molecular weight of the drug is 120 g/mol.

a. Starting with the general differential equation for mass transfer and Fick's flux equation, develop a simple model, in final integrated form, for predicting the flux of the drug from one pore. You may assume that the diffusion process is pseudo-steady state, the stomach fluid serves as an infinite sink for the drug so that $c_A = 0$, and the drug does not chemically degrade inside the pore.

b. From your model, determine the total transfer rate of the drug from the whole pill, W_A, to the body when each 0.2 cm pore is filled to a depth of 0.12 cm with solid drug.

c. How many hours will it take for all of the drug to be released?

26.13 Hydrogen (H_2) gas is commonly used to reduce iron oxide ores to metallic ores. Consider the following heterogeneous reduction reactions on a nonporous flat surface of the following ores:

$$FeO(s) + H_2(g) \rightarrow Fe(s) + H_2O(g)$$

$$TiO_2(s) + 2H_2(g) \rightarrow Ti(s) + 2H_2O(g)$$

$$Fe_2O_3(s) + 3H_2(g) \rightarrow 2Fe(s) + 3H_2O(g)$$

$$Mn_3O_4(s) + 4H_2(g) \rightarrow 3Mn(s) + 4H_2O(g)$$

Write the integrated form of the flux equation, assuming (1) steady-state one-dimensional flux through a gas film of thickness δ along the coordinate z; (2) constant gas composition at the edge of the gas film; (3) instantaneous reaction at the surface; and (4) constant temperature and pressure. Which reactions result in an equimolar-diffusion flux of the gas-phase species? *Hint*: Carefully consider the stoichiometry associated with each reaction.

26.14 As part of the manufacturing process for the fabrication of titanium-oxide-based solar panels, a layer of nonporous titanium oxide must be reduced to metallic titanium, Ti, by hydrogen gas as shown in the following figure.

The reaction at the Ti/TiO$_2$ boundary is given by

$$TiO_2(s) + 2H_2(g) \rightarrow Ti(s) + 2H_2O(g)$$

Pure H_2 gas flows rapidly over the surface of the nanoporous TiO$_2$ slab. As TiO$_2$(s) is reduced to Ti (molecualr weight Ti = 47.9 g/g mol), the path length for mass transfer of H_2 and H_2O gas through the porous slab from the surface to the Ti/TiO$_2$ boundary increases with time. You may assume that (1) the process operates at 1.0 atm and 900 K; (2) the reaction is very fast so that the concentration of H_2 gas at the Ti/TiO$_2$ boundary is zero and the reduction of TiO$_2$(s) is limited by the diffusion of TiO$_2$(g) away from the Ti(s)/TiO$_2$ boundary; (3) the diffusion process is pseudo-steady state along the diffusion path; and (4) the effective gas-phase diffusion coefficient of H_2 within the porous Ti(s) containing a mixture of H_2(g) and H_2O(g) is 0.031 cm^2/s at the temperature and pressure of the process, whereas the effective diffusion coefficient of H_2O(g) within the nanoporous Ti(s) containing a mixture of H_2(g) and H_2O(g) is 0.01 cm^2/s at the temperature and pressure of the process. The density of the nanoporous Ti(s) is 2.6 g/cm^3.

a. Determine the flux of H_2 to the Ti/TiO$_2$ boundary when $\delta = 0.05$ cm, assuming $y_{Ao} = 1.0$.

b. Determine the number of hours necessary for all the TiO$_2$ to be converted to Ti(s).

c. At $z = \delta = 0.05$ cm, determine the concentration profile for H_2(g).

26.15 Two very large tanks, maintained at 323 K and 1 atm total system pressure, are connected by a 0.1-m-diameter circular duct which is 5 m in length. Tank 1 contains a uniform gas of 60 mol% acetone and 40 mol% air, whereas tank 2 contains a uniform gas of 10 mol% acetone and 90 mol% air. Determine the initial rate of acetone transfer between the two tanks. The gas diffusivity of acetone in air at 298 K and 1 atm is 0.093 cm^2/s.

26.16 Consider the novel "nanostructured" catalyst shown in the following figure.

The catalyst support consists of an ordered array of cylindrical "nanowells" of 50nm diameter and 200 nm depth (1 nm = 10^{-9} m). A catalytic surface coats of the bottom of each well. Although gas flows over the catalyst surface, the gas space within each "well" is stagnant, that is, not well mixed. In the present application, the catalyst surface is used to convert unreacted H_2 gas (species A) and O_2 gas (species B) from a fuel cell into water vapor (species C) according to the reaction

$$2H_2(g) + O_2(g) \rightarrow 2H_2O(g)$$

For our first approximation, consider the reaction diffusion limited within the catalyst well. The process is isothermal at 473 K and isobaric at 1.25 atm total system pressure.

a. State reasonable assumptions for this process. What is the simplified form of the general differential equation for the mass transfer of hydrogen? What is the simplified form of Fick's equation for hydrogen? Consider O_2 is the dominant species and both H_2 and H_2O are dilute.

b. What is the effective diffusion coefficient of H_2 gas inside each nanowell if we have a dilute system with bulk gas mole fraction compositions for $y_{H_2} = 0.01$ and $y_{H_2O} = 0.01$. What is the flux of H_2?

26.17 "Microvia" are microscopic passages between two thin films on a microelectronic device. Often, microvia are filled with a conductive metal to make a microscopic conductor for the flow of electrons between the two thin films. In one particular process, tungsten is deposited onto the base of the microvia by the following chemical vapor deposition reaction:

$$WF_6(g) + 3H_2(g) \rightarrow W(s) + 6HF(g)$$

As the tungsten metal forms, it fills the microvia (2.0 µm depth, 0.5 µm diameter) as illustrated in the following figure:

The tungsten metal does not coat the side walls of the microvia; it only grows upward from the base of the microvia where the tungsten was initially seeded. The reactants are significantly diluted in inert Helium (He) gas to lower the deposition rate. The temperature is 700 K, the total system pressure is 75 Pa, and the concentrations of WF_6 and H_2 in the bulk gas space over the microvia are each 0.001 mol. Assume that the tungsten deposition is limited by molecular diffusion. The molecular weight of tungsten, W, is 184 g/mol, the molecular weight of fluorine is 19 g/mol, and the density of solid tungsten is 19.4 g/cm^3.

a. Develop a pseudo-steady-state model to predict the depth of tungsten metal within the microvia as a function of time.

b. Estimate the time required to completely fill the microvia assuming Knudsen diffusion for WF_6 vapor.

26.18 In the distillation of a benzene/toluene mixture, a vapor richer in benzene is produced from a benzene/toluene liquid solution. Benzene is transferred from the liquid to the vapor phase and the less-volatile toluene is transferred in the opposite direction. At the system temperature and pressure, the latent heats of vaporization of benzene and toluene are 30 and 33 kJ/mol, respectively. Both components are diffusing through a gas film of thickness δ. Develop an equation to predict the steady-state flux of benzene through the gas film. The equation must include terms for the bulk gas-phase mole fraction of benzene, the gas-phase mole fraction of benzene in equilibrium with the liquid solution, the diffusion coefficient of benzene/toluene, the diffusion path δ, and the total molar gas concentration. Assume the distillation is an adiabatic process.

26.19 An important step in the purification of uranium isotopes involves the conversion of uranium to uranium hexafluoride, UF_6. In the present process, UF_6 is prepared by exposing uranium pellets, spherical in shape, to fluorine gas at 1000 K and 1 atm pressure. The molecular diffusion of the fluorine gas to the pellet surface is thought to be the controlling step. If the reaction

$$U(s) + 3F_2(g) \rightarrow UF_6(g)$$

occurs irreversibly and instantaneously on the pellet surface, and the diffusivity of fluorine gas through uranium hexafluoride gas is 0.273 cm^2/s at 1000 K and 1 atm, determine the production rate of UF_6 when the pellet diameter is 0.4 cm.

26.20 Consider the "drug patch" shown below. The drug patch looks like a sandwich consisting of a pure drug source mounted on top of a gel diffusion barrier. The gel diffusion barrier has a thickness of 2.0 mm. The gel barrier is in direct contact with the skin. The drug release vs. time profile for a 3.0-cm square patch at 20°C is also shown below. Other experiments showed that the drug was immediately taken up into the body after exiting the patch. The maximum solubility of the drug in the gel diffusion barrier is 0.5 µmol/cm^3. The drug is only slightly soluble in the gel material. The solubility of the drug in the gel diffusion barrier is not affected by the temperature.

a. From the data in the drug release vs. time profile, estimate the effective diffusion coefficient of the drug in the diffusion barrier.

b. When used on the body, heat transfer raises the temperature of the drug patch to about 35°C. What is the new drug delivery rate, W_A, at this temperature in units of µmol/day? You may assume that the gel-like diffusion barrier material approximates the properties of water. The viscosity of liquid water is 0.00993 g/cm s at 20°C and 0.00742 g/cm s at 35°C.

26.21 The permeability of solids by gases is experimentally determined by steady-state diffusion measurements. The diffusing solute is introduced at one side of the membrane and removed from the other side as a gas.

For a diatomic gas, A_2, which dissociates upon dissolving into a solid, Sievert's law relates the concentration of A atoms in the surface layer of the membrane, c_{A_1}, in equilibrium with the applied pressure, p_1, of the diatomic gas by the relation

$$c_{A_1} = k(p_1)^{1/2}$$

This same equation also holds at the other surface of the membrane for the off-gas pressure, p_2. Sievert's law is a variant of Henry's law for gases that dissociate upon dissolving.

a. Prove that the rate of diffusion of a diatomic gas from a high-pressure reservoir, p_1, through a membrane of thickness z, into a low-pressure reservoir, p_2, is

$$J_{A_2} = \frac{D_{A2}k(p_1^{1/2} - p_2^{1/2})}{z}$$

where D_{A2} is the diffusivity of A_2 through the membrane. When standard pressures are employed with a membrane of standard thickness, J_{A_2} is called the *permeability* of A_2.

b. A piece of laboratory equipment operating at 700°C contains hydrogen gas at 8 atm that is separated from a continuously evacuated space by an 8 cm² nickel disk, 2 mm thick. The solubility of hydrogen in nickel at 1 atm pressure and 700°C is approximately 7.0 cm³/100 g of nickel. The diffusivity of hydrogen through nickel at 700°C is 6×10^{-5} cm²/s, and the density of nickel at 700°C is 9.0 g/cm³. Calculate the number of cubic centimeters of hydrogen per hour that diffuses through the nickel.

26.22 Living cells homogeneously distributed (immobilized) with an agarose gel require glucose to survive. An important aspect of the biochemical system design is the effective diffusion coefficient of glucose (species A) into the cell-immobilized gel. Consider the experiment shown below where a slab of the cell-immobilized gel of 1.0 cm thickness is placed within a

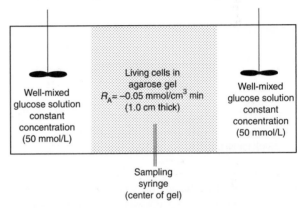

well-mixed aqueous solution of glucose maintained at a concentration of 50 mmol/L. The glucose consumption within the cell-immobilized gel proceeds by a zero-order process given by

$$R_A = -m = -0.05 \text{ mmol/L min} \qquad (26\text{-}70)$$

The solubilities of the glucose in both water and the gel are the same; that is, the concentration of glucose on the water side of the water–gel interface is equal to the concentration of glucose on the gel side of the water–gel interface. A syringe mounted at the center of the gel carefully excises a tiny sample of the gel for glucose analysis.

Develop a model in final integrated form to predict the concentration profile of glucose within the gel. Be careful with specification of boundary conditions, so that your model is truly predictive and is based only on process input parameters, not measured parameters.

26.23 A cylindrical graphite (pure carbon) rod of length 25 cm and initial diameter of 2 cm is inserted into a flowing air stream at 1100 K and 2 atm total pressure. The flowing gas creates a stagnant gas boundary layer 5 mm thick around the external surface of the rod. At this high temperature, the solid carbon

oxidizes to carbon dioxide, CO_2, gas

$$C(s) + O_2(g) \rightarrow CO_2(g)$$

The oxidation reaction is limited by the molecular diffusion of O_2 through the stagnant gas film surrounding the surface of the rod. Outside of the gas film, the bulk composition of the air stream prevails.

a. Estimate the initial rate of CO_2 production from the rod, assuming that the surface reaction is diffusion limited.

b. How long will it take for the rod to disappear? The density of solid graphite is 2.25 g/cm^3.

26.24 Ammonia, NH_3, is selectively removed from an air–NH_3 mixture by absorption into water. In this steady-state process, ammonia is transferred by molecular diffusion through a stagnant gas layer 2 cm thick, and then through a stagnant water layer 1 cm thick as shown in the figure. The concentration of ammonia at the upper boundary of the

gas layer is 3.42 mol%, and the concentration at the lower boundary of the water layer is essentially zero. The temperature of the system is 15°C and the total pressure on the system is 1 atm. The concentration of ammonia (species A) at the interface between the gas and the liquid phase is given by the following equilibrium data:

P_A(mmHg)	5.0	10.0	15.0	20.0	25.0	30.0
c_A(mol/m³)	6.1	11.9	20.0	32.1	53.6	84.8

Determine the flux of ammonia across both the gas and liquid films. At 15°C, the diffusivity of ammonia in air is 0.215 cm^2/s, and the diffusivity of ammonia in liquid water is 1.77×10^{-5} cm^2/s.

26.25 In a combustion chamber, oxygen diffuses through air to the carbon surface where it reacts to make CO and/or CO_2. The mole fraction of oxygen at $z = 0$ is 0.21. The reaction at the surface may be assumed to be instantaneous. No reaction occurs in the gas film.

Determine the rate of oxygen diffusion per hour through one square meter of area if

a. only carbon monoxide, CO, is produced at the carbon surface;

b. only carbon dioxide, CO_2, is produced at the carbon surface;

c. the following instantaneous reaction occurs at the carbon surface:

$$4C(s) + 3O_2(g) \rightarrow 2CO(g) + 2CO_2(g).$$

26.26 The data provided in Figure 26.5 are based on the diffusion of O_2 into SiO_2 formed from the oxidation of (100) crystalline silicon at 1000°C. Estimate the diffusion coefficient of O_2 in SiO_2 formed from the oxidation of (111) crystalline silicon at 1000°C, using the following data provided by Hess[9]:

	Measured SiO^2	film thickness (μm)
Time (h)	(100) Si	(111) Si
1	0.049	0.070
2	0.078	0.105
4	0.124	0.154
7	0.180	0.212
16	0.298	0.339

The maximum solubility of O_2 in the SiO_2 is 9.6×10^{-8} mol O_2/cm^3 solid at 1000°C and 1 atm O_2 gas partial pressure.

26.27 A 20-cm-long, cylindrical graphite (pure carbon) rod is inserted into an oxidizing atmosphere at 1145 K and 1.013×10^5 Pa pressure. The oxidizing process is limited by the diffusion of oxygen counterflow to the carbon monoxide that is formed on the cylindrical surface. Under the conditions of the combustion process, the diffusivity of oxygen in the gas mixture may be assumed to be 1.0×10^{-5} m^2/s.

a. Determine the moles of CO that are produced at the surface of the rod per second at the time when the diameter of the rod is 1.0 cm and the oxygen concentration that is 1.0 cm radial distance from the rod is 40 mol%. Assume a steady-state process.

b. What would be the composition of oxygen 1.0 cm from the center of the rod?

[9]D. W. Hess, *Chem. Eng. Educ.*, **24**, 34 (1990).

26.28   Please refer to Example 4. Consider now that the biofilm is a homogeneous sphere of 2.0 mm diameter. First, verify that the concentration profile of phenol within the sphere is

$$c_A = c_{Ao} \frac{R \sinh(r\sqrt{k_1 D_{AB}})}{r \sinh(R\sqrt{k_1 D_{AB}})}$$

for $0 \leq r \leq R$. From this expression, derive an appropriate expression for the flux N_A at the surface of the sphere ($r = R$). Then calculate N_A using the values for c_{Ao}, k_1, and D_{AB} given in Example 4. Compare your results to the flux obtained in Example 4 and discuss any differences.

Hint: Define a new lumped parameter $y = c_A r$ to reduce the second-order differential equation for $c_A(r)$ from a "variable coefficients" form to a "constant coefficients" form.

26.29   The following problem illustrates the gas-phase diffusion in the neighborhood of a flat catalytic surface. Component A diffuses through a stagnant film containing only A and B. Upon reaching the catalytic surface, it is instantaneously converted into species B by the reaction $A \rightarrow B$. When B diffuses back into the stagnant film, it begins to decompose by the first-order reaction $B \rightarrow A$. The rate of formation of component A with the film is equal to $R_A = k_1 y_B$, moles A produced/(time) (volume), where y_A is the concentration of B expressed in mole fraction.

Determine the rate at which A enters the gas film if this is a steady-state process.

26.30   Reconsider Problem 26.29 and determine the concentration profile of compound A in the stagnant film if in the film B decomposes to form A and if A reacts to form B, both by first-order reactions

$$A \underset{k_1}{\overset{k_1'}{\rightleftharpoons}} B, \quad R_A = k_1 y_B - k_1' y_A$$

Simultaneously, A is instantaneously reacting to form B on the flat catalytic surface.

Chapter 27

Unsteady-State Molecular Diffusion

$\mathbf{I}$n Chapter 26, we restricted our attention to describing the steady-state molecular diffusion where the concentration at a given point was constant with time. In this chapter, we shall consider problems and their solutions that involve the concentration varying with time, thus resulting in *unsteady-state molecular diffusion* or *transient diffusion*. Many common examples of unsteady-state transfer will be cited. These generally fall into two categories: a process that is in an unsteady state only during its initial startup, and a process in which the concentration is continually changing throughout its duration.

The time-dependent differential equations are simple to derive from the general differential equation of mass transfer. The equation of continuity for component A in terms of mass

$$\mathbf{\nabla} \cdot \mathbf{n}_A + \frac{\partial \rho_A}{\partial t} - r_A = 0 \tag{25-5}$$

or in terms of moles

$$\mathbf{\nabla} \cdot \mathbf{N}_A + \frac{\partial c_A}{\partial t} - R_A = 0 \tag{25-11}$$

contains the concentration time-dependent of the unsteady-state accumulation term. It also contains the net rate of mass efflux of species A, which accounts for the variation of the concentration with spatial directions. The solution to the resulting partial differential equations is generally difficult, involving relatively advanced mathematical techniques. We will consider the solutions to some of the less complex mass-transfer processes. A detailed discussion of the mathematics of diffusion is beyond the scope of this book. An excellent reference on the subject is a treatise by Crank.[1]

27.1 UNSTEADY-STATE DIFFUSION AND FICK'S SECOND LAW

Although the differential equations for unsteady-state diffusion are easy to establish, most solutions to these equations have been limited to situations involving simple geometries and boundary conditions, and a constant diffusion coefficient. Many solutions are for one-directional mass transfer as defined by Fick's second "law" of diffusion

$$\frac{\partial c_A}{\partial t} = D_{AB} \frac{\partial^2 c_A}{\partial z^2} \tag{27-1}$$

---

[1] J. Crank, *The Mathematics of Diffusion,* 2nd edition, Oxford University Press, 1975.

This partial differential equation describes a physical situation in which there is no bulk motion contribution, that is, $v = 0$, and no chemical reaction, that is, $R_A = 0$. This situation is encountered when the diffusion takes place in solids, in stationary liquids, or in systems having equimolar counterdiffusion. Due to the extremely slow rate of diffusion within liquids, the bulk motion contribution of Fick's first law (i.e., $y_A \sum N_i$) approaches the value of zero for dilute solutions; accordingly, this system also satisfies Fick's second law of diffusion.

It may be advantageous to express equation (27-1) in terms of other concentration units. For example, the mass density of species A, ρ_A, is equal to $M_A c_A$; by multiplying both sides of equation (27-1) by the constant molecular weight of A, we obtain

$$\frac{\partial \rho_A}{\partial t} = D_{AB} \frac{\partial^2 \rho_A}{\partial z^2} \tag{27-2}$$

If the density of the given phase remains essentially constant during the mass-transfer period, the density of species A can be divided by the total density, ρ_A/ρ; this ratio is the weight fraction of A, w_A, and our equation becomes

$$\frac{\partial w_A}{\partial t} = D_{AB} \frac{\partial^2 w_A}{\partial z^2} \tag{27-3}$$

However, when the phase loses a considerable amount of the solute, the total density, ρ, will not be constant and equation (27-3) cannot be used to explain the transient mass transfer. Under these circumstances, it is preferable to divide by the density of the given phase on a solute-free basis (as an example, during the drying of a piece of wood, the density of the moisture-free solid will be constant). On dividing equation (27-2) by the constant $\rho_{A\text{-free}}$, we obtain

$$\frac{\partial \left(\dfrac{\rho_A}{\rho_{A\text{-free}}} \right)}{\partial t} = D_{AB} \frac{\partial^2 \left(\dfrac{\rho_A}{\rho_{A\text{-free}}} \right)}{\partial z^2}$$

or

$$\frac{\partial w'_A}{\partial t} = D_{AB} \frac{\partial^2 w'_A}{\partial z^2} \tag{27-4}$$

where w'_A is the weight fraction of A divided by one minus the weight fraction of A.

Equations (27-1)–(27-4) are similar in form to Fourier's second "law" of heat conduction

$$\frac{\partial T}{\partial t} = \alpha \frac{\partial^2 T}{\partial z^2} \tag{27-5}$$

thereby establishing an analogy between transient molecular diffusion and heat conduction.

The solution to Fick's second law usually has one of two standard forms. It may involve error functions or related special integrals that are suitable for small values of time, or it may appear in the form of trigonometric series that converges at large values of time. Analytical solutions are commonly obtained by Laplace transform techniques or by the separation of variables. The analytical solution to Fick's second law is described for transient diffusion into a semi-infinite medium and for transient diffusion into a finite-dimensional medium.

27.2 TRANSIENT DIFFUSION IN A SEMI-INFINITE MEDIUM

An important case of transient mass diffusion amenable to analytical solution is the one-dimensional mass transfer of a solute into a semi-infinite stationary liquid or solid medium where surface concentration of the solute is fixed. For example, we might like to describe the absorption of oxygen gas into a deep tank of still water, the "doping" of phosphorous into a silicon wafer, or the solid-phase diffusion process involved in the case-hardening of mild steel within a carburizing atmosphere. Figure 27.1 depicts the concentration profiles as the time increases for a semi-infinite medium that has a uniform initial concentration of c_{Ao} and a constant surface concentration of c_{As}. The differential equation to be solved is

$$\frac{\partial c_A}{\partial t} = D_{AB}\frac{\partial^2 c_A}{\partial z^2}$$

(27-1)

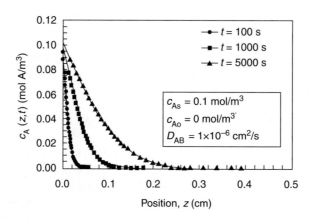

Figure 27.1 Transient diffusion in a semi-infinite medium.

Equation () is subject to the initial condition

$$t = 0, \qquad c_A(z, 0) = c_{Ao} \text{ for all } z$$

Two boundary conditions are required. The first boundary condition at the surface is

$$\text{at } z = 0, \qquad c_A(0, t) = c_{As} \quad \text{for } t > 0$$

A second boundary condition in the z direction must be specified. It is obtained by assuming that the diffusing solute penetrates only a very small distance during the finite time of exposure in comparison to the depth of the medium; this assumption provides the boundary condition

$$\text{at } z = \infty, \qquad c_A(\infty, t) = c_{Ao} \text{ for all } t$$

The analytical solution of equation () under the stated boundary and initial conditions can be obtained by Laplace transform techniques. The analytical solution is facilitated by making the boundary conditions homogeneous using the simple transformation

$$\theta = c_A - c_{Ao}$$

The partial differential equation and its initial and boundary conditions become

$$\frac{\partial \theta}{\partial t} = D_{AB}\frac{\partial^2 \theta}{\partial z^2}$$

(27-6)

with

$$\theta(z, 0) = 0$$
$$\theta(0, t) = c_{As} - c_{Ao}$$
$$\theta(\infty, t) = 0$$

The Laplace transformation of equation (27-6) with respect to time yields

$$s\bar{\theta} - 0 = D_{AB}\frac{d^2\bar{\theta}}{dz^2}$$

which readily transforms into the ordinary differential equation

$$\frac{d^2\bar{\theta}}{dz^2} - \frac{s}{D_{AB}}\bar{\theta} = 0 \tag{27-7}$$

with the transformed boundary conditions of

$$\bar{\theta}(z = 0) = \frac{c_{As} - c_{Ao}}{s}$$

and

$$\bar{\theta}(z = \infty) = 0$$

The general analytical solution of this differential equation is

$$\bar{\theta} = A_1 e^{+\sqrt{s/D_{AB}}z} + B_1 e^{-\sqrt{s/D_{AB}}z}$$

The boundary condition at $z = \infty$ requires integration constant A_1 to be zero. The boundary condition at $z = 0$ requires

$$B_1 = \frac{(c_{As} - c_{Ao})}{s}$$

Therefore, the general analytical solution reduces to

$$\bar{\theta} = \frac{(c_{As} - c_{Ao})}{s} e^{-\sqrt{s/D_{AB}}z} \tag{27-8}$$

The inverse Laplace transform of equation (27-8) can be found in any appropriate Laplace transform table. The result is

$$\theta = (c_{As} - c_{Ao})\text{erfc}\left(\frac{z}{2\sqrt{D_{AB}t}}\right)$$

which can be expressed as the dimensionless concentration change with respect to the initial concentration of species A, c_{Ao}, as

$$\frac{c_A - c_{Ao}}{c_{As} - c_{Ao}} = \text{erf c}\left(\frac{z}{2\sqrt{D_{AB}t}}\right) = 1 - \text{erf}\left(\frac{z}{2\sqrt{D_{AB}t}}\right) \tag{27-9}$$

or with respect to the surface concentration of species A, c_{As}, as

$$\frac{c_{As} - c_A}{c_{As} - c_{Ao}} = \text{erf}\left(\frac{z}{2\sqrt{D_{AB}t}}\right) = \text{erf}(\phi) \tag{27-10}$$

Equation (27-10) is analogous to heat conduction in a semi-infinite wall given by equation (18-20). The argument of the error function, given by the dimensionless quantity

$$\phi = \frac{z}{2\sqrt{D_{AB}t}}$$

contains the independent variables of position (z) and time (t).

In general, the error function is defined by

$$\text{erf}(\phi) = \frac{2}{\sqrt{\pi}} \int_0^{\phi} e^{-\xi^2} d\xi$$

Where ϕ is the argument of the error function and ξ is the dummy variable for ϕ. The error function has the following properties: $\text{erf}(0) = 0$ and $\text{erf}(\infty) = 1.0$. The error function is approximated by

$$\text{erf}(\phi) = \frac{2}{\sqrt{\pi}} \left(\phi - \frac{\phi^3}{3} \right) \qquad \text{if } \phi \leq 0.5$$

and

$$\text{erf}(\phi) = 1 - \frac{1}{\phi\sqrt{\pi}} e^{-\phi^2} \qquad \text{if } \phi > 1.0$$

A short table of the values of $\text{erf}(\phi)$ is presented in Appendix L.

The one-dimensional diffusion flux of species A into the semi-infinite medium at the surface of the medium ($z = 0$) is

$$N_A\Big|_{z=0} = -D_{AB} \frac{dc_A}{dz}\Big|_{z=0}$$

To obtain the diffusive flux of species A into the semi-infinite medium, we need to obtain the derivative of c_A with respect to z. By applying the chain rule of differentiation of the error function to equation (27-10), we obtain

$$\frac{dc_A}{dz}\Big|_{z=0} = -\frac{(c_{As} - c_{Ao})}{\sqrt{\pi D_{AB}}}$$

which is inserted into the flux equation to obtain

$$N_{A,z}\big|_{z=0} = \sqrt{\frac{D_{AB}}{\pi t}} (c_{As} - c_{Ao}) \qquad (27\text{-}11)$$

The total amount of species A transferred with time t can be determined by integrating the flux with time.

$$W_{At} - W_{Ao} = S \int_0^t N_{A,z}\big|_{z=0}\, dt = S \int_0^t \sqrt{\frac{D_{AB}}{\pi t}} (c_{As} - c_{Ao})\, dt = S\sqrt{\frac{4 D_{AB} t}{\pi}} (c_{As} - c_{Ao})$$

27.3 TRANSIENT DIFFUSION IN A FINITE-DIMENSIONAL MEDIUM UNDER CONDITIONS OF NEGLIGIBLE SURFACE RESISTANCE

Solutions for time-dependent mass-transfer processes have been obtained for simple geometrical shapes using the separation-of-variables technique. These bodies, initially possessing a concentration of c_{Ao}, are subjected to a sudden change in the surrounding environment that brings the surface concentration to c_{As}.

To illustrate the analytical technique of separation of variables, consider the molecular diffusion of a solute through a solid slab of uniform thickness L. Due to the slow rate of molecular diffusion within solids, the bulk contribution term of Fick's first law approaches zero, and so our solution for the concentration profile will satisfy the partial differential equation

$$\frac{\partial c_A}{\partial t} = D_{AB} \frac{\partial^2 c_A}{\partial z^2} \qquad (27\text{-}1)$$

with initial and boundary conditions

$$
\begin{aligned}
c_A &= c_{Ao} & \text{at } t = 0 & \quad \text{for } 0 \leq z \leq L \\
c_A &= c_{As} & \text{at } z = 0 & \quad \text{for } t > 0 \\
c_A &= c_{As} & \text{at } t = L & \quad \text{for } t > 0
\end{aligned}
$$

This transport process is analogous to the heating of a body under conditions of negligible surface resistance as discussed in Chapter 18. The boundary conditions above are simplified by expressing the concentrations in terms of the dimensionless concentration change Y, given by

$$
Y = \frac{c_A - c_{As}}{c_{Ao} - c_{As}}
$$

The partial differential equation becomes

$$
\frac{\partial Y}{\partial t} = D_{AB} \frac{\partial^2 Y}{\partial z^2} \tag{27-12}
$$

with the initial and boundary conditions

$$
\begin{aligned}
Y &= Y_o & \text{at } t = 0 & \quad \text{for } 0 \leq z \leq L \\
Y &= 0 & \text{at } z = 0 & \quad \text{for } t > 0 \\
Y &= 0 & \text{at } t = L & \quad \text{for } t > 0
\end{aligned}
$$

Let us assume that there is a product solution to the partial differential equation of the form

$$
Y(z, t) = T(t)Z(z)
$$

where the function $T(t)$ depends only on the time t and the function $Z(z)$ depends only on the coordinate z. The partial derivatives will be

$$
\frac{\partial Y}{\partial t} = Z \frac{\partial T}{\partial t}
$$

and

$$
\frac{\partial^2 Y}{\partial z^2} = T \frac{\partial^2 Z}{\partial z^2}
$$

Substitution into equation (27-12) yields

$$
Z \frac{\partial T}{\partial t} = D_{AB} T \frac{\partial^2 Z}{\partial z^2}
$$

which may be divided by $D_{AB} \cdot T \cdot Z$ to give

$$
\frac{1}{D_{AB} T} \frac{\partial T}{\partial t} = \frac{1}{Z} \frac{\partial^2 Z}{\partial z^2} \tag{27-13}
$$

The left-hand side of this equation depends only on time t and the right-hand side depends only on position z. If t varies, the right-hand side of the equation remains constant, and if z varies, the left-hand side remains constant. Accordingly, both sides must be independent of z and t and are equal to an arbitrary constant, $-\lambda^2$. This produces two separate ordinary differential equations, one for time t

$$
\frac{1}{D_{AB} T} \frac{dT}{dt} = -\lambda^2
$$

with a general solution of

$$T(t) = C_1 e^{-D_{AB}\lambda^2 t}$$

and one for position z

$$-\frac{1}{Z}\frac{d^2 Z}{dz^2} = -\lambda^2$$

with the general solution of

$$Z(z) = C_2\cos(\lambda z) + C_3\sin(\lambda z)$$

Substituting these two solutions into our product solution, we obtain

$$Y = T(t)Z(z) = [C_1'\cos(\lambda z) + C_2'\sin(\lambda z)]e^{-D_{AB}\lambda^2 t} \qquad (27\text{-}14)$$

The constants C_1' and C_2' and the parameter λ are obtained by applying the boundary and initial conditions to the general solution. The first boundary condition, $Y = 0$ at $z = 0$, requires C_1' to be zero. The second boundary condition, $Y = 0$ at $z = L$, stipulates that $\sin(\lambda L) = 0$, because C_2' cannot be equal to zero. Otherwise, if C_2' were equal to zero, the entire equation would be zero, yielding a trivial solution. In this context, $\sin(\lambda L)$ will be equal to zero only when

$$\lambda = \frac{n\pi}{L} \quad \text{for } n = 1, 2, 3, \ldots$$

To obtain C_2', the property of orthogonality must be applied, yielding the complete solution

$$Y = \frac{c_A - c_{As}}{c_{Ao} - c_{As}} = \frac{2}{L}\sum_{n=1}^{\infty}\sin\left(\frac{n\pi z}{L}\right)e^{-(n\pi/2)^2 X_D}\int_0^L Y_o\sin\left(\frac{n\pi z}{L}\right)dz \qquad (27\text{-}15)$$

where L is the thickness of the sheet and X_D is the relative time ratio, given by

$$X_D = \frac{D_{AB}t}{x_1^2}$$

with x_1 being the characteristic length of $L/2$. If the sheet has a uniform initial concentration along z, the final solution is

$$\frac{c_A - c_{As}}{c_{Ao} - c_{As}} = \frac{4}{\pi}\sum_{n=1}^{\infty}\sin\left(\frac{n\pi z}{L}\right)e^{-(n\pi/2)^2 X_D}, \qquad n = 1, 3, 5, \ldots \qquad (27\text{-}16)$$

which is analogous to the heat conduction equation (18-13) obtained for the heating of a body under conditions of negligible surface resistance.

The flux at any position z is

$$N_{Az} = -D_{AB}\frac{\partial c_A}{\partial z}$$

For the concentration profile given by equation (27-16), the flux at any position z and time t within the slab is

$$N_{Az} = \frac{4D_{AB}}{L}(c_{As} - c_{Ao})\sum_{n=1}^{\infty}\cos\left(\frac{n\pi z}{L}\right)e^{-(n\pi/2)^2 X_D}, \qquad n = 1, 3, 5, \ldots \qquad (27\text{-}17)$$

At the center of the slab $(z = L/2), N_A$ is equal to 0. Mathematically, it is equal to zero because the cosine term in equation (27-17) vanishes at intervals of $\pi/2$; physically, it is equal to zero because the *net flux* is equal to zero at the centerline. Consequently, the following boundary condition at $z = L/2$ also holds:

$$\frac{\partial c_A}{\partial z} = 0$$

Recall from Section 25.3 that the mathematical boundary condition also arises when the flux at a boundary is equal to zero because of the presence of barrier that is impermeable to the transfer of diffusing species A. Consequently, equation (27-16) can also be used for the physical situation where a slab of corresponding to a thickness x_1 of has an impermeable barrier at the boundary $z = x_1$.

The following examples illustrate processes that are governed by one-dimensional unsteady-state diffusion of a dilute solute into semi-infinite or finite-dimensional media. The phosphorous doping of silicon wafers illustrates molecular diffusion into a semi-infinite medium, whereas the timed drug release from a spherical capsule illustrates molecular diffusion from a finite-dimensional medium. We take a little extra time at the beginning of each example to describe the interesting technology behind the process.

EXAMPLE 1

In the fabrication of solid-state microelectronic devices, semiconducting thin films can be made by impregnating either phosphorous or boron into a silicon wafer. This process is called *doping*. The doping of phosphorous atoms into crystalline silicon makes an *n-type semiconductor*, whereas the doping of boron atoms into crystalline silicon makes a *p-type semiconductor*. The formation of the semiconducting thin film is controlled by the molecular diffusion of the dopant atoms through crystalline-silicon matrix.

Methods to deliver phosphorous atoms to the silicon wafer surface include chemical vapor deposition and ion implantation. In one typical process, phosphorous oxychloride, $POCl_3$, which has a normal boiling point of 105.3°C, is vaporized. The $POCl_3$ vapors are fed into a chemical vapor deposition (CVD) reactor at elevated temperature and reduced system pressure (e.g., 0.1 atm), where $POCl_3$ decomposes on the silicon surface according to the reaction

$$Si(s) + 2POCl_3(g) \rightarrow SiO_2(s) + 3Cl_2 + 2P(s)$$

A SiO_2 coating rich in molecular phosphorous (P) is formed over the crystalline-silicon surface. The molecular phosphorous then diffuses through the crystalline silicon to form the Si–P thin film. So the coating is the source for mass transfer of phosphorous, and the silicon wafer is the sink for mass transfer of phosphorous.

As one can see in Figure 27.2, the process for making Si–P thin films can be quite complex with many species diffusing and reacting simultaneously. But consider a simplified case where the P-atom

Figure 27.2 Phosphorous doping of a silicon wafer.

concentration is constant at the interface. As the diffusion coefficient of P atoms in crystalline silicon is very low, and only a thin film of Si–P is desired, phosphorous atoms do not penetrate very far into the silicon. Therefore, the phosphorous atoms cannot "see" through the entire thickness of the wafer, and the Si solid serves as a semi-infinite sink for the diffusion process. It is desirable to predict the properties of the Si–P thin film as a function of doping conditions. The concentration profile of the doped phosphorous atoms is particularly important for controlling the electrical conductivity of the semiconducting thin film.

Consider the phosphorous doping of crystalline silicon at 1100°C, a temperature high enough to promote phosphorous diffusion. The surface concentration of phosphorous (c_{As}) in the silicon is 2.5×10^{20} atoms P/cm^3 solid Si, which is relatively dilute, as pure solid silicon is 5×10^{22} atoms Si/cm^3 solid. Furthermore, the phosphorous-rich coating is considered as an infinite source relative to the amount of P atoms transferred, so that c_{As} is constant. Predict the depth of the Si–P thin film after 1 h, if the target concentration is 1% of the surface value (2.5×10^{18} atoms P/cm^3 solid Si), and the concentration profile of P atoms after 1 h.

Based on the assumptions given in the problem statement, the partial differential equation describing the one-dimensional, unsteady-state concentration profile $c_A(z, t)$ of phosphorous (species A) in solid silicon (species B) is

$$\frac{\partial c_A}{\partial t} = D_{AB} \frac{\partial^2 c_A}{\partial z^2} \tag{27-1}$$

For a semi-infinite medium, the initial and boundary conditions are

$$t = 0, \quad c_A(z, 0) = c_{Ao} = 0 \text{ for all } z$$

$$z = 0, \quad c_A(0, t) = c_{As} = 2.5 \times 10^{20} \text{ atoms P/cm}^3 \text{ solid Si, for } t > 0$$

$$z = \infty, \quad c_A(\infty, t) = c_{Ao} = 0 \text{ for all } t$$

If the diffusion coefficient D_{AB} is a constant, then the analytical solution is

$$\frac{c_{As} - c_A}{c_{As} - c_{Ao}} = \text{erf}\left(\frac{z}{2\sqrt{D_{AB}t}}\right) = \text{erf}(\phi)$$

Note that the depth z is imbedded in ϕ, the argument of the error function. The value for erf(ϕ) is calculated from the dimensionless concentration change

$$\frac{c_{As} - c_A}{c_{As} - c_{Ao}} = \frac{2.5 \times 10^{20} \text{ atoms P/cm}^3 - 2.5 \times 10^{18} \text{ atoms P/cm}^3}{2.5 \times 10^{20} \text{ atoms P/cm}^3 - 0} = 0.990 = \text{erf}(\phi)$$

From the table in Appendix L, the argument of the error function at erf(ϕ) = 0.990 is $\phi = 1.82$. From Figure 24.6, the solid-diffusion coefficient of P atoms (species A) in crystalline silicon (species B) is 6.5×10^{-13} cm^2/s at 1100°C (1373 K). The depth z can be backed out from ϕ by

$$z = \phi \cdot 2\sqrt{D_{AB}t} = 1.82 \cdot 2 \cdot \sqrt{\left(6.5 \times 10^{-13} \frac{\text{cm}^2}{\text{s}} \cdot \frac{10^8 \text{ }\mu\text{m}^2}{1 \text{ cm}^2}\right)\left(1 \text{ h} \cdot \frac{3600 \text{ s}}{1 \text{ h}}\right)} = 1.76 \text{ }\mu\text{m}$$

Prediction of the concentration profile after 1 h requires the calculation of ϕ at different values of z, followed by the calculation of erf(ϕ) and finally $c_A(z, t)$. Repetitive calculation of erf(ϕ) is best carried out with the help of a mathematics software package such as Mathcad (MathSoft Inc., Cambridge, MA.). The predicted phosphorous concentration profile is compared with the data of Errana and Kakati[2] obtained under similar process conditions, as shown in Figure 27.3. It is known that the molecular-diffusion coefficient of phosphorous in crystalline silicon is a function of phosphorous concentration. The concentration-dependent diffusion coefficient creates a "dip" in

---

[2] G. Errana and D. Kakati, *Solid State Technol.*, **27** (12), 17 (1984).

Figure 27.3 Phosphorous doping profile in silicon wafer after 1 h at 1100°C.

the observed phosphorous concentration profile. A detailed model of this phenomenon is provided by Middleman and Hochberg.[3]

EXAMPLE 2

One way to deliver a timed dosage of a drug within the human body is to ingest a capsule and allow it to settle in the gastrointestinal system. Once inside the body, the capsule slowly releases the drug by a diffusion-limited process. A suitable drug carrier is a spherical bead of a nontoxic gelatinous material that can pass through the gastrointestinal system without disintegrating. A water-soluble drug (solute A) is uniformly dissolved within the gel and has an initial concentration c_{Ao}. The drug loaded within the bead is the source for mass transfer, whereas the fluid surrounding the bead is the sink for mass transfer. This is an unsteady-state process, as the source for mass transfer is contained within the diffusion control volume itself.

Consider a limiting case where the resistance to film mass transfer of the drug through the liquid boundary layer surrounding the capsule surface to the bulk surrounding the fluid is negligible. Furthermore, assume that the drug is immediately consumed or swept away once it reaches the bulk solution so that in essence the surrounding fluid is an infinite sink. In this particular limiting case, c_{As} is equal to zero, so at a long time the entire amount of drug initially loaded into the bead will be depleted. If radial symmetry is assumed, then the concentration profile is only a function of the r direction (Figure 27.4).

It is desired to design a spherical capsule for the timed release of the drug dimenhydrinate, commonly called Dramamine, which is used to treat motion sickness. A conservative total dosage for one capsule is 10 mg, where 50% of the drug must be released to the body within 3 h. Determine the size of the bead and the initial concentration of Dramamine in the bead necessary to achieve this dosage. The diffusion coefficient of Dramamine (species A) in the gel matrix (species B) is 3×10^{-7} cm^2/s at a body temperature of 37°C. The solubility limit of Dramamine in the gel is 100 mg/cm^3, whereas the solubility of Dramamine in water is only 3 mg/cm^3.

The model must predict the amount of drug released vs. time, bead diameter, initial concentration of the drug within the bead, and the diffusion coefficient of the drug within the gel matrix. The physical system possesses spherical geometry. The development of the differential material balance model and the assumptions associated with it follow the approach presented in Section 25.4.

---

[3] S. Middleman and A. K. Hochberg, *Process Engineering Analysis in Semiconductor Device Fabrication*, McGraw-Hill, New York, 1993.

Figure 27.4 Drug release from a spherical gel bead.

The general differential equation for mass transfer reduces to the following partial differential equation for the one-dimensional unsteady-state concentration profile $c_A(r, t)$:

$$\frac{\partial c_A}{\partial t} = D_{AB}\left(\frac{\partial^2 c_A}{\partial r^2} + \frac{2}{r}\frac{\partial c_A}{\partial r}\right) \tag{27-18}$$

Key assumptions include radial symmetry, dilute solution of the drug dissolved in the gel matrix, and no degradation of the drug inside the bead ($R_A = 0$). The boundary conditions at the center ($r = 0$) and the surface ($r = R$) of the bead are

$$r = 0, \quad \frac{\partial c_A}{\partial r} = 0, \qquad t \geq 0$$

$$r = R, \quad c_A = c_{As} = 0, \quad t > 0$$

At the center of the bead, we note the condition of symmetry where the flux $N_A(0, t)$ is equal to zero. The initial condition is

$$t = 0, \qquad c_A = c_{Ao}, \qquad 0 \leq r \leq R$$

The analytical solution for the unsteady-state concentration profile $c_A(r, t)$ is obtained by separation-of-variables technique described earlier. The details of the analytical solution in spherical coordinates are provided by Crank. The result is

$$Y = \frac{c_A - c_{Ao}}{c_{As} - c_{Ao}} = 1 + \frac{2R}{\pi r}\sum_{n=1}^{\infty}\frac{(-1)^n}{n}\sin\left(\frac{n\pi r}{R}\right)e^{-D_{AB}n^2\pi^2 t/R^2}, \quad r \neq 0, n = 1, 2, 3, \ldots \tag{27-19}$$

At the center of the spherical bead ($r = 0$), the concentration is

$$Y = \frac{c_A - c_{Ao}}{c_{As} - c_{Ao}} = 1 + 2\sum_{n=1}^{\infty}(-1)^n e^{-D_{AB}n^2\pi^2 t/R^2}, \qquad r = 0, n = 1, 2, 3, \ldots \tag{27-20}$$

Once the analytical solution for the concentration profile is known, calculations of engineering interest can be performed, including the rate of drug release and the cumulative amount of drug release over time. The rate of drug release, W_A, is the product of the flux at the surface of the bead ($r = R$) and the surface area of the spherical bead

$$W_A(t) = 4\pi R^2 N_{Ar} = 4\pi R^2\left(-D_{AB}\frac{\partial c_A(R, t)}{\partial r}\right) \tag{27-21}$$

It is not so difficult to differentiate the concentration profile, $c_A(r, t)$, with respect to radial coordinate r, set $r = R$, and then insert back into the above expression for $W_A(t)$ to ultimately obtain

$$W_A(t) = 8\pi R c_{Ao} D_{AB}\sum_{n=1}^{\infty}e^{-D_{AB}n^2\pi^2 t/R^2} \tag{27-22}$$

The above equation shows that the rate of drug release will decrease as time increases until all of the drug initially loaded into the bead is depleted, at which point W_A will go to zero. Initially, the drug is uniformly loaded into the bead. The initial amount of drug loaded in the bead is the product of the initial concentration and the volume of the spherical bead

$$m_{Ao} = c_{Ao} V = c_{Ao} \frac{4}{3} \pi R^3$$

The cumulative amount of drug release from the bead over time is the integral of the drug release rate over time

$$m_{Ao} - m_A(t) = \int_0^t W_A(t) \, dt$$

After some effort, the result is

$$\frac{m_A(t)}{m_{Ao}} = \frac{6}{\pi^2} \sum_{n=1}^{\infty} \frac{1}{n^2} e^{-D_{AB} n^2 \pi^2 t / R^2} \tag{27-23}$$

The analytical solution is expressed as an infinite series summation that converges as "n" goes to infinity. In practice, convergence to a single numerical value can be attained by carrying the series summation out to only a few terms, especially if the dimensionless parameter $D_{AB} t / R^2$ is relatively large. It is a straightforward task to implement the infinite series summation on a spreadsheet program such as Excel (Microsoft Corporation).

Table 27.1 Excel spreadsheet for drug−release profile, Example 2

$m_{Ao} = 10 \, \text{mg}$
$D_{AB} = 3.00\text{E} - 07 \, \text{cm}^2/\text{s}$
$R = 0.326 \, \text{cm}$
$c_{Ao} = 68.9 \, \text{mg/cm}^3$

Time, t (s)	0.0	18	180	1800	3600	7200	10800	14400	18000	21600
Time, t (h)	0.0	0.005	0.05	0.50	1.00	2.00	3.00	4.00	5.00	6.00
$m_A(t)/m_{Ao}$	1.0	0.964	0.925	0.774	0.689	0.578	0.500	0.439	0.389	0.347
$1 - m_A(t)/m_{Ao}$	0.0	0.036	0.075	0.226	0.311	0.422	0.500	0.561	0.611	0.653
Series term $n =$ 1		9.99E−01	9.95E−01	9.51E−01	9.05E−01	8.18E−01	7.40E−01	6.70E−01	6.06E−01	5.48E−01
2		2.49E−01	2.45E−01	2.05E−01	1.67E−01	1.12E−01	7.50E−02	5.02E−02	3.36E−02	2.25E−02
3		1.11E−01	1.06E−01	7.08E−02	4.51E−02	1.83E−02	7.41E−03	3.00E−03	1.22E−03	4.94E−04
4		6.20E−02	5.77E−02	2.80E−02	1.26E−02	2.52E−03	5.07E−04	1.02E−04	2.05E−05	4.11E−06
5		3.95E−02	3.53E−02	1.14E−02	3.26E−03	2.66E−04	2.16E−05	1.76E−06	1.44E−07	1.17E−08
6		2.73E−02	2.32E−02	4.57E−03	7.51E−04	2.03E−05	5.49E−07	1.48E−08	4.01E−10	1.08E−11
7		1.99E−02	1.60E−02	1.75E−03	1.50E−04	1.10E−06	8.07E−09	5.92E−11	4.34E−13	3.19E−15
8		1.51E−02	1.13E−02	6.31E−04	2.55E−05	4.15E−08	6.77E−11	1.10E−13	1.80E−16	2.93E−19
9		1.19E−02	8.22E−03	2.13E−04	3.66E−06	1.08E−09	3.21E−13	9.52E−17	2.82E−20	8.36E−24
10		9.51E−03	6.06E−03	6.64E−05	4.41E−07	1.94E−11	8.56E−16	3.77E−20	1.66E−24	7.33E−29
11		7.78E−03	4.50E−03	1.91E−05	4.43E−08	2.38E−13	1.27E−18	6.84E−24	3.67E−29	1.97E−34
12		6.46E−03	3.37E−03	5.07E−06	3.71E−09	1.98E−15	1.06E−21	5.65E−28	3.02E−34	1.61E−40
13		5.44E−03	2.54E−03	1.23E−06	2.57E−10	1.12E−17	4.87E−25	2.12E−32	9.23E−40	4.02E−47
14		4.62E−03	1.91E−03	2.75E−07	1.48E−11	4.29E−20	1.25E−28	3.61E−37	1.05E−45	3.04E−54
15		3.97E−03	1.44E−03	5.59E−08	7.03E−13	1.11E−22	1.76E−32	2.79E−42	4.41E−52	6.98E−62
16		3.44E−03	1.08E−03	1.04E−08	2.76E−14	1.95E−25	1.38E−36	9.72E−48	6.87E−59	4.85E−70
17		2.99E−03	8.12E−04	1.76E−09	8.93E−16	2.30E−28	5.94E−41	1.53E−53	3.95E−66	1.02E−78
18		2.62E−03	6.08E−04	2.71E−10	2.38E−17	1.83E−31	1.41E−45	1.09E−59	8.41E−74	6.48E−88
19		2.31E−03	4.53E−04	3.80E−11	5.22E−19	9.85E−35	1.86E−50	3.50E−66	6.60E−82	1.24E−97
20		2.05E−03	3.36E−04	4.86E−12	9.43E−21	3.56E−38	1.34E−55	5.06E−73	1.91E−90	7.20E−108

A representative spreadsheet solution is provided in Table 27.1. Note that in Table 27.1 the terms within the series summation rapidly decay to zero after a few terms. The cumulative drug release vs. time profile is shown in Figure 27.5. The drug-release profile is affected by the dimensionless parameter $D_{AB} t/R^2$. If the diffusion coefficient D_{AB} is fixed for a given drug and gel matrix, then the critical engineering-design parameter we can manipulate is the bead radius R. As R increases, the rate of drug release decreases; if it is desired to release 50% of Dramamine from a gel bead within 3 h, a bead radius of 0.326 cm (3.26 mm) is required, as shown in Figure 27.5. Once the bead radius R is specified, the initial concentration of Dramamine required in the bead can be backed out

$$c_{Ao} = \frac{m_{Ao}}{V} = \frac{3m_{Ao}}{4\pi R^3} = \frac{3(10\,\text{mg})}{4\pi(0.326\,\text{cm})^3} = \frac{68.9\,\text{mg}}{\text{cm}^3}$$

In summary, a 6.52-mm-diameter bead with an initial concentration of $68.9\,\text{mg/cm}^3$ will dose out the required 5 mg of Dramamine within 3 h. The concentration profile along the r direction at different points in time is provided in Figure 27.6. The concentration profile was calculated by spreadsheet similar to the format given in Table 27.1. The concentration profile decreases as time increases and then flattens out to zero after the drug is completely released from the bead.

Figure 27.5 Fractional drug release vs. time profiles.

Figure 27.6 Concentration profile of a 0.326-cm-radius bead after 3, 12, and 24 h.

27.4 CONCENTRATION–TIME CHARTS FOR SIMPLE GEOMETRIC SHAPES

In our analytical solutions, the unaccomplished change, Y, was found to be a function of the relative time, X_D. The mathematical solutions, for the unsteady-state mass transfer in several simple shapes with certain restrictive boundary conditions, have been presented in a wide variety of charts to facilitate their use. Two forms of these charts are available in the appendix.

The "Gurney–Lurie" charts present solutions for the flat plate, sphere, and long cylinder. As the defining partial differential equations for heat conduction and molecular diffusion are analogous, these charts may be used to solve either transport phenomenon. For molecular diffusion, the charts are in terms of four dimensionless ratios:

$$Y = \text{unaccomplished concentration change} = \frac{c_{As} - c_A}{c_{As} - c_{A_0}}$$

$$X_D = \text{relative time} = \frac{D_{AB}t}{x_1^2}$$

$$n = \text{relative position} = \frac{x}{x_1}$$

$$m = \text{relative resistance} = \frac{D_{AB}}{k_c x_1}$$

The characteristic length, x_1, is the distance from the point of symmetry. For shapes where the transport takes place from two opposite faces, x_1 is the distance from the midpoint to the surfaces from which the transfer occurs. For shapes where the transport takes place from only one of the faces, the x_1 value in the dimensionless ratios is calculated as if the thickness were twice the true value; that is, for a slab of thickness $2a$, the relative time, X_D, is considered to be $D_{AB}\,t/4a^2$.

The relative resistance, m, is the ratio of the convective mass-transfer resistance to the internal molecular mass-transfer resistance. For a process with negative convective mass-transfer resistance, the convective mass-transfer coefficient, k_c, will be very large relative to D_{AB}, so m will be assumed to be zero in processes where molecular diffusion controls the flux of the diffusing species. For this case of no convection resistance, the concentration of the diffusing species at the surface, c_{As}, will be equal to the concentration in the bulk fluid, $c_{A\infty}$. If m is much greater than zero, then the surface concentration of species A differs from its concentration in the bulk fluid phase.

These charts may be used to evaluate concentration profiles for cases involving molecular mass transfer into as well as out of bodies of the specific shapes if the following conditions are satisfied:

(a) Fick's second law of diffusion is assumed; that is, there is no fluid motion, $v = 0$, no production term, $R_A = 0$, and constant mass diffusivity.

(b) The body has an initial uniform concentration, c_{Ao}.

(c) The boundary is subjected to a new condition that remains constant with time t.

Although the charts were drawn for one-dimensional transport, they may be combined to yield solutions for two- and three-dimensional transfer. In two dimensions, Y_a evaluated with the width, $x_1 = a$ and Y_b evaluated with the depth, $x_1 = b$, are combined to give

$$Y = Y_a Y_b \tag{27-24}$$

A summary of these combined solutions follows:

(1) For transport from a rectangular bar with sealed ends

$$Y_{\text{bar}} = Y_a Y_b \tag{27-25}$$

where Y_a is evaluated with width $x_1 = a$ and Y_b is evaluated with thickness $x_1 = b$.

(2) For transport from a rectangular parallelepiped

$$Y_{\text{parallelepiped}} = Y_a Y_b Y_c \qquad (27\text{-}26)$$

where Y_a is evaluated with width $x_1 = a$, Y_b is evaluated with thickness $x_1 = b$, and Y_c is evaluated with depth $x_1 = c$.

(3) For transport from a cylinder, including both ends

$$Y_{\text{cylinder plus ends}} = Y_{\text{cylinder}} Y_a \qquad (27\text{-}27)$$

The use of these charts will be shown in the following example.

EXAMPLE 3

Recall the drug capsule described in Example 2. The present drug capsule consists of a 0.652-cm-diameter spherical bead (radius of 0.326 cm) containing a uniform initial concentration of 68.9 mg/cm^3 dramamine. (a) what is the residual concentration of dramamine at the center of the spherical bead after 48 h? (b) the capsule is now a cube 0.652 cm on a side. Recalculate part (a) above. (c) the capsule is now a cylindrical tablet of diameter 0.652 cm and thickness 0.3 cm. Recalculate part (a) above. The diffusion coefficient of dramamine (species A) in the gel matrix (species B) is 3×10^{-7} cm^2/s at a body temperature of 37°C. The three capsules are presented in Figure 27.7.

| $R = 0.326$ cm | $x_1 = 0.326$ cm | $x_1 = 0.15$ cm | **Figure 27.7** Three |
| Sphere | Cube | Cylinder | capsule configurations. |

This problem is readily solved using the charts given in Appendix F.

(a) *Spherical capsule*: First, calculate the relative time (X_D), relative position (n), and relative resistance (m) based on the spherical coordinate system

$$X_D = \frac{D_{AB}t}{R^2} = \frac{\left(3 \times 10^{-7}\,\frac{cm^2}{s}\right)\left(48\,\text{h}\,\frac{3600\,\text{s}}{1\,\text{h}}\right)}{(0.326\,\text{cm})^2} = 0.488$$

$$n = \frac{r}{R} = \frac{0\,\text{cm}}{0.326\,\text{cm}} = 0\,(\text{center of sphere})$$

$$m = \frac{D_{AB}}{k_c R} \approx 0$$

From Figure F.1 or Figure 18.3, the value for Y, which in this case is the unaccomplished concentration change at the center of the spherical bead, is about 0.018. We can now calculate c_A

$$Y = 0.018 = \frac{c_{As} - c_A}{c_{As} - c_{Ao}} = \frac{0 - c_A}{0 - 68.9\,\text{mg/cm}^3}$$

The residual Dramamine concentration at the center of the bead after 48 h (c_A) is 1.24 mg/cm^3.

(b) For the cube-shaped capsule the distance from the midpoint of the cube to any of the six faces is 0.652 cm/2. The relative time X_D is now defined as

$$X_D = \frac{D_{AB}t}{x_1^2} = \frac{\left(3 \times 10^{-7} \frac{\text{cm}^2}{\text{s}}\right)\left(48\,\text{h}\frac{3600\,\text{s}}{1\,\text{h}}\right)}{(0.326\,\text{cm})^2} = 0.488$$

Values for n and m are unchanged, with $n = 0$ and $m = 0$. As all of the faces of the cube are of equal dimension, let

$$Y = Y_a Y_b Y_c = Y_a^3$$

From Figure F.4, given $X_D = 0.488$, $m = 0$ and $n = 0$, Y_a is 0.4 for a flat plate of semi-thickness $x_1 = a = 0.326$ cm. Extending this value to a three-dimensional cube using the above relationship, we have

$$Y = Y_a^3 = (0.4)^3 = 0.064$$

Finally,

$$Y = 0.064 = \frac{c_{As} - c_A}{c_{As} - c_{Ao}} = \frac{0 - c_A}{0 - 68.9\,\text{mg/cm}^3}$$

with $c_A = 4.41$ mg/cm^3 after 48 h.

(c) For a cylindrical capsule with exposed ends, $R = 0.652$ cm/2 for the radial coordinate, and $x_1 = a = 0.3$ cm/2 for the axial coordinate. The relative times are

$$X_D = \frac{D_{AB}t}{R^2} = \frac{\left(3.0 \times 10^{-7} \frac{\text{cm}^2}{\text{s}}\right)\left(48\,\text{h}\frac{3600\,\text{s}}{1\,\text{h}}\right)}{(0.326\,\text{cm})^2} = 0.488$$

for the cylindrical dimension and

$$X_D = \frac{D_{AB}t}{x_1^2} = \frac{\left(3.0 \times 10^{-7} \frac{\text{cm}^2}{\text{s}}\right)\left(48\,\text{h}\frac{3600\,\text{s}}{1\,\text{h}}\right)}{(0.15\,\text{cm})^2} = 2.30$$

for the axial dimension. Values for n and m are unchanged, with $n = 0$ and $m = 0$. From Figures F.1 and F.2, respectively, $Y_{\text{cylinder}} = 0.1$ for the cylindrical dimension and $Y_a = 0.006$ for the axial dimension. Therefore,

$$Y = Y_{\text{cylinder}}\, Y_a = (0.1)(0.006) = 0.0006$$

and finally

$$Y = 0.006 = \frac{c_{As} - c_A}{c_{As} - c_{Ao}} = \frac{0 - c_A}{0 - 68.9\,\text{mg/cm}^3}$$

with $c_A = 0.413$ mg/cm^3 after 48 h. As $Y_a \ll Y_{\text{cylinder}}$, the flux directed out of the exposed ends of the cylindrical tablet along the axial dimension dominates.

The above calculations assume that convective mass-transfer resistances associated with external fluid flow over the surface of the capsule are negligible. Problems in Chapter 30 will reconsider the drug release for unsteady-state diffusion and convection in series.

When a phase loses a considerable amount of the solute during a transient mass-transfer process, the total density of the phase will not be constant. This is the case where Fick's second "law" of diffusion must be expressed in terms of a constant solute-free basis as developed in equation (27–4)

$$\frac{\partial w_A'}{\partial t} = D_{AB}\frac{\partial^2 w_A'}{\partial z^2}$$

where w_A' is the weight fraction of the solute A divided by 1.0 minus the weight fraction of A. Our next example illustrates the use of the solute-free basis in explaining a transient mass-transfer drying process.

EXAMPLE 4

An oak pole having an initial uniform moisture content of 45 wt% is placed in a drying kiln where its surface moisture is maintained at 15 wt%. Under the specified drying conditions, the drying was controlled by the internal diffusion of liquid water to the surface. If the maximum moisture content of the pole is set at 25 wt%, how long must the 10-cm-diameter by 45-cm-length pole be dried when the ends of the pole are sealed with a vapor barrier? In comparison, how long must the same cylinder be dried if the cylinder surface is sealed with a vapor barrier and the ends are exposed?

The diffusivity of moisture through the oak is 1.04×10^{-5} cm^2/s under the specified drying conditions.

This problem is readly solved using the charts given in Appendix F. The initial uniform moisture content is 45 wt%, the surface moisture content is 15 wt%, and the maximum moisture content is 25 wt%.

$$Y = \frac{w'_A - w'_{As}}{w_{Ao} - w_{As}} = \frac{\dfrac{0.25}{1.0 - .25} - \dfrac{0.15}{1.0 - 0.15}}{\dfrac{0.45}{1.0 - 0.45} - \dfrac{0.15}{1.0 - 0.15}} = 0.244$$

(a) When the ends of the cylinder are sealed, the moisture (species A) will diffuse to the cylindrical surface, transferring in the r direction only. As the drying was controlled by the internal diffusion of the water, $m = 0$. The 25 wt% moisture content will be at the center of the cylinder.

$$n = \frac{r}{R} = \frac{0\,\text{cm}}{5\,\text{cm}} = 0$$

For $n = 0$, m $= 0$, and $Y = 0.244$, $X = 0.32$, or

$$X = \frac{D_{AB}t}{R^2} = 0.32$$

and

$$t = \frac{0.32R^2}{D_{AB}} = \frac{0.32(5.0\,\text{cm})^2}{1 \times 10^{-5}\,\text{cm}^2/\text{s}} = 7.81 \times 10^4\,\text{s}\ (21.7\text{h})$$

(b) When the cylinder surface is sealed and the ends are exposed, the transfer will be from the two flat ends. With the specified moisture contents, $Y = 0.244$, m will still be 0, and $n = 0$; $X = 0.68$, or

$$X = \frac{D_{AB}t}{R^2} = 0.68$$

and

$$t = \frac{0.68x_1^2}{D_{AB}} = \frac{0.68(22.5\,\text{cm})^2}{1 \times 10^{-5}\,\text{cm}^2/\text{s}} = 3.44 \times 10^7\,\text{s}\ (398\ \text{days})$$

27.5 CLOSURE

In this chapter, we have considered unsteady-state molecular diffusion. The partial differential equations that described the transient processes were obtained from the combination of Fick's equation with the general differential equation for mass transfer. Two approaches for the analytical solution to Fick's second law of diffusion were presented. Charts for solving unsteady-state mass transfer problems were also introduced.

PROBLEMS

27.1 A large tank, filled with water with an initially uniform oxygen concentration of 2 kg/m³, has its surface concentration suddenly raised and maintained at a 9 kg/m³ concentration level. Sketch a concentration level, c_A, as a function of depth z, for a period of

a. 3600 s,

b. 36,000 s,

c. 360,000 s,

if the water is at a temperature of 283 K.

27.2 A silicone polymer film of 4 mm thickness coats a flat, nonporous, inert surface. The polymer film is exposed to stagnant 100% O_2 gas maintained at 1.5 atm. Oxygen gas is sparingly soluble in the silicone polymer film, with the solubility relationship given by

$$c_A = S \cdot p_A$$

with $S = 3.16$ g mol O_2/m³ silicone atm O_2 at 298 K. Initially, there is 0.39 g mol O_2/m³ uniformly dissolved in the film. The diffusion coefficient of O_2 in silicone is 1×10^{-11} m²/s at 298 K.

a. At a very short time of 10 s, O_2 does not penetrate very far into the silicone polymer film. Under this physical situation, what is the flux of O_2 into the surface of the silicone polymer film at this time?

b. At a much longer time, the dissolved O_2 concentration in the silicone polymer film at the polymer/inert interface is now 3.0 g mol/m³. How long will it take to achieve this concentration?

27.3 We are interested in measuring the effective diffusion coefficient of blue dextran dye into an agarose gel. An aqueous, well-mixed solution containing 1.0 g/L of the dye rests over the rigid gel of 2.0 cm thickness, as shown below. The solubilities of the dye in both water and the gel are the same; that is, the concentration of the dye on the water side of the water–gel interface is equal to the concentration of dye on the gel side of the water–gel interface. There is no dye initially in the gel. After 24 h, a tiny section of the gel 2 mm from the surface is very carefully excised with a syringe needle, and the concentration of

the dye within the gel, as measured by a spectrophotometer is 0.203 g dye/L. An aqueous "gel" is generally considered a water-hydrated polymer, where solute molecules diffuse through the hydrated regions.

a. What is the effective diffusion coefficient of blue dextran dye into the gel based on the experimental measurements?

b. State at least three assumptions you had to make to arrive at your answer in part (a).

c. If the system temperature is increased from 20 to 40°C and the experiment is repeated exactly as described above, what would you expect the new measured diffusion coefficient to be? The viscosity of water is 9.93×10^{-4} kg/m · s at 20°C and 6.58×10^{-4} kg/m · s at 40°C.

27.4 Liquid benzene, C_6H_6, a common industrial solvent with a density of 0.88 g/cm³, leaked from a storage tank and seeped into the ground below. As the density of benzene is less than the density of water, the liquid benzene formed a light nonaqueous-phase liquid layer on top of the water-saturated soil, as shown below. At a depth of 1 m below the water-saturated soil layer is a nonporous, impermeable rock layer. There is no groundwater flow through the water-saturated soil layer, it is completely stagnant.

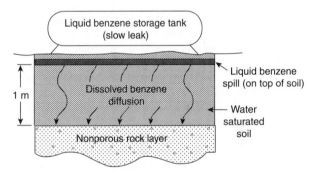

We are interested in the transport of benzene into the water-saturated soil layer, so that we may identify how much soil must be dug up and treated. Although the water-saturated soil is complex mixture, as a medium for benzene diffusion you may assume that it approximates the properties of a homogeneous substance.

At relatively short times or relatively small penetration depths, a finite-dimensional diffusion medium can be approximated by a semi-infinite diffusion medium. Calculate the concentration of the dissolved benzene at a position of 5 cm into the water-saturated soil layer after 72 h of benzene spill. Liquid benzene is sparingly soluble in water, and its solubility limit is 24 mol/m³ at 293 K. The effective diffusion coefficient of benzene in the water-saturated soil is 1×10^{-9} m²/s at 293 K. Initially, there is no benzene dissolved in the water-saturated soil layer. The pure liquid benzene layer resting on top of the soil is essentially a constant source due to the low solubility of benzene in water. Finally, you may assume that benzene does not diffuse into the nonporous rock layer.

27.5 Hydrogen embrittlement weakens the mechanical strength of cast iron. This phenomenon frequently occurs in cast-iron pressure vessels containing 100% hydrogen gas. The H_2 gas dissolves into the iron metal and diffuses into the solid nonporous iron by an interstitial diffusion mechanism. The H_2 gas does not have to penetrate very far into the iron to have a negative effect on the mechanical strength of iron. In the present situation, 100% H_2 gas at 1.0 atm and 100°C is contained within a cylindrical iron vessel of 1.0 m inner diameter and wall thickness of 2.0 cm. The solubility of hydrogen in iron at 100°C is 2.2×10^{-7} mol H atoms/g Fe. The diffusion coefficient of hydrogen atoms in solid iron is 124.0×10^{-9} cm²/s at 100°C. Initially, there are no H atoms in the solid iron. How many hours will it take for the hydrogen level within the iron metal to reach 1.76×10^{-7} mol H atoms/g Fe at a depth of 0.1 cm from the surface exposed to the hydrogen gas?

27.6 A large tanker truck overturns and spills a herbicide over a field. If the mass diffusivity of the fluid in the soil is 1×10^{-3} m²/s and the fluid remains on the soil for 1800 s before evaporating into the air, determine the depth at which plant and insect life is likely to be destroyed if a concentration of 0.1% by weight will destroy most of the life.

27.7 In the fabrication of a p-type semiconductor, elemental boron is diffused a small distance into a solid crystalline silicon wafer. The boron concentration within the solid silicon determines semiconducting properties of the material. A physical deposition process keeps the concentration of elemental boron at the surface of the water equal to 5.0×10^{20} atoms boron/cm³ silicon. In the manufacture of a transistor, it is desired to produce a thin film of silicon doped to a boron concentration of at least 1.7×10^{19} atoms boron/cm³ silicon at a depth of 0.2 μm from the surface of the silicon wafer. It is desired to achieve this target within a 30-min processing time. The density of solid silicon is 5.0×10^{22} atoms Si/cm³ solid. At what temperature must the boron doping process be operated? It is known that the temperature dependence of the diffusion coefficient of boron (A) in silicon (B) is given by

$$D_{AB} = D_0 e^{-Q_0/RT}$$

where $D_0 = 0.019$ cm²/s and $Q_0 = 2.74 \times 10^5$ J/g mol for elemental boron in solid silicon. The thermodynamic constant $R = 8.314$ J/g mol/K.

27.8 A preheated piece of mild steel, having an initial concentration of 0.20% by weight of carbon, is exposed to a carbonizing atmosphere for 1 h. Under the processing conditions, the surface concentration of carbon is 0.70% by weight. If the diffusivity of carbon through steel is 1.0×10^{-11} m²/s at the process temperature, determine the carbon composition at 0.01 cm, 0.02 cm, and 0.04 cm below the surface.

27.9 In the manufacture of semiconducting thin films, a thin film of solid arsenic is laid down onto the surface of a silicon

wafer by the chemical vapor deposition of arsine, AsH_3, as

$$2AsH_3(g) \rightarrow 2As(s) + 3H_2(g)$$

The arsenic atoms then diffuse into the solid silicon to dope the wafer as shown below:

What is the flux of arsenic atoms into the silicon wafer after 1 h, in units of atoms/cm² s? What is arsenic concentration 2 μm into the silicon wafer after 1 h, in units of atoms/cm³? The initial concentration of residual arsenic in the silicon wafer is 10^{12} atoms/cm³. The process temperature is 1050°C. The average diffusivity of arsenic in silicon is 5×10^{-13} cm²/s at this temperature, and the maximum solubility of arsenic in silicon is 2×10^{21} atoms/cm³.

27.10 Please refer to Problem 27.9. The "junction depth" is a critical parameter for establishing the semiconducting properties of doped silicon films. For a particular semiconductor, the junction depth is established when the arsenic concentration in the silicon reaches 2×10^{17} atoms/cm³.

a. Plot out the junction depth with time t during the first 6 h of doping.

b. Now plot the junction depth vs. $t^{1/2}$. Why is the plot linear?

27.11 It is extremely difficult to "aerate" stagnant water. Consider the process shown in the figure below. A sealed tank consists of still, deep liquid water with pure oxygen gas in the headspace. The temperature and pressure of the 100% O_2 gas are 298 K and 2.0 atm, respectively. Initially, the dissolved oxygen concentration of the water within the tank is uniform at 10.0 g/m³, the diffusion coefficient of dissolved O_2 in water is 2.1×10^{-5} cm²/s at 298 K, and the Henry's law constant for dissolved O_2 gas in liquid water is 0.800 atm · m³/g mol at 298 K. How long will it take for dissolved oxygen concentration to reach 20 g/m³ at a depth of 0.3 cm from the water surface?

27.12 A "drug patch" is designed to slowly deliver a drug (species A) through the body tissue to an infected zone of tissue beneath the skin. The drug patch consists of a sealed reservoir containing the drug encapsulated within a porous polymer matrix. The patch is implanted just below the skin. A diffusion barrier attached to the bottom surface of the patch sets the surface concentration of the drug in the body tissue at 2 mol/m^3, which is below the solubility limit. The mean distance from the drug patch to the infected area of tissue is 5 mm. To be effective, the drug concentration must be at least 0.2 mol/m^3 at the top edge of the infected zone. Determine the time it will take in hours for the drug to begin to be effective for treatment. The effective molecular diffusion coefficient of the drug through the body tissue is 1×10^{-6} cm^2/s.

27.13 Please refer to Problem 27.4. How long will it take for the concentration of benzene at the clay barrier (depth of 1 m) to reach 1 g/m^3, the toxicity limit? You might find that the concentration charts are difficult to read accurately at this concentration. Therefore, consider the analytical solution and implement the calculations on a computer spreadsheet. Be aware that the diffusion process may take a long time.

27.14 Please refer to Problem 27.4. Consider a 1 m by 1 m cross section of the water-saturated soil. How many grams of dissolved benzene remain in this 1 m^3 water-saturated soil 2 years after the spill? You may need to analytically integrate the flux given by equation (27–17) at the surface ($z = 0$) over time, and then implement the series solution on a computer spreadsheet.

27.15 The concentration profile resulting from transient diffusion in a slab under conditions of negligible surface resistance is described by equation (27–16). Use this equation to develop an equation for predicting the average concentration, $\bar{c}_A$. Evaluate and plot the dimensionless average profile $(\bar{c}_A - c_{As})/(c_{Ao} - c_{As})$ as a function of the dimensionless relative time ratio, X_D. Use a computer spreadsheet to perform the calculations.

27.16 Living cells immobilized within an agarose gel requires glucose to survive. An important aspect of the biochemical system design is the effective diffusion coefficient of glucose into the agarose gel itself, which you may consider as a homogeneous material that is mostly liquid water. Consider the experiment that is shown below, where a slab of the agarose gel of 1.0 cm thickness is placed within a well-mixed aqueous

Effective diffusion coefficient measurement experiment

Sampling
syringe
(center of gel)

solution of glucose maintained at a concentration of 50 mmol/L, which is relatively dilute. The solubility of the glucose in water is equal to the solubility of glucose in the gel.

A syringe mounted at the center of the gel carefully excises a tiny sample of the gel for glucose analysis. Initially, there is no glucose in the gel. However, after 42 h, the measured concentration of glucose in the gel at the sampling point is 48.5 mmol/L.

a. What is the general differential equation for mass transfer for this process in terms of concentration c_A? Propose reasonable boundary/initial conditions for this equation. List three assumptions made in obtaining this equation.

b. Based on the measurements provided above, what is the effective diffusion coefficient of glucose into the gel?

27.17 Consider cylindrical-shaped absorbent material designed to selectively remove solute A from a solution. The uptake of solute A through the homogeneous material is limited by molecular diffusion. Furthermore, the absorbent has a higher affinity for solute A relative to the surrounding solution as described by the linear equilibrium relationship

$$c_A = Kc'_A$$

where c_A is the molar concentration of A in the absorbent, c'_A is the molar concentration of A in the surrounding fluid, and K is a partition coefficient for solute A between the fluid and the absorbent material. Assume that (1) the surrounding fluid is very well mixed and has a constant concentration of $c'_A = 2.00$ g mol/m^3; (2) the diffusion coefficient of A in the homogeneous absorbent material is 4×10^{-7} cm^2/s and $K = 1.5$ cm^3 fluid/cm^3 absorbent; (3) the cylindrical absorbent pellet is 1.0 cm in diameter and 5.0 cm long; and (4) there is no solute A initially in the absorbent material.

If the "edge effects" associated with the ends of the cylinder can be neglected, how long will take for the solute A concentration to reach 2.94 g mol/m^3 at a depth of 0.4 cm from the surface of the cylinder?

27.18 A small spherical bead is used as a controlled drug-release capsule in the gastrointestinal system. In the particular case, a 0.1-cm-diameter bead has a uniform initial concentration of 0.2 mmol/L of the drug griseofulvin (species A). Determine the time it will take for the concentration of the griseofulvin at the center of the bead to reach 10% of its initial value. The diffusivity of griseofulvin within the bead material is 1.5×10^{-7} cm^2/s. Upon release from the bead, the drug is immediately consumed so that the surface concentration is essentially zero.

27.19 Determine the time necessary to reduce the center moisture content to 10 wt% if a 5-cm-thick slab of clay is placed on a belt passing through a continuous drier, thus restricting the drying to only one of the flat surfaces. The initial moisture content is 15 wt% and the surface moisture content under the constant drying conditions is to be maintained at 4 wt%. The effective diffusivity of water through clay is 1.3×10^{-4} cm^2/s.

27.20 Aluminum is the primary conductor material for fabrication of microelectronic devices. Consider the composite thin film shown in the figure below. A thin film of solid aluminum is sputter-coated onto a wafer surface. Then, a 0.5 μm thin film of silicon is added on top of the aluminum film by chemical vapor deposition of silane. This Si thin film serves as an electrically insulating layer. However, if a high temperature is maintained during processing, the aluminum can diffuse into the Si thin film and change the characteristics of the microelectronic device. Estimate the concentration of Al halfway into the Si thin film if the temperature is maintained at 1250 K for 10 h. Carefully consider whether or not the process represents diffusion within a semi-infinite medium or a finite-dimensional medium. At 1250 K, the maximum solubility of Al in Si is about 1 wt%. Diffusivity data are provided in Figure 24.7.

27.21 A novel material based on the biopolymer polyglucosamine is cast into a homogeneous gel bead. The amine group on the biopolymer has a high affinity for transition metal ions at parts-per-million concentrations. When water containing metal ions contacts the gel, the metal ions selectively partition into the gel at the surface, and then diffuse through the gel. In essence, the gel acts as a second phase that extracts the metal ions from solution and concentrates them within the bead. Consequently, this material can be used to selectively remove toxic transition metal ions from wastewater. Although the real situation is somewhat more complicated, assume as a limiting case that the uptake rate of the metal ions is limited by the passive molecular diffusion of the metal ions through the gel. At relatively low metal ion concentrations in water below 1.0 mol/m^3, the partition constant is defined by the relationship

$$c'_A = Kc_{As}$$

where c'_A is the molar concentration of the metal ion within the gel phase at the surface, c_{As} is the molar concentration of the metal in the aqueous phase at the surface, and K is a dimensionless partition constant, which is dependent on the concentration of biopolymer within the gel. In the present process, the bead diameter is 0.5 cm, and the biopolymer concentration within the

gel sets K equal to 150. How long will it take for the cadmium concentration at the center of the gel bead to reach 12 mol/m^3 if wastewater containing a constant concentration of 0.1 mol/m^3 cadmium (11.2 ppm) rapidly flows over the gel bead? The diffusion coefficient of cadmium within the gel is approximately 2×10^{-6} cm^2/s.

27.22 Consider a rectangular-shaped gel tablet of thickness 0.652 cm and width 1.0 cm. The edges of the gel tablet are sealed. The initial concentration of the drug Dramamine (solute A) in the gel (B) is 64.0 mg/cm^3, to provide a total drug dosage of 41.7 mg. The diffusion coefficient of the drug in the gel is 3.0×10^{-7} cm^2/s. What is the residual concentration of drug at the center of the tablet after 96 h? The concentration of the drug at the exposed surface of the tablet is maintained at zero.

27.23 A porous cylinder, 2.5 cm in diameter and 80 cm long, is saturated with an alcohol. The void space in the solid contains sufficient pores so that molecular diffusion can take place through the liquid in the passage. The cylinder is dropped into a large, well-agitated reservoir of pure water. The agitation maintains a concentration of 1 wt% alcohol at the surface of the cylinder.

If the concentration at the center of the cylinder drops from 30 wt% alcohol to 18 wt% in 10 h, determine the concentration in weight percent at the center after 15 h.

27.24 A common procedure for increasing the moisture content of air is to bubble it through a column of water. The air bubbles are assumed to be spheres having radii of 1 mm and are in thermal equilibrium with water at 298 K. Determine how long the bubble should remain in the water to achieve a vapor concentration at the center that is 90% of the maximum possible (saturation) concentration. Assume that the air is dry when it enters the column of water and that the air inside the small bubble is stagnant. The vapor pressure of water is available from many sources, including the stream tables.

27.25 A rectangular oak pole, 10cm by 10 cm in width and 45 cm in length, having an initial uniform moisture content of 45 wt%, is placed in a drying kiln where its surface moisture is maintained at 15 wt%. The drying rate is controlled by the internal diffusion of the liquid water to the surface. If the maximum moisture content of the dried pole is set at 25 wt%, how long must the pole be dried when the ends of the pole are sealed with a vapor barrier? In comparison, how long must the same pole be dried if the side surfaces are sealed with a vapor barrier and the ends are exposed?

You may assume for the purposes of this problem that wood is an isotropic material. The diffusivity of moisture through the oak is 1.04×10^{-5} cm^2/s under the specified drying conditions.

Chapter 28

Convective Mass Transfer

Convective mass transfer, initially introduced in Section 24.3, involves the transport of material between a boundary surface and a moving fluid or between two immiscible moving fluids separated by a mobile interface. In this chapter, we will discuss the transfer within a single phase where the mass is exchanged between a boundary surface and a moving fluid, and the flux is related to an *individual mass-transfer convective coefficient*. In Chapter 29, we will consider the mass transfer between two contacting phases where the flux is related to an *overall mass-transfer convective coefficient*.

The rate equation for convective mass transfer, generalized in a manner analogous to Newton's law of cooling, is

$$N_A = k_c \, \Delta c_A \tag{24-57}$$

where the mass flux, N_A, is the molar-mass flux of species A, measured relative to fixed spatial coordinates, Δc_A is the concentration difference between the boundary surface concentration and the average concentration of the diffusing species in the moving fluid stream, and k_c is the convective mass-transfer coefficient. Recalling the discussions of the convective heat-transfer coefficient, as defined by Newton's law of cooling, we should realize that the determination of the analogous convective mass-transfer coefficient is not a simple undertaking. Both the heat- and the mass-transfer coefficients are related to the properties of the fluid, the dynamic characteristics of the flowing fluid, and the geometry of the specific system of interest.

In light of the close similarity between the convective heat- and mass-transfer equations used to define these two convective coefficients, we may expect that the analytical treatment of the heat-transfer coefficient in Chapter 19 might be applied to analyze the mass-transfer coefficient. Considerable use will also be made of developments and concepts of Chapters 9–13.

28.1 FUNDAMENTAL CONSIDERATIONS IN CONVECTIVE MASS TRANSFER

When the mass transfer involves a solute dissolving into a moving fluid, the convective mass-transfer coefficient is defined by

$$N_A = k_c(c_{As} - c_A) \tag{28-1}$$

In this equation, the flux N_A represents the moles of solute A leaving the interface per unit time and unit interfacial area. The composition of the solute in the fluid at the interface, c_{AS}, is the composition of the fluid if it were in equilibrium with the solid solute at the temperature and pressure of the system. The quantity c_A represents the composition at some point within the fluid phase. For example, if a concentration boundary layer is defined, c_A can be chosen as the concentration of component A at the edge of the boundary layer and be expressed as $c_{A\infty}$. If the flow were in a closed conduit, the composition c would be the bulk composition or the *mixing-cup concentration*, that is, an average composition of the bulk flow.

Mass transfer at a steady rate from a solid to a gas stream is described by an equation identical to (28-1), with the compositions of the solute in terms of gas-phase concentrations. Equation (28-1) will be used in the following example to evaluate a mass-transfer coefficient.

EXAMPLE 1

Air flows over a solid slab of frozen carbon dioxide (dry ice) with an exposed cross-sectional surface area of 1×10^{-3} m^2. The carbon dioxide sublimes into the 2 m/s flowing stream at a total release rate of 2.29×10^{-4} mol/s. The air is at 293 K and 1.013×10^5 Pa pressure. At that temperature, the diffusivity of carbon dioxide in air is 1.5×10^{-5} m^2/s and the kinematic viscosity of the air is 1.55×10^{-5} m^2/s.

Determine the value of the mass-transfer coefficient of CO_2 subliming into the flowing air under the conditions of the experiment.

By equation (28-1), $N_A = k_c(c_{As} - c_A)$; accordingly,

$$k_c = \frac{N_A}{(c_{As} - c_{A\infty})} = \frac{W_A}{A_x(c_{As} - c_{A\infty})}$$

At 293 K and 1.013×10^5 Pa,

$$c_{As} = \frac{P_A}{RT} = \frac{4.74 \times 10^3 \text{ Pa}}{\left(8.314 \dfrac{\text{Pa} \cdot \text{m}^3}{\text{mol} \cdot \text{K}}\right)(293 \text{ K})} = 1.946 \text{ mol/m}^3$$

If we assume $c_{A\infty} = 0$,

$$k_c = \frac{2.29 \times 10^{-4} \text{ mol/s}}{(1 \times 10^{-3} \text{ m}^2)(1.946 \text{ mol/m}^3)} = 0.118 \text{ m/s}$$

From our early discussions dealing with a fluid flowing past a surface, we may recall that the fluid particles immediately adjacent to the solid boundary are stationary, and a thin layer of fluid close to the surface will be in laminar flow regardless of the nature of the free stream. The mass transfer through this film will involve molecular transport and it will play a role in any convection process. If the fluid flow is laminar, all of the transport between the surface and the moving fluid will be by molecular means. If the flow is turbulent, the mass will be transported by the eddies present within the turbulent core of the stream. As in the case of heat transfer, higher mass-transfer rates are associated with turbulent conditions. The distinction between laminar and turbulent flow will be an important consideration in any convective situation.

The hydrodynamic boundary layer, analyzed in Chapter 12, plays a major role in convective mass transfer. Additionally, we shall define and analyze the concentration boundary layer, which will be vital to the analysis of the convective mass-transfer process. This layer is similar, but not necessary equal in thickness to the thermal boundary layer that was discussed in Chapter 19.

There are four methods of evaluating convective mass-transfer coefficients that will be discussed in this chapter. They are

1. dimensional analysis coupled with experiment;
2. exact laminar boundary-layer analysis;

3. approximate boundary-layer analysis;

4. analogy between momentum, energy, and mass transfer.

Each of these methods will be considered in the following sections.

28.2 SIGNIFICANT PARAMETERS IN CONVECTIVE MASS TRANSFER

Dimensionless parameters are often used to correlate convective transport data. In momentum transfer, we encountered the Reynolds and the Euler numbers. In the correlation of convective heat-transfer data, the Prandtl and the Nusselt numbers were important. Some of the same parameters, along with some newly defined dimensionless ratios, will be useful in the correlation of convective mass-transfer data. In this section, we shall consider the physical interpretation of three such ratios.

The molecular diffusivities of the three transport phenomena have been defined as

$$\textit{momentum diffusivity}, \nu = \mu/\rho$$

$$\textit{thermal diffusivity}, \alpha = \frac{k}{\rho c_p}$$

and

$$\textit{mass diffusivity}, D_{AB}$$

As we have noted earlier, each of the diffusivities has the dimensions L^2/t; thus, a ratio of any two of these must be dimensionless. The ratio of the molecular diffusivity of momentum to the molecular diffusivity of mass is designated the *Schmidt number*

$$\frac{\text{momentum diffusivity}}{\text{mass diffusivity}} = \text{Sc} \equiv \frac{\nu}{D_{AB}} = \frac{\mu}{\rho D_{AB}} \qquad (28\text{-}2)$$

The Schmidt number plays a role in convective mass transfer analogous to that of the Prandtl number in convective heat transfer. The ratio of the thermal diffusivity to the molecular diffusivity of mass is designated the *Lewis number*

$$\frac{\text{thermal diffusivity}}{\text{mass diffusivity}} = \text{Le} \equiv \frac{k}{\rho c_p D_{AB}} \qquad (28\text{-}3)$$

The Lewis number is encountered when a process involves the simultaneous convective transfer of mass and energy. The Schmidt and the Lewis numbers are observed to be combinations of fluid properties; thus, each number may be treated as a property of the diffusing system.

Consider the mass transfer of solute A from a solid to a fluid flowing past the surface of the solid. The concentration profile is depicted in Figure 28.1. For such a case, the mass transfer between the surface and the fluid may be written as

$$N_A = k_c(c_{A,s} - c_{A,\infty}) \qquad (28\text{-}4)$$

Figure 28.1 Concentration and velocity profiles for a fluid flowing past a solid surface.

Since the mass transfer at the surface is by molecular diffusion, the mass transfer may also be described by

$$N_A = -D_{AB} \frac{dc_A}{dy}\bigg|_{y=0}$$

When the boundary concentration, $c_{A,s}$, is constant, this equation simplifies to

$$N_A = -D_{AB} \frac{d(c_A - c_{A,s})}{dy}\bigg|_{y=0} \tag{28-5}$$

Equations () and (28-5) may be equated, as they define the same flux of component A leaving the surface and entering the fluid. This gives the relation

$$k_c(c_{A,s} - c_{A,\infty}) = -D_{AB} \frac{d}{dy}(c_A - c_{A,s})\bigg|_{y=0}$$

which may be rearranged into the following form:

$$\frac{k_c}{D_{AB}} = \frac{-d(c_A - c_{A,s})/dy|_{y=0}}{(c_{A,s} - c_{A,\infty})} \tag{28-6}$$

Multiplying both sides of equation (28-6) by a significant length, L, we obtain the following dimensionless expression:

$$\frac{k_c L}{D_{AB}} = \frac{-d(c_A - c_{A,s})/dy|_{y=0}}{(c_{A,s} - c_{A,\infty})/L} \tag{28-7}$$

The right-hand side of equation (28-7) is the ratio of the concentration gradient at the surface to an overall- or reference-concentration gradient. Accordingly, it may be considered a ratio of the molecular mass-transport resistance to the convective mass-transfer resistance of the fluid. This ratio is referred to as the *Sherwood number*, Sh. Since the development of equation (28-7) parallels the development of equation (19-5) for the Nusselt number encountered in convective heat transfer, the ratio $k_c L/D_{AB}$ has also been referred to as the *mass-transfer Nusselt number*, Nu_{AB}.

These three parameters—Sc, Sh, and Le—will be encountered in the analyses of convective mass transfer in the following sections. The Schmidt number is evaluated in the following example to reveal the relative magnitude of its value in the gas and liquid phases.

EXAMPLE 2

Determine the Schmidt number for methanol in air at 298 K and 1.013×10^5 Pa and in liquid water at 298 K.

At 298 K, the diffusivity of methanol in air can be evaluated from the value reported in Appendix J.1

$$D_{\text{methanol–air}}P = 1.641 \text{ m}^2 \text{ Pa/s}$$

$$D_{\text{methanol–air}} = \frac{1.641 \text{ m}^2 \text{ Pa/s}}{1.013 \times 10^5 \text{ Pa}} = 1.62 \times 10^{-5} \text{ m}^2/\text{s}$$

and the kinematic viscosity of air from Appendix I

$$\nu = 1.569 \times 10^{-5} \text{ m}^2/\text{s}.$$

Accordingly, the Schmidt number of methanol in air is

$$Sc = \frac{\nu}{D_{AB}} = \frac{1.569 \times 10^{-5} \text{ m}^2/\text{s}}{1.62 \times 10^{-5} \text{ m}^2/\text{s}} = 0.968$$

The liquid-phase diffusivity of methanol at 288 K is reported in Appendix J.2 to be $1.28 \times 10^{-9}\ \text{m}^2/\text{s}$. This value can be used to find the liquid-phase diffusivity at 298 K by

$$\left(\frac{D_{AB}\,\mu_{H_2O}}{T}\right)_{298} = \left(\frac{D_{AB}\,\mu_{H_2O}}{T}\right)_{288}$$

$$\left(\frac{D_{AB}\,909\ \text{Pa}\cdot\text{s}}{298\ \text{K}}\right)_{298} = \left(\frac{(1.28 \times 10^{-9}\ \text{m}^2/\text{s})(1193\ \text{Pa}\cdot\text{s})}{288\ K}\right)_{288}$$

$$D_{AB} = 1.738\ 10^{-9}\ \text{m}^2/\text{s}$$

The viscosity values were obtained in Appendix I. The kinematic viscosity value of liquid water at 298 K is also found in Appendix I to be $0.912 \times 10^{-6}\ \text{m}^2/\text{s}$; thus, the Schmidt number for methanol in liquid water is

$$\text{Sc} = \frac{v}{D_{AB}} = \frac{0.912 \times 10^{-6}\ \text{m}^2/\text{s}}{1.738 \times 10^{-9}\ \text{m}^2/\text{s}} = 525$$

28.3 DIMENSIONAL ANALYSIS OF CONVECTIVE MASS TRANSFER

Dimensionless analysis predicts the various dimensionless parameters that are helpful in correlating experimental data. There are two important mass-transfer processes that we shall consider, mass transfer into a stream flowing under forced convection and mass transfer into a phase that is moving under natural-convection conditions.

Transfer into a Stream Flowing Under Forced Convection

Consider the transfer of mass from the walls of a circular conduit to a fluid flowing through the conduit. The transfer is a result of the concentration driving force, $c_{A,s} - c_A$. The important variables, their symbols, and their dimensional representations are listed here.

Variable	Symbol	Dimensions
Tube diameter	D	L
Fluid density	ρ	M/L^3
Fluid viscosity	μ	M/Lt
Fluid velocity	v	L/t
Fluid diffusivity	D_{AB}	L^2/t
Mass-transfer coefficient	k_c	L/t

The above variables include terms descriptive of the system of geometry, the fluid velocity, the fluid properties, and the quantity that is of primary interest, k_c.

By the Buckingham method of grouping the variables as presented in Chapter 11, we can determine that there will be three dimensionless groups. With D_{AB}, ρ, and D as the core variables, the three pi groups to be formed are

$$\pi_1 = D_{AB}^a \rho^b D^c k_c$$

$$\pi_2 = D_{AB}^d \rho^e D^f v$$

and

$$\pi_3 = D_{AB}^g \rho^h D^i \mu$$

Writing π_1 in dimensional form

$$\pi_1 = D_{AB}^a \rho^b D^c k_c$$

$$1 = \left(\frac{L^2}{t}\right)^a \left(\frac{M}{L^3}\right)^b (L)^c \left(\frac{L}{t}\right)$$

equating the exponents of the fundamental dimensions on both sides of the equation, we have for

$$L: \quad 0 = 2a - 3b + c + 1$$
$$t: \quad 0 = -a - 1$$

and

$$M: \quad 0 = b$$

The solution of these equations for the three unknown exponents yields

$$a = -1$$
$$b = 0$$

and

$$c = 1$$

thus $\pi_1 = k_c L / D_{AB}$, which is the Sherwood number, Sh, or its equivalent, the mass-transfer Nusselt number, Nu_{AB}. The two other pi groups could be determined in the same manner, yielding

$$\pi_2 = \frac{Dv}{D_{AB}}$$

and

$$\pi_3 = \frac{\mu}{\rho D_{AB}} = Sc$$

the Schmidt number. Dividing π_2 by π_3, we obtain

$$\frac{\pi_2}{\pi_3} = \left(\frac{Dv}{D_{AB}}\right) \left(\frac{D_{AB}\rho}{\mu}\right) = \frac{Dv\rho}{\mu} = Re$$

the Reynolds number. The result of the dimensional analysis of forced-convection mass transfer in a circular conduit indicates that a correlating relation could be of the form

$$Sh = Nu_{AB} = f(Re, Sc) \tag{28-8}$$

which is analogous to the heat-transfer correlation

$$Nu = f(Re, Pr) \tag{19-7}$$

Transfer into a Phase Whose Motion is Due to Natural Convection

Natural convection currents will develop if there exists any variation in density within a liquid or gas phase. The density variation may be due to temperature differences or relatively large concentration differences.

In the case of natural convection involving mass transfer from a vertical plane wall to an adjacent fluid, the variables will differ from those used in the forced-convection analysis. The important variables, their symbols, and dimensional representations are listed below.

Variable	Symbol	Dimensions
Characteristic length	L	L
Fluid diffusivity	D_{AB}	L^2/t
Fluid density	ρ	M/L^3
Fluid viscosity	μ	M/LT
Buoyant force	$g\,\Delta\rho_A$	$M/L^2 t^2$
Mass-transfer coefficient	k_c	L/t

By the Buckingham theorem, there will be three dimensionless groups. With D_{AB}, L, and μ as the core variables, the three pi groups to be formed are

$$\pi_1 = D_{AB}^a L^b \mu^c k_c$$

$$\pi_2 = D_{AB}^d L^e \mu^f \rho$$

$$\pi_3 = D_{AB}^g L^h \mu^i \Delta\rho_A$$

Solving for the three pi groups, we obtain

$$\pi_1 = \frac{k_c L}{D_{AB}} = \mathrm{Sh}$$

the Sherwood number

$$\pi_2 = \frac{\rho D_{AB}}{\mu} = \frac{1}{\mathrm{Sc}}$$

the reciprocal of the Schmidt number and

$$\pi_3 = \frac{L^3 g\,\Delta\rho_A}{\mu D_{AB}}$$

Multiplying π_2 and π_3, we obtain a parameter that is analogous to the Grashof number in natural-convection heat transfer

$$\pi_2\pi_3 = \left(\frac{\rho D_{AB}}{\mu}\right)\left(\frac{L^3 g\Delta\rho_A}{\mu D_{AB}}\right)$$

$$= \frac{L^3 \rho g\,\Delta\rho_A}{\mu^2} = \frac{L^3 g\,\Delta\rho_A}{\rho v^2} = \mathrm{Gr}_{AB}$$

The result of the dimensional analysis of natural-convection mass transfer suggests a correlating relation of the form

$$\mathrm{Sh} = f(\mathrm{Gr}_{AB}, \mathrm{Sc}) \tag{28-9}$$

For both forced and natural convection, relations have been obtained by dimensional analysis that suggest that a correlation of experimental data may be in terms of three variables instead of the original six. This reduction in variables has aided investigators who have suggested correlations of these forms to provide many of the empirical equations reported in Chapter 30.

28.4 EXACT ANALYSIS OF THE LAMINAR CONCENTRATION BOUNDARY LAYER

Blasius developed an exact solution for the hydrodynamic boundary layer for laminar flow parallel to a flat surface. This solution was discussed in Section 12.5. An extension of the Blasius solution was made in Section 19.4 to explain convective heat transfer. In an exactly analogous manner, we shall also extend the Blasius solution to include convective mass transfer for the same geometry and laminar flow.

The boundary-layer equations considered in the steady-state momentum transfer included the two-dimensional, incompressible continuity equation

$$\frac{\partial v_x}{\partial x} + \frac{\partial v_y}{\partial y} = 0 \qquad (12\text{-}11b)$$

and the equation of motion in the x direction, for constant v and pressure

$$v_x \frac{\partial v_x}{\partial x} + v_y \frac{\partial v_x}{\partial y} = \nu \frac{\partial^2 v_x}{\partial y^2} \qquad (12\text{-}11a)$$

For the thermal boundary layer, the equation describing the energy transfer in a steady, incompressible, two-dimensional, isobaric flow with constant thermal diffusivity was

$$v_x \frac{\partial T}{\partial x} + v_y \frac{\partial T}{\partial y} = \alpha \frac{\partial^2 T}{\partial y^2} \qquad (19\text{-}15)$$

An analogous differential equation applies to mass transfer within a concentration boundary layer if no production of the diffusing component occurs and if the second derivative c_A of with respect to x, $\partial^2 c_A/\partial x^2$, is much smaller in magnitude than the second derivative of c_A with respect to y. This equation written for steady, incompressible two-dimensional flow with constant mass diffusivity is

$$v_x \frac{\partial c_A}{\partial x} + v_y \frac{\partial c_A}{\partial y} = D_{AB} \frac{\partial^2 c_A}{\partial y^2} \qquad (28\text{-}10)$$

The concentration boundary layer is shown schematically in Figure 28.2. The following are the boundary conditions for the three boundary layers:

$$momentum: \quad \frac{v_x}{v_\infty} = 0 \quad \text{at } y = 0 \quad \text{and} \quad \frac{v_x}{v_\infty} = 1 \quad \text{at } y = \infty$$

or, as the velocity in the x direction at the wall, $v_{x,s}$, is zero

$$\frac{v_x - v_{x,s}}{v_\infty - v_{x,s}} = 0 \quad \text{at } y = 0 \quad \text{and} \quad \frac{v_x - v_{x,s}}{v_\infty - v_{x,s}} = 1 \quad \text{at } y = \infty$$

$$thermal: \quad \frac{T - T_s}{T_\infty - T_s} = 0 \quad \text{at } y = 0 \quad \text{and} \quad \frac{T - T_s}{T_\infty - T_s} = 1 \quad \text{at } y = \infty$$

Figure 28.2 The concentration boundary layer for laminar flow past a flat surface.

and

$$concentration: \quad \frac{c_A - c_{A,s}}{c_{A,\infty} - c_{A,s}} = 0 \quad \text{at } y = 0 \quad \text{and} \quad \frac{c_A - c_{A,s}}{c_{A,\infty} - c_{A,s}} = 1 \quad \text{at } y = \infty$$

Equations (28-11), (28-12), and () can be written in terms of the following dimensionless velocity, temperature, and concentration ratios:

$$v_x \frac{\partial\left(\dfrac{v_x - v_{x,s}}{v_\infty - v_{x,s}}\right)}{\partial x} + v_y \frac{\partial\left(\dfrac{v_x - v_{x,s}}{v_\infty - v_{x,s}}\right)}{\partial y} = \nu \frac{\partial^2\left(\dfrac{v_x - v_{x,s}}{v_\infty - v_{x,s}}\right)}{\partial y^2}$$

or if

$$V = \left(\frac{v_x - v_{x,s}}{v_\infty - v_{x,s}}\right)$$

$$v_x \frac{\partial V}{\partial x} + v_y \frac{\partial V}{\partial y} = \nu \frac{\partial^2 V}{\partial^2 y}$$

with the boundary conditions: $V = 0$ at $y = 0$ and $V = 1$ at $y = \infty$ and similarly if

$$\theta = \frac{T - T_s}{T_\infty - T_s}$$

$$v_x \frac{\partial\theta}{\partial x} + v_y \frac{\partial\theta}{\partial y} = \alpha \frac{\partial^2\theta}{\partial y^2}$$

(19-15c)

with the boundary conditions: $\theta = 0$ at $y = 0$ and $\theta = 1$ at $y = \infty$ and if

$$C = \frac{c_A - c_{As}}{c_{A\infty} - c_{As}}$$

$$v_x \frac{\partial C}{\partial x} + v_y \frac{\partial C}{\partial y} = D_{AB} \frac{\partial^2 C}{\partial y^2}$$

(28-10c)

with the boundary conditions: $C = 0$ at $y = 0$ and $C = 1$ at $y = \infty$.

The similarity in the three differential equations, (12-11c), (19-15c), and (28-10c), and the boundary conditions suggests that similar solutions should be obtained for the three transfer phenomena. In Chapter 19, the Blasius solution for equation (28-11) was modified and successfully applied to explain convective heat transfer when the ratio of the momentum to thermal diffusivity $\nu/\alpha = \text{Pr} = 1$. The same type of solution should also describe convective mass transfer when the ratio of the momentum to mass diffusivity $\nu/D_{AB} = \text{Sc} = 1$. Using the nomenclature defined in Chapter 12,

$$f' = 2\frac{v_x}{v_\infty} = 2\frac{v_x - v_{x,s}}{v_\infty - v_{x,s}} = 2\frac{c_A - c_{A,s}}{c_{A,\infty} - c_{A,s}}$$

(28-11)

and

$$\eta = \frac{y}{2}\sqrt{\frac{v_\infty}{\nu x}} = \frac{y}{2x}\sqrt{\frac{xv_\infty}{\nu}} = \frac{y}{2x}\sqrt{\text{Re}_x}$$

(28-12)

the Blasius solution to the momentum boundary layer

$$\frac{df'}{d\eta} = f''(0) = \left.\frac{d[2(v_x/v_\infty)]}{d[(y/2x)\sqrt{\text{Re}_x}]}\right|_{y=0} = 1.328$$

suggests an analogous solution for the concentration boundary layer

$$\frac{df'}{d\eta} = f''(0) = \frac{d[2(c_A - c_{A,s})/(c_{A,\infty} - c_{A,s})]}{d[(y/2x)\sqrt{Re_x}]}\bigg|_{y=0} = 1.328 \qquad (28\text{-}13)$$

Equation (28-13) may be rearranged to obtain an expression for the concentration gradient at the surface

$$\frac{dc_A}{dy}\bigg|_{y=0} = (c_{A,\infty} - c_{A,s})\left[\frac{0.332}{x}Re_x^{1/2}\right] \qquad (28\text{-}14)$$

It is important to recall that the Blasius solution for equation (12–11a) did not involve a velocity in the y direction at the surface. Accordingly, equation (28-14) involves the important assumption that the rate at which mass enters or leaves the boundary layer at the surface is so small that it does not alter the velocity profile predicted by the Blasius solution.

When the velocity in the y direction at the surface, $v_{y,s}$, is essentially zero, the bulk contribution term in Fick's equation for the mass flux in the y direction is also zero. The mass transfer from the flat surface into the laminar boundary layer is described by

$$N_{A,y} = -D_{AB}\frac{\partial c_A}{\partial y}\bigg|_{y=0} \qquad (28\text{-}15)$$

Upon substituting equation (28-14) into equation (28-15), we obtain

$$N_{A,y} = -D_{AB}\left[\frac{0.332\,Re_x^{1/2}}{x}\right](c_{A,\infty} - c_{A,s})$$

or

$$N_{A,y} = D_{AB}\left[\frac{0.332\,Re_x^{1/2}}{x}\right](c_{A,s} - c_{A,\infty}) \qquad (28\text{-}16)$$

The mass flux of the diffusing component was defined in terms of the mass-transfer coefficient by

$$N_{A,y} = k_c(c_{A,s} - c_{A,\infty}) \qquad (28\text{-}4)$$

The right-hand sides of equations (28-16) and (28–4) may be equated to give

$$k_c = \frac{D_{AB}}{x}[0.332Re_x^{1/2}]$$

or

$$\frac{k_c x}{D_{AB}} = Sh_x = 0.332\,Re_x^{1/2} \qquad (28\text{-}17)$$

Equation (28-17) is restricted to systems having a Schmidt number, Sc, of 1 and low mass-transfer rates between the flat plate and the boundary layer.

A graphical presentation of the solution to the concentration boundary-layer equation (28-10) by Hartnett and Eckert[1] is depicted in Figure 28.3. Curves representing positive and negative values of the surface boundary parameter, $(v_{y,s}/v_\infty)(Re_x)^{1/2}$, are shown.

---

[1] J. P. Hartnett and E. R. G. Eckert, *Trans. ASME*, **13**, 247 (1957).

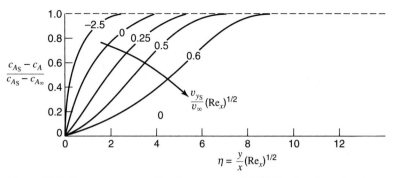

Figure 28.3 Concentration profiles for mass transfer in a laminar boundary layer on a flat plate.

The positive values apply when the mass transfer from the flat plate is into the boundary layer, and the negative values describe mass transfer from the fluid to the plate. As this surface boundary parameter approaches a zero value, the mass-transfer rate diminishes until it is considered to have no effect upon the velocity profile. The slope of the zero line, evaluated at $y = 0$, is 0.332 as predicted by equation (28-13).

In most physical operations involving mass transfer, the surface boundary parameter is negligible, and the low-mass-transfer Blasius type of solution is used to define the transfer into the laminar boundary layer. The vaporization of a volatile material into a gas stream flowing at low pressures is a case in which the low-mass-transfer assumption cannot be made.

For a fluid with a Schmidt number other than unity, similar curves to those shown in Figure 28.3 can be defined. The similarity in differential equations and boundary conditions suggests a treatment for convective mass transfer analogous to Pohlhausen's solution for convective heat transfer. The concentration boundary layer is related to the hydrodynamic boundary layer by

$$\frac{\delta}{\delta_c} = \text{Sc}^{1/3} \tag{28-18}$$

where δ is the thickness of the hydrodynamic boundary layer and δ_c is the thickness of the concentration boundary layer; thus, the Blasius η term must be multiplied by $\text{Sc}^{1/3}$. A plot of the dimensionless concentration vs. $\eta\text{Sc}^{1/3}$ for $v_{y,s} = 0$ is shown in Figure 28.4. The concentration variation given in this form leads to an expression for the convective mass-transfer coefficient similar to equation (28-17). At $y = 0$, the concentration

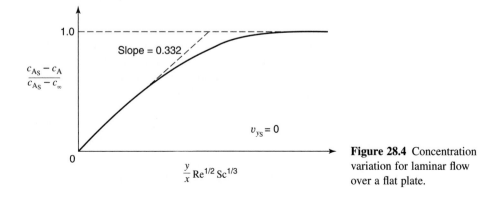

Figure 28.4 Concentration variation for laminar flow over a flat plate.

gradient is

$$\frac{\partial c_A}{\partial y}\bigg|_{y=0} = (c_{A,\infty} - c_{A,s})\left[\frac{0.332}{x}\,\text{Re}_x^{1/2}\text{Sc}^{1/3}\right] \tag{28-19}$$

which, when used with equation (28-15), yields

$$\frac{k_c x}{D_{AB}} = \text{Sh}_x = 0.332\,\text{Re}_x^{1/2}\text{Sc}^{1/3} \tag{28-20}$$

The mean mass-transfer coefficient, $\overline{k}_c$, which applies over a plate of width W and length L, may be obtained by integration. For a plate of these dimensions, the total mass transfer rate, W_A, may be evaluated by

$$\begin{aligned} W_A &= \overline{k}_c A(c_{A,s} - c_{A,\infty}) = \int_A k_c(c_{A,s} - c_{A,\infty})\,dA \\ &= \overline{k}_c WL(c_{A,s} - c_{A,\infty}) \\ &= (c_{A,s} - c_{A,\infty})\int_A \frac{0.332 D_{AB}\,\text{Re}_x^{1/2}\text{Sc}^{1/3}}{x}\,dA \end{aligned}$$

Accordingly,

$$\overline{k}_c WL = 0.332 W D_{AB}\text{Sc}^{1/3}\int_0^L \frac{\text{Re}_x^{1/2}}{x}\,dx$$

$$\overline{k}_c L = 0.332 D_{AB}\text{Sc}^{1/3}\left(\frac{v_\infty \rho}{\mu}\right)^{1/2}\int_0^L x^{-1/2}\,dx$$

$$= 0.664 D_{AB}\text{Sc}^{1/3}\left(\frac{v_\infty \rho}{\mu}\right)^{1/2} L^{1/2}$$

and

$$\frac{\overline{k}_c x}{D_{AB}} = \text{Sh}_L = 0.664\,\text{Re}_L^{1/2}\text{Sc}^{1/3} \tag{28-21}$$

The local Sherwood number at a distance x downstream is related to the mean Sherwood number for the plate by the relation

$$\text{Sh}_L = 2\,\text{Sh}_{x|x=L} \tag{28-22}$$

Equations (28-20) and (28-21) have been experimentally verified.[2] It is interesting to note that this entirely different analysis has produced results of the same form predicted in Section 28.3 by dimensional analysis for forced-convection mass transfer

$$\text{Sh} = f(\text{Re}, \text{Sc}) \tag{28-8}$$

Reconsidering the dimensionless concentration profiles of Harnett and Eckert as presented in Figure 28.3, we can observe that the slope of each curve, when evaluated at $y = 0$, decreases as the positive surface boundary parameter, $(v_{ys}/v_\infty)(\text{Re})^{1/2}$, increases. As the magnitude of the transfer coefficient is directly related to the slope by the relation

$$k_c = D_{AB}\frac{d[(c_{A,s} - c_A)/(c_{A,s} - c_{A,\infty})]}{dy}\bigg|_{y=0} \tag{28-23}$$

---

[2] W. J. Christian and S. P. Kezios, *AIChEJ.* **5**, 61 (1959).

the decrease in slope indicates that the systems having higher values of the surface boundary parameter will have lower mass-transfer coefficients.

When both energy and mass are transferred through the laminar boundary layer, the dimensionless profiles in Figure 28.3 may also represent the dimensionless temperature profiles if the Prandtl and Schmidt numbers for the system are both unity. In the previous paragraph, it was pointed out that the mass-transfer coefficient diminishes in magnitude as mass is transferred into the boundary layer from the surface. Accordingly, we should also expect the heat-transfer coefficient to diminish as mass is transferred into the boundary layer. This may be accomplished by forcing a fluid through a porous plate out into the boundary layer or by sublimating the plate material itself. These simultaneous heat and mass transfer processes, often referred to as *transpiration cooling* and *ablation*, respectively, were used to help reduce the large heat effects during the reentry of a missile into Earth's atmosphere.

EXAMPLE 3

The mass-transfer coefficient for a turbulent boundary layer formed over a flat plate has been correlated in terms of a local Sherwood number by

$$\text{Sh}_x = 0.0292\,\text{Re}_x^{4/5}\text{Sc}^{1/3} \tag{28-24}$$

where x is the distance downstream from the leading edge of the flat plate. The transition from laminar to turbulent flow occurs at $\text{Re}_x = 2 \times 10^5$.

(a) Develop an expression for the mean mass-transfer coefficient for a flat plate of length L. By definition

$$\bar{k}_c = \frac{\int_0^L k_c\,dx}{\int_0^L dx} = \frac{\int_0^{L_t} k_{c,\text{lam}}\,dx + \int_{L_t}^L k_{c,\text{turb}}\,dx}{L} \tag{28-25}$$

where L_t is the measured distance from leading edge to the transition point. $k_{c,\text{lam}}$ is defined by equation (28-20)

$$k_{c,\text{lam}} = 0.332\,\frac{D_{AB}}{x}(\text{Re}_x)^{1/2}(\text{Sc})^{1/3}$$

$k_{c,\text{turb}}$ is defined by equation (28-24)

$$k_{c,\text{turb}} = 0.0292\,\frac{D_{AB}}{x}(\text{Re}_x)^{4/5}(\text{Sc})^{1/3}$$

Upon substitution of these two equations in our equation for the mean mass-transfer coefficient, we obtain

$$\bar{k}_c = \frac{\displaystyle\int_0^{L_t} \frac{0.332 D_{AB}(\text{Re}_x)^{1/2}}{x}(\text{Sc})^{1/3}dx + \int_{L_t}^L \frac{0.0292 D_{AB}(\text{Re}_x)^{4/5}}{x}(\text{Sc})^{1/3}dx}{L}$$

where L_t is the distance from the leading edge of the plane to the transition point where the $\text{Re}_x = 2 \times 10^5$.

$$\bar{k}_c = \frac{0.332 D_{AB}\left(\dfrac{\upsilon}{\nu}\right)^{1/2}(\text{Sc})^{1/3}\displaystyle\int_0^{L_t} x^{-1/2}dx + 0.0292 D_{AB}\left(\dfrac{\upsilon}{\nu}\right)^{4/5}(\text{Sc})^{1/3}\int_{L_t}^L x^{-1/5}dx}{L}$$

$$\bar{k}_c = \frac{0.664 D_{AB}\left(\dfrac{\upsilon}{\nu}\right)^{1/2}(\text{Sc})^{1/3}L_t^{1/2} + 0.0365 D_{AB}\left(\dfrac{\upsilon}{\nu}\right)^{4/5}(\text{Sc})^{1/3}[(L)^{4/5} - (L_t)^{4/5}]}{L} \tag{28-26}$$

$$\bar{k}_c = \frac{0.664 D_{AB}(\text{Re}_t)^{1/2}(\text{Sc})^{1/3} + 0.0365 D_{AB}(\text{Sc})^{1/3}[(\text{Re}_L)^{4/5} - (\text{Re}_t)^{4/5}]}{L}$$

The solution to Problem 28-9, at the end of this chapter, points out the dominating contribution that the turbulent zone has in evaluating the mean mass-transfer coefficient for flow over a flat plate. This is more obvious as the Reynolds number for the plate exceeds its value at the transition point.

(b) In a manufacturing process, an organic solvent (methyl ethyl ketone, MEK) is used to dissolve a thin coating of a polymer away from a nonporous flat surface of length 2.0 m and width 10 cm. The thickness of the polymer film is initially uniform at 0.2 mm. In the process, at the MEK solvent flows a flow rate of 15 cm/s over the open flat plate. You may assume that the concentration of the dissolved polymer in the bulk solvent is essentially zero, $c_{A\infty} = 0$, even though in reality the concentration of dissolved polymer in the solvent increases very slightly as it flows over the pan.

At the temperature and pressure of the process, the following data area available:

- Diffusion coefficient of dissolved polymer (solute) in the MEK solvent $D_{AB} = 3 \times 10^{-6}$ cm^2/s.
- Kinematic viscosity of liquid MEK $\nu_B = 6.0 \times 10^{-3}$ cm^2/s.
- Density of solid polymer film material $\rho_A = 1.05$ g/cm^3.
- Density of liquid MEK $\rho_B = 0.80$ g/cm^3.
- Maximum solubility of dissolved polymer in MEK solvent $\rho_A^* = 0.04$ g/cm^3.

1. Determine the average flux, W_A, from the surface.
2. Determine the mass-transfer coefficient 50 cm downstream from the leading edge.

The Reynolds number for the MEK flow for the pan is evaluated by

$$\text{Re}_L = \frac{v_\infty L}{\nu} = \frac{(15\,\text{cm})(200\,\text{cm})}{6.0 \times 10^{-3}\,\text{cm}^2/\text{s}} = 5 \times 10^5$$

As this is greater than 2×10^5, we recognize that there is a transition point where the boundary layer changes from laminar to turbulent. This transition point can be evaluated from the transition Reynolds number, $\text{Re}_t = 2 \times 10^5$.

$$L = \frac{\text{Re}_t \nu}{v_\infty} = \frac{(2 \times 10^5)(6.0 \times 10^{-3}\,\text{cm}^2/\text{s})}{15\,\text{cm/s}} = 80\,\text{cm}$$

We can evaluate the mean mass-transfer coefficient by using equation (28-26) derived in part (a):

$$\bar{k}_c = \frac{0.664 D_{AB}\text{Re}_t^{1/2}\text{Sc}^{1/3} + 0.0365 D_{AB}\text{Sc}^{1/3}[\text{Re}_L^{4/5} - \text{Re}_t^{4/5}]}{L}$$

The Schmidt number is evaluated by

$$\text{Sc} = \frac{\nu}{D_{AB}} = \frac{6.0 \times 10^{-3}\,\text{cm}^2/\text{s}}{3 \times 10^{-6}\,\text{cm}^2/\text{s}} = 2 \times 10^3 \quad \text{and} \quad \text{Sc}^{1/3} = 12.567$$

The Reynolds number at transition, $\text{Re}_t = 2.0 \times 10^5$, and the Reynolds number at the end of the pan, $\text{Re}_L = 5.0 \times 10^5$, permit evaluating

$$\text{Re}_t^{1/2} = (2.0 \times 10^5)^{1/2} = 447.2$$
$$\text{Re}_t^{4/5} = (2.0 \times 10^5)^{4/5} = 17{,}411$$
$$\text{Re}_L^{4/5} = (5.0 \times 10^5)^{4/5} = 36{,}239$$

Then $[\text{Re}_L^{4/5} - \text{Re}_t^{4/5}] = 18{,}828$, and we have

$$\bar{k}_c = \frac{0.664(3.0 \times 10^{-6}\,\text{cm}^2/\text{s})(447.2)(12.567) + 0.0365(3.0 \times 10^{-6}\,\text{cm}^2/\text{s})(12.567)(18{,}828)}{200\,\text{cm}}$$

$$= 1.854 \times 10^{-4}\,\text{cm/s}$$

The average flux of species A leaving the surface will be

$$w_A = \bar{k}_c(\rho_A^* - \rho_{A\infty}) = (1.855 \times 10^{-4} \text{ cm/s})(0.04 \text{ g/cm}^3 - 0) = 7.42 \times 10^{-6} \text{ g/cm}^2\text{s}$$

2. At $x = 50$ cm the flow will be laminar and

$$\text{Re}_x = \frac{v_\infty x}{\nu} = \frac{(15 \text{ cm/s})(50 \text{ cm})}{6.0 \times 10^{-3} \text{cm}^2/\text{s}} = 1.25 \times 10^5$$

$$k_{cx} = \frac{0.332 D_{AB}(\text{Re}_x)^{1/2}(\text{Sc})^{1/3}}{x} = \frac{0.332(3.0 \times 10^{-6} \text{ cm}^2/\text{s})(1.25 \times 10^5)^{1/2}(12.567)}{50 \text{ cm}}$$

$$k_{cx} = 8.85 \times 10^{-5} \text{ cm/s}$$

28.5 APPROXIMATE ANALYSIS OF THE CONCENTRATION BOUNDARY LAYER

When the flow is other than laminar or the configuration is other than a flat plate, few exact solutions presently exist for mass transport in a boundary layer. The approximate method developed by von Kármán to describe the hydrodynamic boundary layer can be used for analyzing the concentration boundary layer. The use of this approach was discussed in Chapters 12 and 19.

Consider a control volume that is located in the concentration boundary layer as illustrated in Figure 28.5. This volume, designated by the dashed line, has a width of Δ_x, a height equal to the thickness of the concentration boundary layer, δ_c, and a unit depth. A steady-state molar mass balance over the control volume produces the relation

$$W_{A1} + W_{A3} + W_{A4} = W_{A2} \qquad (28\text{-}27)$$

Figure 28.5 The concentration boundary-layer control volume.

where W_A is the molar rate of mass transfer of component A. At each surface, the molar rate is expressed as

$$W_{A1} = \int_0^{\delta_c} c_A v_x dy \bigg|_x$$

$$W_{A2} = \int_0^{\delta_c} c_A v_x dy \bigg|_{x+\Delta x}$$

$$W_{A3} = c_{A\infty} \left[\frac{\partial}{\partial x} \int_0^{\delta_c} v_x \, dy\right] \Delta x$$

and

$$W_{A4} = k_c(c_{A,s} - c_{A\infty})\Delta x$$

In terms of these molar rates, equation (28-27) may be rewritten as

$$\int_0^{\delta_c} c_A v_x dy \bigg|_x + c_{A\infty} \left[\frac{\partial}{\partial x} \int_0^{\delta_c} v_x \, dy\right] \Delta x = k_c(c_{A,s} - c_{A\infty})\Delta x$$

$$= \int_0^{\delta_c} c_A v_x dy \bigg|_{x+\Delta x} \qquad (28\text{-}28)$$

Rearranging, dividing each term by Δx, and evaluating the results in the limit as Δx approaches zero, we obtain

$$\frac{d}{dx}\int_0^{\delta_c} c_A v_x dy = c_{A\infty}\left[\frac{d}{dx}\int_0^{\delta_c} v_x dy\right] + k_c(c_{A,s} - c_{A\infty})$$

or

$$\frac{d}{dx}\int_0^{\delta_c}(c_A - c_{A\infty})v_x dy = k_c(c_{A,s} - c_{A\infty}) \tag{28-29}$$

Equation (28-29) is analogous to equations (12-38) and (19-30). In order to solve equation (28-29), the velocity and the concentration profiles must be known; normally these profiles are unknown and must be assumed. Some of the boundary conditions that must be satisfied by the assumed boundary conditions are

(1) $\qquad\qquad\qquad\qquad v_x = 0 \qquad$ at $y = 0$

(2) $\qquad\qquad\qquad\qquad v_x = v_\infty \qquad$ at $y = \delta$

(3) $\qquad\qquad\qquad\qquad \dfrac{\partial v_x}{\partial y} = 0 \qquad$ at $y = \delta$

and, according to equation (12-33)

(4) $\qquad\qquad\qquad\qquad \dfrac{\partial^2 v_x}{\partial y^2} = 0 \quad$ at $y = 0$

The assumed concentration profile must satisfy the corresponding boundary conditions in terms of concentrations

(1) $\qquad\qquad c_A - c_{A,s} = 0 \qquad\qquad\qquad$ at $y = 0$ $\qquad\qquad$ (28-30)

(2) $\qquad\qquad c_A - c_{A,s} = c_{A,\ \infty} - c_{A,s} \quad$ at $y = \delta_c$ $\qquad\qquad$ (28-31)

(3) $\qquad\qquad \dfrac{\partial}{\partial y}(c_A - c_{A,s}) = 0 \qquad\qquad\quad$ at $y = \delta_c$ $\qquad\qquad$ (28-32)

and

(4) $\qquad\qquad \dfrac{\partial^2}{\partial y^2}(c_A - c_{A,s}) = 0 \qquad\qquad\quad$ at $y = 0$ $\qquad\qquad$ (28-33)

If we reconsider the laminar flow parallel to a flat surface, we can use the von Kármán integral equation (28-29) to obtain an approximate solution. The results can be compared to the exact solution, equation (28-20), and thus indicate how well we have approximated the velocity and the concentration profiles. As our first approximation, let us consider a power-series expression for the concentration variation with y

$$c_A - c_{A,s} = a + by + cy^2 + dy^3$$

Application of the boundary conditions will result in the following expression:

$$\frac{c_A - c_{A,s}}{c_{A,\infty} - c_{A,s}} = \frac{3}{2}\left(\frac{y}{\delta_c}\right) - \frac{1}{2}\left(\frac{y}{\delta_c}\right)^3 \tag{28-34}$$

If the velocity profile is assumed in the same power-series form, then the resulting expression, as obtained in Chapter 12, is

$$\frac{v_x}{v_\infty} = \frac{3}{2}\left(\frac{y}{\delta}\right) - \frac{1}{2}\left(\frac{y}{\delta}\right)^3 \tag{12-40}$$

Upon substituting equations (28-34) and (12-40) into the integral expression (28-29) and solving, we obtain

$$Sh_x = 0.36Re_x^{1/2}Sc^{1/3} \tag{28-35}$$

which is close to the exact solution expressed in equation (28-20).

Although this result is not exact, it is sufficiently close to the exact solution to indicate that the integral method may be used with some degree of confidence in other situations in which an exact solution is unknown. The accuracy of the method depends entirely on the ability to assume good velocity and concentration profiles.

The von Kármán integral equation (28-29) has been used to obtain an approximate solution for the turbulent boundary layer over a flat plate. With the velocity profile approximated by

$$v_x = \alpha + \beta y^{1/7}$$

and the concentration profile approximated by

$$c_A - c_{A,\infty} = \eta + \xi y^{1/7}$$

the local Nusselt number in the turbulent layer is found to be

$$Sh_x = 0.0292\,Re_x^{4/5} \tag{28-24}$$

28.6   MASS, ENERGY, AND MOMENTUM-TRANSFER ANALOGIES

In the previous analyses of convective mass transfer, we have recognized the similarities in the differential equations for momentum, energy, and mass transfer and in the boundary conditions when the transport gradients were expressed in terms of dimensionless variables. These similarities have permitted us to predict solutions for the similar transfer processes. In this section, we shall consider several analogies among transfer phenomena that have been proposed because of the similarity in their mechanisms. The analogies are useful in understanding the transfer phenomena and as a satisfactory means for predicting behavior of systems for which limited quantitative data are available.

The similarity among the transfer phenomena and, accordingly, the existence of the analogies require that the following five conditions exist within the system:

1. There is no energy or mass produced within the system. This, of course, infers that no homogeneous reaction occurs.

2. There is no emission or absorption of radiant energy.

3. There is no viscous dissipation.

4. The velocity profile is not affected by the mass transfer; thus, there is only a low rate of mass transfer.

5. The physical properties are constant. As there may be slight changes in the physical properties due to variations in temperature or concentration, this condition can be approximated by using average concentration and film temperature properties.

Reynolds Analogy

The first recognition of the analogous behavior of momentum and energy transfer was reported by Reynolds.[3] Although this analogy is limited in application, it has served as the catalyst for seeking better analogies, and it has been useful in analyzing complex boundary-layer phenomena of aerodynamics.

Reynolds postulated that the mechanisms for transfer of momentum and energy were identical. We have observed in our earlier discussions on laminar boundary layers that this is true if the Prandtl number, Pr, is unity. From our previous consideration in Section 28.4, we can extend the Reynolds postulation to include the mechanism for the transfer of mass if the Schmidt number, Sc, is also unity. For example, if we consider the laminar flow over a flat plate where Sc $= 1$, the concentration and velocity profiles within the boundary layers are related by

$$\frac{\partial}{\partial y}\left(\frac{c_A - c_{A,s}}{c_{A,\infty} - c_{A,s}}\right)\Bigg|_{y=0} = \frac{\partial}{\partial y}\left(\frac{v_x}{v_\infty}\right)\Bigg|_{y=0} \tag{28-36}$$

Recalling that at the boundary next to the plate, where $y = 0$, we may express the mass flux in terms of either the mass-diffusivity or the mass-transfer coefficient by

$$N_{A,y} = -D_{AB}\frac{\partial}{\partial y}(c_A - c_{A,s})\Bigg|_{y=0} = k_c(c_{A,s} - c_{A,\infty}) \tag{28-37}$$

We can combine equations (28-36) and (28-37) and as $D_{AB} = \mu/\rho$ when Sc $= 1$, achieve an expression that relates the mass-transfer coefficient to the velocity gradient at the surface

$$k_c = \frac{\mu}{\rho v_\infty}\frac{\partial v_x}{\partial y}\Bigg|_{y=0} \tag{28-38}$$

The coefficient of skin friction was related in Chapter 21 to this same velocity gradient by

$$C_f = \frac{\tau_0}{\rho v_\infty^2/2} = \frac{2\mu(\partial v_x/\partial y)|_{y=0}}{\rho v_\infty^2} \tag{12-2}$$

Using this definition, we can rearrange equation (28-38) to obtain the mass-transfer Reynolds analogy for systems with a Schmidt number of 1

$$\frac{k_c}{v_\infty} = \frac{C_f}{2} \tag{28-39}$$

Equation (28-39) is analogous to the energy-transfer Reynolds's analogy for systems with a Prandtl number of 1. This analogy was discussed in Chapter 19 and may be expressed by

$$\frac{h}{\rho v_\infty c_p} = \frac{C_f}{2} \tag{19-36}$$

The Reynolds analogy, equation (28-39), was obtained by using the "exact solution" for the laminar flow over a flat plate equation (28-36), the mass flux equation written at the boundary next to the plate resulting in equation (28-38), and the defining equation for the coefficient of skin friction (12-2). Thus, the Reynolds analogy satisfies the "exact solution"

---

[3] O. Reynolds, *Proc. Manchester Lit. Phil. Soc.*, **8** (1874).

if and only if the Schmidt number equals 1 and the resistance to flow is that due to skin friction (i.e., no form drag is involved). This was experimentally verified by von Kármán for a fully turbulent flow with $Sc = Pr = 1$.

Turbulent-Flow Considerations

In a majority of practical applications, the flow in the main stream is turbulent rather than laminar. Although many investigators have contributed considerably to the understanding of turbulent flow, so far no one has succeeded in predicting convective transfer coefficients or friction factors by direct analysis. This is not too surprising when we recall from our earlier discussions on turbulent flow, in Section 13.1, that the flow at any point is subject to irregular fluctuations in direction and velocity. Accordingly, any particle of the fluid undergoes a series of random movements, superimposed on the main flow. These eddy movements bring about mixing throughout the turbulent core. This process is often referred to as "eddy diffusion." The value of the eddy-mass diffusivity will be very much larger than the molecular diffusivity in the turbulent core.

In an effort to characterize this type of motion, Prandtl proposed the mixing-length hypothesis as discussed in Chapter 12. In this hypothesis, any velocity fluctuation v'_x is due to the y-directional motion of an eddy through a distance equal to the mixing length L. The fluid eddy, possessing a mean velocity, $\bar{v}_x |_y$, is displaced into a stream where the adjacent fluid has a mean velocity, $\bar{v}_x |_{y+L}$. The velocity fluctuation is related to the mean-velocity gradient by

$$v'_x = \bar{v}_x |_{y+L} - \bar{v}_x |_y = \pm L \frac{d\bar{v}_x}{dy} \tag{12-52}$$

The total shear stress in a fluid was defined by the expression

$$\tau = \mu \frac{d\bar{v}_x}{dy} - \rho \overline{v'_x v'_y} \tag{12-51}$$

The substitution of equation (13-10) into (12-51) gives

$$\tau = \rho[\nu + L v'_y] \frac{d\bar{v}_x}{dy} \tag{28-40}$$

or

$$\tau = \rho[\nu + \epsilon_M] \frac{d\bar{v}_x}{dy} \tag{28-41}$$

where $\epsilon_M = L v'_y$ is designated at the eddy momentum diffusivity. It is analogous to the molecular momentum diffusivity, ν.

We may now similarly analyze mass transfer in turbulent flow, as this transport mechanism is also due to the presence of the fluctuations or eddies. In Figure 28.6, the curve represents a portion of the turbulent concentration profile with mean flow in the x direction. The instantaneous rate of transfer of component A in the y direction is

$$N_{A,y} = c'_A v'_y \tag{28-42}$$

where $c_A = \bar{c}_A + c'_A$, the temporal average

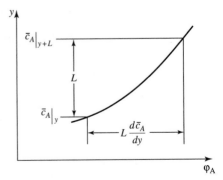

Figure 28.6 Portion of turbulent concentration profile curve, showing the Prandtl mixing length.

or

$$(c_{A,s} - c_{A|\xi}) = \frac{N_{A,y,s}}{D_{AB}} \xi \tag{28-46}$$

Eliminating ξ from these two equations, we obtain

$$\frac{\rho v v_{x|\xi}}{\tau_s} = \frac{D_{AB}}{N_{A,y,s}} (c_{A,s} - c_{A|\xi}) \tag{28-47}$$

The Reynolds analogy, $k_c/v_\infty = C_f/2 = \tau_s/\rho v_\infty^2$, may be used in the turbulent region, from $y = \xi$ to y under bulk conditions. The mass flux in the turbulent region becomes

$$N_{A,y} = k_c(c_{A|\xi} - c_{A,\infty}) = \frac{\tau_s}{\rho(v_\infty - v_{x|\xi})} (c_{A|\xi} - c_{A,\infty}) \tag{28-48}$$

Eliminating $c_{A|\xi}$ between equations (28-47) and (28-48), we obtain

$$\frac{c_{A,s} - c_{A,\infty}}{N_{A,y}} = \frac{\rho}{\tau_s} \left[v_\infty + v_{x|\xi} \left(\frac{v}{D_{AB}} - 1 \right) \right] \tag{28-49}$$

Substituting the defining equations

$$C_f \equiv \frac{\tau_s}{\rho(v_\infty^2/2)}$$

$$k_c \equiv \frac{N_A}{(c_{A,s} - c_{A,\infty})}$$

and

$$Sc \equiv \frac{v}{D_{AB}}$$

into equation (28-49), we may simplify the relation to

$$\frac{1}{k_c} = \frac{2}{C_f v_\infty^2} [v_\infty + v_{x|\xi}(Sc - 1)]$$

or in slightly different form

$$\frac{k_c}{v_\infty} = \frac{C_f/2}{1 + (v_{x|\xi}/v_\infty)(Sc - 1)} \tag{28-50}$$

Note that equation (28-50) simplifies to the Reynolds analogy with the restriction $Sc = 1$. In Chapter 12, the laminar sublayer was defined by $v^+ = y^+ = 5$, where $v^+ = v_{x|\xi}/(v_\infty \sqrt{c_f/2})$; thus,

$$v^+ = \frac{v_{x|\xi}}{v_\infty \sqrt{C_f/2}} = 5$$

or

$$\frac{v_{x|\xi}}{v_\infty} = 5\sqrt{\frac{C_f}{2}} \tag{28-51}$$

Substituting for $v_{x|\xi}/v_\infty$ in equation (28-50), we obtain an analogy for convective mass transfer similar to the *Prandtl analogy* for convective heat transfer

$$\frac{k_c}{v_\infty} = \frac{C_f/2}{1 + 5\sqrt{C_f/2}(Sc - 1)} \tag{28-52}$$

Rearranging and multiplying both sides of equation (28-52) $v_\infty L/D_{AB}$, where L is a characteristic length, we obtain

$$\frac{k_c}{v_\infty}\frac{v_\infty L}{D_{AB}} = \frac{(C_f/2)(v_\infty L/D_{AB})(\rho\mu/\rho\mu)}{1+5\sqrt{C_f/2}(\text{Sc}-1)}$$

or

$$\text{Sh} = \frac{(C_f/2)\text{ReSc}}{1+5\sqrt{C_f/2}(\text{Sc}-1)} \tag{28-53}$$

If the equivalent mass-transfer Nusselt number, Nu_{AB}, were used in place of the Sherwood number, equation (28-53) would be analogous to the Prandtl momentum–energy-transfer analogy, equation (19-57).

von Kármán extended the Prandtl analogy by considering the so-called "buffer layer" in addition to the laminar sublayer and the turbulent core. This led to the development of the *von Kármán analogy*

$$\text{Nu} = \frac{(C_f/2)\text{RePr}}{1+5\sqrt{C_f/2}\{\text{Pr}-1+\ln[(1+5\text{Pr})/6]\}} \tag{19-58}$$

for momentum and energy transfer. The von Kármán analysis for mass transfer yields

$$\text{Sh} = \frac{(C_f/2)\text{ReSc}}{1+5\sqrt{C_f/2}\{\text{Sc}-1+\ln[(1+5\text{Sc})/6]\}} \tag{28-54}$$

or

$$\frac{\text{Sh}}{\text{ReSc}} = \frac{k_c}{v_\infty} = \frac{C_f/2}{1+5\sqrt{C_f/2}\left\{\text{Sc}-1+\ln\left[1+\frac{5}{6}(\text{Sc}-1)\right]\right\}} \tag{28-55}$$

Equation (28-55) is analogous to equation (19-58).

The results of most analogies can be put in a general form, as illustrated in equations (28–52) and (28-55) in which the denominator of the right-hand side is a complex group of terms that serve as a correction to the simple Reynolds analogy.

Chilton–Colburn Analogy

Chilton and Colburn,[4] using experimental data, sought modifications to Reynolds's analogy that would not have the restrictions that Pr and Sc numbers must be equal to 1. They defined the *j factor for mass transfer,*

$$j_D \equiv \frac{k_c}{v_\infty}(\text{Sc})^{2/3}$$

This factor is analogous to the *j* factor for heat transfer that is defined by equation (19-39). Based on the data collected in both laminar and turbulent flow regimes, they found

$$j_D \equiv \frac{k_c}{v_\infty}(\text{Sc})^{2/3} = \frac{C_f}{2} \tag{28-56}$$

[4] A. P. Colburn, *Trans. AIChE*, **29**, 174–210 (1933); T. H. Chilton and A. P. Colburn, *Ind. Eng. Chem.*, **26**, 1183 (1934).

The analogy is valid for gases and liquids within the range $0.6 < Sc < 2500$. Equation (28-56) can be shown to satisfy the "exact solution" for laminar flow over the flat plate

$$Sh_x = 0.332\, Re_x^{1/2} Sc^{1/3} \tag{28-20}$$

If both sides of this equation are divided by $Re_x Sc^{1/3}$, we obtain

$$\frac{Sh_x}{Re_x Sc^{1/3}} = \frac{0.332}{Re_x^{1/2}} \tag{28-57}$$

This equation reduces to the *Chilton–Colburn analogy* when we substitute into the above expression the Blasius solution for the laminar boundary layer

$$\frac{Sh_x}{Re_x Sc^{1/3}} = \frac{Sh_x}{Re_x Sc} Sc^{2/3} = \frac{k_c}{v_\infty} = \frac{C_f}{2}$$

or

$$\left(\frac{k_c x}{D_{AB}}\right)\left(\frac{\mu}{x v_\infty \rho}\right)\left(\frac{\rho D_{AB}}{\mu}\right)(Sc)^{2/3} = \frac{k_c Sc^{2/3}}{v_\infty} = \frac{C_f}{2} \tag{28-58}$$

The complete Chilton–Colburn analogy is

$$j_H = j_D = \frac{C_f}{2} \tag{28-59}$$

which relates all three types of transport in one expression. Equation (28-59) is exact for flat plates and is satisfactory for systems of other geometry provided no form drag is present. For systems where form drag is present, it has been found that neither j_H or j_D is equal to $C_f/2$; however, when form drag is present

$$j_H = j_D \tag{28-60}$$

or

$$\frac{h}{\rho v_\infty c_p}(Pr)^{2/3} = \frac{k_c}{v_\infty}(Sc)^{2/3} \tag{28-61}$$

Equation (28-61) relates convective heat and mass transfer. It permits the evaluation of one unknown transfer coefficient through information obtained for another transfer phenomenon. It is valid for gases and liquids within the ranges $0.6 < Sc < 2500$ and $0.6 < Pr < 100$.

As previously stated, one of the five conditions that should exist if the analogies are to be used requires the physical properties of the fluid stream to be constant. If there are only slight variations in the properties due to the variations in the overall film temperature, one may minimize this restrictive condition by evaluating the physical properties at the mean film temperature.

The Chilton–Colburn analogy for heat and mass transfer has been observed to hold for many different geometries; for example, flow over flat plate, flow in circular pipe and annulus, and flow around cylinders. In the following three examples, we will apply the Chilton–Colburn analogy (1) to determine an unknown convective-transfer coefficient, (2) to predict a correlating equation for mass transfer, and (3) to derive and use the important psychrometry wet-bulb equation.

EXAMPLE 4

Using the statement presented in Example 1 and the mass-transfer coefficient determined in example 1, determine the value of the heat-transfer coefficient, h, for the stated air stream.

In the solution to Example 1, the mass-transfer coefficient was found to be $k_c = 0.118$ m/s and the mass diffusivity of carbon dioxide in air is 1.5×10^{-5} m^2/s and the kinematic viscosity of the air is 1.55×10^{-5} m^2/s. From equation (28-59), we know

$$j_D = j_H$$

$$\frac{k_c}{v_\infty} Sc^{2/3} = \frac{h}{\rho v_\infty c_p} Pr^{2/3}$$

$$h = \rho c_p k_c \left(\frac{Sc}{Pr}\right)^{2/3}$$

The following properties of air at 293 K can be found in Appendix I:

$$\rho_{air} = 1.206 \text{ kg/m}^3$$
$$c_{p\,air} = 1.0061 \text{ J/kg} \cdot \text{K}$$
$$Pr = 0.710$$

The Schmidt number is

$$Sc = \frac{v}{D_{AB}} = \frac{1.55 \times 10^{-5} \text{ m}^2/\text{s}}{1.5 \times 10^{-5} \text{ m}^2/\text{s}} = 1.033$$

Substituting into our equation, we obtain

$$h = \left(1.206 \, \frac{\text{kg}}{\text{m}^3}\right)\left(1.0061 \, \frac{J}{\text{kg K}}\right)\left(0.118 \, \frac{\text{m}}{\text{s}}\right)\left(\frac{1.033}{0.710}\right)^{2/3} = 0.184 \, \frac{J}{\text{m}^2 \, \text{K s}}$$

EXAMPLE 5

Dittus and Boelter proposed the following equation for correlating the heat-transfer coefficient for turbulent flow in a pipe

$$Nu = \frac{hD}{k} = 0.023 \, Re^{0.8} Pr^{1/3}$$

What should be the corresponding equation for the mass-transfer coefficient when the transfer is to a turbulent fluid flowing in a pipe?

According to the Chilton–Colburn relationship (28-61)

$$\frac{h}{\rho v_\infty c_p}(Pr)^{2/3} = \frac{k_c}{v_\infty}(Sc)^{2/3}$$

or

$$h = k_c \rho c_p \left(\frac{Sc}{Pr}\right)^{2/3}$$

Upon substituting this into the Dittus–Boelter equation, we obtain

$$k_c \rho c_p \left(\frac{Sc}{Pr}\right)^{2/3} \frac{D}{k} = 0.023 \, Re^{0.8} Pr^{1/3}$$

or

$$\frac{k_c D}{D_{AB}} \frac{D_{AB}\rho}{\mu} \frac{\mu c_p}{k} \left(\frac{Sc}{Pr}\right)^{2/3} = 0.023 \, Re^{0.8} Pr^{1/3}$$

$$\frac{k_c D}{D_{AB}} \cdot \frac{1}{Sc} \cdot Pr \left(\frac{Sc}{Pr}\right)^{2/3} = 0.023 \, Re^{0.8} Pr^{1/3}$$

This simplifies to

$$Sh = \frac{k_c D}{D_{AB}} = 0.023 \, Re^{0.8} Sc^{1/3}$$

Linton and Sherwood,[5] considering mass transfer into turbulent streams flowing through pipes, correlated their data by

$$\frac{k_c D}{D_{AB}} = 0.023 \, Re^{0.8} Sc^{1/3}$$

for

$$2000 < Re < 70,000$$
$$1000 < Sc < 2260$$

EXAMPLE 6

Dry air at 1.013×10^5 Pa pressure blows across a thermometer whose bulb has been covered with a dampened wick. The classical "wet bulb" thermometer indicates that a steady-state temperature is reached by a small amount of liquid water evaporating into a large reservoir of unsaturated vapor–gas mixture. The thermometer reads 290 K. At this temperature, the following properties were evaluated:

P_A, vapor pressure of water	1.94×10^3 Pa
ρ, density of air	$1.219 \, kg/m^3$
λ_{Ts}, latent heat of vaporization of water	2461 kJ/kg
Pr, Prandtl number	0.71
Sc, Schmidt number	0.61
c_p, specific heat of air	1.006 J/kg K

What is the temperature of the dry air?

Equation (28-1) defines the molar flux of water evaporating

$$N_{H_2O} = k_c(c_{H_2O,s} - c_{H_2O\infty})$$

The energy required to evaporate this water is supplied by convective heat transfer; thus,

$$\frac{q}{A} = h(T_\infty - T_s) = \lambda_{Ts} M_{H_2O} N_{H_2O}$$

where λ_{Ts} is the latent heat of vaporization of water at the surface temperature. This equation may be solved for the bulk temperature

$$T_\infty = \frac{\lambda_{Ts} M_{H_2O} N_{H_2O}}{h} + T_s$$

[5] W. H. Linton and T. K. Sherwood, *Chem. Eng. Prog.*, **46**, 258 (1950).

If we substitute equation (28-1) into this equation, we obtain

$$T_\infty = \lambda_{Ts} M_{H_2O} \frac{k_c}{h}(c_{H_2O,s} - c_{H_2O,\infty}) + T_s$$

Chilton–Colburn j factors give us a relationship for the k_c/h ratio

$$\frac{h}{\rho v_\infty c_p} \mathrm{Pr}^{2/3} = \frac{k_c}{v_\infty} \mathrm{Sc}^{2/3}$$

$$\frac{k_c}{h} = \frac{1}{\rho c_p}\left(\frac{\mathrm{Pr}}{\mathrm{Sc}}\right)^{2/3}$$

When this expression is substituted into our equation for the bulk temperature, we obtain an equation for the psychromatic wet-bulb line

$$T_\infty = \frac{\lambda_{Ts} M_{H_2O}}{\rho c_p}\left(\frac{\mathrm{Pr}}{\mathrm{Sc}}\right)^{2/3}(c_{H_2O,s} - c_{H_2O\infty}) + T_s$$

The concentrations are

$$c_{H_2O,s} = \frac{P_A}{RT} = \frac{1.94 \times 10^3\,\mathrm{Pa}}{\left(8.314\,\dfrac{\mathrm{Pa\,m^3}}{\mathrm{mol\,K}}\right)(290\,\mathrm{K})} = 0.804\,\frac{\mathrm{mol}}{\mathrm{m^3}}$$

$$c_{H_2O,\infty} = 0\,\frac{\mathrm{mol}}{\mathrm{m^3}}$$

Upon substitution of known values, we obtain

$$T_\infty = \frac{\left(2461\,\dfrac{\mathrm{kJ}}{\mathrm{kg}}\right)\left(18\,\dfrac{\mathrm{kg}}{\mathrm{k\,mol}}\right)}{\left(1.219\,\dfrac{\mathrm{kg}}{\mathrm{m^3}}\right)\left(1.006\,\dfrac{J}{\mathrm{kg\,K}}\right)}\left(\frac{0.71}{0.61}\right)^{2/3}\left(0.804\,\frac{\mathrm{mol}}{\mathrm{m^3}}\right) + 290\,\mathrm{K}$$

$$= 322.1\,\mathrm{K}$$

28.7 MODELS FOR CONVECTIVE MASS-TRANSFER COEFFICIENTS

Convective mass-transfer coefficients have been used in the design of mass-transfer equipment for many years. However, in most cases, they have been empirical coefficients that were determined from experimental investigations. A theoretical explanation of the coefficients will require a better understanding of the mechanism of turbulence, as they are directly related to the dynamic characteristics of the flow. In Chapter 26, two possible models for explaining convective mass transfer were introduced. Both the film theory and the penetration theory have been widely applied.

The *film theory* is based upon the presence of a fictitious film of fluid in laminar flow next to the boundary that offers the same resistance to mass transfer as actually exists in the entire flowing fluid. In other words, all resistance to transfer is assumed to exist in a fictitious film in which the transport is entirely by molecular diffusion. The film thickness, δ, must extend beyond the laminar sublayer to include an equivalent resistance encountered as the concentration changes within the buffer layer and the turbulent core. For diffusion through a nondiffusing layer or stagnant fluid, this theory predicts the mass-transfer coefficient to be

$$k_c = \frac{D_{AB}}{\delta}\frac{P}{p_{B,1m}} \tag{26-9}$$

as developed in Chapter 26. For equimolar counterdiffusion, the mass-transfer coefficient was expressed as

$$k_c^0 = \frac{D_{AB}}{\delta} \qquad (26\text{-}24)$$

In both cases, the convective mass-transfer coefficient is directly related to the molecular mass diffusivity. Obviously, the fictitious film thickness, δ, can never be measured, as it does not exist. Because of this and because of its apparent inadequacy in physically explaining convective mass transfer, other theories and models have been postulated to describe this phenomenon.

The *penetration theory* was originally proposed by Higbie[6] to explain the mass transfer in the liquid phase during gas absorption. It has been applied to turbulent flow by Danckwerts[7] and many other investigators when the diffusing component penetrates only a short distance into the phase of interest because of its rapid disappearance through chemical reaction or its relatively short time of contact.

Higbie considered mass to be transferred into the liquid phase by unsteady-state molecular transport. With this concept, the mass flux at the interface between the liquid and the gas phases was expressed as

$$N_A = \sqrt{\frac{D_{AB}}{\pi t_{exp}}}(c_{A,s} - c_{A,\infty}) \qquad (26\text{-}88)$$

Danckwerts applied this unsteady-state concept to the absorption of component A in a turbulent liquid stream. His model assumes that the motion of the liquid is constantly bringing fresh liquid eddies from the interior up to the surface, where they displace the liquid elements previously on the surface. While on the surface, each element of the liquid becomes exposed to the second phase and mass is transferred into the liquid as though it were stagnant and infinitely deep; the rate of transfer is dependent upon the exposure time. Many different assumptions can be made relative to the surface renewal. For instance, each element of the surface may have the same exposure time before being replaced; this infers that the instantaneous mass transfer will occur according to equation (26-88). The total solute penetrating the eddy in an exposure time, t_{exp}, is

$$\int_0^{t_{exp}} N_A dt = (c_{A,s} - c_{A,\infty})\sqrt{\frac{D_{AB}}{\pi}}\int_0^{t_{exp}} t^{-1/2} dt$$

$$= 2(c_{A,s} - c_{A,\infty})\left(\frac{D_{AB}t_{exp}}{\pi}\right)^{1/2}$$

and the average rate of transfer during the exposure is obtained by dividing this equation by the time of exposure

$$N_A = 2(c_{A,s} - c_{A,\infty})\left(\frac{D_{AB}}{\pi t_{exp}}\right)^{1/2} \qquad (28\text{-}62)$$

Danckwerts modified the assumption of constant exposure period by proposing an "infinite" range of ages for the elements at the surface. Surface age distribution functions were introduced to predict the probability of an element of surface being replaced by a fresh eddy. The rate of surface renewal was believed to be constant for a given degree of

---

[6] R. Higbie, *Trans. AIChE*, **31**, 368–389 (1935).
[7] P. V. Danckwerts, *Ind. Eng. Chem.*, **43**, 1460–1467 (1951).

turbulence and equal to a surface-renewal factor s. The rate of mass transfer with random surface renewal is

$$N_A = \sqrt{D_{AB}s}(c_{A,s} - c_{A,\infty}) \tag{28-63}$$

The values of s are currently obtained by experimental investigations. The surface-renewal concept has been very successful in the explanation and analysis of convective mass transfer, particularly when the mass transport is accompanied by chemical reactions in the liquid phase.[8,9] Considerable development and experimental verification are needed to define this model clearly.

A detailed discussion of mass-transfer coefficients for chemically reacting systems is not treated in this text. In our earlier discussions of molecular mass transfer associated with a chemical reaction, in Section 26.2, mass transfer was shown to depend upon the rate constant of the chemical reaction. We should expect a similar dependency for the convective mass-transfer coefficient. Excellent discussions on this subject are available.[8,9]

Toor and Marchello[10] have pointed out that the penetration concept of Danckwerts is valid only when the surface renewal is relatively rapid, thus providing young elements at the surface on a continuous basis. For older elements at the surface, a steady-state concentration gradient is established as predicted by the film theory; accordingly, the rate of mass transfer should be directly proportional to the molecular mass diffusivity. At low Schmidt numbers, a steady concentration gradient is set up very rapidly in any new surface element so that unless the rate of surface renewal is high enough to remove a major fraction of the surface elements before they are penetrated, most of the surface behaves as older elements. As the Schmidt number increases, the time necessary to set up the steady gradient increases rapidly, and accordingly, relatively low surface renewal rates are sufficient to keep most of the elements from being penetrated. When conditions are such that the surface contains both young and older elements, the transfer characteristics are intermediate between the film and penetration models. The convective mass-transfer coefficients will be proportional to a power of the molecular mass diffusivity between 0.5 and 1.0. These conclusions have been supported by experimental data.

In both the film and penetration models, the mass transfer involves an interface between two moving fluids. When one of the phases is a solid, the fluid velocity parallel to the surface at the interface must be zero; accordingly, we should expect the need of a third model, the *boundary-layer model,* for correlating the data involving a solid subliming into a gas or a solid dissolving into a liquid. For diffusion through a laminar boundary layer, the average mass-transfer coefficient was found to be

$$\bar{k}_c = 0.664 \frac{D_{AB}}{L} \text{Re}_L^{1/2} \text{Sc}^{1/3} \tag{28-21}$$

This shows the mass-transfer coefficient varies as $D_{AB}^{2/3}$, which is typical of boundary-layer calculations.

Table 28.1 provides a brief summary of the three models proposed for mass-transfer coefficients. Each model has its own specific diffusion coefficient dependency; this dependency is sometimes used to scale mass-transfer coefficients from one solute to another that are exposed to the same hydrodynamic flows.

---

[8] P. V. Danckwerts, *Gas–Liquid Reactions,* McGraw-Hill, New York, 1970.

[9] G. Astarita, *Mass Transfer with Chemical Reaction,* Elsevier, Amsterdam, 1967.

[10] H. L. Toor and J. M. Marchello, *AIChE J.,* **1,** 97 (1958).

Table 28.1 Models for convective mass-transfer coefficients (dilute systems)

Model	Basic form	$f(D_{AB})$	Notes
Film theory	$k_c = \dfrac{D_{AB}}{\delta}$	$k_c \propto D_{AB}$	δ unknown, may be found when solute has high Sc
Penetration theory	$k_c = \sqrt{\dfrac{D_{AB} v_\infty}{\pi \delta}}$	$k_c \propto D_{AB}^{1/2}$	δ unknown, good model when homogeneous reaction within boundary layer or when solute has low Sc
Boundary-layer theory	$k_c = 0.664 \dfrac{D_{AB}}{L}(\mathrm{Re})^{1/2}(\mathrm{Sc})^{1/3}$	$k_c \propto D_{AB}^{2/3}$	Best way to scale k_c from one solute to another exposed to same hydrodynamic flow

28.8 CLOSURE

In this chapter, we have discussed the principles of mass transfer by forced convection, the significant parameters that help describe convective mass transfer, and the models proposed to explain the mechanism of convective transport. We have seen that the transfer of mass by convection is intimately related to the dynamic characteristics of the flowing fluid, particularly to the fluid in the vicinity of the boundary. Because of the close similarities in the mechanisms of momentum, energy, and mass transfer, we were able to use the same four methods for evaluating convective mass-transfer coefficients that were originally developed to analyze convective heat-transfer coefficients. In all four analyses, the mass-transfer coefficient was correlated by the general equation

$$\mathrm{Nu}_{AB} = f(\mathrm{Re}, \mathrm{Sc})$$

Mass transfer into turbulent streams was discussed, and the eddy mass diffusivity was defined. Analogies were presented for convective mass transfer into turbulent streams.

PROBLEMS

28.1 Determine the Schmidt number for

(a) oxygen in air at 300 K and 1.0 atm; and

(b) oxygen in liquid water at 300 K.

At 300 K, the diffusion coefficient of oxygen in liquid water is $1.5 \times 10^{-9} \text{ m}^2/\text{s}$.

28.2 One step in the manufacture of optical fibers is the chemical vapor deposition of silane, SiH_4, on the inside surface of a hollow glass fiber to form a very thin cladding of solid silicon by the reaction

$$SiH_4(g) \rightarrow Si(s) + 2H_2(g)$$

The process is carried out at temperature of 900 K and a system pressure of 100 Pa. Silane (A) is diluted to 1.0 mol% in the inert helium (B) carrier gas. The binary gas diffusivity of silane in helium at 298 K and 101.3 kPa is 0.518 cm²/s and the kinematic viscosity of the helium gas at 900 K is 5.6 cm²/s.

Determine the Schmidt number of silane (g) in the helium (g) carrier gas.

28.3 Silicon tetrachloride, $SiCl_4$, is a key chemical in the silicon chemical vapor deposition. It is used in the production of silane, SiH_4, which is used as described in Problem 28.2. The purity of $SiCl_4$ is essential to the production of high-quality silicon films. To eliminate trichlorosilane, $SiHCl_3$, within the high-purity silicon tetrachloride, chlorine gas is bubbled through the liquid $SiCl_4$ at 298 K to promote the following reaction:

$$SiHCl_3 + Cl_2 \rightarrow SiCl_4 + HCl$$

The HCl is then easily removed in a stripper, using nitrogen as the stripping gas.

To determine the mass-transfer coefficient of chlorine in liquid $SiCl_4$, a Schmidt number is needed. Evaluate the Schmidt number for chlorine in liquid silicon tetrachloride at 298 K. The following information is available for $SiCl_4$ at 298 K:

$$\rho_L = 1.47 \, \text{g/cm}^3 \quad \text{and} \quad \mu_L = 5.2 \times 10^{-4} \, \text{kg/m} \cdot \text{s}$$

The diffusivity for chlorine in silicon tetrachloride can be evaluated using the Wilke–Chang equation.

28.4 What are the Stanton and Peclet numbers and how are they related to other convective mass-transfer dimensionless numbers?

28.5 In applying dimensional analysis to explain mass-transfer coefficient, one must consider the geometry involved, a variable to explain the flow characteristics of the moving stream, and the properties of the moving stream. Predict the variables that are necessary to explain the mass-transfer coefficient for a gas stream flowing over a flat plate and arrange these variables into dimensionless groups.

28.6 In a mass-transfer spray column, a liquid is sprayed into a gas stream, and mass is transferred between the liquid and gas phases. The mass of the drops that are formed from a spray nozzle is considered a function of the nozzle diameter, acceleration of gravity, surface tension of the liquid against the gas, fluid density, fluid viscosity, fluid velocity, and the viscosity and density of the gas medium. Arrange these variables in dimensionless groups. Should any other variables have been included?

28.7 Mass-transfer data were obtained for the vaporization of naphthalene into a turbulent gas stream flowing in an annulus. Both the inside rod and the outer pipe of the annular conduit were made of naphthalene. To correlate the data, the investigator predicted that the mass-transfer coefficient would depend on the velocity of the flowing stream, v_∞, the diameter of the rod, D_0, the inside diameter of the outer pipe, D_0 the density and viscosity of the gas medium, and the diffusivity of naphthalene in the gas medium. Arrange these variables into dimensionless groups that the investigators might elect to use to correlate their data.

28.8 A long cylinder of porous clay, initially having a uniform concentration of c_{Ao}, is suddenly inserted into an air stream that has a water content of $c_{A\infty}$. If the radius of the cylinder is r_o and the average mass-transfer coefficient from the cylinder into the air stream is k_c, show by means of dimensional analysis that the concentration profile within the cylinder can be expressed in terms of the parameters.

$$\frac{c_A(r) - c_{A\infty}}{c_{Ao} - c_{A\infty}} \quad \frac{r}{r_o} \quad \frac{D_{AB}}{k_c r_o} \quad \text{and} \quad \frac{D_{AB} t}{r_o^2}$$

28.9 The boundary layer solution for a flat plate predicts the following equations:

for laminar flow: $\quad \dfrac{k_c x}{D_{AB}} = 0.332 \, \text{Re}_x^{1/2} \, \text{Sc}^{1/3}$

for turbulent flow: $\quad \dfrac{k_c x}{D_{AB}} = 0.0292 \, \text{Re}_x^{4/5} \, \text{Sc}^{1/3}$

with the transition occurring at $\text{Re}_x = 2 \times 10^5$.

Determine what percentage of the mass transfer occurs in the laminar zone of the flow over the flat plate if the Reynolds number at the end of the plate is $\text{Re}_L = 3 \times 10^6$.

28.10 Consider the chemical vapor deposition (CVD) process for the manufacture of solid silicon thin films as illustrated

above. A dilute mixture of 0.1 mol% silane, SiH_4 (species A), in H_2 gas (species B) enters the chamber and flows over a square Si wafer of 15 cm per side at a bulk velocity of 50 cm/s. At the surface of the Si wafer, the following reaction takes place:

$$SiH_4(g) \rightarrow Si(s) + 2H_2(g)$$

This reaction is very rapid; accordingly, mass transfer controls the Si film formation. Furthermore, the consumption of SiH_4 by the reaction is very small relative to the rate of SiH_4 delivered by the feed gas, so that the bulk gas composition is constant at 0.1 mol% SiH_4 in H_2. The process temperature is maintained at 900 K, and the total system pressure is maintained at 100 Pa. Under these conditions, the mass diffusivity, D_{AB}, equals 4035.5 cm²/s (due to the low system pressure), the viscosity of the hydrogen gas is 1.8×10^{-4} g/cm · s, and the density of the hydrogen gas is 2.67×10^{-8} g/cm³.

(a) What are the Schmidt and Sherwood numbers for this process?

(b) What is the total rate of Si formation on the whole wafer in g mol/min?

(c) Where (at what position x) on the surface would you expect the Si film to be thickest?

28.11 A thin film, 1.0 mm thick, coat of fresh paint has just been sprayed over a 1.5 m by 1.5 m square steel body part, which approximates a flat surface. The paint contains benzene as the solvent. The paint surface exerts a partial pressure of 0.137 atm at the process temperature of 27°C. Sixty cubic meters per minute of air are blown into the rectangular drying chamber as shown above; the chamber measures 1.5 m in length, 1.0 m in height, and 1.5 m in width. The temperature of the air is 27°C and the total system pressure is 1.0 atm.

Determine the average mass-transfer convective coefficient and the solvent evaporation rate from the surface in g/min, assuming that the concentration of the solvent vapor in the bulk gas is very small.

28.12 A thin polymer film contains some residual liquid solvent. It is desired to evaporate the solvent (*n*-hexane, solute A) from the polymer using the process shown below. Both sides of the polymer film

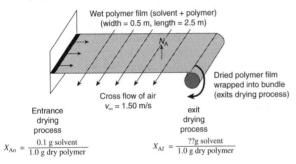

Wet polymer film (solvent + polymer)
(width = 0.5 m, length = 2.5 m)

N_A

Dried polymer film wrapped into bundle (exits drying process)

Cross flow of air
$v_\infty = 1.50$ m/s

Entrance drying process

exit drying process

$X_{Ao} = \dfrac{0.1 \text{ g solvent}}{1.0 \text{ g dry polymer}}$

$X_{Af} = \dfrac{??\text{g solvent}}{1.0 \text{ g dry polymer}}$

$\dot{m}_S = \dfrac{50.0 \text{ g polymer (dry basis)}}{s}$

are exposed to the cross flow of air. The evaporation rate of the solvent from the polymer film is limited by external convection. The dried polymer film is then rolled up into a bundle. During the drying process, the width of the thin polymer film is 0.5 m and the length of the polymer film is 2.5 m. The flowing air has a bulk velocity of 1.5 m/s, a temperature of 20°C, and the total system pressure of 1.0 atm. The wet polymer film is also maintained at 20°C. The vapor pressure of the solvent at 20°C is 0.16 atm, the diffusion coefficient of the solvent in air is 0.080 cm²/s at 20°C and 1.0 atm, and the molecular weight of the solvent is 86 g/g mol. The partial pressure of the solvent in the bulk air flow can be assumed to be near zero. The kinematic viscosity of air is 1.5×10^{-5} m²/s at 20°C.

(a) Evaluate the Sc number and the average Sh number for the solvent evaporation process.

(b) What is the total evaportion rate of the solvent from the 0.5 m by 2.5 m polymer film, recognizing that both sides of the film are exposed to the following air?

(c) The solvent loading in the polymer film at the entrance of the drying process is 0.1 g solvent per gram of dry polymer ($X_{Ao} = 0.1$ g solvent/g dry polymer). The total mass flow rate of the polymer film on a solvent-free dry polymer basis is $m_o = 50.0$ g dry polymer/s. What is the solvent loading in the polymer film exiting the drying process (X_{Af} in g solvent/g dry polymer)?

28.13 A container of acetone was accidently spilled, covering the top, smooth surface of a laboratory bench located in a semiconductor-fabrication building. The exhaust fan for the fabrication building produced a 6 m/s air flow parallel to the 1-m-wide bench surface. The air was maintained at 298 and 1.013×10^5 Pa. The vapor pressure of acetone at 298 K 3.066×10^4 Pa and the diffusivity of acetone in air at 298 K and 1.013×10^5 Pa is 0.93×10^{-5} m²/s.

(a) Determine the mass-transfer coefficient at 0.4 m downstream from the leading edge of the laboratory bench.

(b) Determine the amount of acetone evaporating per 1 m² of surface area each second.

28.14 A bulk gas stream containing 0.9 mol% of CO gas, 0.1 mol% of O_2 gas, and 99.0 mol% of CO_2 gas flows over a flat catalytic surface of length 3.0 m at a bulk velocity of 12.0 m/s. The catalytic surface promotes the reaction.

$$CO(g) + \frac{1}{2}O_2\,(g) \rightarrow CO_2(g)$$

The surface reaction is extremely fast so that the production CO_2 of is limited by the mass transfer of the gaseous reactants CO and O_2 to the catalytic surface. Let A = Co, B = O_2, and C = CO_2. At 27°C and 1.0 atm total system pressure, the following data are available:

- Diffusion coefficients: $D_{AB} = 0.213$ cm²/s, $D_{AC} = 0.155$ cm²/s $D_{BC} = 0.166$ cm²/s.

- Kinematic viscosities: $v_A = 0.158$ cm²/s, $v_B = 0.159$ cm²/s, $v_C = 0.0832$ cm²/s.

(a) What is the Schmidt number associated with O_2 mass transfer?

(b) The mass-transfer process is dominated by which of the following: (1) molecular diffusion; (2) convective mass transfer in the laminar region; or (3) convective mass transfer in the turbulent region?

(c) What is the average mass-transfer coefficient, $\bar{k}_c$, for the flat surface?

28.15 In a manufacturing process, an organic solvent (methyl ethyl ketone, MEK) is used to dissolve a thin coating of a polymer film away from a nonporous flat surface of length 20 cm and width 10 cm as shown in the figure below. The thickness of the polymer film is initially uniform at 0.2 mm. In the process, MEK solvent is added at a flow rate of 30 cm³/s to an open flat pan of length 30 cm and width 10 cm. The depth of the liquid MEK solvent in the pan is maintained at 2.0 cm. You may assume

30 cm³/s liquid MEK

Polymer film
(initially 0.2 mm thick)

2 cm

Exiting MEK

$x = 0$ $x = 20$ cm $x = 30$ cm

Length of pan = 30 cm, width of pan = 10 cm

that the concentration of the dissolved polymer in the bulk solvent is essentially zero, $c_{A\infty} = 0$, even though in reality the concentration of dissolved polymer in the solvent increases very slightly as it flows from the entrance to the exit of the pan. At the temperature and pressure of the process, the following data area available:

- Diffusion coefficient of dissolved polymer (solute A) in the MEK solvent (B) $D_{AB} = 3 \times 10^{-6}$ cm^2/s.
- Kinematic viscosity of liquid MEK $\nu_B = 6.0 \times 10^{-3}$ cm^2/s.
- Density of solid polymer film material $\rho_A = 1.05$ g/cm^3.
- Density of liquid MEK $\rho_B = 0.80$ g/cm^3.
- Maximum solubility of dissolved polymer in the MEK solvent $\rho_A^* = 0.04$ g/cm^3.

(a) Determine the average flux, g A/cm$^2\cdot$ s, from the surface.

(b) How long will it take for the polymer film to completely disappear?

28.16 A well-mixed open pond contains wastewater that is contaminated with a dilute concentration of dissolved methylene

Air flow
$V_\infty = 7.5$ m/s

Pond surface

$L_2 = 100$ m

$L_1 = 500$ m

chloride. The pool is rectangular with dimensions of 500 m by 100 m. Air 27°C at and 1.0 atm blows parallel to the surface of the pond as shown in the figure with a velocity of 7.5 m/s. At 20°C and 1.0 atm, the diffusion coefficient of methylene chloride in air $D_{AB} = 0.085$ cm^2/s and the kinematic viscosity of the air $\nu_B = 0.15$ cm^2/s.

(a) At what position across the pond is the air no longer laminar?

(b) Determine the mean gas-film mass-transfer coefficient associated with the transfer of methylene chloride.

(c) What is the Schmidt number for methylene cholride in the liquid phase if at 20°C the diffusion coefficient of methylene chloride in water is 1.07×10^{-5} cm^2/s and the kinematic viscosity of water is 0.010 cm^2/s?

28.17 A flat steel plate is 2.0 m length and 2.0 m width initially contains a very thin coating of lubricating oil used in a manufacturing process. An engineer is considering the feasibility of using hot forced air convection to remove the lubricating oil from the surface as an alternative to using harmful solvents to rinse the lubricating oil off the surface. In the present process, air at 386 K and 1.0 atm is blown parallel to the surface at a velocity of 50 m/s. At this velocity, the location on the plate where laminar flow ends is 0.097 m from the leading edge. The initial

thickness of the liquid lubricating oil coating the surface is 10.0 μm (1×10^{-5} m). AT 386 K, the viscosity of air is 2.23×10^{-5} kg/m $\cdot$ s, and the mass density of air is 0.917 kg/m^3. At 386 K, the lubricating oil is slightly volatile with a vapor pressure of 0.20 Pa, and the liquid density of the lubricating oil is 1900 kg/m^3. At 386 K and 1.0 atm, the molecular diffusion coefficient of the lubricating oil vapor in air, D_{AB}, is 0.040 cm^2/s.

(a) What is the Schmidt number for the mass transfer of lubricating oil vapor in air?

(b) What is the average mass-transfer coefficient over the entire length of the plate?

(c) Determine the mass-transfer coefficient at 1.2 m downstream from the leading edge of the plate.

28.18 If the local Sherwood number for the laminar layer that is formed over a flat plate is

$$Sh_x = 0.332\, Re_x^{1/2}\, Sc^{1/3}$$

and for the turbulent layer is

$$Sh_x = 0.0292\, Re_x^{4/5}\, Sc^{1/3}$$

evaluate

(a) The value of the mass-transfer coefficient at a point where the Reynolds number is 70,000.

(b) The average film-transfer coefficient, $\bar{k}_c$, for the flat plate from the leading edge to the point where the Reynolds number is 70,000.

(c) The value of the mass-transfer coefficient at a point where the Reynolds number is 700,000.

28.19 In using the von Kármán approximate method for analyzing the turbulent boundary layer over a flat plate, the following velocity and concentration profile were assumed:

$$v_x = \alpha + \beta y^{1/7}$$

and

$$c_A - c_{As} = \eta + \xi y^{1/7}$$

where the four constants α, β, η, and ξ were determined by the appropriate boundary conditions at the surface and at the outer edge of the two boundary layers.

(a) Find the four constants and the resulting velocity and concentration profiles.

(b) Upon the application of the von Kármán momentum integral equation, the following equation for the thickness of the turbulent boundary layer has been determined:

$$\delta = \frac{0.371x}{Re_x^{1/3}}$$

Use this relationship and the solution to von Kármán concentration integral equation when the Schmidt number equals 1 to obtain the following equation for the local mass-transfer coefficient:

$$k_c = 0.0289\, v_\infty (Re_x)^{-1/5}$$

28.20 Assume a linear velocity distribution and a linear concentration profile in the laminar boundary layer over a flat plate. Apply the boundary conditions for a laminar boundary layer and evaluate the velocity and concentration profiles. Are these linear profiles acceptable profiles? Give reasons for your decision.

28.21 A 1×10^{-2} m spherical pellet is sprayed with a very thin coat of paint. The paint contains a volatile solvent. To dry the pellet, a 300 K and 1.013×10^5 Pa air stream flows around it with a bulk velocity of 1 m/s. The estimated loading of the solvent in the wet paint is 0.12 g solvent/cm^3. Physical properties are

vapor pressure of the solvent	$= 1.27 \times 10^4$ Pa
mass diffusivity of solvent in air	$= 9.62 \times 10^{-6}$ m^2/s
kinematic viscosity of air	$= 1.569 \times 10^{-5}$ m^2/s
density of air	$= 1.177$ kg/m^3
thermal conductivity of air	$= 2.624 \times 10^{-2}$ J/m·s·K
thermal diffusivity of air	$= 2.216 \times 10^{-5}$ m^2/s
heat capacity of air	$= 1.006$ J/g·K
molecular weight of the solvent	$= 78$ g/g mole

Use the McAdam's[11] equation

$$Nu = 0.37(\text{Re}_{dp})^{0.6}(\text{Pr})^{1/3}, \quad \text{where} \quad \text{Re}_{dp} = \frac{d_p v_\infty}{\nu}$$

to evaluate

(a) the heat-transfer coefficient, h;

(b) the mass-transfer coefficient, k_c;

(c) the molar flux of the solvent into the air stream.

28.22 Davis[12] investigated heat transfer to fluids flowing in the annular section between concentric tubes. He proposed the following correlation for the heat-transfer coefficient when considering the film on the inner tube:

$$\frac{hd_p}{k} = 0.031(\text{Re}_{d_1})^{0.8}(\text{Pr})^{1/3}\left(\frac{\mu}{\mu_s}\right)^{0.14}\left(\frac{d_2}{d_1}\right)^{0.15}$$

where d_1 and d_2 are the outer diameter of the inner tube and the inner diameter of the outer tube, respectively, μ is the viscosity of the fluid at the bulk temperature of the fluid, and μ_S is the viscosity of the fluid at the heating surface temperature. Use this equation to predict the mass-transfer coefficient from a naphthalene rod used to form the inner tube of an annular duct.

28.23 McAdams presented the heat-transfer equation for the turbulent flow of gases past a single sphere as used in Problem 28.21

$$Nu = \frac{hd_p}{k} = 0.37(\text{Re}_{dp})^{0.6}(\text{Pr})^{1/3}, \quad \text{where} \quad \text{Re}_{dp} = \frac{d_p v_\infty}{\nu}$$

where d_p is the diameter of the sphere.

Predict the equation you might use to correlate the mass-transfer coefficient from a single sphere into a turbulent gas stream. How would you modify your equation to cover the very low Reynolds number range?

28.24 Several thin sheets of naphthalene, 0.25 cm thick and 10 cm square, are arranged parallel to each other with their centers at 1-cm intervals. Air at 273 K and 1.013×10^5 Pa enters this sandwich arrangement with a bulk velocity of 15 m/s. At 273 K, the mass diffusivity for naphthalene in air is 5.14×10^{-6} m^2/s, the Schmidt number is 2.57, and the vapor pressure of naphthalene is 1.0 Pa. Determine the concentration of naphthalene in the air as it leaves the arrangement, evaluating the mass-transfer coefficient using the

(a) Reynolds analogy;

(b) von Kármán analogy;

(c) Chilton–Colburn analogy.

Determine the length of time the sheets must be exposed until half of their mass will be sublimed under these conditions.

28.25 A small droplet of liquid detergent, falling through air in a spray drying tower, has its diameter reduced as water evaporates from the surface. If we assume that the temperature of the liquid within the drop remains at 290 K and the dry air is at 310 K, determine the moisture composition of the drying medium. The following properties are available at the average temperature of the air, 300 K:

kinematic viscosity of air	$= 1.5689 \times 10^{-5}$ m^2/s
thermal diffusivity of air	$= 2.2156 \times 10^{-5}$ m^2/s
mass diffusivity of water in air	$= 2.63 \times 10^{-5}$ m^2/s
density of air	$= 1.177$ kg/m^3
thermal conductivity of air	$= 2.624 \times 10^{-2}$ J/m·s·K
heat capacity of air	$= 1006$ J/g·K
surface temperature of drop, T_s	$= 290$ K
latent heat of vaporization at T_s	$= 2461$ J/g
vapor pressure of water at T_s	$= 1.94 \times 10^3$ Pa

28.26 A "cooling bag," commonly used for storing water in hot, arid environments is made of a thin porous fabric. A small amount of water diffuses through the fabric and evaporates from the surface of the bag. The rate of evaporation is controlled by convective mass transfer from the outer surface of the fabric to the surrounding dry air. The energy for evaporation is supplied by the surrounding hot air. The evaporation of the water cools the remaining liquid water within the bag and a temperature driving force is established. Determine the temperature of the ambient air using mass-transfer considerations, if the following values hold:

surface temperature of the bag	$= 298$ K
heat capacity of air	$= 1$ J/g K
viscosity of air	$= 1.84 \times 10^{-4}$ g/cm·s
density of air	$= 1.17 \times 10^{-3}$ g/cm^3
thermal conductivity of air	$= 2.62 \times 10^{-4}$ J/cm·s·K
mass diffusivity of water in air	$= 3 \times 10^{-5}$ m^2/s
latent heat of vaporization of water at 293 K	$= 2.45$ kJ/g
vapor pressure of water at 293 K	$= 1.3 \times 10^3$ Pa
bulk velocity of air stream	$= 2.2 \times 10^{-1}$ m/s

[11] W. H. McAdams, *Heat Transmission*, 3rd edition, McGraw-Hill, New York, 1949.

[12] E. S. Davis, *Trans. Am. Soc. Mech. Eng.*, **65**, 755 (1943).

28.27 Dry air, flowing at a velocity of 1.5 m/s, enters a 6-m-long, 0.15-m-diameter tube at 310 K and 1.013×10^5 Pa. The inner surface of the tube is lined with a felt material (diameter-to-roughness ratio, D/e, of 10,000) that is continuously saturated with water at 290 K. Assuming constant temperature of the air and the pipe wall, determine the rate at which water must be added to keep the felt continuously saturated. It is important to realize that the bulk composition of the gas stream will be continuously increasing with the length.

28.28 Air passes through a naphthalene tube, that has an inside diameter of 2.5 cm, flowing at a bulk velocity of 15 m/s. The air is at 283 K and an average pressure of 1.013×10^5 Pa. Assuming that the change in pressure along the tube is negligible and that the naphthalene surface is at 283 K, determine the length of tube that is necessary to produce a naphthalene concentration in the exiting gas stream of 4.75×10^{-4} mol/m^3. At 283 K, naphthalene has a vapor pressure of 3 Pa and a diffusivity in air of 5.4×10^{-6} m^2/s.

Chapter 29

Convective Mass Transfer Between Phases

In Chapter 28, convective mass transfer within a single phase was considered; in this case, mass is exchanged between a boundary surface and a moving fluid and the flux is related to an *individual mass-transfer convective coefficient*. Many mass-transfer operations, however, involve the transfer of material between two contacting phases where the flux may be related to an *overall mass-transfer convective coefficient*. These phases may be a gas stream contacting a liquid stream or two liquid streams if they are immiscible. In this chapter, we shall consider the mechanism of steady-state mass transfer between the phases and the interrelations between the individual convective coefficients for each phase and the overall convective coefficient.

Chapter 30 presents empirical equations for the individual mass-transfer convective coefficients involved in the interphase transfer. These equations have been established from experimental investigations. Chapter 31 presents methods of applying interphase concepts to the design of mass-transfer equipment.

29.1 EQUILIBRIUM

The transport of mass within a phase, by either molecular or convective transport mechanisms, has been shown to be directly dependent upon the concentration gradient responsible for the mass transfer. When equilibrium within the system is established, the concentration gradient and, in turn, the net diffusion rate of the diffusing species becomes zero. Transfer between two phases also requires a departure from equilibrium that might exist between the average or bulk concentrations within each phase. As the deviations from equilibrium provides the concentration driving force within a phase, it is necessary to consider interphase equilibrium in order to describe mass transfer between the phases.

Initially, let us consider the equilibrium characteristics of a particular system and then we will generalize the results for other systems. Consider a two-phase system involving a gas contacting a liquid; for example, let the initial system composition include air and ammonia in the gas phase and only water in the liquid phase. When first brought into contact, some of the ammonia will be transferred into the water phase in which it is soluble and some of the water will be vaporized into the gas phase. If the gas–liquid mixture is contained within an isothermal, isobaric container, a dynamic equilibrium between the two phases will eventually be established. A portion of the molecules entering the liquid phase returns to the gas phase at a rate dependent upon the concentration of the ammonia in the liquid phase and the vapor pressure exerted by the ammonia in the aqueous solution. Similarly, a portion of the water vaporizing into the gas phase condenses into the solution. Dynamic equilibrium is

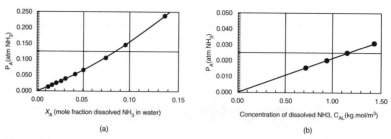

Figure 29.1 Ammonia solubility in water vs. partial pressure of ammonia at 30°C.

indicated by a constant concentration of ammonia in the liquid phase and a constant concentration or partial pressure of ammonia in the gas phase.

This equilibrium condition can be altered by adding more ammonia to the isothermal, isobaric container. After a period of time, a new dynamic equilibrium will be established with a different concentration of ammonia in the liquid and a different partial pressure of ammonia in the gas. Obviously, one could continue to add more ammonia to the system; each time a new equilibrium will be reached. Figure 29.1 (a) and (b) illustrates the equilibrium distribution of ammonia in the gas and liquid phases at 30°C.

Figure 29.1(a) presents the concentrations in terms of the partial pressure of the solute in the gas phase and the mole fraction of the dissolved solute in the liquid phase. Figure 29.1(b) presents the equilibrium distribution as the concentration of ammonia approaches zero, in terms of the partial pressure in the gas phase and the molar concentration in the liquid phase; in this dilute concentration range, the equilibrium distribution is linear and the two concentrations are related by Henry's law, equation (29-4). There are many graphical forms of equilibrium data due to the many ways of expressing concentrations in each of the phases. We will find use for many types of equilibrium plots in Chapter 31.

Equations relating the equilibrium concentrations in the two phases have been developed and are presented in physical chemistry and thermodynamic textbooks. For the case of nonideal gas and liquid phases, the relations are generally complex. However, in cases involving ideal gas and liquid phases, some fairly simple yet useful relations are known. For example, when the liquid phase is ideal, Raoult's law applies

$$p_A = x_A P_A \tag{29-1}$$

where p_A is the equilibrium partial pressure of component A in the vapor phase above the liquid phase, x_A is the mole fraction of A in the liquid phase, and P_A is the vapor pressure of pure A at the equilibrium temperature. When the gas phase is ideal, Dalton's law is obeyed

$$p_A = y_A P \tag{29-2}$$

where y_A is the mole fraction of A in the gas phase and P is the total pressure of the system. When both phases are ideal, the two equations may be combined to obtain a relation between the concentration terms, x_A and y_A, at constant pressure and temperature, the combined Raoult–Dalton equilibrium law stipulates

$$y_A P = x_A P_A \tag{29-3}$$

Another equilibrium relation for gas and liquid phases where dilute solutions are involved is Henry's law. This law is expressed by

$$p_A = H c_A \tag{29-4}$$

where H is the Henry's law constant and c_A is the equilibrium composition of A in the dilute liquid phase. Table 25.1 lists Henry's constant for selected aqueous solutions.

An equation similar to Henry's law relation describes the partition of a solute between two immiscible liquids. This equation, the "distribution-law" equation is

$$c_{A,\text{liquid 1}} = K c_{A,\text{liquid 2}} \tag{29-5}$$

where c_A is the concentration of solute A in the specified liquid phase and K is the partition or distribution coefficient.

A complete discussion of equilibrium relations must be left to physical chemistry and thermodynamic textbooks. However, the following basic concepts common to all systems involving the distribution of a component between two phases are descriptive of interphase mass transfer:

1. At a fixed set of conditions, such as temperature and pressure, Gibbs's phase rule stipulates that a set of equilibrium relations exists, which may be shown in the form of an equilibrium distribution curve.

2. When the system is in equilibrium, there is no net mass transfer between the phases.

3. When a system is not in equilibrium, components or a component of the system will be transported in such a manner as to cause the system composition to shift toward equilibrium. If sufficient time is permitted, the system will eventually reach equilibrium.

The following examples illustrate the application of equilibrium relations for determining equilibrium compositions.

EXAMPLE 1

An exhaust stream from a semiconductor fabrication unit contains 3 mol % acetone and 97 mol % air. In order to eliminate any possible environmental pollution, this acetone-air stream is to be fed to a mass-transfer column in which the acetone will be stripped by a countercurrent, falling 293 K water stream. The tower is to be operated at a total pressure of 1.013×10^5 Pa. If the combined Raoult–Dalton equilibrium relation may be used to determine the distribution of acetone between the air and the aqueous phases, determine

(a) the mole fraction of acetone within the aqueous phase, which would be in equilibrium with the 3 mol % acetone gas mixture.

(b) the mole fraction of acetone in the gas phase, which would be in equilibrium with 20 ppm acetone in the aqueous phase.

At 293 K, the vapor pressure of acetone is 5.64×10^4 Pa.

(a) By Raoult–Dalton law when $y_A = 0.03$

$$y_A P = x_A P_A$$
$$(0.03)(1.013 \times 10^5 \text{ Pa}) = x_A (5.64 \times 10^4 \text{ Pa})$$
$$\text{or } x_A = 0.0539 \text{ mole fraction acetone}$$

(b) 20 ppm acetone in solution

$$= \frac{20 \text{ g acetone}}{999,980 \text{ g water}}$$

$$= \frac{20 \text{ g}/(58 \text{ g/mol})}{999,980 \text{ g}(18 \text{ g/mol})}$$

$$= 6.207 \times 10^{-6} \frac{\text{mol acetone}}{\text{mol water}}$$

For the dilute solution, th mole fraction of acetone will be

$$x_A = \frac{86.027 \times 10^{-6} \, \text{mol acetone}}{1.0 \, \text{mol water} + 6.027 \times 10^{-6} \, \text{mol acetone}} = 6.207 \times 10^{-6}$$

By Raoult-Dalton law

$$y_A P = x_A P_A$$
$$y_A(1.013 \times 10^5 \, \text{Pa}) = (6.207 \times 10^{-6})(5.64 \times 10^4 \, \text{Pa})$$
$$\text{or } y_A = 3.45 \times 10^{-6} \, \text{mole fraction acetone}$$

EXAMPLE 2

The Henry's law constant for oxygen dissolved in water is 4.06×10^9 Pa/(mol of O_2 per total mol of solution) at 293 K. Determine the solution concentration of oxygen in water that is exposed to dry air at 1.013×10^5 Pa and 293 K.

Henry's law can be expressed in terms of the mole fraction units by

$$p_A = H' x_A$$

where H' is 4.06×10^9 Pa/(mol of O_2/total mol of solution).

From example 24.1, we recognize that dry air contains 21 mol% oxygen. By Dalton's law

$$p_A = y_A P = (0.21)(1.013 \times 10^5 \, \text{Pa}) = 2.13 \times 10^4 \, \text{Pa}.$$

The equilibrium mole fraction of the liquid at the interface is computed by Henry's law

$$x_A = \frac{P_A}{H} = \frac{2.13 \times 10^4 \, \text{Pa}}{4.06 \times 10^9 \, \text{Pa/(mol } O_2/\text{mol soln)}}$$
$$= 5.25 \times 10^{-6}(\text{mol } O_2/\text{mol soln})$$

For one cubic meter of very dilute solution, the moles of water in the solution will be approximately

$$n_{\text{water}} = (1 \, \text{m}^3)(1 \times 10^3 \, \text{kg/m}^3)\left(\frac{1}{0.018 \, \text{kg/mol}}\right)$$
$$= 5.56 \times 10^4 \, \text{mol}$$

The total moles in the solution is essentially the moles of water because the concentration of oxygen is quite low. Accordingly, the moles of oxygen in one cubic meter of solution is

$$n_{\text{oxygen}} = (5.25 \times 10^{-6} \, \text{mol } O_2/\text{mol soln})(5.56 \times 10^4 \, \text{mol soln})$$
$$= 0.292 \, \text{mol of } O_2$$

The saturation concentration is

$$(0.292 \, \text{mol/m}^3)(0.032 \, \text{kg/mol}) = 9.34 \times 10^{-3} \, \text{kg } O_2/\text{m}^2 \, (9.34 \, \text{mg/L})$$

29.2 TWO-RESISTANCE THEORY

Many mass-transfer operations involve the transfer of material between two contacting phases. For example, in gas absorption, as illustrated in Figure 29.2, a solute is transferred from the gas phase into a liquid phase. The interphase transfer involves three transfer steps: (1) the transfer of mass from the bulk conditions of one phase to the interfacial surface, (2) the transfer across the interface into the second phase, and (3) the transfer to the bulk conditions of the second phase.

A two-resistance theory, initially suggested by Whitman,[1] is often used to explain this process. The theory has two principal assumptions: (1) the rate of mass transfer between the two phases is controlled by the rates of diffusion through the phases on each side of the interface and (2) no resistance is offered to the transfer of the diffusing component across the interface. The concentration gradient driving force required to produce the mass transfer of component A from the gas phase to the liquid phase, as illustrated in Figure 29.2, is graphically presented in Figure 29.3 with a partial pressure gradient from the bulk gas composition, $p_{A,G}$, to the interfacial gas composition, $p_{A,i}$, and a concentration gradient in the liquid from, $c_{A,i}$, at the interface to the bulk liquid concentration, $c_{A,L}$. On the basis of Whitman's second assumption of no resistance to mass transfer at the interfacial surface, $p_{A,i}$ and $c_{A,i}$, are equilibrium concentrations and are related by thermodynamic relations as discussed in Section 29.1. The interfacial partial pressure, $p_{A,i}$, can be less than, equal to or greater than the value of $c_{A,i}$, according to the equilibrium conditions of the temperature and pressure of the system.

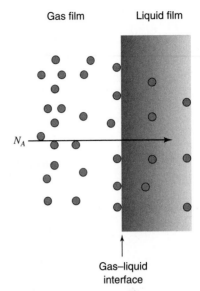

Figure 29.2 Gas absorption with solute A transferred from gas phase to liquid phase.

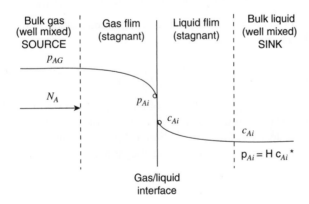

Figure 29.3 Concentration gradients between two contacting phases where solute is transferred from gas to liquid.

When the transfer is from the liquid phase, as in liquid stripping as shown in Figure 29.4, c_{AL}, will be greater than c_{Ai} and p_{A_i} will be greater than p_{AG}. The concentration gradients for this case are graphically presented in Figure 29.5.

Individual Mass-Transfer Coefficients

Restricting our discussion to the steady-state transfer of component A from the gas phase to the liquid phase (for the transfer in the opposite direction, the concentration driving forces

---

[1] W. G. Whitman, *Chem. Met. Engr.*, **29** (4), 197 (1923).

would be reversed; i.e., $p_{A,i} - p_{A,G}$ instead of $p_{A,G} - p_{A,i}$, we can describe the rates of diffusion in the z direction by the equations:

$$N_{A,z} = k_G(p_{A,G} - p_{A,i}). \quad (29\text{-}6)$$

and

$$N_{A,z} = k_L(c_{A,i} - c_{A,L}) \quad (29\text{-}7)$$

where k_G is the *convective mass-transfer coefficient in the gas phase*, in [moles of A transferred/(time)(interfacial area) (Δp_A units of concentration)]; and k_L is the *convective mass-transfer coefficient in the liquid phase*, in [moles of A transferred/(time)(interfacial area) (Δc_A units of concentration)].

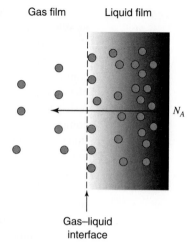

Gas film   Liquid film

N_A

Gas–liquid interface

Figure 29.4 Liquid stripping with solute A transferred from liquid to gas.

The partial pressure difference, $p_{A,G} - p_{A,i}$, is the driving force necessary to transfer component A from the bulk gas conditions to the interface separating the two phases. The concentration difference, $c_{A,i} - c_{A,L}$, is the driving force necessary to continue the transfer of A into the liquid phase.

Under steady-state conditions, the flux of mass in one phase must equal the flux of mass in the second phase. Combining equations (29-6) and (29-7), we obtain

$$N_{A,z} = k_G(p_{A,G} - p_{A,i}) = -k_L(c_{A,L} - c_{A,i}) \quad (29\text{-}8)$$

The ratio of the two convective mass-transfer coefficients may be obtained from equation (29-8) by rearrangement, giving

$$-\frac{k_L}{k_G} = \frac{p_{A,G} - p_{A,i}}{c_{A,L} - c_{A,i}} \quad (29\text{-}9)$$

In Figure 29.6, the application of equation (29-9) for the evaluation of the interfacial compositions for a specific set of bulk compositions as represented by point O is illustrated. The point O, located *above* the equilibrium line, represents conditions found at one plane in a gas absorber where the transfer is from the gas phase to the liquid phase. The bulk conditions

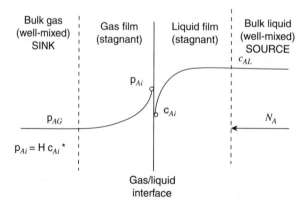

Bulk gas (well-mixed) SINK | Gas film (stagnant) | Liquid film (stagnant) | Bulk liquid (well-mixed) SOURCE

c_{AL}

p_{Ai}

c_{Ai}

p_{AG}

N_A

$p_{Ai} = H\,c_{Ai}{}^{*}$

Gas/liquid interface

Figure 29.5 Concentration gradients between two contacting phases where solute is transferred from liquid to gas.

Figure 29.6 Interfacial compositions as predicted by the two-resistance theory.

at another plane in the gas absorber could be quite different. A similar point representing the bulk conditions found in a liquid stripping tower (where transfer of the solute would be from the liquid phase to the gas phase) would be located below the equilibrium line.

In Table 29.1, the most often encountered individual-phase mass-transfer coefficients are listed and the interrelations between them are noted. A zero superscript on the mass-transfer coefficient for equimolar counterdiffusion is used to designate no net mass transfer into the phase, according to equation (26-24). It is important to realize that there are many other different mass-transfer coefficients for other specific mass-transfer situations; for example, when $N_A = -2N_B$, etc. This table may be helpful in explaining why there are so many different units given for individual coefficients.

Overall Mass-Transfer Coefficients

It is quite difficult to physically measure the partial pressure and the concentration at the interface. It is therefore convenient to employ overall coefficients based on an overall driving force between the bulk compositions, $p_{A,G}$ and $c_{A,L}$. This treatment is similar to the one used in Chapter 15 when the overall heat-transfer coefficient, U, was defined. Obviously, one cannot express the overall driving force as $p_{A,G} - c_{A,L}$ due to the difference in concentration units. In Figure 29.7, one observes the bulk liquid composition $c_{A,L}$ is in

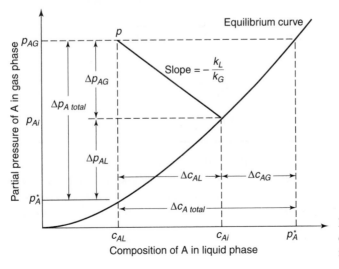

Figure 29.7 Concentration driving forces for the two-resistance theory.

Table 29.1 Individual mass-transfer coefficients

Gas phase

Rate equation		Units of coefficient
Diffusion of A through nondiffusing B	Equimolar counterdiffusion	
$N_A = k_G \Delta p_A$	$N_A = k_G^0 \Delta p_A$	$\dfrac{\text{moles of } A \text{ transferred}}{\text{(time)(area)(pressure)}}$
$N_A = k_c \Delta c_A$	$N_A = k_c^0 \Delta c_A$	$\dfrac{\text{moles of } A \text{ transferred}}{\text{(time)(area)(mol/volume)}}$
$N_A = k_y \Delta y_A$	$N_A = k_y^0 \Delta y_A$	$\dfrac{\text{moles of } A \text{ transferred}}{\text{(time)(area)(mole fraction)}}$
$N_A = k_Y \Delta Y_A$		$\dfrac{\text{moles of } A \text{ transferred}}{\text{(time)(area)(mole } A/\text{mol } B)}$
$n_A = k_{\mathcal{H}} \Delta \mathcal{H}_A$		$\dfrac{\text{mass of } A \text{ transferred}}{\text{(time)(area)(mass } A/\text{mass } B)}$

Gas phase

$k_G = \dfrac{k_y}{P} = \dfrac{k_c}{RT}$	$k_G^0 = \dfrac{k_G \, p_{B,\,lm}}{P}$
$k_G^0 = \dfrac{k_y^0}{P} = \dfrac{k_c^0}{RT}$	$k_y^0 = k_y \dfrac{p_{B,\,lm}}{P}$
$\mathcal{H} = \text{the specific humidity}$	$k_c^0 = k_c \dfrac{p_{B,\,lm}}{P} = k_c \dfrac{c_{B,\,lm}}{c}.$

Liquid phase

Rate equation		Units of coefficient
Diffusion of A through nondiffusing B	Equimolar counterdiffusion	
$N_A = k_L \Delta c_A$	$N_A = k_L^0 \Delta c_A$	$\dfrac{\text{moles of } A \text{ transferred}}{\text{(time)(area)(mole/volume)}}$
$N_A = k_x \Delta x_A$	$N_A = k_x^0 \Delta x_A$	$\dfrac{\text{moles of } A \text{ transferred}}{\text{(time)(area)(mole fraction)}}$
$k_L = \dfrac{k_x}{c}$	$k_L^0 = k_L \dfrac{c_{B,\,lm}}{c} = k_L x_{B,\,lm}$
$k_L^0 = \dfrac{k_x^0}{c}$	$k_x^0 = k_x x_{B,\,lm}$

equilibrium with the partial pressure p_A^*. This is an unique partial pressure at the pressure and temperature of the system; p_A^* is as good a measure of $c_{A,\,L}$ as $c_{A,\,L}$ itself, and it has units consistent with $p_{A,\,G}$. Accordingly, an *overall mass-transfer coefficient*, K_G which includes the resistance to diffusion in both phases in terms of partial pressure driving force, is defined by

$$N_A = K_G (p_{A,\,G} - p_A^*) \tag{29-10}$$

where $p_{A,G}$ is the bulk composition in the gas phase, p_A^* is the partial pressure of A in equilibrium with the bulk composition in the liquid phase, $c_{A,L}$, and K_G is the overall mass-transfer coefficient based on the partial pressure driving force, in mole of A transferred per (time)(interfacial area)(pressure).

Similarly, the overall bulk gas composition, $p_{A,G}$, is in equilibrium with the concentration c_A^*. This is also an unique concentration at the pressure and temperature of the system and c_A^* is as good a measure of $p_{A,G}$ as $p_{A,G}$ itself. An overall mass transfer coefficient, K_L which involves the resistance to diffusion in both phases and is in terms of the liquid phase concentration driving force, is defined by

$$N_A = K_L(c_A^* - c_{A,L}) \tag{29-11}$$

where c_A^* is the concentration of A in equilibrium with $p_{A,G}$, $c_{A,L}$ is the bulk composition in the liquid phase, and K_L is the overall mass-transfer coefficient based on a liquid driving force, in moles of A transferred per (time)(interfacial area)(mole A/volume).

Figure 29.7 illustrates the driving forces associated with each phase and the overall driving forces. The ratio of the resistances in an individual phase to the total resistance may be determined by

$$\frac{\text{resistance in the gas phase}}{\text{total resistance in both phases}} = \frac{\Delta p_{A,\text{gas film}}}{\Delta p_{A,\text{total}}} = \frac{1/k_G}{1/K_G} \tag{29-12}$$

and

$$\frac{\text{resistance in the liquid phase}}{\text{total resistance in both phases}} = \frac{\Delta c_{A,\text{liquid film}}}{\Delta c_{A,\text{total}}} = \frac{1/k_L}{1/K_L} \tag{29-13}$$

A relation between these overall coefficients and the individual phase coefficients can be obtained when the equilibrium relation is linear as expressed by

$$p_{A,i} = mc_{A,i} \tag{29-14}$$

This condition is always encountered at low concentrations, where Henry's law is obeyed; the proportionality constant is then the Henry's law constant, H. Utilizing equation (29-14), we may relate the gas- and liquid-phase concentrations by

$$\begin{aligned} p_{A,G} &= mc_A^* \\ p_A^* &= mc_{A,L} \end{aligned}$$

and

$$p_{A,i} = mc_{A,i}$$

Rearranging equation (29-10), we obtain

$$\frac{1}{K_G} = \frac{p_{A,G} - p_A^*}{N_{A,z}} = \frac{p_{A,G} - p_{A,i}}{N_{A,z}} + \frac{p_{A,i} - p_A^*}{N_{A,z}}$$

or in terms of m

$$\frac{1}{K_G} = \frac{(p_{A,G} - p_{A,i})}{N_{A,z}} + \frac{m(c_{A,i} - c_{A,L})}{N_{A,z}} \tag{29-15}$$

The substitution of equations (29-6) and (29-7) into the above relation relates K_G to the individual phase coefficients by

$$\frac{1}{K_G} = \frac{1}{k_G} + \frac{m}{k_L} \tag{29-16}$$

A similar expression for K_L may be derived as follows:

$$\frac{1}{K_L} = \frac{c_A^* - c_{A,L}}{N_{A,z}} = \frac{(p_{A,G} - p_{A,i})}{mN_{A,z}} + \frac{(c_{A,i} - c_{A,L})}{N_{A,z}}$$

or

$$\frac{1}{K_L} = \frac{1}{mk_G} + \frac{1}{k_L} \tag{29-17}$$

Equations (29-16) and (29-17) stipulate that the relative magnitudes of the individual phase resistances depend on the solubility of the gas, as indicated by the magnitude of the proportionality constant. For a system involving a soluble gas, such as ammonia in water, m is very small. From equation (29-16), we may conclude that the gas-phase resistance is essentially equal to the overall resistance in such a system. When this is true, the major resistance to mass transfer lies in the gas phase, and such a system is said to be *gas-phase controlled*. Systems involving gases of low solubility, such as carbon dioxide in water, have such a large value of m that equation (29-17) stipulates that the gas-phase resistance may be neglected, and the overall coefficient, K_L is essentially equal to the individual liquid-phase coefficient, k_L. This type of system is designated *liquid-phase controlled*. In many systems, both phase resistances are important and must be considered when evaluating the total resistance.

In Chapter 28, the individual phase convective coefficients, k_L and k_G were shown to be dependent on the nature of the diffusing component, on the nature of the phase through which the component is diffusing, and also on the flow conditions of the phase. Even when the individual coefficient, k_G is essentially independent of the concentration, the overall coefficient, K_G, may vary with the concentration unless the equilibrium line is straight. This is also true for the overall coefficient, K_L Accordingly, the overall coefficients should be employed only at conditions similar to those under which they were measured and should not be employed for other concentration ranges unless the equilibrium curve for the system is straight over the entire range of interest.

The two-resistance theory, including the addition of resistances, was proposed by Lewis and Whitman[2] in 1924 as the two-film theory. Although originally proposed in terms of the film model for convective mass transfer, it is equally applicable to the individual phase coefficients evaluated by either the film or the penetration theory. The assumption of negligible interfacial resistance has not been adequately verified; in fact, many investigators have shown that a resistance does exist if dust particles or other foreign particles are carried by the liquid. Nevertheless, most industrial data have been interpreted in terms of the two-resistance theory.

The application of the two-resistance theory for both absorption and stripping of a component will be illustrated in the following two examples.

EXAMPLE 3

In an experimental study of the absorption of ammonia by water in a wetted-wall column, the overall mass-transfer coefficient, K_G was found to be 2.74×10^{-9} kg mol/m$^2 \cdot$ s $\cdot$ Pa. At one point in the column, the gas phase contained 8 mol ammonia and the liquid-phase concentration was 0.064 kg mol ammonia/m^3 of solution. The tower operated at 293 K and 1.013×10^5 Pa. At that temperature, the Henry's law constant is 1.358×10^3 Pa/(kg mol/m^3). If 85% of the total resistance to mass transfer is encountered in the gas phase, determine the individual film mass-transfer coefficients and the interfacial compositions.

---

[2] W. K. Lewis and W. G. Whitman, *Ind. Eng. Chem.*, **16**, 1215 (1924).

The total resistance in both phases, according to equation (29-12), is

$$\frac{1}{K_G} = \frac{1}{2.74 \times 10^{-9} \dfrac{\text{kg mol}}{\text{m}^2 \cdot \text{s} \cdot \text{Pa}}} = 3.65 \times 10^8 \frac{\text{m}^2 \cdot \text{s} \cdot \text{Pa}}{\text{kg mol}}$$

As the resistance in the gas phase, $1/k_G$, is 85% of the total resistance, we may evaluate the individual gas-phase coefficient by

$$\frac{1}{k_G} = 0.85 \left(3.65 \times 10^8 \frac{\text{m}^2 \cdot \text{s} \cdot \text{Pa}}{\text{kg mol}} \right) = 3.10 \times 10^8 \frac{\text{m}^2 \cdot \text{s} \cdot \text{Pa}}{\text{kg mol}}$$

and

$$k_G = \frac{1}{3.10 \times 10^8} = 3.226 \times 10^{-9} \frac{\text{kg mol}}{\text{m}^2 \cdot \text{s} \cdot \text{Pa}}.$$

The liquid-phase coefficient, k_L is evaluated using equation (29-16)

$$\frac{1}{K_G} = \frac{1}{k_G} + \frac{H}{k_L}$$

$$3.65 \times 10^8 = 3.10 \times 10^8 + \frac{1.358 \times 10^3 \,\text{Pa}/(\text{kg mol}/\text{m}^3)}{k_L}$$

$$k_L = 2.47 \times 10^{-5} \,\text{kg mol}/\text{m}^2 \cdot \text{s} \cdot (\text{kg mol}/\text{m}^3)$$

At the stated point in the column

$$p_{A,G} = y_A P = (0.08)(1.013 \times 10^5 \,\text{Pa}) = 8.104 \times 10^3 \,\text{Pa}$$

$$c_{A,L} = 0.064 \,\text{kg mol}/\text{m}^3$$

Upon introducing Henry's law constant, we find the partial pressure, p_A^*, in equilibrium with the bulk liquid concentration

$$p_A^* = H c_{A,L} = \left(1.358 \times 10^3 \frac{\text{Pa}}{\text{kg mol}/\text{m}^3} \right) \left(0.064 \frac{\text{kg mol}}{\text{m}^3} \right) = 87.1 \,\text{Pa}$$

The mass flux, as expressed by equation (29-10), becomes

$$N_A = K_G(p_{A,G} - p_A^*)$$

$$= \left(2.74 \times 10^{-9} \frac{\text{kg mol}}{\text{m}^2 \cdot \text{s} \cdot \text{Pa}} \right) (8.104 \times 10^3 \,\text{Pa} - 87.1 \,\text{Pa})$$

$$= 2.20 \times 10^{-5} \frac{\text{kg mol}}{\text{m}^2 \cdot \text{s}}$$

The interfacial composition can be determined using equation (29-6)

$$N_A = k_G(p_{A,G} - p_{A,i})$$

$$2.20 \times 10^{-5} \frac{\text{kg mol}}{\text{m}^2 \cdot \text{s}} = \left(3.226 \times 10^{-9} \frac{\text{kg mol}}{\text{m}^2 \cdot \text{s} \cdot \text{Pa}} \right)(8.104 \times 10^3 \,\text{Pa} - p_{A,i})$$

$$p_{A,i} = 1284 \,\text{Pa}$$

and using Henry's law

$$p_{A,i} = H c_{A,i}$$

$$(1284 \,\text{Pa}) = \left(1.358 \times 10^3 \frac{\text{Pa}}{\text{kg mol}/\text{m}^3} \right) c_{A,i}$$

$$c_{A,i} = 0.946 \,\text{kg mol}/\text{m}^3$$

EXAMPLE 4

A wastewater stream is introduced to the top of a mass-transfer tower where it flows countercurrent to an air stream. At one point in the tower, the wastewater stream contains 1×10^{-3} g mol A/m^3 and the air is essentially free of any A. At the operating conditions within the tower, the film mass-transfer coefficients are $k_L = 5 \times 10^{-4}$ kg/mol/m$^2 \cdot$ s $\cdot$ (kg mol/m^3) and $k_G = 0.01$ kg mol/m$^2 \cdot$ s $\cdot$ atm. The concentrations are in the Henry's law region where $p_{A,i} = Hc_{A,i}$ with $H = 10$ atm/(kg mol/m^3). Determine

   (a) the overall mass flux of A.

   (b) the overall mass-transfer coefficients, K_L and K_G.

At the specified plane,

$$c_{A,L} = 1.0 \times 10^{-6} \frac{\text{kg mol A}}{\text{m}^3}$$

and

$$p_{A,G} = 0$$

A sketch of the partial pressure of A vs. concentration of A reveals this is a stripping operation.

By equation (29-17)

$$\frac{1}{K_L} = \frac{1}{Hk_G} + \frac{1}{k_L}$$

$$= \frac{1}{\left(10 \dfrac{\text{atm}}{\text{kg mol/m}^3}\right)\left(0.01 \dfrac{\text{kg mol}}{\text{m}^2 \cdot \text{s} \cdot \text{atm}}\right)} + \frac{1}{5 \times 10^{-4} \dfrac{\text{kg mol}}{\text{m}^2 \cdot \text{s} \cdot \text{kg mol/m}^3}}$$

$$K_L = 4.97 \times 10^{-4} \frac{\text{kg mol}}{\text{m}^2 \cdot \text{s} \cdot \text{kg mol/m}^3}$$

The equilibrium concentrations are

$$c_A^* = \frac{p_{A,G}}{H} = \frac{0 \text{ atm}}{10 \dfrac{\text{atm}}{\text{kg mol/m}^3}} = 0 \frac{\text{kg mol}}{\text{m}^3}$$

$$p_A^* = Hc_{A,L} = \left(10 \frac{\text{atm}}{\text{kg mol/m}^3}\right)\left(1 \times 10^{-6} \frac{\text{kg mol}}{\text{m}^3}\right)$$

$$= 1 \times 10^{-5} \text{ atm}$$

The flux of A for this stripping operation is

$$N_A = K_L(c_{A,L} - c_A^*) = \left(4.97 \times 10^{-4} \frac{\text{m}}{\text{s}}\right)\left(1.0 \times 10^{-6} \frac{\text{kg mol}}{\text{m}^3}\right)$$

$$= 4.97 \times 10^{-10} \frac{\text{kg mol}}{\text{m}^2 \cdot \text{s}}$$

The overall mass-transfer coefficient, K_G can be determined in two ways.

$$K_G = \frac{N_A}{p_A^* - p_{A,G}} = \frac{4.97 \times 10^{-10} \dfrac{\text{kg mol}}{\text{m}^2 \cdot \text{s}}}{1 \times 10^{-5} \text{ atm} - 0}$$

$$= 4.97 \times 10^{-5} \frac{\text{kg mol}}{\text{m}^2 \cdot \text{s} \cdot \text{atm}}$$

If we multiply both sides of equation (29-17) by Henry's law constant, H, and relate the results to equation (29-16), we obtain

$$\frac{H}{K_L} = \frac{1}{k_G} + \frac{H}{k_L} = \frac{1}{K_G}$$

then

$$K_G = \frac{K_L}{H} = \frac{4.97 \times 10^{-4} \dfrac{\text{kg mol}}{\text{m}^2 \cdot \text{s} \cdot (\text{kg mol/m}^3)}}{10 \, \text{atm/(kg mol/m}^3)}$$

$$= 4.97 \times 10^{-5} \frac{\text{kg mol}}{\text{m}^2 \cdot \text{s} \cdot \text{atm}}.$$

29.3 CLOSURE

In this chapter, we have considered the mechanism of steady-state mass transfer between phases. The two-resistance theory was presented. This theory defines the mass transfer in each phase as a function of the concentration driving force and the individual mass-transfer coefficient according to the equations

$$N_{A,z} = k_G(p_{A,G} - p_{A,i})$$

and

$$N_{A,z} = k_L(c_{A,i} - c_{A,L})$$

The overall mass-transfer coefficients were defined by

$$N_{A,z} = K_G(p_{A,G} - p_A^*)$$
$$N_{A,z} = K_L(c_A^* - c_{A,L})$$

and related to the individual coefficients by the relations

$$\frac{1}{K_G} = \frac{1}{k_G} + \frac{m}{k_L}$$

and

$$\frac{1}{K_L} = \frac{1}{mk_G} + \frac{1}{k_L}$$

PROBLEMS

29.1 Determine the value of Henry's law constant, in Pa/(kg/m^3) of chlorine for the chlorine–water system. The following equilibrium data at 293 K were reported in the *Chemical Engineering Handbook.*[3]

partial pressure of Cl_2, in Pa	666	1330	4000	6660	13,200
solubility, kg Cl_2/m^3	0.438	0.575	0.937	1.210	1.773

---

[3] J. H. Perry, *Chemical Engineering Handbookas*, Fifth Edition, McGraw-Hil Book Company, New York, 1973.

29.2 Plot the following experimental equilibrium values for trichloroethylene, TCE, in water at 20°C. Determine the Henry's law constant for this TCE–water system, with H in atm/(g mol/m^3).

partial pressure of TCE, atm	0.000	0.050	0.150	0.200
solubility, g mol/m^3	0.00	5.00	15.00	20.00

29.3 A liquid solution containing 49 moles of benzene and 21 moles of toluene is slowly heated to 363 K and a pressure of 1.013×10^5 Pa in a closed container.

a. What would be the composition of the liquid phase and the gas phase in equilibrium at 363 K and 1.013×10^5 Pa? At 363 K, the vapor pressure of benzene is 1.344×10^5 Pa and the vapor pressure of toluene is 5.38×10^5 Pa.

b. Determine the amount of liquid that would be present after this mixture is heated to 363 K.

29.4 In the aeration of wastewater, liquid-gas contact systems are designed to raise the concentration toward equilibrium levels. This goal is accomplished by dispersing air bubbles into the water. An aqueous solution, initially containing 2×10^{-3} kg $O_2/100$ kg H_2O is brought into contact with a large volume of ordinary air at 293 K and a total pressure of 1.013×10^5 Pa. At 293 K, the Henry's law constant for the oxygen–water system equals 4.06×10^9 Pa/mol fraction of oxygen in the liquid.

a. Will the solution gain or lose oxygen?

b. What will be the concentration of oxygen in the final equilibrium solution in kg O_2 per 100 kg H_2O?

29.5 To raise the oxygen concentration level in wastewater, air is injected through spargers located near the bottom of a water-holding, aeration tank. Oxygen is transferred from the released air bubbles into the surrounding aqueous phase. Determine the overall liquid mass-transfer coefficient, K_L, and the percent resistance encountered in the liquid phase if the individual mass-transfer coefficients of oxygen transferring from air into 293 K water are $k_L = 2.15 \times 10^{-5}$ kg mol/$m^2 \cdot s \cdot$ (kg mol/m^3) and $k_G = 9.28 \times 10^{-8}$ kg mol/$m^2 \cdot s \cdot$ Pa. The Henry's law coefficient for oxygen in water at 293 K is 4.06×10^9 Pa/(moles of oxygen/total moles of solution).

29.6 Consider the interphase mass-transfer process for the chlorine dioxide, ClO_2–air–water system. ClO_2 gas (solute A) is sparingly soluble in water. The Henry's law constant for the dilute solution of ClO_2 in water is 7.7×10^{-4} atm/(g mol/m^3). At the current conditions of operation, the mole fraction of ClO_2 in the bulk gas phase is $y_A = 0.040$ and the mole fraction of ClO_2 in the bulk liquid phase is $x_A = 0.00040$. The mass density of the liquid phase is 992.3 kg/m^3 and is not dependent on the very small amount of ClO_2, dissolved in it. The total system pressure is 1.5 atm.

a. Is the process gas absorption or liquid stripping?

b. If the ClO_2 partial pressure in the bulk gas phase is maintained at 0.06 atm, what is the maximum possible dissolved ClO_2 concentration, g mol/m^3, in the liquid phase that could possibly exit the process?

c. If $k_x = 1.0$ g mol/$m^2 \cdot s$ and $k_G = 0.010$ g mol/$m^2 \cdot s \cdot$ atm, what is K_y, the overall mass-transfer coefficient based upon the overall gas phase driving force?

d. Based upon the bulk gas and liquid phase compositions, what is the mass transfer flux for ClO_2 in units g mol/$m^2 \cdot s$?

29.7 In the absorption of component A from an air stream into an aqueous stream, the composition of the two adjacent streams were analyzed to be $c_{A,L} = 4.0$ kg mol/m^3 and $p_{A,G} = 1.013 \times 10^4$ Pa. The Henry's law constant for this system is 1.674×10^3 Pa/(kg mol/m^3). The overall liquid coefficient, K_L, was equal to 1.26×10^{-6} kg mol/$m^2 \cdot s \cdot$ (kg mol/m^3). If 53% of the total resistance of mass transfer is encountered in the liquid film, determine

a. the liquid-film coefficient, k_c;

b. the gas-film coefficient, k_G;

c. the concentration on the liquid side of the interface, $c_{A,L}$;

d. the mass flux of A.

29.8 In a stripping process for the removal of TCA from wastewater (very dilute, 50 mg TCA/l), air is bubbled into the water to transfer the TCA from the liquid phase to the gas phase. This process is carried out at 20°C and 1.25 atm. At 20°C, $H' = 400$ atm when Henry's law is defined by $p_A = H' x_A$ for the TCA–water system.

a. What would be the Henry's constant based on the mole fractions of TCA in gas and liquid?

b. What is the Henry's constant based on the partial pressure of TCA in the gas phase and the molar concentration of TCA (k mol/m^3) in the liquid phase?

c. If k_c in the gas film is 0.01 m/s, what is k_G? What is k_y?

d. If k_L in the liquid film is 0.01 m/s, what is k_x?

29.9 At a particular location in a countercurrent stripper for the removal of solute A from a liquid stream, the mole fraction of the transferring species A in the bulk gas phase is 0.01, and the mole fraction of solute A in the bulk liquid phase is 0.035. The total system pressure is 2.0 atm, and the temperature is 300 K. Eighty percent of the resistance to mass transfer is in the liquid phase. The equilibrium concentrations are given in the following x–y diagram:

a. Determine the interfacial compositions, $x_{A,i}$ and $y_{A,i}$.

b. If $k_y = 1.25$ g mol/$m^2 \cdot s \cdot$ (mole fraction), calculate the overall coefficient, K_y, for the gas phase at the operating point of the process.

c. Calculate the overall gas phase coefficient K_G and K_c.

29.10 At a particular location in a countercurrent absorber used to remove solute A from a gas stream, the mole fraction of the transferring species A in the bulk gas stream is 0.035 and the mole fraction of solute A in the bulk liquid phase is 0.01. The total system pressure is 2.0 atm and the temperature is 300 K. If 60% of the resistance is in the gas phase

a. Determine the interfacial compositions, $x_{A,i}$ and $y_{A,i}$. The gas–liquid equilibrium relationship is linear, as given by $y_A = 0.3\,x_A$.

b. If $k_y = 1.25\,\text{g mole/m}^2 \cdot \text{s} \cdot (\text{mole fraction})$, calculate the overall coefficient, K_y for the gas phase at the operating point of the process.

c. Calculate the overall coefficient K_x for the liquid phase at the operating point of the process.

29.11 A wetted-wall tower is used to "aerate" water using air at 2.0 atm total system pressure and 20°C. The molar composition of air is 21% O_2, 78% N_2, and 1% other gases. At 20°C, the Henry's law constant for dissolution of oxygen in water is 40,100 atm, and the mass density of liquid water is $1000\,\text{kg/m}^3$

a. What is the maximum mole fraction of oxygen that could be dissolved in the water?

b. What is the maximum molar concentration of oxygen that could be dissolved in the water?

c. If the total system pressure increases, will the dissolved oxygen concentration in the water (1) increase; (2) decrease; or (3) stay the same?

29.12 An interphase convective mass-transfer process involves the transfer of the industrial contaminant, methylene chloride (species A), between air and water at 20°C and 2.20 atm total system pressure. Air is the inert carrier gas, and water is the inert solvent. The equilibrium mole fraction of methylene chloride dissolved in water vs. the mole fraction of methylene chloride in air at 20°C and 2.2 atm total pressure is shown below.

Mole fraction of methylene chloride dissolved in water, x_A

At the present operating conditions, the bulk phase mole fraction of methylene chloride is 0.10 in the gas phase and 0.0040 in the liquid phase. The fluid flow rate associated with each case predicts a gas film mass transfer coefficient, k_y, to be $0.010\,\text{g mol/m}^2 \cdot \text{s}$ and a liquid film convective mass transfer

coefficient, k_x, to be $0.125\,\text{g mol/m}^2 \cdot \text{s}$. At 20°C, the density of liquid water is $992.3\,\text{kg/m}^3$.

a. Is the process (1) a liquid stripping process or (2) a gas absorption process?

b. Determine the Henry's law constant for methylene chloride dissolved in water, according to the definition $p_A = H\,c_A$.

c. Determine the overall coefficient K_L.

d. What is the flux of species A across the gas and liquid phases?

e. Determine the interface mole fractions $x_{A,i}$ and $y_{A,i}$.

29.13 It is desired to recover hexane vapor from air using an absorption process. The absorption solvent is a nonvolatile mineral oil, which has a mass density of $0.80\,\text{g/cm}^3$ and a molecular weight of 180 g/g mole. In the dilute concentration range, the equilibrium relationship for the dissolution of hexane vapor in the mineral oil is described by

$$p_A = 0.15\,x_A \quad \text{where} \quad p_A \text{ has the units of atm.}$$

At the present conditions of operation, the hexane partial pressure in the bulk gas stream is 0.015 atm, and the dissolved hexane in the bulk absorption solvent is 5.0 mol%. The total system pressure is 1.5 atm and the temperature is 20°C. The liquid film mass transfer coefficient, k_x, is $0.01\,\text{kg mol/m}^2 \cdot \text{s}$, and the gas film mass transfer coefficient, k_y, is $0.02\,\text{kg mol/m}^2 \cdot \text{s}$.

a. What is the overall mass-transfer coefficient based on the liquid phase, K_L?

b. What is the composition of hexane at the gas–liquid interface, in terms of $p_{A,i}$ and $x_{A,i}$?

29.14 An absorption tower, operating at 20°C and 1.0 atm, is used to absorb sulfur dioxide from an air mixture into water. At one point in the absorber, the partial pressure of the SO_2 in the gas stream is $4 \times 10^3\,\text{Pa}$ and the concentration of the contacting liquid stream is $0.55\,\text{kg mol/m}^3$. The individual gas film mass-transfer coefficient is $k_G = 3.95 \times 10^{-9}\,\text{kg mol/m}^2 \cdot \text{s} \cdot Pa$ and the individual liquid film mass-transfer coefficient is $k_L = 1.1 \times 10^{-4}\,\text{kg mol/m}^2 \cdot \text{s} \cdot (\text{kg mol/m}^3)$. Equilibrium data at 20°C are follows:

partial pressure SO_2, Pa	67	426	1132	3466	7864
concentration, SO_2 kg mol/m^3		0.306	1.458	2.780	6.208	10.896

a. Evaluate the interfacial concentrations, $c_{A,i}$ and $p_{A,i}$.

b. Determine the values for the following quantities

Coefficient	Driving Force
$k_G = \rule{2cm}{0.4pt}$	$p_{AG} - p_{Ai} = \rule{2cm}{0.4pt}$
$k_L = \rule{2cm}{0.4pt}$	$c_{Ai} - c_{AL} = \rule{2cm}{0.4pt}$
$K_G = \rule{2cm}{0.4pt}$	$p_{AG} - p_A^* = \rule{2cm}{0.4pt}$
$K_L = \rule{2cm}{0.4pt}$	$c_A^* - c_{AL} = \rule{2cm}{0.4pt}$

c. Determine the percentage of the overall mass-transfer resistance in the gas film.

29.15 An engineer at a pulp plant is considering the feasibility of sparging a waste gas stream containing a small concentration of chlorine gas into a chlorine-bleaching unit in order to augment the chlorine requirements for the unit. In this processing unit, the waste gas stream at 1.013×10^5 Pa containing 0.2 mol% chlorines at the flow rate of the streams is bubbled countercurrent to the absorbing water stream. The gas film coefficient, k_y, is $1 \, \text{kg mol/m}^2 \cdot \text{h} \cdot (\Delta y \, \text{mole fraction})$ and the liquid-film coefficient, k_x, is $10 \, \text{kg mol/m}^2 \cdot \text{h} \cdot (\Delta x \, \text{mole fraction})$. The Henry's law constant, H, is 6.13×10^4 Pa/ (kg mol/m³). The gas stream, containing 0.2 mol% chlorine gas, is in contact with the aqueous stream containing $2.6 \times 10^{-3} \, \text{kg mol Cl}_2/\text{m}^3$. Determine

a. the overall coefficient, K_x;

b. the chlorine molar flux;

c. the interfacial compositions;

d. the percent resistance to mass transfer in the liquid phase.

29.16 A packed tower has been designed to strip component A from an aqueous stream into a counter-flowing air stream. At a given plane in the tower, the concentrations of the two adjacent streams are $p_{A,G} = 4 \times 10^3$ Pa and $c_{A,L} = 4 \, \text{kg mol/m}^3$ of solution. Under the given flow conditions, the overall gas mass-transfer coefficient, K_G is equal to $2.46 \times 10^{-8} \, \text{kg mol/m}^2 \cdot \text{s} \cdot \text{Pa}$ and 60% of the resistance to mass transfer is encountered in the gas phase. At the tower's operating condition of 290 K and 1.013×10^5 Pa, the system satisfies Henry's law with a Henry's law constant of $1400 \, \text{Pa/(kg mol/m}^3)$. Determine

a. the individual gas-film coefficient, k_G;

b. the individual liquid-film coefficient, k_L;

c. the interfacial gas concentration, p_{ai};

d. the overall liquid mass-transfer coefficient, K_L.

29.17 Chlorinated water for pulp bleaching is being prepared by absorbing chlorine gas into water within a packed tower operating at 293 K and 1.013×10^5 Pa pressure. At one point in the tower, the partial pressure of chlorine in the gas is 4.0×10^4 Pa and the concentration of chlorine in the aqueous stream is $1 \, \text{kg/m}^3$. Data on the solubility of chlorine in water at 293 K are given in Problem 29.1. If 75% of the resistance to mass transfer lies in the liquid phase, determine the interfacial concentrations.

29.18 Wastewater containing solute A at a concentration of $1.0 \times 10^{-3} \, \text{g mol/m}^3$ enters an open tank at a volumetric flow rate of $0.2 \, \text{m}^3/\text{min}$, and exits at the same rate. Solute A transfers from the wastewater in the open tank to the surrounding air which is essentially free of the solute. The total system pressure is 1.0 atm. The diameter of the cylindrical tank is 4 m, and the depth of the liquid in the tank is 1.0 m. At these conditions the individual liquid-film mass-transfer coefficient, k_L, is $5 \times 10^{-4} \, \text{kg mol/m}^2 \cdot \text{s} \cdot (\text{kg mol/m}^3)$ and the individual gas-film mass-transfer coefficient, k_G is $0.01 \, \text{kg mol/m}^2 \cdot \text{s} \cdot \text{atm}$. The concentrations are in the Henry's law region where $p_{Ai} = H c_{Ai}$ with H equal to

10 atm/ (kg mol/m³). What is the percent resistance to mass transfer in the liquid film?

Open tank
$p_A \approx 0$ atm, $P = 1.0$ atm

Waste water
$v_o = 0.2 \, \text{m}^3/\text{min}$
$C_{AL,o} = 0.001 \, \text{g mol/m}^3$

C_{AL} well mixed

Treated waste water
$v_o = 0.2 \, \text{m}^3/\text{min}$
$C_{AL} = ?$

29.19 Jasmone, $C_{11}H_{16}O$, a valuable intermediate in the fine chemical industry, is obtained from the jasmine plant. A common method of manufacture is to extract the plant material in water, and then use benzene to concentrate the jasmone in a simple liquid–liquid extraction process. Jasmone (species A) is 170 times more soluble in benzene than in water; thus,

$$c_A'(\text{benzene phase}) = 170 \, c_A''(\text{aqueous phase})$$

In a proposed extraction unit, the benzene phase is well mixed, with the film mass-transfer coefficient, $k_L' = 3.5 \times 10^{-6} \, \text{kg mol/ m}^2 \cdot \text{s} \cdot (\text{kg mol/m}^3)$ and the aqueous phase is also well mixed with its film mass-transfer coefficient, $k_L'' = 2.5 \times 10^{-5} \, \text{kg mol/m}^2 \cdot \text{s} \cdot (\text{kg mol/m}^3)$. Determine

a. the overall liquid coefficient, K_L', based on the benzene side;

b. the overall liquid coefficient, K_L'', based on the aqueous side;

c. the percent resistance to mass transfer encountered in the aqueous liquid film.

29.20 Consider the waste treatment operation proposed in the figure below. In this process, wastewater conraining a TCE concentration of $50 \, \text{g mol/m}^3$ enters a clarifier, which is essentially a shallow, well-mixed tank with an exposed liquid surface. The overall diameter is 20.0 m and the maximum depth of the liquid in the tank is 4.0 m. The clarifier is enclosed to contain the gases (often quite odorous) that are emitted from the wastewater. Fresh air is blown into this enclosure to sweep away the gases emitted from the clarifier and is then sent to an incinerator. The TCE content in the effluent gas is 4.0 mol%, whereas the TCE content in the effluent liquid phase is $10 \, \text{g mol TCE}/\text{m}^3$ liquid. The clarifier operates at 1.0 atm and a constant temperature of 20°C. In independent pilot

$D = 20$ m

Fresh air → 4.0 mol% TCE → Effluent air + 4.0 mol% TCE

$C_{AL,o} = 50 \, \text{g mole TCE}/\text{m}^3$
$v_o = ? \, \text{m}^3/\text{hr}$

$C_{AL,o} = 10 \, \text{gmol TCE}/\text{m}^3$

Effluent liquid
$C_{AL,o} = 10 \, \text{g mol TCE}/\text{m}^3$

plant studies for TCE, the liquid film mass-transfer coefficient for the clarifier was, $k_x = 200 \, \text{g mol/m}^2 \cdot \text{s}$, whereas the gas film mass-transfer coefficient for the clarifier was $k_y = 0.1\,\text{g mol/}$ $\text{m}^2 \cdot \text{s}$. Equilibrium data for the air–TCE–water system at 20°C are represented by Henry's law in the form $p_A = H' x_A$ with $H' = 550$ atm. The molar density of the effluent liquid is $66\,\text{g mol/m}^3$.

a. What is the overall mass-transfer coefficient based on the liquid phase, K_L?

b. What is the flux of TCE from the clarifier liquid surface?

c. What is the inlet volumetric flow rate of wastewater, in units of m^3/h, needed to ensure that the liquid effluent TCE concentration is $10\,\text{g mol TCE/m}^3$?

29.21 Ammonia, NH_3, and hydrogen sulfide, H_2S, must both be stripped from wastewater before it can be treated for reuse. Individual mass-transfer coefficients for ammonia transfer within a packed tower are

$$k_G = 3.20 \times 10^{-9}\,\text{kg mol/m}^2 \cdot \text{s} \cdot \text{Pa}$$

and

$$k_L = 1.73 \times 10^{-5}\,\text{kg mol/m}^2 \cdot \text{s} \cdot (\text{kg mol/m}^3).$$

As the molecular weight of ammonia and hydrogen sulfide are approximately the same, the value of the individual mass-transfer coefficients for hydrogen sulfide will be similar. With this assumption, evaluate and compare the overall mass-transfer coefficient for each gas, recognizing that Henry's law constant for ammonia is $1.36 \times 10^3\,\text{Pa/(kg mol/m}^3)$ and for hydrogen sulfide is $8.81 \times 10^5\,\text{Pa/(kg mol/m}^3)$.

29.22 In a wetted-wall tower where ammonia, NH_3, is stripped from an ammonia-water solution into an air stream, the overall gas coefficient, k_g is $3.12 \times 10^{-9}\,\text{kg mol/m}^2 \cdot \text{s} \cdot \text{Pa}$. At a plane in the tower, the bulk concentration of the falling aqueous stream is $4\,\text{kg mol/m}^3$ of solution and the partial pressure of ammonia in the rising gas stream is 3.04×10^3 Pa. For dilute solutions of ammonia in water the equilibrium partial pressure may be evaluated by Henry's law:

$$p_{Ai} = 1360\,\text{Pa/(kg mol/m}^3)c_{Ai}$$

If the gas phase comprises 75% of the total resistance to mass transfer, calculate

a. the individual gas-film coefficient, k_G;

b. the individual liquid-film coefficient, k_L;

c. the overall liquid-film coefficient, K_L;

d. the interfacial concentrations, p_{Ai} and c_{Ai}.

29.23 A mass-transfer process is used to remove ammonia, NH_3, solute A, from a mixture of NH_3 and air, using water as the solvent. The partial pressure of ammonia in the bulk gas phase is 0.2 atm, and the mole fraction of dissolved ammonia in the water is 0.04. The total system pressure is 2.0 atm, and the temperature is 30°C. The equilibrium distribution at 30°C is shown in the figure:

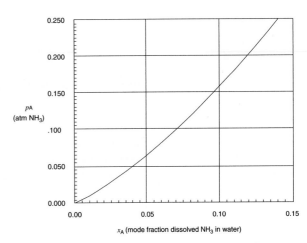

x_A (mode fraction dissolved NH_3 in water)

At 30°C, the molar solution density is $55.6\,\text{kg mol/m}^3$. The film mass-transfer coefficient are $k_G = 1.0\,\text{kg mol/m}^2 \cdot \text{s} \cdot \text{atm}$ and $k_L = 0.045$ m/s.

a. Evaluate k_x.

b. Determine the interfacial concentrations, c_{Ai} and p_{Ai}, What are x_{Ai} and y_{Ai}?

c. Evaluate the overall mass-transfer coefficient, K_G

d. What is the flux, N_A, for this process?

29.24 A packed absorption tower was used to absorb compound A from a gas mixture into solvent B. At one point in the tower, the partial pressure of A in the gas stream was 1.519×10^4 Pa and the concentration of A in the contacting liquid stream was $1.0 \times 10^{-3}\,\text{kg mol/m}^3$. The mass transfer between the gas stream and the liquid stream at that point in the tower was $4 \times 10^{-5}\,\text{kg mol/m}^2 \cdot \text{s}$.

The individual gas-film transfer coefficient, k_G was found to be $3.95 \times 10^{-9}\,\text{kg mol/m}^2 \cdot \text{s} \cdot \text{Pa}$. A laboratory experiment verified that the system satisfied Henry's law and that the liquid composition, $1 \times 10^{-3}\,\text{kg mol/m}^3$ was in equilibrium with a partial pressure of 3.04×10^3 Pa.

a. Determine values for the following table:

Coefficient		Driving Force	
$k_G =$		$p_{AG} - p_{Ai} =$	
$k_L =$		$c_{Ai} - c_{AL} =$	
$K_G =$		$p_{AG} - p_A^* =$	
$K_L =$		$c_A^* - c_{AL} =$	

b. Determine the percentage of the overall mass-transfer resistance in the gas film.

29.25 Ammonia, NH_3, in air is being absorbed into water within the enclosed tank shown in the figure below. The liquid and gas phases are both well mixed, and mass transfer occurs only at the exposed gas–liquid interface. The diameter of the cylindrical tank is 4 m, total liquid volume inside the tank is constant. The bulk gas pressure of ammonia is maintained at 0.02 atm, and the total gas pressure is constant at 1.0 atm. The system is isothermal at 20°C. The inlet volumetric flow rate of water is 200 L/h and enters NH_3 free. You may assume that

Henry's law, $p_{Ai} = Hc_{Ai}$ describes the equilibrium distribution of NH_3 between the gas and the liquid phases where $H = 0.02$ atm/(kg mol/m^3). The film mass-transfer coefficients are $k_G = 1.25$ kg mol/m$^2 \cdot$ h $\cdot$ atm and $k_L = 0.05$ kg mol/m$^2 \cdot$ h $\cdot$ (kg mol/m^3).

a. Determine the overall mass transfer coefficient, K_G.

b. Determine the partial pressure of NH_3 at the gas–liquid interface, p_{Ai}.

c. Determine W_A the total molar rate of ammonia transfer.

d. Develop a material balance for NH_3; then determine c_{Ai} the concentration of NH_3 in the outlet liquid stream.

Chapter 30

Convective Mass-Transfer Correlations

Thus far, we have considered convective mass transfer from an analytical viewpoint, and from relations developed for the analogous transport of momentum or convective heat. Although these considerations have given an insight into the mechanisms of convective mass transport, the validity of the analysis must be proven by comparison with experimental data. In this chapter, we will present dimensionless correlations for mass-transfer coefficients based upon experimental results. There will be no attempt to review all of the mass-transfer investigations, as reviews are available in several excellent references.[1] However, correlations will be presented to show that the forms of many of these equations are indeed predicted by the analytical expressions derived in Chapter 28. Additional correlations will be given for those situations that have not been successfully treated analytically.

In Chapter 28, several dimensionless numbers are introduced as important parameters used in correlating convective-transport data. Before presenting the convective mass-transfer correlations, let us summarize in Table 30.1 the dimensionless variables that have been used frequently in reported correlations. The Sherwood and Stanton numbers are dimensionless parameters involving the mass-transfer coefficient. The Schmidt, Lewis, and Prandtl numbers are required when two separate convective-transport processes are simultaneously involved, and the Reynolds, Peclet, and Grashof numbers are used to describe flow. The two *j*-factors are included in Table 30.1 because they are often used to develop a new correlation for mass transfer based on a previously established heat-transfer coefficient as discussed in Chapter 28.

30.1 MASS TRANSFER TO PLATES, SPHERES, AND CYLINDERS

Extensive data have been obtained for the transfer of mass between a moving fluid and certain shapes, such as flat plates, spheres, and cylinders. The techniques employed include sublimation of a solid, vaporization of a liquid into air, and the dissolution of a solid into

---

[1] W. S. Norman, *Absorption, Distillation and Cooling Towers,* John Wiley & Sons, New York, 1961; C. J. Geankoplis, *Mass Transfer Phenomena,* Holt, Rinehart, and Winston, New York, 1972; A. H. P. Skelland, *Diffusional Mass Transfer,* John Wiley & Sons, New York, 1974; T. K. Sherwood, R. L. Pigford, and C. R. Wilke, *Mass Transfer,* McGraw-Hill Inc., New York, 1975; L. J. Thibodeaux, *Chemodynamics—Environmental Movement of Chemicals in Air, Water and Soil,* John Wiley & Sons, New York, 1979; R. E. Treybal, *Mass Transfer Operations,* McGraw-Hill Book Company, New York, 1980; E. L. Cussler, *Diffusion-Mass Transfer in Fluid Systems,* Second Edition, Cambridge University Press, 1997; S. Middleman, *An Introduction to Heat and Mass Transfer,* John Wiley & Sons, New York, 1998.

Table 30.1 Dimensionless numbers used in correlating mass-transfer data
(L = Characteristic Length)

Name	Symbol	Dimensionless Group
Reynolds number	Re	$\dfrac{v_\infty \rho L}{\mu} = \dfrac{v_\infty L}{\nu}$
Sherwood number	Sh	$\dfrac{k_c L}{D_{AB}}$
Schmidt number	Sc	$\dfrac{\mu}{\rho D_{AB}} = \dfrac{\nu}{D_{AB}}$
Lewis number	Le	$\dfrac{\alpha}{D_{AB}} = \dfrac{k}{\rho c_p D_{AB}}$
Prandtl number	Pr	$\dfrac{\nu}{\alpha} = \dfrac{\mu c_p}{k}$
Peclet number	Pe_{AB}	$\dfrac{v_\infty L}{D_{AB}} = \text{Re Sc}$
Stanton number	St_{AB}	$\dfrac{k_c}{v_\infty}$
Grashof number	Gr	$\dfrac{L^3 \rho g \, \Delta\rho}{\mu^2}$
Mass transfer j-factor	j_D	$\dfrac{k_c}{v_\infty} (\text{Sc})^{2/3}$
Heat transfer j-factor	j_H	$\dfrac{h}{\rho c_p v_\infty} (\text{Pr})^{2/3}$

water. By correlating the data in terms of dimensionless parameters, these empirical equations can be extended to other moving fluids and geometrically similar surfaces.

Flat Plate

Several investigators have measured the evaporation from a free liquid surface or sublimation from a flat, volatile solid surface into a controlled air stream. Mass-transfer coefficients obtained from these experiments compare favorably with the mass-transfer coefficients theoretically predicted for laminar and turbulent boundary layers. The appropriate correlations are

$$\text{Sh}_L = \frac{k_c L}{D_{AB}} = 0.664 \, \text{Re}_L^{1/2} \, \text{Sc}^{1/3} \qquad \text{(laminar)} \, \text{Re}_L < 2 \times 10^5 \tag{28-21}$$

$$\text{Sh}_L = \frac{k_c L}{D_{AB}} = 0.0365 \, \text{Re}_L^{0.8} \, \text{Sc}^{1/3} \qquad \text{(turbulent)} \, \text{Re}_L > 2 \times 10^5 \tag{28-26}$$

with Re_L defind as

$$\text{Re}_L = \frac{\rho v_\infty L}{\mu}$$

where L is the characteristic length of the flat plate in the direction of flow. At a distance x from the leading edge of the flat plate, the exact solution to the laminar boundary layer problem resulting in the theoretical prediction for the local Sherwood number, given by

$$\text{Sh}_x = \frac{k_c x}{D_{AB}} = 0.332 \, \text{Re}_x^{1/2} \, \text{Sc}^{1/3} \tag{28-20}$$

also agrees with experimental data, where the local Reynolds number, Re_x, is defined as

$$Re_x = \frac{\rho v_\infty x}{\mu}$$

The above equations may also be expressed in terms of the j-factor by recalling

$$j_D = \frac{k_c}{v_\infty} Sc^{2/3} = \frac{k_c L}{D_{AB}} \cdot \frac{\mu}{Lv_\infty \rho} \cdot \frac{D_{AB}\rho}{\mu} \cdot \left(\frac{\mu}{\rho D_{AB}}\right)^{2/3} = \frac{Sh_L}{Re_L\, Sc^{1/3}} \tag{30-1}$$

Upon rearranging equations (28-21) and (28-26) into the form of equation (30-1), we obtain

$$j_D = 0.664\, Re_L^{-1/2} \quad \text{(laminar)}\, Re_L < 2 \times 10^5 \tag{30-2}$$

and

$$j_D = 0.0365\, Re_L^{-0.2} \quad \text{(turbulent)}\, Re_L > 2 \times 10^5 \tag{30-3}$$

These equations may be used if the Schmidt number is in the range $0.6 < Sc < 2500$. The j-factor for mass transfer is also equal to the j-factor for heat transfer in the Prandtl number range of $0.6 < Pr < 100$ and is equal to $C_f/2$.

Example 28.3 illustrates the use of the boundary-layer equations for evaluating the point value and the average value mass-t⸺ctive coefficients for a flow over a flat plate. In most situations, the hydrodynar⸺ntration boundary layers both start at the same position along x, the direction⸺. However, there are some situations where the hydrodynamic and the conc⸺ndary layers have different starting points, as illustrated in Figure 30.1. The⸺er a portion of an inert surface *before* flowing over a surface that can also ser⸺e or sink for mass transfer. Consequently, the hydrodynamic boundary la⸺ develop before the concentration boundary layer, so that the boundary con⸺oncentration of transferring species A becomes

$$0 \le x < ?$$
$$X \le x < ?$$

Also, consider a situation for flow over a fl⸺ there is an unheated starting length prior to the heated zone. In this case, the temperatures at the wall surface are

$$0 \le x < X, \qquad T_{wall} = T_\infty$$
$$X \le x < \infty, \qquad T_{wall} = T_s$$

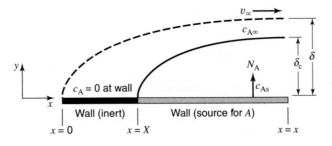

Figure 30.1 Laminar flow over a flat plate with the initiation of the hydrodynamic boundary layer before the concentration boundary layer.

and the hydrodynamic and thermal boundary layers have different starting points. The local Nusselt number for heat transfer is

$$Nu_x = 0.332 \, Re_x^{1/2} \left(\frac{Pr}{1 - \left(\dfrac{X}{x} \right)^{3/4}} \right)^{1/3} \qquad (30\text{-}4)$$

using the analytical approach outlined in Problem 19.7. From this result, the local Sherwood number for the analogous mass-transfer phenomenon is

$$Sh_x = 0.332 \, Re_x^{1/2} \left(\frac{Sc}{1 - \left(\dfrac{X}{x} \right)^{3/4}} \right)^{1/3} \qquad (30\text{-}5)$$

The following example illustrates the application of boundary-layer analysis to evaluate the mass-transfer coefficient for laminar flow over a flat plate, where the transport zone occurs at a short distance downstream from the leading edge of the flat plate.

EXAMPLE 1

A horizontal chemical vapor deposition (CVD) reactor for growth of gallium arsenide (GaAs) thin films is shown in Figure 30.2. In this process, arsine (AsH_3), trimethyl gallium ($Ga(CH_3)_3$), and H_2 gases are fed into the reactor. Inside the reactor, the silicon wafer rests on a heated plate called a susceptor. The reactant gases flow parallel to the surface of the wafer and deposit a GaAs thin film according to the simplified CVD reactions

$$2AsH_3(g) \rightarrow 2As(s) + 3H_2(g) \text{ and } 2Ga(CH_3)_3(g) + 3H_2(g) \rightarrow 2Ga(s) + 6CH_4(g)$$

If the process is considerably diluted in H_2 gas, then the mass transfer of each species in the H_2 carrier gas can be treated separately. The surface reaction is very rapid, and so the mass transfer of the gaseous reactants to the surface of the wafer limits the rate of GaAs thin film formation.

In the present process, the edge of a 10-cm silicon wafer is positioned 4 cm downstream of the leading edge of the susceptor plate. The wafer is inset within this plate so that a contiguous flat surface is maintained. The process temperature is 800 K, and the total system pressure 101.3 kPa (1 atm). Consider a limiting case where the flow rate of the H_2-rich feed gas to the reactor results in a bulk linear velocity of 100 cm/s, where trimethylgallium is present in dilute concentration. Determine the local mass-transfer coefficient (k_c) for trimethylgallium in H_2 gas at the center of the wafer using (a) boundary-layer theory and (b) film theory. The binary gas phase diffusion coefficient of trimethylgallium in H_2 is 1.55 cm²/s at 800 K and 1 atm.

At sufficiently high flowrate within the reactor, the physical system represents convective mass transfer over a flat plate, where the hydrodynamic boundary layer develops before the concentration

Figure 30.2 Horizontal CVD reactor.

boundary layer. Trimethylgallium diluted in H_2 gas serves as the source for mass transfer, and the reaction at the boundary surface is the sink for trimethylgallium gas.

(a) First, the local Reynolds number is evaluated at $x = 9$ cm, the distance from the edge of the susceptor (4 cm) to the middle of the wafer (5 cm). The kinematic viscosity of the feed gas is approximated by the properties of the H_2 gas, as trimethylgallium is present in only dilute concentration; from Appendix I, $\nu = 5.686$ cm^2/s at 800 K and 1 atm for H_2 gas. Therefore

$$Re_x = \frac{v_\infty x}{\nu} = \frac{(100 \text{ cm/s})(9 \text{ cm})}{5.686 \text{ cm}^2/\text{s}} = 158.3$$

and so the flow is laminar. The Schmidt number is

$$Sc = \frac{\nu}{D_{AB}} = \frac{5.686 \text{ cm}^2/\text{s}}{1.55 \text{ cm}^2/\text{s}} = 3.67$$

For the present system where the hydrodynamic boundary layer starts to develop before the concentration boundary layer, the local Sherwood number for laminar flow is

$$Sh_x = 0.332\, Re_x^{1/2}\left(\frac{Sc}{1 - \left(\dfrac{X}{x}\right)^{3/4}}\right)^{1/3} = 0.332(158.3)^{1/2}\left(\frac{3.67}{1 - \left(\dfrac{4 \text{ cm}}{9 \text{ cm}}\right)^{3/4}}\right)^{1/3} = 8.375$$

Finally, the local mass-transfer coefficient predicted by boundary-layer theory is

$$k_c = \frac{Sh_x}{x} D_{AB} = \left(\frac{8.375}{9 \text{ cm}}\right)\left(1.55 \frac{\text{cm}^2}{\text{s}}\right) = 1.44 \frac{\text{cm}}{\text{s}}$$

(b) For film theory, recall that the local mass-transfer coefficient is defined as

$$k_c = \frac{D_{AB}}{\delta_c}$$

For laminar flow, the hydrodynamic boundary-layer thickness is

$$\delta = \frac{5x}{\sqrt{Re_x}} = \frac{5(9 \text{ cm})}{\sqrt{158.3}} = 3.58 \text{ cm}$$

Now recall that the relationship between the hydrodynamic boundary-layer thickness and the concentration boundary-layer thickness is

$$\frac{\delta}{\delta_c} = Sc^{1/3} \tag{28-18}$$

for laminar flow over a flat plate where the hydrodynamic and concentration boundary layers have the same starting point. For laminar flow over a flat plate where the hydrodynamic boundary layer is initiated before the concentration boundary layer, the combination of equations (28-21), (28-26), and (30-5) yields

$$\frac{\delta}{\delta_c} = \left(\frac{Sc}{1 - \left(\dfrac{X}{x}\right)^{3/4}}\right)^{1/3} \tag{30-6}$$

Substituting in the appropriate values, we obtain

$$\frac{\delta}{\delta_c} = \left(\frac{Sc}{1 - \left(\dfrac{X}{x}\right)^{3/4}}\right)^{1/3} = \left(\frac{3.67}{1 - \left(\dfrac{4 \text{ cm}}{9 \text{ cm}}\right)^{3/4}}\right)^{1/3} = 2.004$$

and so

$$\delta_c = \frac{3.58\,\text{cm}}{2.004} = 1.79\,\text{cm}$$

Finally

$$k_c = \frac{D_{AB}}{\delta_c} = \frac{1.55\,\text{cm}^2/\text{s}}{1.79\,\text{cm}} = 0.868\,\text{cm/s}$$

We note that k_c predicted by film theory is 37% lower than k_c predicted by boundary-layer theory.

The preceding analysis is valid only in the limiting case where the thickness of the hydrodynamic boundary layer is significantly less than the height of the enclosure housing the susceptor plate. However, if the Reynolds number is sufficiently low, then the hydrodynamic boundary-layer thickness approaches the height of the enclosure. In this situation, the CVD reactor housing itself becomes the conduit for flow, and the analyses provided by Middleman and Hochberg[2] and Middleman[3] are recommended.

Single Sphere

The Sherwood number (Sh) and Reynolds number (Re) for a sphere are defined as

$$\text{Sh} = \frac{k_c D}{D_{AB}}$$

and

$$Re = \frac{\rho v_\infty D}{\mu}$$

where D is the diameter of the sphere, D_{AB} is the diffusion coefficient of the transferring species A in gaseous or liquid species B, v_∞ is the bulk fluid velocity flowing over the sphere, and ρ and μ are the density and viscosity of the fluid mixture, respectively, usually approximated as species B at dilute concentration of A. Mass-transfer correlations for single spheres consider the sum of the molecular diffusion and forced convection contributions

$$\text{Sh} = \text{Sh}_o + C\,\text{Re}^m\,\text{Sc}^{1/3}$$

where C and m are correlating constants. If there is no forced convection, then the Sherwood number is 2. This value can be derived theoretically by considering the molecular diffusion flux of species A from a sphere into an infinite sink of stagnant fluid B. Accordingly, the generalized equation becomes

$$\text{Sh} = 2 + C\,\text{Re}^m\,\text{Sc}^{1/3}$$

For mass transfer into liquid streams, the equation of Brian and Hales[4]

$$\text{Sh} = \frac{k_L D}{D_{AB}} = \left(4 + 1.21\,\text{Pe}_{AB}^{2/3}\right)^{1/2} \tag{30-7}$$

[2] S. Middleman and A. K. Hochberg, *Process Engineering Analysis in Semiconductor Device Fabrication*, McGraw-Hill Inc., New York, 1993.

[3] S. Middleman, *An Introduction to Heat and Mass Transfer*, John Wiley & Sons, New York, 1998.

[4] P. L. T. Brian and H. B. Hales, *A.I.Ch.E. J.*, **15**, 419 (1969).

correlates data where the mass-transfer Peclet number, Pe_{AB}, is less than 10,000. From Table 30.1, recall Pe_{AB} is the product of the Reynolds and Schmidt numbers, $Re \cdot Sc$. For Peclet numbers greater than 10,000, Levich[5] recommends the simpler relationship

$$Sh = \frac{k_L D}{D_{AB}} = 1.01 \, Pe_{AB}^{1/3} \qquad (30\text{-}8)$$

For mass transfer into gas streams, the Fröessling equation[6]

$$Sh = \frac{k_c D}{D_{AB}} = 2 + 0.552 \, Re^{1/2} \, Sc^{1/3} \qquad (30\text{-}9)$$

correlates the data at Reynolds numbers ranging from 2 to 800 and Schmidt numbers ranging from 0.6 to 2.7. Data of Evnochides and Thodos[7] have extended the Fröessling equation to a Reynolds number range of 1500 to 12 000 under a Schmidt number range of 0.6 to 1.85.

Equations (30-7)–(30-9) can be used to describe forced convection mass-transfer coefficients only when the effects of free or natural convection are negligible; that is

$$Re \geq 0.4 \, Gr^{1/2} \, Sc^{-1/6} \qquad (30\text{-}10)$$

The following correlation of Steinberger and Treybal[8] is recommended when the transfer occurs in the presence of natural convection

$$Sh = Sh_o + 0.347(Re \, Sc^{1/2})^{0.62} \qquad (30\text{-}11)$$

where Sh_o is dependent on $Gr \, Sc$

$$Sh_o = 2 + 0.569(GrSc)^{0.25} \qquad Gr \, Sc \leq 10^8 \qquad (30\text{-}12)$$

$$Sh_o = 2 + 0.0254(GrSc)^{1/3} \, (Sc)^{0.244} \quad Gr \, Sc \geq 10^8 \qquad (30\text{-}13)$$

From Table 30.1, the Grashof number is defined as

$$Gr = \frac{D^3 \rho g \, \Delta\rho}{\mu^2}$$

where density, ρ, and viscosity, μ, are taken at the bulk conditions of the flowing fluid, and $\Delta\rho$ is the positive density difference between the two phases in contact. The prediction for Sh is valid when $2 \leq Re \leq 3 \times 10^4$ and $0.6 \leq Sc \leq 3200$.

EXAMPLE 2	Estimate the distance a spherical drop of liquid water, originally 1 mm in diameter, must fall in quiet, dry air at 323 K in order to reduce its volume by 50%. Assume that the velocity of the drop is its terminal velocity evaluated at its mean diameter and that the water temperature remains at 293 K. Evaluate all gas properties at the average gas film temperature of 308 K.

The physical system requires a combined analysis of momentum and mass transport. The liquid water droplet is the source for mass transfer, the surrounding air serves as an infinite sink, and water vapor (species A) is the transferring species. The rate of evaporation is sufficiently small so that the water droplet is considered isothermal at 293 K; otherwise, a combined analysis of momentum, mass,

[5] V. G. Levich, *Physicochemical Hydrodynamics,* Prentice-Hall, Englewood Cliffs, NJ, 1962.

[6] N. Fröessling, *Gerlands Beitr. Geophys.,* **52,** 170 (1939).

[7] S. Evnochides and G. Thodos, *A.I.Ch.E. J.,* **5,** 178 (1960).

[8] R. L. Steinberger and R. E. Treybal, *A.I.Ch.E. J.,* **6,** 227 (1960).

and heat transport would be required! By considering a force balance on a spherical particle falling in a fluid medium, we can show that the terminal velocity of the particle is

$$v_o = \sqrt{\frac{4d_p(\rho_w - \rho_{air})g}{3C_D\rho_{air}}}$$

where d_p is the diameter of the particle, ρ_w is the density of the water droplet, ρ_{air} is the density of the surrounding fluid (air), g is the acceleration due to gravity, and C_D is the drag coefficient, which is a function of the Reynolds number of the spherical particle as illustrated in Figure 12.4. The arithmetic mean droplet diameter is evaluated by

$$\bar{d}_p = \frac{d_{p|t_1} + d_{p|t_2}}{2} = \frac{d_{p|t_1} + \left(\frac{1}{2}\right)^{1/3} \cdot d_{p|t_1}}{2}$$
$$= 0.897d_{p|t_1} = (0.897)(1 \times 10^{-3}\,\text{m}) = 8.97 \times 10^{-4}\,\text{m}$$

Hence, the arithmetic mean radius is equal to 4.48×10^{-4} m. At 293 K, the density of the water droplet (ρ_w) is 9.95×10^2 kg/m^3. At 308 K, the density of the air is 1.14 kg/m^3 and the viscosity of air is 1.91×10^{-5} Pa $\cdot$ s. Substitution of these values into the terminal velocity equation yields

$$v_o = \sqrt{\frac{(4)(8.97 \times 10^{-4}\,\text{m})(9.95 \times 10^2\,\text{kg/m}^3 - 1.14\,\text{kg/m}^3)(9.8\,\text{m/s}^2)}{(3)(1.14\,\text{kg/m}^3)C_D}} = \sqrt{\frac{10.22\,\text{m}^2/\text{s}^2}{C_D}}$$

By trial and error, guess a value for v_o, calculate a Reynolds number, and read C_D from Figure 12.4. Then, check the guessed value of v_o by the above equation. Guess $v_o = 3.62$ m/s. The Reynolds number is

$$\text{Re} = \frac{d_p v_o \rho_{air}}{\nu_{air}} = \frac{(8.97 \times 10^{-4}\,\text{m})(3.62\,\text{m/s})(1.14\,\text{kg/m}^3)}{1.19 \times 10^{-5}\,\text{Pa} \cdot \text{s}\left(\frac{\text{kg/m} \cdot \text{s}}{\text{Pa} \cdot \text{s}}\right)} = 194$$

and Figure 12.4, $C_D = 0.78$. Now recalculate v_o

$$v_o = \sqrt{\frac{10.22\,\text{m}^2/\text{s}^2}{C_D}} = \sqrt{\frac{10.22\,\text{m}^2/\text{s}^2}{0.78}} = 3.62\,\text{m/s}$$

Therefore, the guessed value for v_o is correct. The Schmidt number must now be calculated. From Appendix J.1, the gas diffusivity (D_{AB}) for water vapor in air at 298 K is 2.60×10^{-5} m^2/s, which is corrected to the desired temperature by

$$D_{AB} = (2.60 \times 10^{-5}\,\text{m}^2/\text{s})\left(\frac{308\,\text{K}}{298\,\text{K}}\right)^{3/2} = 2.73 \times 10^{-5}\,\text{m}^2/\text{s}$$

The Schmidt number is

$$\text{Sc} = \frac{\mu_{air}}{\rho_{air}D_{AB}} = \frac{(1.91 \times 10^{-5}\,\text{Pa} \cdot \text{s})\left(\frac{\text{kg/m} \cdot \text{s}}{\text{Pa} \cdot \text{s}}\right)}{(1.14\,\text{kg/m}^3)(0.273 \times 10^{-4}\,\text{m}^2/\text{s})} = 0.61$$

The Fröessling equation (30-9) can now be used to evaluate the mass-transfer coefficient for transfer of water vapor from the surface of the droplet to the surrounding air

$$\frac{k_c d_p}{D_{AB}} = 2 + 0.552\,\text{Re}^{1/2}\text{Sc}^{1/3}$$

or

$$k_c = \frac{D_{AB}}{d_p}(2 + 0.552\,Re^{1/2}Sc^{1/3})$$

$$= \frac{(0.273 \times 10^{-4}\,m^2/s)}{8.97 \times 10^{-4}\,m}(2.0 + 0.552(194)^{1/2}(0.61)^{1/3}) = 0.276\,m/s$$

The average rate of water evaporation from the droplet is

$$W_A = 4\pi\,\bar{r}_p^2\,N_A = 4\pi\,\bar{r}_p^2 k_c(c_{As} - c_{A\infty})$$

The dry-air concentration, $c_{A\infty}$, is zero, and the surrounding is assumed to be an infinite sink for mass transfer. The surface concentration is evaluated from the vapor pressure of water at 293 K

$$c_{As} = \frac{P_A}{RT} = \frac{2.33 \times 10^3\,Pa}{\left(8.314\dfrac{Pa \cdot m^3}{mol \cdot K}\right)(293\,K)} = 0.956\,\frac{mol}{m^3}$$

When we substitute the known values into the rate of evaporation equation, we obtain

$$W_A = 4\pi(4.48 \times 10^{-4}\,m)^2(0.276\,m/s)(0.956\,mol/m^3 - 0) = 6.65 \times 10^{-7}\,mol/s$$

or 1.2×10^{-8} kg/s on a mass basis. The amount of water evaporated is

$$m_A = \rho_w \Delta V = \rho_w(V_{t,1} - V_{t,2}) = \rho_w(V_{t,1} - 0.5V_{t,1}) = \frac{\rho_w V_{t,1}}{2}$$

$$= \frac{\rho_w}{2}\frac{4\pi}{3}r_p^{-3} = \frac{4\pi}{6}(9.95 \times 10^2\,kg/m^3)(4.48 \times 10^{-4}\,m)^3 = 1.87 \times 10^{-7}\,kg$$

The time necessary to reduce the volume by 50% is

$$t = \frac{m_A}{W_A} = \frac{1.87 \times 10^{-7}\,kg}{1.20 \times 10^{-8}\,kg/s} = 15.6\,s$$

and the distance of the fall is equal to $v_o t$ or 56.5 m.

Spherical Bubble Swarms

Consider a process where a gas is bubbled into a column of liquid. Usually the spherical gas bubbles are produced in swarms or clusters by the orifice that introduces the gas into the liquid.

Unlike the single, rigid sphere previously described, viscous circulation also occurs within the deformable bubble as it rises through the liquid. Consequently, a single-sphere correlation fails to describe the mass transport accurately in the vicinity of the gas–liquid interface of a rising bubble. Calderbank and Moo-Young[9] recommend the following two-

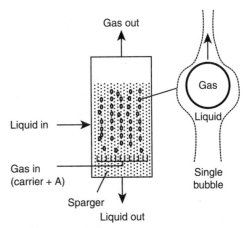

Figure 30.3 Gas dispersed in liquid column.

[9] P. H. Calderbank and M. Moo-Young, *Chem. Eng. Sci.*, **16**, 39 (1961).

point correlation for the mass-transfer coefficient associated with the transfer of a sparingly soluble gaseous solute A into solvent B by a swarm of gas bubbles in a natural convection process.

For gas bubble diameters (d_b) less than 2.5 mm, use

$$\text{Sh} = \frac{k_L d_b}{D_{AB}} = 0.31\,\text{Gr}^{1/3}\,\text{Sc}^{1/3} \tag{30-14a}$$

For bubble diameters greater or equal to 2.5 mm, use

$$\text{Sh} = \frac{k_L d_b}{D_{AB}} = 0.42\,\text{Gr}^{1/3}\,\text{Sc}^{1/2} \tag{30-14b}$$

In the above correlations, the Grashof number is defined as

$$\text{Gr} = \frac{d_b^3 \rho_L g\, \Delta\rho}{\mu_L^2}$$

where $\Delta\rho$ is the difference of the density of the liquid and the density of the gas inside the bubble, with density (ρ_L) and viscosity (μ_L) determined at the bulk average properties of the liquid mixture. For dilute solutions, the fluid properties of the solvent approximate the fluid properties of the liquid mixture. The diffusion coefficient D_{AB} is with respect to dissolved gaseous solute A in solvent B.

In order to relate flux N_A to transfer rate W_A using the above correlations, the gas holdup ratio must be known. The gas holdup ratio, ϕ_g, is defined as the volume of gas bubbles (V_g) per unit volume of liquid. Consequently, the interphase mass-transfer area per unit volume for bubbles of average diameter d_b is

$$\frac{A_i}{V} = \frac{V_g}{V} \cdot \frac{\text{bubble area}}{\text{bubble volume}} = \frac{6\phi_g}{d_b} \tag{30-15}$$

For gas-sparged vessels with no mechanical agitation, the gas holdup is proportional to the ratio of the superficial gas velocity and the terminal velocity of the rising bubble in the liquid, where the superficial gas velocity is volumetric gas flow rate per cross-sectional area of the empty vessel. The gas holdup is typically less than 0.2 for most sparging operations. Correlations for gas holdup are beyond the scope of this text, and the reader is encouraged to consult Treybal[10] for further information.

Single Cylinder

Several investigators have studied the sublimation from a solid cylinder into air flowing normal to its axis. Additional results on the dissolution of solid cylinders into a turbulent water stream have been reported. Bedingfield and Drew[11] correlated the available data to obtain

$$\frac{k_G P(\text{Sc})^{0.56}}{G_M} = \frac{k_c(\text{Sc})^{0.56}}{v_\infty} = 0.281(\text{Re}_D)^{-0.4} \tag{30-16}$$

which is valid for $400 < \text{Re}_D < 25{,}000$ and $0.6 < \text{Sc} < 2.6$. In this correlation, P is the system total pressure and G_M is the superficial molar velocity of the gas flowing normal to the cylinder in units of $\text{kg mol/m}^2 \cdot \text{s}$. The Reynolds number for flow normal to a solid

[10] R. E. Treybal, *Mass Transfer Operations*, McGraw-Hill Book Company, New York, 1980.

[11] C. H. Bedingfield and T. B. Drew, *Ind. Eng. Chem.*, **42**, 1164 (1950).

cylinder, Re_D, is defined as

$$Re_D = \frac{\rho v_\infty D}{\mu}$$

where D is cylinder diameter, v_∞ is the fluid velocity normal to the solid cylinder, and ρ and μ for the gas stream evaluated at the film average temperature.

The full analogy among momentum, heat, and mass transfer breaks down when the flow is around bluff bodies, such as spheres and cylinders. The total drag force includes the form drag in addition to the skin friction and so the j-factor will not equal $c_f/2$. However, the analogy between heat and mass transfer, $j_H = j_D$, still holds. Accordingly, the mass-transfer coefficient for a single cylinder that does not satisfy the specified ranges for equation (30-16) can be evaluated by using the Chilton–Colburn analogy and the appropriate heat-transfer relations described in Section 20.3.

EXAMPLE 3

In a humidification apparatus, liquid water flows in a thin film down the outside of a vertical, circular cylinder. Dry air at 310 K and 1.013×10^5 Pa (1 atm) flows at right angles to the 0.076-m diameter, 1.22-m-long vertically aligned cylinder at a velocity of 4.6 m/s. The liquid film temperature is 290 K. Calculate the rate at which liquid must be supplied to the top of the cylinder if the entire surface of the cylinder is to be used for the evaporating process and no water may drip off from the bottom of the cylinder.

The liquid film on the outside of the cylinder represents the source for mass transfer, and the air stream flowing normal to the cylinder represents an infinite sink. The properties of the air stream are evaluated at the film-average temperature of 300 K. The properties of air may be obtained from Appendix I, with $\rho = 1.1769$ kg/m^3 and $\nu = 1.5689 \times 10^{-5}$ m^2/s at 300 K and 1 atm. The Reynolds number is

$$Re_D = \frac{Dv_\infty}{\nu_{air}} = \frac{(0.076 \text{ m})(4.6 \text{ m/s})}{1.5689 \times 10^{-5} \text{ m}^2/\text{s}} = 22283$$

From Appendix Table J.1, the diffusivity of water in air at 298 K and 1 atm is 2.60×10^{-5} m^2/s, which corrected for temperature becomes

$$D_{AB} = (2.60 \times 10^{-5} \text{ m}^2/\text{s}) \left(\frac{300 \text{ K}}{298 \text{ K}}\right)^{3/2} = 2.63 \times 10^{-5} \text{ m}^2/\text{s}$$

The Schmidt number is

$$Sc = \frac{\nu_{air}}{D_{AB}} = \frac{1.5689 \times 10^{-5} \text{ m}^2/\text{s}}{2.63 \times 10^{-5} \text{ m}^2/\text{s}} = 0.6$$

The superficial molar velocity of the air normal to the cylinder is

$$G_M = \frac{v_\infty \rho_{air}}{M_{air}} = \frac{(4.6 \text{ m/s})(1.1769 \text{ kg/m}^3)}{29 \text{ kg/kg mol}} = 0.187 \frac{\text{kg mol}}{\text{m}^2 \cdot \text{s}}$$

Upon substitution of the known values into equation (30-16), we can solve for the gas-phase film mass-transfer coefficient

$$\frac{k_G P \, Sc^{0.56}}{G_M} = 0.281(Re_D)^{-0.4}$$

or

$$\frac{k_G(1.013 \times 10^5 \text{ Pa})(0.60)^{0.56}}{0.187 \dfrac{\text{kg mol}}{\text{m}^2 \cdot \text{s}}} = \frac{0.281}{(22283)^{0.4}}$$

Finally

$$k_G = 1.26 \times 10^{-8} \frac{\text{kg mol}}{\text{m}^2 \cdot \text{s} \cdot \text{Pa}}$$

The flux of water can be evaluated by

$$N_A = k_G(p_{A,i} - p_{A\infty})$$

The vapor pressure of water at 290 K is 1.73×10^3 Pa, and the partial pressure of the dry air ($p_{A\infty}$) is zero, as the surrounding air stream is assumed to be an infinite sink for mass transfer. Consequently,

$$N_A = \left(1.26 \times 10^{-8} \frac{\text{kg mol}}{\text{m}^2 \cdot \text{s} \cdot \text{Pa}}\right)(1.73 \times 10^3 \text{ Pa} - 0) = 2.18 \times 10^{-5} \frac{\text{kg mol}}{\text{m}^2 \cdot \text{s}}$$

Finally, the mass-feed rate of water for a single cylinder is the product of the flux and the external surface area of the cylinder is

$$W_A = N_A M_A(\pi D L) = \left(2.18 \times 10^{-5} \frac{\text{kg mol}}{\text{m}^2 \cdot \text{s}}\right)\left(18 \frac{\text{kg}}{\text{kg mol}}\right)(\pi)(0.076 \text{ m} \cdot 1.22 \text{ m})$$

$$= 1.14 \times 10^{-4} \text{ kg/s}$$

30.2 MASS TRANSFER INVOLVING FLOW THROUGH PIPES

Mass transfer from the inner walls of a tube to a moving fluid has been studied extensively. Most of the data have been obtained for vaporization of liquids into air, although some data have also been obtained for mass transfer of a soluble solid into a moving liquid, where the solid coats the inner surface of the tube. Gilliland and Sherwood[12] studied the vaporization of nine different liquids into air flowing through the inside of a tube and obtained the correlation

$$\frac{k_c D}{D_{AB}} \frac{p_{B,lm}}{P} = 0.023 \, \text{Re}^{0.83} \text{Sc}^{0.44} \qquad (30\text{-}17)$$

where

$$\text{Re} = \frac{\rho v_\infty D}{\mu}$$

and D is now the inner diameter of the pipe, $p_{B,lm}$ is the log mean composition of the carrier gas B, P is the total system pressure, D_{AB} is the mass diffusivity of the diffusing component A in the flowing carrier gas B. The Reynolds and Schmidt numbers are evaluated at the bulk conditions of the gas inside the pipe; for dilute solutions, the density and viscosity of the carrier gas can be assumed. The correlation is valid for gases where $2000 < \text{Re} < 35,000$ and $0.6 < \text{Sc} < 2.5$.

In a subsequent study, Linton and Sherwood[13] extended the range of the Schmidt number when they investigated the dissolution of benzoic acid, cinnamic acid, and β-naphthol in various solvents flowing through a tube. The combined experimental results of Gilliland and Sherwood and Linton and Sherwood were correlated by the relation

$$\text{Sh} = \frac{k_L D}{D_{AB}} = 0.023 \, \text{Re}^{0.83} \text{Sc}^{1/3} \qquad (30\text{-}18)$$

which is valid for liquids where $2000 < \text{Re} < 35000$ and $1000 < \text{Sc} < 2260$. The Reynolds and Schmidt numbers are evaluated at the bulk conditions of the liquid inside the pipe.

---

[12] E. R. Gilliland and T. K. Sherwood, *Ind. Eng. Chem.*, **26**, 516 (1934).
[13] W. H. Linton and T. K. Sherwood, *Chem. Eng. Prog.*, **46**, 258 (1950).

Again, for dilute solutions, the density and viscosity of the fluid approximates those properties of the solvent carrier B.

We note here that the Sherwood number for mass transfer (Sh) is analogous to the Nusselt number for heat transfer (Nu). The similarity between equation (30-18) and the Dittus–Boelter equation for energy transfer (20-26) illustrates the analogous behavior of these two transport phenomena.

For *laminar* flow of a fluid through a tube, with a Reynolds number range of $10 < \text{Re} < 2000$, the appropriate mass-transfer correlation is

$$\text{Sh} = 1.86\left(\frac{v_\infty D^2}{L D_{AB}}\right)^{1/3} = 1.86\left(\frac{D}{L}\frac{v_\infty D}{\nu} \cdot \frac{\nu}{D_{AB}}\right)^{1/3} = 1.86\left(\frac{D}{L}\text{Re Sc}\right)^{1/3} \quad (30\text{-}19)$$

where L the length of the pipe and v_∞ is the bulk average velocity. Equation (30-19) is analogous to the Sieder–Tate equation (20-25) for laminar flow heat transfer inside a tube. Furthermore, the Graetz solution for mass transfer inside a tube with laminar fluid flow is analogous to the Graetz solution for heat transfer provided in Section 20.2.

30.3 MASS TRANSFER IN WETTED-WALL COLUMNS

Much of the data for interphase mass transfer of a solute between gas and liquid carrier streams have been obtained using wetted-wall columns. In a wetted-wall column, gas flows into the bottom of the tube and moves upward, as shown in Figure 30.4. Liquid is loaded into the top of the column, and a weir evenly distributes the flow of liquid around the inner perimeter of the tube, forming a falling liquid film that evenly wets the inner surface of the tube down its length. The liquid film is often somewhat thin, often less than a few millimeters, and the liquid velocity is relatively high due to gravitational acceleration. There are two principal reasons for using wetted-wall columns in interphase mass-transfer investigations. First, the contacting area between the two phases can be accurately measured. Second, the experiments can be readily set up for steady-state operation.

Figure 30.4 Wetted-wall column for interphase mass transfer of gas and liquid.

$$\frac{k_L z}{D_{AB}} = 0.433(\text{Sc})^{1/2}\left(\frac{\rho_L^2 g z^3}{\mu_L^2}\right)^{1/6}(\text{Re}_L)^{0.4} \quad (30\text{-}20)$$

The convective mass-transfer coefficient for the gas film based on either turbulent or laminar flow is defined by the correlating equaions (30-18) and (30-19), respectively, with the Reynolds number of the gas flowing through the tube is

$$\text{Re}_t = \frac{\rho v_\infty D}{\mu}$$

where v_∞ and μ/ρ refer to the gas stream. However, for interphase mass-transfer analysis, the liquid film mass-transfer coefficient is also needed. A suitable correlation proposed by Vivian and Peaceman[14] for convective mass transfer of a gaseous solute into a falling liquid

[14] L. E. Scriven and R. L. Pigford, *A.I.Ch.E. J.*, **4**, 439 (1958).

film evenly wetting the inner surface of a tube is where z is the length of contact down the falling film, D_{AB} is the mass diffusivity of the diffusing component A into liquid solvent B, ρ_L is the density of the liquid, μ_L is the viscosity of liquid, g is the acceleration due to gravity, and Sc is the Schmidt number for the solute dissolved in the liquid evaluated at the liquid film temperature. The Reynolds number of the liquid flowing down the tube is defined as

$$\mathrm{Re}_L = \frac{4\Gamma}{\mu_L} = \frac{4w}{\pi D \mu_L}$$

where w is the mass flow rate of liquid, D is the inner diameter of the cylindrical column, and Γ is the mass flow rate of liquid per unit wetted perimeter of the column.

The liquid-film mass-transfer coefficients predicted by equation (30-20) were found to be 10–20% lower than the theoretical equation for absorption of a gaseous solute into a laminar falling liquid film, as discussed in Section 26.4. This may have been due to ripples along the liquid surface or to disturbances in the liquid flow at the two ends of the wetted-wall column. These discrepancies between the theoretical and measured rates of mass transfer have often led to the suggestion that a resistance to the mass transfer exists at the gas–liquid interface. However, investigations by Scriven and Pigford and others have substantiated that the interfacial resistance is negligible in normal interphase mass-transfer operations.

EXAMPLE 4

Trichloroethylene (TCE), a common industrial solvent, is often found at low concentrations in industrial waste waters. Stripping is a common process for removing sparingly soluble, volatile organic solutes such as TCE from aqueous solution. A wetted-wall column is used to study the stripping of TCE from water to air at a constant temperature of 293 K and total system pressure of 1 atm. The column inner diameter is 4 cm and the height is 2 m. In the present process, the volumetric air flow rate into the column is 2000 cm^3/s (2.0×10^{-3} m^3/s) and the volumetric flow rate of water is 50 cm^3/s (5×10^{-5} m^3/s). Estimate K_L, the overall liquid phase mass transfer coefficient for TCE across the liquid and gas film. Assume that water loss by evaporation is negligible.

Relevant physical property data are provided below. The process is very dilute so that the bulk gas has the properties of air and the bulk liquid has the properties of water. The equilibrium solubility of TCE in water is described by Henry's law of the form

$$p_A = H \cdot x_A$$

where H is 550 atm at 293 K. The binary gas phase diffusivity of TCE in air is 8.08×10^{-6} m^2/s at 1 atm and 293 K, as determined by the Fuller–Shettler–Giddings correlation. The binary liquid-phase diffusivity of TCE in water at 293 K is 8.9×10^{-10} m^2/s, as determined by the Hayduk–Laudie correlation.

With this physical property information in hand, our strategy is to estimate the gas film coefficient k_G, the liquid film coefficient k_L, and then the overall mass-transfer coefficient K_L. First, the bulk velocity of the gas is

$$v_\infty = \frac{4Q_g}{\pi D^2} = \frac{4 \cdot \left(2.0 \times 10^{-3} \dfrac{\mathrm{m}^3}{\mathrm{s}}\right)}{\pi (0.04 \,\mathrm{m})^2} = 1.59 \frac{\mathrm{m}}{\mathrm{s}}$$

The Reynolds number for air flow through the inside of the wetted wall column is

$$Re = \frac{\rho_{\mathrm{air}} v_\infty D}{\mu_{\mathrm{air}}} = \frac{\left(1.19 \dfrac{\mathrm{kg}}{\mathrm{m}^3}\right) \left(1.59 \dfrac{\mathrm{m}}{\mathrm{s}}\right) (0.04 \,\mathrm{m})}{1.84 \times 10^{-5} \dfrac{\mathrm{kg}}{\mathrm{m} \cdot \mathrm{s}}} = 4113$$

and the Schmidt number for TCE in air is

$$Sc = \frac{\mu_{\mathrm{air}}}{\rho_{\mathrm{air}} D_{\mathrm{TCE-air}}} = \frac{1.84 \times 10^{-5} \dfrac{\mathrm{kg}}{\mathrm{m} \cdot \mathrm{s}}}{\left(1.19 \dfrac{\mathrm{kg}}{\mathrm{m}^3}\right) \left(8.08 \times 10^{-6} \dfrac{\mathrm{m}^2}{\mathrm{s}}\right)} = 1.91$$

where the properties of air are found from Appendix I. As the gas flow is not laminar (Re > 2000), equation (30-17) is appropriate for estimation of k_c. Therefore

$$k_c = \frac{D_{AB}}{D} 0.023 \, \text{Re}^{0.83} \text{Sc}^{0.44} = \left(\frac{8.08 \times 10^{-6} \, \text{m}^2/\text{s}}{0.04 \, \text{m}} \right)$$

$$(0.023)(4113)^{0.83}(1.91)^{0.44} = 6.17 \times 10^{-3} \, \text{m/s}$$

for a dilute solution where $p_{B,lm}/P$ is essentially 1. The conversion to k_G is

$$k_G = \frac{k_c}{RT} = \frac{6.17 \times 10^{-3} \frac{\text{m}}{\text{s}}}{\left(0.08206 \frac{\text{m}^3 \cdot \text{atm}}{\text{kg mol} \cdot \text{K}} \right)(293 \, \text{K})} = 2.57 \times 10^{-4} \frac{\text{kg mol}}{\text{m}^2 \cdot \text{s} \cdot \text{atm}}$$

The liquid-film coefficient is now estimated. The Reynolds number for the falling liquid film is

$$\text{Re}_L = \frac{4\rho_L \dot{V}_L}{\pi D \mu_L} = \frac{4 \cdot \left(998.2 \frac{\text{kg}}{\text{m}^3} \right) \left(5 \times 10^{-5} \frac{\text{m}^3}{\text{s}} \right)}{\pi \cdot (0.04 \, \text{m}) \left(9.93 \times 10^4 \frac{\text{kg}}{\text{m} \cdot \text{s}} \right)} = 1600$$

and the Schmidt number is

$$\text{Sc} = \frac{\mu_L}{\rho_L D_{\text{TCE-H}_2\text{O}}} = \frac{\left(9.93 \times 10^{-4} \frac{\text{kg}}{\text{m} \cdot \text{s}} \right)}{\left(998.2 \frac{\text{kg}}{\text{m}^3} \right) \left(8.90 \times 10^{-10} \frac{\text{m}^2}{\text{s}} \right)} = 1118$$

where the properties of liquid water at 293 K are found in Appendix I. Equation (30-20) is appropriate for estimation of k_L:

$$k_L = \frac{D_{AB}}{z} 0.433(\text{Sc})^{1/2} \left(\frac{\rho_L^2 g z^3}{\mu_L^2} \right)^{1/6} (\text{Re}_L)^{0.4}$$

$$= \frac{8.9 \times 10^{-10} \frac{\text{m}^2}{\text{s}}}{2 \, \text{m}} 0.433 \cdot (1118)^{1/2}$$

$$\left(\frac{\left(998.2 \frac{\text{kg}}{\text{m}^3} \right)^2 \left(\frac{9.8 \, \text{m}}{\text{s}^2} \right)(2 \, \text{m})^3}{\left(9.93 \times 10^{-4} \frac{\text{kg}}{\text{m} \cdot \text{s}} \right)^2} \right)^{1/6} (1600)^{0.4} = 2.55 \times 10^{-5} \, \text{m/s}$$

Since the process is dilute, Henry's law constant in units consistent with k_L and k_G is

$$H = 550 \, \text{atm} \frac{M_{\text{H}_2\text{O}}}{\rho_{L,\text{H}_2\text{O}}} = (550 \, \text{atm}) \left(\frac{18 \, \text{kg}}{\text{kg mol}} \right) \left(\frac{1}{993.2 \, \text{kg/m}^3} \right) = 9.968 \frac{\text{atm} \cdot \text{m}^3}{\text{kg mol}}$$

The overall liquid mass-transfer coefficient, k_L, is

$$\frac{1}{K_L} = \frac{1}{k_L} + \frac{1}{Hk_G} = \frac{1}{2.55 \times 10^{-5} \frac{\text{m}}{\text{s}}} + \frac{1}{\left(9.96 \frac{\text{atm} \cdot \text{m}^3}{\text{kg mol}} \right) \left(2.57 \times 10^{-4} \frac{\text{kg mol}}{\text{m}^2 \cdot \text{s} \cdot \text{atm}} \right)}$$

or $k_L = 2.52 \times 10^{-5}$ m/s. As $K_L \cong k_L$, the process is liquid film mass-transfer controlling, which is characteristic of interphase mass-transfer processes involving a large value for Henry's law constant.

30.4   MASS TRANSFER IN PACKED AND FLUIDIZED BEDS

Packed and fluidized beds are commonly used in industrial mass-transfer operations, including adsorption, ion exchange, chromatography, and gaseous reactions that are catalyzed by solid surfaces. Numerous investigations have been conducted for measuring mass-transfer coefficients in packed beds and correlating the results. In general, the agreement among the investigators is poor, which is to be expected when one realizes the experimental difficulties. Sherwood, Pigford, and Wilke[15] presented a graphical representation of most of the data for mass transfer in packed beds with single-phase fluid and gas flows. They found that a single straight line through the experimental points did a fair job of representing all the data; this line is represented by a fairly simple equation

$$j_D = 1.17\,\text{Re}^{-0.415} \qquad 10 < \text{Re} < 2500 \tag{30-21}$$

where

$$\text{Re} = \frac{d_p u_{\text{ave}} \rho}{\mu}$$

u_{ave} = superficial fluid velocity

d_p = diameter of sphere having the same surface or volume as the particle

This equation may be employed for engineering estimates.

Most of the earlier correlations for packed beds failed to account for variations in the void fraction of the beds, ϵ, which in beds of spheres and pellets can range from 0.3 to 0.5. Mass transfer between liquids and beds of spheres was investigated by Wilson and Geankoplis[16] who correlated their data by

$$\epsilon\, j_D = \frac{1.09}{\text{Re}'''} \tag{30-22}$$

for $0.0016 < \text{Re}''' < 55$, $165 < \text{Sc} < 70\,600$ and $0.35 < \epsilon < 0.75$, and by

$$\epsilon\, j_D = \frac{0.25}{(\text{Re}''')^{0.31}} \tag{30-23}$$

for $55 < \text{Re}''' < 1500$ and $165 < \text{Sc} < 10\,690$. The Reynolds number, Re''', is defined in terms of the diameter of the spheres, d_p, and the superficial mass velocity of the fluid, G, in mass per unit time per unit cross section of the tower without packing. The void fraction in the packed bed is designated as ϵ, the volume void space between the solid particles divided by the total volume of void space plus the solid particles. These values range from about 0.3 to 0.5 in most packed beds. The correlation of Gupta and Thodos[17]

$$\epsilon\, j_D = \frac{2.06}{(\text{Re}''')^{0.575}} \tag{30-24}$$

[15] T. K. Sherwood, R. L. Pigford, and C. R. Wilke, *Mass Transfer,* McGraw-Hill Book Company, New York, 1975.

[16] E. J. Wilson and C. J. Geankoplis, *Ind. Eng. Chem. Fund.,* **5,** 9 (1966).

[17] A. S. Gupta and G. Thodos, *A.I.Ch.E. J.,* **9,** 751 (1963).

is recommended for mass transfer between gases and beds of spheres in the Reynolds number range $90 < \text{Re}''' < 4000$. Data above this range indicate a transitional behavior and are reported in a graphical form by Gupta and Thodos.[18]

Mass transfer in both gas and liquid fluidized beds of spheres has been correlated by Gupta and Thodos[19] with the equation

$$\epsilon\, j_D = 0.010 + \frac{0.863}{(\text{Re}''')^{0.58} - 0.483} \tag{30-25}$$

A detailed discussion of heat and mass transfer in fluidized beds is provided in the book by Kunii and Levenspiel.[20]

30.5   GAS–LIQUID MASS TRANSFER IN STIRRED TANKS

An important industrial process is the aeration of water, which is used in wastewater treatment and aerobic fermentation operations. Air is bubbled into the bottom of a vessel containing liquid water. Oxygen gas inside the air bubble absorbs into the water, where it is sparingly soluble. Usually, the air bubbles are produced in swarms or clusters by the gas sparger. In many gas–liquid mass-transfer operations of this type, a gas is sparged into a tank filled with liquid that is mixed with a rotating impeller. The stirred tank promotes gas–liquid contact by breaking up the rising gas bubbles released at the bottom of the tank and dispersing them throughout the liquid volume. Due to the continual collision of the gas bubbles resulting from gas sparging and mechanical agitation of the submerged impeller, the interfacial area for mass transfer is impossible to measure. Consequently, measured mass-transfer coefficients for aerated stirred tanks are reported as capacity coefficients, for example, $k_L a$, where the mass-transfer coefficient is lumped together with the parameter a, which is defined as

$$a = \frac{A_i}{V} = \frac{\text{area available for interphase mass transfer (m}^2)}{\text{liquid volume (m}^3)} \tag{30-26}$$

Capacity coefficients based on a ''concentration driving force'' mass-transfer coefficient (e.g., k_L) have units of reciprocal time, with a typical units conversion of the following:

$$k_L a = k_L \frac{A_i}{V} = \left(\frac{\text{m}}{\text{s}}\right)\left(\frac{\text{m}^2}{\text{m}^3}\right) = \text{s}^{-1}$$

Consequently, capacity coefficients cannot be used to calculate flux N_A directly. Instead, they are used to compute the total rate of interphase transfer W_A by

$$W_A = N_A \cdot \frac{A_i}{V} \cdot V = k_L a \cdot V (c_A^* - c_A) \tag{30-27}$$

Van't Riet[21] reviewed many studies of gas–liquid mass-transfer processes associated with oxygen transfer to low-viscosity liquids in agitated vessels. The following correlations for the liquid-film capacity coefficients are valid for the interphase mass transfer of oxygen into liquid water. For a stirred vessel of coalescing air bubbles, a suitable correlation is

$$(k_L a)_{\text{O}_2} = 2.6 \times 10^{-2} \left(\frac{P_g}{V}\right)^{0.4} (u_{gs})^{0.5} \tag{30-28}$$

[18] A. S. Gupta and G. Thodos, *Ind. Eng. Chem. Fund.*, **3**, 218 (1964).
[19] A. S. Gupta and G. Thodos, *A.I.Ch.E. J.*, **8**, 608 (1962).
[20] D. Kunii and O. Levenspiel, *Fluidization Engineering*, Wiley, New York (1969).
[21] K. Van't Riet, *Ind. Eng. Chem. Proc. Des. Dev.*, **18**, 357 (1979).

valid for $V < 2.6\,\text{m}^3$ of liquid and $500 < P_g/V < 10{,}000\,\text{W/m}^3$. For a stirred vessel of noncoalescing air bubbles, a suitable correlation is

$$(k_L a)_{O_2} = 2 \times 10^{-3} \left(\frac{P_g}{V}\right)^{0.7} (u_{gs})^{0.2} \qquad (30\text{-}29)$$

valid for $V < 4.4\,\text{m}^3$ of liquid and $500 < P_g/V < 10{,}000\,\text{W/m}^3$. In both correlations, the following units must be strictly followed: $(k_L a)_{O_2}$ is the liquid-phase film-capacity coefficient for O_2 in water in units of s^{-1}, P_g/V is the power consumption of the aerated vessel per unit *liquid* volume in units of W/m^3, and u_{gs} is the superficial velocity of the gas flowing through the *empty* vessel in units of m/s, which can be obtained by dividing the volumetric flow rate of the gas by the cross-sectional area of the vessel. These correlations agree with experimental data at ± 20 to 40% accuracy for equations (30-28) and (30-29), respectively, regardless of the type of impeller used (e.g., paddle, marine, or flat-blade disk turbine impeller).

The power input per unit liquid volume (P/V) is a complex function of impeller diameter (d_i), impeller rotation rate (N, revolutions per time), impeller geometry, liquid viscosity, liquid density, and aeration rate. A correlation shown in Figure 30.5 for the nonaerated, nonvortexing agitation of a Newtonian fluid provides a reasonable approximation for estimating the nonaerated power input P. In Figure 30.5, the impeller Reynolds number is defined as

$$\text{Re}_i = \frac{d_i^2 N \rho_L}{\mu_L} \qquad (30\text{-}30)$$

and the Power number P_o is defined as

$$P_o = \frac{P g_c}{\rho N^3 d_i^5} \qquad (30\text{-}31)$$

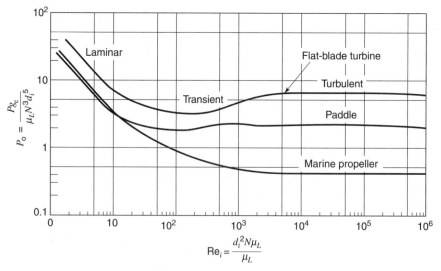

Figure 30.5 Power number vs. Reynolds number for impellers immersed in single-phase liquids.[22]

---

[22] J. H. Ruston, E. W. Costich, and H. J. Everett, *Chem. Eng. Prog.*, **46**, 467 (1950).

Aerating a stirred tank of liquid lowers the impeller power input. Nagata[23] suggests the following correlation for estimation of gased power input (P_g) as a function of gas volumetric flow rate (Q_g) for a flat-blade disk turbine impeller:

$$\log_{10}\left(\frac{P_g}{P}\right) = -192\left(\frac{d_i}{d_T}\right)^{4.38}\left(\frac{d_i^2 N \rho_L}{\mu_L}\right)^{0.115}\left(\frac{d_i N^2}{g}\right)^{1.96\left(\frac{d_i}{d_T}\right)}\left(\frac{Q_g}{d_i^3 N}\right) \tag{30-32}$$

where d_T is the diameter of the vessel. An alternative means to obtain P_g/V is to measure the power input into the stirred tank, but this is often not practical for equipment-design purposes.

30.6   CAPACITY COEFFICIENTS FOR PACKED TOWERS

Although the wetted-wall column has a definite interfacial surface area, the corresponding area in other types of equipment, which is described in Chapter 31, is virtually impossible to measure. For this reason, an engineering factor a must be introduced to represent the interfacial surface area per unit volume of the mass-transfer equipment. Both a and the mass-transfer coefficient depend on the physical geometry of the equipment and on the flow rates of the two contacting, immiscible streams; accordingly, they are normally correlated together as the *capacity coefficient, $k_c a$*. The units of $k_c a$ are moles of A transferred/ (h)(volume)(moles of A/volume). The capacity coefficient is encountered in the basic design equations of Chapter 31.

Empirical equations for capacity coefficients must be experimentally obtained for each type of mass-transfer operation. Such a correlation was obtained by Sherwood and Holloway[24] in the first comprehensive investigation of liquid-film mass-transfer coefficients in packed absorption towers. The experimental results for a variety of packings were represented by

$$\frac{k_L a}{D_{AB}} = \alpha\left(\frac{L}{\mu}\right)^{1-n}\left(\frac{\mu}{\rho D_{AB}}\right)^{0.5} \tag{30-33}$$

where $k_L a$ is the mass-transfer capacity coefficient, in lb mol/h ft^3 (lb mol/ft^3); L is the liquid rate, in lb/h ft^3; μ is the viscosity of the liquid, in lb/h ft; ρ is the density of the liquid, in lb/ft^3; and D_{AB} is the liquid mass diffusivity of component A in liquid B, in ft^2/h. The values of the constant α and the exponent n for various packing are given in Table 30.2.

Further correlations for capacity coefficients can be found in treatises on mass-transfer operations in the discussion of each specific operation and each specific type of tower.[25]

[23] S. Nagata, *Mixing: Principles and Application,* John Wiley & Sons, 1975.

[24] T. K. Sherwood and F. A. Holloway, *Trans. A.I.Ch.E.,* **36,** 21, 39 (1940).

[25] T. K. Sherwood, R. L. Pigford and C. R. Wilke, *Mass Transfer,* McGraw-Hill Book Company, New York, 1975; R. E. Treybal, *Mass Transfer Operations,* McGraw-Hill Book Company, New York, 1980; C. J. King, *Separation Processes,* McGraw-Hill Book Company, New York, 1971; W. S. Norman, *Absorption, Distillation and Cooling Towers,* Wiley, 1961; A. H. P. Skelland, *Diffusional Mass Transfer,* Wiley, New York, 1974.

Table 30.2 Packing coefficients for equation (30-33)

Packing	α	n
2-in. rings	80	0.22
$1\frac{1}{2}$-in. rings	90	0.22
1-in. rings	100	0.22
$\frac{1}{2}$-in. rings	280	0.35
$\frac{3}{8}$-in. rings	550	0.46
$1\frac{1}{2}$-in. saddles	160	0.28
1-in. saddles	170	0.28
$\frac{3}{8}$-in. saddles	150	0.28
3-in. spiral tiles	110	0.28

30.7 STEPS FOR MODELING MASS-TRANSFER PROCESSES INVOLVING CONVECTION

In many real processes, the flux is coupled to a material balance on the control volume of the physical system. Processes of this type are modeled similarly to the five-step procedure described earlier in Section 25.4.

Step 1: Draw a picture of the physical system. Label the important features, including the boundary surface where convective mass transfer occurs. Decide where the source and the sink for mass transfer are located.

Step 2: Make a "list of assumptions" based on your consideration of the physical system. Assumptions can be added as the model develops.

Step 3: Formulate material balances on the species undergoing mass transfer, and then incorporate the appropriate mass-transfer correlation(s) into the material balance. Processes dominated by convective mass transfer generally fall into two types: (1) the well-mixed control volume of uniform concentration of the transferring species, i.e., a stirred tank or (2) the differential control volume with a one-dimensional variation in concentration of the transferring species, i.e., flow through a conduit.

Once the material balance is set up, substitute the convective mass-transfer relationship, $N_A = k_c \Delta c_A$, into the material-balance model, and carefully define Δc_A, taking into consideration the concentrations of the transferring species in the fluid at the boundary surface and in the bulk phase. Finally, specify the appropriate correlation for k_c, keeping in mind restrictions on Re, Sc, geometry, and the phase of the transferring species.

Step 4: Recognize and specify the process boundary and initial conditions. These are distinct from the concentration values at the boundary surface for convective mass transfer, which should be specified in *Step 3*.

Step 5: Solve the algebraic or differential equation(s) resulting from the material balance(s) to obtain the concentration profile, flux, or other parameters of engineering interest. In many cases, k_c can be estimated beforehand.

The following examples illustrate how convective mass-transfer relationships are integrated into process-material balances.

EXAMPLE 5

One step in the manufacture of microelectronic devices is microlithography, which traces a microscopic-circuit pattern on the silicon wafer. In one typical process, a thin polymer film, typically less than a thickness 2 μm, is coated over the surface of the silicon wafer. A microscopic template, called a mask, is placed over the surface, and irradiated. Radiation that passes through the very tiny holes in the mask hits the photoresist. For a negative photoresist, the radiation initiates reactions that greatly increase the molecular weight of polymer, rendering the photoresist insoluble in an organic solvent. The unreacted photoresist is then dissolved away from the silicon wafer with an organic solvent, and the circuit pattern is revealed by the reacted, insoluble photoresist.

We are interested in using a "spinning disk" mass-transfer device shown in Figure 30.6 to study the photoresist-dissolution process within a closed tank of organic solvent. Consider a limiting case where all of the photoresist on surface of the wafer is soluble in the organic solvent. The negative photoresist is polystyrene (species A), and the organic solvent is methyl ethyl ketone (MEK, species B). The initial thickness of the photoresist coating (l_o) is 2 μm, the diameter of the wafer (d) is 10 cm, and the volume of the solvent in the tank (V) is 500 cm^3. If the dissolution process is controlled by the convective mass-transfer rate at the polymer-solvent interface, determine the time required to completely dissolve the photoresist if the disk rotates at 0.5 rev/s (30 rpm). The solubility limit of the developed photoresist in the solvent (c_A^*) is 0.04 g/cm^3, and diffusivity of the photoresist in the solvent at infinite dilution (D_{AB}) is 2.93×10^{-7} cm^2/s at a molecular weight of 7×10^5 g/mol, as reported by Tu and Ouano.[26] The viscosity of the solvent (μ) is 5×10^{-3} g/cm·s, the density of the solvent (ρ) is 0.805 g/cm^3, and the density of the solid polymer ($\rho_{A,\text{solid}}$) is 1.05 g/cm^3. All physical properties are valid at the process temperature of 298 K. The mass-transfer correlation for a spinning disk is given by

$$\frac{k_c d}{D_{AB}} = 0.62 \, \text{Re}^{1/2} \text{Sc}^{1/3}$$

with

$$\text{Re} = \frac{d^2 \omega}{\nu} \tag{30-34}$$

where ω is the angular rotation rate (radians/time) of the disk.

The strategy for solving this problem is to develop a material-balance model for the process, and then incorporate the appropriate mass-transfer correlation(s) into the material-balance calculations.

Figure 30.6 Dissolution of photoresist coating on spinning silicon wafer disk into methyl ethyl ketone (MEK) solvent.

---

[26] Y. O. Tu and A. C. Ouano, *IBM J. Res. Dev.,* **21**, 131 (1977).

The physical system represents a closed process where the polymer coating on the surface of the disk is the source for mass transfer, and the surrounding well-mixed organic solvent of constant volume is the sink for mass transfer. Under these assumptions, the unsteady-state material balance on the dissolved photoresist in the solvent phase of the well-mixed tank is

$$\begin{pmatrix} \text{rate of photoresist} \\ \text{added to solvent} \end{pmatrix} + \begin{pmatrix} \text{rate of photoresist} \\ \text{exiting solvent tank} \end{pmatrix} = \begin{pmatrix} \text{rate of accumulation of} \\ \text{photoresist within solvent} \end{pmatrix}$$

or

$$N_A \frac{\pi d^2}{4} - 0 = \frac{d(c_A V)}{dt}$$

where c_A represents the concentration of dissolved photoresist in the solvent at time t. The only input term is convective mass-transfer flux from the surface of the spinning disk (c_{As}) to the solvent (c_A)

$$N_A = k_c(c_{As} - c_A)$$

At the polymer-solvent interface, the dissolved concentration of the photoresist is at its solubility limit. Furthermore, as the source is a pure component, $c_A^* = c_{As}$ remains constant. The material balance reduces to

$$k_c(c_{As} - c_A) \frac{\pi d^2}{4} = V \frac{dc_A}{dt}$$

Separation of the dependent variable c_A from the independent variable t, followed by integration from the initial condition, $t = t_o$, $c_A = c_{Ao}$, to the final condition where all the photoresist is dissolved, $t = t_f$, $c_A = c_{Af}$, yields

$$-\int_{c_{Ao}}^{c_{Af}} \frac{-dc_A}{c_{As} - c_A} = \frac{k_c \pi d^2}{4V} \int_{t_o}^{t_f} dt$$

and finally

$$t_f - t_o = \frac{4V}{\pi d^2 k_c} \ln\left(\frac{c_{As} - c_{Ao}}{c_{As} - c_{Af}}\right)$$

The final concentration c_{Af} and convective mass-transfer coefficient, k_c, must now be determined. First, c_{Af} is estimated by an overall material balance for the photoresist on the solid wafer and dissolved in the solution, i.e.,

$$c_{Af} V - c_{Ao} V = m_{Ao} - m_{Af}$$

where m_A is the remaining mass of solid photoresist on the wafer, and m_{Ao} is the initial mass, given by

$$m_{Ao} = \rho_{A,\text{solid}} \frac{\pi d^2}{4} l_o$$

When all the photoresist is dissolved, $m_{Af} = 0$. If there is initially no photoresist dissolved in the solvent, then $c_{Ao} = 0$ and c_{Af} is

$$c_{Af} = \frac{m_{Ao}}{V} = \frac{\rho_{A,\text{solid}} \pi d^2 l_o}{4V} = \frac{\left(1.05 \frac{g}{cm^2}\right) \pi (cm)^2 \left(2.0 \, \mu m \frac{1 \, cm}{10^4 \, \mu m}\right)}{4 \cdot (500 \, cm^3)} = 3.3 \times 10^{-5} \frac{g}{cm^3}$$

The final concentration is well below the solubility limit of 0.04 g/cm^3, and so all of the photoresist will dissolve. In order to calculate k_c, Sc, and Re are needed. For a dilute system, the fluid properties are essentially the properties of the solvent, and so Sc and Re are

$$Sc = \frac{\mu}{\rho D_{AB}} = \frac{5 \times 10^{-5} \frac{g}{cm \cdot s}}{\left(0.805 \frac{g}{cm^3}\right)\left(2.93 \times 10^{-7} \frac{cm^2}{s}\right)} = 21\,199$$

$$Re = \frac{d^2 \omega \rho}{\mu} = \frac{(10\,cm)^2 \left(\frac{0.5\,rev}{s} \frac{2\pi\,rad}{rev}\right)\left(0.805 \frac{g}{cm^3}\right)}{5 \times 10^{-3} \frac{g}{cm \cdot s}} = 50{,}580$$

Consequently

$$\frac{k_c d}{D_{AB}} = 0.62\,Re^{1/2}Sc^{1/3} = 0.62(50\,580)^{1/2}(21\,199)^{1/3} = 3859$$

or

$$k_c = \frac{3859 \cdot \left(2.93 \times 10^{-7} \frac{cm^2}{s}\right)}{10\,cm} = 1.13 \times 10^{-4}\,cm/s$$

Finally, the time required to completely dissolve the photoresist is

$$t_f = \frac{(4)(500\,cm^3)}{\pi(10\,cm)^2\left(1.13 \times 10^{-4} \frac{cm}{s}\right)} \ln\left(\frac{0.04 - 0}{0.04 - 3.3 \times 10^{-5}}\right) = 46\,s$$

Notice that the concentration difference $c_{As} - c_A$ is relatively constant because c_A is very small. It is left to the reader to show that

$$W_A = k_c c_{As} \frac{\pi d^2}{4}$$

and

$$t_f = \frac{4m_{Ao}}{k_c c_{As} \pi d^2}$$

for the limiting case where $c_{As} \gg c_A$, i.e., the surrounding solvent represents an *infinite sink* for mass transfer.

EXAMPLE 6

Consider the remediation trench shown in Figure 30.7, a very simple process to treat contaminated wastewater before discharge to a lake or river. The remediation trench consists of a narrow outdoor open channel with an air sparger aligned along the bottom of the trench. Wastewater containing a volatile contaminant dissolved in the water enters one end of the trench. As the wastewater flows down the trench, the aeration gas strips out the dissolved volatile solute and transfers it to the surrounding atmosphere by an interphase mass-transfer process. Consequently, the concentration of the solute in the wastewater decreases down the length of the trench. Remediation trenches can be long, and may extend from a holding pond to the discharge point.

We wish to design an aerated remediation trench to treat wastewater contaminated with trichloroethylene (TCE) at a concentration of 50 mg/L (50 g TCE/m^3 wastewater). The trench is open duct of width (W) 1 m and depth (H) 2 m, and the volumetric flow rate of wastewater added to the

Figure 30.7 Aerated remediation trench for stripping TCE from wastewater.

trench is 0.1 m³/s. Air is sparged into the bottom of the duct at a rate that provides a gas holdup of 0.02 m³ of gas per 1 m³ of water, and the average bubble diameter is 1 cm (0.01 m). Determine the length of the trench necessary to reduce the effluent TCE concentration to 0.05 mg/L. The process temperature is 293 K and the total system pressure is 1 atm.

Again, the strategy is to develop a material-balance model for the process, and then incorporate the appropriate mass-transfer correlation(s) into the material-balance calculations. The physical system represents a steady-state, continuous flow process where TCE is transferred from the wastewater to the aeration gas. As the aeration gas locally mixes the liquid, assume that the concentration profile is one-dimensional along axial coordinate z. Referring to Figure 30.7, a steady-state material balance for TCE (species A) in the liquid phase of the trench within the differential volume element $W \cdot H \Delta z$ for $R_A = 0$ is

$$
\begin{pmatrix} \text{rate of TCE into} \\ \text{volume element} \\ \text{carried by water} \end{pmatrix} - \begin{pmatrix} \text{rate of TCE exiting} \\ \text{volume element} \\ \text{carried by water} \end{pmatrix} - \begin{pmatrix} \text{rate of TCE transferred} \\ \text{from water to air by} \\ \text{interphase mass transfer} \end{pmatrix} = 0
$$

or

$$
W \cdot H (v_\infty c_A)|_z - W \cdot H (v_\infty c_A)|_{z+\Delta z} - N_A \cdot \frac{A_i}{V} W \cdot H \Delta z = 0
$$

where v_∞ is the bulk average velocity of the wastewater and $W \cdot H$ is the cross-sectional area of the trench for fluid flow. Dividing by $W \cdot H \Delta z$ and taking the limit as Δz goes to zero yields

$$
v_\infty \frac{dc_A}{dz} + N_A \frac{A_i}{V} = 0
$$

For interphase mass transfer of TCE *from* water *to* the gas bubble, N_A must be

$$
N_A = K_L (c_A - c_A^*)
$$

where K_L is the overall interphase mass-transfer coefficient based on the overall liquid phase driving force and c_A^* is the concentration of TCE in the liquid that is in equilibrium with the partial pressure of TCE vapor in the air bubble. This equilibrium relationship is described by Henry's law

$$
c_A^* = \frac{p_A}{H}
$$

where H is Henry's law constant for TCE in water, which is equal to 9.97 atm · m³/kg mol at 293 K. As the rate of TCE transfer will be very small relative to the air flow rate bubbled into the trench, p_A for TCE inside the air bubble is essentially zero, and so c_A^* will be essentially zero. Furthermore, as the value of H for TCE in water is very high, TCE is only sparingly soluble in water, and so the interphase mass-transfer process will be liquid-phase controlling. Consequently, the liquid-film mass-transfer coefficient, k_L, will suffice for the overall mass-transfer coefficient k_L. Therefore, the flux equation reduces to

$$N_A = k_L c_A$$

The material balance is now

$$v_\infty \frac{dc_A}{dz} + k_L c_A \frac{A_i}{V} = 0$$

Separation of dependent variable c_A from independent variable z followed by integration from the entrance of the trench, $z = 0$, $c_A = c_{Ao}$, to the exit of the trench, $c_A = c_{AL}$, $z = L$ yields

$$-\int_{c_{Ao}}^{c_{AL}} \frac{dc_A}{c_A} = \frac{k_L}{v_\infty} \cdot \frac{A_i}{V} \int_o^L dz$$

The final design equation is

$$L = \frac{\ln\left(\frac{c_{Ao}}{c_{AL}}\right)}{\frac{k_L}{v_\infty} \cdot \frac{A_i}{V}}$$

The bulk average velocity of water through the open channel is

$$v_\infty = \frac{0.1 \frac{m^3}{s}}{W \cdot H} = \frac{0.1 \frac{m^3}{s}}{1\,m \cdot 2\,m} = 0.05 \frac{m}{s}$$

As the bulk velocity is relatively slow, we assume that the natural convection of the rapidly rising bubbles will dominate the convective mass-transfer process. The term $k_L A_i/V$ must now be evaluated. For a nonagitated bubble swarm of average bubble diameter d_b, the interphase mass-transfer area per unit liquid volume is

$$\frac{A_i}{V} = \frac{V_g}{V} \cdot \frac{6}{d_b} = \left(\frac{0.02\,m^3\,air}{1\,m^3\,water}\right) \frac{6}{0.01\,m} = 12 \frac{m^2}{m^3}$$

where V_g/V is the aeration gas volume per unit volume of liquid, commonly called the gas holdup ratio. This value for A_i/V is much higher than the open surface area per unit volume, which is $0.5\,m^2/m^3$ of liquid. Equation (30-14a) for mass transfer of gas bubble swarms into a nonagitated liquid is appropriate

$$Sh = \frac{k_L d_b}{D_{AB}} = 0.42\,Gr^{1/3}\,Sc^{1/2} \text{ for } d_b > 2.5\,mm \tag{30-15}$$

with

$$Gr = \frac{d_b^3 \rho_L g \Delta\rho}{\mu_L^2} = \frac{(0.01\,m)^3 \left(998.2 \frac{kg}{m^3}\right)\left(9.8 \frac{m}{s^2}\right)(998.2 - 1.19)\frac{kg}{m^3}}{\left(9.93 \times 10^{-4} \frac{kg}{m \cdot s}\right)^2} = 9.89 \times 10^6$$

and

$$Sc = \frac{\mu_L}{\rho_L D_{AB}} = \frac{9.93 \times 10^{-4} \frac{kg}{m \cdot s}}{\left(998.2 \frac{kg}{m^3}\right)\left(8.9 \times 10^{-10} \frac{m^2}{s}\right)} = 1118$$

where values for ρ_L and μ_L for water and ρ for air at 1 atm and 293 K were obtained from Appendix I. The molecular diffusion coefficient of TCE in water (D_{AB}) at 293 K was determined by the Hayduk–Laudie correlation. Finally

$$k_L = \frac{D_{AB}}{d_b} 0.42 \, \text{Gr}^{1/3} \, \text{Sc}^{1/2} = \frac{8.9 \times 10^{-10} \frac{\text{m}^2}{\text{s}}}{0.01 \, \text{m}} \, 0.42(9.89 \times 10^6)^{1/3}(1118)^{1/2}$$

$$= 2.683 \times 10^{-4} \frac{\text{m}}{\text{s}}$$

The channel length necessary to reduce the dissolved TCE concentration from $c_{Ao} = 50$ mg/L to $c_{AL} = 0.05$ mg/L is

$$L = \frac{\ln\left(\dfrac{c_{Ao}}{c_{AL}}\right)}{\dfrac{k_L}{v_\infty} \cdot \dfrac{A_i}{V}} = \frac{\ln\left(\dfrac{50 \, \text{mg/L}}{0.05 \, \text{mg/L}}\right)}{\dfrac{\left(2.683 \times 10^{-4} \dfrac{\text{m}}{\text{s}}\right)\left(12 \dfrac{\text{m}^2}{\text{m}^3}\right)}{0.05 \dfrac{\text{m}}{\text{s}}}} = 107.3 \, \text{m}$$

EXAMPLE 7

The dissolution of lead from lead-soldered joints in household piping is a possible health problem. If the water flowing through the pipe is mildly acidic, then some of the lead will dissolve from the inner surface of the lead-soldered pipe section into the water. New drinking water requirements stipulate that soluble lead (Pb) concentration levels should not exceed 0.015 g/m^3. At present, water flows through a 0.025-m I.D. pipe at a bulk velocity of 0.2 m/s. The water chemistry results in a soluble lead concentration of 10 g/m^3 at the surface of the soldered pipe section. Determine the soluble lead concentration in the water after it passes across three lead-soldered joints that are each 0.05 m in length. The diffusivity of soluble lead in water is 1×10^{-9} m^2/s, and the kinematic viscosity of water is 1×10^{-6} m^2/s.

It is important to realize that the concentration of soluble lead will be continuously increasing as water flows past the lead-soldered sections of pipe. A steady-state mass balance for soluble lead (species A) on a differential volume element of pipe along axial coordinate z is

$$\text{Input (moles Pb/time)} = \text{Output(moles Pb/time)}$$

$$\left. c_A v_\infty \frac{\pi D^2}{4} \right|_z + k_L(c_{As} - c_A)\pi D \, \Delta z = \left. c_A v_\infty \frac{\pi D^2}{4} \right|_{z+\Delta z}$$

where c_A represents the bulk concentration of soluble lead in the water stream. Upon rearrangement, we obtain

$$\frac{v_\infty D}{4}\left(\frac{c_A|_{z+\Delta z} - c_A|_z}{\Delta z}\right) = k_L(c_{As} - c_A)$$

At the limit of $\Delta z \to 0$, the resulting differential equation is

$$\frac{v_\infty D}{4} \frac{dc_A}{dz} = k_L(c_{As} - c_{A\infty})$$

Separation of variables gives

$$\int_{c_{Ao}}^{c_{AL}} \frac{dc_A}{c_{As} - c_A} = \frac{4k_L}{Dv_\infty} \int_0^L dz$$

As c_{As} is constant, the integral is

$$\ln\left(\frac{c_{As} - c_{Ao}}{c_{As} - c_{AL}}\right) = \frac{4L}{D}\frac{k_L}{v_\infty}$$

The mass-transfer coefficient (k_L) will be determined by both correlation and by analogy. Both approaches require the Reynolds and Schmidt numbers. For flow through a pipe, the Reynolds number is

$$Re = \frac{v_\infty D}{\nu} = \frac{(0.2\text{ m/s})(0.025\text{ m})}{1 \times 10^{-6}\text{ m}^2/\text{s}} = 5000$$

The Schmidt number is

$$Sc = \frac{\nu}{D_{AB}} = \frac{1 \times 10^{-6}\text{m}^2/\text{s}}{1 \times 10^{-9}\text{m}^2/\text{s}} = 1000$$

If $Re > 2000$, for liquids, the Linton–Sherwood correlation is appropriate for estimation of k_L

$$Sh = \frac{k_L D}{D_{AB}} = 0.023\,Re^{0.83}\,Sc^{1/3} = 0.023(5000)^{0.83}(1000)^{1/3} = 270$$

or

$$k_L = Sh\frac{D_{AB}}{D} = 270\frac{1 \times 10^{-9}\text{ m/s}}{0.025\text{ m}} = 1.08 \times 10^{-5}\frac{\text{m}}{\text{s}}$$

The Chilton–Colburn analogy can also be used to obtain k_L. From Figure 14.1, the friction factor is 0.0095 for flow through a smooth pipe at Re equal to 5000. From the Chilton–Colburn analogy, we obtain

$$\frac{k_L}{v_\infty} = \frac{C_f}{2}\frac{1}{Sc^{2/3}} = \frac{0.0095}{2}\frac{1}{(1000)^{2/3}} = 4.75 \times 10^{-5}$$

From this result, $k_L = 9.5 \times 10^{-6}$ m/s at v_∞ equal to 0.2 m/s, which agrees with the Linton–Sherwood correlation to within 12%.

As there is no transfer of soluble lead into the flowing water stream between the lead-soldered pipe sections, the total transfer length (L) for the three 0.05-m sections (L) is 0.15 m. The outlet concentration of soluble lead, c_{AL}, is

$$c_{AL} = c_{As} - (c_{As} - c_{Ao})\exp\left(-\frac{4L}{D}\frac{k_L}{v_\infty}\right)$$

$$= 10\frac{\text{g}}{\text{m}^3} - (10-0)\frac{\text{g}}{\text{m}^3}\exp\left(-\frac{4 \cdot 0.15\text{ m}}{0.025\text{ m}}\frac{1.08 \times 10^{-5}\frac{\text{m}}{\text{s}}}{0.2\frac{\text{m}}{\text{s}}}\right)$$

or $c_{AL} = 0.013$ g/m^3, using k_L obtained by the Linton–Sherwood correlation. Based on this calculation, the soluble-lead concentration is just below drinking water standards.

30.8 CLOSURE

In this chapter, we have presented correlating equations for convective mass-transfer coefficients obtained from experimental investigations. The correlations have verified the validity of the analysis of convective transport as presented in Chapter 23. In Chapter 31, methods have been developed for applying the capacity coefficient correlations to the design of mass-transfer equipment.

PROBLEMS

30.1 A very thin polymeric coating of thickness 0.1 mm uniformly coats a rectangular surface. The rectangular surface has a length of 20 cm and a width of 10 cm. the coating contains a solvent that must be evaporated away from the coating in order to cure the coating. Initially, there is 0.001 mole of solvent per cm^3 of coating loaded in the coating. A heated plate just beneath the surface maintains the coating at a uniform temperature of 40°C, and the vapor pressure exerted by the solvent is 0.05 atm at 40°C. Air gently flows parallel to the surface at a velocity of 5.0 cm/s. The surrounding air at 1.0 atm total system pressure and 20°C represents an "infinite sink" for mass transfer. You may neglect any molecular diffusion of the solvent through the very thin polymeric film and focus only on the convection aspects of the problem. The diffusion coefficient of species in air at 20°C is $0.1 \, cm^2/s$.

a. Determine the Reynolds, Schmidt, and Sherwood numbers associated with this process.

b. What is the film mass-transfer coefficient, k_y, (mole fraction based driving force) associated with this process?

c. How long will it take for the solvent to completely evaporate from the coating?

30.2 A 1-m square thin plate of solid naphthalene is oriented parallel to a stream of air flowing at 20 m/s. The air is at 310 K and 1.013×10^5 Pa. The naphthalene remains at 290 K; at this temperature, the diffusivity of naphthalene in air is $5.61 \times 10^{-6} \, m^2/s$ and its vapor pressure is 26 Pa. Determine

a. the value of the mass-transfer coefficient at a point 0.3 m downstream from the leading edge.

b. the moles of naphthalene per hour lost from the section of the plate 0.5–0.75 m downstream from the leading edge.

30.3 Ethanol, C_2H_5OH, flows in a thin film down the outside surface of an inclined plane, 2 m wide and 4 m long. The liquid temperature is 289 K. Ethanol-free air at 303 K and 1 atm flows across the width of the plate parallel to the surface. At the average temperature of the gas film, the diffusivity of ethanol vapor in air is $1.32 \times 10^{-5} \, m^2/s$, The vapor pressure of ethanol is 6.45×10^{-2} atm at 289 K. If the air velocity is 3 m/s, determine the rate at which the ethanol should be supplied to the top of the plate so that evaporation will prevent it from reaching the very bottom of the plate.

30.4 (Please refer to Problem 27.1) Gasoline from an under-storage storage tank leaked down onto an impermeable clay barrier and collected into a liquid pool 2 cm deep. Directly over this underground pool of liquid gasoline (*n*-octane) is a layer of gravel 1 m thickness (*L*) and width (*W*) of 10 m. The volatile *n*-octane vapors diffuse through the highly porous gravel layer of void fraction 0.4, through a gas film formed by flow of air over the flat surface, and then out to the bulk atmosphere where the *n*-octane is instantaneously diluted to below detectable levels. A simplified picture of the situation was provided in Problem 27.1.

There is no adsorption of *n*-octane vapor onto the porous gravel layer, and *n*-octane vapor concentration inside the porous gravel layer is dilute. Assume that the mass-transfer process is allowed to achieve a steady state. The temperature is constant at 15°C. At this temperature, liquid *n*-octane exerts a vapor pressure of 1039 Pa. The void spaces between the rocks are large enough so that the molecular-diffusion coefficient can be used for the gas phase. The system is at atmospheric pressure.

a. What is the average mole fraction of *n*-octane vapor at the surface ($z = L$) of the rock layer if the air velocity is only 2 cm/s? What is the flux of *n*-octane vapor across both the porous gravel layer and the concentration boundary layer at the surface?

b. What would be the average mole fraction of *n*-octane vapor at the surface of the rock layer if the air velocity is 50 cm/s? What would the new flux be?

c. The Biot number associated with a mass-transfer process involving diffusion and convection in series is defined as

$$Bi_{AB} = \frac{k_c L}{D_{AB}}$$

where *L* refers to the path length for molecular diffusion within the porous gravel layer and D_{AB} refers to the diffusion coefficient of species A within medium B for molecular diffusion. Determine the Biot number for parts (a) and (b) and then assess the relative importance of convective mass transfer in determining the *n*-octane vapor emissions rate.

30.5 (Please refer to example 1, this chapter) In the presence process, arsine and trimethylgallium vapors are diluted in hydrogen gas and fed to the horizontal CVD reactor shown in Figure 30.2. The composition of arsine and trimethylgallium in the feed gas are both 0.1 mol %, which is very dilute. You may assume that the amount of arsine and trimethylgallium delivered with the feed gas is much higher than the amount of arsine and trimethylgallium consumed by the reaction, so that the concentration of these reactants in the bulk gas phase is essentially constant down the length of the reactor. You may also assume that the surface-reaction rates are instantaneous relative to the rates of mass transfer, so that the gas-phase concentrations of both arsine vapor and trimethylgallium vapor at the surface of the wafer are equal to zero.

a. What are the local molar fluxes of arsine and trimethylgallium along the surface of the 10-cm Si wafer at $x = 4 \, cm$ (left edge), $x = 9 \, cm$ (center), and $x = 14 \, cm$ (right edge), assuming that the surface reactions are controlled by the rate of convective mass transfer? From this result, what is the composition of the *GaAs* composite thin film; e.g., the molar composition of gallium (Ga) and arsenic (As) in the solid?

b. How could the feed-gas composition be adjusted so that the molar ratio of Ga to As within the solid thin film is 1:1?

c. Based on your results in part (a), calculate the thickness of the GaAs thin film (in units of μm) after 2 min at each position x, assuming that the density of the GaAs composite thin film is $5.8\,\text{g/cm}^3$.

30.6 Spherical pellets of 1.0 cm diameter are spray painted with a very thin coat of paint. The paint contains a volatile solvent. The vapor pressure of the solvent at 298 K is 1.17×10^4 Pa, and the diffusivity of the solvent vapor in air at 296 K is $0.0962\,\text{cm}^2\text{/s}$. The amount of solvent in the wet paint on the pellet is 0.12 g solvent per cm^2 of pellet surface area. The molecular weight of the solvent is 78 kg/kg mol.

a. Determine the minimum time to dry the painted pellet if still air at 298 K and 1.0 atm surrounds the pellet.

b. Determine the minimum time to dry the painted pellet if air at 298 K and 1.0 atm pressure flows around the pellet at a bulk velocity of 1.0 m/s.

30.7 Investigators studying the mass transfer from single spheres recommended the generalized equation

$$Sh = 2.0 + CRe^m Sc^{1/3}$$

As discussed in section 30.1, the value of 2 can be derived theoretically by considering the molecular diffusion from a sphere into a large volume of stagnant fluid. Prove this is the correct value and state what assumptions must be made to obtain it.

30.8 A spherical pellet containing pure solid A is suspended in a flowing liquid stream at 20°C. The initial diameter of the pellet is 1.0 cm, and the bulk liquid velocity is 5 cm/s. Component A is soluble in the liquid, and as time progresses, the diameter of the pellet decreases. The flowing liquid effectively serves as an "infinite sink" for the solute, so that the dissolved solute concentration in the bulk liquid is essentially zero. This process represents a "pseudo steady state" convective mass transfer system. All relevant physical properties of the system are provided below

ρ_A, the density of solid A, $= 2.0\,\text{g/cm}^3$;

M_A, the molecular weight of solute A, $= 110\,\text{g/g mol}$;

c_A^*, the equilibrium solubility of solute A, at liquid-solid interface, $= 7 \times 10^{-4}\,\text{g mole/cm}^3$;

ν, the kinematic viscosity of the bulk fluid at 20°C $= 9.95 \times 10^{-3}\,\text{cm}^2\text{/s}$;

D_{AB}, the diffusion coefficient of solute A in the liquid at 20°C $= 1.2 \times 10^{-5}\,\text{cm}^2\text{/s}$.

a. Estimate the film transfer coefficient at the initial pellet 1.0 cm diameter.

b. Estimate the rate of pellet shrinkage, dR/dt, in cm/h, when the pellet diameter is 1.0 cm.

c. If the pellet diameter decreases from 1.0 to 0.5 cm, the total mass transfer rate, W_A, will change by what factor?

30.9 Determine the mass transfer coefficient, k_c, for the transfer from a glucose sphere, 0.3 cm in diameter, which is sub-

merged in a 25°C aqueous stream flowing at a bulk velocity of 0.15 m/s. The mass diffusivity of glucose in water at 25°C equals $6.9 \times 10^{-10}\,\text{m}^2\text{/s}$.

a. Estimate the mass transfer coefficient, k_L.

b. Determine whether increasing the diameter of the glucose sphere or increasing the velocity of the water stream effects the magnitude of the mass transfer coefficient, k_c. Which of these variables has the larger effect on the coefficient?

30.10 Silicon tetrachloride, $SiCl_4$, is an important chemical used in the manufacture of microelectronic devices. A common impurity in silicon tetrachloride is trichlorosilane, $SiHCl_3$. High purity $SiCl_4$ is manufactured by bubbling chlorine Cl_2, gas into unpurified liquid $SiCl_4$. In the presence of ultraviolet light, the dissolved Cl_2 reacts with residual trichlorosilane in the unpurified liquid to form silicon tetrachloride according to the homogeneous reaction;

$$SiHCl_3 + Cl_2 \rightarrow SiCl_4 + HCl$$

Estimate the liquid-film mass-transfer coefficient for the transfer of Cl_2 (species A) into liquid $SiCl_4$, assuming pure Cl_2 gas at 298 K and 1.0 atm is bubbled into the liquid $SiCl_4$ with a mean bubble diameter of 2 mm. At 298 K, the density of liquid $SiCl_4$ is $1.47\,\text{g/cm}^3$, the viscosity of liquid $SiCl_4$ is 0.52 cp, and the liquid diffusivity of Cl_2 in $SiCl_4$ is $5.6 \times 10^{-5}\,\text{cm}^2\text{/s}$. The dissolution of Cl_2 gas in $SiCl_4$ liquid is defined by $p_A = H x_A^*$ where H is 6.76 atm at 298 K.

What is the flux of Cl_2, assuming that the Cl_2 is instantaneously consumed once it reaches the bulk liquid phase? You may assume that the total molar concentration of the mixture approximates the molar concentration of pure liquid $SiCl_4$.

30.11 An open trench containing $200\,\text{m}^3$ of wastewater is contaminated with a small concentration of trichloroethylene, TCE, a common industrial solvent. The trench is 2.0 m deep and 1.0 m in width. The figure below illustrates the open trench.

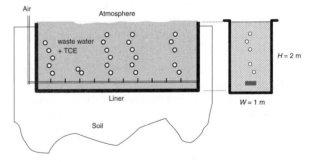

There is no inflow or out of flow of water to or from the trench. Air is sparged into the bottom of the trench to strip out the dissolved TCE. The inlet air flow rate is $0.15\,\text{m}^3$ per $1.0\,\text{m}^3$ of water per minute. At this air flow rate, the gas holdup is calculated to be $0.015\,\text{m}^3$ air per $1.0\,\text{m}^3$ aerated water. The air sparger provides an average bubble diameter of 5.0 mm.

The temperatures of the air and water are 293 K, with a liquid viscosity, $\mu_L = 9.93 \times 10^{-4}\,\text{kg/m} \cdot \text{s}$ and a liquid density, $\rho_L = 998.2\,\text{kg/m}^3$ and a gas density. $\rho_G = 1.19\,\text{kg/m}^3$.

TCE is only sparingly soluble in water, and the Henry's law constant for TCE in liquid water is 9.97 atm/(kg mol/m^3). The molecular weight for TCE is 131.4 g/g mol. Assume that the air stream serves as an infinite sink for TCE transfer, so that the partial pressure of TCE in the gas bubble is essentially zero. Furthermore, assume that the interphase mass transfer process is liquid film controlling. The mass diffusivity of TCE in water at 293 K is $D_{AB} = 8.9 \times 10^{-10}$ m^2/s.

a. Develop a well-mixed, unsteady state material balance model for dissolved TCE in the liquid phase of the trench.

b. Determine the required time to reduce the dissolved TCE concentration from 50 to 0.005 g TCE/m^3.

30.12 Consider now the open trench illustrated below where wastewater containing a volatile contaminate dissolved in the water enters one end of the trench. As the wastewater flows down the

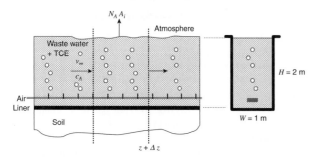

trench, the aeration gas strips out the dissolved volatile solute, TCE, and transfers it to the surrounding atmosphere by an interphase mass-transfer process. Consequently, the concentration of the solute in the wastewater changes down the length of the trench. Remediation trenches can be long, and may extend from a holding pond to the discharge point.

a. Develop a material balance model for the dissolved TCE profile in the liquid phase of the trench.

b. Determine the length required to reduce the dissolved TCE concentration from 50 to 0.005 g TCE/m^3 if the volumetric flow rate of liquid into the trench is 0.1 m^3/s.

30.13 A stent is used to "prop up" a clogged artery to allow blood to pass through it. But stents can also be loaded with drugs to facilitate the timed release of the drug into the body, especially if the drug is not very soluble in body fluids. Consider the very simple stent design as illustrated below. The pole of the stent, which is 0.2 cm in diameter and 1.0 cm in length, is coated with the anticancer drug Taxol. The thickness of the coating is 0.01 cm and 5.0 mg total Taxol is loaded. Blood flows through the 1.0 cm diameter cylindrical blood vessel at a volumetric flow rate of 10.0 cm^3/s. Taxol is not very soluble in aqueous environments; the maximum solubility of Taxol in blood is 2.5×10^{-4} mg/cm^3. The viscosity of blood is 0.040 g/cm · s and its density is 1.05 g/cm^3. Blood is a complex fluid but you may assume that its average molecular weight is close to that of liquid water. The molecular diffusion coefficient of Taxol in fluid blood is $D_{AB} = 1.0 \times 10^{-6}$ cm^2/s.

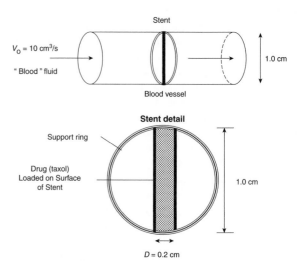

a. What is the convective mass transfer coefficient, k_c around the outer surface of the cylindrical portion of the stent, for the solute Taxol?

b. What is the minimum time it will take for all the Taxol to be completely discharged from the stent?

30.14 The air freshener shown in the figure below is used to deliver a pleasant vanilla fragrance in a room. The device consists of a rod 1 cm in diameter and 5 cm in length. The rod is composed of a solid mixture containing a volatile aromatic substance that smells like vanilla. A heating element in the center of the rod, 0.5 cm in diameter, maintains a constant surface temperature of 40°C to evaporate the aromatic substance. Air at 25°C flows perpendicular to the rod at a velocity of 0.2 m/s, and the aroma vapors are immediately diluted to a very low concentration. The density of the gel is 1.1 g/cm^3, and the initial loading of the aromatic compound in the solid is 40 mol%. The diffusion coefficient of the volatile substance generating the aroma, which approximates the properties of benzaldehyde, is 0.08 cm^2/s in air at 40°C. The vapor pressure of this substance is

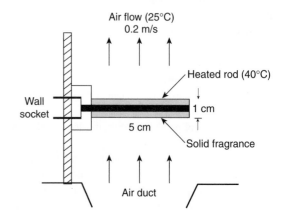

428 Pa at 40°C, and its molecular weight is 106 g/gmol. Assume that the release of the aromatic compound is controlled by the convective transport across the gas film surrounding the rod, and that the diameter of the rod remains constant. Neglect internal diffusion of the aroma substance within the rod itself.

a. What is the initial rate of aroma delivery to the surrounding in grams/hour?

b. How long can the air freshener provide fragrance?

30.15 An entrance section of process piping is fouled with calcium carbonate scale. Specifically, the inner surface of the pipe is coated with 0.1 cm film of solid calcium carbonate (species A) that is sparingly soluble in water. The length of the pipe containing the calcium carbonate is 1 m, and the inner diameter of the pipe is 2 cm. In an attempt to dissolve the scale from the pipe surface, "clean" water containing no soluble calcium carbonate flows through the pipe at a volumetric flow rate of 314 cm^3/s and a temperature of 293 K. Determine the time in hours that it will take for the calcium carbonate coating to completely dissolve from the inner surface of the tube, assuming that bulk concentration of solute A in the liquid stream is essentially zero throughout the 1 m length of the scaled section. The following potentially useful data are available:

M_A,   the molecular weight of solute A = 100 kg/kg mol.
ρ_A,   the density of solid A = $2.7 \times 10^3 \text{ kg/m}^3$.
D_{AB},   the diffusivity of solute A in water = $1.2 \times 10^{-9} \text{ m}^2$/s.
c_A^*,   the solubility of solute A in water = 0.14 mol/m^3.

30.16 The "bubbleless" shell-and-tube membrane aeration system shown below is used to transfer oxygen gas to liquid water. Water completely free of dissolved oxygen is added to the tube side at the entrance. Pure oxygen gas, maintained at a constant pressure of 1.0 atm, flows through the annular space. The inner diameter of the tube is 1.0 cm. The tube wall is made of silicone, a polymer that is highly permeable to O_2 gas but not to water vapor. The O_2 transfer properties associated with the thin tube wall are neglected, so that the concentration of dissolved O_2 in the water at the inner surface of the wall is established by the O_2 partial pressure on the inside of the tube. The tube length is 500 cm, and the bulk velocity of liquid through the tube is 50 cm/s. The process is maintained at 25°C, where the Henry's law constant for O_2 gas dissolved in water is 0.78 atm (mol/m^3) and the diffusion coefficient for dissolved O_2 in water is $2.1 \times 10^{-5} \text{ cm}^2$/s, and the kinematic viscosity of water is $9.12 \times 10^{-3} \text{ cm}^2$/s.

a. What are the Sherwood and Schmidt numbers inside the tube?

b. What is the predicted exit concentration of dissolved oxygen? It will be wise to develop a material balance model on O_2 for the shell-and-tube mass exchanger.

c. What would be the required length of tubing to achieve 60% of dissolved O_2 saturation?

30.17 The experiment shown below is used to estimate mass transfer coefficients. In the present experiment, two portions of the inside of the tube are coated with naphthalene, a volatile solid. The tube is maintained at 100°C and 1.0 atm total system pressure. At 100°C, the sublimation vapor pressure of solid naphthalene is 1.0 mmHg. The inner diameter of the tube is 2.0 cm. The total length of the tube is 100 cm, and each section of the tube bearing the naphthalene is 2.0 cm. Air enters the tube at a

mass flow rate of 4.5 g/min. The measured outlet mole fraction of naphthalene vapor in the exit of the tube is 0.0066, as determined by gas chromatography. The diffusion coefficient of naphthalene in air is 0.086 cm^2/sat 100°C and 1.0 atm total system pressure. At this temperature and pressure, the kinematic viscosity of air is 0.25 cm^2/s and the density of air is $9.5 \times 10^{-4} \text{ g/cm}^3$.

a. Develop a material balance model to predict the mole fraction of naphthalene vapor exiting the tube.

b. Based on your model and the data given in the problem statement, estimate k_c.

c. Now, evaluate the mass-transfer coefficient using the appropriate established correlation and compare your results to part (b)

30.18 Fatty deposits in blood vessels pose a health risk. Chemical agents added to the blood can increase the solubility of the fatty deposits in blood, thereby providing a means to dissolve the fatty deposits *in situ*. To explore the mass-transfer characteristics of this phenomenon, a biological engineer set up the experiment shown in the figure. The inner walls of a smooth, 0.8 cm inner diameter tube are uniformly coated with a thin layer, 0.01 cm, of lipophilic material called fatty compound A (FA for short). The length of the tubing containing the FA layer is 50 cm. A synthetic solution is pumped through the tube at a volumetric flow rate of 35.0 cm^3/s. The maximum solubility of FA in this synthetic solution is 20 mg FA/cm^3 solution. At the conditions of the experiment, the kinematic viscosity of the synthetic solution is 0.02 cm^3/s and the density of the synthetic solution is 1.04 g/cm^3. The density of the solid FA is 1.10 g/cm^3.

a. Develop a material balance to predict the concentration of FA in the solution exiting the tube.

b. Experiments were conducted with the test unit shown above, with a measured dissolved concentration of FA in the synthetic solution exiting the tube of 0.1 mg FA/cm^3. This concentration was constant over time. From this data, using your model, back out the molecular diffusion coefficient of FA in the solution.

30.19   Consider the device shown below that will be used to humidity an air stream. In the current process, air containing 1.0 mol% water vapor (species A) at 40°C and 1.0 atm enters the tube of 3.0 cm inner diameter at a velocity of 300 cm/s. A water-adsorbent material saturated with liquid water lines the inner surface of the tube. Water evaporates from this liner material as air passes over it. There is no diffusion resistance of water vapor

Water adsorbent liner
(1.0 mm thick, 0.6 g H$_2$O/cm^3 liner initially)

Gas stream
$v_\infty = 300$ cm/sec
$C_{A,1}$ $D = 3.0$ cm (tube) 1.0 atm, 40°C $C_{A,2}$

$L = 8.0$ m

through the liner material itself. The process is maintained at 40°C. The saturation vapor pressure of water is 55.4 mmHg at 40°C, At this temperature and 1.0 atm, the viscosity of air is 1.91×10^{-4} g/cm · s and the density of air is 1.13×10^{-3} g/cm^3.

a. What is the Schmidt number for this process?

b. Develop a model, in final integrated form, to predict the concentration of water vapor exiting the tube.

c. The thickness of the adsorbent material lining the inner surface of the tube is 1.0 mm and it initially contains 0.6 g of water per cm^3. If the tube is 8.0 m long and the humidification occurs at 40°C and 1.0 atm system pressure, determine how long this process can operate before the liquid water is depleted. You will need to determine the outlet concentration of water vapor and then perform an overall material

balance to estimate the net overall rate of water vapor transferred.

30.20   Fatty deposits in blood vessels pose a health risk. Chemical agents added to the blood can increase the solubility of the fatty deposits in blood, thereby providing a means to dissolve the fatty deposits *in situ*. To explore the mass-transfer characteristics of this phenomenon, a biological engineer set up the experiment shown below. The inner walls of a smooth, 0.8 cm inner diameter tube are uniformly coated with a thin layer, 0.01 cm, of lipophilic material called fatty compound A (FA for short). A synthetic solution (B) is pumped through the tube at a

volumetric flow rate of 3.0 cm^3/s. The maximum solubility of FA in this synthetic solution is 20 mg FA/cm^3 solution (c_A^*). At the conditions of the experiment, the kinematic viscosity of the synthetic solution is 0.02 cm^2/s, the density of the synthetic solution is 1.04 g/cm^3 and the diffusion coefficient is 4.0×10^{-5} cm^2/s. The density of the solid FA is 1.10 g/cm^3. As the solution flows down the length of the tube, the concentration of FA changes with axial position, z, according to the material balance model:

$$\ln\left(\frac{c_A^* - c_{AL,o}}{c_A - c_{AL,z}}\right) = \frac{4k_L z}{v_\infty D}$$

How long should the tubing be if the desired outlet concentration of FA is 0.010 mg/cm^3?

30.21   The wetted-well column shown in Figure 30.4 is used to humidify a pressurized air stream. Bone-dry air (containing no water vapor) at 2.0 atm and 27°C is introduced to the bottom of the column at a volumetric flow rate of 4000 cm^3/s. Liquid water is added to the top of the column. The liquid water uniformly coats the inner surface of the tube as a thin liquid film flowing downward. The falling liquid film of water evaporates into the air flowing upward. The wetted-wall column has an inner diameter of 5.0 cm and a length of 6.0 m. The column temperature is maintained at 27°C, and the total system pressure is maintained at 2.0 atm. At 27°C, the kinematic viscosity of air is 0.157 cm^2/s and the vapor pressure of liquid water is 0.035 atm. The gas is ideal with $R = 82.06$ cm^3 · atm/g mol · K.

a. What is the diffusion coefficient of water vapor in air at 2.0 atm and 27°C?

b. Determine the Sherwood number for the gas phase.

c. Evaluate the gas film coefficient, k_{cG}, based on the partial pressure driving force.

d. What is the maximum possible concentration of water vapor in the air?

e. Evaluate the concentration of water vapor in the air stream exiting the wetted-wall column. You will need to do a material balance on water vapor within a differential volume element of the column.

30.22 A process is being developed to produce carbonated beverages. As part of this process, a wetted-wall absorption column of 2.0 m length will be used to dissolve carbon dioxide, CO_2, gas into water. Pure mountain spring water, containing no dissolved CO_2 enters the top of the column at a flow rate of 2.0 g mol/s. Pure 100% CO_2 gas at 2.54 atm is fed to the bottom of the column at a flow rate of 0.5 g mol/s. As the liquid flows down the wetted-wall column, CO_2 gas dissolves into the water. The carbonated water exits the bottom of the column, and the unused CO_2 gas leaves the top of the column. The inner column diameter is 6.0 cm. The temperature is maintained at 20°C. At this temperature, the Henry's law constant for the dissolution of CO_2 gas in water is 25.4 atm/(kg mol/m^3). At 20°C, the molar density of liquid water is 55.5 kg mol/m^3, the mass density of liquid water is 998.2 kg/m^3, and the viscosity of liquid water is 993×10^{-6} kg/m · s.

Water has a finite vapor pressure at 20°C. However, for this problem you may assume that the water solvent is essentially nonvolatile, so that 100% CO_2 gas composition is always maintained down the length of the column.

a. What is the maximum possible concentration of dissolved CO_2 in water at 2.54 atm CO_2 partial pressure?

b. What is the liquid-phase mass-transfer coefficient for this process?

c. What is the exit concentration of dissolved CO_2 in the carbonated water if the wetted-wall column is 2.0 m in length?

30.23 Tetraethoxysilane, also called TEOS or $Si(OC_2H_5)_4$, is a liquid chemical used in the semiconductor industry to produce thin films of silicon dioxide by chemical vapor deposition (CVD). In order to deliver the TEOS vapor to the CVD reactor, liquid TEOS is fed to a wetted-wall column. The TEOS liquid uniformly coats the inner surface of the tube as a thin liquid film as it flows downward. The falling liquid film of TEOS evaporates into an inert helium carrier gas flowing upwards at a volumetric flow rate of 2000 cm^3/s. The wetted-wall column has an inner diameter of 5 cm and a length of 2 m. The column temperature is maintained at 333 K, and the total system pressure is 1 atm. At 333 K, the kinematic viscosity of helium gas is 1.47 cm^2/s, the diffusion coefficient of TEOS vapor in helium gas is 1.315 cm^2/s, and the vapor pressure of liquid TEOS is 2,133 Pa.

a. What is the gas mass-transfer coefficient, k_G?

b. What is the mole fraction of TEOS vapor exiting the column?

c. What is the required mass flow rate of liquid TEOS flowing into the column if all of the liquid TEOS evaporates by the time it reaches the bottom of the column?

30.24 A wetted-wall column, as illustrated in Figure 30.4, is used as a mass-transfer device for transferring ethyl acetate liquid to an air stream by a mass-transfer limited evaporation

process. At the current conditions of operation, the inlet air stream bulk velocity is 0.2 m/s, the temperatures of the gas and the liquid are maintained at 27°C, the total system pressure is 1.0 atm, the column diameter is 0.05 m, the column length is 10 m, and the vapor pressure of ethyl acetate is 0.080 atm.

a. Determine the mass-transfer coefficient for the process at the current conditions of operation. You may assume that the gas phase represents a dilute system.

b. As the gas moves up the tube, it picks up ethyl acetate vapor. Develop a mathematical expression to predict the ethyl acetate vapor concentration in the gas stream exiting the top of the tube. Then, calculate the ethyl acetate concentration in the gas stream exiting the tube. At the specified conditions of operation, what is the overall evaporation rate of ethyl acetate vapor in g mol/h if the inlet air stream contains no ethyl acetate vapor?

30.25 Ozone, O_3, dissolved in water is used in many wastewater treatment applications. Pure 100% ozone gas at 1.0 atm and 20°C is continuously bubbled into a tank of liquid water. The rising bubbles keeps the liquid well mixed. There is no inflow or outflow of water. Initially, there is no dissolved O_3 in the water, but as time goes on, the dissolved O_3 concentration increases. The total volume of liquid in the tank is 2.0 m^3. The Henry's law constant, H, for O_3–water system is 6.67×10^{-2} atm/ (g mol/m^3).

It is desired to achieve a dissolved O_3 concentration of 4.0 g mol/m^3 within 10.0 min. What is the required capacity coefficient, k_La for O_3 needed to meet this constraint?

30.26 Wilke and Hougan[27] studied the mass-transfer characteristics of packed beds containing granular solids. In their experimental investigations, hot air was blown through a packed bed of porous celite pellets saturated with liquid water. The water evaporated under adiabatic conditions, and the rate of water transfer was determined by material balance using humidity measurements. From this data, they calculated the gas-film mass-transfer coefficient at a given flow rate. In one run, the following data were reported:

Gas film coefficient, k_G	4.42×10^{-3} kg mol/m^2 · s · atm
Effective particle diameter	0.571 cm
Packed bed void fraction	0.75
Gas stream mass velocity	0.816 kg/m^2 · s
Temperature at particle surface	311 K
Total system pressure	9.77×10^4 Pa

Estimate the gas-film mass-transfer coefficient by two appropriate correlations, one that includes the packed bed void fraction, and a simpler correlation that does not account for the void fraction. Compare these estimates to the measured gas-film coefficient above.

30.27 Often aerobatic fermentations give off CO_2 gas because the respiring cells produce CO_2 as glucose nutrient is converted

---

[27] C.R. Wilke and O.A. Hougan. *Trans., A.I.Ch.E.*, 41, 445 (1945).

to chemical energy. The conditions used to set $k_L a$ for O_2 transfer in the fermenter also establish the same hydrodynamic conditions for CO_2 transfer. Consequently, the transfer rate of CO_2 from the liquid to the aeration gas is also described by this volumetric mass-transfer coefficient, $k_L a$. Consider that the $k_L a$ for O_2 in the fermenter is $300\ h^{-1}$. Scale the $k_L a$ for O_2 to $k_L a$ for CO_2 transfer using film theory, boundary layer theory, and penetration theory.

30.28 A process is being developed to produce carbonated beverages. As part of this process, a packed-bed absorption tower will be used to dissolve carbon dioxide, CO_2, gas into water. Pure mountain spring water, containing no dissolved CO_2 enters the top of the column at a flow rate of 5 kg mol/min. Pure CO_2 gas at 2 atm is also fed to the top of the tower at a flow rate of 1 kg mol/min. As the liquid flows down the tower, CO_2 gas absorbs into the water, and the dissolved CO_2 concentration increases down the length of the bed. The carbonated water and the unused CO_2 gas exit the bottom of the tower. The absorption process is liquid-film controlling because only pure CO_2 is present in the gas phase. The tower is packed with 1 in. ceramic rings and the inner tower diameter is 0.25 m. The temperature is maintained at 20°C. At this temperature, the Henry's law constant for the dissolution of CO_2 gas in water is 25.4 atm/ $(kg\ mol/m^3)$. At 20°C, the molar density of liquid water is 55.5 kg mol/m^3, the mass density of liquid water is 998.2 kg/m^3, and the viscosity of liquid water is 993×10^{-6} kg/m · s.

a. What is the liquid phase mass transfer coefficient, $k_L a$, for CO_2 in water flowing through the packed bed?

b. Estimate the depth of packing if the desired concentration of dissolved CO_2 in the outlet liquid is 95% of the saturation value for dissolved CO_2 in water under CO_2 partial pressure of 2 atm and 20°C.

30.29 The "soil venting" shown in the figure below is used to treat soil contaminated with volatile, toxic liquids. In the present situation, the porous soil particles are saturated with liquid TCE, a common industrial solvent. The contaminated soil is dug up at the waste site and loaded into a rectangular trough. The soil consists of coarse, porous mineral particles with an average diameter of 3 mm, loosely compacted into a packed bed with a void fraction 0.5. Air is introduced at the bottom of the trough through a distributor and flows upward around the soil particles. Liquid TCE satiating the pores of the soil particle evaporate into the air stream. Consequently, the TCE concentration in the air stream increases from the bottom to the top of the trough. Usually, the rate of TCE evaporation is slow enough so that the liquid TCE within the soil particle is a constant source for mass transfer, at least until 80% of the volatile TCE soaked within the soil is removed. Under these conditions, the transfer of TCE from the soil particle to the air stream is limited by convective mass transfer across the gas film surrounding the soil particles. The mass flow rate of air per unit cross section of the empty bed is $0.1\ kg/m^3$ · s. The process is carried out at 293 K. At this temperature, the vapor pressure of TCE is 58 mmHg. The molecular diffusion coefficient of TCE vapor in air is given in example 4 of this chapter.

Air + TCE vapors
diluited to atmosphere

Air blower

TCE contaminated packed soil bed

Raised trough for soil venting

$z = L$

$z = 0$

Air flow distributor
(0.1 kg air/m² · s)

a. What is the gas-film mass-transfer coefficient for TCE vapor in air?

b. At what position in the bed will the TCE vapor in the air stream reach 90% of its saturated vapor pressure? In your solution, you may want to consider a material balance on TCE in the gas phase within a differential volume element of the bed. Assume the convective mass-transfer resistances associated with air flowing over the top surface of the bed are negligible and that there is no pressure drop of the gas stream through the bed so that the total system pressure remains constant at 1 atm.

Chapter 31

Mass-Transfer Equipment

In earlier chapters, the theory currently used to explain the mechanism of convective mass transfer between phases was introduced, and correlations for the interphase convective mass-transfer coefficients were listed. In this chapter, we will develop methods for applying the transport equations to the design of commercial mass-transfer equipment. It is important to realize that design procedures are not restricted to the design of new equipment, for they may also be applied in analyzing existing equipment for possible improvement in performance.

The presentation or development of mass transfer from the defining equations to the final design equations, which is presented in this chapter, is completely analogous to our earlier treatment of energy transfer. Convective mass-transfer coefficients are defined in Chapter 28. These definitions and the methods of analysis are similar to those presented in Chapter 19 for convective heat-transfer coefficients. An overall driving force and an overall transfer coefficient expressed in terms of the individual coefficients were developed to explain the transfer mechanisms of both mass- and energy-transport processes. By integrating the appropriate energy-transfer relation in Chapter 23, we were able to evaluate the area of a heat exchanger. Accordingly, we should expect to find similar mass-transfer relations that can be integrated to yield the total transfer contact area within a mass exchanger.

31.1 TYPES OF MASS-TRANSFER EQUIPMENT

A substantial number of industrial operations in which the compositions of solutions and/or mixtures are changed involve interphase mass transfer. Typical examples of such operations could include (1) the transfer of a solute from the gas phase into a liquid phase, as encountered in absorption, dehumidification, and distillation; (2) the transfer of a solute from the liquid phase into a gas phase, as encountered in desorption or stripping and humidification; (3) the transfer of a solute from one liquid phase into a second immiscible liquid phase (such as the transfer from an aqueous phase to a hydrocarbon phase), as encountered in liquid–liquid extraction; (4) the transfer of a solute from a solid into a fluid phase as encountered in drying and leaching; and (5) the transfer of a solute from a fluid onto the surface of a solid as encountered in adsorption and ion exchange.

Mass-transfer operations are commonly encountered in towers or tanks that are designed to provide intimate contact of the two phases. This equipment may be classified into one of the four general types according to the method used to produce the interphase contact. Many varieties and combinations of these types exist or are possible; we will restrict our discussion to the major classifications.

Bubble towers consist of large open chambers through which the liquid phase flows and into which the gas is dispersed into the liquid phase in the form of fine bubbles. The small gas

Figure 31.1 Bubble tower.

bubbles provide the desired contact area. Mass transfer takes place both during the bubble formation and as the bubbles rise up through the liquid. The rising bubbles create mixing action within the liquid phase, thus reducing the liquid-phase resistance to mass transfer. Bubble towers are used with systems in which the liquid phase normally controls the rate of mass transfer; for example, it is used for the absorption of relatively insoluble gases, as in the air oxidation of water. Figure 31.1 illustrates the contact time and the direction of phase flow in a bubble tower. As one would expect, the contact time, as well as the contact area, plays an important role in determining the amount of mass transferred between the two phases. The basic mass-transfer mechanism involved in bubble towers is also encountered in *batch bubble tanks or ponds* where the gas is dispersed at the bottom of the tanks. Such equipment is commonly encountered in biological oxidation and in wastewater-treatment operations.

Exactly opposite in principle to the bubble tower is the *spray tower.* In the spray tower, the gas phase flows up through a large open chamber and the liquid phase is introduced by spray nozzles or other atomizing devices. The liquid, introduced as fine droplets, falls countercurrent to the rising gas stream. The spray nozzle is designed to subdivide the liquid into a large number of small drops; for a given liquid flow rate, smaller drops provide a greater interphase contact area across which mass is transferred. However, as also encountered in bubble towers, care in design must be exercised to avoid producing drops so fine that they become entrained in the exiting, countercurrent stream. Figure 31.2 illustrates the contact time and the direction of phase flow in the spray tower. Resistance to transfer within the gas phase is reduced by the swirling motion of the falling liquid droplets.

Figure 31.2 Spray tower.

Figure 31.3 Countercurrent packed tower.

Spray towers are used for the mass transfer of highly soluble gases where the gas-phase resistance normally controls the rate of mass transfer.

Packed towers are the third general type of mass-transfer equipment, which involves a continuous countercurrent contact of two immiscible phases. These towers are vertical columns that have been filled with packing as illustrated in Figure 31.3. A variety of packing materials is used, ranging from specially designed ceramic and plastic packing, as illustrated in Figure 31.4, to crushed rock. The chief purpose of the packing is to provide a large contact area between the two immiscible phases. The liquid is distributed over the packing and flows down the packing surface as thin films or subdivided streams. The gas generally flows upward, countercurrent to the falling liquid. Both phases are well agitated. Thus, this type of equipment may be used for gas–liquid systems in which either of the phase resistances controls or in which both resistances are important.

Raschig ring   Berl saddle   Lessing ring   Pall ring

Figure 31.4 Common industrial tower packing.

Special types of packed towers are used to cool water so that it can be recirculated as a heat-transfer medium. These structures are made of wood-slat decks, having louver construction so that air can flow across each deck. The water is sprayed above the top deck and then trickles down through the various decks to a bottom collection basin. Cooling towers may be classified as natural draft when sufficient natural wind is available to carry away the humid air or as forced or induced draft when a fan is used. In the forced-draft towers, air is pulled into louvers at the bottom of the structure and then flows up through the decks countercurrent to the water flow.

Bubble-plate and *sieve-plate towers* are commonly used in industry. They represent the combined transfer mechanisms observed in the spray and the bubble towers. At each plate, bubbles of gas are formed at the bottom of a liquid pool by forcing the gas through small holes drilled in the plate or under slotted caps immersed in the liquid. Interphase mass transfer occurs during the bubble formation, and as the bubbles rise up through the agitated liquid pool. Additional masstransfer takes place above the liquid pool because of spray carryover produced by the active mixing of the liquid and gas on the plate. Such plates are arranged one above the other in a cylindrical shell as schematically illustrated in Figure 31.5. The liquid flows downward, crossing first the upper plate and then the plate below. The vapor rises through each plate. As Figure 31.5 illustrates, the contact of the two phases is stepwise. Such towers cannot be designed by equations that are obtained by integrating over a continuous area of interphase contact. Instead, they are designed by stagewise calculations that are developed and used in design courses of stagewise operations. We shall not consider the design of plate towers in this book; our discussions will be limited to continuous-contact equipment.

31.2 GAS–LIQUID MASS-TRANSFER OPERATIONS IN WELL-MIXED TANKS

Aeration is a common gas–liquid contacting operation where compressed air is introduced to the bottom of a tank of liquid water through small-orifice dispersers, such as

Figure 31.5 Plate towers.

perforated pipes, porous sparger tubes, or porous plates. These dispersers produce small bubbles of gas that rise through the liquid. Often, rotating impellers break up the bubble swarms and disperse the bubbles throughout the liquid volume. Gas–liquid mass-transfer processes induced by aeration include *absorption* and *stripping*. In gas absorption, a solute in the aeration gas is transferred from the gas to the liquid. Often, the solute is the oxygen gas in air, which is sparingly soluble in water. The absorption of oxygen into water underlies many processes important to biochemical engineering. In liquid stripping, the volatile dissolved solute is transferred from the liquid to the aeration gas. Stripping underlies many wastewater-treatment processes important to environmental engineering.

The gas–liquid contacting pattern in aeration processes is gas dispersed, meaning that the gas is dispersed within a continuous liquid phase. Consequently, the material balances for solute mass transfer are based on the liquid phase. Recall from Section 29.2 that the interphase mass-transfer flux for solute A across the gas- and liquid-phase films, based on the overall liquid phase driving force, is

$$N_A = K_L(c_A^* - c_A) \qquad (29\text{-}11)$$

and the transfer rate for solute A is

$$W_A = K_L \frac{A_i}{V} V(c_A^* - c_A) = K_L a \cdot V(c_A^* - c_A) \qquad (30\text{-}27)$$

with

$$c_A^* = \frac{p_A}{H}$$

where P_A is the partial pressure of solute A in the bulk gas phase. Recall from Section 30.5 that the interphase mass-transfer area per unit volume is hard to measure, and so capacity coefficients, e.g., $K_L a$, are used.

Well-mixed gas–liquid contacting processes can be either continuous or batch with respect to the liquid phase. A continuous process is shown in Figure 31.6. For a batch

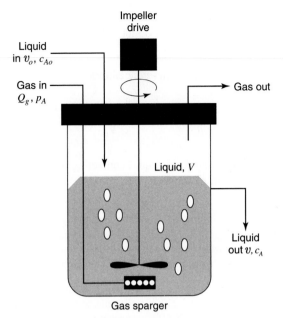

Liquid
in v_o, c_{Ao}

Impeller
drive

Gas in
Q_g, p_A

Gas out

Liquid, V

Liquid
out v, c_A

Gas sparger

Figure 31.6 Aerated stirred tank.

process, the liquid flow is turned off, but the gas flow remains on. In this case, the unsteady-state material balance on solute A in the liquid phase is

$$0 - 0 + N_A \cdot A_i + R_A \cdot V = \frac{d(c_A V)}{dt}$$

which is subject to the initial condition $t = 0$, $c_A = c_{Aa}$. If the liquid volume V is constant, then

$$K_L a \cdot V(c_A^* - c_A) + R_A \cdot V = V \frac{dc_A}{dt}$$

Furthermore, if (1) the partial pressure of solute A is constant so that C_A^* is constant and (2) there is no homogeneous reaction of dissolved solute A in the liquid phase so that $R_A = 0$, then

$$\int_{c_{A0}}^{c_A} \frac{-dc_A}{c_A^* - c_A} = -K_L a \int_0^t dt$$

which upon integration yields

$$\ln\left(\frac{c_A^* - c_{Aa}}{c_A^* - c_A^i}\right) = K_L a \cdot t$$

or

$$c_A = c_A^* - (c_A^* - c_{Ao})e^{-k_L a \cdot t} \tag{31-1}$$

In equation (31-1), the concentration of A in the liquid (c_A) exponentially approaches c_A^* as time t goes to infinity.

For a continuous process with one liquid inflow stream and one liquid outflow stream, the steady-state material balance is

$$c_{Ao}\dot{V}_o + N_A A_i - c_A \dot{V} + R_A \cdot V = 0$$

where c_{Ao} is now the inlet concentration of solute A. For a dilute process, the inlet liquid volumetric flow rate v_0 approaches outlet volumetric flow rate v. Consequently

$$\dot{V}_o(c_{Ao} - c_A) + K_L a \cdot V(c_A^* - c_A) + R_A \cdot V = 0 \qquad (31\text{-}2)$$

If $R_A = 0$, then the predicted outlet concentration is

$$c_A = \frac{\dfrac{\dot{V}_o}{V} c_{Ao} + K_L a \cdot c_A^*}{\dfrac{\dot{V}_o}{V} + K_L a} \qquad (31\text{-}3)$$

An application of well-mixed gas–liquid contacting operations to the design of an aerobic fermenter is provided in example 1.

EXAMPLE 1	The design of aeration systems for aerobic-fermentation processes is based on gas–liquid mass transfer. Microorganisms grow in a liquid suspension and feed on dissolved nutrients such as glucose and mineral salts. *Aerobic* microorganisms in liquid suspension also require dissolved oxygen for growth. If oxygen is not supplied at a rate sufficient to support cell growth, the cells will die.

In the present process, *Aerobacter aerogenes* is being cultivated within a continuous flow fermenter of 3 m³ liquid volume (V) and tank diameter (d_T) of 1.5 m. Fresh nutrient medium containing a trace amount of dissolved O_2 at concentration 0.01 mole O_2/m³ enters the fermenter at a flow rate of 1.8 m³/h. At steady-state conditions, the aerobic fermenter operates at a cell concentration (c_X) of 5 kg dry mass m³ of liquid culture. The cell concentration is determined by the specific growth rate of the organism and the nutrient composition of the liquid medium, details of which will not be presented here. The liquid cell suspension consumes oxygen proportional to the cell concentration according to the rate equation

$$R_A = -q_o \cdot c_X$$

where q_o is the specific oxygen consumption rate of the cells, equal to 20 mole O_2/kg cells $\cdot$ h, which is assumed to be constant. Determine the $K_L a$ value necessary to ensure that the dissolved oxygen concentration in the liquid culture (c_A) is at least 0.05 mol/m³. Also, determine the power input into a 3 m³ fermenter if the gas flow rate into the fermenter is 1 m³ of air per minute at the process conditions of 298 K and 1 atm. Assume that the bubbles are noncoalescing. At 298 K, Henry's law constant for dissolution of O_2 in the liquid nutrient medium is 0.826 atm $\cdot$ m³/mol.

The required $K_L a$ is backed out from a material balance on dissolved oxygen (species A) within the well-mixed liquid phase of the fermenter. Recall equation (31-2)

$$\dot{V}_o(c_{Ao} - c_A) + K_L a \cdot V(c_A^* - c_A) + R_A \cdot V = 0$$

Inserting $R_A = -q_o c_X$ and solving for the required $K_L a$ yields

$$K_L a = \frac{q_o \cdot c_X - \dfrac{\dot{V}_o}{V}(c_{Ao} - c_A)}{c_A^* - c_A} \qquad (31\text{-}4)$$

The saturation concentration of dissolved oxygen is determined by Henry's law

$$c_A^* = \frac{p_A}{H} = \frac{0.21 \text{ atm}}{0.826 \dfrac{\text{atm} \cdot \text{m}^3}{\text{mol}}} = 0.254 \frac{\text{mol } O_2}{\text{m}^3}$$

The partial pressure of oxygen (p_A) is presumed constant, as the rate of O_2 transferred to the sparingly soluble liquid is very small in comparison to the molar flow rate of O_2 in the aeration gas. Finally

$$K_L a = \frac{\left(20 \dfrac{\text{mol } O_2}{\text{kg cells} \cdot \text{h}} \cdot 5 \dfrac{\text{kg cells}}{\text{m}^3} - \dfrac{1.8 \text{ m}^3/\text{h}}{3.0 \text{ m}^3}(0.01 - 0.05) \dfrac{\text{mol } O_2}{\text{m}^3}\right) \cdot \dfrac{1 \text{ h}}{3600 \text{ s}}}{(0.254 - 0.05) \dfrac{\text{mol } O_2}{\text{m}^3}} = 0.136 \text{ s}^{-1}$$

In the limiting case of $c_A = 0$ and $c_{Ao} = 0$, the minimum $K_L a$ for O_2 transfer is defined as

$$(K_L a)_{min} = \frac{q_o \cdot c_X}{c_A^*} = \frac{\left(\dfrac{\text{mol } O_2}{\text{kg cells} \cdot \text{h}} \cdot 5 \dfrac{\text{kg cells}}{\text{m}^3} \right) \dfrac{1 \text{ h}}{3600 \text{ s}}}{0.254 \dfrac{\text{mol } O_2}{\text{m}^3}} = 0.109 \text{ s}^{-1}$$

From the above equation, it is evident that the biological oxygen consumption most strongly determines the required $K_L a$.

The power input to the aerated tank is backed out from the correlation

$$(k_L a)_{O_2} = 2 \times 10^{-3} \left(\frac{P_g}{V} \right)^{0.7} (u_{gs})^{0.2} \tag{30-29}$$

where $K_L a$ has units of s^{-1}, P_g/V has units of W/m^3, and u_{gs} has units of m/s. The superficial velocity of the gas through the empty tank is

$$u_{gs} = \frac{4Q_g}{\pi d_T^2} = \frac{(4) \left(1 \dfrac{\text{m}^3}{\text{min}} \cdot \dfrac{1 \text{min}}{60 \text{ s}} \right)}{\pi (1.5 \text{ m})^2} = 0.0094 \frac{\text{m}}{\text{s}}$$

If the gas is sparingly soluble in the liquid, the interphase mass-transfer process is liquid-phase controlling so that $K_L a \cong k_L a$. Therefore

$$0.136 = 2 \times 10^{-3} \left(\frac{P_g}{V} \right)^{0.7} (0.0094)^{0.2}$$

or

$$\frac{P_g}{V} = 1572 \frac{\text{W}}{\text{m}^3}$$

The total required power input (Pg) for the 3 m^3 aerated fermenter is 4716 W.

Eckenfelder[1] developed a general correlation for the transfer of oxygen from air bubbles rising in a column of still water

$$K_L \frac{A}{V} = \frac{\theta_g Q_g^{1+n} h^{0.78}}{V} \tag{31-5}$$

where θ_g is a correlating constant dependent on the type of disperser, Q_g is the gas flow rate in standard cubic feet per minute, n is a correlating constant that is dependent on the size of the small orifices in the disperser, and h is the depth below the liquid surface at which the air is introduced to the aeration tank. Typical data for a sparger-aeration unit, correlated according to equation (31-5), are presented in Figure 31.7.

Figure 31.7 Oxygen transfer factor of a single sparger unit in an aeration tank.

[1]W. W. Eckenfelder, Jr., *J. Sanit. Engr. Div.*, Amer. Soc. Civ. Engr., **85**, SA4, 89 (1959).

EXAMPLE 2

A 20,000-ft^3(566 m^3) aeration pond is aerated with 15 spargers, each using compressed air at a rate of 15 standard cubic feet per minute (7.08 × 10^{-3} m^3/s). The spargers will be located 15 ft (4.57 m) below the surface of the pond. Find the time required to raise the dissolved oxygen from 2 to 5 mg/L if the water temperature is 293 K.

From Figure 3.16, the transfer factor, $K_L(A/V)V$, for a single sparger is 1200 ft^3/h (9.44 × 10^{-3} m^3/s) and for the system

$$K_L\left(\frac{A}{V}\right) = \frac{(9.44 \times 10^{-3} \text{ m}^3/\text{s})(15 \text{ spargers})}{566 \text{ m}^3} = 2.50 \times 10^{-4} s^{-1}.$$

The average hydrostatic pressure of the rising air bubble is equal to the arithmetic mean of the pressure at the top and the bottom of the pond.

$$P_{\text{bottom}} = 1 \text{ atm} + (15 \text{ ft H}_2\text{O})(0.0295 \text{ atm/ft H}_2\text{O})$$

$$= 1.44 \text{ atm } (1.459 \times 10^5 \text{ Pa})$$

$$P_{\text{mean}} = \frac{1 \text{ atm} + 1.44 \text{ atm}}{2} = 1.22 \text{ atm}(1.236 \times 10^5 \text{ Pa})$$

As the mole fraction of oxygen in air is 0.21, the partial pressure of oxygen within the bubble will equal $y_{o_2}P = (0.21)(1.236 \times 10^5 \text{ Pa}) = 2.60 \times 10^4 \text{ Pa}$. The equilibrium concentration of a slightly soluble gas is related to its partial pressure by Henry's law. At 293 K, this law stipulates for oxygen

$$p_{O_2} = (4.06 \times 10^9 \text{ Pa/mole fraction}) \cdot x_{O_2}$$

Accordingly

$$x_{O_2} = \frac{2.60 \times 10^4 \text{ Pa}}{4.06 \times 10^9 \text{ Pa/mol fraction}} = 6.40 \times 10^{-6}$$

For 1 L of solution, which is essentially pure water, the equilibrium concentration in milligrams per liter can be calculated

$$\text{moles of water} = \frac{(1000 \text{ cm}^3 \text{ of water})(1 \text{ g/cm}^3 \text{ water})}{18 \text{ g water/mol}}$$

$$= 55.6 \text{ mol}.$$

In the liter of water, the moles of oxygen equal

$$x_{O_2} \cdot (\text{moles of solution}) = (6.4 \times 10^{-6})(55.6 \text{ mol}) = 3.56 \times 10^{-4} \text{ mol}$$

The grams of oxygen per liter equal

$$(3.56 \times 10^{-4} \text{ mol})(32 \text{g/mol}) = 1.139 \times 10^{-2} \text{ g/L} = 11.39 \quad \text{mg/L}.$$

Using equation (31-1), we can solve for the required time

$$t = \ln\left(\frac{c_A^* - c_{Ao}}{c_A^* - c_{At}}\right)\left(\frac{1}{K_L\left(\frac{A}{V}\right)}\right) = \ln\left(\frac{11.39 - 2}{11.39 - 5}\right)\left(\frac{1}{2.50 \times 10^{-4}/\text{s}}\right)$$

$$t = 1540 \text{ s}.$$

31.3   MASS BALANCES FOR CONTINUOUS CONTACT TOWERS: OPERATING-LINE EQUATIONS

There are four important fundamentals that constitute the basis for continuous-contact equipment design:

1. Material and enthalpy balances, involving the equations of conservation of mass and energy;
2. Interphase equilibrium;
3. Mass-transfer equations;
4. Momentum-transfer equations.

Interphase-equilibrium relations are defined by laws of thermodynamics as discussed in Section 29.1. Momentum-transfer equations, as presented in Section 9.3, are used to define the pressure drop within the equipment. We shall not treat this subject in this chapter, as it was previously discussed. *The material and enthalpy balances are important, as they provide expressions for evaluating the bulk compositions of the two contacting phases at any plane in the tower as well as the change in bulk compositions between two planes in the tower.* The mass-transfer equations will be developed in differential form, combined with a differential material balance, and then integrated over the area of interfacial contact to provide the length of contact required in the mass exchanger.

Countercurrent Flow

Consider any steady-state mass-transfer operation that involves the countercurrent contact of two insoluble phases as schematically shown in Figure 31.8. The two insoluble phases will be identified as phase G and phase L.

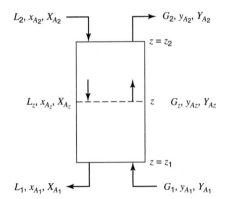

Figure 31.8 Steady-state countercurrent process.

At the bottom of the mass-transfer tower, the flow rates and concentrations are defined as follows:

G_1 is the total moles of phase G entering the tower per hour per cross-sectional area of the tower;

L_1 is the total moles of phase L leaving the tower per hour per cross-sectional area of the tower;

y_{A_1} is the mole fraction of component A in G_1, expressed as moles of A per total moles in phase G;

x_{A_1} is the mole fraction of component A in L_1, expressed as moles of A per total moles in phase L.

Similarly at the top of the tower, or plane z_2, the total moles of each phase will be G_2 and L_2, and the compositions of each stream will be y_{A2} and x_{A2}. An overall, macroscopic mass balance for component A around the steady-state mass exchanger, in which there is no chemical production or disappearance of A, requires

$$\begin{bmatrix} \text{moles of } A \text{ entering} \\ \text{the tower} \end{bmatrix} = \begin{bmatrix} \text{moles of } A \text{ leaving} \\ \text{the tower} \end{bmatrix}$$

or

$$G_1 y_{A_1} + L_2 x_{A_2} = G_2 y_{A_2} + L_1 x_{A_1} \tag{31-6}$$

A mass balance for component A around plane $z = z_1$ and the arbitrary plane z stipulates

$$G_1 y_{A_1} + L_z x_{A_z} = G_z y_{A_z} + L_1 x_{A_1} \tag{31-7}$$

Simpler relations, and certainly easier equations to use, may be expressed in terms of *solute-free concentration units*. The concentration of each phase will be defined as follows:

Y_A is the moles of A in G per mole of A-free G; that is,

$$Y_A = \frac{y_A}{1 - y_A} \tag{31-8}$$

and X_A is the moles of A in L per mole of A-free L; that is

$$X_A = \frac{x_A}{1 - x_A} \tag{31-9}$$

The flow rates to be used with the solute-free concentration units are L_S and G_S, where L_S is the moles of phase L on a solute-free basis, that is, the moles of carrier solvent in phase L per hour per cross-sectional area of the tower, and is equal to $L(1 - x_A)$ where both L and x_A are evaluated at the same plane in the tower, that is, $L_1(1 - x_{A_1})$ or $L_2(1 - x_{A_2})$. G_S is the moles of phase G on a solute-free basis; that is, the moles of carrier solvent in phase G per hour per cross-sectional area of the tower and is equal to $G(1 - x_A)$ where both G and y_A are evaluated at the same plane in the tower. The overall balance on component A may be written using the solute-free terms as

$$G_S Y_{A_1} + L_S X_{A_2} = G_S Y_{A_2} + L_S X_{A_1}$$

or

$$G_S(Y_{A_1} - Y_{A_2}) = L_S(X_{A_1} - X_{A_2}) \tag{31-10}$$

Rearranging we obtain

$$\frac{L_S}{G_S} = \frac{Y_{A_1} - Y_{A_2}}{X_{A_1} - X_{A_2}}$$

Equation (31-10) is an equation of a straight line that passes through points (x_{A_1}, y_{A_1}) and (x_{A_2}, y_{A_2}) with a slope of L_S/G_S. A mass balance of component A around plane z_1 and an arbitrary plane $z = z$ in solute-free terms is

$$G_S Y_{A_1} + L_S X_{Az} = G_S Y_{Az} + L_S X_{A_1}$$

or

$$G_S(Y_{A_1} - Y_{Az}) = L_S(X_{A_1} - X_{Az}) \tag{31-11}$$

Rearranging, we obtain

$$\frac{L_S}{G_S} = \frac{Y_{A_1} - Y_{Az}}{X_{A_1} - X_{Az}}$$

Equation (31-11) is an equation of a straight line, one which passes through the points (X_{A_1}, Y_{A_1}) and (X_{A_2}, Y_{A_2}) with a slope of L_S/G_S. Two straight lines having the same slope and a point in common lie on the same straight line. Equation 31.40, is therefore, a general expression relating the bulk compositions of the two phases at any plane in the mass exchanger. As it defines operating conditions within the equipment, it is designated the *operating line for countercurrent operations*. In our earlier discussions on interphase transfer in Section 29.2, point O of Figure 29.6 is one of many points that lie on the operating line.

It is important that the student recognizes the difference between equations (31-7) and (31-11). Although both equations describe the mass balance for component A, only equation (31-11) is an equation of a straight line. When written in the solute-free units, X and Y, the operating line is straight because the mole-ratio concentrations are based on both the constant quantities, L_s and G_s. When written in mole-fraction units, x and y, the total moles in a phase, L or G, change as the solute is transferred into or out of the phase; this produces a curved operating line on the x–y coordinates.

Figure 31.9 illustrates the location of the operating line relative to the equilibrium line when the transfer of the solute is from phase G to phase L as in the case of absorption. The bulk equilibrium, located on the operating line, must be greater than the equilibrium concentration in order to provide the driving forces, $Y_{AG} - Y_{Ai}$ or its equivalent or $p_{AG} - p_{Ai}$ or $Y_{AG} - Y_A^*$ or the equivalent $p_{AG} - p_A^*$, needed for transfer from G phase to L phase.

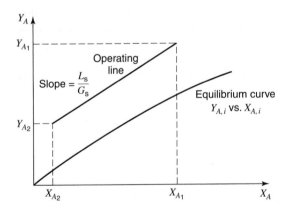

Figure 31.9 Steady-state countercurrent process, transfer from phase G to L.

Figure 31.10 illustrates the location of the operating line relative to the equilibrium line when the transfer of the solute is from phase L to phase G as in desorption or stripping. The location of the operating line below the equilibrium line assures correct driving forces, $x_{Ai} - x_{AL}$ or its equivalent $c_{Ai} - c_{AL}$ or $x_A^* - x_{AL}$ or its equivalent $c_A^* - c_{AL}$.

A mass balance for component A over the differential length, dz, is easily obtained by differentiating equation (31-11). This differential equation

$$L_S \, dX_A = G_S \, dY_A$$

relates the moles transferred in countercurrent operations per time per cross-sectional area available in the length dz.

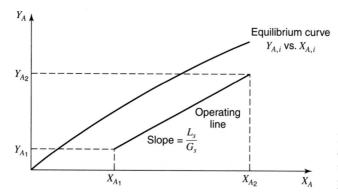

Figure 31.10 Steady-state countercurrent process, transfer from phase L to phase G.

In the design of mass-transfer equipment, the flow rate of at least one phase and three of the four entering and exiting compositions must be fixed by the process requirements. The necessary flow rate of the second phase is often a design variable. For example, consider the case in which phase G, with a known G_S, changes in composition from Y_{A_1} to Y_{A_2} by transferring solute to a second phase that enters the tower with composition X_{A_2}. According to equation (31-10), the operating line must pass through point (X_{A_2}, Y_{A_2}) and must end at the ordinate Y_{A_1}. Three possible operating lines are shown in Figure 31.11. Each line has a different slope, L_S/G_S, and since G_S is fixed by the process requirement, each line represents a different quantity, L_S, of the second phase. In fact, as the slope decreases, L_S decreases. The *minimum, L_S which may be used*, corresponds to the operating line ending at point P_3. This quantity of the second phase corresponds to an operating line that touches the equilibrium line. If we recall from Chapter 29 the definition of driving forces, we should immediately recognize that the closer the operating line is to the equilibrium curve, the smaller will be the driving force for overcoming any mass-transfer resistance. At the point of tangency, the diffusional driving force is zero; thus, mass transfer between the two phases cannot occur. This then represents a limiting condition, the *minimum L_S/G_S ratio for mass transfer*. In the case of equilibrium curves that are concave upward, the minimum L_S/G_S ratio corresponds to the second phase L_1 leaving in equilibrium with the entering phase G_1; that is, point (X_1, Y_1) lies on the equilibrium curve.

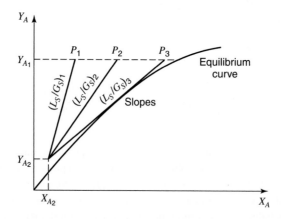

Figure 31.11 Operating-line locations.

EXAMPLE 3

Ammonia is to be absorbed from an air mixture at 293 K and 1.013×10^5 Pa pressure in a countercurrent packed tower, using water at 293 K as the absorbent. An inlet gas rate of 1.21×10^{-2} m³/s and an ammonia-free water rate of 9.46×10^{-3} kg/s will be used. If the ammonia, NH_3, concentration is reduced from 3.52 to 1.29% by volume, determine the ratio of $(L_S/G_S)_{actual}/(L_S/G_S)_{min}$. Equilibrium data for the system at 293 K and 1.013×10^5. Pa are as follows:

$X \dfrac{\text{kg mol NH}_3}{\text{kg mol H}_2\text{O}}$	0.0164	0.0252	0.0349	0.0455	0.0722
$Y \dfrac{\text{kg mol NH}_3}{\text{kg mol H}_2\text{O}}$	0.021	0.032	0.042	0.053	0.08

The total moles of gas entering the tower per hour may be evaluated using the ideal gas law.

$$G_1 A = \text{moles entering gas/s} = \frac{P\dot{V}}{RT} = \frac{(1.013 \times 10^5 \text{Pa})(1.21 \times 10^2 \text{ m}^2/\text{s})}{\left(8.314 \dfrac{\text{Pa} \cdot \text{m}^3}{\text{mol} \cdot \text{K}}\right)(293 \text{ K})}$$

$$= 0.503 \quad \text{mol/s}$$

The gas enters the tower with a mole fraction of ammonia, Y_{A_1}, equal to 0.0352. Choosing a cross-sectional area for the tower of A m^2, we may evaluate the moles of G on a solute-free basis as

$$G_S = G_1(1 - y_{A1}) = \left(0.503 \frac{\text{mol}}{\text{s}}\right)\left(\frac{0.9648}{A \text{ m}^2}\right) = \frac{0.483}{A} \frac{\text{mol}}{\text{s} \cdot \text{m}^2}$$

The moles of phase L on a solute-free basis are

$$L_S = \left(9.46 \times 10^{-3} \frac{\text{kg}}{\text{s}}\right)\left(\frac{\text{kg mol}}{18 \text{ kg}}\right)\left(\frac{1}{A \text{ m}^2}\right) = \frac{5.26 \times 10^{-4}}{A} \frac{\text{kg mol}}{\text{s} \cdot \text{m}^2} = \frac{0.526}{A} \frac{\text{mol}}{\text{s} \cdot \text{m}^2}$$

The ratio of the actual L_S to G_S is evaluated as

$$\left(\frac{L_S}{G_S}\right)_{actual} = \frac{0.526}{A} \frac{A}{0.483} = 1.09 \frac{\text{mole NH}_{3\text{-free}}L \text{ phase}}{\text{mole NH}_{3\text{-free}}G \text{ phase}}$$

The composition of the known streams, G_1, G_2, and L_2 on a solute-free basis, are evaluated from the known mole fraction as

$$Y_{\text{NH}_3,1} = \frac{y_{\text{NH}_3,1}}{1 - y_{\text{NH}_3,1}} = \frac{0.0352}{0.9648} = 0.0365$$

$$Y_{\text{NH}_3,2} = \frac{y_{\text{NH}_3,2}}{1 - y_{\text{NH}_3,2}} = \frac{0.0129}{0.9871} = 0.0131$$

$$X_{\text{NH}_3,2} = \frac{x_{\text{NH}_3,2}}{1 - x_{\text{NH}_3,2}} = 0$$

The exiting composition, $X_{\text{NH}_3,1}$, can be evaluated by

$$G_S(Y_{A_1} - Y_{A_2}) = L_S(X_{A_1} - X_{A_2})$$

$$\frac{0.483}{A}(0.0365 - 0.0131) = \frac{0.526}{A}(X_{A_1} - 0)$$

$$X_{A_1} = 0.0215$$

The actual and minimum operating lines are shown in Figure 31.12. The composition of the solution in equilibrium with $Y_{\text{NH}_3,1} = 0.0365$ is obtained from the equilibrium curve as $X_{\text{NH}_3, \text{ equil.}} = 0.0296$. The slope of the minimum operating line is

$$\left(\frac{L_S}{G_S}\right)_{minimum} = \frac{\Delta Y}{\Delta X} = \frac{0.0365 - 0.0131}{0.0296 - 0} = 0.79 \frac{\text{mole NH}_{3\text{-free}} \text{ liquid phase}}{\text{mole NH}_{3\text{-free}} \text{ gas phase}}$$

The desired ratio, $(L_S/G_S)_{actual}/(L_S/G_S)_{minimum}$ is then a ratio of the two values, 1.09/0.79 or 1.38.

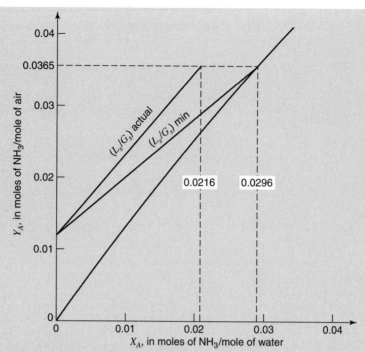

Figure 31.12 Solution to example 3.

EXAMPLE 4

In Problem 30.10, the production of ultra pure silicon tetrachloride for the chemical vapor deposition of silicon was described. Hydrogen chloride, HCl, was produced in the elimination of trichlorosilane, SiHCl$_3$, and the HCl was stripped from the liquid silicon tetrachloride into an N$_2$ carrier gas stream.

(a) If the HCl concentration was reduced from 1.8 to 0.2 mol% by its transfer into the 298 K and 1.013×10^5 Pa nitrogen gas carrier stream, determine the $(L_s/G_s)_{min}$ ratio. At the involved low concentration levels, the HCl–H$_2$ system obeys Henry's law with the Henry's constant, H', equal to 47.5 atm/mol fraction HCl in the liquid.

(b) Determine the HCl concentration in the exiting N$_2$ stream when 40 times the minimum G_s is used in the stripper.

The compositions of the known streams, L_2, L_1, and G_1, on a solute-free basis are evaluated from the known mole fractions as

$$X_{HCl,2} = \frac{x_{HCl,2}}{1 - x_{HCl,2}} = \frac{0.018}{0.982} = 0.0183$$

$$X_{HCl,1} = \frac{x_{HCl,1}}{1 - x_{HCl,1}} = \frac{0.002}{0.998} = 0.002$$

$$Y_{HCl,1} = \frac{y_{HCl,1}}{1 - y_{HCl,1}} = 0$$

Note that the solute-free compositions are essentially equal to the mole-fraction compositions for low concentrations.

The equilibrium concentration is evaluated with Henry's law

$$p_{HCl}(atm) = \left(47.5 \frac{atm}{mole\ fraction}\right)(x_{HCl})$$

Figure 31.13 Solution to example 4.

At end 2 of the tower

$$p_{HCl,2} = \left(47.5 \frac{atm}{mole\ fraction}\right)(0.018) = 0.855\ atm$$

and

$$Y^*_{HCl,2} = \frac{p_{HCl,2}}{P - p_{HCl,2}} = \frac{0.855}{0.145} = 5.90$$

Figure 31.13 illustrates the operating lines for both $G_{S,min}$ and $G_{S,actual}$ for this example. As one notes, the operating lines for the stripping process are below the equilibrium line.

$$\left(\frac{L_S}{G_S}\right)_{min} = \frac{Y^*_2 - Y_1}{X_2 - X_1} = \frac{5.90 - 9}{0.0183 - 0.002} = 362$$

and

$$\left(\frac{L_S}{G_S}\right)_{actual} = \frac{L_S}{(G_S)(40)} = \frac{362}{40} = 9.05$$

$$\left(\frac{L_S}{G_S}\right)_{actual} = 9.05 = \frac{Y_2 - Y_1}{X_2 - X_1} = \frac{Y_2 - 0}{0.0183 - 0.002}$$

$$Y_2 = 0.147$$

The mole fraction of the exiting gas stream is

$$y_{HCl,2} = \frac{Y_2}{1 + Y_2} = \frac{0.147}{1.147} = 0.128.$$

Cocurrent Flow

For steady-state mass-transfer operations involving cocurrent contact of two immiscible phases as shown in Figure 31.14, the overall mass balance for component A is

$$L_S X_{A_2} + G_S Y_{A_2} = L_S X_{A_1} + G_S Y_{A_1}$$

or

$$L_S(X_{A_2} - X_{A_1}) = G_S(Y_{A_1} - Y_{A_2}) \tag{31-12}$$

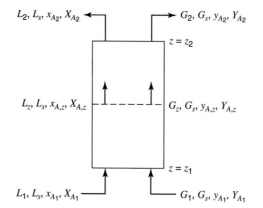

Figure 31.14 Steady-state cocurrent process.

The mass balance on component A around planes 1 and an arbitrary plane z stipulates that

$$L_S X_{A,z} + G_S Y_{A,z} = L_S X_{A_1} + G_S Y_{A_1}$$

or

$$L_S(X_{A,z} - X_{A_1}) = G_S(Y_{A_1} - Y_{A,z}) \tag{31-13}$$

Equations (31-12) and (31-13) are both straight lines that pass through a common point (X_{A_1}, Y_{A_1}) and have the same slope, $-L_S/G_S$, Equation (31-13) is the general expression that relates to the composition of the two contacting phases at any plane in the equipment. It is designated as the *operating-line equation for cocurrent operations*. Figures 31.15 and 31.16 illustrate the location of the operating line relative to the equilibrium line. A mass balance for component A over the differential length, dz, for cocurrent flow:

$$L_S \, dX_A = -G_S \, dY_A$$

verifies the slope of the operating line for cocurrent operation is $-L_s/G_s$.

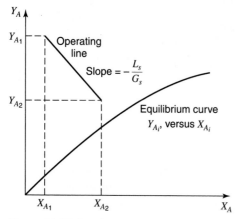

Figure 31.15 Steady-state cocurrent process, transfer from phase G to phase L.

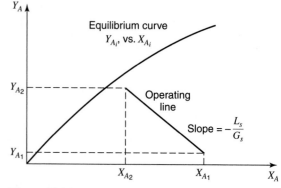

Figure 31.16 Steady-state cocurrent process transfer from phase L to phase G.

As in the case of countercurrent flow, there is a minimum L_s/G_s ratio for cocurrent mass-transfer operations established from the fixed process variables: $G_{S_1}, Y_{A_1}, Y_{A_2}$, and X_{A_1}. Its evaluation involves the same procedure as discussed for countercurrent flow.

EXAMPLE 5

The ammonia–air feed stream described in example 3 is fed cocurrently with an ammonia–free water stream. The ammonia concentration is to be reduced from 3.52 to 1.29% by volume, using a water stream 1.37 times the minimum. Determine (a) the minimum L_s/G_s ratio, (b) the actual water rate, and (c) the concentration in the exiting aqueous stream.

In example 3, the following compositions were evaluated:

$$\begin{array}{lll} \text{entering} & Y_{\text{NH}_3,1} = 0.0365 \\ \text{exiting} & Y_{\text{NH}_3,2} = 0.0131 \\ \text{entering} & X_{\text{NH}_3,1} = 0.0 \end{array}$$

The moles of G on a solute-free basis were evaluated to be $\frac{0.483}{A} \frac{\text{mol}}{\text{m}^2 \cdot \text{s}}$. In Figure 31.17, the minimum and actual operating lines are shown. For these operating lines

$$\left(\frac{L_s}{G_s}\right)_{\min} = \frac{Y_{\text{NH}_3,1} - Y_{\text{NH}_3,2}}{X_{\text{NH}_3,2} - X_{\text{NH}_3,2}} = \frac{0.0365 - 0.0131}{0.01 - 0}$$

$$= 2.34 \frac{\text{moles} \quad \text{NH}_3\text{-free } L \text{ phase}}{\text{moles NH}_3\text{-free } G \text{ phase}}$$

and

$$\left(\frac{L_S}{G_S}\right)_{\text{actual}} = 1.37 \left(\frac{L_S}{G_S}\right)_{\min} = 1.37(2.34) = 3.21 \frac{\text{moles NH}_3\text{-free } L \text{ phase}}{\text{moles NH}_3\text{-free } G \text{ phase}}$$

The composition of the exiting stream can be evaluated with the slope of the actual operating line by

$$\left(\frac{L_S}{G_S}\right)_{\text{actual}} = 3.21 = \frac{Y_{\text{NH}_3,1} - Y_{\text{NH}_3,2}}{X_{\text{NH}_3,2} - X_{\text{NH}_3,1}} = \frac{0.0365 - 0.0131}{X_{\text{NH}_3,2} - 0}$$

Figure 31.17 Solution to example 5.

or

$$X_{NH_3,2} = \frac{0.0234}{3.21} = 0.0073 = \frac{\text{mol NH}_3}{\text{mol NH}_3\text{-free water}}$$

The moles of NH_3-free water fed to the tower, L_s, is also evaluated using the value of

$$\left(\frac{L_S}{G_S}\right)_{\text{actual}} = 3.21 \frac{\text{mol NH}_3 \text{ free } L \text{ phase}}{\text{mol NH}_3 \text{ free } G \text{ phase}}$$

Then

$$L_S = 3.21 G_S = 3.21\left(\frac{0.481}{A}\right) = \frac{1.55}{A}\frac{\text{mol}}{\text{m}^2 \cdot \text{s}}$$

31.4 ENTHALPY BALANCES FOR CONTINUOUS-CONTACTS TOWERS

Many mass-transfer operations are isothermal. This is especially true when we are dealing with dilute mixtures. However, when large quantities of solute are transferred, the heat of mixing can produce a temperature rise in the receiving phase. If the temperature of the phase changes, the equilibrium solubility of the solute will be altered, and in turn, the diffusion driving forces will be altered.

Consider the steady-state countercurrent process illustrated in Figure 31.18. An enthalpy balance around the planes $z = z_2$ and z is

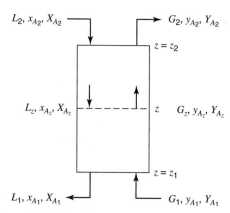

Figure 31.18 Steady-state countercurrent process.

$$L_2 H_{L_2} + G H_G = G_2 H_{G_2} + L H_L \tag{31-14}$$

where H is the molal enthalpy of the stream at its particular temperature, pressure, and concentration. The enthalpies are normally based upon a reference of pure solute-free carrier solvent and pure solute at a chosen base temperature, T_0. The normal enthalpy of a liquid mixture is evaluated above this base temperature by the relation

$$H_L = c_{pL}(T_L - T_0)M_{\text{avg}} + \Delta H_S \tag{31-15}$$

where H_L is the enthalpy of the liquid stream, in kJ/mol of L; c_{pL} is the heat capacity of the mixture on a mass basis, kJ/kg · K; T_L is the temperature of the mixture in K; M_{avg} is the average molecular weight of the mixture; and ΔH_S is the integral heat of solution, evaluated at the base temperature, T_0, and at the concentration of the mixture in kJ/mol.

The molal enthalpy of a gas mixture, with the same base temperature and standard state of the solute is expressed as

$$H_G = [y_{\text{solute}}c_{p,G \text{ solute}}M_{\text{solute}} + (1 - y_{\text{solute}})(c_{p,G \text{ solute-free } G\text{-phase}})$$
$$(M_{\text{solute-free } G\text{-phase}})](T_G - T_0) + y_{\text{solute}}h_{f,g \text{ solute}}M_{\text{solute}} \tag{31-16}$$

where H_G is the enthalpy of the gas stream, in kJ/mol of G; c_p, g is the heat capacity in the gas phase in kJ/kg·K; T_G is the temperature of the gas mixture in K; M is the molecular weight; and $h_{f,g\,\text{solute}}$ is the heat of vaporization of the solute in kJ/kg. The integral heat of solution, ΔH_S, is zero for ideal solutions and essentially zero for gas mixtures. For nonideal solutions, it is a negative quantity if heat is evolved on mixing and a positive quantity if heat is absorbed on mixing.

Equation (31-14) may be used to compute the temperature of a given phase at any plane within the mass-transfer equipment. The calculations involve the simultaneous application of the mass balance in order to know the flow rate of the stream associated with the particular enthalpy term.

31.5   MASS-TRANSFER CAPACITY COEFFICIENTS

The individual mass-transfer coefficient, k_G, was defined by the expression

$$N_A = k_G(p_{A,G} - p_{A,i}) \tag{29-6}$$

and the overall mass-transfer coefficient was defined by a similar equation in terms of the overall driving force in partial pressure units

$$N_A = K_G(p_{A,G} - p_A^*) \tag{29-10}$$

In both expressions, the interphase mass transfer was expressed as moles of A transferred per unit time per unit area per unit driving force in terms of partial pressure. In order to use these equations in the design of mass exchangers, the interphase contact area must be known. Although the wetted-wall column, as described in Chapter 26, has a definite interfacial surface area, the corresponding area in other types of equipment is virtually impossible to measure. For this reason the factor a must be introduced to represent the interfacial surface area per unit volume of the mass-transfer equipment. The mass transfer within a differential height, dz, per unit cross-sectional area of the mass exchanger is

$$N_A \left[\frac{\text{moles of } A \text{ transferred}}{(\text{time})(\text{interfacial area})}\right] \left[a\left(\frac{\text{interfacial area}}{\text{volume}}\right)\right] dz(\text{length})$$

$$= \frac{\text{moles of } A \text{ transferred}}{(\text{time})(\text{cross-sectional area})}$$

or, in terms of the mass-transfer coefficients

$$N_A a \, dz = k_G a(p_{A,G} - p_{A,i}) \, d_z \tag{31-17}$$

and

$$N_A a \, dz = K_G a(p_{A,G} - p_A^*) \, d_z \tag{31-18}$$

As both the factor a and the mass-transfer coefficients depend on the geometry of the mass-transfer equipment and on the flow rates of the two contacting, immiscible streams, they are commonly combined as a product. The *individual capacity coefficient, $k_G a$*, and the *overall capacity coefficient, $k_G a$*, are each experimentally evaluated as a combined process variable. The units of the gas phase capacity coefficient are

$$k_G a \left[\frac{\text{moles of } A \text{ transferred}}{(\text{time})\,(\text{interfacial area})(\text{pressure})}\right] \left[\frac{\text{interfacial area}}{\text{volume}}\right]$$

$$= \frac{\text{moles of } A \text{ transferred}}{(\text{time})(\text{volume})(\text{pressure})}$$

the most often encountered units are g moles of $A/s \cdot m^3 \cdot Pa$. The capacity coefficients in terms of liquid concentration driving forces are similarly defined by

$$N_A a \, dz = k_L a(c_{A,i} - c_{A,L}) \, dz \qquad (31\text{-}19)$$

and

$$N_A a \, dz = K_L a(c_A^* - c_{A,L}) \, dz \qquad (31\text{-}20)$$

The most common units for the liquid-phase capacity coefficients are g moles of $A/s \cdot m^3 \cdot$ g moles of A/m^3 of solution. Mass-transfer capacity equations in terms of $k_y a$, $k_x a$, $k_Y a$, and $k_X a$ are similarly defined.

31.6   CONTINUOUS-CONTACT EQUIPMENT ANALYSIS

The moles of the diffusing component A transferred per time per cross-sectional area have been defined by two entirely different concepts, the material balance and the mass-transfer equations. For equipment involving the continuous-contact between the two immiscible phases, these two equations may be combined and the resulting expression integrated to provide a defining relation for the unknown height of the mass exchanger.

Constant Overall Capacity Coefficient

Consider an isothermal, countercurrent mass exchanger used to achieve a separation in a system that has a constant overall mass-transfer coefficient $K_Y a$ through the concentration range involved in the mass-transfer operations. The mass balance for component A over the differential length dz is described by

$$\frac{\text{moles of } A \text{ transferred}}{(\text{time})(\text{cross-sectional area})} = L_S \, dX_A = G_S \, dY_A \qquad (31\text{-}11)$$

The mass transfer of component A in the differential length dz is defined by

$$\frac{\text{moles of } A \text{ transferred}}{(\text{time})(\text{cross-sectional area})} = N_A a \, dz = K_Y a(Y_A - Y_A^*) \, dz \qquad (31\text{-}21)$$

As equation (31-21) involves the mass flux of component A, N_A, it is a vectorial equation that not only defines the quantity of A transferred per time and cross-sectional area but it also indicates the direction of mass transfer. If the driving force, $Y_A - Y_A^*$, is positive, the transfer of A must be from the bulk composition in the G phase to the bulk composition in the L phase. The two differential quantities in equation (31-11), $L_S dX_A$ and $G_S dY_A$, stipulate only the quantity of A transferred per time and cross-sectional area; each term must have either a positive or a negative sign assigned to it to indicate the direction of the A transfer.

Let us consider the transfer of A from the G phase to the L phase; $-G_s \, dY_A$ indicates that the G phase is losing A. The mass transfer of component A in the differential length dz may be defined by combining equation (31-21) and the modified component of equation (31-11)

$$-G_S \, dY_A = K_Y a(Y_A - Y_A^*) \, dz$$

or

$$dz = \frac{-G_S}{K_Y a} \cdot \frac{dY_A}{Y_A - Y_A^*} \qquad (31\text{-}22)$$

This equation may be integrated over the length of the mass exchanger, with the assumption of a constant overall capacity coefficient

$$\int_{z_1}^{z_2} dz = \frac{-G_S}{K_Y a} \cdot \int_{Y_{A_2}}^{Y_{A_1}} \frac{dY_A}{Y_A - Y_A^*}$$

(31-23)

$$z = (z_2 - z_1) = \frac{G_S}{K_Y a} \cdot \int_{Y_{A_2}}^{Y_{A_1}} \frac{dY_A}{Y_A - Y_A^*}$$

The evaluation of the right-hand side of this equation often requires a numerical or a graphical integration. As discussed in Section 31.3, we may evaluate $Y_A - Y_A^*$ from the plot of Y_A vs. X_A as illustrated in Figure 31.19.

The vertical distance between the operating line and the equilibrium line represents the overall driving force in Y-units. It may be determined for each value of the bulk concentration, Y_A, and its reciprocal, $1/(Y_A - Y_A^*)$, may then be plotted vs. Y_A, as illustrated in Figure 31.20.

After obtaining the area under the curve in Figure 31.19, we may evaluate the length of the mass exchanger by Equation (31-23).

When the transfer is from the L phase to the G phase, as in the case of desorption or stripping, G, dy_A will be positive; the mass transfer of component A in the differential length dz will be

$$G_S \, dY_A = K_Y a (Y_A^* - Y_A) \, dz$$

or

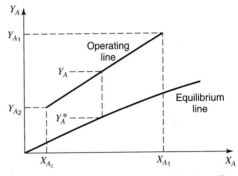

Figure 31.19 Evaluation of $Y_A - Y_A^*$ the overall driving force.

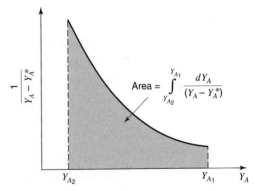

Figure 31.20 Evaluation of the integral

$$dz = \frac{G_S}{K_Y a} \cdot \frac{dY_A}{Y_A^* - Y_A} = \frac{-G_S}{K_Y a} \cdot \frac{dY_A}{Y_A - Y_A^*}$$

and the same design equation (31-23) will be obtained.

A similar but more complex equation can be derived in terms of the constant overall capacity coefficient, $K_G a$

$$z = \frac{G_S}{K_G a} \cdot \int_{Y_{A_2}}^{Y_{A_1}} \frac{dY_A}{p_{AG} - p_A^*}$$

(31-24)

As two different units for gas concentrations are involved, this equation is a little more difficult to evaluate.

The length of the mass exchanger can also be determined by an equation written in terms of the overall liquid capacity coefficient, $K_X a$. For the transfer of A from the G phase into the L phase,

$$-L_S \, dX_A = K_X a(X_A^* - X_A) \, dz$$

or

$$dz = \frac{-L_S}{K_X a} \cdot \frac{dX_A}{X_A^* - X_A}$$

If the capacity coefficient is constant over the concentration range involved in the mass transfer operation

$$\int_{z_1}^{z_2} dz = \frac{-L_S}{K_X a} \cdot \int_{X_{A_1}}^{X_{A_2}} \frac{dX_A}{X_A^* - X_A}$$

or   (31-25)

$$z = \frac{L_S}{K_X a} \cdot \int_{X_{A_2}}^{X_{A_1}} \frac{dX_A}{X_A^* - X_A}$$

The overall driving force, $X_A^* - X_A$, is the horizontal difference between the operating line and the equilibrium line values on a plot similar to Figure 31.19.

Variable Overall Capacity Coefficient: Allowance for Resistance in Both Gas and Liquid Phase

In Chapter 29, the overall coefficient was found to vary with concentration unless the equilibrium line was straight. Accordingly, we should expect that the overall capacity coefficient will also vary when the slope of the equilibrium line varies within the region that includes the bulk and interfacial concentrations. With slightly curved equilibrium lines, one may safely use the design equations (31-23)–(31-25). However, in the case of equilibrium lines with more pronounced curvature, the exact calculations should be based on one of the individual capacity coefficients.

The mass balance for component A over the differential length dz is

$$L_S \, dX_A = G_S \, dY_A \tag{31-11}$$

Differentiating Equation (31-8), we obtain

$$dY_A = \frac{dy_A}{(1 - y_A)^2}$$

This relation may be substituted into equation (31-11) to give

$$L_S \, dX_A = G_S \frac{dy_A}{(1 - y_A)^2} \tag{31-26}$$

The mass transfer of component A in the differential length, dz, is defined in terms of the individual gas-phase capacity coefficient by

$$N_A a \, dz = k_G a(p_{AG} - p_{Ai}) \, dz \tag{31-27}$$

Combining equation (31-27) and a modified equation (31-26) that accounts for the transfer from the G phase, we obtain upon rearranging

$$dz = \frac{-G_S \, dy_{AG}}{k_G a(p_{AG} - p_{Ai})(1 - y_{AG})^2}$$

or

$$dz = \frac{-G_S \, dy_{AG}}{k_G a P(y_{AG} - y_{Ai})(1 - y_{AG})^2} \qquad (31\text{-}28)$$

As discussed in Chapter 29, the interfacial compositions, y_{Ai} and x_{Ai} may be found for each point on the operating line by drawing a line from the point toward the equilibrium line. The slope of this line is $-k_L/k_G$ on a p_A vs. c_A plot, or is $-c k_L/k_G P$ on a y_A vs. x_A plot, where c is the molar concentration in the liquid phase. In Figure 31.21, the location of the interfacial compositions on both phase plots are illustrated. It is important to recall from the discussion in Section 31.3 that the operating line is not straight on plots of y_A vs. x_A and p_A vs. c_A, except when we are dealing with relatively dilute gas and liquid mixtures. Knowing the interfacial composition, y_{Ai}, for every bulk composition y_{AG} in the gas stream, we may numerically or graphically integrate equation (31-28) to obtain the length of the mass exchanger.

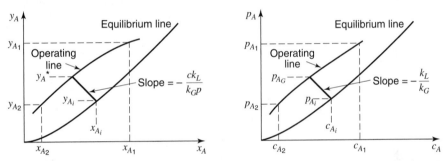

Figure 31.21 Determination of interfacial composition for transfer from phase G to phase L.

Logarithmic-Mean Driving Force

Although the graphical integration procedure must be employed in most practical design calculations, it is sometimes possible to use a much simpler equation based upon a logarithmic-mean driving force. When the two contacting streams are relatively dilute, the equilibrium curve and the operating line may both be linear in terms of the mole fractions over the range of concentration involved in the mass-transfer operation. Under these conditions, $G_1 \approx G_2 \approx G$ and $L_1 \approx L_2 \approx L$. The mass balance for component A may be approximated by

$$L(x_{A_1} - x_A) = G(y_{A_1} - y_A) \qquad (31\text{-}29)$$

or

$$L \, dx_A = G \, dy_A \qquad (31\text{-}30)$$

The rate of interphase transfer may be expressed in terms of the overall gas-phase capacity coefficient by

$$N_A a \, dz = K_G a(p_{A,G} - p_A^*) \, dz$$

or

$$N_A a \, dz = K_G a P (y_A - y_A^*) \, dz \tag{31-31}$$

As the operating and equilibrium lines are straight, the difference in the ordinates of the two lines must vary linearly in composition. Designating the difference $y_A - y_A^*$ by Δ, we see that this linearity stipulates

$$\frac{d\Delta}{dy_A} = \frac{\Delta_{\text{end}_1} - \Delta_{\text{end}_2}}{y_{A_1} - y_{A_2}} = \frac{\Delta_1 - \Delta_2}{y_{A_1} - y_{A_2}} \tag{31-32}$$

For the transfer of A from the G phase into the L phase, we can combine the rate of mass transfer from equation (31-31) and the mass balance for A using a modified component of equation (31-30) to obtain

$$dz = \frac{-G}{K_G a P} \cdot \frac{dy_A}{y_A - y_A^*} = \frac{-G}{K_G a P} \cdot \frac{dy_A}{\Delta}$$

$$dz = \frac{-G}{K_G a P} \cdot \frac{(y_{A_1} - y_{A_2})}{\Delta_1 - \Delta_2} \cdot \frac{d\Delta}{\Delta} \tag{31-33}$$

Integrating over the length of the mass exchanger, we obtain

$$z = \frac{G}{K_G a P} \cdot \frac{(y_{A_1} - y_{A_2})}{\Delta_1 - \Delta_2} \times \ln \frac{\Delta_1}{\Delta_2}$$

or

$$z = \frac{G}{K_G a P} \frac{y_{A_1} - y_{A_2}}{(y_A - y_A^*)_{lm}} \tag{31-34}$$

where

$$\Delta_{lm} = \frac{\Delta_1 - \Delta_2}{\ln \Delta_1 / \Delta_2} = (y_A - y_A^*)_{lm}$$

$$= \frac{(y_A - y_A^*)_1 - (y_A - y_A^*)_2}{\ln[(y_A - y_A^*)_1 / (y_A - y_A^*)_2]} \tag{31-35}$$

A similar expression in terms of the overall liquid-phase capacity coefficient is

$$z = \frac{L(x_{A_1} - x_{A_2})}{K_L a c (x_A^* - x_A)_{lm}} \tag{31-36}$$

where

$$(x_A^* - x_A)_{lm} = \frac{(x_A^* - x_A)_1 - (x_A^* - x_A)_2}{\ln \left[(x_A^* - x_A)_1 / (x_A^* - x_A)_2 \right]} \tag{31-37}$$

EXAMPLE 6

Ammonia is to be absorbed from air at 293 K and 1.013×10^5 Pa pressure in a countercurrent, packed tower, 0.5 m in diameter, using ammonia-free water as the absorbent. The inlet gas rate will be 0.2 m³/s and the inlet water rate will be 203 kg/s. Under these conditions, the overall capacity coefficient, K_Y, may be assumed to be $80 \dfrac{\text{mole}}{\text{m}^3 \cdot \text{s} \cdot \Delta Y_A}$. The ammonia concentration will be reduced from 0.0825 to 0.003 mol fraction. The tower will be cooled, the operation thus taking place essentially at 293 K; the equilibrium data of example 3 may be used. Determine the length of the mass exchanger.

The concentrations of three of the streams were given; these may be expressed on a NH_3-free basis by

$$Y_{NH_3,1} = \frac{y_{NH_3,1}}{1 - y_{NH_3,1}} = \frac{0.0825}{0.9175} = 0.09$$

$$Y_{NH_3,2} = \frac{y_{NH_3,2}}{1 - y_{NH_3,2}} = \frac{0.003}{0.997} = 0.003$$

and

$$X_{NH_3,2} = 0.0$$

The cross-sectional area of the tower is equal to $\frac{\pi D^2}{4} = \frac{\pi (0.5 \text{ m})^2}{4} = 0.196 \text{ m}^2$. The gas enters the exchanger at plane 1, and its flow

$$G = \frac{VP}{RT} \cdot \frac{1}{A} = \frac{(0.20 \text{ m}^3/\text{s})(1.013 \times 10^5 \text{ Pa})}{\left(8.314 \dfrac{\text{Pa} \cdot \text{m}^3}{\text{mol} \cdot \text{K}}\right)(293 \text{ K})} \cdot \frac{1}{(0.196 \text{ m}^2)} = 42.4 \frac{\text{mol}}{\text{m}^2 \cdot \text{s}}$$

The NH_3-free gas flow rate is

$$G_s = G_1(1 - y_{NH_3,1}) = 42.4 \frac{\text{mol}}{\text{m}^2 \cdot \text{s}} (0.9175) = 38.9 \frac{\text{mol}}{\text{m}^2 \cdot \text{s}}$$

The NH_3-free water flow rate is evaluated using the flow into the exchanger at plane 2.

$$L_s = (203 \text{ g/s}) \left(\frac{\text{mol}}{18 \text{ g}}\right) \left(\frac{1}{0.196 \text{ m}^2}\right) = 57.5 \frac{\text{mol}}{\text{m}^2 \cdot \text{s}}$$

The liquid leaves the exchanger at plane 1; its concentration is evaluated, using the countercurrent material balance equation

$$G_s(Y_{NH_3,1} - Y_{NH_3,2}) = L_s(X_{NH_3,1} - X_{NH_3,2})$$
$$38.9(0.09 - 0.003) = 57.5(X_{NH_3,1} - 0)$$
$$X_{NH_3,1} = 0.059$$

The operating and equilibrium lines are shown in Figure 31.22.

Figure 31.22 The operating line for example 6.

As this is not a case in which the gas and liquid concentrations are dilute enough for us to assume straight equilibrium and operating lines on the plot of y vs. x, the height of the tower must be evaluated by the graphical integration procedure, using

$$z = \frac{G_S}{K_Y a} \int_{Y_{A_2}}^{Y_{A_1}} \frac{dY_A}{Y_A - Y_A^*}$$

$$z = \frac{38.9 \dfrac{\text{mol}}{\text{m}^2 \cdot \text{s}}}{80 \dfrac{\text{mol}}{\text{m}^3 \cdot \text{s} \cdot \Delta Y_A}} \int_{Y_{NH_3,2}}^{Y_{NH_3,1}} \frac{dY_{NH_3}}{Y_{NH_3} - Y_{NH_3}^*}$$

In Table 31.1, Y_{NH_3} is the composition at a point on the operating line, and $Y_{NH_3}^*$ is the composition on the equilibrium line directly below the Y_{NH_3} value. An example of these compositions is illustrated in Figure 31.21. The integral

$$\int_{Y_{NH_3,2}}^{Y_{NH_3,1}} \frac{dY_{NH_3}}{Y_{NH_3} - Y_{NH_3}^*}$$

is graphically evaluated in Figure 31.23.

Table 31.1 Example 3 gas compositions

Y_A	Y_A^*	$Y_A - Y_A^*$	$1/(Y_A - Y_A^*)$
0.003	0	0.003	333.3
0.01	0.0065	0.0035	296
0.02	0.0153	0.0047	212.5
0.035	0.0275	0.0075	133.3
0.055	0.0425	0.0125	80.0
0.065	0.0503	0.0147	68.0
0.075	0.058	0.017	58.9
0.09	0.0683	0.0217	47.6

Figure 31.23 Graphical evaluation of the integral.

The area under the curve is the numerical value of the integral. The resulting length of the mass exchanger is

$$Z = \frac{38.9 \, \frac{mol}{m^2 \cdot s}}{80 \, \frac{mol}{m^3 \cdot s \cdot \Delta Y_A}} (10.95) = 5.32 \text{ m}$$

The integral $\int_{Y_{NH_3,2}}^{Y_{NH_3,1}} \frac{dY_{NH_3}}{Y_{NH_3} - Y_{NH_3}^*}$ can also be evaluated numerically using either a spreadsheet or a mathematical software program. With the values for G_s, L_s, $y_{NH_3,1}$ and $X_{NH_3,1}$ already determined, the operating line equation becomes

$$G_S(Y_{NH_3,1} - Y_{NH_3,z}) = L_S(X_{NH_3,1} - X_{NH_3,z})$$

$$38.9 \, \frac{mol}{m^2 \cdot s} (0.09 - Y_{NH_3,z}) = 57.7 \, \frac{mol}{m^2 \cdot s} (0.059 - X_{NH_3,z})$$

which upon rearrangement yields

$$X_{NH_3,z} = 0.6765 \quad Y_{NH_3,z} - 0.0020$$

The equilibrium data for example 3, when fitted to a polynomial of the second order, can be represented by

$$Y_{NH_3}^* = -2.6903 \, X_{NH_3}^2 + 1.2953 \, X_{NH_3} + 0.0003$$

These two equations are used to develop the following table:

Y_{NH_3}	X_{NH_3}	$Y_{NH_3}^*$	$Y - Y^*$	$\frac{1}{Y - Y^*}$	$\left(\frac{1}{Y - Y^*}\right)_{avg}$	ΔY_{NH_3}	$\left(\frac{1}{Y - Y^*}\right)_{avg} \Delta Y_{NH_3}$
0.003	0	0	0.003	333.3			
					305.95	0.007	2.14
0.01	0.00476	0.06411	0.003589	278.6			
					236.95	0.01	2.37
0.02	0.01153	0.01488	0.00512	195.3			
					170.15	0.01	1.70
0.03	0.01830	0.02310	0.006897	145.0			
					128.5	0.01	1.28
0.04	0.02506	0.03107	0.008929	112.0			
					100.6	0.01	1.01
0.05	0.03182	0.03879	0.01121	89.2			
					81.03	0.01	0.81
0.06	0.03859	0.04628	0.01372	72.9			
					66.78	0.01	0.67
0.07	0.04536	0.05351	0.01648	60.66			
					55.98	0.01	0.56
0.08	0.05212	0.06050	0.01950	51.29			
					47.62	0.01	0.48
0.09	0.05889	0.06725	0.02275	43.96			
							Total area = 11.02

The integral $\int_{Y_{NH_3,2}}^{Y_{NH_3,1}} \frac{dY_{NH_3}}{Y_{NH_3} - Y_{NH_3}^*}$ equals 11.02 and the height of the mass exchanger is

$$Z = \frac{G_S}{K_Y a} \int_{Y_{NH_3,2}}^{Y_{NH_3,1}} \frac{dY_{NH_3}}{Y_{NH_3} - Y_{NH_3}^*}$$

$$Z = \frac{38.9 \, \frac{mol}{m^2 \cdot s}}{80 \, \frac{mol}{m^3 \cdot s}} (11.02) = 5.36 \text{ m}$$

EXAMPLE 6

Recently enacted legislation requires that the effectiveness of the absorption system, currently employed to reduce the level of a pollutant in a process off-gas, must be improved to meet new state requirements. Under current operating conditions, 13.6 mol/s· m² of off-gas flows countercurrent to 27.2 mol/s · m² of water in the 12-m tall absorber. The concentration of A is reduced from 2 to 0.5 mol%.

It has been suggested that the required concentration level in the off-gas may be obtained by doubling the water rate in the existing tower. Its diameter is large enough to permit this higher flow rate. If it is assumed that the process is gas-film controlled and that the overall gas capacity coefficient, $K_G a$, is proportional to the mass velocity of the solvent raised to the 0.4 power, determine the concentration of A in the effluent gas when the same 12-m tower and the same gas mass velocity are used with the doubled liquid flow rate.

Equilibrium for the system is defined by a modified Henry's law; $y_A^* = 1.5_{x_A}$.

As the concentrations are in the dilute range where the equilibrium curve is a straight line, the height of the tower may be evaluated by equation (31-34)

$$Z = \frac{G(y_{A_1} - y_{A_2})}{K_G a P(y_A - y_A^*)_{lm}}$$

An overall material balance on the existing system (system I) establishes the composition of its exiting liquid stream.

$$G(y_{A_1} - y_{A_2}) = L(x_{A_1} - x_{A_2})$$
$$(13.6\,\text{mol/s} \cdot \text{m}^2)(0.02 - 0.005) = (27.2\,\text{mol/s} \cdot \text{m}^2)(x_{A_1} - 0)$$
$$x_{A_1} = 0.0075$$

The compositions at each end of the tower for the existing system are

$$\text{Bottom}: y_{A_1} = 0.02$$
$$x_{A_1} = 0.0075$$
$$y_{A_1}^* = 1.5 x_{A_1} = 1.5(0.0075) = 0.0113$$
$$\text{Top}: y_{A_2} = 0.005$$
$$x_{A_2} = 0.0$$
$$y_{A_2}^* = 1.5_{x_{A_2}} = 0.0$$

For the existing tower, $(y_A - y_A^*)_{lm}$ is evaluated by equation (31-35)

$$(y_A - y_A^*)_{lm} = \frac{(y_A - y_A^*)_{end_1} - (y_A - y_A^*)_{end_2}}{\ln\left[\dfrac{(y_A - y_A^*)_{end_1}}{(y_A - y_A^*)_{end_2}}\right]}$$

$$= \frac{(0.02 - 0.0113) - (0.005 - 0)}{\ln\dfrac{(0.02 - 0.0113)}{(0.005 - 0.0)}} = 0.0067$$

Upon substituting the known values into equation (31-34), we obtain

$$12\,\text{m} = \frac{(13.6\,\text{mol/s} \cdot \text{m}^2)(0.02 - 0.005)}{K_G a_I P(0.0067)}$$

$$12\,\text{m} = \frac{(13.6\,\text{mol/s} \cdot \text{m}^2)(2.24)}{K_G a_I P} \tag{31-38}$$

When we consider the proposed system (system II), we obtain the following relationship with equation (31-34)

$$12 = \frac{(13.6\,\text{mol/s} \cdot \text{m}^2)(0.02 - y_{A_2})}{K_G a_{II} P(y_A - y_A^*)_{lm}} \tag{31-39}$$

Upon equating and simplifying equations (31-38) and (31-39), we obtain

$$\frac{2.24}{K_G a_{\mathrm{I}}} = \frac{(0.02 - y_{A_2})}{K_G a_{\mathrm{II}} (y_A - y_A^*)_{lm}}\bigg|_{\mathrm{II}}$$

or

$$\frac{K_G a_{\mathrm{II}}(2.24)}{K_G a_{\mathrm{I}}} = \frac{(0.02 - y_{A_2})}{(y_A - y_A^*)_{lm}}\bigg|_{\mathrm{II}} \tag{31-40}$$

As the capacity coefficient is proportional to the mass velocity of the solvent raised to the 0.4 power, the ratio of the capacity coefficients is

$$\frac{K_G a_{\mathrm{II}}}{K_G a_{\mathrm{I}}} = \left[\frac{L_{\mathrm{II}}}{L_{\mathrm{I}}}\right]^{0.4} = \left[\frac{(54.4 \text{ mol/s} \cdot \text{m}^2)(0.018 \text{ kg/mol})}{(27.2 \text{ mol/s} \cdot \text{m}^2)(0.018 \text{ kg/mol})}\right]^{0.4} = 1.32$$

Accordingly, equation (31-40) becomes

$$\frac{0.02 - y_{A_2}}{(y_A - y_A^*)_{lm}}\bigg|_{\mathrm{II}} = (1.32)(2.24) = 2.95 \tag{31-41}$$

This equation requires a trial-and-error solution. With a guess of $y_{A_2} = 0.0021$, we can make an overall balance to establish x_{A_1} for the proposed system.

$$G(y_{A_1} - y_{A_2}) = L(x_{A_1} - x_{A_2})$$

$$(13.6 \text{ mol/s} \cdot \text{m}^2)(0.02 - 0.0021) = (54.4 \text{ mol/s} \cdot \text{m}^2)(x_{A_1} - 0.0)$$

$$x_{A_1} = 0.0045$$

The compositions at each end of the proposed tower (system II) are

Bottom: $y_{A_1} = 0.02$

$$x_{A_1} = 0.0045$$

$$y_{A_1}^* = 1.5 x_{A_1} = 1.5(0.0045) = 0.0067$$

Top: $y_{A_2} = 0.0021$ (estimated value)

$$x_{A_2} = 0.0$$

$$y_{A_2}^* = 1.5 x_{A_2} = 1.5(0.0) = 0.0$$

The $(y_A - y_A^*)_{lm}$ for the proposed tower is evaluated by equation (31-35)

$$(y_A - y_A^*)_{lm} = \frac{(y_A - y_A^*)_{\mathrm{end}_1} - (y_A - y_A^*)_{\mathrm{end}_2}}{\ln\left[\dfrac{(y_A - y_A^*)_{\mathrm{end}_1}}{(y_A - y_A^*)_{\mathrm{end}_2}}\right]}$$

$$(y_A - y_A^*)_{lm} = \frac{(0.02 - 0.0067) - (0.0021 - 0.0)}{\ln\left[\dfrac{(0.02 - 0.0067)}{(0.0021 - 0.0)}\right]} = 0.00607$$

$$\frac{0.02 - y_{A_2}}{(y_A - y_A^*)_{lm}} = \frac{0.02 - 0.0021}{0.00607} = 2.95$$

This satisfies equation (31-41); as a result of the trial-and-error solution, the concentration of A in the effluent gas is 0.21%.

Packed-Tower Diameter

The packed tower is the most commonly encountered continuous-contacting equipment in gas–liquid operations. A variety of packing material is used, ranging from specially designed ceramic or plastic packing, as illustrated in Figure 31.4, to crushed rock. The packing is chosen to promote a large area of contact between the phases, with a minimum resistance to the flow of the two phases. Table 31.2 lists some of the properties of packing frequently used in the industry.

Table 31.2   Tower Packing Characteristics[†]

Packing	Nominal size, in. (mm)					
	$\frac{1}{4}$ (6)	$\frac{1}{2}$ (13)	$\frac{3}{4}$ (19)	1 (25)	$1\frac{1}{2}$ (38)	2 (50)
Raschig rings						
Ceramic						
ϵ	0.73	0.63	0.73	0.73	0.71	0.74
c_f	1600	909	255	155	95	65
a_p ft^2/ft^3	240	111	80	58	38	28
Metal						
ϵ	0.69	0.84	0.88	0.92		
c_f	700	300	155	115		
a_p ft^2/ft^3	236	128	83.5	62.7		
Berl saddles						
Ceramic						
ϵ	0.60	0.63	0.66	0.69	0.75	0.72
c_f	900	240	170	110	65	45
a_p ft^2/ft^3	274	142	82	76	44	32
Intalox saddles						
Ceramic						
ϵ	0.75	0.78	0.77	0.775	0.81	0.79
c_f	725	200	145	98	52	40
a_p ft^2/ft^3	300	190	102	78	59.5	36
Plastic						
ϵ				0.91		0.93
c_f				33		56.5
a_p ft^2/ft^3				63		33
Pall rings						
Plastic						
ϵ				0.90	0.91	0.92
c_f				52	40	25
a_p ft^2/ft^3				63	39	31
Metal						
ϵ				0.94	0.95	0.96
c_f				48	28	20
a_p ft^2/ft^3				63	39	31

[†]R. E. Treybal, *Mass-Transfer Operations*, McGraw-Hill Book Company, New York, 1980.

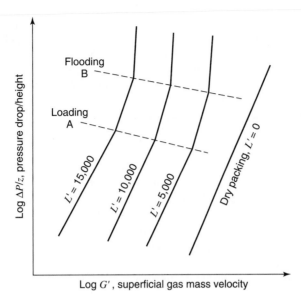

Figure 31.24 Typical gas pressure drop for countercurrent, packed tower.

We have previously established that the height of a continuous-contact tower is determined by the rate of mass transfer. The diameter of the tower is established to handle the flow rates of the two phases to be treated.

As illustrated in Figure 31.24, the pressure drop encountered by the gas phase as it flows through the packing is influenced by the flow rates of both phases. This is to be expected because both phases will be competing for the free cross section that is available for the streams to flow through. Let us consider a tower operating with a fixed liquid flow rate, L'; below the region marked A, the quantity of liquid retained in the packed bed will remain

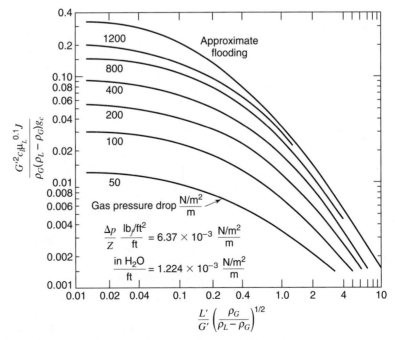

Figure 31.25 Flooding and pressure drop in random-packed towers.

reasonably constant with changing gas velocities. As the gas flow rate increases, the interphase friction increases and a greater quantity of liquid is held up in the packing. This is known as *loading*. Finally, at a certain value of the gas flow rate, G', the holdup is so high that the tower starts to fill with liquid. The tower cannot be operated above this *flooding velocity*, which is a function of the liquid velocity, the fluid properties, and the characteristics of the packing.

In Figure 31.25, a correlation is given for the flooding velocity in a random packed tower. Absorbers and desorbers are designed to operate well below the pressure drop that is associated with flooding. Typically, they are designed for gas pressure drops of 200–400 N/m^2 per meter of packed depth. The abscissa on this figure involves a ratio of the superficial liquid and gas-mass flow rates and the densities of the gas and liquid phases. The ordinate involves the superficial gas-mass flow rate, the liquid-phase viscosity, the liquid and gas densities, a packing characteristic, c_f, which can be obtained from Table 31.2, and two constants. For SI units, g_c equals 1, and J equals 1; for U.S. English units, μ_L is in centipoise, densities are in lb$_m$/ft^3, mass flow rates are in lb$_m$/(ft^2)(h), g_c equals 4.18×10^8, and J equals 1.502.

Example 8 will illustrate how we may evaluate the diameter of a packed tower by using Figure 31.25.

EXAMPLE 8

A packed tower is to be used to reduce the ammonia concentration in the gas stream from 4 to 0.3% by volume. A water stream is fed to the top of the tower at a rate of 231 g/s and the gas is fed countercurrently at a rate of 0.2 m^3/s. The tower, which is packed with 2.54-cm (1-inch) Raschig rings, is operated isothermally at 293 K and 1.013×10^5 Pa.

Determine (a) the composition of the exiting liquid stream; (b) the diameter of the absorption tower if the gas pressure drop is limited to 200 N/m^2 per meter of packing.

The concentration of the three known streams may be expressed on a NH$_3$-free basis

$$Y_{\mathrm{NH_3},1} = \frac{y_{\mathrm{NH_3},1}}{1 - y_{\mathrm{NH_3},1}} = \frac{0.04}{0.96} = 0.0417$$

$$Y_{\mathrm{NH_3},2} = \frac{y_{\mathrm{NH_3},2}}{1 - y_{\mathrm{NH_3},2}} = \frac{0.003}{0.997} = 0.003$$

$$X_{\mathrm{NH_3},2} = 0.0$$

The gas enters the tower at plane 1, and its molar flow rate is

$$G_1 = \frac{\dot{V}P}{RT} \cdot \frac{1}{A} = \frac{\left(0.20\,\dfrac{\mathrm{m^3}}{\mathrm{s}}\right)(1.013 \times 10^5\,\mathrm{Pa})}{\left(8.314\,\dfrac{\mathrm{Pa \cdot m^3}}{\mathrm{mol \cdot K}}\right)(293\,\mathrm{K})} \cdot \frac{1}{A} = 8.317\,\frac{\mathrm{mol}}{\mathrm{s}} \cdot \frac{1}{A}$$

$$G_S = G_1(1 - y_{\mathrm{NH_3},1}) = 8.317\,\frac{\mathrm{mol}}{\mathrm{s}} \cdot \frac{1}{A}(0.96) = 7.98\,\frac{\mathrm{mol}}{\mathrm{s}} \cdot \frac{1}{A}$$

The NH$_3$-free water flow rate is

$$L_S = L_2(1 - x_{\mathrm{NH_3},2}) = 231\,\frac{\mathrm{g}}{\mathrm{s}} \cdot \frac{\mathrm{mol}}{18\,\mathrm{g}} \cdot \frac{1}{A} = 12.83\,\frac{\mathrm{mol}}{\mathrm{s}} \cdot \frac{1}{A}$$

By an overall balance for ammonia, the concentration of the exiting liquid stream can be established.

$$L_S(X_{\mathrm{NH_3},1} - X_{\mathrm{NH_3},2}) = G_S(Y_{\mathrm{NH_3},1} - Y_{\mathrm{NH_3},2})$$

$$12.83\,\frac{\mathrm{mol}}{\mathrm{s}} \cdot \frac{1}{A}(X_{\mathrm{NH_3},1} - 0) = 7.98\,\frac{\mathrm{mol}}{\mathrm{s}} \cdot \frac{1}{A}(0.0417 - 0.003)$$

$$X_{\mathrm{NH_3},1} = 0.024$$

or

$$x_{NH_3,1} = \frac{X_{NH_3,1}}{1 + X_{NH_3,1}} = \frac{0.024}{1.024} = 0.0234$$

The liquid flow rate at end 1 is

$$L_1 = \frac{L_S}{1 - x_{NH_3,1}} = \frac{12.83 \frac{mol}{s} \cdot \frac{1}{A}}{1 - 0.0234} = 13.13 \frac{mol}{s} \cdot \frac{1}{A}$$

The maximum mass flow rates for both phases will occur at end 1. Accordingly, we will use these flow rates to determine the diameter of the tower.

On a unit-area basis, the liquid stream at end 1 will contain 12.83 mol/s and H_2O 13.13 − 12.83 = 0.3 mol/s NH_3. The total mass flow rate is

$$12.83 \frac{mol\ H_2O}{s} \cdot \frac{1}{A} \cdot \left(\frac{18\,g}{mol}\right) + 0.3 \frac{mol\ NH_3}{s} \cdot \frac{1}{A} \cdot \left(\frac{17\,g}{mol}\right) = 236.1 \frac{g}{s} \cdot \frac{1}{A}$$

The average molecular weight of the gas is

$$\left(0.04 \frac{mol\ NH_3}{mol}\right)\left(\frac{17\,g}{mol}\right) + \left(0.96 \frac{mol\ air}{mol}\right)\left(\frac{29\,g}{mol}\right) = 28.5 \frac{g}{mol}$$

The total gas-mass flow rate is

$$\left(8.317 \frac{mol}{s} \cdot \frac{1}{A}\right)\left(28.5 \frac{g}{mol}\right) = 237 \frac{g}{s} \cdot \frac{1}{A}$$

The ratio of the two mass flow rates is

$$\frac{L'}{G'} = \frac{\left(236.1 \frac{g}{s}\right)\left(\frac{1}{A}\right)}{\left(237 \frac{g}{s}\right)\left(\frac{1}{A}\right)} = 0.996$$

Note that this ratio can be evaluated without knowing either the diameter or the cross-sectional area. The density of the gas stream entering the tower is

$$\rho_G = \frac{n}{V} M_W = \frac{P}{RT} M_W = \frac{1.013 \times 10^5\ Pa}{\left(8.314 \frac{Pa \cdot m^3}{mol \cdot K}\right)(293\ K)} \cdot 28.5 \frac{g}{mol}$$
$$= 1185\ g/m^3 = 1.185\ kg/m^3$$

The density of the dilute aqueous stream will be essentially that of water at 293 K that is 998.2 kg/m³. The abscissa for Figure 31.25 becomes

$$\frac{L'}{G'} \left(\frac{\rho_G}{\rho_L - \rho_G}\right)^{1/2} = 0.996 \left(\frac{1.185}{998.2 - 1.185}\right)^{1/2} = 0.034$$

At a pressure drop of 200 N/m² per meter of packing, this abscissa value indicates an ordinate value of 0.049; consequently

$$0.049 = \frac{(G')^2 c_f (\mu_L)^{0.1} J}{\rho_G (\rho_L - \rho_G) g_c}$$

Upon rearrangement

$$(G')^2 = \frac{0.049 \rho_G (\rho_L - \rho_G) g_c}{c_f (\mu_L)^{0.1} J}$$

From Table 31.2, we evaluate the c_f for 1-in. Raschig rings to be 155. The viscosity of the aqueous stream at 293 K is given in Appendix Table I to be 993×10^{-6} Pa· s.

$$(G')^2 = \frac{0.049\left(1.185\frac{kg}{m^3}\right)\left(998.2 - 1.185\frac{kg}{m^3}\right)(1.0)}{(155)(993 \times 10^{-6}\,Pa \cdot s)(1.0)} = 0.745$$

$$G' = 0.863\frac{kg}{m^2 \cdot s} = 863\frac{g}{m^2 \cdot s}$$

As the gas feed rate, G', to the tower is equal to $237\,\frac{g}{s} \cdot \frac{1}{A}$ and $863\,\frac{g}{m^2 \cdot s}$, the cross-sectional area of the tower is

$$A = \frac{237\frac{g}{s}}{863\frac{g}{m^2 \cdot s}} = 0.275\ m^2$$

The area is $\dfrac{\pi D^2}{4}$; accordingly, the diameter is

$$D = \left[\frac{(0.275\ m^2)(4)}{\pi}\right]^{1/2} = 0.59\ m$$

31.7 CLOSURE

Continuous-contact mass exchangers are designed by integrating an equation that relates the mass balance and the mass-transfer relations for a differential area of interfacial contact. In this chapter, we have described the four major types of mass-transfer equipment. The fundamental equations for the design of continuous-contact equipment have been developed. A mass balance for the diffusing component A in terms of solute-free concentration units produced the following important operating line equations:

steady-state countercurrent operations

$$G_S(Y_{A_1} - Y_{A_2}) = L_S(X_{A_1} - X_{A_2})$$

and

$$G_S\,dY_A = L_S\,dX_A$$

steady-state cocurrent operations

$$G_S(Y_{A_1} - Y_{A_2}) = L_S(X_{A_2} - X_{A_1})$$

and

$$G_S\,dY_A = -L_S\,dX_A$$

Because of the difficulty in measuring the interphase contact area within most mass-transfer equipment, the factor a, the interphase surface area per unit volume of the mass exchanger, was introduced. The product of the mass-transfer convective coefficient and the factor a was designated the mass-transfer capacity coefficient. The mass transferred in a differential length per cross-sectional area was expressed in empirical equations, one of which was

$$N_A a\,dz = K_G a(p_{A,G} - p_A^*)\,dz$$

For equipment involving the continuous-contact between two immiscible phases, the differential mass balance and the differential mass-transfer equations were combined to produce the following design equations:

Constant overall capacity coefficient $K_Y a$

$$z = \frac{G_S}{K_Y a} \int_{Y_{A_2}}^{Y_{A_1}} \frac{dY_A}{(Y_A - Y_A^*)}$$

The integral evaluation must be accomplished graphically.

Variable overall capacity coefficient-allowance for resistance in both gas and liquid phases

$$z = \int_{Y_{A_2}}^{Y_{A_1}} \frac{G_S \, dy_A}{k_G a P(y_A - y_{Ai})(1 - y_A)^2}$$

This integral is also evaluated graphically.

Straight equilibrium and operating lines on the x–y coordinates—the log mean driving force

$$z = \frac{G(y_{A_1} - y_{A_2})}{k_G a P(y_A - y_A^*)_{lm}}$$

where

$$(y_A - y_A^*)_{lm} = \frac{(y_A - y_A^*)_1 - (y_A - y_A^*)_2}{\ln[(y_A - y_A^*)_1/(y_A - y_A^*)_2]}$$

The similarity between mass and energy transfer was further emphasized in this chapter. Using a combined term representing the total resistance, $k_G a$ as compared to UA, and a total resistance, $(p_{A,G} - p_A^*)$ as compared to $\Delta T_{overall}$, we have obtained design equations for mass exchangers by integrating over the area of contact.

PROBLEMS

31.1 A 283-m^3(10,000 ft^3) basin is to be aerated with six spargers, each sparger using air at the rate of 7.08×10^{-3} m^3/s. If 283 K wastewater, initially containing a dissolved oxygen level of 5×10^{-3} mmol/L and filling the basin to a depth of 4.55 m, is aerated for 9000 s, evaluate the final dissolved oxygen level. The dissolved solids content will be low enough so that Henry's law will be obeyed, with a Henry's law constant of 3.27×10^4 atm/mole fraction.

31.2 An ozone, O$_3$, sparger system has been proposed for a water-treatment plant. The incoming water in a 80 m^3 pond has been stripped of all ozone-demanding substances, and the pH is sufficiently low such that the ozone decomposition will be insignificant. The ozone bubbler will have eight spargers, each using a compressed gas feed containing 4 mol % ozone. The spargers will be located 3.2 m below the surface of the pond and each sparger's gas rate will be 17.8 m^3/h. Henry's law constant for the ozone–water system will be 0.0677 atm/(mol/m^3)

a. Use the Eckenfelder plot for oxygen transfer factor, $k_L(A/V)$, and evaluate the $k_L(A/V)$ for ozone, recalling that

the penetration theory predicts, for mass transfer from gas bubbles into liquids, that $k_L(A/V)$ will be proportional to the liquid diffusivity raised to the power of 1/2. The diffusivity of oxygen in water is 2.14×10^{-5} cm^2/s and the diffusivity of ozone into water is 1.7×10^{-5} cm^2/s.

b. Determine the time required to raise the initially ozone-free water in the pond to a concentration level of 0.15 mol/m^3.

31.3 A 425-m^3(15,000 ft^3) basin filled with wastewater is to be aerated with 10 spargers, each sparger using air at the rate of 7.08×10^{-3} m^2/s. Find the time that is necessary to raise the dissolved oxygen level in the wastewater from 8×10^{-2} to 2×10^{-1} mmol/L if the temperature of the water is 283 K and the depth of the water above the spargers is 3.2 m. The dissolved solids content will be low enough so that Henry's law will be obeyed, with a Henry's law constant of 3.27×10^4 atm/mole fraction.

31.4 A 425-m^3(15,000 ft^3) basin, equipped with 10 gas dispersers of a new design, is to be installed for stripping hydrogen sulfide, H_2S, from wastewater. The dispensers, each to use air at a rate of 7.08×10^{-3} m^3/s, will be located at a depth of 3.2 m (10.5 ft) below the liquid surface. A preliminary analysis of the 283 K wastewater indicates an initial H_2S concentration of 0.3 mmol/L.

In a pilot investigation of the proposed new dispenser, the designated flow rate of 7.08×10^{-3} m^3/s was discharged from a single dispenser located at 3.2 m below the surface of the water held in a 28.3-m^3 (1000-ft^3) tank. This water was maintained at 283 K. After 4 h of aeration, the dissolved oxygen concentration increased from 0.04 to 0.25 mmol/L. Based upon this pilot investigation, determine

a. the K_La for oxygen for this new dispenser when it is operated 3.2 m below the liquid surface with the air flow rate of 7.08×10^{-3} m^3/s. At 283 K, Henry's law constant for the oxygen–water system is 3.27×10^4 atm/mole fraction.

b. the K_La for H_2S for this new dispenser when it is operated 3.2 m below the liquid surface with the air flow rate of 7.08×10^{-3} m^3/s. If the penetration-theory model holds, the capacity coefficient, K_La, is proportional to the liquid mass diffusivity raised to the power of 1/2. The liquid diffusivity of oxygen in water at 283 K is 2.14×10^{-5} cm^2/s and of H_2S in water at 283 K is 1.4×10^{-5} cm^2/s.

c. the concentration of H_2S in the 425-m^3 basin after 4h of operation if the wastewater's initial H_2S concentration is 0.3 m mole/L. Henry's law constant for H_2S–water system at 283 K is 0.0367×10^4 atm/mole fraction.

31.5 An exhaust stream leaving a semiconductor fabrication plant contains 4 mol % volatile organic chemical, VOC, pollutant. Twenty-five cubic meters per minute enter the bottom of a packed absorption tower and flows countercurrent to an initially VOC-free solvent stream that was fed to the top of the tower. At the given temperature and pressure of the system, the equilibrium of the system will obey Henry's law as represented by

$$Y_{VOC} = 0.236X_{VOC}$$

a. Determine the moles of solvent required per mole of VOC-free carrier gas if the exiting gas stream is to contain only 0.2 mol % VOC and if 1.5 times the minimum solvent is used.

b. Determine the mole fraction composition of the exiting liquid.

31.6 It is desired to strip trichloroethylene, O_2HCl_3, from a contaminated groundwater stream. The groundwater, which can be treated as a dilute water stream, will be fed to the top of the tower operating at 288 K and 1.013×10^5 Pa. An air stream, flowing countercurrent to the aqueous stream, is to reduce the TCE concentration from 40 to 5μg/L. At 288 K, the Henry's law constant for TCE–air system is 11.7×10^{-3} atm/(mol/m^3).

a. Determine the moles of wastewater per mole of air if three times the minimum air stream is used.

b. Determine the concentration of the exiting air stream.

31.7 The absorber for Problem 31.5 would have been excessively tall; accordingly, various schemes are being considered for using two shorter absorbers side by side. Make freehand sketches of operating diagrams, showing the operating line for each tower and the equilibrium line relative to the operating lines, for each scheme. Indicate, in general, where the concentrations of each stream will be with respect to each other, but do not compute the actual concentration.

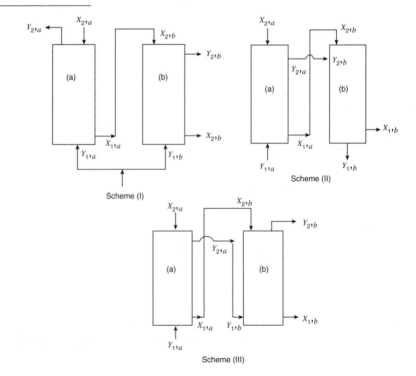

Scheme (I)

Scheme (II)

Scheme (III)

31.8 Another designer was asked to propose a design for the stripping of a pollutant from a liquid solution. He decided to use the same schemes of two towers side by side as shown in Problem 31.7. Make freehand sketches of operating diagrams, showing the operating line for each tower and the equilibrium line relative to the operating lines, for each scheme. Indicate, in general, where the concentrations of each stream will be with respect to each other, but do not compute the actual concentration.

31.9 An absorber is being used by an industrial company as a gas scrubber. A dilute air–ammonia mixture, containing 5.0 mol % NH_3 is drawn through the tower countercurrent to an initially ammonia-free water stream. The inlet gas stream rate is 136 mol/m^2· s and the inlet NH_3 -free water rate is 3.4 kg/m^2· s. At the temperature and pressure of the absorber, the equilibrium concentrations are related by

$$y_{NH_3} = 1.075\, x_{NH_3}$$

If the exiting gas stream is to contain only 0.2 mol % ammonia, determine

a. the composition of the exiting liquid stream;

b. the ratio of the actual L_s/G_s is larger than the minimum L_s/G_s.

c. the composition of the liquid stream flowing by the gas stream when the gas mixture contains 2.0 mol % ammonia.

31.10 The absorber used in Problem 31.9 is now going to be used by the same industrial company to reduce a dilute air–ammonia mixture from 4.93 mol % NH_3 to the desired 0.2 mol % NH_3. The inlet gas rate will be 136 mol/m^2· s and the inlet NH_3 -free water rate will flow countercurrent to the rising gas stream, flowing at an actual L_s/G_s ratio that is 1.4 times the minimum L_s/G_s ratio. At that flow rate, the overall mass-transfer capacity coefficient, K_Ya, is estimated to be equal to 107 mol/m^2· s · ΔY_A. At the temperature and pressure of the absorber, the equilibrium concentrations are related by

$$y_{NH_3} = 1.075\, x_{NH_3}$$

Determine

a. the kg of ammonia absorbed per second per square meters of cross-sectional area;

b. the height of packing required for the absorption process.

31.11 A scheme for the removal of H_2S from a gas stream by scrubbing with water at 293 K and 1.013×10^5 Pa is being considered. The liquid will initially contain 3.7 mol % H_2S. It is desired that the exiting gas stream contains only 0.2 mol % H_2S. The absorbing water will enter at the top of the tower that is H_2S free. At the given temperature and pressure of the system, the equilibrium of H_2S–water will obey Henry's law according to the following relationship:

$$Y_{H_2S} = 48.3\, X_{H_2S}$$

a. For a countercurrent absorber, determine the moles of water that are required per mole of H_2S-free carrier gas if 1.5 times the minimum ratio will be used.

b. Determine the composition of the exiting liquid.

31.12 Benzene freed from coal gas in a scrubber is recovered by passing the benzene–wash oil solution through a tower in contact with steam. The entering liquid stream contains 7 mol % benzene and the steam is benzene free. It is desired to recover 85% of the benzene by using a steam rate 1.4 times the minimum steam rate. A wash-oil (liquid solvent-benzene free) flow rate of 6.94 mol/s will be used. Determine the required moles of steam if

a. a countercurrent tower is used;

b. a cocurrent tower is used.

Equilibrium data for the benzene–wash oil–steam system are as follows

$X \dfrac{\text{mole benzene}}{\text{mole wash oil}}$	0.00	0.02	0.04	0.06	0.08	0.10	0.12	0.14
$Y \dfrac{\text{mole benzene}}{\text{mole steam}}$	0.00	0.07	0.14	0.22	0.31	0.405	0.515	0.65

31.13 An absorber is to be designed to remove a VOC pollutant from an exhaust gas stream. Fifteen cubic meters per minute of gas at 289 K and 1.013×10^5 Pa containing 5.0 mol % VOC is fed to the bottom of the absorption tower. By feeding a VOC-free solvent stream to the top of the tower, the VOC concentration is reduced to 0.3 mol %. The solvent stream leaves the tower containing 3.65 mol % VOC. With the specified stream flow rates, the overall mass-transfer capacity coefficient, K_Ya, equals 52.0 mol/ s · m^2 · ΔY_A. The cross-sectional area of the tower is 0.2 m^2.

At 298 K, the equilibrium for the VOC-solvent system may be represented by

$$Y_A = 0.8\, X_A$$

Determine the hight of the tower.

31.14 A tower, 15 cm in diameter, is to be used to lower the ammonia, NH_3, concentration in a gas stream from 3.6 to 0.3 mol %. Water is fed to the top of the tower at a rate of 14.5 mol/s and the gas stream is fed to the bottom at a rate of 8 mol/s. The tower operates isothermally at 293 K and 1.013×10^5 Pa. The overall mass-transfer capacity coefficient, K_Ya, may be assumed to equal 71 mol/m^2· s · ΔY_A. The equilibrium data at 293 K are as follows:

$X \dfrac{\text{mole } NH_3}{\text{mole } NH_3\text{-free water}}$	0.00	0.0164	0.0252	0.0349	0.0445	0.0722
$Y \dfrac{\text{mole } NH_3}{\text{mole } NH_3\text{-free air}}$	0.00	0.021	0.032	0.042	0.053	0.080

Evaluate

a. the ratio $(L_S/G_S)_{\text{actual}}$ to $(L_S/G_S)_{\text{minimum}}$;

b. the height of tower required, by graphical integration.

31.15 A tower, 15 cm in diameter, is to be used to lower the ammonia, NH_3, concentration in a gas stream from 3.6 to

0.3 mol %. Water is fed to the top of the tower at a rate of 14.5 mol/s and the gas stream is fed to the bottom at a rate of 8 mol/s. The tower operates isothermally at 293 K and 1.013×10^5 Pa. The overall mass-transfer capacity coefficient, $K_Y a$, may be assumed to equal 71 mol/m$^2 \cdot s \cdot \Delta Y_A$. The equilibrium data at 293 K are as follows:

$X \dfrac{\text{mole NH}_3}{\text{mole NH}_3\text{-free water}}$	0.00	0.0164	0.0252	0.0349	0.0445	0.0722
$Y \dfrac{\text{mole NH}_3}{\text{mole NH}_3\text{-free air}}$	0.00	0.021	0.032	0.042	0.053	0.080

Determine the height of the tower using a modified Equation (31-34).

31.16 A mass-transfer tower is to be designed for reducing a VOC concentration from 0.0394 to 0.0131 mg mol/L. Five thousand gallons per hour of wastewater will be fed to the 0.6 m diameter tower, countercurrent to a stripping air stream that will enter the tower free of any VOC. If a volumetric flow rate $Q_L(\text{m}^3/\text{h})$ to a gas flow rate $G(\text{mol/h})$ of 1.5 is used, the overall mass-transfer capacity coefficient, $K_L a$, will equal 0.01/s. The equilibrium data at the pressure and temperature of the tower are

c_A, moles VOC/m^3	0.014	0.0240	0.0349	0.0498
y_A,VOC	0.018	0.030	0.042	0.053

Determine
a. the minimum gas flow rate that might be used;
b. the height of the tower.

31.17 The absorption of water in sulfuric acid is an exothermic process. Describe the effect this would have on the required height of a mass-transfer column as compared to the height that is evaluated if isothermal conditions are assumed.

31.18 An absorber is to be designed to reduce the concentration level of a mercaptan pollutant in an exhaust stream from 5.0 to 0.3 mol %. A nonvolatile, mercaptan-free solvent stream will be fed to the top of the tower and will flow countercurrent to 0.236 m^3/s of exhaust gas stream. The tower is to operate at 293 K and 1.013×10^5 Pa; at these conditions, the equilibrium data for the mercaptan-solvent system are as follows:

$X \dfrac{\text{moles mercaptan}}{\text{mole mercaptan-free solvent}}$	0.00	0.01	0.02	0.03	0.04
$Y \dfrac{\text{moles mercaptan}}{\text{mole mercaptan-free air}}$	0.00	0.0045	0.0145	0.0310	0.0545

Determine the height of packing required if the solvent leaves the bottom of the tower containing 3.05 mol % mercaptan, the overall mass-transfer capacity coefficient, $K_Y a$, equals 40.0 mol/m^2.s.ΔY_A, and the cross-sectional area of the tower is 0.2 m^2.

31.19 An absorber packed to the height of 4.5 m with 1.0-in. Raschig rings has been designed to reduce the concentration level of a mercaptan pollutant in an exhaust stream from 5.0 to 0.3 mol %. A nonvolatile, mercaptan-free solvent stream will be fed to the top of the tower and will flow countercurrent to 0.236 m^3/s of the exhaust gas stream. The solvent leaves the bottom of the tower containing 3.05 mol % mercaptan. At the pressure and temperature of the tower, 293 K and 1.013×10^5 Pa, the equilibrium data for the mercaptan-solvent system are as follows:

$X \dfrac{\text{moles mercaptan}}{\text{mole mercaptan-free solvent}}$	0.00	0.01	0.02	0.03	0.04
$X \dfrac{\text{moles mercaptan}}{\text{mole mercaptan-free air}}$	0.00	0.0045	0.0145	0.0310	0.0545

Determine the diameter of the tower if the gas pressure drop within the packing is limited to 300 N/m^3 of packing. The gas enters the tower with an average molecular weight of 30.1 and the liquid stream, leaving the bottom of the tower, has an average molecular weight of 180, a specific gravity of 0.81 and a viscosity of 3.9×10^{-3} Pa· s.

31.20 An absorber is to be designed for reducing a process off-gas, initially containing 12.5% to 0.4% HCl by volume. HCl-free water at 293 K and 1.013×10^5 Pa pressure will be fed to the top of a packed, 0.6-m-diameter column, flowing countercurrent to the rising gas stream. Preliminary design calculations indicate that an operating L_s/G_s ratio should be 1.64 times the minimum L_s/G_s ratio. With these flow rates, the overall mass-transfer capacity coefficient, $K_G a$, may be taken as 8.8×10^{-8} kg mol/m$^3 \cdot$ s· Pa. The equilibrium relation for solutions of HCl in water at 293 K is given by the following data:

x_{HCl}	0.210	0.243	0.287	0.330	0.353	0.375	0.400	0.425
y_{HCl}	0.0023	0.00956	0.0215	0.0523	0.0852	0.135	0.203	0.322

Determine
a. the percentage of HCl to be removed by the absorption process;
b. the exit concentration of the acid stream;
c. If 5 m^3/min of process off-gas, measured at 293 K and 1.013×10^5 Pa pressure, are blown into the tower, determine the height of ceramic packing required in the absorber.

Nomenclature

a	interphase mass-transfer area per unit volume; ft^2/ft^3, m^2/m^3.
$\mathbf{a}$	acceleration; ft/s^2, m/s^2.
a_p	packing characteristic; ft^2/ft^3, m^2/m^3.
A	area; ft^2, m^2.
A_i	interphase mass-transfer area; ft^2, m^2.
A_p	projected area of surface; ft^2, m^2; equation (12-3).
c	total molar concentration; lb mol/ft^3; mol/m^3.
c_A	concentration of A in equilibrium with the bulk composition of gas phase, $P_{A.G}$; lb mol/ft^3; mol/m^3.
c_{A_o}	concentration of A at time $t = 0$; lb mol/ft^3; mol/m^3.
$c_{A,i}$	liquid molar concentration of A at the interface; lb mol/ft^3, mol/m^3; Section 29.2.
$c_{A,L}$	liquid molar concentration of A in the bulk stream; lb mol/ft^3, mol/m^3; Section 29.2.
$c_{A,s}$	concentration of A at the surface; lb mol/ft^3, mol/m^3.
$c_{A,\infty}$	concentration of A in the bulk stream; lb mol/ft^3, mol/m^3.
c_i	molar concentration of species i; lb mol/ft^3, mol/m^3; equation (24-4).
c_p	heat capacity; Btu/lb °F, J/kg K.
$\underline{C}$	dimensionless concentration; dimensionless.
C	average random molecular velocity; m/s; Sections 7.3 and 15.2.
C_C	capacity rate of cold fluid stream; Btu/h °F, kW/K; equation (22-1).
C_D	drag coefficient; dimensionless; equation (12-3).
C_f	coefficient of skin friction; dimensionless; equation (12-2).
C_f	packing characteristic; dimensionless.
C_H	capacity rate of hot fluid stream; Btu/h °F, kW/K; equation (22-1)
C_{sf}	correlating coefficient for nucleate boiling; dimensionless; Table 21.1.
d_c	diameter of cylinder; ft, m.
d_p	diameter of spherical particle; ft, m.
D	tube diameter; ft, m.
D_{AB}	mass diffusivity or diffusion coefficient for component A diffusing through component B; ft^2/h, m^2/s; equation (24-15).
D_{Ae}	effective diffusion coefficient of species A within straight pores; ft^2/s, m^2/s.
D'_{Ae}	effective diffusion coefficient of species A within random pores; ft^2/s, m^2/s.
$D_{A,mix}$	diffusion coefficient of species A in a multicomponent mixture; ft^2/s, m^2/s.
$D_{eq,}$	equivalent diameter; ft, m; equation (13-18).
D_{KA}	Knudsen diffusion coefficient of species A; ft^2/s, m^2/s.
D^o_{AB}	diffusion coefficient of solute A in solvent B at infinite dilution; ft^2/s, m^2/s.
d_{pore}	pore diameter; Å, nm.
d_s	molecular diameter; Å, nm.
e	pipe roughness; in., mm; Section 13.1.

e	specific energy or energy per unit mass; Btu/lb$_m$, J/kg; Section 6.1.
E	total energy of system; Btu, J; Section 6.1.
E	total emissive power; Btu/h ft^2, W/m^2; equation (23-2).
E	electrical potential; V; Section 15.5.
E_b	blackbody emissive power; Btu/h ft^2, W/m^2; equation (23-12).
f	dependent variable used in the Blasius solution of boundary layer; dimensionless; equation (12-13).
f'	similarity parameter for convective analysis of boundary layer, prime denotes derivative with respect to η; dimensionless; equation (19-16).
f_D	Darcy friction factor; dimensionless; equation (13-4).
f_f	Fanning friction factor; dimensionless; equation (13-3).
F	force; lb$_f$, N; Section 1.2.
F	correction factor for compact heat-exchanger configurations; dimensionless; equation (22-14).
F_{ii}	view factor for radiant heat transfer; dimensionless; Section 23.7.
$\overline{F}_{ij}$	reradiating view factor; dimensionless; Section 23.9.
$\mathbf{g}$	acceleration due to gravity; ft/s^2, m/s^2.
g_c	dimensional conversion factor; 32.2 ft lb$_m$/lb$_f$ s^2, 1 kg $\cdot$ m/s$^2 \cdot$ N.
G	irradiation; Btu/h ft^2, W/m^2; Section 23.7.
G	mass velocity; lb$_m$/ft^2 h, g/m$^2 \cdot$ s.
G	total moles of the gas phase per time per cross-sectional area; lb mol/ft^2 h, g mol/m$^2 \cdot$ s.
G'	superficial gas-mass flow rate; lb$_m$/h ft^2; Section 31.6.
G_b	mass velocity of bubbles; lb$_m$/ft^2 s, kg/m$^2 \cdot$ s; equation (21-4).
G_M	molar velocity; lb mol/ft^2 h, g mol/m$^2 \cdot$ s.
G_s	moles of gas stream on a solute-free basis per time per cross-sectional area; lb mol/ft^2 h, g mol/m$^2 \cdot$ s.
h	convective heat-transfer coefficient; Btu/h ft^2 °F, W/m$^2 \cdot$ K; equation (15-11).
$h_{fg,solute}$	heat of vaporization of solute; Btu/lb$_m$, kJ/kg.
h_L	head loss, $\Delta P/\rho$; ft lb$_f$/lb$_m$, Pa/(kg/m$_3$)); Section 13.1.
h_r	radiation heat-transfer coefficient; Btu/h ft^2 °F, W/m$^2 \cdot$ K; Section 23.12.
H	Henry's law constant; concentration of gas phase/concentration of liquid phase.
$\mathbf{H}$	moment of momentum; lb$_m$ ft^2/s, kg $\cdot$ m^2/s; equation (5-7).
H_i	enthalpy of species i; Btu, J.
ΔH_s	integral heat of solution; Btu/lb mol of solute; J/g mol of solute.
$\Delta H_{v,A}$	enthalpy of vaporization for species A; Btu/lb mol, J/g mol.
$\bar{H}_i$	partial molar enthalpy of species i; Btu/lb mol, J/mol.
I	intensity of radiation; Btu/h ft^2, W/m^2; Section 23.3.
j'	j factor for heat transfer with tube bundles; dimensionless; Figures 20.12 and 20.13.
j_D	j factor for mass transfer, Chilton-Colburn analogy; dimensionless.
j_H	j factor for heat transfer, Colburn analogy; dimensionless; equation (19-38).
$\mathbf{j}_i$	mass flux relative to the mass-average velocity; lb$_m$/ft^2 h, kg/m$^2 \cdot$ s; equation (24-17).
J	radiosity; Btu/h ft^2, W/m^2; Section 23.10.
$\mathbf{J}_i$	molar flux relative to the molar-average velocity; lb mol/h ft^2, mol/m$^2 \cdot$ s; equation (24-15).
k	thermal conductivity; Btu/h ft °F, W/m $\cdot$ K; equation (15-1).
k	rate constant for chemical reaction, used to define r_A and R_A; Section 25.1.

k^0	mass-transfer coefficient with no net mass transfer into film; lb mol/ft^2 s Δ_{c_A}; mol/m$^2 \cdot$ s $\cdot$ **Δc_A**.
k_c	convective mass-transfer coefficient; lb mol/ft^2 h Δc_A, mol/m$^2 \cdot$ s $\cdot$ mol/m^3.
$\overline{k}_c$	mean convective mass-transfer coefficient; lb mol/ft^2 h Δ_{c_A}, mol/m$^2 \cdot$ s $\cdot$ mol/m^3.
k_G	convective mass-transfer coefficient in the gas phase; lb mol/ft^2 h atm, mol/m$^2 \cdot$ s $\cdot$ Pa.
k_L	convective mass-transfer coefficient in the liquid phase; lb mol/ft^2 h lb mol/ft^3, mol/m$^2 \cdot$ s $\cdot$ mol/m^3.
$k_G a$	individual gas-capacity coefficient; lb mol/h ft^3 atm, mol/s $\cdot$ m$^3 \cdot$ Pa.
$k_L a$	individual liquid-capacity coefficient; lb mol/h ft$^3 \Delta c_A$, mol/s $\cdot$ m$^3 \cdot$ mol/m^3.
K_G	overall mass-transfer coefficient in the gas phase; lb mol/h ft^2 atm, mol/s $\cdot$ m$^2 \cdot$ Pa.
K_L	overall mass-transfer coefficient in the liquid phase; lb mol/h ft$^2 \Delta c_A$, mol/s $\cdot$ m$^2 \cdot$ mol/m^3.
$K_G a$	overall gas-capacity coefficient; lb mol/h ft^3 atm, mol/s $\cdot$ m$^3 \cdot$ Pa.
$K_L a$	overall liquid-capacity coefficient; lb mol/h ft$^3 \Delta c_A$, mol/s $\cdot$ m$^3 \cdot$ mol/m^3.
$K_X a$	overall liquid-capacity coefficient based on ΔX_A driving force; lb mol/h ft$^3 \Delta X_A$, mol/s $\cdot$ m$^3 \cdot \Delta X_A$.
$K_Y a$	overall gas-capacity coefficient based on ΔY_A driving force; lb mol/h ft$^3 \Delta Y_A$, mol/s $\cdot$ m$^3 \cdot \Delta Y_A$.
L	mixing length; equations (12-52), (19-41), and (28-43).
L	characteristic length; ft, m.
L	total moles of liquid phase per time per cross-sectional area; lb mol/h ft^2, mol/s $\cdot$ m^2.
L_{eq}	equivalent length; ft, m; equation (13-17).
L_m	molar liquid-mass velocity; lb mol/h ft^2, mol/s $\cdot$ m^2.
L_s	moles of liquid phase on a solute-free basis per time per cross-sectional area; lb mol/h ft^2, mol/s $\cdot$ m^2.
m	mass of molecule; Section 7.3.
m	relative resistance $= D_{AB}/k_c x_1$; dimensionless; Section 27.4.
m	slope of the equilibrium line; units of gas concentration per units of liquid concentration.
M	moment; lb$_m$ ft^2/s^2, kg $\cdot$ m^2/s^2.
M_i	molecular weight of species i; lb/lb mol, kg/kg mol.
n	packed bed constant; dimensionless; equation (30-33).
n	number of species in a mixture; equations (24-1), (24-3), and (24-6).
n	relative position $= x/x_1$; dimensionless; Section 27.4.
N	molecules per unit volume; Section 7.3.
n	outward directed unit normal vector; Sections 4.1, 5.1, 6.1.
n_i	number moles of species i.
n$_i$	mass flux relative to a set of stationary axes; lb$_m$/h ft^2, kg/s $\cdot$ m^2.
N$_i$	molar flux relative to a set of stationary axes; lb mol/h ft^2, mol/s $\cdot$ m^2.
NTU	number of transfer units; dimensionless; Section 22.4.
p_A^*	partial pressure of A in equilibrium with bulk composition in liquid phase, $c_{A,L}$; atm, Pa.
$p_{A,G}$	partial pressure of component A in the bulk gas stream; atm, Pa; Section 29.2.
$p_{A,i}$	partial pressure of component A at the interface; atm, Pa; Section 29.2.
p_i	partial pressure of species i; atm, Pa.
$p_{B,lm}$	log mean of partial pressure of the nondiffusing gas; atm, Pa.

P	total pressure; atm, Pa.
P	power input for stirred tank of liquid; N · m/s.W.
P_A	vapor pressure of pure volatile liquid species A; $lb_f/in.^2$, Pa.
$\mathbf{P}$	total linear momentum of system; lb_m ft/s, kg · m/s; equation (5-1).
P_c	critical pressure; atm, Pa.
P_i	vapor pressure of species i; atm, Pa.
P_g	power input for aerated stirred tank of liquid; N · m/s., W.
q	heat flow rate; Btu/h, W; equation (15-1).
q	volumetric energy generation rate; Btu/h ft^3, W/m^3; equation (16-1).
Q	heat transfer; Btu, J; Section 6.1.
Q	activation energy for solid diffusion coefficient; J/g mol.
r	radial distance in both cylindrical and spherical coordinates; ft, m.
r	radius; ft, m.
r_{crit}	critical radius of insulation; ft, m; equation (17-13).
R	radius of sphere, ft, m.
R	gas constant; 0.73 atm ft^3/lb mol °F, 8.314 Pa · m^3/mol · K.
R_t	thermal resistance; h °F/Btu, K/W; equation (15-16).
r_A	rate of the production of mass A within the control volume; lb_m/ft^3 h, kg/m^3 · s.
R_A	rate of production of moles A within control volume; lb mol/ft^3 h, mol/m^3 · s.
s	surface renewal factor.
S	shape factor; ft or m; equation (15-19).
S	partition coefficient for dissolution of a gas into a solid; kg mol/m^3 · Pa.
t	time; h, s.
t_{exp}	time of exposure; s.
T	absolute temperature; °R, K.
T	dimensionless temperature; dimensionless.
T_b	normal boiling temperature; K.
T_c	critical temperature; K.
T_f	film temperature; °F, K; equation (19-28).
T_{sat}	temperature of saturated liquid-vapor mixtures; °F, K; Figure 21.1.
u	mean molecular speed; ft/s, m/s.
U	overall heat-transfer coefficient; Btu/h ft^2 °F, W/m^2 · K; equation (15-17).
$\dot{V}$	volumetric flow rate of fluid; ft^3/s, m^3/s.
v_x	x component of velocity, $\mathbf{v}$; ft/s, m/s.
v_y	y component of velocity, $\mathbf{v}$; ft/s, m/s.
v_z	z component of velocity, $\mathbf{v}$; ft/s, m/s.
v_∞	free stream velocity of flowing fluid; ft/s, m/s.
v^+	dimensionless velocity.
v	velocity, ft/s, m/s.
$\mathbf{v}$	mass-average velocity for multicomponent mixture; ft/s, m/s equation (24-13).
$\mathbf{v}_i$	velocity of species i; ft/s, m/s.
$\mathbf{v}_i - \mathbf{v}$	diffusion velocity of species i relative to mass-average velocity; ft/s, m/s; Section 24.1.
$\mathbf{v}_i - \mathbf{V}$	diffusion velocity of species i relative to the molar-average velocity; ft/s, m/s; Section 24.1.
V	volume; ft^3, m^3.
V_b	molecular volume at the normal boiling point, cm^3/g mol.
V_c	critical molecular volume; cm^3/g mol.
$\mathbf{V}$	molar-average velocity; ft/s, m/s equation (24-14).
W	work done; Btu, J; Section 6.1.

w_A	mass rate of flow of species A; lbm/h, g/s.
W_s	shaft work; Btu, J; Section 6.1.
W_δ	normal stress work; Btu, J; Section 6.1.
W_τ	shear work; Btu, J; Section 6.1.
x	rectangular coordinate.
x_A	mole fraction in either liquid or solid phase; dimensionless; equation (24-7).
X_A	mole of A/mole of A-free liquid.
X_D	relative time, $D_{AB}t/x_1^2$; dimensionless; Section 27.3.
y	rectangular coordinate.
y^+	dimensionless distance; equation (12-60).
y_A	mole fraction in the gas phase.
$y_{B,\text{lm}}$	log-mean mole fraction of the carrier gas.
y_n'	log fraction of component n in a gas mixture on species i-free basis; equation (24-49).
Y	parameter in heat exchanger analysis; dimensionless; equation (22-12).
Y	unaccomplished change; dimensionless; Section 27.4.
Y_A	mole of A/mole of A-free gas.
z	distance in the z direction; ft, m.
z	rectangular coordinate.
Z	wall collision frequency; equation (7-8).
Z	parameter in heat-exchanger analysis; dimensionless; equation (22-13).
α	absorptivity; dimensionless; Section 23.2.
α	thermal diffusivity; ft^2/h, m^2/s; equation (16-17).
α	ratio of fluxes, N_B/N_A; dimensionless.
α	packed-bed constant.
β	bulk modulus of elasticity; lb$_f$/ft, N/m equation (1-11a).
β	coefficient of thermal expansion; 1/ °F, 1/K; equation (19-10).
δ	boundary layer thickness; ft, m; equation (12-28).
δ	thickness of stagnant or laminar layer; ft, m.
δ_c	concentration boundary layer thickness; ft, m.
δ_i	thermal boundary layer thickness; ft, m; equation (19-22).
Δ_{lm}	log-mean concentration difference, $(y_A - y_A^*)_{lm}$; dimensionless; equation (31-35).
ϵ	emissivity; dimensionless; equation (23-2).
ϵ	volume void.
ϵ	packing characteristic; dimensionless.
ϵ_{AB}	a Lennard-Jones parameter; erg.
ϵ_D	eddy mass diffusivity, ft^2/h, m^2/s.
ϵ_H	eddy thermal diffusivity; ft^2/h, m^2/s; Section 19.7.
ϵ_i	a Lennard-Jones parameter; erg.
ϵ_M	eddy momentum diffusivity or eddy viscosity; ft^2/h, m^2/s; equation (12-52).
η	dependent variable used by Blasius in solution of boundary layer; dimensionless; equation (12-12).
η	similarity parameter for convection analysis; dimensionless; equation (19-17).
η_F	fin efficiency; dimensionless; Section 17.3, Figure 17.11.
θ	temperature parameter $= T - T_\infty$; °F, K; Section 17.3.
θ	fractional void space of a catalyst.
θ	angle in cylindrical or spherical coordinates; rad.
θ_g	correlating constant.
φ	reduced pore diameter; dimensionless.

κ	Boltzmann constant; 1.38×10^{-16} erg/K.
λ	molecular mean free path; Sections 7.3, 15.2, and 24.2.
λ	wave length of thermal radiation; μm; Section 23.4.
λ	ionic conductance; $(A/cm^2)(V/cm)(g\ equivalent/cm^3)$.
λ_{Ts}	latent heat of vaporization; Btu/lb mol, J/g mol.
μ	viscosity; lb_m/ft s, Pa $\cdot$ s; equation (7-4).
μ_B	viscosity of solvent B; cp.
μ_c	chemical potential of given species; Btu/mol, J/mol.
ν	frequency; Hz; Section 23.1.
ν	kinematic viscosity, μ/ρ; ft^2/s, m^2/s.
π	pi groups in dimensional analysis; Sections 11.3, 13.1, 19.3, and 28.3.
ρ	density of a fluid; lb_m/ft^3, kg/m^3; Section 1.2.
ρ	mass density of mixture; lb_m/ft^3, kg/m^3.
ρ	reflectivity; dimensionless; Section 23.2.
ρ_i	mass concentration of species i, lb_m/ft^3, kg/m^3.
σ	surface tension; lb_f/ft, N/m.
σ	Stefan-Boltzmann constant; 0.1714×10^{-8} Btu/h ft^2 °F^4, 5.672×10^{-8} W/$m^2 \cdot$ K^4; equation (15-13).
σ_A	Lennard-Jones molecular diameter of species A; Å, nm.
σ_{AB}	Lennard-Jones parameter; Å.
σ_i	a Lennard-Jones parameter; Å.
σ_{ii}	normal stress; lb_f/$in.^2$, N/m^2; Section 1.2.
τ	transmissivity; dimensionless; Section 23.2.
τ_{ij}	shear stress; lb_f/$in.^2$, N/m^2; Section 1.2.
τ_0	shear stress at the surface; lb_f/$in.^2$, N/m^2; equation (12-30).
ϕ	velocity potential; Section 10.4.
ϕ	angle in spherical coordinates; rad.
ϕ	argument of error function; dimensionless.
ω	angular velocity; 1/s.
ω_i	mass fraction of species i; dimensionless.
$\omega/2$	vorticity; equation (10-4).
Γ	flow rate of condensate film per width; lb_m/ft s, kg/m $\cdot$ s; equation (21-13).
Δ	$y_A - y_A^*$ dimensionless; equation (31-31).
ΔT_{lm}	logarithmic-mean temperature difference; °F, K; equation (22-9).
$\mathscr{E}$	heat-exchanger effectiveness; dimensionless; equation (22-17).
Φ_B	association parameter.
Ψ	stream function; Section 10.2.
Ω	solid angle; rad; Section 23.3.
Ω_D	collision integral: Appendix K.
Ω_k	Lennard-Jones collision integral; equation (15-7) and Appendix K.
Ω_μ	Lennard-Jones collision integral; equation (7-10) and Appendix K.

DIMENSIONLESS PARAMETERS

Bi	Biot number, $(hV/A)/k$; equation (18-7).
Eu	Euler number, $P/\rho v^2$; equation (11-5).
Fo	Fourier number, $\alpha t/(V/A)^2$; equation (18-8).
Fr	Froude number, v^2/gL; equation (11-4).
Gr	Grashof number, $\beta g \rho^2 L^3 \Delta T/\mu^2$; equation (19-12).
Gr_{AB}	mass-transfer Grashof number, $L^3 g \Delta \rho_A/\rho v^3$; equation (28-9).

Gz	Graetz number, $(\pi/4)(D/x)$ Re Pr; Section 20.2.
Le	Lewis number, $k/\rho c_p D_{AB}$; equation (28-3).
Nu	Nusselt number, hL/k; equation (19-6).
Nu_{AB}	mass-transfer Nusselt number, $k_c L/D_{AB}$; equation (28-7).
Pe	Peclet number, $Dv\rho c_p/k = Re$ Pr; Section 20.2.
Pe_{AB}	mass-transfer Peclet number, $Dv/D_{AB} = Re$ Sc; Table 30.1.
Pr	Prandtl number, $\nu/\alpha = \mu c_p/k$; equation (19-1).
Re	Reynolds number, $Lv\rho/\mu = Lv/\nu$; equation (11-7).
Sc	Schmidt number, $\mu/\rho D_{AB}$; equation (28-2).
Sh	Sherwood number, $k_c L/D_{AB}$; Section 28.3.
St	Stanton number, $h/\rho v c_p$; equation (19-8).
St_{AB}	mass-transfer Stanton number, k_c/v_∞.

MATHEMATICAL OPERATIONS

D/Dt	substantial derivative; equation (9-4).
div **A**	or $\nabla \cdot \mathbf{A}$, divergence of a vector.
erf ϕ	the error function of ϕ; Appendix L.
exp x	or e^x, exponential function of x.
ln x	logarithm of x to the base e.
$\log_{10} x$	logarithm of x to base 10.

$$\nabla = \frac{\partial}{\partial x}\mathbf{e}_x + \frac{\partial}{\partial y}\mathbf{e}_y + \frac{\partial}{\partial z}\mathbf{e}_z.$$

Appendix A

Transformations of the Operators ∇ and ∇² to Cylindrical Coordinates

THE OPERATOR ∇ IN CYLINDRICAL COORDINATES

In Cartesian coordinates, ∇ is written as

$$\nabla = \mathbf{e}_x \frac{\partial}{\partial x} + \mathbf{e}_y \frac{\partial}{\partial y} + \mathbf{e}_z \frac{\partial}{\partial z} \tag{A-1}$$

When transforming this operator into cylindrical coordinates, both the unit vectors and the partial derivatives must be transformed.

A cylindrical coordinate system and a Cartesian coordinate system are shown in Figure A.1. The following relations are observed to exist between the Cartesian and cylindrical coordinates:

$$z = z, \quad x^2 + y^2 = r^2, \quad \tan\theta = \frac{y}{x} \tag{A-2}$$

Thus,

$$\left(\frac{\partial}{\partial z}\right)_{cyl} = \left(\frac{\partial}{\partial z}\right)_{cart} \tag{A-3}$$

whereas, from the chain rule

$$\left(\frac{\partial}{\partial x}\right) = \frac{\partial}{\partial r}\frac{\partial r}{\partial x} + \frac{\partial}{\partial \theta}\frac{\partial \theta}{\partial x}$$

As

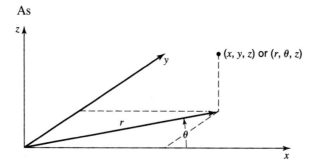

Figure A.1 Cylindrical and Cartesian coordinates.

thus

$$\left(\frac{\partial}{\partial x}\right) = \cos\theta\left(\frac{\partial}{\partial r}\right) - \frac{\sin\theta}{r}\left(\frac{\partial}{\partial\theta}\right) \tag{A-4}$$

In a similar manner,

$$\frac{\partial}{\partial y} = \frac{\partial}{\partial r}\frac{\partial r}{\partial y} + \frac{\partial}{\partial\theta}\frac{\partial\theta}{\partial y}$$

where

$$\frac{\partial r}{\partial y} = \frac{y}{r} = \sin\theta \quad \text{and} \quad \frac{\partial\theta}{\partial y} = \frac{1}{x\sec^2\theta} = \frac{\cos\theta}{r}$$

Thus, $(\partial/\partial y)$ becomes

$$\left(\frac{\partial}{\partial y}\right) = \sin\theta\left(\frac{\partial}{\partial r}\right) + \frac{\cos\theta}{r}\left(\frac{\partial}{\partial\theta}\right) \tag{A-5}$$

The unit vectors must also be transformed. Resolving the unit vectors into their x-, y-, and z-direction components, we obtain

$$\mathbf{e}_z = \mathbf{e}_z \tag{A-6}$$
$$\mathbf{e}_x = \mathbf{e}_r\cos\theta - \mathbf{e}_\theta\sin\theta \tag{A-7}$$
$$\mathbf{e}_y = \mathbf{e}_r\sin\theta + \mathbf{e}_\theta\cos\theta \tag{A-8}$$

Substituting the above relations into equation (A-1), we obtain

$$\mathbf{e}_x\frac{\partial}{\partial x} = \mathbf{e}_r\cos^2\theta\frac{\partial}{\partial r} - \mathbf{e}_r\frac{\sin\theta\cos\theta}{r}\frac{\partial}{\partial\theta} - \mathbf{e}_\theta\sin\theta\cos\theta\frac{\partial}{\partial r} + \mathbf{e}_\theta\frac{\sin^2\theta}{r}\frac{\partial}{\partial\theta}$$

$$\mathbf{e}_y\frac{\partial}{\partial y} = \mathbf{e}_r\sin^2\theta\frac{\partial}{\partial r} + \mathbf{e}_r\frac{\sin\theta\cos\theta}{r}\frac{\partial}{\partial\theta} + \mathbf{e}_\theta\sin\theta\cos\theta\frac{\partial}{\partial r} + \mathbf{e}_\theta\frac{\cos^2\theta}{r}\frac{\partial}{\partial\theta}$$

and

$$\mathbf{e}_z\frac{\partial}{\partial z} = \mathbf{e}_z\frac{\partial}{\partial z}$$

Adding the above relations, we obtain, after noting that $\sin^2\theta + \cos^2\theta = 1$,

$$\mathbf{\nabla} = \mathbf{e}_r\left(\frac{\partial}{\partial r}\right) + \frac{\mathbf{e}_\theta}{r}\left(\frac{\partial}{\partial\theta}\right) + \mathbf{e}_z\left(\frac{\partial}{\partial z}\right) \tag{A-9}$$

THE OPERATOR ∇^2 IN CYLINDRICAL COORDINATES

A unit vector may not change magnitude; however, its direction may change. Cartesian unit vectors do not change their absolute directions, but in cylindrical coordinates both $\mathbf{e}_r$ and $\mathbf{e}_\theta$ depend upon the angle θ. As these vectors change direction, they have derivatives with respect to θ. As $\mathbf{e}_r = \mathbf{e}_x \cos\theta + \mathbf{e}_y \sin\theta$ and $\mathbf{e}_\theta = -\mathbf{e}_x \sin\theta + \mathbf{e}_y \cos\theta$ it may be seen that

$$\frac{\partial}{\partial r}\mathbf{e}_r = 0, \quad \frac{\partial}{\partial r}\mathbf{e}_\theta = 0$$

whereas

$$\frac{\partial}{\partial\theta}\mathbf{e}_r = \mathbf{e}_\theta \tag{A-10}$$

and

$$\frac{\partial}{\partial\theta}\mathbf{e}_\theta = -\mathbf{e}_r \tag{A-11}$$

Now the operator $\nabla^2 = \nabla \cdot \nabla$ and thus

$$\nabla \cdot \nabla = \nabla^2 = \left(\mathbf{e}_r\frac{\partial}{\partial r} + \frac{\mathbf{e}_\theta}{r}\frac{\partial}{\partial\theta} + \mathbf{e}_z\frac{\partial}{\partial z}\right) \cdot \left(\mathbf{e}_r\frac{\partial}{\partial r} + \frac{\mathbf{e}_\theta}{r}\frac{\partial}{\partial\theta} + \mathbf{e}_z\frac{\partial}{\partial z}\right)$$

Performing the indicated operations, we obtain

$$\mathbf{e}_r\frac{\partial}{\partial r} \cdot \nabla = \frac{\partial^2}{\partial\theta}$$

$$\frac{\mathbf{e}_\theta}{r}\frac{\partial}{\partial\theta} \cdot \nabla = \frac{\mathbf{e}_\theta}{r} \cdot \frac{\partial}{\partial\theta}\left(\mathbf{e}_r\frac{\partial}{\partial r}\right) + \frac{\mathbf{e}_\theta}{r} \cdot \frac{\partial}{\partial\theta}\left(\frac{\mathbf{e}_\theta}{r}\frac{\partial}{\partial\theta}\right)$$

or

$$\frac{\mathbf{e}_\theta}{r}\frac{\partial}{\partial\theta} \cdot \nabla = \frac{1}{r}\frac{\partial}{\partial r} + \frac{1}{r^2}\frac{\partial^2}{\partial\theta^2}$$

and

$$\mathbf{e}_z\frac{\partial}{\partial z} \cdot \nabla = \frac{\partial^2}{\partial z^2}$$

Thus, the operator ∇^2 becomes

$$\nabla^2 = \frac{\partial^2}{\partial r^2} + \frac{1}{r}\frac{\partial}{\partial r} + \frac{1}{r^2}\frac{\partial^2}{\partial\theta^2} + \frac{\partial^2}{\partial z^2} \tag{A-12}$$

or

$$\nabla^2 = \frac{1}{r}\frac{\partial}{\partial r}\left(r\frac{\partial}{\partial r}\right) + \frac{1}{r^2}\frac{\partial^2}{\partial\theta^2} + \frac{\partial^2}{\partial z^2} \tag{A-13}$$

Appendix B

Summary of Differential Vector Operations in Various Coordinate Systems

CARTESIAN COORDINATES

Coordinate system

Figure B.1 Unit vectors at the point (x, y, z).

Gradient

$$\nabla P = \frac{\partial P}{\partial x}\mathbf{e}_x + \frac{\partial P}{\partial y}\mathbf{e}_y + \frac{\partial P}{\partial z}\mathbf{e}_z \qquad \text{(B-1)}$$

Divergence

$$\nabla \cdot \mathbf{v} = \frac{\partial v_x}{\partial x} + \frac{\partial v_y}{\partial y} + \frac{\partial v_z}{\partial z} \qquad \text{(B-2)}$$

Curl

$$\nabla \times \mathbf{v} = \left\{ \begin{array}{c} \left(\dfrac{\partial v_z}{\partial y} - \dfrac{\partial v_y}{\partial z}\right)\mathbf{e}_x \\[2mm] \left(\dfrac{\partial v_x}{\partial z} - \dfrac{\partial v_z}{\partial x}\right)\mathbf{e}_y \\[2mm] \left(\dfrac{\partial v_y}{\partial x} - \dfrac{\partial v_x}{\partial y}\right)\mathbf{e}_z \end{array} \right\} \qquad \text{(B-3)}$$

Laplacian of a scalar

$$\nabla^2 T = \frac{\partial^2 T}{\partial x^2} + \frac{\partial^2 T}{\partial y^2} + \frac{\partial^2 T}{\partial z^2} \qquad \text{(B-4)}$$

CYLINDRICAL COORDINATES

Coordinate system

Figure B.2   Unit vectors at the point (r, θ, z).

Gradient

$$\nabla P = \frac{\partial P}{\partial r}\mathbf{e}_r + \frac{1}{r}\frac{\partial P}{\partial \theta}\mathbf{e}_\theta + \frac{\partial P}{\partial z}\mathbf{e}_z \tag{B-5}$$

Divergence

$$\nabla \cdot \mathbf{v} = \frac{1}{r}\frac{\partial}{\partial r}(rv_r) + \frac{1}{r}\frac{\partial v_\theta}{\partial \theta} + \frac{\partial v_z}{\partial z} \tag{B-6}$$

Curl

$$\nabla \times \mathbf{v} = \left\{ \begin{aligned} &\left(\frac{1}{r}\frac{\partial v_z}{\partial \theta} - \frac{\partial v_\theta}{\partial z}\right)\mathbf{e}_r \\ &\left(\frac{\partial v_r}{\partial z} - \frac{\partial v_z}{\partial r}\right)\mathbf{e}_\theta \\ &\left\{\frac{1}{r}\left[\frac{\partial}{\partial r}(rv_\theta) - \frac{\partial v_r}{\partial \theta}\right]\right\}\mathbf{e}_z \end{aligned} \right\} \tag{B-7}$$

Laplacian of a scalar

$$\nabla^2 T = \frac{1}{r}\frac{\partial}{\partial r}\left(r\frac{\partial T}{\partial r}\right) + \frac{1}{r^2}\frac{\partial^2 T}{\partial \theta^2} + \frac{\partial^2 T}{\partial z^2} \tag{B-8}$$

SPHERICAL COORDINATES

Coordinate system

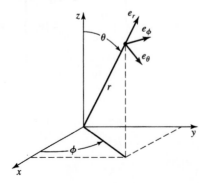

Figure B.3   Unit vectors at the point (r, θ, ϕ).

Gradient

$$\nabla P = \frac{\partial P}{\partial r}\mathbf{e}_r + \frac{1}{r}\frac{\partial P}{\partial \theta}\mathbf{e}_\theta + \frac{1}{r \sin \theta}\frac{\partial P}{\partial \phi}\mathbf{e}_\phi \tag{B-9}$$

Divergence

$$\nabla \cdot \mathbf{v} = \frac{1}{r^2}\frac{\partial}{\partial r}(r^2 v_r) + \frac{1}{r \sin \theta}\frac{\partial}{\partial \theta}(v_\theta \sin \theta) + \frac{1}{r \sin \theta}\frac{\partial v_\phi}{\partial \phi} \tag{B-10}$$

Curl

$$\nabla \times \mathbf{v} = \left\{ \begin{array}{l} \dfrac{1}{r \sin \theta}\left[\dfrac{\partial}{\partial \theta}(v_\phi \sin \theta) - \dfrac{\partial v_\theta}{\partial \phi}\right]\mathbf{e}_r \\[2ex] \left[\dfrac{1}{r \sin \theta}\dfrac{\partial v_r}{\partial \phi} - \dfrac{1}{r}\dfrac{\partial}{\partial r}(r v_\phi)\right]\mathbf{e}_\theta \\[2ex] \dfrac{1}{r}\left[\dfrac{\partial}{\partial r}(r v_\theta) - \dfrac{\partial v_r}{\partial \theta}\right]\mathbf{e}_\phi \end{array} \right\} \tag{B-11}$$

Laplacian of a scalar

$$\nabla^2 T = \frac{1}{r^2}\frac{\partial}{\partial r}\left(r^2 \frac{\partial T}{\partial r}\right) + \frac{1}{r^2 \sin \theta}\frac{\partial}{\partial \theta}\left(\sin \theta \frac{\partial T}{\partial \theta}\right) + \frac{1}{r^2 \sin^2 \theta}\frac{\partial^2 T}{\partial \phi^2} \tag{B-12}$$

Appendix C

Symmetry of the Stress Tensor

The shear stress τ_{ij} can be shown to be equal to $\tau_{j,i}$ by the following simple argument. Consider the element of fluid shown in Figure C.1. The sum of the moments on the element will be related to the angular acceleration by

$$\sum M = I\dot{\omega} \qquad \text{(C-1)}$$

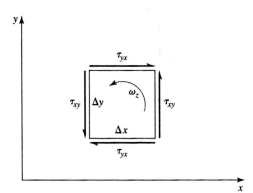

Figure C.1 Free body of element

where I is the mass moment of inertia of the element. Substituting into equation (C-1)

$$-(\tau_{yx}\,\Delta x\,\Delta z)\Delta y + (\tau_{xy}\,\Delta y\,\Delta z)\Delta x = \rho\,\Delta x\,\Delta y\,\Delta z\frac{(\Delta x^2 + \Delta y^2)}{12}\dot{\omega}_z$$

where the moment of inertia of a rectangular prism has been used for the element.

The volume of the element $\Delta x\,\Delta y\,\Delta z$ may be canceled to yield

$$\rho\left(\frac{\Delta x^2 + \Delta y^2}{12}\right)\dot{\omega}_z = \tau_{xy} - \tau_{yx} \qquad \text{(C-2)}$$

Now the difference in shear stress is seen to depend upon the size of the element. As the element shrinks to a point. Δx and Δy approach zero independently, and we obtain, in the limit,

$$\tau_{xy} = \tau_{yx}$$

or, as this can be done about any axis,

$$\tau_{ij} = \tau_{ji}$$

Another way to look at equation (C-2) is to determine the angular acceleration ω_z as the element shrinks to a point. The angular acceleration at a point must be finite; hence, τ_{yx} and τ_{xy} must be equal.

Appendix D

The Viscous Contribution to the Normal Stress

The normal stress, σ, may be divided into two parts: the pressure contribution, $-P$, and a viscous contribution, σ_v. The viscous contribution to the normal stress is obtained by analogy with Hooke's law for an elastic solid. In Hooke's law for three-dimensional stress, the normal stress, $\sigma_{x,x}$ in the x direction is related to the strains in the x, y, and z directions by[1]

$$\sigma_{x,x} = 2G\epsilon_x + \frac{2G\eta}{1 - 2\eta}(\epsilon_x + \epsilon_y + \epsilon_z) \tag{D-1}$$

where G is the share modulus, η is Poisson's ratio, and ϵ is the axial strain.

When Newton's viscosity relation was discussed, the shear strain in a solid was seen to be analogous to the rate of shear strain in a fluid. Accordingly, the axial strain in a solid, ϵ_x, is taken to be analogous to the axial strain rate in a fluid, $\partial v_x/\partial x$.

When the velocity derivatives are substituted for the strains in equation (D-1), and the viscosity is used in place of the shear modulus, we obtain

$$(\sigma_{x,x})_{\text{viscous}} = 2\mu \frac{\partial v_x}{\partial x} + \lambda \nabla \cdot \mathbf{v} \tag{D-2}$$

Here the sum of the strain-rate derivatives is observed to be equal to $\nabla \cdot \mathbf{v}$, and the second coefficient has been designated λ and is called the bulk viscosity or second viscosity coefficient. The total normal stress in the x direction becomes

$$\sigma_{x,x} = -P + 2\mu \frac{\partial v_x}{\partial x} + \lambda \nabla \cdot \mathbf{v} \tag{D-3}$$

If the corresponding normal stress components in the y and z directions are added together, we obtain

$$\sigma_{x,x} + \sigma_{y,y} + \sigma_{z,z} = -3P + (2\mu + 3\lambda)\nabla \cdot \mathbf{v}$$

so that the average normal stress $\bar{\sigma}$ is given by

$$\bar{\sigma} = -P + \left(\frac{2\mu + 3\lambda}{3}\right)\nabla \cdot \mathbf{v}$$

---

[1] A more familiar form is

$$\sigma_{x,x} = \frac{E}{(1 + \eta)(1 - 2\eta)}[(1 - \eta)\epsilon_x + \eta(\epsilon_y + \epsilon_z)]$$

The shear modulus G has been replaced with its equivalent, $E/2(1 + \eta)$.

Thus, unless $\lambda = -\dfrac{2}{3}\mu$, the average stress will depend upon the flow properties rather than the fluid property, P. Stokes assumed that $\lambda = -\dfrac{2}{3}\mu$, and experiments have indicated that λ is of the same order of magnitude as μ of air. As $\boldsymbol{\nabla} \cdot \mathbf{v} = 0$ in an incompressible flow, the value of λ is of no concern except for compressible fluids.

The resulting expressions for normal stress in a Newtonian fluid are

$$\sigma_{x,x} = -P + 2\mu\frac{\partial v_x}{\partial x} - \frac{2}{3}\mu\boldsymbol{\nabla} \cdot \mathbf{v} \tag{D-4}$$

$$\sigma_{y,y} = -P + 2\mu\frac{\partial v_y}{\partial y} - \frac{2}{3}\mu\boldsymbol{\nabla} \cdot \mathbf{v} \tag{D-5}$$

$$\sigma_{z,z} = -P + 2\mu\frac{\partial v_z}{\partial z} - \frac{2}{3}\mu\boldsymbol{\nabla} \cdot \mathbf{v} \tag{D-6}$$

Appendix E

The Navier–Stokes Equations for Constant ρ and μ in Cartesian, Cylindrical, and Spherical Coordinates

CARTESIAN COORDINATES

x direction

$$\rho\left(\frac{\partial v_x}{\partial t} + v_x\frac{\partial v_x}{\partial x} + v_y\frac{\partial v_x}{\partial y} + v_z\frac{\partial v_x}{\partial z}\right) = -\frac{\partial P}{\partial x} + \rho g_x + \mu\left(\frac{\partial^2 v_x}{\partial x^2} + \frac{\partial^2 v_x}{\partial y^2} + \frac{\partial^2 v_x}{\partial z^2}\right) \quad \text{(E-1)}$$

y direction

$$\rho\left(\frac{\partial v_y}{\partial t} + v_x\frac{\partial v_y}{\partial x} + v_y\frac{\partial v_y}{\partial y} + v_z\frac{\partial v_y}{\partial z}\right) = -\frac{\partial P}{\partial y} + \rho g_y + \mu\left(\frac{\partial^2 v_y}{\partial x^2} + \frac{\partial^2 v_y}{\partial y^2} + \frac{\partial^2 v_y}{\partial z^2}\right) \quad \text{(E-2)}$$

z direction

$$\rho\left(\frac{\partial v_z}{\partial t} + v_x\frac{\partial v_z}{\partial x} + v_y\frac{\partial v_z}{\partial y} + v_z\frac{\partial v_z}{\partial z}\right) = -\frac{\partial P}{\partial z} + \rho g_z + \mu\left(\frac{\partial^2 v_z}{\partial x^2} + \frac{\partial^2 v_z}{\partial y^2} + \frac{\partial^2 v_z}{\partial z^2}\right) \quad \text{(E-3)}$$

CYLINDRICAL COORDINATES

r direction

$$\rho\left(\frac{\partial v_r}{\partial t} + v_r\frac{\partial v_r}{\partial r} + \frac{v_\theta}{r}\frac{\partial v_r}{\partial \theta} - \frac{v_\theta^2}{r} + v_z\frac{\partial v_r}{\partial z}\right)$$
$$= -\frac{\partial P}{\partial r} + \rho g_r + \mu\left[\frac{\partial}{\partial r}\left(\frac{1}{r}\frac{\partial}{\partial r}(rv_r)\right) + \frac{1}{r^2}\frac{\partial^2 v_r}{\partial \theta^2} - \frac{2}{r^2}\frac{\partial v_\theta}{\partial \theta} + \frac{\partial^2 v_r}{\partial z^2}\right] \quad \text{(E-4)}$$

θ direction

$$\rho\left(\frac{\partial v_\theta}{\partial t} + v_r\frac{\partial v_\theta}{\partial r} + \frac{v_\theta}{r}\frac{\partial v_\theta}{\partial \theta} + \frac{v_r v_\theta}{r} + v_z\frac{\partial v_\theta}{\partial z}\right) = -\frac{1}{r}\frac{\partial P}{\partial \theta} + \rho g_\theta$$
$$+ \mu\left[\frac{\partial}{\partial r}\left(\frac{1}{r}\frac{\partial}{\partial r}(rv_\theta)\right) + \frac{1}{r^2}\frac{\partial^2 v_\theta}{\partial \theta^2} + \frac{2}{r^2}\frac{\partial v_r}{\partial \theta} + \frac{\partial^2 v_\theta}{\partial z^2}\right] \quad \text{(E-5)}$$

z direction

$$\rho\left(\frac{\partial v_z}{\partial t} + v_r\frac{\partial v_z}{\partial r} + \frac{v_\theta}{r}\frac{\partial v_z}{\partial \theta} + v_z\frac{\partial v_z}{\partial z}\right)$$

$$= -\frac{\partial P}{\partial z} + \rho g_z + \mu\left[\frac{1}{r}\frac{\partial}{\partial r}\left(r\frac{\partial v_z}{\partial r}\right) + \frac{1}{r^2}\frac{\partial^2 v_z}{\partial \theta^2} + \frac{\partial^2 v_z}{\partial z^2}\right] \quad \text{(E-6)}$$

SPHERICAL COORDINATES[1]

r direction

$$\rho\left(\frac{\partial v_r}{\partial t} + v_r\frac{\partial v_r}{\partial r} + \frac{v_\theta}{r}\frac{\partial v_r}{\partial \theta} + \frac{v_\phi}{r\sin\theta}\frac{\partial v_r}{\partial \phi} - \frac{v_\phi^2}{r} - \frac{v_\theta^2}{r}\right)$$

$$= -\frac{\partial P}{\partial r} + \rho g_r + \mu\left[\nabla^2 v_r - \frac{2}{r^2}v_r - \frac{2}{r^2}\frac{\partial v_\theta}{\partial \theta} - \frac{2}{r^2}v_\theta\cot\theta - \frac{2}{r^2\sin\theta}\frac{\partial v_\phi}{\partial \phi}\right] \quad \text{(E-7)}$$

θ direction

$$\rho\left[\frac{\partial v_\theta}{\partial t} + v_r\frac{\partial v_\theta}{\partial r} + \frac{v_\theta}{r}\frac{\partial v_\theta}{\partial \theta} + \frac{v_\phi}{}\frac{\partial v_\theta}{} + \frac{v_r v_\theta}{} \quad \frac{\partial v_\phi^2\cot\theta}{}\right]$$

$$= -\frac{1}{r}\frac{\partial P}{\partial \theta} + \rho g_\theta \quad \text{(E-7)}$$

φ direction

$$\rho\left(\frac{\partial v_\phi}{\partial t} + v_r\frac{\partial v_\phi}{\partial r} + \right.$$

$$= -\frac{1}{r\sin\theta}\frac{\partial P}{\partial \phi} + \rho \quad \text{(E-7)}$$

---

[1] In the above equations,

$$\nabla^2 = \frac{1}{r^2}\frac{\partial}{\partial r}\left(r\frac{\partial}{\partial r}\right) + \frac{1}{r^2\sin\theta}\frac{\partial}{\partial \theta}\left(\sin\theta\frac{\partial}{\partial \theta}\right) + \frac{1}{r^2\sin\theta}\frac{\partial^2}{\partial \phi^2}$$

Charts for Solution of Unsteady Transport Problems

Table F.9 Symbols for unsteady-state charts

	Parameter symbol	Molecular mass transfer	Heat conduction
Unaccomplished change, a dimensionless ratio	Y	$\dfrac{c_{A_1} - c_A}{c_{A_1} - c_{A_0}}$	$\dfrac{T_\infty - T}{T_\infty - T_0}$
Relative time	X	$\dfrac{D_{AB}t}{x_1^2}$	$\dfrac{\alpha t}{x_1^2}$
Relative position	n	$\dfrac{x}{x_1}$	$\dfrac{x}{x_1}$
Relative resistance	m	$\dfrac{D_{AB}}{k_c x_1}$	$\dfrac{k}{h x_1}$

T = temperature
c_A = concentration of component A
x = distance from center to any point
t = time
k = thermal conductivity
h, k_c = convective-transfer coefficients
α = thermal diffusivity
D_{AB} = mass diffusivity

Subscripts:
 0 = initial condition at time $t = 0$
 1 = boundary
 A = component A
 ∞ = reference condition for temperature

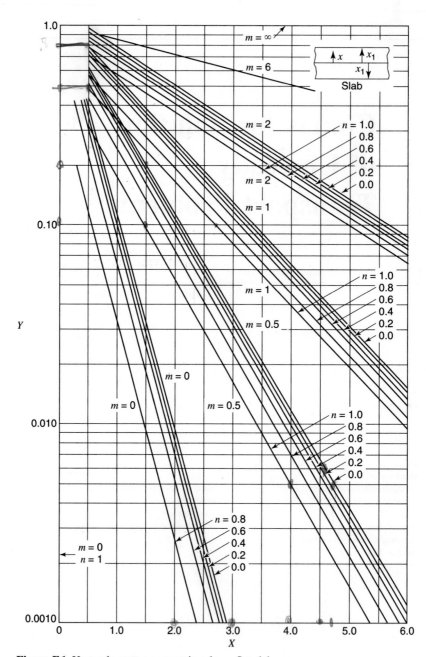

Figure F.1 Unsteady-state transport in a large flat slab.

Figure F.2 Unsteady-state transport in a long cylinder.

Figure F.3 Unsteady-state transport in a sphere.

Figure F.4 Center temperature history for an infinite plate.

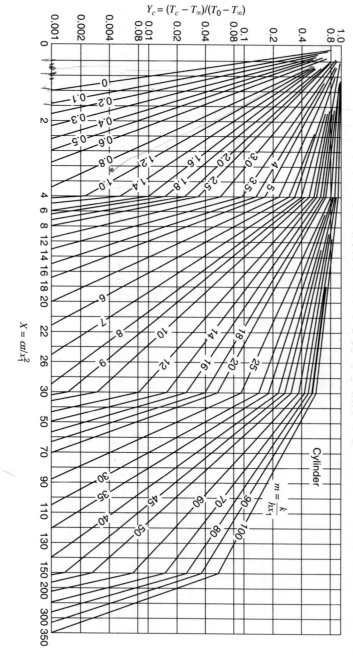

Figure F.5 Center temperature history for an infinite cylinder.

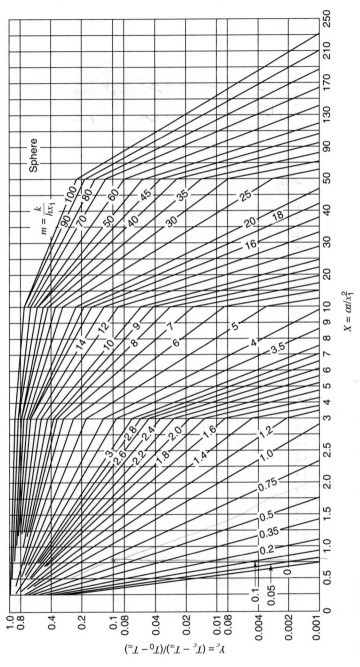

Figure F.6 Center temperature history for a sphere.

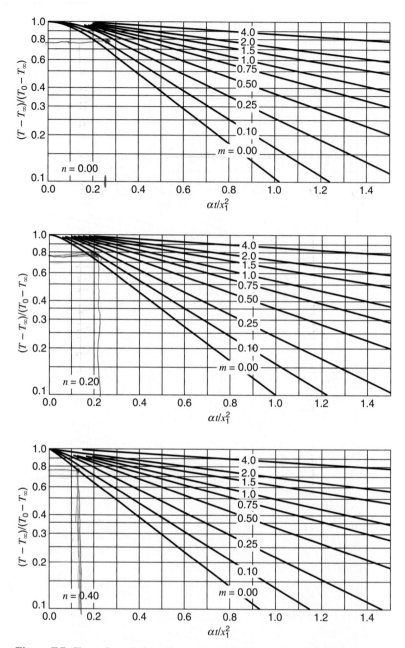

Figure F.7 Charts for solution of unsteady transport problems: flat plate.

Figure F.7 Continued.

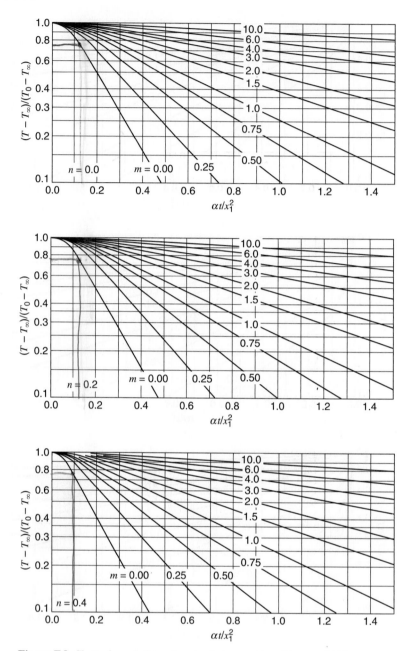

Figure F.8 Charts for solution of unsteady transport problems: cylinder.

Figure F.8 Continued.

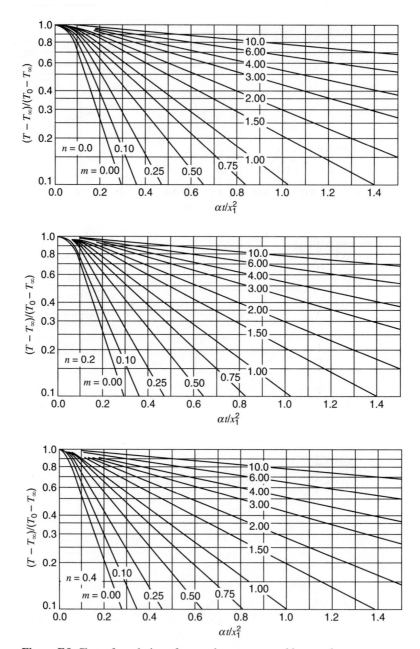

Figure F.9 Charts for solution of unsteady transport problems: sphere.

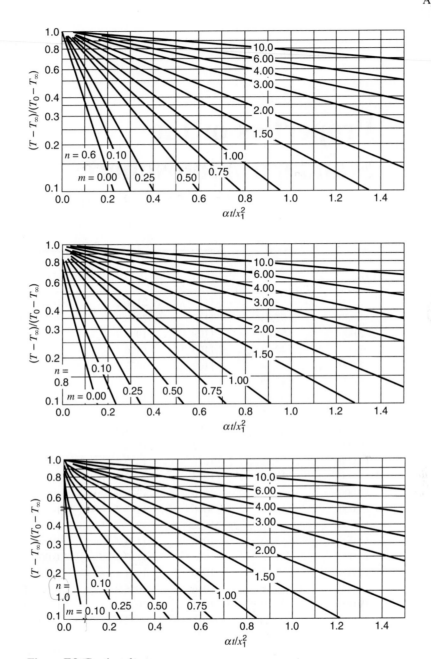

Figure F.9 Continued.

Appendix **G**

Properties of the Standard Atmosphere[1]

Table G.1 English units

h (ft)	T (°F)	a (fps)	P (lb/ft^2)	P (slug/ft^3)	$\mu \times 10^7$ (slug/ft s)
0	59.00	1117	2116.2	0.002378	3.719
1,000	57.44	1113	2040.9	0.002310	3.699
2,000	51.87	1109	1967.7	0.002242	3.679
3,000	48.31	1105	1896.7	0.002177	3.659
4,000	44.74	1102	1827.7	0.002112	3.639
5,000	41.18	1098	1760.8	0.002049	3.618
6,000	37.62	1094	1696.0	0.001988	3.598
7,000	34.05	1090	1633.0	0.001928	3.577
8,000	30.49	1086	1571.9	0.001869	3.557
9,000	26.92	1082	1512.8	0.001812	3.536
10,000	23.36	1078	1455.4	0.001756	3.515
11,000	19.80	1074	1399.8	0.001702	3.495
12,000	16.23	1070	1345.9	0.001649	3.474
13,000	12.67	1066	1293.7	0.001597	3.453
14,000	9.10	1062	1243.2	0.001546	3.432
15,000	5.54	1058	1194.3	0.001497	3.411
16,000	1.98	1054	1147.0	0.001448	3.390
17,000	−1.59	1050	1101.1	0.001401	3.369
18,000	−5.15	1046	1056.9	0.001355	3.347
19,000	−8.72	1041	1014.0	0.001311	3.326
20,000	−12.28	1037	972.6	0.001267	3.305
21,000	−15.84	1033	932.5	0.001225	3.283
22,000	−19.41	1029	893.8	0.001183	3.262
23,000	−22.97	1025	856.4	0.001143	3.240
24,000	−26.54	1021	820.3	0.001104	3.218
25,000	−30.10	1017	785.3	0.001066	3.196
26,000	−33.66	1012	751.7	0.001029	3.174
27,000	−37.23	1008	719.2	0.000993	3.153
28,000	−40.79	1004	687.9	0.000957	3.130
29,000	−44.36	999	657.6	0.000923	3.108

(*Continued*)

Table G.1   Continued

h (ft)	T (°F)	a (fps)	P (lb/ft^2)	P (slug/ft^3)	$\mu \times 10^7$ (slug/ft s)
30,000	−47.92	995	628.5	0.000890	3.086
31,000	−51.48	991	600.4	0.000858	3.064
32,000	−55.05	987	573.3	0.000826	3.041
33,000	−58.61	982	547.3	0.000796	3.019
34,000	−62.18	978	522.2	0.000766	2.997
35,000	−65.74	973	498.0	0.000737	2.974
40,000	−67.6	971	391.8	0.0005857	2.961
45,000	−67.6	971	308.0	0.0004605	2.961
50,000	−67.6	971	242.2	0.0003622	2.961
60,000	−67.6	971	150.9	0.0002240	2.961
70,000	−67.6	971	93.5	0.0001389	2.961
80,000	−67.6	971	58.0	0.0000861	2.961
90,000	−67.6	971	36.0	0.0000535	2.961
100,000	−67.6	971	22.4	0.0000331	2.961
150,000	113.5	1174	3.003	0.00000305	4.032
200,000	159.4	1220	0.6645	0.00000062	4.277
250,000	−8.2	1042	0.1139	0.00000015	3.333

Table G.2 SI units—Properties of the standard atmosphere

h (m)	T (K)	a (m/s)	P (Pa)	ρ (kg/m^3)	$\mu \times 10^5$ (Pa · s)
0	288.2	340.3	1.0133×10^5	1.225	1.789
500	284.9	338.4	0.95461	1.167	1.774
1,000	281.7	336.4	0.89876	1.111	1.758
1,500	278.4	334.5	0.84560	1.058	1.742
2,000	275.2	332.5	0.79501	1.007	1.726
2,500	271.9	330.6	0.74692	0.9570	1.710
3,000	268.7	328.6	0.70121	0.9093	1.694
3,500	265.4	326.6	0.65780	0.8634	1.678
4,000	262.2	324.6	0.61660	0.8194	1.661
4,500	258.9	322.6	0.57753	0.7770	1.645
5,000	255.7	320.5	0.54048	0.7364	1.628
5,500	252.4	318.5	0.50539	0.6975	1.612
6,000	249.2	316.5	0.47218	0.6601	1.595
6,500	245.9	314.4	0.44075	0.6243	1.578
7,000	242.7	312.3	0.41105	0.5900	1.561
7,500	239.5	310.2	0.38300	0.5572	1.544
8,000	236.2	308.1	0.35652	0.5258	1.527
8,500	233.0	306.0	0.33154	0.4958	1.510
9,000	229.7	303.8	0.30801	0.4671	1.493
9,500	226.5	301.7	0.28585	0.4397	1.475
10,000	223.3	299.5	0.26500	0.4135	1.458
11,000	216.8	295.2	0.22700	0.3648	1.422
12,000	216.7	295.1	0.19399	0.3119	1.422
13,000	216.7	295.1	0.16580	0.2666	1.422
14,000	216.7	295.1	0.14170	0.2279	1.422
15,000	216.7	295.1	0.12112	0.1948	1.422
16,000	216.7	295.1	0.10353	0.1665	1.422
17,000	216.7	295.1	8.8497×10^3	0.1423	1.422
18,000	216.7	295.1	7.5652	0.1217	1.422
19,000	216.7	295.1	6.4675	0.1040	1.422
20,000	216.7	295.1	5.5293	0.08891	1.422
25,000	221.5	298.4	2.5492	0.04008	1.448
30,000	226.5	301.7	1.1970	0.01841	1.475
35,000	236.5	308.3	0.57459	0.008463	1.529
40,000	250.4	317.2	0.28714	0.003996	1.601
45,000	264.2	325.8	0.14910	0.001966	1.671
50,000	270.7	329.8	7.9779×10^1	0.001027	1.704
55,000	265.6	326.7	4.27516	0.0005608	1.678
60,000	255.8	320.6	2.2461	0.0003059	1.629
65,000	239.3	310.1	1.1446	0.0001667	1.543
70,000	219.7	297.1	5.5205×10^0	0.00008754	1.438
75,000	200.2	283.6	2.4904	0.00004335	1.329
80,000	180.7	269.4	1.0366	0.00001999	1.216

Appendix H

Physical Properties of Solids

Material	ρ (lb_m/ft^3) (68°F)	ρ (kg/m^3) (293 K)	c_p ($Btu/lb_m°F$) (293 K)	c_p ($J/kg \cdot 1K$) $\times 10^{-2}$ (293K)	α (ft^2/h) (68°F)	α ($m^2/s) \cdot 10^5$ (293k)	k ($Btu/h\,ft°F$) (68)	(212)°F	(572)	k ($W/m \cdot K$) (293)	K (373)	(573)
Metals												
Aluminum	168.6	2,701.1	0.224	9.383	3.55	9.16	132	133	133	229	229	230
Copper	555	8,890	0.092	3.854	3.98	10.27	223	219	213	386	379	369
Gold	1206	19,320	0.031	1.299	4.52	11.66	169	170	172	293	294	298
Iron	492	7,880	0.122	5.110	0.83	2.14	42.3	39	31.6	73.2	68	54
Lead	708	11,300	0.030	1.257	0.80	2.06	20.3	19.3	17.2	35.1	33.4	29.8
Magnesium	109	1,750	0.248	10.39	3.68	9.50	99.5	96.8	91.4	172	168	158
Nickel	556	8,910	0.111	4.560	0.87	2.24	53.7	47.7	36.9	93.0	82.6	63.9
Platinum	1340	21,500	0.032	1.340	0.09	0.23	40.5	41.9	43.5	70.1	72.5	75.3
Silver	656	10,500	0.057	2.388	6.42	16.57	240	237	209	415	410	362
Tin	450	7,210	0.051	2.136	1.57	4.05	36	34	—	62	59	—
Tungsten	1206	19,320	0.032	1.340	2.44	6.30	94	87	77	160	150	130
Uranium α	1167	18,700	0.027	1.131	0.53	1.37	16.9	17.2	19.6	29.3	29.8	33.9
Zinc	446	7,150	0.094	3.937	1.55	4.00	65	63	58	110	110	100
Alloys												
Aluminum 2024	173	2,770	0.230	9.634	1.76	4.54	70.2			122		
Brass (70% Cu, 30% Ni)	532	8,520	0.091	3.812	1.27	3.28	61.8	73.9	85.3	107	128	148
Constantan (60% Cu, 40% Ni)	557	8,920	0.098	4.105	0.24	0.62	13.1	15.4		22.7	26.7	
Iron, cast	455	7,920	0.100	4.189	0.65	1.68	29.6	26.8		51.2	46.4	
Nichrome V	530	8,490	0.106	4.440	0.12	0.31	7.06	7.99	9.94	12.2	13.8	17.2
Stainless steel	488	7,820	0.110	4.608	0.17	0.44	9.4	10.0	13	16	17.3	23
Steel, mild (1% C)	488	7,820	0.113	4.733	0.45	1.16	24.8	24.8	22.9	42.9	42.9	39.0

Nonmetals									
Asbestos	36				0.092	0.11	0.125	0.159	0.190	0.21
Brick (fire clay)	144					0.65			1.13	
Brick (masonry)	106				0.38			0.66		
Brick (chrome)	188					0.67			1.16	
Concrete	144				0.70			1.21		
Corkboard	10				0.025			0.043		
Diatomaceous earth, powdered	14				0.03			0.05		
Glass, window	170				0.45			0.78		
Glass, Pyrex	140				0.63	0.67	0.84	1.09	1.16	1.45
Kaolin firebrick	19						0.052			0.09
85% Magnesia	17				0.038	0.041		0.066	0.071	
Sandy loam, 4% H_2O	104				0.54			0.94		
Sandy loam, 10% H_2O	121				1.08			1.87		
Rock wool	10	160			0.023	0.033		0.040	0.057	
Wood, oak ⊥ to grain	51	820	0.57	23.9	0.12			0.21		
Wood, oak ‖ to grain	51	820	0.57	23.9	0.23			0.40		

Appendix I

Physical Properties of Gases and Liquids[1]

[1] All gas properties are for atmospheric pressure.

					Gases				
T ($°F$)	ρ (lb_m/ft^3)	c_p ($Btu/lb_m°F$)	$\mu \times 10^5$ ($lb_m/ft\,s$)	$\nu \times 10^3$ (ft^2/s)	k ($Btu/h\,ft\,°F$)	α (ft^2/h)	Pr	$\beta \times 10^3$ ($1/°F$)	$g\beta\rho^2/\mu^2$ ($1/°F \cdot ft^3$)
					Air				
0	0.0862	0.240	1.09	0.126	0.0132	0.639	0.721	2.18	4.39×10^6
30	0.0810	0.240	1.15	0.142	0.0139	0.714	0.716	2.04	3.28
60	0.0764	0.240	1.21	0.159	0.0146	0.798	0.711	1.92	2.48
80	0.0735	0.240	1.24	0.169	0.0152	0.855	0.708	1.85	2.09
100	0.0710	0.240	1.28	0.181	0.0156	0.919	0.703	1.79	1.76
150	0.0651	0.241	1.36	0.209	0.0167	1.06	0.698	1.64	1.22
200	0.0602	0.241	1.45	0.241	0.0179	1.24	0.694	1.52	0.840
250	0.0559	0.242	1.53	0.274	0.0191	1.42	0.690	1.41	0.607
300	0.0523	0.243	1.60	0.306	0.0203	1.60	0.686	1.32	0.454
400	0.0462	0.245	1.74	0.377	0.0225	2.00	0.681	1.16	0.264
500	0.0413	0.247	1.87	0.453	0.0246	2.41	0.680	1.04	0.163
600	0.0374	0.251	2.00	0.535	0.0270	2.88	0.680	0.944	79.4×10^3
800	0.0315	0.257	2.24	0.711	0.0303	3.75	0.684	0.794	50.6
1000	0.0272	0.263	2.46	0.906	0.0337	4.72	0.689	0.685	27.0
1500	0.0203	0.277	2.92	1.44	0.0408	7.27	0.705	0.510	7.96

T (K)	ρ (kg/m^3)	$c_p \times 10^{-3}$ ($J/kg \times K$)	$\mu \times 10^5$ ($Pa \times s$)	$\nu \times 10^5$ (m^2/s)	$k \times 10^2$ ($W/m \times K$)	$\alpha \times 10^5$ (m^2/s)	Pr	$g\beta\rho^2/\mu^2$ ($1/K \cdot m^3$)
				Air				
250	1.4133	1.0054	1.5991	1.1315	2.2269	1.5672	0.722	4.638×10^8
260	1.3587	1.0054	1.6503	1.2146	2.3080	1.6896	0.719	2.573
280	1.2614	1.0057	1.7503	1.3876	2.4671	1.9448	0.713	1.815
300	1.1769	1.0063	1.8464	1.5689	2.6240	2.2156	0.708	1.327
320	1.1032	1.0073	1.9391	1.7577	2.7785	2.5003	0.703	0.9942
340	1.0382	1.0085	2.0300	1.9553	2.9282	2.7967	0.699	0.7502
360	0.9805	1.0100	2.1175	2.1596	3.0779	3.1080	0.695	0.5828
400	0.8822	1.0142	2.2857	2.5909	3.3651	3.7610	0.689	0.3656
440	0.8021	1.0197	2.4453	3.0486	3.6427	4.4537	0.684	0.2394
480	0.7351	1.0263	2.5963	3.5319	3.9107	5.1836	0.681	0.1627
520	0.6786	1.0339	2.7422	4.0410	4.1690	5.9421	0.680	0.1156
580	0.6084	1.0468	2.9515	4.8512	4.5407	7.1297	0.680	7.193×10^6
700	0.5040	1.0751	3.3325	6.6121	5.2360	9.6632	0.684	3.210
800	0.4411	1.0988	3.6242	8.2163	5.7743	11.9136	0.689	1.804
1000	0.3529	1.1421	4.1527	11.1767	6.7544	16.7583	0.702	0.803

T (°F)	ρ (lb$_m$/ft^3)	c_p (Btu/lb$_m$°F)	$\mu \times 10^5$ (lb$_m$/ft s)	$\nu \times 10^3$ (ft^2/s)	k (Btu/h ft °F)	α (ft^2/h)	Pr	$\beta \times 10^3$ (1/°F)	$g\beta\rho^2/\mu^2$ (1/°F·ft^3)
				Steam					
212	0.0372	0.493	0.870	0.234	0.0145	0.794	1.06	1.49	0.873×10^6
250	0.0350	0.483	0.890	0.254	0.0155	0.920	0.994	1.41	0.698
300	0.0327	0.476	0.960	0.294	0.0171	1.10	0.963	1.32	0.493
400	0.0289	0.472	1.09	0.377	0.0200	1.47	0.924	1.16	0.262
500	0.0259	0.477	1.23	0.474	0.0228	1.85	0.922	1.04	0.148
600	0.0234	0.483	1.37	0.585	0.0258	2.29	0.920	0.944	88.9×10^3
800	0.0197	0.498	1.63	0.828	0.0321	3.27	0.912	0.794	37.8
1000	0.0170	0.517	1.90	1.12	0.0390	4.44	0.911	0.685	17.2
1500	0.0126	0.564	2.57	2.05	0.0580	8.17	0.906	0.510	3.97

T (K)	ρ (kg/m^3)	$c_p \times 10^{-3}$ (J/kg·K)	$\mu \times 10^5$ (Pa·s)	$\nu \times 10^5$ (m^2/s)	$k \times 10^2$ (W/m·K)	$\alpha \times 10^5$ (m^2/s)	Pr	$g\beta\rho^2/\mu^2$ (1/K·m^3)
				Steam				
380	0.5860	2.0592	12.70	2.1672	2.4520	2.0320	1.067	5.5210×10^7
400	0.5549	2.0098	13.42	2.4185	2.6010	2.3322	1.037	4.1951
450	0.4911	1.9771	15.23	3.1012	2.9877	3.0771	1.008	2.2558
500	0.4410	1.9817	17.03	3.8617	3.3903	3.8794	0.995	1.3139
550	0.4004	2.0006	18.84	4.7053	3.8008	4.7448	0.992	0.8069
600	0.3667	2.0264	20.64	5.6286	4.2161	5.6738	0.992	0.5154
650	0.3383	2.0555	22.45	6.6361	4.6361	6.6670	0.995	0.3415
700	0.3140	2.0869	24.25	7.7229	5.0593	7.7207	1.000	0.2277
750	0.2930	2.1192	26.06	8.8942	5.4841	8.8321	1.007	0.1651
800	0.2746	2.1529	27.86	10.1457	5.9089	9.9950	1.015	0.1183

T (°F)	ρ (lb$_m$/ft^3)	c_p (Btu/lb$_m$°F)	$\mu \times 10^5$ (lb$_m$/ft s)	$\nu \times 10^3$ (ft^2/s)	k (Btu/h ft °F)	α (ft^2/h)	Pr	$\beta \times 10^3$ (1/°F)	$g\beta\rho^2/\mu^2$ (1/°F·ft^3)
				Nitrogen					
0	0.0837	0.249	1.06	0.127	0.0132	0.633	0.719	2.18	4.38×10^6
30	0.0786	0.249	1.12	0.142	0.0139	0.710	0.719	2.04	3.29
60	0.0740	0.249	1.17	0.158	0.0146	0.800	0.716	1.92	2.51
80	0.0711	0.249	1.20	0.169	0.0151	0.853	0.712	1.85	2.10
100	0.0685	0.249	1.23	0.180	0.0154	0.915	0.708	1.79	1.79
150	0.0630	0.249	1.32	0.209	0.0168	1.07	0.702	1.64	1.22
200	0.0580	0.249	1.39	0.240	0.0174	1.25	0.690	1.52	0.854
250	0.0540	0.249	1.47	0.271	0.0192	1.42	0.687	1.41	0.616
300	0.0502	0.250	1.53	0.305	0.0202	1.62	0.685	1.32	0.457
400	0.0443	0.250	1.67	0.377	0.0212	2.02	0.684	1.16	0.263
500	0.0397	0.253	1.80	0.453	0.0244	2.43	0.683	1.04	0.163
600	0.0363	0.256	1.93	0.532	0.0252	2.81	0.686	0.944	0.108
800	0.0304	0.262	2.16	0.710	0.0291	3.71	0.691	0.794	0.0507
1000	0.0263	0.269	2.37	0.901	0.0336	4.64	0.700	0.685	0.0272
1500	0.0195	0.283	2.82	1.45	0.0423	7.14	0.732	0.510	0.00785

T (K)	ρ (kg/m^3)	$c_p \times 10^{-3}$ (J/kg·K)	$\mu \times 10^5$ (Pa·s)	$\nu \times 10^5$ (m^2/s)	$k \times 10^2$ (W/m·K)	$\alpha \times 10^5$ (m^2/s)	Pr	$g\beta\rho^2/\mu^2$ (1/K·m^3)
				Nitrogen				
250	1.3668	1.0415	1.5528	1.1361	2.2268	1.5643	0.729	3.0362×10^8
300	1.1383	1.0412	1.7855	1.5686	2.6052	2.1981	0.713	1.3273
350	0.9754	1.0421	2.0000	2.0504	2.9691	2.9210	0.701	0.6655
400	0.8533	1.0449	2.1995	2.5776	3.3186	3.7220	0.691	0.3697
450	0.7584	1.0495	2.3890	3.1501	3.6463	4.5811	0.688	0.2187
500	0.6826	1.0564	2.5702	3.7653	3.9645	5.4979	0.684	0.1382
600	0.5688	1.0751	2.9127	5.1208	4.5549	7.4485	0.686	6.237×10^6
700	0.4875	1.0980	3.2120	6.5887	5.0947	9.5179	0.691	3.233
800	0.4266	1.1222	3.4896	8.1800	5.5864	11.6692	0.700	1.820
1000	0.3413	1.1672	4.0000	11.7199	6.4419	16.1708	0.724	0.810

T (°F)	ρ (lb$_m$/ft^3)	c_p (Btu/lb$_m$°F)	$\mu \times 10^5$ (lb$_m$/ft s)	$\nu \times 10^3$ (ft^2/s)	k (Btu/h ft °F)	α (ft^2/h)	Pr	$\beta \times 10^3$ (1/°F)	$g\beta\rho^2/\mu^2$ (1/°F·ft^3)
					Oxygen				
0	0.0955	0.219	1.22	0.128	0.0134	0.641	0.718	2.18	4.29×10^6
30	0.0897	0.219	1.28	0.143	0.0141	0.718	0.716	2.04	3.22
60	0.0845	0.219	1.35	0.160	0.0149	0.806	0.713	1.92	2.43
80	0.0814	0.220	1.40	0.172	0.0155	0.866	0.713	1.85	2.02
100	0.0785	0.220	1.43	0.182	0.0160	0.925	0.708	1.79	1.74
150	0.0720	0.221	1.52	0.211	0.0172	1.08	0.703	1.64	1.19
200	0.0665	0.223	1.62	0.244	0.0185	1.25	0.703	1.52	0.825
250	0.0168	0.225	1.70	0.276	0.0197	1.42	0.700	1.41	0.600
300	0.0578	0.227	1.79	0.310	0.0209	1.60	0.700	1.32	0.442
400	0.0511	0.230	1.95	0.381	0.0233	1.97	0.698	1.16	0.257
500	0.0458	0.234	2.10	0.458	0.0254	2.37	0.696	1.04	0.160
600	0.0414	0.239	2.25	0.543	0.0281	2.84	0.688	0.944	0.103
800	0.0349	0.246	2.52	0.723	0.0324	3.77	0.680	0.794	49.4×10^3
1000	0.0300	0.252	2.79	0.930	0.0366	4.85	0.691	0.685	25.6
1500	0.0224	0.264	3.39	1.52	0.0465	7.86	0.696	0.510	7.22

T (K)	ρ (kg/m^3)	$c_p \times 10^{-3}$ (J/kg·K)	$\mu \times 10^5$ (Pa·s)	$\nu \times 10^5$ (m^2/s)	$k \times 10^2$ (W/m·K)	$\alpha \times 10^5$ (m^2/s)	Pr	$g\beta\rho^2/\mu^2$ (1/K·m^3)
				Oxygen				
250	1.5620	0.9150	1.7887	1.1451	2.2586	1.5803	0.725	2.9885×10^8
300	1.3007	0.9199	2.0633	1.5863	2.6760	2.2365	0.709	1.2978
350	1.1144	0.9291	2.3176	2.0797	3.0688	2.9639	0.702	0.6469
400	0.9749	0.9417	2.5556	2.6214	3.4616	3.7705	0.695	0.3571
450	0.8665	0.9564	2.7798	3.2081	3.8298	4.6216	0.694	0.2108
500	0.7798	0.9721	2.9930	3.8382	4.1735	5.5056	0.697	0.1330
550	0.7089	0.9879	3.1966	4.5092	4.5172	6.4502	0.700	8.786×10^6
600	0.6498	1.0032	3.3931	5.2218	4.8364	7.4192	0.704	5.988

T (°F)	ρ (lb$_m$/ft^3)	c_p (Btu/lb$_m$°F)	$\mu \times 10^5$ (lb$_m$/ft s)	$\nu \times 10^3$ (ft^2/s)	k (Btu/h ft °F)	α (ft^2/h)	Pr	$\beta \times 10^3$ (1/°F)	$g\beta\rho^2/\mu^2$ (1/°F · ft^3)
				Carbon dioxide					
0	0.132	0.193	0.865	0.0655	0.00760	0.298	0.792	2.18	16.3×10^6
30	0.124	0.198	0.915	0.0739	0.00830	0.339	0.787	2.04	12.0
60	0.117	0.202	0.965	0.0829	0.00910	0.387	0.773	1.92	9.00
80	0.112	0.204	1.00	0.0891	0.00960	0.421	0.760	1.85	7.45
100	0.108	0.207	1.03	0.0953	0.0102	0.455	0.758	1.79	6.33
150	0.100	0.213	1.12	0.113	0.0115	0.539	0.755	1.64	4.16
200	0.092	0.219	1.20	0.131	0.0130	0.646	0.730	1.52	2.86
250	0.0850	0.225	1.32	0.155	0.0148	0.777	0.717	1.41	2.04
300	0.0800	0.230	1.36	0.171	0.0160	0.878	0.704	1.32	1.45
400	0.0740	0.239	1.45	0.196	0.0180	1.02	0.695	1.16	1.11
500	0.0630	0.248	1.65	0.263	0.0210	1.36	0.700	1.04	0.485
600	0.0570	0.256	1.78	0.312	0.0235	1.61	0.700	0.944	0.310
800	0.0480	0.269	2.02	0.420	0.0278	2.15	0.702	0.794	0.143
1000	0.0416	0.280	2.25	0.540	0.0324	2.78	0.703	0.685	75.3×10^3
1500	0.0306	0.301	2.80	0.913	0.0340	4.67	0.704	0.510	19.6

T (K)	ρ (kg/m^3)	$c_p \times 10^{-3}$ (J/kg · K)	$\mu \times 10^5$ (Pa · s)	$\nu \times 10^5$ (m^2/s)	$k \times 10^2$ (W/m · K)	$\alpha \times 10^5$ (m^2/s)	Pr	$g\beta\rho^2/\mu^2$ (1/K · m^3)
				Carbon dioxide				
250	2.1652	0.8052	1.2590	0.5815	1.2891	0.7394	0.793	1.1591×10^9
300	1.7967	0.8526	1.4948	0.8320	1.6572	1.0818	0.770	0.4178
350	1.5369	0.8989	1.7208	1.1197	2.0457	1.4808	0.755	0.2232
400	1.3432	0.9416	1.9318	1.4382	2.4604	1.9454	0.738	0.1186
450	1.1931	0.9803	2.1332	1.7879	2.8955	2.4756	0.721	6.786×10^7
500	1.0733	1.0153	2.3251	2.1663	3.3523	3.0763	0.702	4.176
550	0.9756	1.0470	2.5073	2.5700	3.8208	3.7406	0.685	2.705
600	0.8941	1.0761	2.6827	3.0004	4.3097	4.4793	0.668	1.814

T (°F)	ρ (lb_m/ft^3)	c_p ($\text{Btu}/\text{lb}_m\,°F$)	$\mu \times 10^5$ ($\text{lb}_m/\text{ft s}$)	$\nu \times 10^3$ (ft^2/s)	k (Btu/h ft °F)	α (ft^2/h)	Pr	$\beta \times 10^3$ (1/°F)	$g\beta\rho^2/\mu^2$ (1/°F · ft^3)
				Hydrogen					
0	0.00597	3.37	0.537	0.900	0.092	4.59	0.713	2.18	87,000
30	0.00562	3.39	0.562	1.00	0.097	5.09	0.709	2.04	65,700
60	0.00530	3.41	0.587	1.11	0.102	5.65	0.707	1.92	50,500
80	0.00510	3.42	0.602	1.18	0.105	6.04	0.705	1.85	42,700
100	0.00492	3.42	0.617	1.25	0.108	6.42	0.700	1.79	36,700
150	0.00450	3.44	0.653	1.45	0.116	7.50	0.696	1.64	25,000
200	0.00412	3.45	0.688	1.67	0.123	8.64	0.696	1.52	17,500
250	0.00382	3.46	0.723	1.89	0.130	9.85	0.690	1.41	12,700
300	0.00357	3.46	0.756	2.12	0.137	11.1	0.687	1.32	9,440
400	0.00315	3.47	0.822	2.61	0.151	13.8	0.681	1.16	5,470
500	0.00285	3.47	0.890	3.12	0.165	16.7	0.675	1.04	3,430
600	0.00260	3.47	0.952	3.66	0.179	19.8	0.667	0.944	2,270
800	0.00219	3.49	1.07	4.87	0.205	26.8	0.654	0.794	1,080
1000	0.00189	3.52	1.18	6.21	0.224	33.7	0.664	0.685	571
1500	0.00141	3.62	1.44	10.2	0.265	51.9	0.708	0.510	158

T (K)	ρ (kg/m^3)	c_p (J/kg · K)	$\mu \times 10^6$ (Pa · s)	$\nu \times 10^6$ (m^2/s)	$k \times 10^2$ (W/m · K)	$\alpha \times 10^4$ (m^2/s)	Pr	$g\beta\rho^2/\mu^2 \times 10^{-6}$ (1/K · m^3)
				Hydrogen				
50	0.5095	10.501	2.516	4.938	0.0362	0.0633	0.78	
100	0.2457	11.229	4.212	17.143	0.0665	0.2410	0.711	333.8
150	0.1637	12.602	5.595	34.178	0.0981	0.4755	0.719	55.99
200	0.1227	13.504	6.813	55.526	0.1282	0.7717	0.719	15.90
250	0.0982	14.059	7.919	80.641	0.1561	1.131	0.713	6.03
300	0.0818	14.314	8.963	109.57	0.182	1.554	0.705	2.72
350	0.0702	14.436	9.954	141.79	0.206	2.033	0.697	1.39
400	0.0613	14.491	10.864	177.23	0.228	2.567	0.690	0.782
450	0.0546	14.499	11.779	215.73	0.251	3.171	0.680	0.468
500	0.0492	14.507	12.636	256.83	0.272	3.811	0.674	0.297
600	0.0408	14.537	14.285	350.12	0.315	5.311	0.659	0.134
700	0.0349	14.574	15.890	455.30	0.351	6.901	0.660	0.0677
800	0.0306	14.675	17.40	568.63	0.384	8.551	0.665	0.0379
1000	0.0245	14.968	20.160	822.86	0.440	11.998	0.686	0.0145
1200	0.0205	15.366	22.75	1109.80	0.488	15.492	0.716	0.00667

T (°F)	ρ (lbm/ft³)	c_p (Btu/lbm°F)	$\mu \times 10^5$ (lbm/ft s)	$\nu \times 10^3$ (ft²/s)	k (Btu/h ft °F)	α (ft²/h)	Pr	$\beta \times 10^3$ (1/°F)	$g\beta\rho^2/\mu^2$ (1/°F·ft³)
				Carbon monoxide					
0	0.0832	0.249	1.05	0.126	0.0128	0.620	0.749	2.18	4.40×10^6
30	0.0780	0.249	1.11	0.142	0.0134	0.691	0.744	2.04	3.32
60	0.0736	0.249	1.16	0.157	0.0142	0.775	0.740	1.92	2.48
80	0.0709	0.249	1.20	0.169	0.0146	0.828	0.737	1.85	2.09
100	0.0684	0.249	1.23	0.180	0.0150	0.884	0.735	1.79	1.79
150	0.0628	0.249	1.32	0.210	0.0163	1.04	0.730	1.64	1.19
200	0.0580	0.250	1.40	0.241	0.0174	1.20	0.726	1.52	0.842
250	0.0539	0.250	1.48	0.275	0.0183	1.36	0.722	1.41	0.604
300	0.0503	0.251	1.56	0.310	0.0196	1.56	0.720	1.32	0.442
400	0.0445	0.253	1.73	0.389	0.0217	1.92	0.718	1.16	0.248
500	0.0399	0.256	1.85	0.463	0.0234	2.30	0.725	1.04	0.156
600	0.0361	0.259	1.97	0.545	0.0253	2.71	0.723	0.944	0.101
800	0.0304	0.266	2.21	0.728	0.0288	3.57	0.730	0.794	48.2×10^3
1000	0.0262	0.273	2.43	0.929	0.0324	4.54	0.740	0.685	25.6
1500	0.0195	0.286	3.00	1.54	0.0410	7.35	0.756	0.510	6.93

T (K)	ρ (kg/m³)	$c_p \times 10^{-3}$ (J/kg·K)	$\mu \times 10^5$ (Pa·s)	$\nu \times 10^5$ (m²/s)	$k \times 10^2$ (W/m·K)	$\alpha \times 10^5$ (m²/s)	Pr	$g\beta\rho^2/\mu^2$ (1/K·m³)
				Carbon monoxide				
250	1.3669	1.0425	1.5408	1.1272	2.1432	1.5040	0.749	3.0841×10^8
300	1.1382	1.0422	1.7854	1.5686	2.5240	2.1277	0.737	1.3273
350	0.9753	1.0440	2.0097	2.0606	2.8839	2.8323	0.727	0.6590
400	0.8532	1.0484	2.2201	2.6021	3.2253	3.6057	0.722	0.3623
450	0.7583	1.0550	2.4189	3.1899	3.5527	4.4408	0.718	0.2133
500	0.6824	1.0642	2.6078	3.8215	3.8638	5.3205	0.718	0.1342
550	0.6204	1.0751	2.7884	4.4945	4.1587	6.2350	0.721	8.843×10^6
600	0.5687	1.0870	2.9607	5.2061	4.4443	7.1894	0.724	6.025

T (°F)	ρ (lbm/ft³)	c_p (Btu/lbm°F)	$\mu \times 10^6$ (lbm/ft s)	$\nu \times 10^3$ (ft²/s)	k (Btu/h ft °F)	α (ft²/h)	Pr	$\beta \times 10^3$ (1/°F)	$g\beta\rho^2/\mu^2 \times 10^{-6}$ (1/°F·ft³)
				Chlorine					
0	0.211	0.113	8.06	0.0381	0.00418	0.175	0.785	2.18	48.3
30	0.197	0.114	8.40	0.0426	0.00450	0.201	0.769	2.04	36.6
60	0.187	0.114	8.80	0.0470	0.00480	0.225	0.753	1.92	28.1
80	0.180	0.115	9.07	0.0504	0.00500	0.242	0.753	1.85	24.3
100	0.173	0.115	9.34	0.0540	0.00520	0.261	0.748	1.79	19.9
150	0.159	0.117	10.0	0.0629	0.00570	0.306	0.739	1.64	13.4

T (°F)	ρ (lb$_m$/ft^3)	c_p (Btu/lb$_m$°F)	$\mu \times 10^7$ (lb$_m$/ft s)	$\nu \times 10^3$ (ft^2/s)	k (Btu/h ft °F)	α (ft^2/h)	Pr	$\beta \times 10^3$ (1/°F)	$g\beta\rho^2/\mu^2$ (1/°F · ft^3)
				Helium					
0	0.0119	1.24	122	1.03	0.0784	5.30	0.698	2.18	66,800
30	0.0112	1.24	127	1.14	0.0818	5.89	0.699	2.04	51,100
60	0.0106	1.24	132	1.25	0.0852	6.46	0.700	1.92	40,000
80	0.0102	1.24	135	1.32	0.0872	6.88	0.701	1.85	33,900
100	0.00980	1.24	138	1.41	0.0892	7.37	0.701	1.79	29,000
150	0.00900	1.24	146	1.63	0.0937	8.36	0.703	1.64	20,100
200	0.00829	1.24	155	1.87	0.0977	9.48	0.705	1.52	14,000
250	0.00772	1.24	162	2.09	0.102	10.7	0.707	1.41	10,400
300	0.00722	1.24	170	2.36	0.106	11.8	0.709	1.32	7,650
400	0.00637	1.24	185	2.91	0.114	14.4	0.714	1.16	4,410
500	0.00572	1.24	198	3.46	0.122	17.1	0.719	1.04	2,800
600	0.00517	1.24	209	4.04	0.130	20.6	0.720	0.994	1,850
800	0.00439	1.24	232	5.28	0.145	27.6	0.722	0.794	915
1000	0.00376	1.24	255	6.78	0.159	35.5	0.725	0.685	480
1500	0.00280	1.24	309	11.1	0.189	59.7	0.730	0.510	135

T (°F)	ρ (lb$_m$/ft^3)	c_p (Btu/lb$_m$°F)	$\mu \times 10^5$ (lb$_m$/ft s)	$\nu \times 10^3$ (ft^2/s)	k (Btu/h ft °F)	α (ft^2/h)	Pr	$\beta \times 10^3$ (1/°F)	$g\beta\rho^2/\mu^2$ (1/°F · ft^3)
				Sulfur dioxide					
0	0.195	0.142	0.700	3.59	0.00460	0.166	0.778	2.03	50.6×10^6
100	0.161	0.149	0.890	5.52	0.00560	0.233	0.854	1.79	19.0
200	0.136	0.157	1.05	7.74	0.00670	0.313	0.883	1.52	8.25
300	0.118	0.164	1.20	10.2	0.00790	0.407	0.898	1.32	4.12
400	0.104	0.170	1.35	13.0	0.00920	0.520	0.898	1.16	2.24
500	0.0935	0.176	1.50	16.0	0.00990	0.601	0.958	1.04	1.30
600	0.0846	0.180	1.65	19.5	0.0108	0.711	0.987	0.994	0.795

				Liquids					
T (°F)	ρ (lb$_m$/ft^3)	c_p (Btu/lb$_m$°F)	$\mu \times 10^3$ (lb$_m$/ft s)	$\nu \times 10^5$ (ft^2/s)	k (Btu/h ft °F)	$\alpha \times 10^3$ (ft^2/h)	Pr	$\beta \times 10^4$ (1/°F)	$g\beta\rho^2/\mu^2 \times 10^{-6}$ (1/°F · ft^3)
				Water					
32	62.4	1.01	1.20	1.93	0.319	5.06	13.7	−0.350	
60	62.3	1.00	0.760	1.22	0.340	5.45	8.07	0.800	17.2
80	62.2	0.999	0.578	0.929	0.353	5.67	5.89	1.30	48.3
100	62.1	0.999	0.458	0.736	0.364	5.87	4.51	1.80	107
150	61.3	1.00	0.290	0.474	0.383	6.26	2.72	2.80	403
200	60.1	1.01	0.206	0.342	0.392	6.46	1.91	3.70	1,010
250	58.9	1.02	0.160	0.272	0.395	6.60	1.49	4.70	2,045
300	57.3	1.03	0.130	0.227	0.395	6.70	1.22	5.60	3,510
400	53.6	1.08	0.0930	0.174	0.382	6.58	0.950	7.80	8,350
500	49.0	1.19	0.0700	0.143	0.349	5.98	0.859	11.0	17,350
600	42.4	1.51	0.0579	0.137	0.293	4.58	1.07	17.5	30,300

T (K)	ρ (kg/m^3)	c_p (J/kg $\times$ K)	$\mu \times 10^6$ (Pa $\times$ s)	$\nu \times 10^6$ (m^2/s)	k (W/m $\times$ K)	$\alpha \times 10^6$ (m^2/s)	Pr	$g\beta\rho^2/\mu^2 \times 10^{-9}$ (1/K $\cdot$ m^3)
					Water			
273	999.3	4226	1794	1.795	0.558	0.132	13.6	
293	998.2	4182	993	0.995	0.597	0.143	6.96	2.035
313	992.2	4175	658	0.663	0.633	0.153	4.33	8.833
333	983.2	4181	472	0.480	0.658	0.160	3.00	22.75
353	971.8	4194	352	0.362	0.673	0.165	2.57	46.68
373	958.4	4211	278	0.290	0.682	0.169	1.72	85.09
473	862.8	4501	139	0.161	0.665	0.171	0.94	517.2
573	712.5	5694	92.2	0.129	0.564	0.139	0.93	1766.0

T ($^\circ$F)	ρ (lb$_m$/ft^3)	c_p (Btu/lb$_m$ $^\circ$F)	$\mu \times 10^5$ (lb$_m$/ft s)	$\nu \times 10^5$ (ft^2/s)	k (Btu/h ft $^\circ$F)	$\alpha \times 10^3$ (ft^2/h)	Pr	$\beta \times 10^3$ (1/$^\circ$F)	$g\beta\rho^2/\mu^2 \times 10^{-6}$ (1/$^\circ$F $\cdot$ ft^3)
					Aniline				
60	64.0	0.480	305	4.77	0.101	3.29	52.3		
80	63.5	0.485	240	3.78	0.100	3.25	41.8		
100	63.0	0.490	180	2.86	0.100	3.24	31.8	0.45	17.7
150	61.6	0.503	100	1.62	0.0980	3.16	18.4		
200	60.2	0.515	62	1.03	0.0962	3.10	12.0		
250	58.9	0.527	42	0.714	0.0947	3.05	8.44		
300	57.5	0.540	30	0.522	0.0931	2.99	6.28		

T ($^\circ$F)	ρ (lb$_m$/ft^3)	c_p (Btu/lb$_m$ $^\circ$F)	$\mu \times 10^5$ (lb$_m$/ft s)	$\nu \times 10^5$ (ft^2/s)	k (Btu/h ft $^\circ$F)	$\alpha \times 10^3$ (ft^2/h)	Pr	$\beta \times 10^3$ (1/$^\circ$F)	$g\beta\rho^2/\mu^2 \times 10^{-7}$ (1/$^\circ$F $\cdot$ ft^3)
					Ammonia				
-60	43.9	1.07	20.6	0.471	0.316	6.74	2.52	0.94	132
-30	42.7	1.07	18.2	0.426	0.317	6.93	2.22	1.02	265
0	41.3	1.08	16.9	0.409	0.315	7.06	2.08	1.1	467
30	40.0	1.11	16.2	0.402	0.312	7.05	2.05	1.19	757
60	38.5	1.14	15.0	0.391	0.304	6.92	2.03	1.3	1130
80	37.5	1.16	14.2	0.379	0.296	6.79	2.01	1.4	1650
100	36.4	1.19	13.5	0.368	0.287	6.62	2.00	1.5	2200
120	35.3	1.22	12.6	0.356	0.275	6.43	2.00	1.68	3180

T (°F)	ρ (lb$_m$/ft^3)	c_p (Btu/lb$_m$°F)	$\mu \times 10^5$ (lb$_m$/ft s)	$\nu \times 10^5$ (ft^2/s)	k (Btu/h ft °F)	$\alpha \times 10^3$ (ft^2/h)	Pr	$\beta \times 10^3$ (1/°F)	$g\beta\rho^2/\mu^2 \times 10^{-6}$ (1/°F · ft^3)
				Freon-12					
-40	94.5	0.202	125	1.32	0.0650	3.40	14.0	9.10	168
-30	93.5	0.204	123	1.32	0.0640	3.35	14.1	9.60	179
0	90.9	0.212	116	1.28	0.0578	3.00	15.4	11.4	225
30	87.4	0.221	108	1.24	0.0564	2.92	15.3	13.1	277
60	84.0	0.230	99.6	1.19	0.0528	2.74	15.6	14.9	341
80	81.3	0.238	94.0	1.16	0.0504	2.60	16.0	16.0	384
100	78.7	0.246	88.4	1.12	0.0480	2.48	16.3	17.2	439
150	71.0	0.271	74.8	1.05	0.0420	2.18	17.4	19.5	625

T (°F)	ρ (lb$_m$/ft^3)	c_p (Btu/lb$_m$°F)	$\mu \times 10^5$ (lb$_m$/ft s)	$\nu \times 10^5$ (ft^2/s)	k (Btu/h ft °F)	$\alpha \times 10^3$ (ft^2/h)	Pr	$\beta \times 10^3$ (1/°F)	$g\beta\rho^2/\mu^2 \times 10^{-6}$ (1/°F · ft^3)
				n-Butyl Alcohol					
60	50.5	0.55	225	4.46	0.100	3.59	44.6		
80	50.0	0.58	180	3.60	0.099	3.41	38.0	0.25	6.23
100	49.6	0.61	130	2.62	0.098	3.25	29.1	0.43	2.02
150	48.5	0.68	68	1.41	0.098	2.97	17.1		

T (°F)	ρ (lb$_m$/ft^3)	c_p (Btu/lb$_m$°F)	$\mu \times 10^5$ (lb$_m$/ft s)	$\nu \times 10^5$ (ft^2/s)	k (Btu/h ft °F)	$\alpha \times 10^3$ (ft^2/h)	Pr $\times 10^{-2}$	$\beta \times 10^4$ (1/°F)	$g\beta\rho^2/\mu^2 \times 10^{-6}$ (1/°F · ft^3)
				Benzene					
60	55.2	0.395	44.5	0.806	0.0856	3.93	7.39		
80	54.6	0.410	38	0.695	0.0836	3.73	6.70	7.5	498
100	53.6	0.420	33	0.615	0.0814	3.61	6.13	7.2	609
150	51.8	0.450	24.5	0.473	0.0762	3.27	5.21	6.8	980
200	49.9	0.480	19.4	0.390	0.0711	2.97	4.73		

T (°F)	ρ (lb$_m$/ft^3)	c_p (Btu/lb$_m$°F)	$\mu \times 10^5$ (lb$_m$/ft s)	$\nu \times 10^5$ (ft^2/s)	k (Btu/h ft °F)	$\alpha \times 10^3$ (ft^2/h)	Pr	$\beta \times 10^3$ (1/°F)	$g\beta\rho^2/\mu^2 \times 10^{-4}$ (1/°F · ft^3)
				Hydraulic fluid (MIL-M-5606)					
0	55.0	0.400	5550	101	0.0780	3.54	1030	0.76	2.39
30	54.0	0.420	2220	41.1	0.0755	3.32	446	0.68	13.0
60	53.0	0.439	1110	20.9	0.0732	3.14	239	0.60	44.1
80	52.5	0.453	695	13.3	0.0710	3.07	155	0.52	95.7
100	52.0	0.467	556	10.7	0.0690	2.84	136	0.47	132
150	51.0	0.499	278	5.45	0.0645	2.44	80.5	0.32	346
200	50.0	0.530	250	5.00	0.0600	2.27	79.4	0.20	258

T (°F)	ρ (lb$_m$/ft^3)	c_p (Btu/lb$_m$°F)	μ (lb$_m$/ft s)	$\nu \times 10^2$ (ft^2/s)	k (Btu/h ft °F)	$\alpha \times 10^3$ (ft^2/h)	Pr $\times 10^{-2}$	$\beta \times 10^3$ (1/°F)	$g\beta\rho^2/\mu^2$ (1/°F · ft^3)
				Glycerin					
30	79.7	0.540	7.2	9.03	0.168	3.91	832		
60	79.1	0.563	1.4	1.77	0.167	3.75	170		
80	78.7	0.580	0.6	0.762	0.166	3.64	75.3	0.30	166
100	78.2	0.598	0.1	0.128	0.165	3.53	13.1		

T (°F)	ρ (lb$_m$/ft^3)	c_p (Btu/lb$_m$°F)	$\mu \times 10^5$ (lb$_m$/ft s)	$\nu \times 10^5$ (ft^2/s)	k (Btu/h ft °F)	$\alpha \times 10^3$ (ft^2/h)	Pr	$\beta \times 10^3$ (1/°F)	$g\beta\rho^2/\mu^2$ (1/°F · ft^3)
				Kerosene					
30	48.8	0.456	800	16.4	0.0809	3.63	163		
60	48.1	0.474	600	12.5	0.0805	3.53	127	0.58	120
80	47.6	0.491	490	10.3	0.0800	3.42	108	0.48	146
100	47.2	0.505	420	8.90	0.0797	3.35	95.7	0.47	192
150	46.1	0.540	320	6.83	0.0788	3.16	77.9		

T (°F)	ρ (lb$_m$/ft^3)	c_p (Btu/lb$_m$°F)	$\mu \times 10^5$ (lb$_m$/ft s)	$\nu \times 10^5$ (ft^2/s)	k (Btu/h ft °F)	$\alpha \times 10^3$ (ft^2/h)	Pr	$\beta \times 10^3$ (1/°F)	$g\beta\rho^2/\mu^2 \times 10^{-4}$ (1/°F · ft^3)
				Liquid hydrogen					
−435	4.84	1.69	1.63	0.337	0.0595	7.28	1.67		
−433	4.77	1.78	1.52	0.319	0.0610	7.20	1.59		
−431	4.71	1.87	1.40	0.297	0.0625	7.09	1.51	7.1	2.59
−429	4.64	1.96	1.28	0.276	0.0640	7.03	1.41		
−427	4.58	2.05	1.17	0.256	0.0655	6.97	1.32		
−425	4.51	2.15	1.05	0.233	0.0670	6.90	1.21		

T (°F)	ρ (lbm/ft³)	c_p (Btu/lbm°F)	$\mu \times 10^5$ (lbm/ft s)	$\nu \times 10^5$ (ft²/s)	$k \times 10^3$ (Btu/h ft °F)	$\alpha \times 10^5$ (ft²/h)	Pr	$\beta \times 10^3$ (1/°F)	$g\beta\rho^2/\mu^2 \times 10^{-8}$ (1/°F·ft³)
				Liquid oxygen					
-350	80.1	0.400	38.0	0.474	3.1	9.67	172		
-340	78.5	0.401	28.0	0.356	3.4	10.8	109		
-330	76.8	0.402	21.8	0.284	3.7	12.0	85.0		
-320	75.1	0.404	17.4	0.232	4.0	12.2	63.5	3.19	186
-310	73.4	0.405	14.8	0.202	4.3	14.5	50.1		
-300	71.7	0.406	13.0	0.181	4.6	15.8	41.2		

T (°F)	ρ (lbm/ft³)	c_p (Btu/lbm°F)	$\mu \times 10^3$ (lbm/ft s)	$\nu \times 10^6$ (ft²/s)	k (Btu/h ft °F)	α (ft²/h)	Pr	$\beta \times 10^3$ (1/°F)	$g\beta\rho^2/\mu^2 \times 10^{-9}$ (1/°F·ft³)
				Bismuth					
600	625	0.0345	1.09	1.75	8.58	0.397	0.0159		
700	622	0.0353	0.990	1.59	8.87	0.405	0.0141	0.062	0.786
800	618	0.0361	0.900	1.46	9.16	0.408	0.0129	0.065	0.985
900	613	0.0368	0.830	1.35	9.44	0.418	0.0116	0.068	1.19
1000	608	0.0375	0.765	1.26	9.74	0.427	0.0106	0.071	1.45
1100	604	0.0381	0.710	1.17	10.0	0.435	0.00970	0.074	1.72
1200	599	0.0386	0.660	1.10	10.3	0.446	0.00895	0.077	2.04
1300	595	0.0391	0.620	1.04	10.6	0.456	0.00820		

T (°F)	ρ (lbm/ft³)	c_p (Btu/lbm°F)	$\mu \times 10^3$ (lbm/ft s)	$\nu \times 10^6$ (ft²/s)	k (Btu/h ft °F)	α (ft²/h)	Pr	$\beta \times 10^3$ (1/°F)	$g\beta\rho^2/\mu^2 \times 10^{-9}$ (1/°F·ft³)
				Mercury					
40	848	0.0334	1.11	1.31	4.55	0.161	0.0292		1.57
60	847	0.0333	1.05	1.24	4.64	0.165	0.0270		1.76
80	845	0.0332	1.00	1.18	4.72	0.169	0.0252		1.94
100	843	0.0331	0.960	1.14	4.80	0.172	0.0239		2.09
150	839	0.0330	0.893	1.06	5.03	0.182	0.0210		2.38
200	835	0.0328	0.850	1.02	5.25	0.192	0.0191		2.62
250	831	0.0328	0.806	0.970	5.45	0.200	0.0175		2.87
300	827	0.0328	0.766	0.928	5.65	0.209	0.0160		3.16
400	819	0.0328	0.700	0.856	6.05	0.225	0.0137	0.084	3.70
500	811	0.0328	0.650	0.803	6.43	0.243	0.0119		4.12
600	804	0.0328	0.606	0.754	6.80	0.259	0.0105		4.80
800	789	0.0329	0.550	0.698	7.45	0.289	0.0087		5.54

T (°F)	ρ (lb$_m$/ft^3)	c_p (Btu/lb$_m$°F)	$\mu \times 10^3$ (lb$_m$/ft s)	$\nu \times 10^6$ (ft^2/s)	k (Btu/h ft °F)	α (ft^2/h)	Pr	$\beta \times 10^3$ (1/°F)	$g\beta\rho^2/\mu^2 \times 10^{-6}$ (1/°F · ft^3)
					Sodium				
200	58.1	0.332	0.489	8.43	49.8	2.58	0.0118		68.0
250	57.6	0.328	0.428	7.43	49.3	2.60	0.0103		87.4
300	57.2	0.324	0.378	6.61	48.8	2.64	0.00903		110
400	56.3	0.317	0.302	5.36	47.3	2.66	0.00725		168
500	55.5	0.309	0.258	4.64	45.5	2.64	0.00633	0.15	224
600	54.6	0.305	0.224	4.11	43.1	2.58	0.00574		287
800	52.9	0.304	0.180	3.40	38.8	2.41	0.00510		418
1000	51.2	0.304	0.152	2.97	36.0	2.31	0.00463		548
1300	48.7	0.305	0.120	2.47	34.2	2.31	0.00385		795

Appendix J

Mass-Transfer Diffusion Coefficients in Binary Systems

Table J.1 Binary mass diffusivities in gases[†]

System	T (K)	$D_{AB}P(\text{cm}^2\ \text{atm/s})$	$D_{AB}P(\text{m}^2\ \text{Pa/s})$
Air			
Ammonia	273	0.198	2.006
Aniline	298	0.0726	0.735
Benzene	298	0.0962	0.974
Bromine	293	0.091	0.923
Carbon dioxide	273	0.136	1.378
Carbon disulfide	273	0.0883	0.894
Chlorine	273	0.124	1.256
Diphenyl		0.160	1.621
Ethyl acetate			0.718
Ethanol			1.337
Ethyl ether			0.908
Iodine			0.845
Methanol			1.641
Mercury			4.791
Naphthalene			0.619
Nitrobenzene			0.879
n-Octane			0.610
Oxygen			1.773
Propyl acet			0.932
Sulfur diox			1.236
Toluene			0.855
Water			2.634
Ammonia			
Ethylene			1.793
Argon			
Neon	293		3.333
Carbon dioxide			
Benzene	318	0.0715	0.724
Carbon disulfide	318	0.0715	0.724
Ethyl acetate	319	0.0666	0.675

(continued)

Table J.1   (*Continued*)

System	T (K)	$D_{AB}P$(cm^2 atm/s)	$D_{AB}P$(m^2 Pa/s)
Ethanol	273	0.0693	0.702
Ethyl ether	273	0.0541	0.548
Hydrogen	273	0.550	5.572
Methane	273	0.153	1.550
Methanol	298.6	0.105	1.064
Nitrogen	298	0.165	1.672
Nitrous oxide	298	0.117	1.185
Propane	298	0.0863	0.874
Water	298	0.164	1.661
Carbon monoxide			
Ethylene	273	0.151	1.530
Hydrogen	273	0.651	6.595
Nitrogen	288	0.192	1.945
Oxygen	273	0.185	1.874
Helium			
Argon	273	0.641	6.493
Benzene	298	0.384	3.890
Ethanol	298	0.494	5.004
Hydrogen	293	1.64	16.613
Neon	293	1.23	12.460
Water	298	0.908	9.198
Hydrogen			
Ammonia	293	0.849	8.600
Argon	293	0.770	7.800
Benzene	273	0.317	3.211
Ethane	273	0.439	4.447
Methane	273	0.625	6.331
Oxygen	273	0.697	7.061
Water	293	0.850	8.611
Nitrogen			
Ammonia	293	0.241	2.441
Ethylene	298	0.163	1.651
Hydrogen	288	0.743	7.527
Iodine	273	0.070	0.709
Oxygen	273	0.181	1.834
Oxygen			
Ammonia	293	0.253	2.563
Benzene	296	0.0939	0.951
Ethylene	293	0.182	1.844

[†]R. C. Reid and T. K. Sherwood, *The Properties of Gases and Liquids,* McGraw-Hill, New York, 1958, Chapter. 8.

Table J.2 Binary mass diffusivities in liquids[†]

Solute A	Solvent B	Temperature (K)	Solute concentration (g mol/L or kg mol/m^3)	Diffusivity (cm^2/s × 10^5 or m^2/s × 10^9)
Chlorine	Water	289	0.12	1.26
Hydrogen chloride	Water	273	9	2.7
			2	1.8
		283	9	3.3
			2.5	2.5
		289	0.5	2.44
Ammonia	Water	278	3.5	1.24
		288	1.0	1.77
Carbon dioxide	Water	283	0	1.46
		293	0	1.77
Sodium chloride	Water	291	0.05	1.26
			0.2	1.21
			1.0	1.24
			3.0	1.36
			5.4	1.54
Methanol	Water	288	0	1.28
Acetic acid	Water	285.5	1.0	0.82
			0.01	0.91
		291	1.0	0.96
Ethanol	Water	283	3.75	0.50
			0.05	0.83
		289	2.0	0.90
n-Butanol	Water	288	0	0.77
Carbon dioxide	Ethanol	290	0	3.2
Chloroform	Ethanol	293	2.0	1.25

[†]R. E. Treybal, *Mass Transfer Operations*, McGraw-Hill, New York, 1955, p. 25.

Table J.3 Binary diffusivities in solids[†]

Solute	Solid	Temperature (K)	Diffusivity (cm^2/s or m^2/s ×10^4)	Diffusivity (ft^2/h)
Helium	Pyrex	293	4.49×10^{-11}	1.74×10^{-10}
		773	2.00×10^{-8}	7.76×10^{-8}
Hydrogen	Nickel	358	1.16×10^{-8}	4.5×10^{-8}
		438	1.05×10^{-7}	4.07×10^{-7}
Bismuth	Lead	293	1.10×10^{-16}	4.27×10^{-16}
Mercury	Lead	293	2.50×10^{-15}	9.7×10^{-15}
Antimony	Silver	293	3.51×10^{-21}	1.36×10^{-20}
Aluminum	Copper	293	1.30×10^{-30}	5.04×10^{-30}
Cadmium	Copper	293	2.71×10^{-15}	1.05×10^{-14}

[†]R. M. Barrer, *Diffusion In and Through Solids*, Macmillan, New York, 1941.

Appendix K

Lennard–Jones Constants

Table K.1 The collision integrals, Ω_μ and Ω_D based on the Lennard–Jones potential[†]

$\kappa T/\epsilon$	$\Omega_\mu = \Omega_k$ (for viscosity and thermal conductivity)	Ω_D (for mass diffusivity)	kT/ϵ	$\Omega_\mu = \Omega_k$ (for viscosity and thermal conductivity)	Ω_D (for mass diffusivity)
			1.75	1.234	1.128
0.30	2.785	2.662	1.80	1.221	1.116
0.35	2.628	2.476	1.85	1.209	1.105
0.40	2.492	2.318	1.90	1.197	1.094
0.45	2.368	2.184	1.95	1.186	1.084
0.50	2.257	2.066	2.00	1.175	1.075
0.55	2.156	1.966	2.10	1.156	1.057
0.60	2.065	1.877	2.20	1.138	1.041
0.65	1.982	1.798	2.30	1.122	1.026
0.70	1.908	1.729	2.40	1.107	1.012
0.75	1.841	1.667	2.50	1.093	0.9996
0.80	1.780	1.612	2.60	1.081	0.9878
0.85	1.725	1.562	2.70	1.069	0.9770
0.90	1.675	1.517	2.80	1.058	0.9672
0.95	1.629	1.476	2.90	1.048	0.9576
1.00	1.587	1.439	3.00	1.039	0.9490
1.05	1.549	1.406	3.10	1.030	0.9406
1.10	1.514	1.375	3.20	1.022	0.9328
1.15	1.482	1.346	3.30	1.014	0.9256
1.20	1.452	1.320	3.40	1.007	0.9186
1.25	1.424	1.296	3.50	0.9999	0.9120
1.30	1.399	1.273	3.60	0.9932	0.9058
1.35	1.375	1.253	3.70	0.9870	0.8998
1.40	1.353	1.233	3.80	0.9811	0.8942
1.45	1.333	1.215	3.90	0.9755	0.8888
1.50	1.314	1.198	4.00	0.9700	0.8836
1.55	1.296	1.182	4.10	0.9649	0.8788
1.60	1.279	1.167	4.20	0.9600	0.8740
1.65	1.264	1.153	4.30	0.9553	0.8694

(continued)

Table K.1 *(continued)*

$\kappa T/\epsilon$	$\Omega_\mu = \Omega_k$ (for viscosity and thermal conductivity)	Ω_D (for mass diffusivity)	kT/ϵ	$\Omega_\mu = \Omega_k$ (for viscosity and thermal conductivity)	Ω_D (for mass diffusivity)
1.70	1.248	1.140	4.40	0.9507	0.8652
4.50	0.9464	0.8610	10.0	0.8242	0.7424
4.60	0.9422	0.8568	20.0	0.7432	0.6640
4.70	0.9382	0.8530	30.0	0.7005	0.6232
4.80	0.9343	0.8492	40.0	0.6718	0.5960
4.90	0.9305	0.8456	50.0	0.6504	0.5756
5.0	0.9269	0.8422	60.0	0.6335	0.5596
6.0	0.8963	0.8124	70.0	0.6194	0.5464
7.0	0.8727	0.7896	80.0	0.6076	0.5352
8.0	0.8538	0.7712	90.0	0.5973	0.5256

Table K.2   Lennard–Jones force constants calculated from viscosity data[†]

Compound	Formula	ϵ_A/κ, in (K)	σ, in Å
Acetylene	C_2H_2	185	4.221
Air		97	3.617
Argon	A	124	3.418
Arsine	AsH_3	281	4.06
Benzene	C_6H_6	440	5.270
Bromine	Br_2	520	4.268
i-Butane	C_4H_{10}	313	5.341
n-Butane	C_4H_{10}	410	4.997
Carbon dioxide	CO_2	190	3.996
Carbon disulfide	CS_2	488	4.438
Carbon monoxide	CO	110	3.590
Carbon tetrachloride	CCl_4	327	5.881
Carbonyl sulfide	COS	335	4.13
Chlorine	Cl_2	357	4.115
Chloroform	$CHCl_3$	327	5.430
Cyanogen	C_2N_2	339	4.38
Cyclohexane	C_6H_{12}	324	6.093
Ethane	C_2H_6	230	4.418
Ethanol	C_2H_5OH	391	4.455
Ethylene	C_2H_6	205	4.232
Fluroine	F_2	112	3.653
Helium	He	10.22	2.576
n-Heptane	C_7H_{16}	282[‡]	8.88[3]
n-Hexane	C_6H_{14}	413	5.909
Hydrogen	H_2	33.3	2.968
Hydrogen chloride	HCl	360	3.305

[†] R. C. Reid and T. K. Sherwood, *The Properties of Gases and Liquids*, McGraw-Hill, New York, 1958.
[‡] Calculated from virial coefficients.[1]

Table K.2 *(continued)*

Compound	Formula	ϵ_A/k, in (K)	σ, in Å
Hydrogen iodide	HI	324	4.123
Iodine	I_2	550	4.982
Krypton	Kr	190	3.60
Methane	CH_4	136.5	3.822
Methanol	CH_3OH	507	3.585
Methylene chloride	CH_2Cl_2	406	4.759
Methyl chloride	CH_3Cl	855	3.375
Mercuric iodide	HgI_2	691	5.625
Mercury	Hg	851	2.898
Neon	Ne	35.7	2.789
Nitric oxide	NO	119	3.470
Nitrogen	N_2	91.5	3.681
Nitrous oxide	N_2O	220	3.879
n-Nonane	C_9H_{20}	240	8.448
n-Octane	C_8H_{18}	320	7.451
Oxygen	O_2	113	3.433
n-Pentane	C_5H_{12}	345	5.769
Propane	C_3H_8	254	5.061
Silane	SiH_4	207.6	4.08
Silicon tetrachloride	$SiCl_4$	358	5.08
Sulfur dioxide	SO_2	252	4.290
Water	H_2O	356	2.649
Xenon	Xe	229	4.055

The Error Function[1]

ϕ	erf ϕ	ϕ	erf ϕ
0	0.0	0.85	0.7707
0.025	0.0282	0.90	0.7970
0.05	0.0564	0.95	0.8209
0.10	0.1125	1.0	0.8427
0.15	0.1680	1.1	0.8802
0.20	0.2227	1.2	0.9103
0.25	0.2763	1.3	0.9340
0.30	0.3286	1.4	0.9523
0.35	0.3794	1.5	0.9661
0.40	0.4284	1.6	0.9763
0.45	0.4755	1.7	0.9838
0.50	0.5205	1.8	0.9891
0.55	0.5633	1.9	0.9928
0.60	0.6039	2.0	0.9953
0.65	0.6420	2.2	0.9981
0.70	0.6778	2.4	0.9993
0.75	0.7112	2.6	0.9998
0.80	0.7421	2.8	0.9999

[1] J. Crank, *The Mathematics of Diffusion*, Oxford University Press, London, 1958.

Appendix M

Standard Pipe Sizes

Nominal pipe size (in.)	Outside diameter (in.)	Schedule no.	Wall thickness (in.)	Inside diameter (in.)	Cross-sectional area of metal (in.^{2})	Inside sectional area (ft^2)
$\frac{1}{3}$	0.405	40	0.068	0.269	0.072	0.00040
		80	0.095	0.215	0.093	0.00025
$\frac{1}{4}$	0.540	40	0.088	0.364	0.125	0.00072
		80	0.119	0.302	0.157	0.00050
$\frac{3}{8}$	0.675	40	0.091	0.493	0.167	0.00133
		80	0.126	0.423	0.217	0.00098
$\frac{1}{2}$	0.840	40	0.109	0.622	0.250	0.00211
		80	0.147	0.546	0.320	0.00163
		160	0.187	0.466	0.384	0.00118
$\frac{3}{4}$	1.050	40	0.113	0.824	0.333	0.00371
		80	0.154	0.742	0.433	0.00300
		160	0.218	0.614	0.570	0.00206
1	1.315	40	0.133	1.049	0.494	0.00600
		80	0.179	0.957	0.639	0.00499
		160	0.250	0.815	0.837	0.00362
$1\frac{1}{2}$	1.900	40	0.145	1.610	0.799	0.01414
		80	0.200	1.500	1.068	0.01225
		160	0.281	1.338	1.429	0.00976
2	2.375	40	0.154	2.067	1.075	0.02330
		80	0.218	1.939	1.477	0.02050
		160	0.343	1.689	2.190	0.01556
$2\frac{1}{2}$	2.875	40	0.203	2.469	1.704	0.03322
		80	0.276	2.323	2.254	0.02942
		160	0.375	2.125	2.945	0.02463
3	3.500	40	0.216	3.068	2.228	0.05130
		80	0.300	2.900	3.016	0.04587
		160	0.437	2.626	4.205	0.03761

(continued)

Nominal pipe size (in.)	Outside diameter (in.)	Schedule no.	Wall thickness (in.)	Inside diameter (in.)	Cross-sectional area of metal (in.^{2})	Inside sectional area (ft^2)
4	4.500	40	0.237	4.026	3.173	0.08840
		80	0.337	3.826	4.407	0.07986
		120	0.437	3.626	5.578	0.07170
		160	0.531	3.438	6.621	0.06447
5	5.563	40	0.258	5.047	4.304	0.1390
		80	0.375	4.813	6.112	0.1263
		120	0.500	4.563	7.963	0.1136
		160	0.625	4.313	9.696	0.1015
6	6.625	40	0.280	6.065	5.584	0.2006
		80	0.432	5.761	8.405	0.1810
		120	0.562	5.501	10.71	0.1650
		160	0.718	5.189	13.32	0.1469
8	8.625	20	0.250	8.125	6.570	0.3601
		30	0.277	8.071	7.260	0.3553
		40	0.322	7.981	8.396	0.3474
		60	0.406	7.813	10.48	0.3329
		80	0.500	7.625	12.76	0.3171
		100	0.593	7.439	14.96	0.3018
		120	0.718	7.189	17.84	0.2819
		140	0.812	7.001	19.93	0.2673
		160	0.906	6.813	21.97	0.2532
10	10.75	20	0.250	10.250	8.24	0.5731
		30	0.307	10.136	10.07	0.5603
		40	0.365	10.020	11.90	0.5475
		60	0.500	9.750	16.10	0.5158
		80	0.593	9.564	18.92	0.4989
		100	0.718	9.314	22.63	0.4732
		120	0.843	9.064	26.34	0.4481
		140	1.000	8.750	30.63	0.4176
		160	1.125	8.500	34.02	0.3941
12	12.75	20	0.250	12.250	9.82	0.8185
		30	0.330	12.090	12.87	0.7972
		40	0.406	11.938	15.77	0.7773
		60	0.562	11.626	21.52	0.7372
		80	0.687	11.376	26.03	0.7058
		100	0.843	11.064	31.53	0.6677
		120	1.000	10.750	36.91	0.6303
		140	1.125	10.500	41.08	0.6013
		160	1.312	10.126	47.14	0.5592

Appendix N

Standard Tubing Gages

Outside diameter (in.)	Wall thickness B.W.G. and Stubs's gage	(in.)	Inside diameter (in.)	Cross-sectional area (in.²)	Inside sectional area (ft²)
$\frac{1}{2}$	12	0.109	0.282	0.1338	0.000433
	14	0.083	0.334	0.1087	0.000608
	16	0.065	0.370	0.0888	0.000747
	18	0.049	0.402	0.0694	0.000882
	20	0.035	0.430	0.0511	0.001009
$\frac{3}{4}$	12	0.109	0.532	0.2195	0.00154
	13	0.095	0.560	0.1955	0.00171
	14	0.083	0.584	0.1739	0.00186
	15	0.072	0.606	0.1534	0.00200
	16	0.065	0.620	0.1398	0.00210
	17	0.058	0.634	0.1261	0.00219
	18	0.049	0.652	0.1079	0.00232
1	12	0.109	0.782	0.3051	0.00334
	13	0.095	0.810	0.2701	0.00358
	14	0.083	0.834	0.2391	0.00379
	15	0.072	0.856	0.2099	0.00400
	16	0.065	0.870	0.1909	0.00413
	17	0.058	0.884	0.1716	0.00426
	18	0.049	0.902	0.1463	0.00444
$1\frac{1}{4}$	12	0.109	1.032	0.3907	0.00581
	13	0.095	1.060	0.3447	0.00613
	14	0.083	1.084	0.3042	0.00641
	15	0.072	1.106	0.2665	0.00677
	16	0.065	1.120	0.2419	0.00684
	17	0.058	1.134	0.2172	0.00701
	18	0.049	1.152	0.1848	0.00724
$1\frac{1}{2}$	12	0.109	1.282	0.4763	0.00896
	13	0.095	1.310	0.4193	0.00936
	14	0.083	1.334	0.3694	0.00971

(*continued*)

Outside diameter (in.)	Wall thickness		Inside diameter (in.)	Cross-sectional area (in.^{2})	Inside sectional area (ft^2)
	B.W.G. and Stubs's gage	(in.)			
	15	0.072	1.358	0.3187	0.0100
	16	0.065	1.370	0.2930	0.0102
	17	0.058	1.384	0.2627	0.0107
	18	0.049	1.402	0.2234	0.0109
$1\frac{3}{4}$	10	0.134	1.482	0.6803	0.0120
	11	0.120	1.510	0.6145	0.0124
	12	0.109	1.532	0.5620	0.0128
	13	0.095	1.560	0.4939	0.0133
	14	0.083	1.584	0.4346	0.0137
	15	0.072	1.606	0.3796	0.0141
	16	0.065	1.620	0.3441	0.0143
2	10	0.134	1.732	0.7855	0.0164
	11	0.120	1.760	0.7084	0.0169
	12	0.109	1.782	0.6475	0.0173
	13	0.095	1.810	0.5686	0.0179
	14	0.083	1.834	0.4998	0.0183
	15	0.072	1.856	0.4359	0.0188
	16	0.065	1.870	0.3951	0.0191

Author Index

Subject Index